CONTEMPORARY CARBENE CHEMISTRY

Wiley Series of Reactive Intermediates in Chemistry and Biology

Steven E. Rokita, Series Editor

Quinone Methides
Edited by Steven E. Rokita

Radical and Radical Ion Reactivity in Nucleic Acid Chemistry
Edited by Marc Greenberg

Carbon-Centered Free Radicals and Radical Cations
Edited by Malcolm D.E. Forbes

Copper-Oxygen Chemistry
Edited by Kenneth D. Karlin and Shinobu Ihoh

Oxidation of Amino Acids, Peptides, and Proteins: Kinetics and Mechanism
By Virender K. Sharma

Nitrenes and Nitrenium Ions
Edited by Daniel E. Falvey and Anna D. Gudmundsdottir

Contempopary Carbene Chemistry
Edited by Robert A. Moss and Michael P. Doyle

CONTEMPORARY CARBENE CHEMISTRY

Edited by

ROBERT A. MOSS
MICHAEL P. DOYLE

WILEY

Published by John Wiley & Sons, Inc., Hoboken, New Jersey
Published simultaneously in Canada

***Library of Congress Cataloging-in-Publication Data*:**

Moss, Robert A., author.
Contemporary carbene chemistry / Robert A. Moss, Michael P. Doyle.
pages cm. – (Wiley series of reactive intermediates in chemistry and biology)
Includes bibliographical references and index.
ISBN 978-1-118-23795-3 (hardback)
1. Carbenes (Methylene compounds) 2. Carbon compounds. I. Doyle, Michael P., author. II. Title.
QD305.H5M67 2014
547′.01–dc23

2013023529

Printed in the United States of America

10 9 8 7 6 5 4 3 2 1

CONTENTS

PREFACE TO SERIES

Most stable compounds and functional groups have benefited from numerous monographs and series devoted to their unique chemistry, and most biological materials and processes have received similar attention. Chemical and biological mechanisms have also been the subject of individual reviews and compilations. When reactive intermediates are given center stage, presentations often focus on the details and approaches of one discipline despite their common prominence in the primary literature of physical, theoretical, organic, inorganic, and biological disciplines. The *Wiley Series on Reactive Intermediates in Chemistry and Biology* is designed to supply a complementary perspective from current publications by focusing each volume on a specific reactive intermediate or target and endowing it with the broadest possible context and outlook. Individual volumes may serve to supplement an advanced course, sustain a special topics course, and provide a ready resource for the research community. Readers should feel equally reassured by reviews in their specialty, inspired by helpful updates in allied areas, and intrigued by topics not yet familiar.

This series revels in the diversity of its perspectives and expertise. Where some books draw strength from their focused details, this series draws strength from the breadth of its presentations. The goal is to illustrate the widest possible range of literature that covers the subject of each volume. When appropriate, topics may span theoretical approaches for predicting reactivity, physical methods of analysis, strategies for generating intermediates, utility for chemical synthesis, applications in biochemistry and medicine, impact on the environment, occurrence in biology, and more. Experimental systems used to explore these topics may be equally broad and range from simple models to complex arrays and mixtures such as those found in the final frontiers of cells, organisms, earth, and space.

Advances in chemistry and biology gain from a mutual synergy. As new methods are developed for one field, they are often rapidly adapted for application in the other. Biological transformations and pathways often inspire analogous development of new procedures in chemical synthesis, and likewise, chemical characterization and identification of transient intermediates often provide the foundation for understanding the biosynthesis and reactivity of many new biological materials. While individual chapters may draw from a single expertise, the range of contributions contained within each volume should collectively offer readers with a multidisciplinary analysis and exposure to the full range of activities in the field. As this series grows, individualized compilations may also be created through electronic access to highlight a particular approach or application across many volumes that, together, cover a variety of different reactive intermediates.

Interest in starting this series came easily, but the creation of each volume of this series required vision, hard work, enthusiasm, and persistence. I thank all of the contributors and editors who graciously accepted the challenge.

Steven E. Rokita
John Hopkins University

PREFACE

The origins of modern carbene chemistry can be found in Jack Hine's incisive mechanistic studies of chloroform hydrolysis, which implicated dichlorocarbene (or "carbon dichloride," as Hine termed it) as the key intermediate. Published in 1950, Hine's work initiated a myriad of studies, which shows no sign of abating even after 60 years.

Along with the continuing stream of primary research reports, pertinent monographs have appeared regularly, summarizing progress and solidifying our increasing understanding of the physical organic chemistry and synthetic potential of divalent carbon compounds. It is instructive to examine the contents and foci of the earliest texts, not only to chart the developmental history of carbene chemistry, but also to trace the temporal succession and evolution of those problems that have been of central concern.

Hine's *Divalent Carbon* (1964) focused on the formation and reactions of key carbene types, e.g., methylene, halocarbenes, and alkoxycarbenes. The formation of methylene and its derivatives from diazo compounds and ketenes was also reviewed. Significantly, Hine recognized isonitriles as "double-bonded divalent carbon," relatives of today's stable carbenes. Kirmse's *Carbene Chemistry* (1964) adopted a functional group approach to carbene structure, considering methylene, alkyl- and arylcarbenes, as well as carboalkoxy-, alkoxy-, amino-, halo-, and ketocarbenes. Importantly, Gaspar and Hammond's now-classic chapter on carbene spin states was included. A second edition of *Carbene Chemistry* (1971) added chapters on structure and reactivity, structural theory, spectroscopy, carbene analogs, and reaction types. The two volumes of *Carbenes* by Jones and Moss (1973, 1975) featured essays on carbenes from diazo compounds, carbene–alkene addition reactions, reactions of carbon atoms and unsaturated carbenes, organometallic carbene precursors, physical methods applied to carbenes in solution, and an updated version of Gaspar and Hammond's spin state chapter. *Carbene* (*Carbenoide*), a landmark, multiauthored, two-volume set edited by Manfred Regitz (1989), summarized in exceptional depth, scope, and detail nearly 40 years of modern divalent carbon chemistry. Finally, *Advances in Carbene Chemistry,* a three volume set edited by Udo Brinker (1994, 1998, 2001), exemplified the breadth and depth of twentieth century carbene chemistry, while *Carbene Chemistry: From Fleeting Intermediates to Powerful Reagents*, edited by Guy Bertrand (2002), heralded the new country's focus on selectivity.

Now, some 10 years later, carbene chemistry has experienced a remarkable renaissance, entering a new era, driven by advances in spectroscopy—especially fast and ultrafast laser flash photolysis, continuing progress in theory and computational methodology, and signal developments in *synthetic* chemistry. The latter feature an ever-expanding arsenal of *stable* carbenes and *metal* carbenes, useful both as catalysts and as reagents. Spectroscopy now allows the visualization of transient carbenes, together with their excited state precursors, on the nanosecond to femtosecond time scales, while theory enables increasingly accurate calculation of the carbenes' energies and spectra, as well as many features of the energy surfaces upon which their reactions occur. Additionally, developments in reaction dynamics help to characterize the optimum reaction channels for transient carbenes generated on relatively flat energy surfaces.

At the same time, carbene chemistry now provides widely used, highly selective synthetic methodologies that were only nascent 20 years ago. Chemical transformations of carbenes, including metathesis and carbon–hydrogen insertion reactions, have claimed their place in the synthetic chemistry lexicon, controlling carbene reactivity via transition metal-bound intermediates. Moreover, the role of carbenes in these catalytic processes is well-understood in both mechanistic and synthetic terms. From "unselective" as a characteristic description in the 1960s, carbenes, tamed by association with transition metals, are now paragons of catalytic selectivity, while stable carbenes themselves are important catalysts and ligands in an ever-widening spectrum of reactions.

Contemporary Carbene Chemistry focuses on the most innovative and promising aspects of carbene research over the past decade. We do not dwell on those classical mechanistic concerns, which, for the most part, are now resolved. Rather, we explore newer structural, catalytic, and organometallic aspects of carbene chemistry, with a particular emphasis on synthetic applications. The essays that follow are not intended to be exhaustive reviews. Instead, they highlight the most interesting, fruitful, and promising initiatives in their topical areas. In Part 1, Properties and Reactions of Carbenes, we consider new paradigms of carbene stability, acid–base behavior, and catalysis. Aspects of carbenic structure and reactivity are highlighted by chapters on stable carbenes, carbodicarbenes, carbenes as guests in supramolecular hosts, tunneling in the reactions of carbenes and oxacarbenes, and the ultrafast kinetics of carbenes and their excited state precursors. Theoretical concerns are explored in chapters on computational methods, and dynamics applied to the reactions of carbenes. In Part 2, Metal Carbenes, our attention turns to the synthetic dimension, particularly the reactions and catalytic properties of metal carbenes. The metals considered include, lithium, rhodium, ruthenium, chromium, molybdenum, tungsten, cobalt, and gold. In addition, each chapter features a concluding section summarizing the current situation, highlighting near-term problems, and pointing to likely research directions. A list of key reviews and suggestions for further reading also accompanies each chapter.

Any book that bears "contemporary" in its title challenges the future. We welcome this challenge. If *Contemporary Carbene Chemistry* aids the research that renders it redundant, then it will have served its purpose; its authors and editors will be more than content.

New Brunswick ROBERT A. MOSS
College Park MICHAEL P. DOYLE

CONTRIBUTORS

Christopher W. Bielawski*, Department of Chemistry and Biochemistry, The University of Texas at Austin, Austin, TX, USA

Jeffrey W. Bode*, Laboratorium fur Organische Chemie, Department of Chemistry and Applied Biosciences, Swiss Federal Institute of Technology, ETH Zürich, Zürich, Switzerland

Udo H. Brinker*, Department of Chemistry, Institute of Organic Chemistry, University of Vienna, Vienna, Austria; Department of Chemistry, The State University of New York at Binghamton, Binghamton, NY, USA

Gotard Burdzinski*, Quantum Electronics Laboratory, Faculty of Physics, Adam Mickiewicz University, Poznan, Poland

Vito Capriati*, Dipartimento di Farmacia, Università degli Studi di Bari "Aldo Moro," Consorzio Interuniversitario Nazionale Metodologie e Processi Innovativi di Sintesi, Bari, Italy

Xin Cui, Department of Chemistry, University of South Florida, Tampa, FL, USA

Huw M. L. Davies*, Department of Chemistry, Emory University, Atlanta, GA, USA

Steven T. Diver*, Department of Chemistry, University at Buffalo, The State University of New York, Amherst, NY, USA

Jonathan M. French, Department of Chemistry, University at Buffalo, The State University of New York, Amherst, NY, USA

Gernot Frenking*, Fachbereich Chemie, Philipps-Universität Marburg, Marburg, Germany

Dennis Gerbig, Institute of Organic Chemistry, Justus-Liebig University, Giessen, Germany

Scott Gronert*, Department of Chemistry, Virginia Commonwealth University, Richmond, VA, USA

Christopher M. Hadad*, Department of Chemistry and Biochemistry, The Ohio State University, Columbus, OH, USA

K. N. Houk*, Department of Chemistry and Biochemistry, University of California, Los Angeles, CA, USA

James R. Keeffe, Department of Chemistry and Biochemistry, San Francisco State University, San Francisco, CA, USA

Hoi Ling Luk, Department of Chemistry and Biochemistry, The Ohio State University, Columbus, OH, USA

Jessada Mahatthananchai, Laboratorium fur Organische Chemie, Department of Chemistry and Applied Biosciences, Swiss Federal Institute of Technology, ETH Zürich, Zürich, Switzerland

Richard S. Massey, Department of Chemistry, Durham University, Durham, UK

Dina C. Merrer*, Department of Chemistry, Barnard College, New York, NY, USA

Jean-Luc Mieusset, Department of Chemistry, Institute of Organic Chemistry, University of Vienna, Vienna, Austria

Jonathan P. Moerdyk, Department of Chemistry and Biochemistry, The University of Texas at Austin, Austin, TX, USA

Rory A. More O'Ferrall*, School of Chemistry and Chemical Biology, University College Dublin, Dublin, Ireland

AnnMarie C. O'Donoghue*, Department of Chemistry, Durham University, Durham, UK

Brendan T. Parr, Department of Chemistry, Emory University, Atlanta, GA, USA

Mathew S. Platz*, Department of Chemistry, The Ohio State University, Columbus, OH, USA

Murray G. Rosenberg, Department of Chemistry, The State University of New York at Binghamton, Binghamton, NY, USA

Peter R. Schreiner*, Institute of Organic Chemistry, Justus-Liebig University, Giessen, Germany

Ralf Tonner, Fachbereich Chemie, Philipps-Universität Marburg, Marburg, Germany

Zachary J. Tonzetich*, Department of Chemistry, University of Texas at San Antonio, San Antonio, TX, USA

Shubham Vyas, Department of Chemistry and Biochemistry, The Ohio State University, Columbus, OH, USA

Lai Xu, Department of Chemistry and Biochemistry, Texas Tech University, Lubbock, TX, USA

Liming Zhang*, Department of Chemistry and Biochemistry, University of California, Santa Barbara, CA, USA

X. Peter Zhang*, Department of Chemistry, University of South Florida, Tampa, FL, USA

*Denotes lead author.

PART 1

PROPERTIES AND REACTIONS OF CARBENES

1

CARBENE STABILITY

SCOTT GRONERT
Department of Chemistry, Virginia Commonwealth University, Richmond, VA, USA

JAMES R. KEEFFE
Department of Chemistry and Biochemistry, San Francisco State University, San Francisco, CA, USA

RORY A. MORE O'FERRALL
School of Chemistry and Chemical Biology, University College Dublin, Belfield, Dublin, Ireland

1.1 INTRODUCTION

Assessing the stabilities of carbenes has been a richly rewarding, but frustrating endeavor in organic chemistry over the past several decades.[1–9] The investigations have required heroic efforts in the synthesis of precursors, validation of carbene formation, and extraction of thermodynamic parameters. The field is studded with examples of elegant and sophisticated approaches to avoid the complications presented by these often highly reactive intermediates. The pursuit of some carbenes has been so extensive, controversial, and fraught with difficulty, that it offers material possibly better suited for a novel than a scientific review. In this chapter, space limitations do not allow for a full historical review and accounting of all the major contributions that have been made in the study of carbene stability. The review will focus on the latest or most

This chapter is dedicated to the memory of our coauthor, Professor Rory A. More O'Ferrall, who passed away during its composition.

Contemporary Carbene Chemistry, First Edition. Edited by Robert A. Moss and Michael P. Doyle.

accurate values that are available for selected species of fundamental interest (multiple values will be given if a consensus value is not available). As a result, some classical and seminal work may not be directly referenced in this chapter, but instead be found in the cited references. We will begin with a brief discussion of strategies for assessing carbene stability, and the experimental and computational methods that have been used, and then move directly to the stabilities of various classes of carbenes.

1.2 BACKGROUND

1.2.1 Measures of Carbene Stability

Although stability is a familiar concept in all of organic chemistry, describing it precisely can be problematic in practical applications because appropriate, universal reference states cannot always be defined. In some cases, such as carbanions, a single reaction process, protonation in this case, provides a stability measure, proton affinity (PA), which has been broadly embraced by the organic chemistry community. For carbocations, hydride affinities have also been widely used to characterize stability. This has not been true for carbenes, in part because they have varied reaction patterns, but also because they have not been amenable to some physical measurements. As a result, carbene stability has been described in a number of ways, including singlet–triplet energy gaps, reaction energetics such as hydrogenation energies, and kinetic reactivity. Each of these approaches probes a somewhat different aspect of carbene stability and generally correlates with a different reference state. In this chapter, the goal will be to provide a broad overview of these measures of stability, along with a very detailed analysis of the stability of fundamental species on the basis of computed hydrogenation energies.

When choosing a reference state for judging carbene stability, there are generally two considerations that drive the decision process. In the first approach, the decision is driven by a chemical process of interest. This is usually straightforward and can be determined on the basis of relative reaction energies or kinetics. However, this approach is specific to the process and may or may not answer the question of stability in the most general sense. For example, the singlet–triplet gap is a well-defined measure of the relative stability of electronic states of carbenes and offers insight into their spectral properties, but it has only indirect relevance to the ease of formation or the bond-forming reactivity of the carbene.

The second approach is to define a universal reference state that offers no strain or special stabilization. Heats of formation are a simple example of this approach, but they offer limited utility for comparative work because they scale with the molecular formula and therefore direct comparisons are only valid with isomers. A more practical application of this strategy is to relate stability to a model that lacks any of the strain or stabilization that is being probed in the target. These approaches often refer back to alkyl groups and/or employ isodesmic/homodesmotic reactions to extract the stabilization or strain energies.

In this chapter, we will focus on two measures of carbene stability. The primary measure will be hydrogenation energies for the conversion of the carbene center to a tetravalent carbon (Eq. 1.1). The motivations for this choice are: (i) the conversion generally eliminates special orbital interactions that potentially stabilize or destabilize the carbene; (ii) stabilization of singlet and triplet carbenes can be determined separately; and (iii) the reference states are often stable species with well-established thermochemistry.[10]

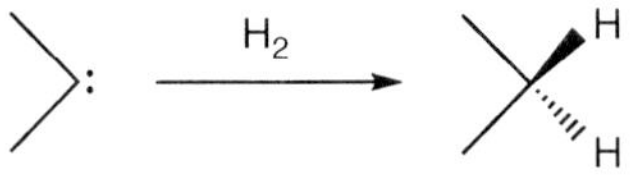

(1.1)

To conveniently express these stabilizations as relative values, Equation (1.1) can be recast as a homodesmotic process with methylene as the standard reference (Eq. 1.2).

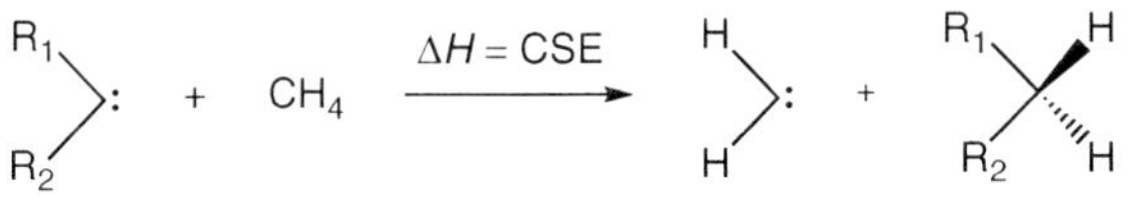

(1.2)

Here, the *carbene stabilization energy* (CSE) is defined as the enthalpy of Equation (1.2) and can be calculated with the carbene species in either the singlet or triplet state (although the product is taken as a singlet so that the reaction violates spin conservation for the triplet). By this definition, a positive CSE value indicates that the substituent provides greater stabilization to the carbene than a hydrogen. This general approach has been taken in the past to assess carbene stability. In 1980, Rondan et al. introduced a similar term based on computed data, referred to as ΔE_{stab}, which gave good correlations with measures of carbene reactivity.[10a] Boehme and Frenking also employed a similar approach when assessing carbene, silylene, and germylene stability with computed data.[10b]

Although CSE values offer insight into both singlet and triplet stabilizations, they do not directly reveal the singlet–triplet gap, ΔE_{ST}, because the former is an ensemble rather than a molecular term. For that reason, we will report both ΔE_{ST} and CSE values in cases where both are available. Some kinetic data will also be presented to construct correlations between carbene stability and kinetic reactivity.

Finally, the chapter will present extensive computational data along with the experimental data. The rationale for this approach is that experimental data for carbene stability are still far from comprehensive, are derived from experiments in a variety of media, and often suffer from significant uncertainties in the data needed to satisfy Equations (1.1) and (1.2). Computational values, despite their inherent and indefinable uncertainties, allow for the construction of structure/stability relationships that offer powerful qualitative, if not quantitative, insights into the structural features that modulate the stabilities of singlet and triplet carbenes.

1.2.2 Bonding and Orbital Interactions in Carbenes

In following chapters, many aspects of carbene structure and reactivity will be explored. Therefore, only a brief overview of critical orbital interactions will be provided here as a foundation for explaining trends in carbene stability. A logical place to begin is to examine the valence orbitals of a carbene in a linear or a bent geometry. The contrast between them gives insight into factors that will control the preferences for singlet and triplet ground states. Qualitative orbital diagrams for a linear and a bent carbene are given in Figure 1.1. The bending leads to rehybridization and a number of effects on several molecular orbitals, but the key changes are in the *p*-orbitals that are nonbonding in the linear geometry (dashed box). The degeneracy in these orbitals is broken when a bend is introduced and the energy gap between the pure *p*-orbital and the one that is rehybridizing (moving towards sp^2-like) grows as the angle at the carbene center is contracted. The carbene carbon has six valence electrons and four will occupy the two, low-energy sigma-bonding orbitals. The remaining two electrons are then distributed into one or both of the orbitals shown as nonbonding in Figure 1.1. In the linear geometry one expects, by analogy to Hund's Rule, that the electrons will occupy each of the degenerate orbitals with a preference for a triplet ground state. In the bent geometry, the situation is not so clear because the gap in energy between the two nonbonding orbitals may be sufficient to overcome the spin-pairing energy and a singlet can become the preferred ground state.

The orbital occupancies for a triplet and a singlet in a bent geometry are outlined in Figure 1.2. The ΔE_{nb} is the important energy term and as noted earlier, it is expected to increase as the angle at the carbenic carbon is reduced. So far, orbital interactions with the X and Y groups on the carbene center (Fig. 1.1) have been ignored. Clearly, these groups can alter ΔE_{nb} by stabilizing or destabilizing the operative orbitals. Much of the discussion in this chapter will focus on the orbital interactions of the groups attached to the carbene center.

Before discussing specific cases in the following sections, it is useful to consider generalizations about factors that affect the preferences for singlet versus triplet ground states. As a starting point, CH_2 (X,Y = H), is a ground-state triplet with a 9 kcal/mol gap between the lowest triplet and singlet states.[11] Consequently, carbene substituents must provide at least 9 kcal/mol of preferential stabilization to the singlet to tip the balance to a singlet ground state. Although it is a crude model, one can view a singlet carbene as possessing the key characteristics of both a carbocation and a carbanion, namely, an empty *p*-orbital and a carbon-centered lone pair. Therefore, X and Y groups that provide orbital interactions that can stabilize a carbocation or carbanion, relative to a radical (the key feature of the triplet orbital occupancy), could potentially stabilize the singlet state. It is well-known that substituent effects are considerably larger for carbocations than carbanions and the stabilization of singlet carbenes by substituents gener-

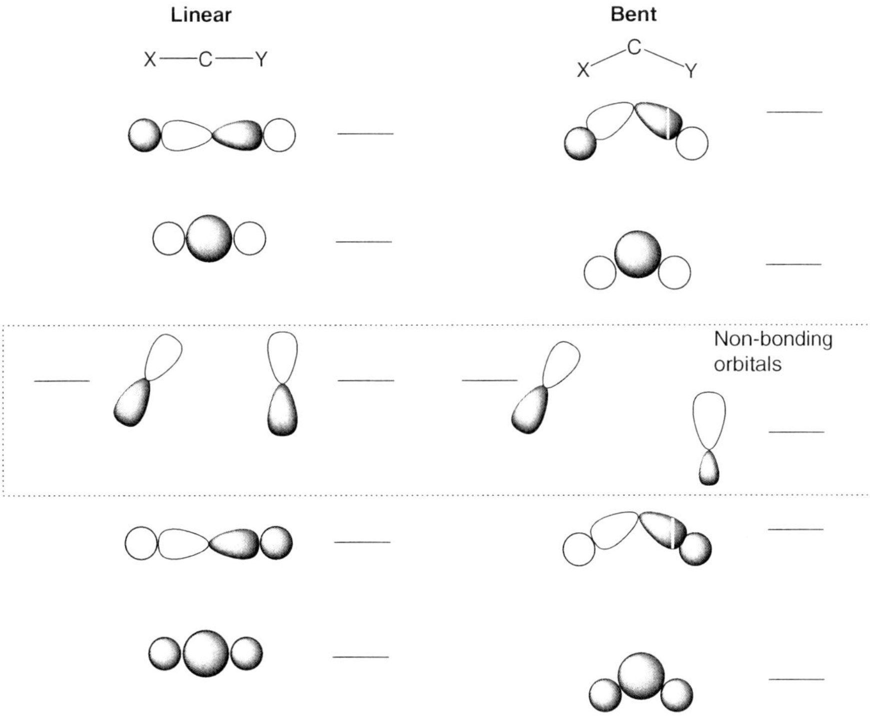

Figure 1.1. Molecular orbitals for linear and bent carbenes.

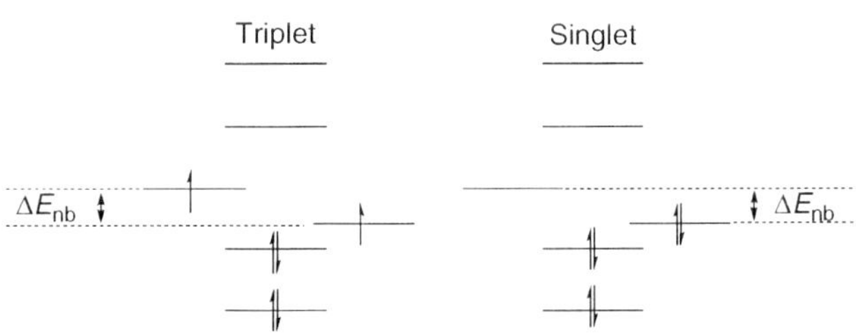

Figure 1.2. Orbital occupancies for triplet and singlet carbenes.

ally parallels trends found for carbocations. With that in mind, the most obvious stabilizing interaction for a singlet carbene would be electron donation to the unoccupied, nonbonding orbital by either conjugation with a lone pair/π-bond or hyperconjugation to a saturated carbon. The situation is illustrated for a fluoro-substituted carbene in Figure 1.3. Interaction of the fluorine lone pair with the unoccupied $2p$ orbital on the carbene stabilizes the overall singlet system by reducing the n_F orbital energy, but the unoccupied molecular orbital is shifted by this interaction to a higher energy than the starting $2p$ orbital, causing a net increase in ΔE_{nb} which, in turn, disfavors the triplet state. In fact, the singlet of

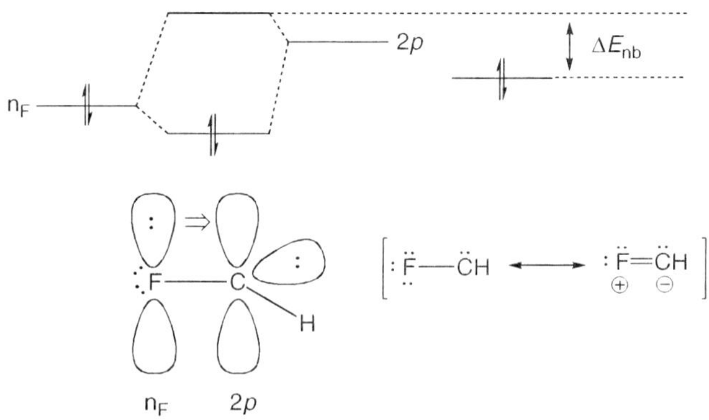

Figure 1.3. Interaction of a fluorine lone pair with a carbene.

FCH is favored by 15 kcal/mol over the triplet,[12] a net shift of 24 kcal/mol in the singlet–triplet gap relative to unsubstituted CH_2. The valence bonding from these orbital interactions can be illustrated by the introduction of an ylide resonance form. The contribution of the ylide resonance depends on the nature of the substituents and when it dominates, the carbene is expected to behave like an ylide and display little of the reactivity that is typically seen in singlet carbenes (i.e., insertion reactions and dimerizations).

In summary, molecular orbital considerations suggest two general factors that should modulate preferences for singlets and triplets. (i) Larger X–C–Y angles will favor the triplet by reducing the hybridization difference and therefore the energy gap between the nonbonding orbitals. (ii) Electron-donating groups favor the singlet by raising the energy of carbene's pure 2*p* orbital. These factors will be discussed in more detail in the sections that follow. Finally, the analogy to carbocations suggests that singlet carbenes can gain significant stabilization via conjugative interactions, possibly diminishing or eliminating their characteristic carbene reactivity.

1.2.3 Determining Carbene Stability

Experimental measures of carbene stability offer several challenges. In many simple examples, such alkyl carbenes, the barrier to rearrangements, often to alkenes via a 1,2 hydrogen shift, may be negligible and therefore carbene lifetimes can be too short for direct analysis even in an inert environment (however, they may be probed via spectroscopy on their corresponding radical anions, see section 1.2.3.1 "Anionic Photoelectron Spectroscopy"). For these short-lived carbenes, the options for characterizing their stability are generally limited to fast kinetic methods or computational modeling. For species that are stable in inert environments or have significant lifetimes, there are other experimental techniques that can provide quantitative stability data. In some cases, such as *N*-heterocylic carbenes, the carbene is indefinitely stable and its properties can be probed by conventional approaches such as pK_a determinations (see section "Other Experimental Measures of Stability" under Section 1.3.4.4). Here, brief descriptions are provided for three approaches that are applicable to the study of relatively reactive carbenes.

1.2.3.1 Anionic Photoelectron Spectroscopy. The most accurate method for determining singlet–triplet gaps has been anionic photoelectron spectroscopy.[13] In this technique, a radical anion precursor to the carbene is formed, thermalized in the gas phase, and then subjected to electron detachment via a laser pulse (Eq. 1.3). The kinetic energy distribution of the resulting electron provides information about the electron-binding energies of the singlet and triplet carbene electronic states, as well as their vibrational states. If thresholds for electron detachment to the singlet and triplet states of the resulting carbene can be identified, the singlet–triplet gap can be determined. Although the spectra can be challenging to assign in some cases, and can be complicated by the presence of vibrational "hot" bands in the precursor ion, the data are capable of providing exceptionally accurate singlet–triplet gaps. This technique has provided the singlet–triplet gap for the prototypical carbene, CH_2. Although powerful, the technique has several limitations. First, there must be a convenient source for producing the carbanion in the gas phase in its vibrational ground state. The electron-binding energy must be within a range that is accessible by common lasers, and the Franck–Condon factors must allow for a confident assignment of the thresholds for detachment to the singlet and triplet. These conditions are met for a number of small, fundamental systems, and data from anionic photoelectron spectroscopy are presented in Table 1.1 and Table 1.3 in the following sections.

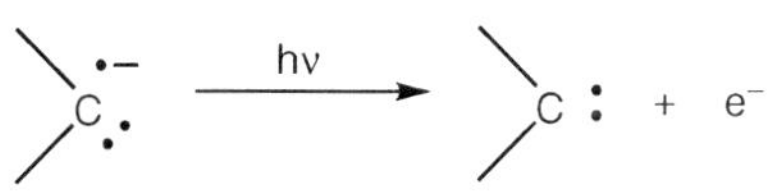

(1.3)

1.2.3.2 Fragmentation Energies. Carbenes can be formed in the gas phase via the fragmentation of an anionic precursor. One approach has been the fragmentation of an α-halo anion via collision-induced dissociation (CID) to produce a halide anion and a carbene (Eq. 1.4).[14] If the heat of formation of the starting anion is known (via its proton affinity and the heat of formation of its conjugate acid), the threshold dissociation energy can be used to calculate the heat of formation of the resulting carbene. Like anionic photoelectron spectroscopy, the technique is limited to ions that can be formed and thermalized. However, it lacks the accuracy of anionic photoelectron spectroscopy because it does not generally offer resolution at the vibrational level, and therefore dissociation measurements rely on difficult estimates of the threshold as well as corrections for kinetic shifts in the threshold. Moreover, the data are only for the electronic state in which the carbene is formed, which is generally assumed to be the ground state.

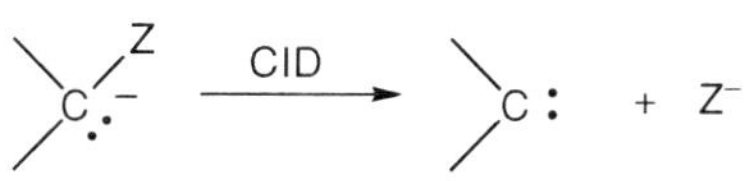

(1.4)

*1.2.3.3 **Addition/Rearrangement Rates**.* Typically, kinetic measures of reactivity only provide qualitative information about carbene thermodynamic stability.[15] Of course, they provide direct measures of kinetic stability with respect to the particular process that is being studied. Some kinetic studies of stability will be summarized later in this chapter. In cases where singlet and triplet states are in rapid equilibrium, addition rates have been used, in conjunction with a number of assumptions, to estimate the equilibrium constant between the singlet and triplet state in solution (Eq. 1.5). The working principle is that only the singlet will undergo the reaction to produce the ylide. This method has been successfully applied in a limited number of cases including phenylcarbene.[16]

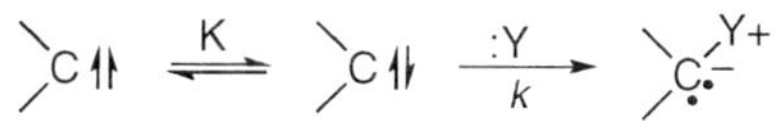

(1.5)

*1.2.3.4 **Computational Modeling**.* Given the instability of many carbenes, as well as their high reactivity, computational modeling has been used extensively in efforts to assess carbene stability. However, carbenes present special challenges to computational methods. First, the species needed to assess the stability have very different electronic properties. The singlet carbene possesses a lone pair and an unoccupied *p*-orbital, the triplet has two unpaired electrons, and the reference compounds are often simple, closed-shell species. Therefore, the chosen computational method must handle all of these situations equally well, whereas the advantageous cancellation of errors that often occurs when similar species are compared computationally is much less likely when addressing carbene stability. Second, the electronic states of carbenes can be close in energy and mixing is possible. Finally, even the assumption that electrons are paired in the singlet is not universally valid (singlet diradicals are possible).

There are three general computational approaches that have been routinely applied to carbenes. Traditional *ab initio* methods with varying levels of electron correlation have been widely used. The second approach is an extension of the first and involves incorporating multiconfiguration wave functions to deal with the mixing of electronic structures. The third is density functional theory (DFT), where a number of functionals are employed including the very popular B3LYP functional. It is safe to say that none of these approaches has been universally accurate and applicable to carbene systems, though very accurate results have been obtained in many cases. Nonethleless, many methods, including B3LYP, cannot accurately characterize the singlet–triplet gap in CH_2 without empirical corrections. For example, the computed singlet–triplet gaps for CH_2 are 12.2 kcal/mol at the B3LYP/6-311++G(*d*,*p*) level,[17] 12.5 kcal/mol at the RCCSD(T)/6-311G(*d*,*p*) level,[17] 9.4 kcal/mol at the G3MP2 level,[18] 12.1 kcal/mol at the CASSCF(8,8)/6-311++G(*d*,*p*) level,[17] and 11.6 kcal/mol at the MRCI + Q(6,6)6-311++G(*d*,*p*) level.[17] The experimental value is 9.0 kcal/mol.[11] The computational demands of these methods vary greatly. Multireference approaches incorporating extensive electron correlation are difficult to extend to moderate-sized carbene systems, whereas DFT calculations on smaller systems can be completed in a few minutes on a personal computer. In this chapter, the main computational data will be from a previous study involving the G3MP2 composite method.[18] This approach should have good general accuracy in systems where a single configuration is adequate. Some data from other computational methods are also included for comparison.

1.3 CARBENE STABILITY

1.3.1 Hydrocarbon-Substituted Carbenes

1.3.1.1 Singlets. In Table 1.1, data are given for representative cyclic and acyclic alkyl-substituted carbenes. In the footnotes to Table 1.1, a sampling of the many other computed values for carbene ΔE_{ST} are listed. As noted earlier, a positive CSE value indicates that the substituent provides greater stabilization to the carbene than hydrogen. A quick survey of Table 1.1 reveals that CSE is positive for all of the singlets—this is not surprising because with the structural duality of an unoccupied *p*-orbital and a lone pair, electron-withdrawing and electron-donating groups have means of stabilizing the singlet carbene. A clear, simple trend is seen in the CSE(singlet) values for the acyclic carbenes, **1–4**. Each additional alkyl group at the carbene center provides about 10 kcal/mol of stabilization relative to the hydrogen substituents in CH_2. A convenient explanation is that the alkyl groups act as electron-donors via hyperconjugation to the unoccupied *p*-orbital on the carbene carbon. The comparison between the methyl (**2**) and dimethyl (**4**) systems suggests that the effect is nearly, but not completely additive. The cyclopropyl-substituted system (**5**) is an interesting case which experiences 10 kcal/mol more stabilization than the analogous isopropyl-substituted system (**3**). This is consistent with the exalted electron-donating ability that is typically seen for the cyclopropyl substituent. For example, the σ_p values for cyclopropyl and isopropyl are −0.21 and −0.15, respectively.[19]

The cyclic carbenes in the table are all analogs of the dimethyl-substituted carbene (**4**). For the four- and five-membered rings, the CSE(singlet) values are reasonably close to that of the acyclic analog, but somewhat higher. In interpreting this result, one must also take into account the reference states that are used in calculating the CSE values, namely the hydrogenation products. For the three-membered to five membered rings, the hydrogenation products are strained and, therefore, the CSE values also contain the changes in strain that accompany hydrogenation. For the four- and five-membered rings, issues with the reference compounds do not appear to be important because there is nothing unusual in the CSE(triplet) values for these species, which share the same reference (see in the following). A likely explanation is that the four- and five-membered ring carbenes benefit from conformational effects that enhance hyperconjugation relative to acyclic systems, and thus a small increase in CSE(singlet) is found. This effect is not seen in the three-membered ring where the CSE(singlet) value is similar to that of dimethylcarbene;

TABLE 1.1. Computed CSE Values, and Computed and Experimental Singlet–Triplet Gaps for Hydrocarbon-Substituted Carbenes[a]

Carbene	CSE (singlet)	CSE (triplet)	ΔE_{ST} (G3MP2)	ΔE_{ST} (exp)
Acyclic				
1 CH_2	0	0	9.4	9.0[11]
2 CH_3CH	12.1	5.6	3.0[b]	
3 $(CH_3)_2CH–CH$	13.2	4.4	0.6	
4 $(CH_3)_2C$	21.0	9.7	−1.0[c]	
5 (cyclopropyl)-CH	23.0	7.1	−6.5[d]	
Cyclic				
6 Cyclopropylidene	21.0	−5.1	−16.8[e]	
7 Cyclobutylidene	26.5	6.7	−10.5[f]	
8 Cyclopentylidene	27.6	8.3	−10.0[g]	
9 Cyclohexylidene	19.8	8.4	−2.1	
Conjugated				
10 $H_2C=CH–CH$	18.0	16.8	8.2[h]	
11 PhCH	18.7	11.4	2.3[i]	2.3[16]
12 Ph_2C	31.7	19.7	−2.6[j]	2.6[24]
13 HC≡C–CH	22.0	24.0	11.4[k]	
Cyclic conjugated				
14 Cyclopropenylidene	69.1	6.8	−52.9[l]	
15 Cyclobut-1-en-3-ylidene	45.6	12.1	−24.1	
16 Cyclopentadienylidene	21.7	17.8	5.5[m]	
17 Fluorenylidene	25.0	13.0	−2.6[n]	$\Delta G = 0.2 - 1$[25]
18 Cycloheptatrienylidene	39.9	27.6	−2.9	

[a]In kcal/mol. Unless noted otherwise, values are from Reference 20. The cycloheptatrienylidene system was estimated based on G3MP2 data for phenylcarbene and a thermodynamic cycle linking phenylcarbene to computed data from Matzinger et al. for cycloheptatrienylidene and phenylcarbene. Their computed ΔE_{ST} value is −2.1. This carbene easily converts to a more stable allenic structure.[26]
[b]Values of 5,[27] 3,[28] and 7.9 (DFT)[29] have also been reported.
[c]Values of −1.4,[30] −1.6,[31] and 0.3 (DFT)[32] have also been reported.
[d]Values of −3.7 and −1.7 (DFT) have been reported.[33]
[e]Values of −12.9,[34] −13.8,[35] and −14.2 (DFT)[21] have been reported.
[f]Values of −5.9 and −4.4 (DFT) have been reported.[35]
[g]Values of −8.7[36] and −10.2 (DFT)[37] have been reported.
[h]A value of 11.7 has been reported.[38]
[i]A value of 2.5 has been reported by two groups,[17,39] and a value of 7.0 (DFT)[32] by another.
[j]A value of −2.9 (DFT) has been reported.[39]
[k]Values of 12.9[40] and 14.1 (DFT)[41] have been reported.
[l]Values of −70[42] and −49.8 (DFT)[43] have been reported.
[m]Values of 6.3[44] and 5.6 (DFT)[32] have been reported.
[n]A value of −2.8 (DFT) has been reported.[32]

however, there is more to this system. The enhanced *s*-character in the carbene lone pair of the three-membered ring should offer some stabilization (it is the factor that enhances the acidity of cyclopropanes), but this must be balanced by the disadvantages of introducing an unoccupied *p*-orbital into a three-membered ring (cyclopropyl cations have unusually high hydride affinities)[20]—the net effect is a CSE(singlet) value near that of dimethylcarbene. As expected, the six-membered ring system gives a hydrogenation product without strain and the CSE(singlet) value is similar to the dimethylcarbene. Not surprisingly, an unsaturated substituent can provide more stabilization to a carbene center than a simple alkyl group. Both vinyl and phenyl substituents provide an extra 6 kcal/mol of stabilization to the singlet relative to a methyl group. The effect appears to be additive and singlet diphenylcarbene (**12**) experiences 10 kcal/mol of added stabilization relative to dimethylcarbene.

The final entries in Table 1.1 are a group of cyclic, conjugated carbenes. These are fascinating species because aromaticity can play important roles in the stabilities of singlet states. In these systems, the singlet carbene could fundamentally complete the π-system of the ring either through its unoccupied *p*-orbital or its lone pair. The smallest in the series is cyclopropenylidene. This species can form a two-electron aromatic π-system if the carbene completes the π-system with its unoccupied *p*-orbital and aligns its lone pair in the plane of the ring. It can be viewed as the deprotonation product of the highly stable cyclopropenylium cation. The CSE(singlet) value for cyclopropenylidene is extremely large for a hydrocarbon system, 69.1 kcal/mol, giving it a value that is over 50 kcal/mol greater than the stabilization from a simple vinyl group. Aromaticity in cyclopropenylidene is also supported by its calculated nucleus-independent chemical shift (NICS). This is a common theoretical approach to evaluating ring current effects and aromaticity. A negative value is associated with aromaticity. Cyclopropenylidene has an NICS value of −16.8.[21] As a reference point, the cyclopentadienyl anion has an NICS of −14.3.[22] Although aromaticity must play a role in the large CSE(singlet) for cyclopropenylidene, it is insufficient to explain a stabilization of this magnitude. The other factor to consider is the carbene lone pair. In cyclopropenylidene, it benefits from the three-membered ring structure, which increases the *s*-character in the lone pair orbital and enhances the effective electronegativity of the orbital. The combination of these effects, aromaticity and high *s*-character in the carbene's lone pair, combine to provide the observed stabilization of the singlet carbene.

Cyclobut-1-en-3-ylidene (**15**) has a much larger CSE(singlet) value than vinylcarbene, but not as large as cyclopropenylidene. This system can potentially benefit from homoaromaticity via a cross ring interaction, which appears to be the case given the exalted CSE(singlet) value. With the next largest carbene in this series, cyclopentadienylidene, the electron need of the cyclic π-system is reversed—to form an aromatic π-system, the carbene must donate two electrons to it. Here, the CSE(singlet) value is only 21.7 kcal/mol, a value that is only about 4 kcal/mol greater than that of vinylcarbene. Clearly, singlet cyclopentadienylidene does not experience any special stabilization and, in fact, the cyclic structure appears to reduce the stabilization by the π-system. Computational modeling, including NICS analysis,[23] confirms that cyclopentadienylidene does not possess an aromatic π-system; the carbene completes the π-system with its unoccupied *p*-orbital, leading to an antiaromatic π-system. This electronic structure can be rationalized by considering the cost of completing the π-system with the lone pair. It would require the carbene to have an unoccupied sp^2-hybrid orbital in the plane of the ring, which is energetically very unfavorable (analogous to a vinylic cation) and apparently outweighs the benefits of an aromatic π-system.

Fluorenylidene (**17**) provides a closely related π framework, but its more extended π-system adds only marginal stability to the singlet and a CSE(singlet) value of 25 kcal/mol is obtained. Comparison of this value with the one obtained for diphenylcarbene gives insight into the impact of antiaromaticity in fluorenylidene. By forming the five-membered ring (fluorenylidene can be viewed as diphenylcarbene bridged through the ortho carbons), the CSE(singlet) drops by nearly 7 kcal/mol. Although a number of other factors are at play, this simple comparison highlights the fact that cyclopentadienyl-based carbenes suffer destabilization in the singlet state due to antiaromaticity in their π-systems.

The last entry in Table 1.1 is cycloheptatrienylidene (**18**). Like cyclopropenylidene, this carbene can form an aromatic π-system by incorporating its unoccupied orbital in the π-system. In this case, a six-electron, aromatic π-system results and cycloheptatrienylidene has a CSE(singlet) value of 39.9 kcal/mol. Although not as exceptional as that of cyclopropenylidene, this CSE(singlet) is unusually large and significantly exceeds that of diphenylcarbene. Of course in cycloheptatrienylidene, the lone pair does not have the benefit of occupying an orbital with exalted *s*-character, a key feature in stabilizing singlet cyclopropenylidene.

1.3.1.2 Triplets. The CSE(triplet) values are much smaller in magnitude and the variation in the acyclic systems is limited. This is not surprising given that the magnitude of hyperconjugative stabilization possible for interactions with a singly occupied orbital is modest at best. Analogies could be made to alkyl-substituted radicals, where alkyl substituents reduce C–H bond strengths by only a few kcal/mol (currently, the magnitude of hyperconjucative stabilization in radicals is a controversial topic[45–51]). The only system that stands out is the cyclopropylidene system (**6**), where CSE(triplet) is negative, indicating that this triplet is less stable than CH_2 with respect to the relevant reference species. As noted earlier, the triplet prefers large X–C–Y angles and should be destabilized by the small angle in the cyclopropylidene system. The effect is very significant and amounts to almost 15 kcal/mol of destabilization relative to the acyclic analog, dimethylcarbene. The impact of an unsaturated substituent on triplet stability is significant, particularly for a vinyl group. Relative to methyl, a phenyl substituent provides over 5 kcal/mol of stabilization to the triplet and a vinyl substituent provides about 11 kcal/mol, qualitatively resembling benzyl and allyl radicals in this respect. Again, the effect is additive and diphenylcarbene experiences about twice the stabilization of its triplet compared with phenylcarbene.

Cyclic, conjugated systems offer a wider range of CSE(triplet) values. Cyclopropenylidene has a value similar to that of methylcarbene, but a more useful comparison is to vinylcarbene. From this perspective, triplet cyclopropenylidene suffers 10 kcal/mol of destabilization due to its cyclic structure. The reason for the destabilization is likely the same as in cyclopropylidene, and the magnitude is roughly similar. In the cyclopentadienyl-based carbenes, there is significant stabilization of the triplet. As in the acyclic systems, vinyl groups (e.g., cyclopentadienylidene, **16**) provide more stabilization than phenyl groups (e.g., fluorenylidene, **17**). Finally, the largest of the cyclic π-systems, cycloheptatrienylidene, provides exceptional stabilization to the triplet and yields a CSE(triplet) value of over 25 kcal/mol. This is only about 12 kcal/mol smaller than the CSE(singlet) for this carbene, which as noted earlier, experiences some degree of aromatic stabilization. Clearly, extensive delocalization is a powerful stabilizing force in the triplet.

1.3.1.3 Singlet–Triplet Gap. In the parent CH_2 system, there is a 9 kcal/mol preference for the triplet. Alkyl substituent groups preferentially stabilize the singlet because hyperconjugation has a greater impact on an unoccupied orbital (singlet) than on a singly occupied one (triplet). Therefore, the addition of alkyl groups to the carbene center reduces the singlet–triplet gap and dimethylcarbene is a ground-state singlet. In strained ring systems, the preference for the singlet is more extreme and is greater than 10 kcal/mol. In fact, the singlet–triplet gap for the three-membered ring system is dominated by the severe destabilization of the triplet. In the four- and five-membered rings, the changes in the singlet–triplet gap are dominated by the stabilization of the singlet carbenes. This comparison points out the value of having independent measures of singlet and triplet stability rather than just the difference in their stability, ΔE_{ST}. The six-membered ring system has an ΔE_{ST} value similar to the acyclic analog, **4**, and mirrors it in each of the measures in Table 1.1.

Unsaturated substituents provide stabilization to both the singlet and triplet, so the impact on the singlet–triplet gap is moderate. Phenyl and vinyl carbenes are ground-state triplets, with the latter having a triplet preference nearly as large as CH_2. The computed singlet–triplet gap for phenylcarbene is the same as that reported in solution. The computations suggest that diphenylcarbene is a ground-state singlet, but experiments indicate a small preference for the triplet—in acetonitrile it is favored by 2.6 kcal/mol.[24] Of the cyclic, conjugated systems, cyclopropenylidene stands out, and its singlet is over 50 kcal/mol more stable than its triplet. As noted earlier, this is driven by aromatic stabilization of the singlet and geometric destabilization of the triplet. The computed value, −52.9 kcal/mol, is close to the experimental ΔE_{ST} for this carbene, −53.8 kcal/mol. For the rest of the cyclic, conjugated carbenes, the computed singlet–triplet gaps are small and although a singlet ground state is suggested for fluorenylidene, experimental data indicate that it is a ground-state triplet by a small margin.[25]

1.3.1.4 Other Experimental Measures of Stability. Much interest has been focused on methylene, the parent carbene, and its singlet–triplet gap. After some controversy, it is now established experimentally as 9 kcal/mol.[11] The G3MP2 value listed in Table 1.1 is in good agreement with this value. The triplet's heat of formation has been determined to be 93 kcal/mol[52] and this leads to a heat of hydrogenation of −111 kcal/mol (Eq. 1.2). For comparison, the G3MP2 calculations indicate a heat of hydrogenation of −109 kcal/mol for the triplet. In addition, the experimental data suggest a heat of formation of 102 kcal/mol for the singlet state. Reliable thermochemical data are not available for most alkyl-substituted carbenes. This is partly because the corresponding anions cannot be formed in the gas phase, so that anionic photoelectron spectroscopy is not an option for probing relative spin-state energies. Values are available in the Russian literature for methylcarbene (triplet) and cyclopropylidene (singlet). Takhistov[53] has reported heats of formation of 88 and 95 kcal/mol, respectively. These values imply a CSE(triplet) of 1 kcal/mol for methylcarbene and a CSE(singlet) value of 23 kcal/mol for cyclopropylidene. Data are available for vinylcarbene and phenylcarbene from the gas-phase dissociation of α-halo carbanions (i.e., step-wise α-elimination).[54] Heat of formation values of 93 kcal/mol for vinylcarbene and 103 kcal/mol for phenylcarbene were reported for what are assumed to be triplet ground states. These correspond to CSE(triplet) values of 21 kcal/mol for vinylcarbene and 18 kcal/mol for phenylcarbene. The phenylcarbene value is significantly higher than the value listed in Table 1.1, and other computational data suggest that the experimental heat of formation may be too low by 8 kcal/mol.[17] Clauberg and Chen have reported a heat of formation of 114 kcal/mol for cyclopropenylidene.[55] This implies a CSE(singlet) of 70 kcal/mol, which is very close to the calculated value. From the value for cyclopropenylidene, they have used computational data to estimate a heat of formation for singlet HC≡C–CH of 140 kcal/mol.[56] This gives a CSE(singlet) value of 22 kcal/mol.

1.3.2 Substituted Phenylcarbenes

1.3.2.1 Singlets. In Table 1.2, data are given for a series of phenylcarbenes. These compounds provide a good illustration of the subtler impacts of electron-donating and electron-withdrawing groups on carbene stability. In the singlets, the trend in stabilization energy is clearly linked to electron-donating ability. The nitro-substituted phenylcarbenes have the lowest CSE(singlet) values, whereas the amino-substitution, particularly at the paraposition, gives the highest CSE(singlet) values for the neutral substituents. However, the impact is attenuated by the π-system and is only moderate. For example, methylation at the carbene center, $PhCCH_3$ (**19**), provides nearly as much stabilization as a para-amino group, and considerably more than any of the other ring substituents in Table 1.2, with the exception of an oxygen anion (carbene **30**). For **30**, the structure can also be viewed as a deprotonated quinone methide and this resonance form should dominate; see Scheme 1.1. This explains the large

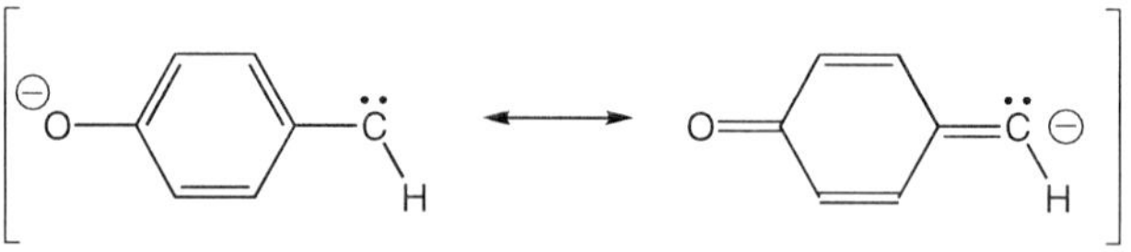

Scheme 1.1.

TABLE 1.2. Computed CSE Values and Singlet–Triplet Gaps for Substituted Phenylcarbenes[a]

Carbene	CSE(singlet)	CSE(triplet)	ΔE_{ST} (G3MP2)
11 PhCH	18.7	11.4	2.3
19 $PhCCH_3$	24.9	16.3	1.0
20 3-O_2N–PhCH	15.7	9.7	3.6
21 4-O_2N–PhCH	15.1	10.7	5.1
22 3-NC-PhCH	16.0	9.4	3.0
23 4-NC-PhCH	15.7	10.5	4.4
24 3-F-PhCH	17.4	10.5	2.7
25 4-F-PhCH	19.9	10.5	0.2
26 3-HO-PhCH	19.0	11.3	1.9
27 4-HO-PhCH	19.6	11.5	−1.3
28 3-H_2N–PhCH	19.2	11.3	1.7
29 4-H_2N–PhCH	25.6	12.1	−3.5
30 4-(O^-)-PhCH	53.6	19.9	−24.1

[a]Data are from Reference 20.

CSE(singlet) value and highlights the need to consider alternate resonance forms when assessing the stabilization of carbenes.

1.3.2.2 Triplets. The effect of ring substituents on the stabilities of the triplets is even subtler. From 3-nitro to 4-amino, the variation in the CSE(triplet) values is only from 9.7 to 12.1 kcal/mol. Once again we see what little impact substituents generally have on triplet carbene stability. Here, the triplet is generally stabilized by electron-donating groups and destabilized by electron-withdrawing groups, but the pattern is not perfectly consistent.

1.3.2.3 Singlet–Triplet Gap. Phenylcarbene has a small preference for a triplet ground state (2.3 kcal/mol), which can be overcome by the effects of ring substituents. The impact of substituents is greatest in the singlet and factors related to the singlet dominate the singlet–triplet gap. With a strong-electron donor, 4-amino, the singlet is favored by over 3 kcal/mol. With a 4-nitro group, the triplet phenylcarbene is favored by over 5 kcal/mol. However, for most of the substituted phenylcarbenes, the singlet–triplet gap is computed to be small so that these systems offer potential to tune the singlet–triplet gap, tipping the balance between singlet and triplet ground states.

1.3.2.4 Hammett Correlations with Phenylcarbenes. A summary of the data in Table 1.2 is presented in Figure 1.4 as a Hammett plot. Each of the terms is plotted separately. The fits are only moderately good, but give ρ values of −3.8, −0.7, and 2.9 for CSE(singlet), CSE(triplet), and ΔE_{ST}, respectively. To put these values in context, a comparison can be made to related carbenium ions

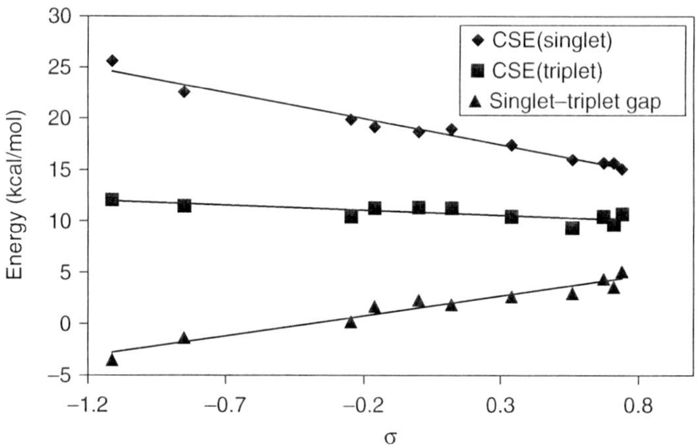

Figure 1.4. Correlation of carbene properties with σ parameters. For para, σ^+ is used and for meta, σ_m is used.

(protonated carbenes). For the analogous set of benzyl cations, a ρ value of about −20 is found for the correlation to hydride ion affinity, which is a general measure of carbocation stability and shares the same reference state as CSE values. It is not surprising that the carbenium ions are much more sensitive to substituent effects than the carbenes—they experience greater electron deficiency at the hypovalent carbon and utilize the aromatic ring as a means of delocalizing charge.

1.3.2.5 Other Measures of Stability. This series also allows for an exploration of how electronic factors impact the kinetic stability of singlet carbenes. Singlet phenylcarbene is known to readily rearrange to bicyclo[4.1.0]hepta-2,4,6-triene (Scheme 1.2). The barriers to this process have been computed with DFT for many substituted phenylcarbenes, including the set in Table 1.2.[57]

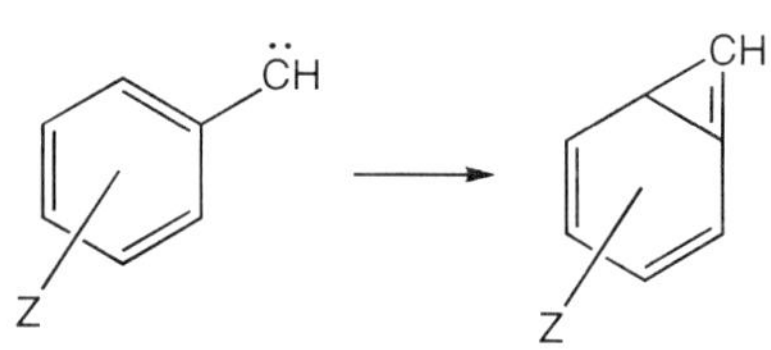

Scheme 1.2.

The assumption here is that the more stabilization provided by the substituent, the larger the barrier to the rearrangement process. The data are plotted in Figure 1.5. In this case, two distinct correlations are seen, one for *para*- and one for *meta*-substituted phenylcarbenes. For the *para*-substituents, there is a significant impact on the reaction barrier, with electron-donating groups enhancing the barrier. This is consistent with the idea that carbene stability

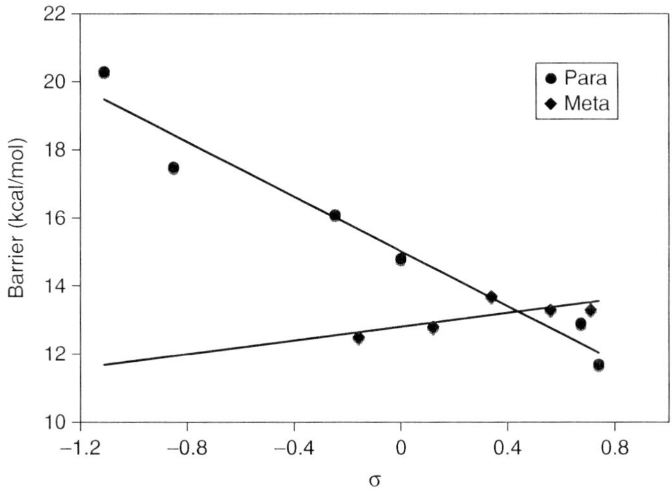

Figure 1.5. Hammett correlation of computed rearrangement barriers with σ parameters for substituted phenylcarbenes. For para, σ^+ is used and for meta, σ_m is used. Data from Hadad and Geise.[57]

controls the rearrangement rate. The ρ value is −3.0 for the *para*-substituents, nearly as large as the ρ value obtained for the correlation of CSE(singlet) with σ^+ (−3.8). This suggests that most of the substituent's stabilizing effect is lost in the transition state, and therefore there is close to a 1:1 relationship (75% to be more exact) between stabilization of the *para*-substituted carbene and its kinetic stability toward rearrangement. In this case, carbene stability, as measured by the CSE(singlet), is a good proxy for the corresponding kinetic stability of the carbene. *Meta*-substituted phenylcarbenes present a different picture. Here, the ρ value is moderately positive (0.7), which indicates that electron-donating groups reduce the barrier to rearrangement. This is in contrast to the correlation with CSE(singlet), where the *meta*-substituents followed a similar trend line as the *para*-substituents (the slope in Figure 1.4 is only modestly altered if the *para*-substituents are removed). The shift in slope found for meta substituents in Figure 1.4 and Figure 1.5 indicates that, in the meta position, electron-donating substituents provide more stabilization to the rearrangement transition state than to the starting carbene. This is a curious result, but it can be rationalized. If we assume that the rearrangement mechanism involves the carbene center in an electrophilic attack at an ortho carbon, then the *meta*-substituent becomes ortho or para to the site of attack on the ring, and can have a significant stabilizing effect on the transition state (a situation similar to an electrophilic aromatic substitution transition state). This may be viewed as a special case; the data for the *para*-substituted phenylcarbenes illustrate the more general correlation between the thermodynamic stability of carbenes and their kinetic stability with respect to rearrangement reactions.

1.3.3 Heteroatom-Substituted Carbenes

1.3.3.1 Singlets. In Table 1.3, data are given for carbenes containing a heteroatom as part of the substituent, and directly attached to the carbene center. Again, a sampling of alternative computational values for ΔE_{ST} is given in the table notes. The first group of entries in the table is for N, O, and F substituents. Each of these elements can stabilize the carbene by donation of a lone pair to give an ylide resonance form (Fig. 1.3). Given the electronegativities and the known electron-donating capabilities, one expects that singlet stabilization should fall in the order N > O > F. This is borne out by the data. The CSE(singlet) values drop from 58.1 to 46.3 to 28.5 kcal/mol for NH_2, OH, and F as the substituents. Alkylation of the heteroatom (e.g., OCH_3 vs. OH) has little impact on the CSE(singlet value). Adding a second heteroatom provides a large amount of additional stabilization, and the CSE(singlet) value for diaminocarbene (**32**) reaches over 80 kcal/mol. CSE(singlet) values in this range suggest very strong π-bonding with the substituent and a substantial/dominant contribution from an ylide resonance form. This is discussed in more detail in Section 1.3.5. Finally, bis(trifluorylmethyl)carbene (**39**) is interesting in that the strong electron-withdrawing groups have only a small impact on the singlet stability relative to CH_2. There must be a rough balance between the CF_3 effect on the carbon lone pair (stabilizing) and on its unoccupied *p* orbital (destabilizing), which is slightly tilted toward the latter.

Table 1.3 also lists several period III elements as substituent groups. The pattern here is more complicated than with the period II elements. Along the series SiH_3, PH_2, SH, and Cl, the CSE(singlet) values are 0.7, 34.2, 42.7, and 24.4 kcal/mol, respectively. The SiH_3 group provides much less stabilization than a methyl group, and is actually similar to a hydrogen in its impact on the carbene. It stands out in this series because it lacks a lone pair and cannot conjugate with the singlet carbene's unoccupied orbital. In addition, Si–H bonds are not very effective in hyperconjugation. All of the other period III elements provide substantial stabilization. The factors that make PH_2 a weaker stabilizing group than SH are not very clear, but gas-phase hydride affinity data for carbocations suggest that sulfur is better able to stabilize adjacent carbocations, presumably by electron donation to the unoccupied *p*-orbital. The computational data indicate other factors are also in play. For example, there is significant elongation of the S–H and P–H bonds if they are anti to the carbene lone pair, indicating negative hyperconjugation to the carbon lone pair and enhanced bonding in the C–S and C–P linkages.[18] There is only one example given in Table 1.3, but it appears that the impact of 3rd row elements is additive. Thus, dichlorocarbene has a CSE(singlet) value that is nearly 20 kcal/mol greater than that of chlorocarbene. Although the cyano group does not link to the carbene through a heteroatom, it has been included here because it is often grouped with the other substituents of Table 1.3. Its impact on the CSE(singlet) value is almost the same as a methyl group, though the mode of stabilization must be much different (i.e., hyperconjugation will not be operative).

TABLE 1.3. Computed CSE Values, and Computed and Experimental Singlet–Triplet Gaps for Heteroatom-Substituted Carbenes

Carbene	CSE(singlet)	CSE(triplet)	ΔE_{ST} (G3MP2)	ΔE_{ST} (exp)
31 H_2NCH	58.1	14.2	−34.3[a]	
32 $(NH_2)_2C$	82.9	16.1	−57.1[b]	
33 HOCH (anti)	46.3	11.2	−25.4[c]	
34 CH_3OCH	47.9	10.4	−27.8	
35 CH_3OCCH_3	53.5	11.4	−32.6	
36 $(CH_3O)_2C$	77.3	8.1	−59.8[d]	
37 FCH	28.5	4.5	−14.6	−14.9[12]
38 F_2C	58.7	−6.9	−56.2	−54[58]
39 F_3CCCF_3	−4.7	−5.7	8.4[e]	
40 SiH_3CH	0.7	9.4	18.1[f]	
41 $(SiH_3)_2C$	4.4[31]	17.0[31]	22.1[g]	
42 H_2PCH	34.2	11.4	−13.3[h]	
43 HSCH (anti)	42.7	12.7	−20.4[i]	
44 CH_3SCH	48.4	14.1	−24.8	
45 ClCH	24.4	8.1	−6.9	−4.3[12]
46 Cl_2C	43.7	12.5	−21.7	−20.8[59]
47 NCCH	11.5	12.0	10.0	11.1[38]
48 $(NC)_2C$	23.6	25.9	12.2	12[60]
49 H_2NCCN	65.3	23.5	−31.9[j]	
50 H_2NCNO_2	60.1	9.2	−41.5	

Unless noted otherwise, values are from Reference 20.
[a]Values of −28.8[61] and −31.5 (DFT)[62] have been reported.
[b]Values of −51.8[21] and −52.1 (DFT)[62] have been reported.
[c]Values of −25.3[63] and −25.7 (DFT)[29] have been reported.
[d]A value of −54.0 (DFT)[32] has been reported.
[e]A value of 11.2 has been reported.[8]
[f]Values of 19.1[64] and 23.3 (DFT)[9] have been reported.
[g]Values of 19.6[64] and 24.7 (DFT)[8] have been reported.
[h]Values of −7.9[65] and −6.6 (DFT)[62] have been reported.
[i]Values of −18.2[66] and −13.8 (DFT)[62] have been reported.
[j]Values of −31.2 and −30.3 (DFT) have been reported.[67]

Here, the cyano group will be a poor donor to the carbene's unoccupied *p*-orbital, but could act as an acceptor of its lone pair. The net effect is modest apparently, resulting in a CSE(singlet) close to that of a simple alkyl group. The last two entries in the table are carbenes in which an electron-donating (NH_2) and an electron-withdrawing group (CN or NO_2) are both attached to the carbene center. The "push–pull" substituents might be expected to provide unusual stability to the carbene because they address each of its reactive elements, the unoccupied *p*-orbital via the electron-donor and the carbon lone pair via the electron acceptor. In each of these cases, the carbene has a large CSE(singlet), but it is only marginally larger than the carbene with only one NH_2 substituent, and much less than the CSE(singlet) resulting from diamino substitution on the carbene. The impact is not additive—the sum of the CSE(singlet) values for an NH_2 and a CN group is 69.6 kcal/mol, but H_2NCCN

has a CSE(singlet) value of only 65.3 kcal/mol. The limited effectiveness of the "push–pull" stabilization of carbenes can be rationalized on the basis of the hybridization of the carbene center. The cross-conjugation needed for the "push–pull" interaction requires sp hybridization at the carbene center, which typically is unfavorable. In fact the carbenic angle in singlet aminocyanocarbene is only 110.8°. Overall, the conclusion is that singlet carbene stability is much more sensitive to groups that stabilize the unoccupied *p*-orbital; hence the exalted stability of carbenes substituted with two electron-donating groups, rather than an electron-donating and an electron-withdrawing group.

1.3.3.2 Triplets. As with the hydrocarbon substituents, heteroatom substituent effects are also less pronounced for the triplet carbenes. For the period II elements, the CSE(triplet) values range from 14.2 kcal/mol for nitrogen to 4.5 kcal/mol for fluorine (about the same as a methyl group, 5.6 kcal/mol). The pattern is generally consistent with expectations about the ability of these elements to stabilize an unpaired electron on an adjacent atom (i.e., patterns in bond dissociation energies).[68] The effect of adding multiple period II heteroatoms to the carbene is notable, particularly with fluorine. Here the CSE(triplet) is negative, indicating that the fluorines are destabilizing relative to the hydrogens in the reference CH_2 species. This effect is not observed in the monofluoro species (it has a positive CSE(triplet) value), so it is a consequence of the double substitution. Systems with F–C–F components are known to prefer small bond angles (Bent's Rule), whereas triplet carbenes prefer wide X–C–Y bond angles. This forces the difluorocarbene to compromise between these contradictory geometric preferences. The reference species needed for Equation (1.2), CH_4 and CF_2H_2, do not need to make such compromises and, therefore, triplet difluorocarbene is characterized as a destabilized carbene. This effect is also seen to a lesser extent for dimethoxycarbene, where the second CH_3O group reduces the CSE(triplet) value.

For bis(trifluoromethyl)carbene (**39**), the triplet also experiences a small destabilization relative to the parent CH_2 system. The period III elements give regular and moderate stabilizations to the triplets, with CSE(triplet) values ranging from 9.4 to 12.7 kcal/mol of stabilization for single substitutions. In contrast to the fluorine system, there is no penalty for a second chlorine on the carbene, but the positive impact is small. Chlorine is much less electronegative than fluorine and the energetic impact of Bent's Rule is expected to be less important. The cyano group provides considerable stabilization to the triplet and this is additive, leading to a CSE(triplet) of over 25 kcal/mol for dicyanocarbene (**48**). The cyano group is known to be an excellent stabilizer of adjacent radical centers. In the "push–pull" systems, the impact of the substituents on triplet stability appears to be somewhat less than additive. Summing the effects of a NH_2 and a CN group suggests a combined CSE(triplet) value of 26.2 kcal/mol, but the H_2NCCN system has a CSE(triplet) value of 23.5 kcal/mol.

1.3.3.3 Singlet–Triplet Gap. Heteroatom substituents tend to have significant CSE(singlet) values and moderate to low CSE(triplet) values. The net effect is a strong shift toward the singlet state. Of all the carbenes listed in Table 1.3, only the carbenes with CN, CF_3, and SiH_3 substituents are computed to be ground-state triplets—the rest are singlets. Many have large singlet–triplet gaps with preferences for the singlet reaching nearly 60 kcal/mol for systems doubly substituted with nitrogen, oxygen, or fluorine groups. Difluorocarbene stands out in this set because its large preference for a singlet is mainly driven by the unusual instability of its triplet rather than the stability of its singlet. In the period III, the silyl system prefers a triplet ground state by a large margin. Here it is a large stabilization of the triplet combined with limited stabilization of the singlet that leads to the large magnitude of the triplet preference. These examples point out again the advantage of using CSE values rather than singlet–triplet gaps alone. The other period III elements provide substantial stabilization to the singlet and push the systems to be ground-state singlets. Finally, the cyano and CF_3 groups are the sole substituents in Table 1.3 that provide virtually equal stabilization to the singlet and triplet states. As a result, their carbenes have singlet–triplet gaps that are very close to the parent system, CH_2. The CF_3 group is also unusual in that it has only small impacts on the singlet and triplet stabilities, so that despite its strong electron-withdrawing ability, it is a relatively benign carbene substituent. With the push–pull systems, H_2NCCN has a large, negative singlet–triplet gap, but its magnitude is smaller than that of H_2NCH because the CN group stabilizes the triplet. The opposite is true for H_2NCNO_2 which has a larger singlet–triplet gap than H_2NCH, albeit by a small amount, because the NO_2 group destabilizes triplets (see in the following).

1.3.3.4 Other Experimental Measures of Stability. A number of the species in Table 1.3 have been the subject of experimental studies and data are available for them. The halogen-containing carbenes are well-known species and accurate, experimental singlet–triplet gaps are listed in Table 1.3. Heats of formation have been reported for FCH (30 kcal/mol), CF_2 (−43.5 kcal/mol), ClCH (80 kcal/mol), and CCl_2 (57 kcal/mol).[69] These data lead to CSE(singlet) values of 31, 55.5, 20, and 40 kcal/mol, respectively. Pau and Hehre[70] have found that HCOH is 54.2 kcal/mol less stable than formaldehyde, giving it an estimated heat of formation of 28 kcal/mol (CSE(singlet) = 43 kcal/mol). Liu et al.[71] give CH_3COH a heat of formation of 16 kcal/mol, which suggests CSE(singlet) = 48 kcal/mol. Du et al.[72] give a heat of formation of −55 kcal/mol for $(CH_3O)_2C$ (CSE(singlet) = 92 kcal/mol). The experimental value of CSE(singlet) for $(CH_3O)_2C$ is about 15 kcal/mol above the computed value and also exceeds the computed CSE(singlet) of $(H_2N)_2C$. This is the only system where the experimental value differs so significantly from the computed value. The calculation of the CSE(singlet) of $(CH_3O)_2C$ from experimental data involves the heats of formation of $(CH_3O)_2CH_2$ and $(CH_3O)_3CH$, and they could be the root of the discrepancy. In any case, some caution should be exercised with the experimental heat of formation of $(CH_3O)_2C$ because it would be unlikely for it to have a CSE(singlet) greater than that of $(H_2N)_2C$. Finally, the heat of formation of HCCN has been reported by Poutsma, Upshaw, Squires, and Wenthold to be 116 kcal/mol, giving a CSE(triplet) of 13 kcal/mol.[38] In all these cases (except $(CH_3O)_2C$), the experimentally derived CSE values are in good accord with the computed values, which lends confidence to the other computed CSE values presented here.

1.3.4 Conjugated Heterocyclic Carbenes

1.3.4.1 Singlets. In these systems, cyclic conjugation and therefore aromatic properties are possible. Data are given in Table 1.4 and include saturated analogs for comparison (Scheme 1.3). In each of these systems, the carbene center is flanked by atoms with lone pairs and strong stabilization of the singlet carbene is anticipated. The heteroatom systems containing nitrogen fall into an important class of metal ligands that have been referred to as *N*-heterocyclic carbenes (NHCs).[73–76] They offer unique properties, in part, because they can act as σ-donors and π-acceptors through the singlet carbene's lone pair and formally unoccupied orbital, respectively. The strong stabilization of the NHC systems is due to a large contribution from an ylide resonance form. Based on their stability and reactivity, some have argued that they are not true carbenes and should be viewed as ylides instead. This will be discussed in more detail in Section 1.3.6.

51 X = NH, Y = NH
53 X = N–CH_3, Y = N–CH_3
55 X = O, Y = O
57 X = NH, Y = O
58 X = PH, Y = PH
60 X = S, Y = S
62 X = N–CH_3, Y = S

52 X = NH, Y = NH
54 X = N–CH_3, Y = N–CH_3
56 X = O, Y = O
59 X = PH, Y = PH
61 X = S, Y = S

Scheme 1.3.

Imidazol-2-ylidene (**51**) and *N*,*N*'-dimethylimidazol-2-ylidene (**53**) are prototypical NHCs. The CSE(singlet) values exceed 100 kcal/mol, an energy that is well beyond the expected contribution from a fully formed π-bond. For comparison, saturated analogs, such 4,5-dihydroimidazol-2-ylidene (**52**) and diaminocarbene (**32**), have CSE(singlet) values of about 80 kcal/mol. This suggests that the presence of the cyclic π-system affords the carbene an additional 20 kcal/mol of stabilization. In the corresponding oxygen system, 1,3-dioxol-2-ylidene (**55**), the CSE(singlet) value is 84.4 kcal/mol, which is only slightly higher than the saturated analog, 1,3-dioxolan-2-ylidene (**56**). Here, there is high stabilization, but the cyclic conjugation plays a much smaller role. The mixed oxygen and nitrogen species, oxazol-2-ylidene (**57**) gives intermediate

TABLE 1.4. Computed CSE Values and Singlet–Triplet Gaps for Conjugated Heterocyclic Carbenes

Carbene	CSE(singlet)	CSE(triplet)	ΔE_{ST} (G3MP2)
51 Imidazol-2-ylidene	103.1	8	−85.5
52 4,5-Dihydroimidazol-2-ylidene	81.9	2.3	−72.3
53 *N*,*N*′-Dimethylimidazol-2-ylidene	103.2	6.2	−87.5
54 4,5-Dihydro-*N*,*N*′-dimethylimidazol-2-ylidene	101.4	8.4	−83.5
55 1,3-Dioxol-2-ylidene	84.4	−11.6	−86.1
56 1,3-Dioxolan-2-ylidene	80.0	−2.1	−72.5
57 Oxazol-2-ylidene	91.8	−4.4	−86.3
58 1,3-Diphosphol-2-ylidene	48.1	12.6	−26.2
59 1,3-Diphospholan-2-ylidene	43.2	11.8	−21.1
60 1,3-Dithiol-2-ylidene	82.5	16.2	−56.5
61 1,3-Thiolan-2-ylidene	71.2	14.6	−47
62 *N*-methylthiazol-2-ylidene[a]	92.7	−1.6	−84.5

Unless noted otherwise, values are from Reference 20.
[a]Value from present work.

behavior. As is the case in the simple period III systems (Table 1.3), sulfur provides much more stabilization than phosphorous in the cyclic systems. In addition, the effect of cyclic conjugation is more important in the sulfur system and the saturated analog of 1,3-dithiol-2-ylidene (**61**) has a CSE(singlet) value that is over 10 kcal/mol smaller than the unsaturated system (**60**). The mixed nitrogen/sulfur heterocycle, *N*-methylthiazol-2-ylidene (**62**), has an intermediate CSE(singlet) value that is close to that of the nitrogen/oxygen heterocycle, oxazol-2-ylidene (**57**). For this group of carbenes, the impact of cyclic conjugation on the CSE(singlet) values follows the order N > S > O, P for the various heteroatoms (i.e., the difference in CSE(singlet) between unsaturated and saturated cyclic systems). The location of oxygen in this series is somewhat surprising and may be related to the unsaturated system forcing three adjacent lone pairs (oxygen, carbene, oxygen) into the same plane. Also, the π-donation requires formal positive charges on the heteroatoms, which will be more destabilizing for oxygen than the other atoms.

The most obvious explanation for the added stabilization of the singlet in the cyclic, conjugated systems is aromaticity (Scheme 1.4). The C–C π-bond and the two heteroatom lone pairs combine with the unoccupied *p*-orbital on the carbene carbon to give an aromatic, 5-atom, 6-electron ring system. From the ylide resonance form in Scheme 1.4, it is easy to see the analogy to related aromatic systems such as imidazole and pyrrole. By the CSE(singlet) measure,

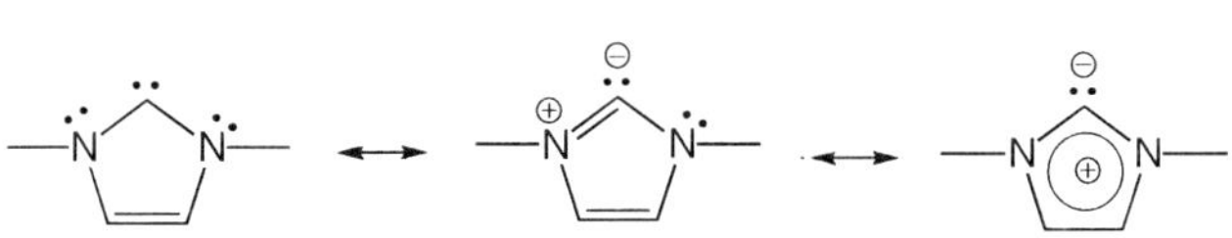

Scheme 1.4.

the added advantage of aromaticity appears to be relatively modest in the NHCs (up to 20 kcal/mol), but charge separation in systems of this type can limit the full manifestation of aromaticity. However, the NICS value for **51** (−14.7)[18] is near that of the cyclopentadienyl anion (−14.3).[22]

1.3.4.2 Triplets. The situation of the triplets in conjugated heterocyclic carbenes is very different. With the exception of the sulfur and phosphorous heterocycles, the CSE(triplet) values tend to be small, and in some cases, negative. Useful comparisons can be made to acyclic analogs. Good comparison systems are $(H_2N)_2C$ and $(CH_3O)_2C$, with CSE(triplet) values of 16.1 and 8.1 kcal/mol, respectively. In the nitrogen systems, imidazol-2-ylidene (**51**) and its partially saturated analog (**52**), the CSE(triplet) values are only 8.0 and 2.3 kcal/mol, respectively. The corresponding oxygen systems, **55** and **56**, have CSE(triplet) values of −11.6 and −2.1 kcal/mol, respectively. These comparisons suggest that the heterocyclic structures destabilize the triplets by 10–20 kcal/mol. However, the destabilization of the triplet carbene is not strongly linked with cyclic conjugation because in the nitrogen system, destabilization (relative to the acyclic analog) is greatest in the saturated species, while the opposite is true in the oxygen system.

Clues to the destabilization can be seen in the geometries of the triplet heterocyclic carbenes. Instead of adopting planar or nearly planar ring structures like the singlets, the triplets give puckered conformations with pyramidalization at carbon and nitrogen, even in the fully unsaturated system (**51**). These data suggest that the rings are forcing the triplet carbene into unfavorable orbital interactions with the adjacent heteroatom. When the heteroatom is from the period III, these interactions do not seem to be significant and routine CSE(triplet) values are seen for the sulfur and phosphorous systems, though the mixed sulfur/nitrogen system has a slightly negative CSE(triplet) value.

1.3.4.3 Singlet–Triplet Gap. Given the strong stabilization of singlets in some systems, and the generally weak stabilization of triplets, the conjugated heterocycles produce enormous singlet–triplet gaps that reach nearly 90 kcal/mol. The *N*- and *O*-heterocycles are clearly locked into singlet states and their chemistry is unlikely to be influenced by the triplet state. This is another factor that separates these species from conventional carbenes. The phosphorous heterocycles are the only systems that have singlet–triplet gaps less than 30 kcal/mol and therefore have electronic states that are comparatively close in energy.

1.3.4.4 Other Experimental Measures of Stability. Species in Table 1.4 can in some cases be formed by the deprotonation of stable cations (i.e., they can be viewed as ylides as well as carbenes, see Section 1.3.5 for more discussion of this duality, as well as Chapter 3). As a result, the pK_a of the cationic precursor offers insight into the stability of the resulting carbene. Richard and Amyes have provided data for the imidazolium precursors to **51** and **53**. They report pK_a values of 23.8 and 23.0, respectively, in water.[77] The most comprehensive set of measurements come from O'Donoghue[78] and Chu et al.[79] Their work also suggests pK_a's of about 23 for 1,3-dialkylimidazolium ions, although bulky alkyl groups push the pK_a's up to 25. An interesting result from their work is that unsaturation of the ring has little impact on the pK_a. 1,3-Bis(2,4,6-trimethylphenyl) imidazolium and its 4,5-dihydro analog, have p*K*a's of 20.8 and 21.3, respectively, indicating that cyclic conjugation has little effect on the free energy of deprotonation. This suggests that the aromatic stabilization seen in species such as **51** and **53** is equal to that in the imidazolium precursor. Related species have also been studied in the gas phase. A recent report by Lee and coworkers[80] suggests that values in earlier work may have been in error.[81,82] The latest value for the gas-phase proton affinity of **53** is 259 kcal/mol. This value is consistent with the computed proton affinities (PA) for this species.[20,80] For context, this places **53** near the top of the gas-phase basicity scale (for neutral species) and it is slightly more basic than the superbase, ETBD (7-ethyl-1,5,7-triazabicyclo[4.4.0] dec-5-ene).[83]

1.3.5 Carbenes with Other Electron-Withdrawing Groups

When the carbene bears an electron-withdrawing group with lone pairs on atoms other than the point of connection to the carbene, the singlets are prone to facile rearrangements which appear to have small or zero barriers in the gas phase. Examples include carbonyl groups, nitro groups, and sulfonyl groups. This is not surprising given the electron-deficiency of the carbene, and analogous rearrangements are seen in carbocations. Computational studies suggest two different modes of rearrangement for the singlet carbenes.[20c] In the first, the atom with the lone pair coordinates with the carbene center to produce a cyclic, bridged structure. In this case, the interaction can be viewed as another mechanism for a substituent to stabilize a carbene. However, if the bridging process is accompanied by a bond cleavage, then a complete rearrangement is achieved and the carbene center is converted into another functional group. A generalized pathway is shown in Scheme 1.5. In the case of intramolecular

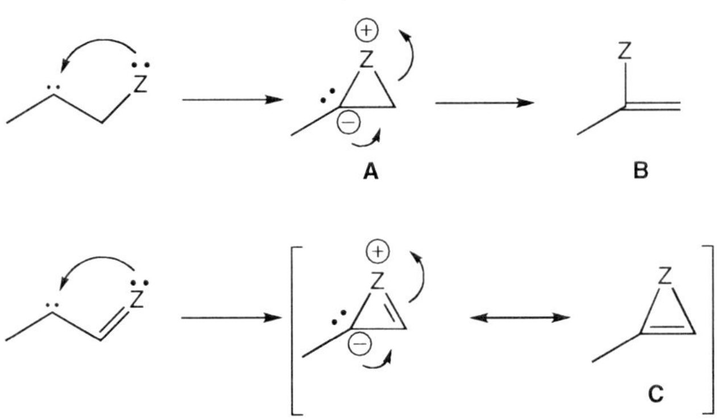

Scheme 1.5.

TABLE 1.5. Computed CSE Values and Singlet–Triplet Gaps for Carbenes with Other Electron-Withdrawing Groups

Carbene	CSE(singlet)	CSE(triplet)	ΔE_{ST} (G3MP2)
63 HCC(O)H	[a]	7.7	[a]
64 HCC(O)CH_3	18.1	7.0	−1.9
65 O_2NCH	[a]	0.7	[a]
66 $(O_2N)_2C$	25.0	−4.1	−20.0
67 $HCSOCH_3$	43.2	6.2	−27.4
68 $HCS(O_2)CH_3$	[a]	0.2	[a]
69 $HCS(O_2)CF_3$	6.2	−1.6	1.6

Values are from Reference 20.
[a]Singlet is unstable and rearranges.

ylide formation (A in Scheme 1.5), one might still view the system as a carbene, albeit highly altered by the mode of stabilization, and a CSE(singlet) as well as a ΔE_{ST} could be calculated. If the neighboring group contains a double bond, the ylide also has an unsaturated, heterocyclic resonance form (**C**). Again one can calculate a CSE(singlet) value, but it hardly reflects the intrinsic carbene stability. Finally, if rearrangement occurs to give a new species where there are no important resonance forms with carbene character (B in Scheme 1.5), then it is not possible to assess carbene stability in the absence of applying arbitrary geometric constraints in a computational chemistry approach.

Formylcarbene (**63**) is calculated to be unstable as a singlet and rearranges to ketene without a barrier. The triplet gives a stationary point computationally and the CSE(triplet) value is small, which suggests little conjugative stabilization by the carbonyl group (Table 1.5). Acetylcarbene (**64**) does not rearrange to a ketene, but gives a bridged oxirene structure (C in Scheme 1.5). The calculated CSE(singlet) is modest despite the bridging (18.1 kcal/mol) and makes the acetyl group about as stabilizing as a vinyl or phenyl group. It should be pointed out that the CSE(singlet) values computed for the bridging species should represent upper bounds on the potential stabilization of a true carbene structure—bridging would only occur if it were exothermic and enhanced the apparent CSE(singlet) value. The triplet has a conventional structure and like formylcarbene, experiences modest stabilization from the substituent. The net effect is a computed ΔE_{ST} that is close to zero.

Nitrocarbene (**65**) is also computed to be unstable as a singlet and rearranges to nitrosoformaldehyde by a pathway akin to **A** ⇒ **B** in Scheme 1.5.[84] The triplet does not rearrange and has a very small CSE value of 0.7 kcal/mol. Computations of singlet dinitrocarbene (**66**) indicate that it will cyclize, but not fragment. The CSE(singlet) associated with the cyclization product is 25 kcal/mol. The triplet is unusual in that the CSE(triplet) value is negative. The only other example of a significantly negative CSE(triplet) in the data reviewed here is CF_2. Like CF_2, dinitrocarbene is subject to Bent's rule and has a rather small N–C–N angle for a triplet (123°). The sulfoxide-substituted carbene (**67**) does not cyclize, but it has strong π-donation from the sulfur with a very short S–CH bond (1.560 Å vs. 1.807 Å for DMSO). The CSE(singlet) value is slightly smaller than that of CH_3SCH, indicating that the oxygen limits some of the π-donation to the carbene center. A methylsulfone-substituted carbene (**68**) rearranges as a singlet via oxygen transfer to give a sulfur ester of the carbene, $HCOS(O)CH_3$. However, a trifluoromethylsulfone-substituted carbene (**69**) gives a stable singlet. The data indicate that, like a CF_3 group, the sulfone functional group has a limited impact on the singlet and triplet.

1.3.6 Carbenes versus Ylides

The most stable of the carbenes are species where an ylide resonance form is expected to be a major contributor. Here the unoccupied orbital of the carbene is engaged in a strong π-bond with an adjacent heteroatom. Although a carbene resonance form can be drawn, are these species best described as carbenes? There are two obvious factors that might be considered in answering this question. First is the electron distribution consistent with a carbene—is there evidence of significant hypovalency at the carbenic carbon? Second, does the species undergo reactions that are characteristic of carbenes such as insertions and dimerizations? For the species to be viewed as a carbene, it seems reasonable that at least one of these criteria must be met.

Let us consider the prototypical NHC, *N,N′*-dimethylimidazol-2-ylidene (**53**) as an example. The CSE(singlet) value for this carbene is in excess of 100 kcal/mol and its singlet–triplet gap is more than 80 kcal/mol. As noted earlier, this CSE(singlet) value is greater than typical π-bond strengths. To put the singlet–triplet gap in perspective, it is greater than the singlet–triplet gap in ethene ($\pi \Rightarrow \pi^*$ transition).[85] Each of these observations suggests a very strong π-bond to the carbenic carbon and likely full valency at that carbon. This is supported by the aromaticity of this species (based on energetic and ring-current data, section "Singlets" under Section 1.3.3.1). This outcome is not a surprise. This carbene is derived from the deprotonation of the *N,N′*-dimethylimidazolium cation, and this species is not characterized as a stabilized carbocation (i.e., containing a hypovalent carbon). Therefore, bonding perspectives and their underlying electronic foundations do not support a carbene resonance form. The NHC species also generally do not undergo typical carbene reactions (i.e., rapid insertions and dimerizations) so they mostly fail on both factors that might characterize them as carbenes. Their novel behavior, particularly as metal ligands, highlights the fact that they have unusual properties, but NHCs would have been better labeled as ylides, or more specifically, vinylic ylides than as carbenes.

So what is a reasonable cutoff for distinguishing between a true carbene and an ylide. For the present purposes, a distinction based on the stabilization of the carbene will be proposed. A key factor in carbene reactivity is the relatively small spacing between the singlet and triplet states. This is true because if the carbenic carbon is truly hypovalent, promotion of the lone pair electron into the "unoccupied" orbital should be relatively facile or (even) favorable.

When the singlet–triplet gap becomes large, it suggests strong bonding to the "unoccupied" orbital and potentially the absence of carbene character. As an arbitrary, but reasonable cutoff, one could propose a singlet–triplet gap of 65 kcal/mol, the approximate strength of a C–C π-bond. The assumption is that conversion to the triplet disrupts the π-bonding to the carbene and therefore the singlet–triplet gap gives a rough measure of π-bonding in the carbene. This is a purely arbitrary cutoff, but it does cleanly separate the NHCs from all the other carbenes considered in this chapter.

1.4 CORRELATIONS INVOLVING CARBENE STABILITY

1.4.1 CSE versus Singlet–Triplet Gap

Because it can be obtained directly from spectroscopic measurements, the singlet–triplet gap has often been invoked in discussions of carbene stability. One aspect of the singlet–triplet gap is that it embraces factors that impact the stability of each spin state. Therefore, it can be difficult to rationalize substituent effects because it is not possible link their impact to a particular spin state. To illustrate this challenge, the computed singlet–triplet gaps for the species in Table 1.1, Table 1.3, and Table 1.4 have been plotted against their computed CSE(singlet) and CSE(triplet) values in Figure 1.6 and Figure 1.7. When plotted against CSE(singlet), one sees a reasonable correlation between the two properties of the carbene. In contrast, the correlation of the singlet–triplet gap with CSE(triplet) gives essentially a scatter plot with no discernible pattern. Another evident difference in the two plots is the very different ranges of values for CSE(singlet) and CSE(triplet). The former has values from 0 to 100 kcal/mol whereas the latter is limited to the range from about −12 to 26 kcal/mol.

The much wider range for the singlet is expected. The singlet, like a carbocation, has a vacant *p*-orbital that can directly conjugate with substituents bearing lone pairs or π-bonds, or hyperconjugate with alkyl substituents. Carbocations are very sensitive to substituent effects and the same is expected of singlet carbenes. A good correlation has been demonstrated between CSE(singlet) values, cast in terms of hydrogenation energies, and the hydride affinities of analogous carbocations (protonated carbenes).[20] With its unpaired electrons, the triplet carbene is much more akin to a radical and an analogy to bond dissociation energies would suggest that CSE(triplet) values should be much less sensitive to substituent effects. The net result is that the singlet–triplet gap is generally, but not always, dominated by variations in the singlet and can therefore be a proxy for singlet stability. However, one must take care in using this interpretation of singlet–triplet gaps. For example, cyclopropylidene (**6**) and difluorocarbene (**38**) have large singlet–triplet gaps that are driven to a great extent by their unusually low (negative) CSE(triplet) values rather than high CSE(singlet) values. These species are outliers in Figure 1.6, but the effect is obscured to some extent by the large scale of the plot.

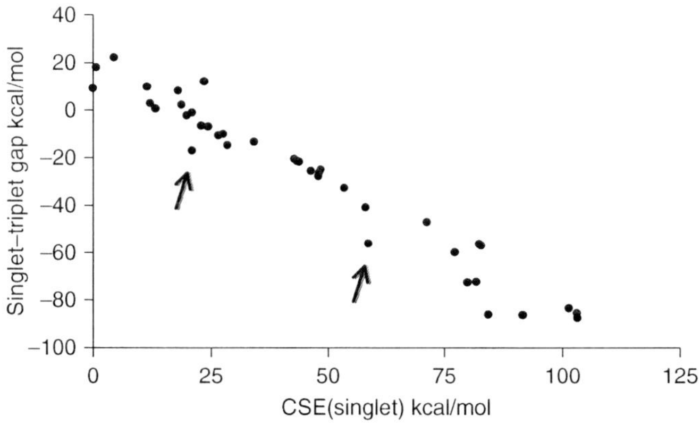

Figure 1.6. Plot of singlet–triplet gap versus CSE(singlet) for species in Table 1.1, Table 1.3, and Table 1.4. Arrows indicate cyclopropylidene and difluorocarbene, respectively.

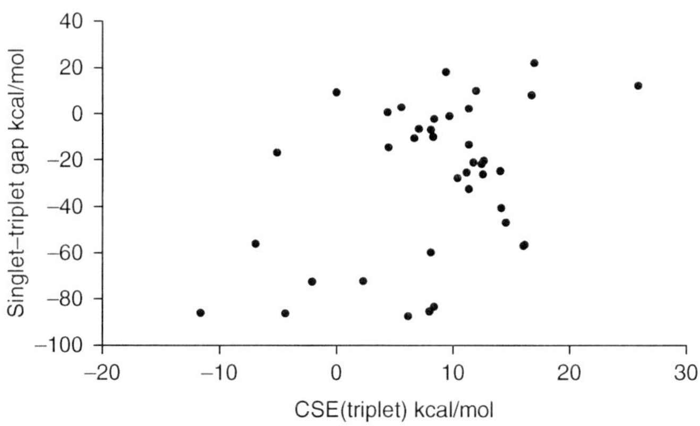

Figure 1.7. Plot of singlet–triplet gap versus CSE(triplet) for species in Table 1.1, Table 1.3, and Table 1.4.

1.4.2 CSE(Singlet) versus Proton Affinities

A possible reaction process for singlet carbenes is protonation, which results in the formation of a carbocation. This process eliminates the carbon lone pair, but retains the empty *p*-orbital. As a result, strong correlations between CSE(singlet) and PA reveal features of carbene stability related to stabilization of the empty *p*-orbital, whereas deviations from this correlation highlight factors that stabilize the carbene's lone pair. A plot based on computed data for a few fundamental species is shown in Figure 1.8.[20] On the plot, a 1:1 correlation line has been included. There is a rough correlation between the CSE(singlet) and PA, but many species are well below or above the 1:1 correlation line. For species below the line, the substituent(s) provide more relative stabilization to the singlet carbene than its protonation product. These include all the heteroatom-substituted carbenes. Conjugating hydrocarbon substituent

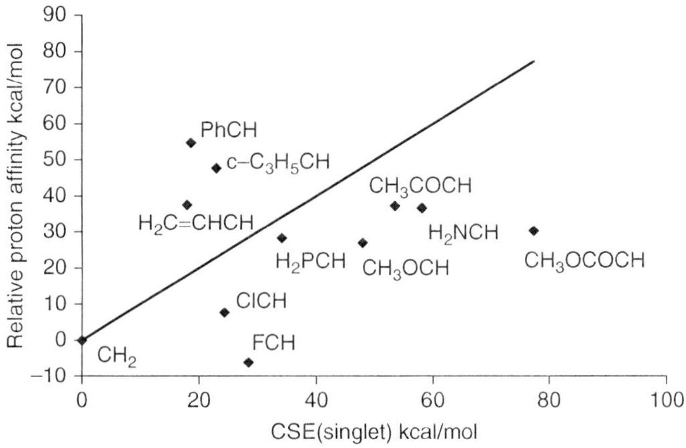

Figure 1.8. Plot of relative gas-phase proton affinity versus CSE(singlet) for selected species.[20]

groups are found above the line; these substituents provide more relative stabilization to the carbocation (carbene protonation product). One naturally expects that atoms or groups with higher electronegativity will be weaker in their stabilization of a carbocation, but the effect here is striking. Although fluorocarbene, chlorocarbene, and cyclopropylcarbene have quite similar CSE(singlet) values, their PA's vary by about by 60 kcal/mol. This result highlights one of the special features of carbenes: the ability of fluorine to stabilize a singlet carbene and enhance the singlet–triplet gap is conceptually linked to its ability to act as a π-donor. However, the data in Figure 1.8 indicate that the donating ability is unique to the carbene and likely driven by a push–pull interaction (π-donation vs. σ-withdrawal) that is facilitated by the carbene's lone pair. Another interesting comparison involves CH_3OCH and $(CH_3O)_2C$. The addition of the second methoxy group has little impact on the PA, but a large impact on the CSE. This again points to the unusual electronic structure of carbenes that allows for potent electron-donation to an empty *p*-orbital by electronegative substituents without paying the price for introducing a σ electron-withdrawing group at a center with a positive charge. Linear-free energy fits to related data indicate that CSE(singlet) values are considerably more sensitive to resonance effects than are the PA's of the carbenes.[20]

1.4.3 CSE(Singlet) versus Carbene Reaction Rates

In the previous sections, the stability of carbenes has been assessed exclusively in thermodynamic or spectroscopic terms. Kinetic stability is also a critical factor when considering reactive intermediates such as carbenes. Several subsequent chapters in the book will focus on the reactivity of carbenes, so here only representative bimolecular reactivity of a few fundamental species will be presented and contrasted with the other measures of stability presented previously.

1.4.3.1 Reaction Rates with Pyridine. The reaction of a carbene with pyridine to give an ylide has been a common method of trapping carbenes as well as assessing their intrinsic reactivity. Data are available for a number of simple carbenes and the range of rate constants extends over several orders of magnitude. For the least stabilized carbenes, the rates are at or near the diffusion-controlled limit (10^9–$10^{10}\,M^{-1}s^{-1}$) and offer little insight into the relative reactivity of carbenes. The majority of the data in Table 1.6 are from Moss et al., and the carbenes in those studies were formed by laser flash photolysis of diazirenes.[86] The first two entries in Table 1.6 illustrate the impact of electron-withdrawing groups on the ability of an oxygen substituent to stabilize a carbene. The added CF_3 group decreases the stabilization from the carbene's alkoxy substituent and increases the rate of reaction with pyridine by a factor of 20. A similar effect is seen in the comparison of stabilization by a methoxy versus an acetoxy group (entries 3 and 4). Here the inclusion of the carbonyl in the substituent enhances the reactivity with pyridine by a factor of over 200. Modest stabilization is possible from bridging in the acetoxycarbene.[20c] With just phenyl, halide, or alkyl substituents, the rates rise into the $10^9\,M^{-1}\,s^{-1}$ range.

1.4.3.2 Reaction Rates with Alkenes. There are also some data available on the kinetic reactivity of stabilized carbenes with alkenes. These reactions can also have small barriers and high rate constants, but a wide range of reactivity is observed depending on the stabilization of the carbene. The data in Table 1.7 illustrate some general trends. CF_2 and CCl_2 are very reactive with the alkenes, with the former giving somewhat slower rates (by about a factor of 5), suggesting that CF_2 is more stabilized by its substituents. This is consistent with the respective CSE(singlet) values for these carbenes (58.7 and 43.7 kcal/mol), although the difference in kinetic reactivity is much smaller than the difference

TABLE 1.6. Rate Constants for the Reactions of Carbenes with Pyridine in Hydrocarbon Solvents

Carbene	k ($M^{-1}s^{-1}$)
CH_3COCH_3	6.6×10^5
$CH_3COCH_2CF_3$	1.6×10^7
$PhCOCH_3$	1.2×10^5
$PhCOC(O)CH_3$	3.5×10^7
PhCF	1.5×10^9
PhCCl	1.5×10^9
PhCBr	1.2×10^9
CBr_2[a]	5.0×10^9
CFCl[b]	8.0×10^9
Adamantyl[c]	5.0×10^9

Data from Reference 86 unless noted otherwise.
[a]Cyclohexane data from Reference 87.
[b]Reference 88.
[c]Reference 89.

TABLE 1.7. Rate Constants for the Reactions of Carbenes with Alkenes in Hydrocarbon Solvents

Carbene	k ($M^{-1}s^{-1}$)		
	$(CH_3)_2C{=}C(CH_3)_2$	c-C_6H_{10}	$CH_2{=}CClCN$
CF_2[a]	6.4×10^8	1.4×10^7	
CCl_2[a]	4.7×10^9	6.4×10^7	8.1×10^6
$C(OCH_3)_2$[a]			5.0×10^5
CH_3OCCl[b]			5.6×10^5
CH_3COCH_3[c]			4.9×10^7
PhCCl[d]	2.8×10^8		2.1×10^8
PhCF[d]	1.6×10^8		1.2×10^8
$PhCOCH_3$[c,e]			3.4×10^7
C_6F_5CF[d]	1.8×10^9		

[a]Reference 90.
[b]Reference 91.
[c]Reference 92.
[d]Reference 93.
[e]Reference 94.

in CSE(singlet) would suggest. A comparison of CCl_2 and $C(OCH_3)_2$ is possible, and the latter is 15 times less reactive with α-chloroacrylonitrile. Again this is consistent with the respective CSE(singlet) values of 43.7 and 77.3 kcal/mol.

The relationship between CSE(singlet) and relative reactivity with alkenes is not universal and CH_3COCH_3 provides a simple counterexample. The CSE(singlet) for CH_3COCH_3 is 53.5 kcal/mol, which places it between CCl_2 (43.7 kcal/mol) and $C(OCH_3)_2$ (77.3 kcal/mol). However, it is more reactive than CCl_2 with α-chloroacrylonitrile. The failure of the correlation in this case illustrates an interesting aspect of carbene reactivity. If we make the same comparison, but use 2-methyl-2-butene as the alkene, a very different result is seen. The rate with CCl_2 is $2.5 \times 10^9\ M^{-1}s^{-1}$, whereas the rate with CH_3COCH_3 is $4.8 \times 10^3 M^{-1}\ s^{-1}$—this is over a million-fold shift in relative reaction rates. Clearly, rates of addition to alkenes are driven by more than the stability of the carbene as measured by terms such as the CSE(singlet). The key factor here is the nature of the carbene reaction with the alkene. Because the carbene has both electrophilic and nucleophilic characteristics, its mode of attack with an alkene is linked to the nature of the alkene. With α-chloroacrylonitrile, the alkene is electron-poor and it is prone to nucleophilic attack by the carbene. It consequently experiences enhanced reactivity with electron-rich, nucleophilic carbenes such as CH_3COCH_3. In contrast, 2-methyl-2-butene is an electron-rich alkene and the reactivity swings in favor of the more electron-poor carbene, CCl_2. This effect has been discussed in great detail in the past and is a well-known and tested aspect of carbene reactivity.[3]

The relevance in the present context is that this effect highlights the danger of attempting to link thermodynamic properties to kinetic data. The reactions with pyridine (previous section) focus only on the electrophilic character of the carbenes, and the data in Table 1.6 are consistent with the CSE(singlet) values. When comparing similar carbenes (like the first examples of this section), the alkene addition kinetics seem to reflect the intrinsic stability. For example, PhCCl is slightly more reactive than PhCF with α-chloroacrylonitrile, which is consistent with the observation that fluorine is more effective than chlorine in stabilizing carbenes, as measured by CSE(singlet) values. The extension of this series to phenylmethoxycarbene leads to the expected drop in reactivity with the introduction of the more powerful carbene stabilizing group. As expected, the greater electrophilicity of pentafluorophenylfluorocarbene, relative to phenylfluorocarbene, is manifested in reactions with an electron-rich alkene (e.g., tetramethylethene).

1.5 CONCLUSION AND OUTLOOK

The future offers many new opportunities in the study of carbene stability. Experimentally, new ionization sources for mass spectrometers will allow the formation of more complex precursor anions for the study of carbene singlet–triplet gaps via anion photoelectron spectroscopy. Analysis of the data from these experiments will be aided with advances in computational modeling, particularly of vibrational spectra. New fast kinetic methods allow for the identification of short-lived carbenes in solution and the ability to probe their properties, including their stability. Recent equilibrium measurements by Moss highlight some of these new capabilities,[95] but there is also mounting evidence that singlet carbenes interact strongly with solvent, potentially giving weakly bound ylides, and this can affect interpretations of their stability in solution.[96–98] Computational chemistry has become a real workhorse in the study of carbene structure, reactivity, and stability. Advances in computer technology have allowed larger and larger carbene systems to be studied with the most advanced theoretical methods. However, carbenes still offer significant challenges to computational modeling; high accuracy for both singlet and triplet states has not been possible in the general case. Even with the highest levels of theory, empirical corrections are still required to match some experimental singlet–triplet gaps. As a result, the development of methods, possibly focused on DFT, that could provide a general solution to evenly addressing both spin systems would be a significant advance in the field.

SUGGESTED READING

Bertrand, G. "Stable Singlet Carbenes." In *Reactive Intermediate Chemistry*. Moss, R. A., Platz, M. S., Jones, M., Jr., Eds. Wiley-Interscience: New York, **2004**.

Bourissou, D., Guerret, O., Gabbai, F. P., Bertrand, G. "Stable Carbenes," *Chem. Rev.* **2000**, *100*, 39.

Gronert, S., Keeffe, J. R., O'Ferrall, R. A. M. "Stabilities of Carbenes: Independent Measures for Singlets and Triplets," *J. Am. Chem. Soc.* **2011**, *133*, 3381.

Hirai, K., Itoh, T., Tomioka, H. "Persistent Triplet Carbenes," *Chem. Rev.* **2009**, *109*, 3275.

Jones, M., Jr., Moss, R. A. "Singlet Carbenes." In *Reactive Intermediate Chemistry*. Moss, R. A., Platz, M. S., Jones, M., Jr., Eds. Wiley-Interscience: New York, **2004**.

Nefedov, O. M., Egorov, M. P., Ioffe, A. I., Menchikov, L. G., Zuev, P. S., Minkin, V. I., Simkin, B. Y., Glukhovtsev, M. N. "Critical Compilation of Physical Properties of Short-Lived Intermediates – Carbenes and Carbene Analogues," *Pure Appl. Chem.* **1992**, *64*, 265.

Tomioka, H. "Triplet Carbenes." In *Reactive Intermediate Chemistry*. Moss, R. A., Platz, M. S., Jones, M., Jr., Eds. Wiley-Interscience: New York, **2004**.

Vignolle, J., Cattoen, X., Bourissou, D. "Stable Noncyclic Singlet Carbenes," *Chem. Rev.* **2009**, *109*, 3333.

REFERENCES

1. Baird, N. C., Taylor, K. F. *J. Am. Chem. Soc.* **1978**, *100*, 1333.
2. Mueller, P. H., Rondan, N. G., Houk, K. N., Harrison, J. F., Hooper, D., Willen, B. H., Liebman, J. F. *J. Am. Chem. Soc.* **1981**, *103*, 5049.
3. Jones, M., Jr., Moss, R. A. "Singlet Carbenes." In *Reactive Intermediate Chemistry*. Moss, R. A., Platz, M. S., Jones, M., Jr., Eds. Wiley-Interscience: New York, **2004**.
4. Bertrand, G. "Stable Singlet Carbenes." In *Reactive Intermediate Chemistry*. Moss, R. A., Platz, M. S., Jones, M., Jr., Eds. Wiley-Interscience: New York, **2004**.
5. Tomioka, H. "Triplet Carbenes." In *Reactive Intermediate Chemistry*. Moss, R. A., Platz, M. S., Jones, M., Jr., Eds. Wiley-Interscience: New York, **2004**.
6. Hirai, K., Itoh, T., Tomioka, H. "Persistent Triplet Carbenes," *Chem. Rev.* **2009**, *109*, 3275.
7. Vignolle, J., Cattoen, X., Bourissou, D. "Stable Noncyclic Singlet Carbenes," *Chem. Rev.* **2009**, *109*, 3333.
8. Nemirowski, A., Schreiner, P. R. *J. Org. Chem.* **2007**, *72*, 9533.
9. Mendez, F., Garcia-Garibay, M. A. *J. Org. Chem.* **1999**, *64*, 7061.
10. (a) Rondan, N. G., Houk, K. N., Moss, R. A. *J. Am. Chem. Soc.* **1980**, *102*, 1770. (b) Boehme, C., Frenking, G. *J. Am. Chem. Soc.* **1996**, *118*, 2039.
11. Leopold, D. G., Murray, K. K., Miller, A. E. S., Lineberger, W. C. *J. Chem. Phys.* **1985**, *83*, 4849.
12. Gilles, M. K., Ervin, K. M., Ho, J., Lineberger, W. C. *J. Phys. Chem.* **1992**, *96*, 1130.
13. Wenthold, P. G., Lineberger, W. C. *Acc. Chem. Res.* **1999**, *32*, 597.
14. Paulino, J. A., Squires, R. R. *J. Am. Chem. Soc.* **1991**, *113*, 1845.
15. Jackson, J. E., Soundararajan, N., Platz, M. S., Liu, M. T. H. *J. Am. Chem. Soc.* **1988**, *110*, 5595.
16. Admasu, A., Gudmundsdottir, A. D., Platz, M. S. *J. Phys. Chem. A* **1997**, *101*, 3832.
17. Nguyen, T. L., Kim, G. S., Mebel, A. M., Nguyen, M. T. *Chem. Phys. Lett.* **2001**, *349*, 571.
18. Gronert, S., Keeffe, J. R., O'Ferrall, R. A. M. *J. Am. Chem. Soc.* **2011**, *133*, 3381.
19. Hansch, C., Leo, A., Taft, R. W. *Chem. Rev.* **1991**, *91*, 165.
20. (a) Gronert, S., Keeffe, J. R., O'Ferrall, R. A. M. *J. Org. Chem.* **2009**, *74*, 5250. (b) Keeffe, J. R., O'Ferrall, R. A. M. *Arkivoc* **2008**, *10*, 183. (c) Gronert, S., Keefe, J. R. *J. Phys. Org. Chem.* **2013**, *26*, in press.

21 Kassaee, M. Z., Ghambarian, M., Musavi, S. M., Shakib, F. A., Momeni, M. R. *J. Phys. Org. Chem.* **2009**, *22*, 919.
22 Goldfuss, B., Schleyer, *P. v. R. Organometallics* **1997**, *16*, 1543.
23 Martin-Santamaria, S., Rzepa, H. S. *J. Chem. Soc. Perkin Trans. 2* **2000**, 2372.
24 Eisenthal, K. B., Moss, R. A., Turro, N. J. *Science* **1984**, *225*, 1439.
25 Wang, J., Kubicki, J., Hilinski, E. F., Mecklenburg, S. L., Gustafson, T. L., Platz, M. S. *J. Am. Chem. Soc.* **2007**, *129*, 13683.
26 Matzinger, S., Bally, T., Patterson, E. V., McMahon, R. J. *J. Am. Chem. Soc.* **1996**, *118*, 1535.
27 Gallo, M. M., Schaefer, H. F. *J. Phys. Chem.* **1992**, *96*, 1515.
28 Khodabandeh, S., Carter, E. A. *J. Phys. Chem.* **1993**, *97*, 4360.
29 Chamorro, E., Perez, P., De Proft, F., Geerlings, P. *J. Chem. Phys.* **2006**, *124*, 044105.
30 Richards, C. A., Kim, S. J., Yamaguchi, Y., Schaefer, H. F. *J. Am. Chem. Soc.* **1995**, *117*, 10104.
31 Matzinger, S., Fulscher, M. P. *J. Phys. Chem.* **1995**, *99*, 10747.
32 Goedecke, C., Leibold, M., Siemeling, U., Frenking, G. *J. Am. Chem. Soc.* **2011**, *133*, 3557.
33 Muck-Lichtenfeld, C. *J. Org. Chem.* **2000**, *65*, 1366.
34 Sulzbach, H. M., Bolton, E., Lenoir, D., Schleyer, P. v. R., Schaefer, H. F. *J. Am. Chem. Soc.* **1996**, *118*, 9908.
35 Stracener, L. L., Halter, R. J., McMahon, R. J., Castro, C., Karney, W. L. *J. Org. Chem.* **2000**, *65*, 199.
36 Xu, G. P., Chang, T. M., Zhou, J. L., McKee, M. L., Shevlin, P. B. *J. Am. Chem. Soc.* **1999**, *121*, 7150.
37 Nicolaides, A. *J. Am. Chem. Soc.* **2003**, *125*, 9070.
38 Poutsma, J. C., Upshaw, S. D., Squires, R. R., Wenthold, P. G. *J. Phys. Chem. A* **2002**, *106*, 1067.
39 Woodcock, H. L., Moran, D., Brooks, B. R., Schleyer, P. v. R., Schaefer, H. F. *J. Am. Chem. Soc.* **2007**, *129*, 3763.
40 Taatjes, C. A., Klippenstein, S. J., Hansen, N., Miller, J. A., Cool, T. A., Wang, J., Law, M. E., Westmoreland, P. R. *Phys. Chem. Chem. Phys.* **2005**, *7*, 806.
41 Kassaee, M. Z., Hossaini, Z. S., Haerizade, B. N., Sayyed-Alangi, S. Z. *J. Mol. Struct.: Theochem.* **2004**, *681*, 129.
42 Lee, T. J., Bunge, A., Schaefer, H. F. *J. Am. Chem. Soc.* **1985**, *107*, 137.
43 Schoeller, W. W., Frey, G. D., Bertrand, G. *Chem.—Eur. J.* **2008**, *14*, 4711.
44 Collins, C. L., Davy, R. D., Schaefer, H. F. *Chem. Phys. Lett.* **1990**, *171*, 259.
45 Gronert, S. *J. Org. Chem.* **2006**, *71*, 1209.
46 Gronert, S. *Chem.—Eur. J.* **2009**, *15*, 5372.
47 Poutsma, M. L. *J. Org. Chem.* **2008**, *73*, 8921.
48 Wodrich, M. D., Schleyer, P. v. R. *Org. Lett.* **2006**, *8*, 2135.
49 Wodrich, M. D., Wannere, C. S., Mo, Y., Jarowski, P. D., Houk, K. N., Schleyer, P. v. R. *Chem.—Eur. J.* **2007**, *13*, 7731.
50 Kemnitz, C. R., Mackey, J. L., Loewen, M. J., Hargrove, J. L., Lewis, J. L., Hawkins, W. E., Nielsen, A. E. *Chem.—Eur. J.* **2010**, *16*, 6942.
51 Mitoraj, M., Zhu, H. J., Michalak, A., Ziegler, T. *J. Org. Chem.* **2006**, *71*, 9208.
52 Boehland, T., Temps, F., Wagner, H. G. *J. Phys. Chem.* **1987**, *91*, 1205.
53 Takhistov's data are cited by Orlov, Y. D., Lebedev, Y. A., Menchikov, L. G., Nefedov, O. M. *Russ. Chem. Bull.* **1997**, *46*, 52.

54 Poutsma, J. C., Nash, J. J., Paulino, J. A., Squires, R. R. *J. Am. Chem. Soc.* **1997**, *119*, 4686.
55 Clauberg, H., Chen, P. *J. Am. Chem. Soc.* **1991**, *113*, 1445.
56 Clauberg, H., Minsek, D. W., Chen, P. *J. Am. Chem. Soc.* **1992**, *114*, 99.
57 Geise, C. M., Hadad, C. M. *J. Org. Chem.* **2002**, *67*, 2532.
58 Schwartz, R. L., Davico, G. E., Ramond, T. M., Lineberger, W. C. *J. Phys. Chem. A* **1999**, *103*, 8213.
59 Wren, S. W., Vogelhuber, K. M., Ervin, K. M., Lineberger, W. C. *Phys. Chem. Chem. Phys.* **2009**, *11*, 4745.
60 Goebbert, D. J., Pichugin, K., Khuseynov, D., Wenthold, P. G., Sanov, A. *J. Chem. Phys.* **2010**, *132*.
61 Sumathi, R. *Theochem—J. Mol. Struct.* **1996**, *364*, 97.
62 Olah, J., De Proft, F., Veszpremi, T., Geerlings, P. *J. Phys. Chem. A* **2004**, *108*, 490.
63 Matus, M. H., Nguyen, M. T., Dixon, D. A. *J. Phys. Chem. A* **2006**, *110*, 8864.
64 Irikura, K. K., Goddard, W. A., Beauchamp, J. L. *J. Am. Chem. Soc.* **1992**, *114*, 48.
65 Nguyen, M. T., VanKeer, A., Vanquickenborne, L. G. *J. Organomet. Chem.* **1997**, *529*, 3.
66 Ochsenfeld, C., Kaiser, R. I., Lee, Y. T., Head-Gordon, M. *J. Chem. Phys.* **1999**, *110*, 9982.
67 Freeman, F., Gomarooni, M. *Int. J. Quantum Chem.* **2006**, *106*, 2379.
68 McMillen, D. F., Golden, D. M. *Annu. Rev. Phys. Chem.* **1982**, *33*, 493.
69 Afeefy, H. Y., Liebman, J. F., Stein, S. E. "Neutral Thermochemical Data." In *NIST Standard Reference Database Number 69*. Mallard, W. G., Linstrom, P. J., Eds. National Institute of Standards and Technology (http://webbook.nist.gov): Gaithersburg, MD, **2012**.
70 Pau, C. F., Hehre, W. J. *J. Phys. Chem.* **1982**, *86*, 1252.
71 Liu, X. P., Gross, M. L., Wenthold, P. G. *J. Phys. Chem. A* **2005**, *109*, 2183.
72 Du, X. M., Fan, H., Goodman, J. L., Kesselmayer, M. A., Krogh-Jespersen, K., La Villa, J.A., Moss, R. A., Shen, S., Sheridan, R.S. *J. Am. Chem. Soc.* **1990**, *112*, 1920.
73 Bugaut, X., Glorius, F. *Chem. Soc. Rev.* **2012**, *41*, 3511.
74 Lin, J. C. Y., Huang, R. T. W., Lee, C. S., Bhattacharyya, A., Hwang, W. S., Lin, I. J. B. *Chem. Rev.* **2009**, *109*, 3561.
75 Marion, N., Nolan, S. P. *Chem. Soc. Rev.* **2008**, *37*, 1776.
76 Samojlowicz, C., Bieniek, M., Grela, K. *Chem. Rev.* **2009**, *109*, 3708.
77 Amyes, T. L., Diver, S. T., Richard, J. P., Rivas, F. M., Toth, K. *J. Am. Chem. Soc.* **2004**, *126*, 4366.
78 Higgins, E. M., Sherwood, J. A., Lindsay, A. G., Armstrong, J., Massey, R. S., Alder, R. W., O'Donoghue, A. C. *Chem. Commun.* **2011**, *47*, 1559.
79 Chu, Y., Deng, H., Cheng, J. P. *J. Org. Chem.* **2007**, *72*, 7790.
80 Liu, M., Chen, M., Zhang, S. S., Yang, I., Buckley, B., Lee, J. K. *J. Phys. Org. Chem.* **2011**, *24*, 929.
81 Liu, M., Yang, I., Buckley, B., Lee, J. K. *Org. Lett.* **2010**, *12*, 4764.
82 Chen, H., Justes, D. R., Cooks, R. G. *Org. Lett.* **2005**, *7*, 3949.
83 Hunter, E. P., Lias, S. G. "Proton Affinity Data." In *NIST Standard Reference Database Number 69*. Mallard, W. G., Linstrom, P. J., Eds. National Institute of Standards and Technology (http://webbook.nist.gov): Gaithersburg, MD, **2012**.
84 Tao, Y. G., Ding, Y. H., Li, Z. S., Huang, X. R., Sun, C. C. *J. Phys. Chem. A* **2001**, *105*, 3388.
85 Ge, Y. B., Shore, T. C. *Comput. Theor. Chem.* **2011**, *978*, 57.

86 Ge, C. S., Jang, E. G., Jefferson, E. A., Liu, W. G., Moss, R. A., Wlostowska, J., Xue, S. *J. Chem. Soc., Chem. Comm.* **1994**, *1479*.
87 Robert, M., Snoonian, J. R., Platz, M. S., Wu, G. X., Hong, H., Thamattoor, D. M., Jones, M. *J. Phys. Chem. A* **1998**, *102*, 587.
88 Tippmann, E. M., Platz, M. S. *J. Phys. Chem. A* **2003**, *107*, 8547.
89 Bonneau, R., Hellrung, B., Liu, M. T. H., Wirz, J. *J. Photochem. Photobiol. A* **1998**, *116*, 9.
90 Zhang, M., Moss, R. A., Thompson, J., Krogh-Jespersen, K. *J. Org. Chem.* **2012**, *77*, 843.
91 Moss, R. A., Zhang, M. *Org. Lett.* **2008**, *10*, 4045.
92 Moss, R. A., Ge, C. S., Wlostowska, J., Jang, E. G., Jefferson, E. A., Fan, H. *Tetrahedron Lett.* **1995**, *36*, 3083.
93 Moss, R. A., Shen, Y. M., Wang, L., Krogh-Jespersen, K. *Org. Lett.* **2011**, *13*, 4752.
94 Moss, R. A., Shen, S., Hadel, L. M., Kmiecik-Lawrynowicz, G., Wlostowska, J., Krogh-Jespersen, K. *J. Am. Chem. Soc.* **1987**, *109*, 4341.
95 Wang, L., Moss, R. A., Krogh-Jespersen, K. *J. Am. Chem. Soc.* **2012**, *134*, 17459.
96 Hoijemberg, P. A., Moss, R. A., Krogh-Jespersen, K. *J. Phys. Chem. A* **2012**, *116*, 4745.
97 Hoijemberg, P. A., Moss, R. A., Krogh-Jespersen, K. *J. Phys. Chem. A* **2012**, *116*, 358.
98 Moss, R. A. *J. Phys. Org. Chem.* **2011**, *24*, 866.

2

STABLE CARBENES

Jonathan P. Moerdyk and Christopher W. Bielawski
Department of Chemistry and Biochemistry, The University of Texas at Austin, Austin, TX, USA

2.1 INTRODUCTION

Although the existence of divalent carbon was postulated before the tetravalency of carbon was established and pursued through several different synthetic routes (beginning with Dumas' unsuccessful dehydration of methanol in 1845[1]), carbenes—a term coined by Doering, Winstein, and Woodward on a nocturnal Chicago taxi ride in the 1950s[2]—were not realized as intermediary species until the last half of the nineteenth and the beginning of the twentieth centuries.[3,4] Due to their high reactivities, free carbenes were largely viewed as structural curiosities and/or transient intermediates. That perspective changed, however, with the isolation of the first stable carbene in 1988 by Bertrand[5] and the first crystalline carbene a few years later by Arduengo[6] (Fig. 2.1). These reports catalyzed an explosion of research activities and facilitated the use of carbenes in a broad range of applications. While the definition of "stable" is subjective, this chapter views a stable carbene as one that can be isolated in its pure form at or above room temperature. The term "persistent" is reserved for carbenes that may be spectroscopically observed, but not isolated. It should be noted that although many classes of isolable carbenes are known, not every derivative within a class may be stable, or even persistent. As summarized in an excellent review by Tomioka and coworkers,[7] triplet carbenes, wherein the two electrons are distributed across degenerate or nearly degenerate orbitals on

Contemporary Carbene Chemistry, First Edition. Edited by Robert A. Moss and Michael P. Doyle.

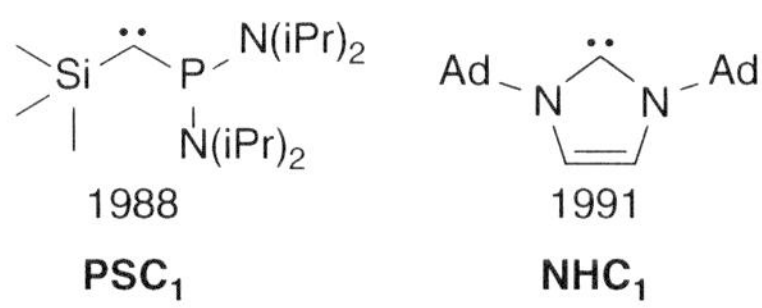

Figure 2.1. The first isolable (**PSC$_1$**) and crystalline (**NHC$_1$**) carbenes. iPr = isopropyl; Ad = adamantyl.

the carbene nucleus, have seen enormous gains in stability over the past several years; however, these compounds will not be the focus of this chapter due to the current inability to isolate triplet carbenes in the solid form. Instead, this chapter will focus exclusively on the chemistry of stable, singlet carbenes with the aim of providing readers with an overview and appreciation for the rich history, key developments, and future directions of this exciting and growing field.

2.2 TYPES OF STABLE CARBENES

As the fundamentals of carbene stabilization were covered in the preceding chapter, only a brief overview will be given here. In general, stable carbenes contain at least one heteroatom covalently bonded to the carbene nucleus that is capable of π-donation. Acceptance of electron density either through π-effects or induction also provides stabilization and helps maintain electroneutrality[8] at the carbene center. Moreover, as will be described in more detail below, the resulting ylidic resonance structure resulting from π-donation is often used to explain the nucleophilic characteristics displayed by many stable carbenes. Regardless, based on the dominant means of stabilization provided by the carbene substituents, the terms push–push, push–pull, and pull–pull are generally used. Strong donation of electron density, usually via lone pairs, is considered a "push" (e.g., NR_2, OR, PR_3) while "pull" refers to functional groups with π-electron accepting or strongly inductive withdrawing characteristics (e.g., SiR_3, $^+PR_3$, CF_3). For example, the phosphinosilyl carbene **PSC$_1$** is considered a push–pull system because the phosphorus π-donates while the silyl π-accepts electron density. *N*-heterocyclic carbene **NHC$_1$**, in comparison, is a push–push scaffold because both nitrogen atoms primarily stabilize the carbene center via π-donation, although inductive withdrawing effects are also operative. Additionally, carbene scaffolds, such as the alkylaminocarbenes (AACs), may contain a "spectator" group (e.g., CR_3) which is neither strongly withdrawing nor donating. Stabilizing effects also influence the energy differential between the highest occupied and lowest unoccupied molecular orbitals (HOMO and LUMO, respectively). The difference in energy is defined as the singlet–triplet gap (ΔS–T) and lower ΔS–T energies generally equate to higher chemical reactivity (or at least the potential for increased reactivity).

The aforementioned classes of stable carbene scaffolds will be discussed below with attention directed toward comparing the collective structural and electronic properties; the chemistry of these compounds follows.

2.2.1 Push–Push Carbenes

2.2.1.1 Diaminocarbenes. A classic push–push carbene, cyclic diaminocarbenes, commonly called *N*-heterocyclic carbenes (NHCs),[9] represent the most studied class of isolable carbenes[10] to date. The attention may be attributed, at least partially, to their readily modifiable core (*N*-substituents, backbone) and prior interest in enetetramines (c.f., diaminocarbene dimers) which react with similar substrates as their corresponding free carbenes.[11] Most importantly, NHCs have found exceptional utility for many applications, particularly as ligands for metals[12] or as organocatalysts,[13] wherein the high stability, excellent electron-donating abilities and variable steric properties[14] have yielded remarkable results (see Chapter 9 and Part II). Regardless, a plethora of NHC scaffolds have been synthesized including imidazolylidenes, imidazolinylidenes, triazolylidenes, benzimidazolylidenes, and various fused and ring-expanded systems (Fig. 2.2).[9]

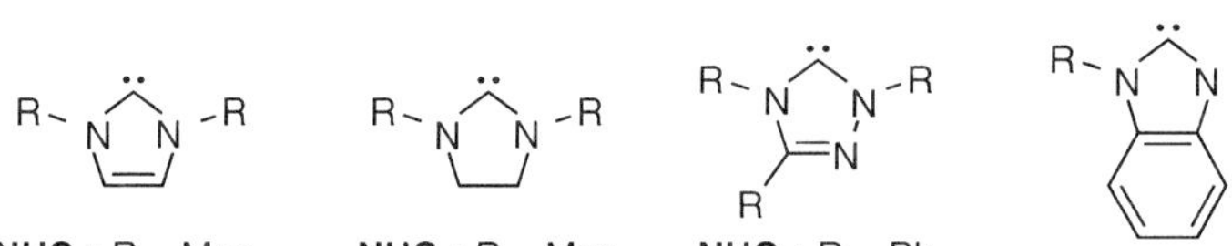

NHC$_2$: R = Mes Imidazolylidene; **NHC$_3$**: R = Mes Imidazolinylidene; **NHC$_4$**: R = Ph Triazolylidene; Benzimidazolylidene

Figure 2.2. Selected classes of stable NHCs.

The large array of commercially available amines from which NHCs may be constructed facilitates the diversification of the steric as well as electronic properties at the *N*-aryl substituents. Moreover, methods for substituting the backbones of five-membered NHCs with various electron withdrawing or donating groups have also been developed, though generally only for unsaturated motifs.[15] Incorporation of various heteroatoms into the backbone from commercially available triazolylidenes[16] to various boron[17] and phosphorus[18] containing scaffolds have also been successful.

The majority of NHCs are synthesized by exposing their respective protonated precursors to basic conditions (Fig. 2.3).[9b] However, other methods, such as the extrusion of volatile small molecules (e.g., alcohols, $CHCl_3$, NH_3, C_6F_5H) at high temperature and/or under reduced pressure are useful when carbene formation needs to be latent (e.g., polyurethane synthesis[19]) or when deprotonation is challenging.[20] Similarly, the loss of CO_2[21] from carbene–carboxylate adducts or the thermal reversion of other zwitterions (e.g., NHC-isothiocyanate adducts[22]), the reduction of chloroiminiums with bis(trimethylsilyl)mercury,[23] or the reduction of

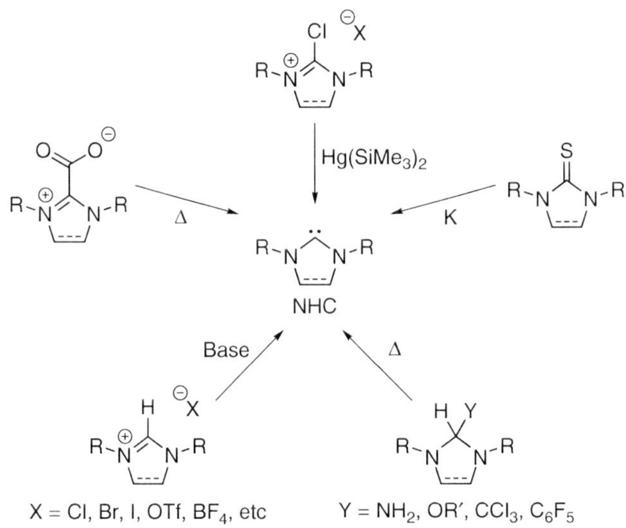

Figure 2.3. Selected methods for generating stable diaminocarbenes (NHCs) from various precursors.

thioureas with elemental potassium[24] have also been utilized, though the latter method is much less common today.

2.2.1.2 Acyclic Diaminocarbenes (ADCs). Although most diaminocarbenes are cyclic, acyclic variants are known, with the first example being reported by Alder in 1996. Combining diisopropylformamide with phosphorus oxychloride generated α-chloroiminium **1** *in situ* which, upon addition of diisopropylamine followed by deprotonation, afforded acyclic diaminocarbene **ADC**$_1$ (see Fig. 2.4, top).[25] Since that time, several general methods for preparing ADCs have been reported,[26] including the dialkylation of an aryl or

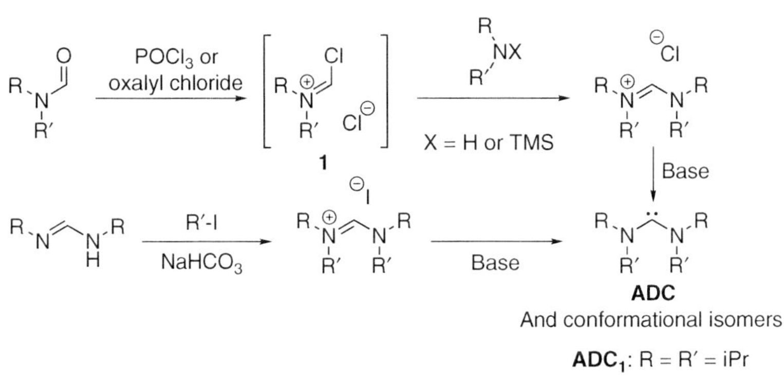

Figure 2.4. Selected syntheses of ADCs from formamide (top) or formamidine (bottom) precursors.

alkyl formamidine followed by deprotonation[27] as well as a variant on Alder's synthesis wherein a silylated, rather than a secondary amine, is added to the α-chloroiminium intermediate.[28] The latter method, driven by the formation of the Si–Cl bond, is particularly useful for the synthesis of ADC precursors wherein both *N*-substituents are sterically demanding (e.g., adamantyl or 2,6-diisopropylphenyl, DIPP).

On account of the relaxed geometric constraints, the N–C–N bond angles displayed by ADCs are significantly wider than NHCs.[29] Combined with the increased conformational flexibility and enhanced steric properties, the strong donating ability[30] of the ADCs has endowed this class of carbenes with potential for use as ligands for transition metals.[31] Moreover, the increased rotational freedom about the N–C–N bonds disrupt π-conjugation and lowers the $\Delta S\text{–}T$ for ADCs when compared to typical NHCs (ca.40 kcal/mol vs. >65 kcal/mol, respectively).[32] The associated decrease in $\Delta S\text{–}T$, which often correlates with increased reactivity, results in a higher sensitivity to atmospheric conditions and enables reactions not observed with typical cyclic diaminocarbenes, such as the coupling of **ADC**$_1$ with carbon monoxide (Section 2.4.4).[32]

2.2.1.3 N,N'-Diamidocarbene (DAC). The diamidocarbenes (DACs; Fig. 2.5) are an emerging class of stable carbenes that utilize amides rather than amines to stabilize the carbene nucleus. Prepared via the condensation of a formamidine and a diacid chloride followed by deprotonation, five-,[33] six-,[34] and seven-membered[35] derivatives have been synthesized, though dimerization of the five-membered DAC has precluded its isolation. Density functional theory (DFT) calculations revealed that, due to the weakly donating amido group, the LUMO of the DAC is substantially lower-lying than that of typical NHCs although the HOMO levels are similar. As a result, the $\Delta S\text{–}T$ of the former (ca. 45 kcal/mol for **DAC**$_1$) is significantly smaller than that of the latter (>65 kcal/mol for **NHC**$_2$ and **NHC**$_3$).[36] This has endowed DACs with a broad reactivity profile. For example, DACs coordinate various transition metals[37] and condense with electrophiles, but also exhibit reactivities reminiscent of transient carbenes. In addition to DACs, monoaminoamidocarbenes (MAACs), hybrids both in structure, electronic properties, and chemical reactivity between DACs and NHCs, as well as other related derivatives have also been isolated.[38]

2.2.1.4 Diphosphinocarbene (PPC). Given the high stability of diaminocarbenes, one might expect that the PPCs (Fig. 2.5) may be similarly stable.

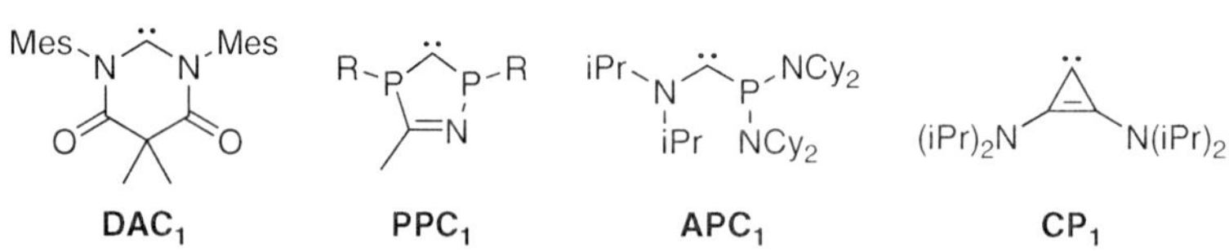

Figure 2.5. Selected crystalline examples of DACs, PPCs, APCs, and CPs. Mes=2,4,6-trimethylphenyl; R=2,4,6-tri-tertbutylphenyl; iPr=isopropyl; Cy=cyclohexyl.

However, while nitrogen readily adopts a planar configuration, phosphorus tends to be strongly pyramidalized which significantly reduces π-donation and stabilization. To overcome this drawback, Bertrand reported the synthesis of **PPC_1** in which the ring system and bulky P-aryl groups limit inversion of the phosphorus atoms and enforce a greater degree of planarity in addition to sterically shielding the carbene center.[39] Moreover, **PPC_1**, which was prepared via deprotonation of the salt formed from the [3+2] cycloaddition of acetonitrile and an *in situ* generated diphosphallylic cation, was shown to ligate to metals and was calculated to be a strong σ-donor compared to its structurally related triazolylidene (**NHC_4**) and other NHCs.[39] Although the reactivity of free PPCs with organic substrates is still under investigation,[40] the DFT calculated singlet-triplet gap for PPCs is significantly lower than for the analogous imidazolylidene derivatives (21 kcal/mol vs. 79 kcal/mol, respectively) which would be expected to enhance the reactivity of the carbene toward organic substrates.[41]

2.2.1.5 ***Aminophosphinocarbene (APC).*** As a hybrid of diamino- and PPCs, the aminophosphinocarbenes (APCs, Fig. 2.5) readily undergo intramolecular C–H insertion and other decomposition pathways above −20 °C. Inspection of the crystallographic data of **APC_1** revealed a strongly pyramidalized environment around the phosphorus center ($\Sigma_{C-P-N,N-P-N}$=304.5°), a long P–$C_{carbene}$ bond distance (1.856 Å), and a short N–$C_{carbene}$ bond distance (1.296 Å); collectively, these metrics were consistent with the nitrogen atom being the dominant π-stabilizer of the carbene nucleus.[42] Interestingly, APCs react with isonitriles and BF_3 at the carbene center; however, treatment of APCs with a single equivalent of elemental sulfur or BH_3 induces adduct formation at the phosphorus atom.

2.2.1.6 ***Cyclopropenylidene (CP).*** Contrasting all other isolable singlet carbene scaffolds, CPs contain no heteroatoms directly adjacent to the divalent carbon atom. A two π-electron Hückel aromatic derivative of the transient CP observed in interstellar space, **CP_1** (Fig. 2.5) is stabilized by strong π-donation of the amino substituents on the cyclopropene ring.[43] The geometric constraints of **CP_1**, which may be synthesized via deprotonation of the corresponding air-stable salt, affords a sterically unencumbered carbene. Because of their relatively small size, 2 equiv of **CP_1** can ligate to Pd(TMEDA)Cl, $Ni(COD)_2$, or $[Rh(COD)Cl]_2$ (yielding the cationic biscarbene complex Rh(COD)(**CP_1**)$_2$ and anionic $Rh(COD)Cl_2$). For comparison, the same metal complexes typically coordinate only one NHC ligand (COD=1,5-cyclooctadiene; TMEDA=tetramethylethylenediamine).[44]

2.2.1.7 ***Aminooxy (AOC) and Aminothio (ATC) Carbenes.*** In addition to their role in the biocatalytic activity of thiamine, a coenzyme of Vitamin B_1,[45] ATCs were utilized in the earliest organocatalytic applications of singlet carbenes dating back to the benzoin condensations reported by Ugai in 1943.[46] However, while sulfur is a π-donor, its divalency provides significantly less steric shielding and thus less kinetic stabilization, when compared to trivalent nitrogen or phosphorus. Isolation of the first stable ATC, *N*-(2,6-DIPP)-4,5-dimethylthiazol-2-ylidene (**ATC$_1$**), was achieved by Arduengo in 1997; notably, changing the 2,6-DIPP substituent to a 2,4,6-trimethylphenyl (Mes) resulted in dimerization.[47] Similar to the imidazolinylidenes, the dimerization of ATCs appears to be catalyzed by Lewis acids.[47]

Analogous to ATCs, the limited steric bulk at oxygen can challenge the isolation of AOCs, although Alder has isolated several acyclic variants (as well as acyclic **ATC$_2$**) through the use of bulky *O*-aryl and *N*-alkyl groups.[48] In short, diiminium **2** was prepared from triflic anhydride and the appropriate formamide. Subsequent nucleophilic attack of **2** by an alkoxide (or thiolate in the case of **ATC$_2$**) followed by deprotonation of the resultant salt afforded **AOC$_{1-3}$** (Fig. 2.6).

2.2.1.8 Phosphinoarylcarbene (PAC). Building upon the PSC$_1$ scaffold wherein the silyl group serves as a π-electron acceptor and σ-donor, Bertrand exchanged the silyl group of PSCs for CF_3 to elucidate the capability of a single heteroatom center to stabilize a free carbene.[49] Prepared from the diazo precursor, dimerization was observed when generated in nonpolar solvents or upon concentration while the free carbene was only stable in polar solvents such as THF up to −30 °C. However, replacing CF_3 with 2,6-bis(trifluoromethyl) phenyl significantly increased stability and facilitated the isolation of **PAC$_1$** (Fig. 2.7). Crystallographic analysis of **PAC$_1$** revealed an essentially planar environment at the phosphorus atom ($\Sigma_{N-P-C,N-P-N}$=359.9°) consistent with π-donation of the phosphorus lone pair, while the aryl group resided perpendicular to the P–C–C plane and therefore out of conjugation, at least in the solid state.[49] More electron-rich aryl groups as well as the absence of ortho-substitution destabilized the PAC, underscoring the importance of electronic as well as steric effects in carbene stability.[50]

Similar to the APCs, exposing the PACs to metals did not always result in coordination at the carbene center.[51] The addition of $[Rh(COD)Cl]_2$, for example, lead to a mixture of traditional carbene–metal bonding and η^2-ligation involving coordination of Rh by the phosphorus as well as the carbene carbon (Fig. 2.7, right). In addition to coupling to isonitriles upon which the resulting ketenimines decomposed to cyclic dihydroazaphosphetes, the migration of a phosphorus dialkylamino group occurred upon exposure of **PAC$_1$** to 312 nm light.[52] Collectively, the observed chemistry may reflect a low-lying LUMO capable of being accessed either thermally or photochemically.

2.2.1.9 Aminoaryl (AA). Like their heavier-atom counterparts (i.e., the PACs), the lone nitrogen in aminoarylcarbenes (AAs, Fig. 2.7) is sufficient to provide electronic stabilization of the carbene center given the appropriate choice of substituents. Readily synthesized via the condensation of an amine and an aryl aldehyde followed by alkylation and deprotonation, *N*-(*o*-*tert*-butylphenyl) derivatives were found to be unstable, undergoing intramolecular C–H insertion into the tert-butyl methyl group over several hours at

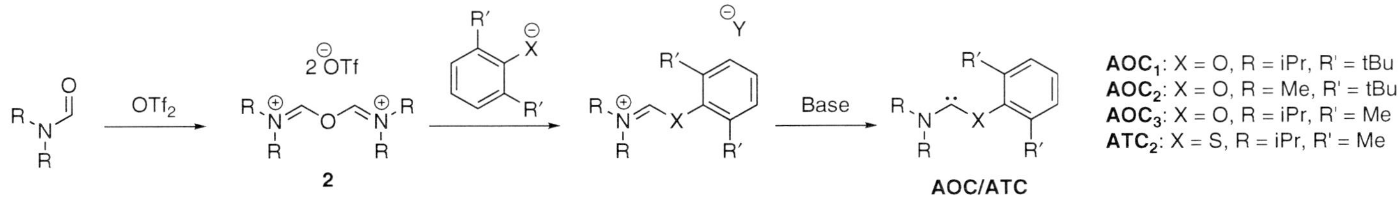

Figure 2.6. Syntheses of various acyclic AOCs and an acyclic ATC. Y = OTf or BF_4.

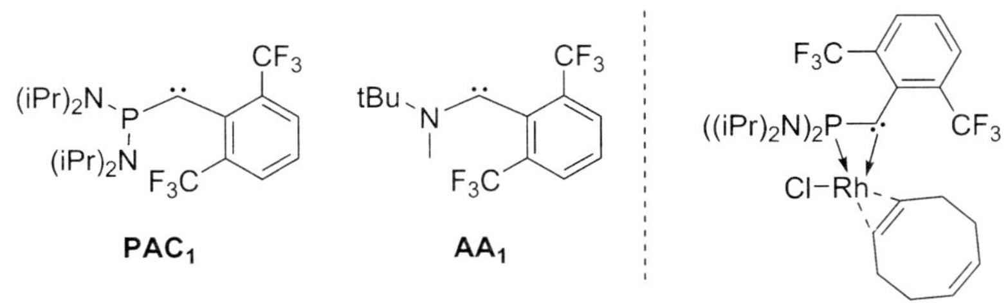

Figure 2.7. (left) Stable phophinoaryl- (PACs) and AAs. (right) η^2-coordination of a PAC to rhodium.

room temperature.[53] However, the use of aryl substituents with *o*-CF_3 groups that preclude C–H insertion were readily isolable. Electronic differences between the substituents are likely less important in determining stability since a relatively long C–C bond (1.453 Å) observed in the solid state structure of $\mathbf{AA_1}$ indicated minimal π-delocalization into the aryl group (again located perpendicular to the N–C–C plane in the solid state). While $\mathbf{AA_1}$ was not found to ligate to metals, less encumbered AAs undergo ligation as well as reactions with hindered alcohols and BF_3.[54] Importantly, the PACs and AAs demonstrated that carbenes may be stabilized with a single heteroatom.

2.2.1.10 Alkylaminocarbenes (AACs). AACs represent examples of a push-spectator system as the alkyl substituent is expected to only be a weak σ-donor. Both cyclic and acyclic stable AACs are known and may be synthesized using a variety of methods.[55] Initially, cyclic AAC precursors were prepared by the ring opening of an epoxide and followed by triflic anhydride promoted ring closure (Fig. 2.8). More recently, an alternate synthesis involving the nucleophilic substitution of 3-chloro-2-methylpropene with a lithiated imine followed by ring closure upon exposure to HCl has been reported.[56] DFT calculations have indicated that the AACs are more nucleophilic than their NHC counterparts, a result that has been supported experimentally through the analysis of metal complexes supported by the aforementioned carbenes. AACs were also calculated to be more electrophilic than NHCs which, in conjunction with the enhanced nucleophilicity, has enabled numerous transformations not observed with many other scaffolds. For example, cyclic and acyclic AACs are the only stable carbenes to date to split dihydrogen (see Section 2.4.5).

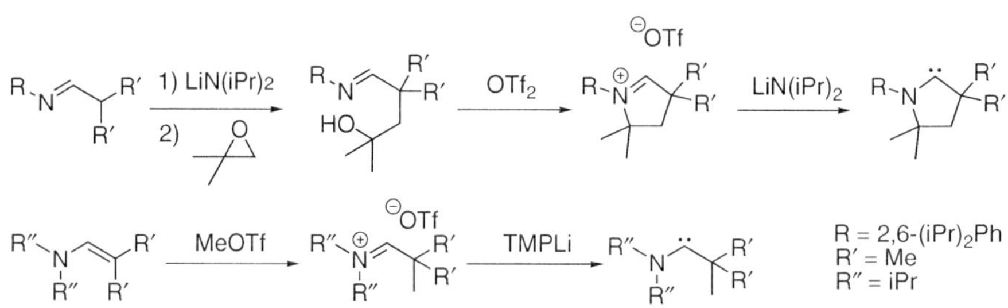

Figure 2.8. Selected syntheses of stable cyclic (top) and acyclic (bottom) AACs.

2.2.2 Push–Pull Carbenes

2.2.2.1 Phosphinosilyl Carbene (PSC). The PSCs are typically prepared by treating a phosphorus chloride (**3**) with a silyldiazomethane salt (**4**) followed by pyrolytic or photolytic loss of N_2 (Fig. 2.9). Stabilization of the free carbene was accomplished through the π-donation of the phosphorus as well as π-accepting properties of silicon. Unlike other singlet carbenes, PSCs are much less nucleophilic and are reluctant to bind transition metals. However, PSCs do exhibit many characteristics of more transient carbene species such as an ability to undergo

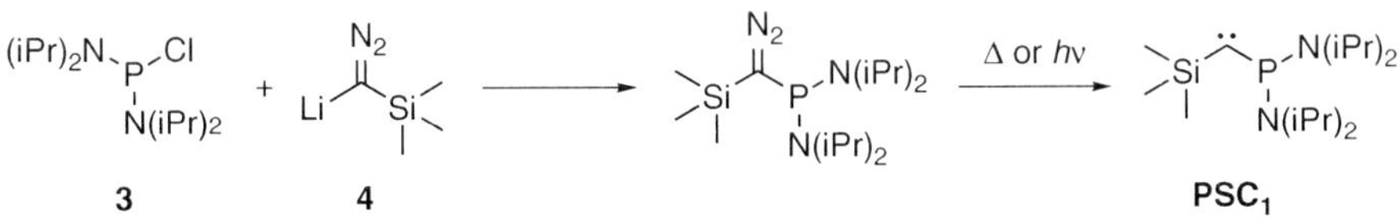

Figure 2.9. Synthesis of **PSC**$_1$.

intramolecular C–H insertion upon pyrolysis and [2+1] cycloaddition chemistry.[57] Indeed, the more electrophilic character of PSCs may account for much of the chemical reactivity observed with these species (see below).

2.2.2.2 Aminophosphonio- (APC⁺) and Phosphinophosphoniocarbenes (PPC⁺s). While most carbenes are charge neutral, two classes of cationic, isolable carbenes have been reported. Methylation at the phosphorus centers of free APCs with methyl triflate at −30°C has afforded cationic APC⁺s, which are stable up to −10°C in solution or room temperature in the solid state (Fig. 2.10).[58] In addition to the stabilization afforded by its π-accepting character, the phosphonium represents an excellent leaving group. As such, APC⁺s are subject to nucleophilic displacement of phosphine by alkoxides, lithium sulfides, or lithium phosphides and provide a unique method for preparing a variety of neutral AOCs, ATCs, or APCs.

Surprisingly, phosphinophosphoniocarbenes (PPC⁺s; Fig. 2.10), rather than their charged precursors, are obtained upon treatment of the corresponding neutral diphosphinodiazo compounds with triflic acid at −40°C.[59] The neutral phosphorus–carbon bond (P–C=1.548 Å) of a crystalline PPC⁺ is reminiscent of a P–C triple bond; however, the cationic phosphorus–carbon distance (P⁺–C=1.605 Å) is too short for a single bond. Based on these observations, the dominant resonance structure was concluded to involve a P–C=P interaction with a formal positive charge at both phosphorus atoms and a negative charge at the carbene center (Fig. 2.10, right).

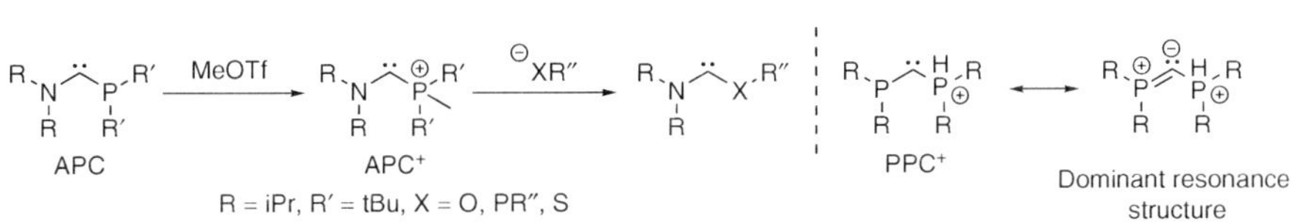

Figure 2.10. (left) Preparation of APC⁺s and subsequent displacement reactions. (right) General PPC⁺ structure and dominant resonance form.

2.2.3 Abnormal Carbenes

While the deprotonation of an imidazolium salt typically affords the corresponding imidazol-2-ylidene, another carbene, where loss of a proton from the C4 or C5 position, may also form to afford a so-called abnormal or mesoionic *N*-heterocyclic carbene (abNHC). Compared to their "normal" NHC isomers, abNHCs are significantly more basic ($pK_a \approx 33$)[60] and less stable (ca. 20 kcal/mol

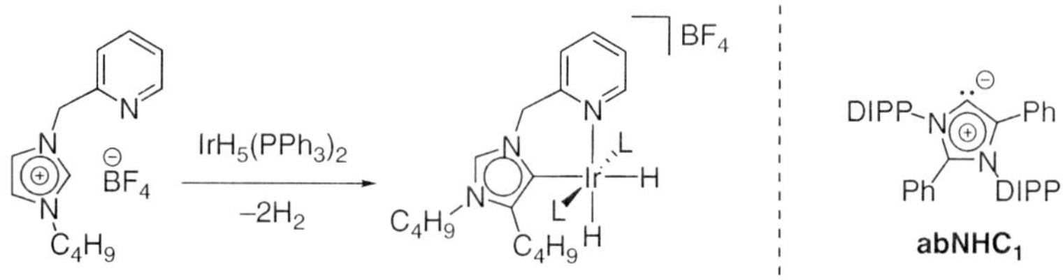

Figure 2.11. (left) First experimental observation of an abnormal NHC (abNHC) and (right) an example of an isolable abNHC.

higher in energy than typical NHCs).[61] The first abNHCs were obtained as metal ligands through C–H bond activation at the C4-position rather than deprotonation (Fig. 2.11).[62] Most abNHCs are still primarily isolated as metal ligands but, through judicious choice of substitution, free abNHCs may be isolated. For example, monosubstitution of the C2 position between the nitrogens prevents the "normal" NHC from forming and the incorporation of bulky *N*- and C4-substituents kinetically stabilize the resultant free $\mathbf{abNHC_1}$. Ligated abnormal triazolylidenes,[63] thiazolylidenes,[64] oxazolylidenes,[65] and dithiolylidenes[66] have also been reported from stable, sometimes masked, precursors, though the corresponding free carbenes are often not isolable. Regardless, the strongly donating properties of abnormal carbenes (on the order of acyclic NHCs) has made abNHCs of interest for use as ligands for a variety of transition metal complexes.[67]

2.3 SPECTROSCOPIC CHARACTERISTICS

As many carbenes are utilized for their electron-donating properties, several methods have been developed to evaluate the "donicity" of carbenes, namely through the comparison of ^{13}C NMR shifts,[68] Tolman Electronic Parameters (TEPs), or the Lever Method. In principle, downfield NMR shifts of the carbene nucleus may be attributed to a less electron-rich scaffold and lower donicity; however, care should be taken when comparing different scaffolds. For example, the AACs are considered excellent donors (on the order of acyclic NHCs) but possess a far downfield ^{13}C carbene shift (319 ppm) compared to the less donating $\mathbf{NHC_2}$ (219.7 ppm). In lieu of NMR spectroscopy, which can be challenging for highly unstable free carbenes, the TEP[69] may be used. This method is based on the infrared (IR) stretching frequency of the carbonyl ligands in $(L)M(CO)_2Cl$ complexes where L is the ligand of interest and M = Rh or Ir. Though originally employed to evaluate the electronics of phosphine ligands in nickel tricarbonyl complexes, adaptation to carbenes as well as formulas for comparing Rh, Ir, and Ni complexes have been reported.[70] The convenience of comparing multiple metal systems and the relative stability of the metal complexes are several factors that have resulted in the TEP method being commonly used to measure the electron-donicity (and roughly by extension nucleophilicity) of carbenes.

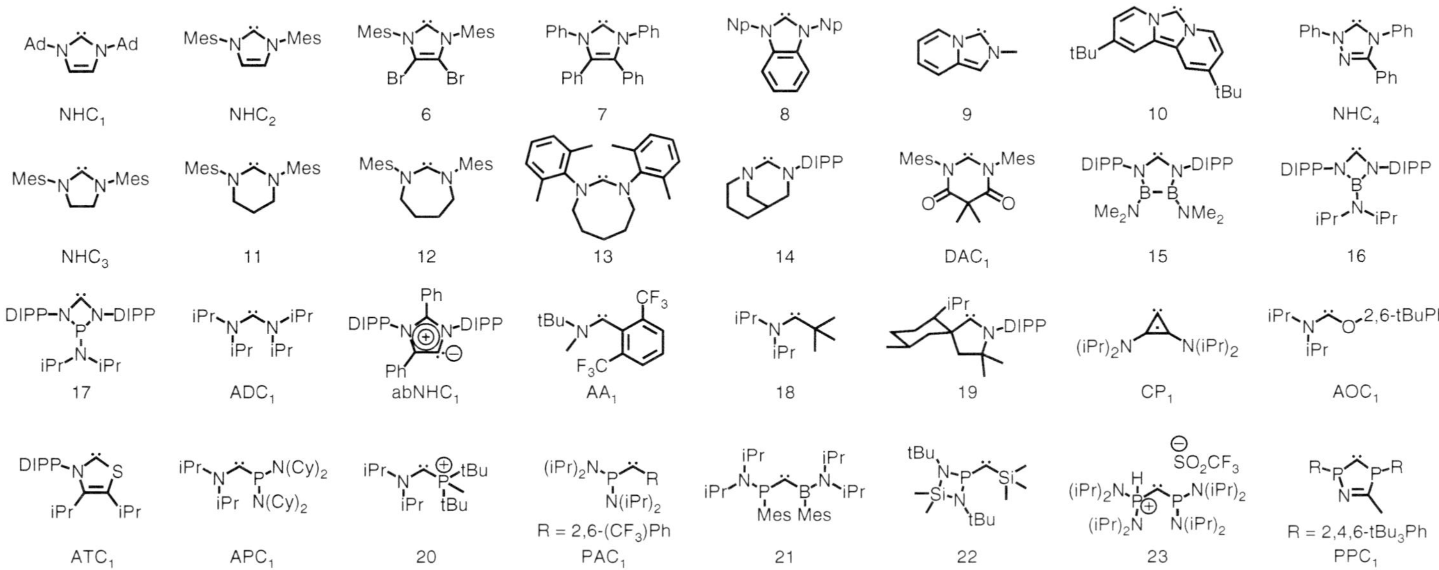

Figure 2.12. Selected examples of crystalline carbenes.

A disadvantage to TEP measurements is the indirect nature of the technique which analyzes the ligated carbon monoxides rather than the metal center. One alternative is electrochemistry. Most transition metals are redox-active at accessible potentials, and electrochemical measurements directly probe the metal center, obviating the need for carbonyl ligands. Moreover, the sensitivity in terms of measuring electron density at the metal center via electrochemical measurements is significantly higher than that of IR spectroscopy.[71] Lever developed a system based on the Ru^{III}/Ru^{II} redox couple whereby every ligand is assigned a Lever Electronic Parameter (LEP) independent of the other ligands present.[72] However, few electrochemical studies have been performed on NHC-ligated ruthenium complexes and LEPs are not widely used in the carbene community as of yet.[37,73]

A summary of the experimental, spectroscopic, and crystallographic data for various crystalline carbenes (Fig. 2.12) is given in Table 2.1 and inspection of these data reveal several trends in the ^{13}C NMR shifts, carbene bond angles, and TEPs:

1. Acyclic carbenes trend to lower TEP values (i.e., better donors), larger bond angles, and downfield shifted ^{13}C NMR chemical shifts (Entries 18, 20, and 21 vs. 2, 9, and 22). More obtuse bond angles are particularly accentuated for acyclic carbenes that include a π-accepting moiety and no nitrogen adjacent to the carbene nucleus (Entries 28–31).
2. As ring size increases, the TEP values, bond angles, and ^{13}C NMR chemical shifts tend to increase (Entries 9–12).
3. Saturated NHCs have slightly wider N–C–N bonds, downfield shifted ^{13}C NMR signals, and higher TEP values than their unsaturated analogues (e.g., Entries 1, 3 vs. 9).
4. Additional heteroatoms (B, P, N) in the ring structure decrease TEP values and increase the ^{13}C NMR chemical shift (Entries 8, 15–17).
5. Phosphorus substitution of the carbene center shifts the ^{13}C NMR signal upfield compared to nitrogen derivatives (Entry 8 vs. 32).
6. Based on N–C bond distances, nitrogen atoms dominate π-donation into the carbene carbon.

2.4 CHEMICAL REACTIVITY

Every known stable carbene contains at least one heteroatom capable of π-donating into the carbene center. The donation raises the energy of the LUMO and renders these scaffolds less electrophilic. Indeed, the archetypal carbenes commonly introduced in undergraduate organic courses, such as methylene and dichlorocarbene, exhibit significant electrophilic character, while the reactivity of stable carbenes has largely revolved around their strong nucleophilic properties. These properties have frequently endowed many stable carbenes with exceptional catalytic properties when used as ligands for

TABLE 2.1. Selected Spectroscopic Data for Various Crystalline Carbenes

Entry	Carbene	R1	R2	R1–C1 (Å)	R2–C1 (Å)	R1–C1–R2	^{13}C NMR (solvent)	TEP (cm^{-1})[a]
1[6]	**NHC$_1$**	N	N	1.367(2)	1.373(2)	102.2(1)	212.4 (thf-d_8)	2049.5[70b]
2[74]	**NHC$_2$**	N	N	1.365(4)	1.371(4)	101.4(2)	219.7 (thf-d_8)	2050.8[70b]
3[75]	**6**	N	N	1.375(8)	1.388(7)	102.4(5)	221.2 (C_6D_6)	—
4[76]	**7**	N	N	1.369(3)	1.369(3)	102.1(3)	219.59 (thf-d_8)	2045.1[77]
5[78]	**8**	N	N	1.361(2)	1.374(2)	103.49(13)	231.5 (thf-d_8)	—
6[79]	**9**	N	N	1.355(6)	1.389(6)	100.4(4)	196.3 (thf-d_8)	—
7[79]	**10**	N	N	1.380(4)	1.374(4)	99.6(2)	209.7(thf-d_8)	—
8[16]	**NHC$_4$**	N	N	1.351(3)	1.373(4)	100.6(2)	214.6 (C_6D_6)	2057.2[70b]
9[80]	**NHC$_3$**	N	N	1.352(5)	1.345(5)	104.7(3)	244.5 (thf-d_8)	2052.0[70b]
10[81]	**11**	N	N	1.3464(12)	1.3464(12)	114.65(13)	244.9 (C_6D_6)	2043.5[b,82]
11[81]	**12**	N	N	1.349(5)	1.346(5)	116.6(4)	257.3 (C_6D_6)	2042.7[b,82]
12[83]	**13**	N	N	1.3475(16)	1.3475(16)	120.11(18)	251.1 (C_6D_6)	—
13[84]	**14**	N	N	1.349(2)	1.407(2)	112.9(1)	282 (C_6D_6)	2048.2[84]
14[34c]	**DAC$_1$**	N	N	1.372(3)	1.369(3)	114.83(19)	277.7 (C_6D_6)	2057.0[37]
15[17a]	**15**	N	N	1.352(1)	1.404(1)	108.45(8)	303.6 (C_6D_6)	—
16[17b]	**16**	N	N	1.395(2)	1.385(3)	94.02(16)	312.6 (C_6D_6)	—
17[18]	**17**	N	N	1.373(2)	1.387(2)	96.72(13)	285.0 (C_7D_8)	2050.3[b,85]
18[25]	**ADC$_1$**	N	N	1.363(6)	1.381(6)	121.0(5)	255.5 (C_6D_6)	2036.7[b,30]
19[60]	**abNHC$_1$**	N	C	1.417(2)	1.383(3)	101.03(17)	201.9 (thf-d_8)	2039.6[67b]
20[53]	**AA$_1$**	N	C	1.283(3)	1.453(3)	121.0(2)	302.2 (thf-d_8)	—
21[55a]	**18**	N	C	1.2976(17)	1.5392(18)	120.50(12)	326.27 (thf-d_8)	2043.9[b,55a]
22[55b]	**19**	N	C	1.315(3)	1.516(3)	106.54(18)	319.0 (thf-d_8)	2042.2[55b]

(Continued)

TABLE 2.1. (Continued)

Entry	Carbene	R1	R2	R1–C1 (Å)	R2–C1 (Å)	R1–C1–R2	^{13}C NMR (solvent)	TEP (cm^{-1})[a]
23[43]	**CP$_1$**	C	C	1.404(3)	1.405(3)	57.19(16)	185.51 (C_6D_6)	2045.1[b,44]
24[48]	**AOC$_1$**	N	O	1.319(5)	1.379(5)	109.3(3)	267.3 (C_6D_6)	—
25[47]	**ATC$_1$**	N	S	1.344(3)	1.720(3)	104.2(2)	254.3 (thf-d_8)	(2053.7)[86]
26[42]	**APC$_1$**	N	P	1.296(4)	1.856(3)	116.5(2)	329.7 (thf-d_8)	—
27[58]	**20**	N	P	1.287(6)	1.770(5)	127.3(4)	302.0 ($CDCl_3$)	—
28[49]	**PAC$_1$**	P	C	1.544(3)	1.390(4)	162.1(3)	146.1 (C_6D_6)	—
29[87]	**21**	P	B	1.563(13)	1.54(2)	151.6(13)	208.9 (thf-d_8)	—
30[88]	**22**	P	Si	1.532(2)	1.795(3)	152.6(3)	77.6 (C_6D_6)	—
31[59]	**23**	P^+	P	1.605(5)	1.548(4)	165.1(4)	98.9 ($CDCl_3$)	—
32[39]	**PPC$_1$**	P_N	P_C	1.673(6)	1.710(6)	98.2(3)	184.4 (thf-d_8)	2037.9[b,39]

[a] The TEP values were calculated from the average carbonyl stretching frequency ($\nu_{CO}^{av/Ir}$) of the corresponding $LIr(CO)_2Cl$ complex using the linear regression TEP (cm^{-1}) = $0.8475 \cdot \nu_{CO}^{av/Ir} + 336.2$.[9b]

[b] The $\nu_{CO}^{av/Ir}$ values were calculated from the corresponding $LRh(CO)_2Cl$ complex using the linear regression $\nu_{CO}^{av/Ir}$ (cm^{-1}) = $0.9441 \cdot \nu_{CO}^{av/Rh} + 98.9$.[9b]

transition metal catalysis[12] or as organocatalysts[13] (see Chapter 9 and Part II). Herein, we discuss the chemical reactivity of stable carbenes with small molecules and main group elements.

2.4.1 Heteroallenes/Electrophilic Traps

The classic methods for trapping singlet carbenes when the stability is not yet known or as proof of formation for nonisolable derivatives are the condensations of the carbene with elemental sulfur, selenium, or CS_2. The latter can be particularly informative as, in addition to a diagnostic infrared absorbance, complexation of CS_2 provides an additional ^{13}C NMR handle that is often useful for characterization purposes, is frequently accompanied by an intense color change, and affords an inner salt for which the difference in solubility may facilitate isolation.[89] As with CS_2, the addition of carbon dioxide, isothiocyanates, isocyanates, and ketenes results in a new C–C bond and the formation of an inner salt with the positive charge at the carbene carbon (Fig. 2.13). For reactions involving CO_2, CS_2, and isothiocyanates, the corresponding zwitterions are frequently isolable though in some instances, especially for less bulky carbenes, the addition of a second equivalent of heteroallene via a [3+2] dipolar cycloaddition may occur.[89, 90] In contrast, 1:1 adducts between carbenes and isocyanates or ketenes are rarely isolable as they readily participate in [3+2] cycloaddition chemistry with another equivalent of heteroallene.[91] Moreover, several NHCs catalyze the trimerization of isocyanates to their corresponding isocyanurates.[92] Thus, heteroallene–carbene coupling reactions are useful methods for the synthesis of a variety of functionalized heterocycles.[90]

In contrast, mixing nucleophilic carbenes and azides or diazo compounds results in the formation of a double bond between the terminal nitrogen and carbene carbon.[93] One attractive feature of the linear products formed is the retention of conjugation which has been exploited in the formation of

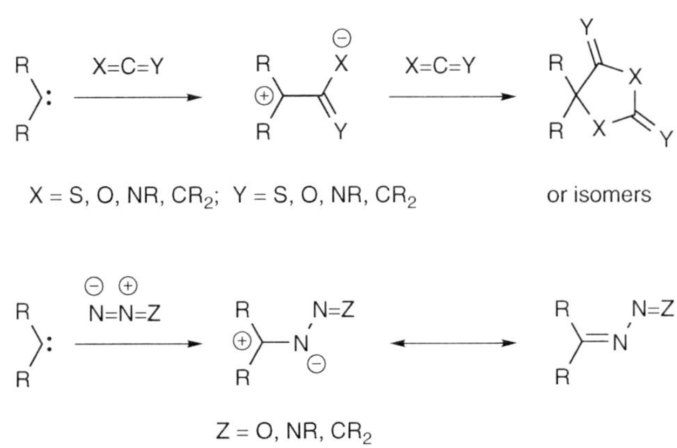

Figure 2.13. Generalized representation of typical heteroallene–carbene adducts.

conjugated polymers via the condensation of bis(NHCs) and bisazides.[94] Recently, Severin extended the coupling of NHCs and heteroallenes to include nitrous oxide,[95] a greenhouse gas 300 times more potent than CO_2.[96] The resultant, linear heterodiene with zwitterionic resonance structures, may be cleaved with a vanadium complex at the N–N and N–O bonds without the release of N_2 or NO typical for bond cleavage of nitrous oxide by metal complexes.[97] Phosphorus-containing carbenes provide different products with heteroallenes as interactions of the substrate with the π-accepting phosphorus can arise.[5] In all, heteroallene–carbene chemistry had been well-studied before carbenes were isolated but continues to be a field of interest as new substrates, reactions, and applications are uncovered.

2.4.2 Lewis Acids: Boryl Compounds

The formation of a Lewis acid–base pair between $\mathbf{NHC_1}$ and AlH_3 revealed the high Lewis basicity of NHCs.[98] Similarly, many isolable carbenes coordinate various boron species such as BF_3 or BH_3.[42, 99] In particular, the Lewis acid–base pair formed between NHCs and boranes has been a topic of recent interest. Work by Curran and others has demonstrated that NHC-borane adducts may be functionalized at the boron center through the addition of various acids or alkyl halides or via deprotonation followed by introduction of common electrophiles.[100] Additionally, NHC-boranes may be oxidized to NHC-boryl radicals and used to accelerate Type I radical photopolymerizations,[101] catalyze the radical reduction of alkyl- and arylhalides,[102] and facilitate Barton McCombie deoxygenations of xanthates.[103] NHC-boron adducts have even been employed as chiral reducing agents[104] and to enhance the electron-donating and electron-accepting character of conjugated systems that hold potential for use in electronic applications.[105]

Beyond NHCs, phosphinosilyl-, aminophosphino-, and alkylamino carbenes also coordinate to various boron-containing compounds, although differing reactivity may be observed. For example, treating an AAC with pinacolborane results in B–H bond insertion across the carbene center.[106] Remarkably, boron, a classic Lewis acid, may be rendered Lewis basic when coordinated by two AACs. In short, reduction of the Lewis acid–base adduct of BBr_3 and an AAC with potassium on graphite afforded a trivalent $B(AAC)_2H$ adduct containing a formal lone pair (Fig. 2.14).[107] Both protonation with triflic acid and one electron oxidation to the radical cation were observed and supported by theoretical calculations.

2.4.3 Alkenes, Alkynes, and Nitriles

The [2+1] cycloaddition of an alkene and carbene to form a cyclopropane ring is a well-known reaction. Given the familiarity and fundamental nature of such chemistry, it may be surprising that relatively few isolable carbenes are known to participate in cyclopropanation chemistry. This may be due to the high

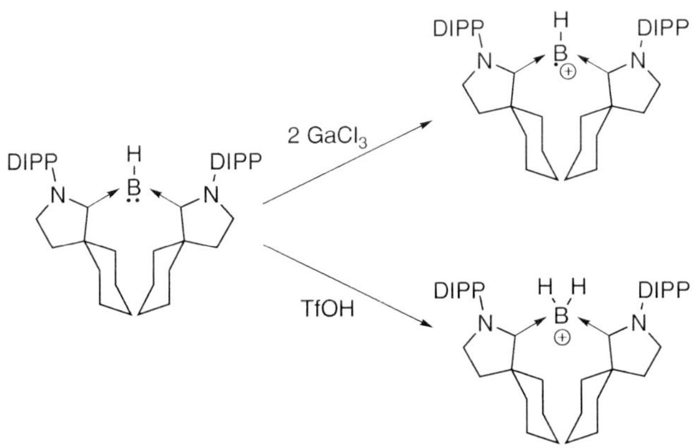

Figure 2.14. One electron oxidation or protonation of a carbene-stabilized, tricoordinate boron center.

nucleophilicity and low electrophilicity of highly stabilized carbenes. For example, **NHC$_4$** was reported to react with diethylmaleate but afforded the corresponding exocyclic olefin rather than a cyclopropane (Fig. 2.15).[16] The result may be explained by nucleophilic attack of the carbene onto the electron-deficient olefin to form a zwitterionic intermediate susceptible to a 1,2-proton shift that results in the olefinic product. A cyclopropane in equilibrium with the charge-separated intermediate may be involved in this process but has not yet been observed spectroscopically. The absence of cyclopropane formation may be exploited for certain applications. For example,

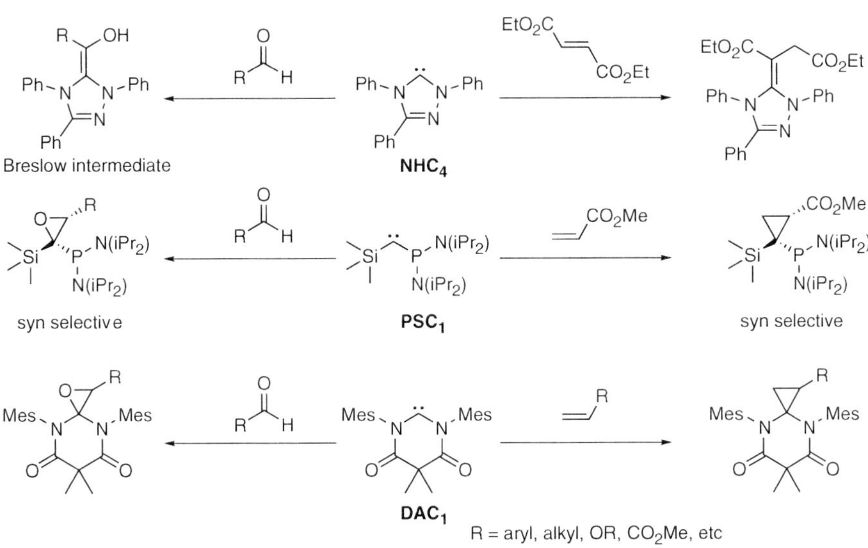

Figure 2.15. Chemical reactivity comparison of various NHCs, PSCs, and DACs.

Bazan and coworkers have demonstrated that, unlike more transient carbenes, NHCs form Lewis acid–base adducts with C_{60} and C_{70} in lieu of cyclopropanes. The results reflect the potential applications of NHC–fullerene interactions in organic semiconductors and photovoltaics.[108] Regardless, only ferrocenophanes,[32] phosphinosilyl-,[57, 109] alkylamino-,[55a] PAC-,[52, 110] and DACs[111, 112] have been reported to date as suitable partners in cyclopropanation reactions.

Generally, nucleophilic carbenes react more readily with electron-deficient alkenes, though latent electrophilic properties cannot be discounted when considering carbene–olefin interactions.[113] However, only the **DAC**$_1$ has been reported to cyclopropanate olefins more electron-rich than 4-methoxystyrene (e.g., vinyl ethers). Mechanistic studies into the cyclopropanation of olefins have supported an asynchronous mechanism that is typically initiated by the nucleophilic approach of the carbene.[114] The concerted nature of the cyclopropanations involving other singlet carbenes typically results in the retention of stereochemistry from the olefinic starting material,[115] as opposed to the stepwise cyclopropanation of radical carbenes that often results in scrambling. Still, exceptions have been observed where the thermodynamically favored trans-cyclopropane was formed from *cis*-olefins and singlet carbenes.[111, 116] In addition, Bertrand has demonstrated syn-selectivity with respect to the phosphanyl group when using phosphinosilylcarbenes (Fig. 2.15). Initially, the selectivity was attributed to secondary orbital interactions between the phosphorus and olefin substituent but later assigned to steric and electrostatic effects.[114b] Regardless, PSCs with chiral *N*-substituents at the phosphorus have further enabled enantioselective cyclopropanations[117] which has typically required a transition metal-based carbenoid species.

Compared to cyclopropanations, the [2+1] cycloaddition of a carbene and aldehyde are much less common, which may be at least in part due to the propensity of transient electrophilic carbenes to form ylides rather than oxiranes in the presence of aldehydes.[118] The only singlet carbenes active in oxiranation chemistry have been the phosphinosilylcarbenes and the DACs which both react with a variety of alkyl as well as aryl derivatives.[57, 111, 119] Stereoselectivity is observed with PSC-derived oxiranes in a manner similar to the aforementioned cyclopropanation chemistry and may be rendered enantioselective for some systems.[57, 119] The DAC-derived oxiranes are readily reversible under mild conditions as demonstrated by scrambling studies,[111] and in this manner are reminiscent of NHC-aldehyde interactions although different products are involved. NHCs attack the carbonyl carbon to form an inner salt that rapidly undergoes a 1,2-proton shift to the enolate form that is commonly referred to as Breslow's intermediate.[120] Indeed, homoenolate formation plays a crucial role in many of the organocatalytic applications involving NHCs[13] and may not be obtained from scaffolds that form oxiranes.

Beyond cyclopropanes and oxiranes, [2+1] cycloadditions may be an effective method for generating the smallest unsaturated cycles (e.g., azirines or cyclopropenes). Phosphinosilyl carbenes have been demonstrated to react with a variety of nitriles or phosphalkynes to afford azirines and phosphirenes,

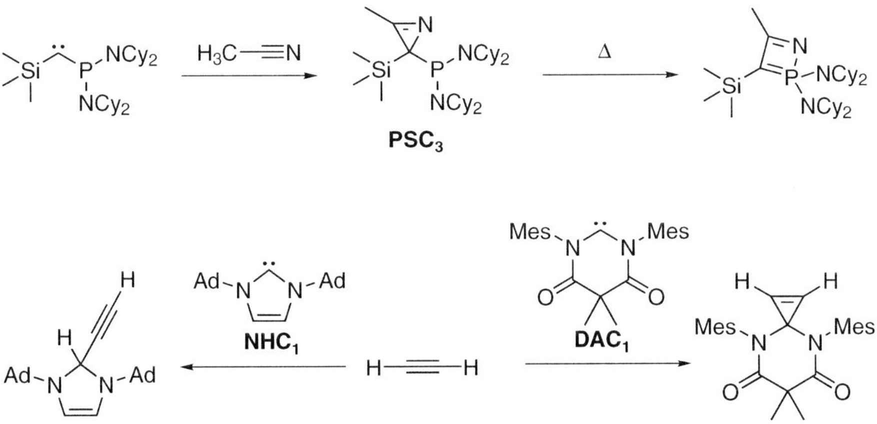

Figure 2.16. (top) C–H insertion between **NHC**$_1$ and acetylene compared to cyclopropenation with acetylene and **DAC**$_1$. (bottom) [2+1] cycloaddition of **PSC**$_3$ and acetonitrile followed by isomerization to the azaphosphete.

respectively, although ring-expansion to the corresponding azaphosphetes or diphospetes also may occur (Fig. 2.16).[121] Upon exposure to nitriles, **DAC**$_1$ also affords the corresponding azirines although the reaction is reversible (K_{eq}=4−72 M^{-1}).[122] Moreover, the **DAC**$_1$ is the only stable carbene reported to react with alkynes via a [2+1] cycloaddition. The cyclopropenation reaction is general for terminal, internal, alkyl as well as aryl alkynes, with a preference for more electron-deficient and less bulky alkyne substrates.[122]

The rarity of [2+1] cycloadditions between stable carbenes and substrates with triple bonds is related to the high proton affinities (NHCs=229.9–270.6 kcal/mol, **abNHC**$_1$=287.0 kcal/mol)[123] and high basicity of many isolable carbenes (pK_a=16–27 for typical NHCs in DMSO).[124] In lieu of cycloaddition chemistry, formal C–H insertions through acid–base type processes may occur as observed with **NHC**$_1$ and acetylene or acetonitrile (Fig. 2.16).[125] However, the lack of cyclizations with internal alkynes indicates that other factors such as the singlet–triplet gap energy and/or steric effects may also influence the reactivity. In total, [2+1] cycloadditions present a challenge for the development of new scaffolds and highlight the importance of accessing isolable carbenes with relatively small HOMO-LUMO gaps to facilitate the formation of unsaturated, three-membered rings.

2.4.4 Carbon Monoxide and Isonitriles

Whereas the coupling of carbenes with electrophilic partners such as CS_2 and other heteroallenes is a classic hallmark of nucleophilic carbenes and particularly well-studied for NHCs as well as AACs and DACs, the coupling of isonitriles has been utilized to study electrophilic characteristics. Due to their strongly nucleophilic and π-accepting properties, isonitriles may attack the

LUMO of a carbene followed by formation of the corresponding ketenimine. Carbenes with inaccessible LUMOs would therefore be expected to be unreactive toward isonitriles as is observed for many NHCs. However, various isolable scaffolds with lower-lying LUMOs such as phosphinosilyl-,[57] PAC-,[52] AA-,[53] aminophosphino-,[42] diamido-,[33,34c,35,36] and ferrocenophane[32] carbenes all have been shown to react with isonitriles. More recently, the LUMO of diaminocarbene was successfully lowered by incorporating a nitrogen atom into the bridgehead position of a bicycle which enabled coupling with various isonitriles.[84]

Though isonitriles and carbon monoxide are isolable, the formation of ketenes from stable carbenes has been relatively scarce. Several years after Arduengo's synthesis of **NHC**$_1$, Lyashchuk claimed that **NHC**$_1$ reacted with CO to afford the corresponding crystalline ketene (Fig. 2.17).[126] However, the result was refuted by Arduengo who found discrepancies in the crystal structure and that computationally the ketene was not located at an energy minimum.[127] Only the AACs and DACs have been found to couple carbon monoxide to afford isolable ketenes.[34a,c,35,128] In contrast, ferrocenophanes react with CO in a 2:1 fashion to form a zwitterion, presumably through the coupling of CO and ferrocenophane to afford an intermediate diaminoketene that undergoes attack by a second equivalent of carbene to form the oxyanion-containing inner salt.[32] Additionally, an acyclic alkylamino carbene known as Alder's carbene (**aNHC**$_1$), in the presence of CO yielded a beta-lactam derivative, presumably through decomposition of a ketene intermediate.[32] All of these scaffolds contain relatively small singlet–triplet gaps which may in part account for their reactivity with CO. The higher entropic penalty for coupling gaseous CO versus a liquid or solid isonitrile is also likely related to the large number of carbene coupling partners for isonitriles compared to CO.

2.4.5 X–H Activation

C–H insertions represent a valuable method for transforming relatively inert bonds into useful functional groups. Many transient and stable carbenes are known to insert into benzylic, tertiary, and other activated C–H groups. For example, **NHC**$_5$ and **DAC**$_2$ intramolecularly insert into the aryl methyl or isopropyl groups at elevated temperatures (50–70 °C; Fig. 2.18).[34a,129] Pyrolysis of **PSC**$_1$ above 300 °C results in decomposition through intramolecular insertion into the *N*-isopropyl tertiary carbon.[5] For entropic reasons, the formation of five-membered rings is favored over other ring sizes. This propensity is nicely illustrated in the synthesis of **AA**$_2$ which resulted in insertion into the *t*-butyl methyl group at room temperature even though the primary carbon is not activated toward insertion.[53] Beyond alkyl C–H scission, biphenyl-substituted **AA**$_3$ was reported to intramolecularly insert into an aryl C–H bond, precluding the isolation of the corresponding free carbene.[130]

Except in the case of **NHC**$_5$ for which hydrolysis of the resultant aminal affords an *N*-substituted indole,[129] intramolecular C–H insertions of stabilized

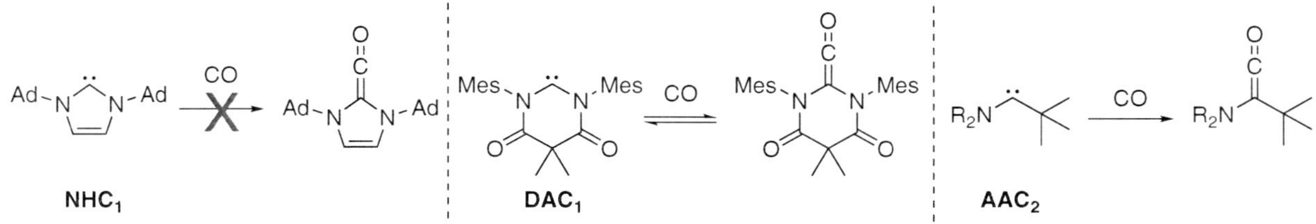

Figure 2.17. A summary of the chemistry observed between various stable NHCs and carbon monoxide.

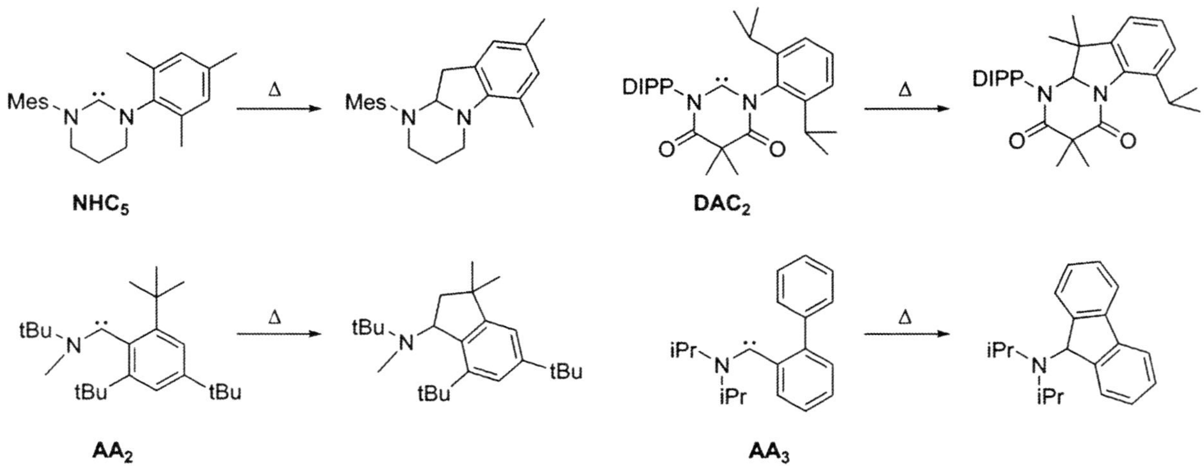

Figure 2.18. Intramolecular C–H insertions involving various carbenes.

carbenes are largely utilized to study carbene characteristics or intermediates rather than as precursors to more complex products. In contrast, intermolecular C–H insertions should facilitate the incorporation of a wide variety of structures and functionality and lead to new synthetic methodologies. However, since intramolecular reactions are favored kinetically and intermolecular C–H insertions disfavored entropically, the former dominates. Although acid–base-driven, formal C–H insertions such as those between **NHC**$_1$ and acetylene or acetonitrile have been realized,[125, 131] this area of carbene chemistry holds great potential. Only one isolable carbene, a six-membered NHC, has been reported to insert into a nonacidic C–H bond in an intramolecular fashion. In this case, the insertion of a ^{13}C-labeled NHC into a variety of toluene derivatives was reported to be catalyzed by residual potassium hexamethyldisilazide base.[132]

When compared to their lighter C–H congeners, compounds with Si–H bonds react more readily with carbenes and may undergo subsequent rearrangements. For example, heating various imidazolylidenes with Ph_3SiH resulted in ring-expansion concomitant with incorporation of the silicon atom into the carbene ring; this process is believed to proceed through an H–C–Si intermediate (Fig. 2.19).[133] Likewise, AACs also split the Si–H bonds of primary, secondary, and tertiary silanes as do other NHCs.[106] The broader activation of Si–H bonds is likely due to the low-lying orbitals of silicon susceptible to nucleophilic attack by the carbene lone pair, a view consistent with the coordination of tetrachloro- and tetrabromosilane by NHCs. Potassium metal reduction of the resultant Lewis acid–base pairs enabled the isolation of NHC-supported Si(II)Cl_2 or Si(II)Br_2, two compounds not previously isolable at room temperature.[134, 135]

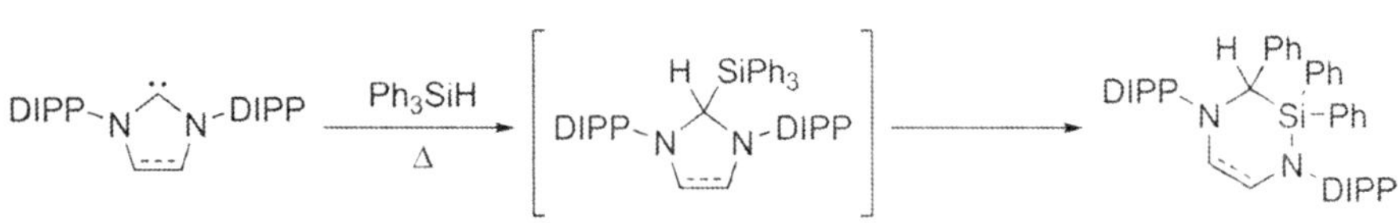

Figure 2.19. Example of the insertion of NHCs into a Si–H bond and subsequent ring-expansion.

The propensity of carbenes to split O–H bonds accounts for their often high sensitivity toward water and may limit the reaction conditions under which they may be used. In some cases, the addition of volatile alcohols such as methanol across the carbene carbon may be reversible upon heating under reduced pressure, a property that has been exploited to generate triazolylidenes.[16] The imidazol-2-ylidenes appear to show the greatest stability toward water. Indeed, 1,3-dimesityl-4,5-dichloroimidazol-2-ylidene has been reported to be stable toward air when kept in the solid state for several days or in solution overnight.[136] Still, reactivity with alcohols and water is generally rapid and can challenge the purification of free carbenes.

Despite the ubiquity of alcohol additions to carbenes, the addition of compounds containing P–H or N–H bonds are relatively rare, and may be attributed to the decreased acidity of these functional groups. Both AACs and

NHCs have been shown to insert into P–H bonds.[106] Insertion into secondary amines has been reported for triazolylidenes[16] as well as enetetramines formed by the dimerization of sterically unencumbered imidazolinylidenes,[11a] but a few other examples that involved stable NHCs have been reported. Recently, AACs, DACs, and ferrocenophanes were shown to activate ammonia, although, except for the latter, only the ring-opened isomer of the presumed initial N–H insertion product was observed.[32,34c,137] Ammonia activation is particularly noteworthy as many metals typically form Werner-type complexes rather than inducing N–H bond scission with this abundant gas.

2.4.6 Homodiatomic Bond Activation

Based on the similar bond dissociation energies of ammonia and dihydrogen, it may be possible for carbenes to activate dihydrogen, analogous to many transition metal complexes. Indeed, Bertrand demonstrated that exposing the AACs to an atmosphere of hydrogen at 35 °C reduced the AACs by adding H_2 across the carbene center (Fig. 2.20).[137] Computational studies indicated the process was initiated by nucleophilic attack of one hydrogen atom by the AAC, resulting in polarization of the H–H bond and subsequent bond scission. To date, no other carbene has been shown to activate dihydrogen, but this remains an area of intense interest as other nonmetallic systems such as frustrated Lewis pairs (FLPs) activate hydrogen and facilitate the hydrogenation of unsaturated substrates.[138] Together, the activation of ammonia, coupling of carbon monoxide (sometimes reversibly), and splitting of H_2 suggest that carbenes may hold potential as alternatives for catalytically active transition metals.

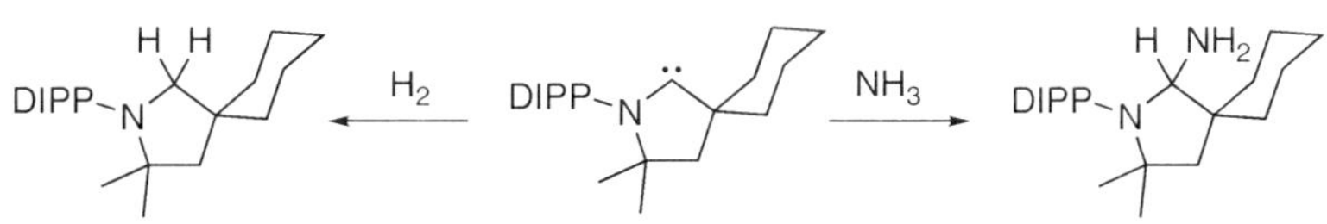

Figure 2.20. Activation of H_2 and ammonia by a cyclic AAC.

The activation of dihydrogen by the AACs also reflects the ability of isolable carbenes to activate other homonuclear bonds. Indeed, the cleavage of various homonuclear bonds such as O–O, S–S, C–C, and Br–Br have recently been reported either through the use of isolable carbenes or FLPs with a carbene as the Lewis base component. While FLPs typically are based on compounds which feature N–B or P–B connectivity, Alcarazo revealed that a completely carbogenic FLP comprised of a fluorene-derived allene and a bulky NHC may heterolytically cleave aryl disulfides to form imidazolium salts and vinylsulfides (Fig. 2.21, top).[139]

Given the strongly nucleophilic properties of NHCs and the excellent leaving group ability of bromides, the formation of 2-bromoimidazolium bromides has been observed upon mixing bromine with imidazolylidenes.[75] For stronger homonuclear bonds, the DACs have been shown to activate peroxides, disulfides, cyclopropenones, and various 1,2-diones (Fig. 2.21, bottom).[140] Collectively, the

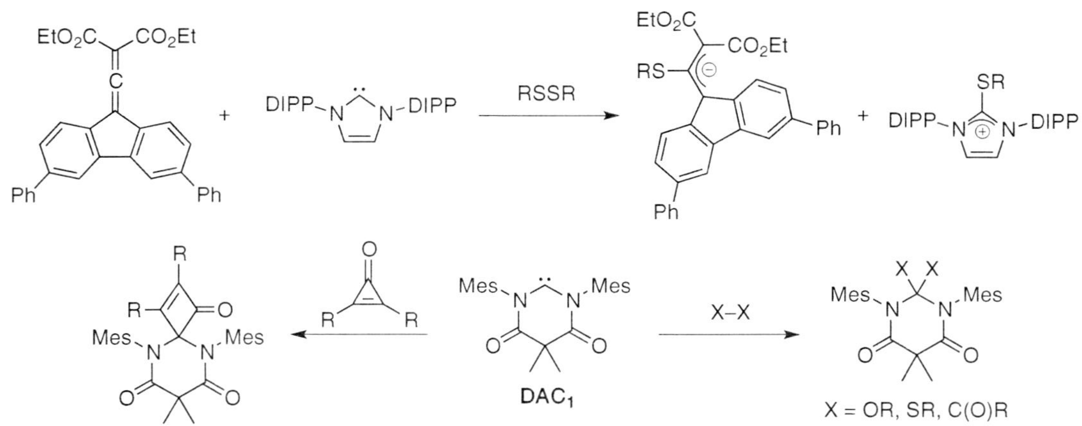

Figure 2.21. (top) Activation of disulfides by an NHC-containing FLP. (bottom) Insertion of $\mathbf{DAC_1}$ into the C–C bond of cyclopropenones or various homonuclear bonds.

results demonstrate the ability of the isolable carbenes to insert into a wide variety of homonuclear bonds and future work may generalize the reactivity to include other classes of carbenes as well as more inert bonds.

2.4.7 Elemental Allotropes and Other Reactive Species

Carbenes are now often employed as stabilizers and solubilizers of various reactive intermediates and elemental allotropes. This area has seen particular growth in the past five years as the use of the electron-donating properties of singlet carbenes for stabilizing a variety of reactive Lewis acids and other electron-deficient species has been uncovered. Highly pyrophoric in air, white phosphorus (P_4) is the starting material for almost all phosphine-containing compounds, and its activation remains of high interest. Various NHCs and AACs react with P_4 to give stabilized allotropes of phosphorus including P_2,[141] P_4,[142] and P_{12} (Fig. 2.22).[143] Further modification to crystalline P_2 radicals and dications was also reported,[144] while combining **CP**$_1$ and white phosphorus results in stabilized P_1.[145] Beyond P(0), crystalline phosphinyl radical cations,[146] phosphorus mononitride and its radical cation,[147] and P(III) cations[148] have also been stabilized using similar methods.

In addition to the Lewis acid–base interaction of carbenes and boron compounds mentioned earlier, similar interactions have also been utilized to form a B–B triple bond,[149] a boryl anion,[150] and a transient boraborylene.[151] Additionally, an NHC-beryllium borohydride was isolated following the reduction of the corresponding NHC-$BeCl_2$ adduct with lithium borohydride.[152] A variety of main group elements in the zero oxidation state also have been stabilized by the strong donating ability of NHCs and/or AACs, including Si_2,[153] digermanium(0),[154] and $[Fe_nS_n]^0$ clusters.[155] Moreover, Frenking and Tonner proposed through computational studies that carbenes may be used to stabilize C(0) that possesses two lone pairs of electrons.[156] This concept was realized through the subsequent isolation of crystalline **24** (Fig. 2.22)[157] and a related acyclic NHC-based allene[158] realized. Single crystal X-ray diffraction analysis revealed a C–C–C angle of 134.8°, indicating a large deviation from linearity consistent with a divalent C(0) rather than an allene structure. Beyond the stabilization of various elemental allotropes and reactive intermediates, the supporting organic carbene scaffolds solubilize what are or would likely be insoluble allotropes and/or reactive intermediates, facilitating their characterization and study of their chemistry.[159] Together, the positive stabilization derived from strongly donating carbene scaffolds is expected to continue to enhance our understanding of reactive intermediates derived from various elemental allotropes.

2.5 CONCLUSIONS AND OUTLOOK

Over the past 25 years, the views of carbenes have progressed from transient intermediates to stable compounds that have found tremendous utility in catalysis, small molecule activation, and the stabilization of reactive species.

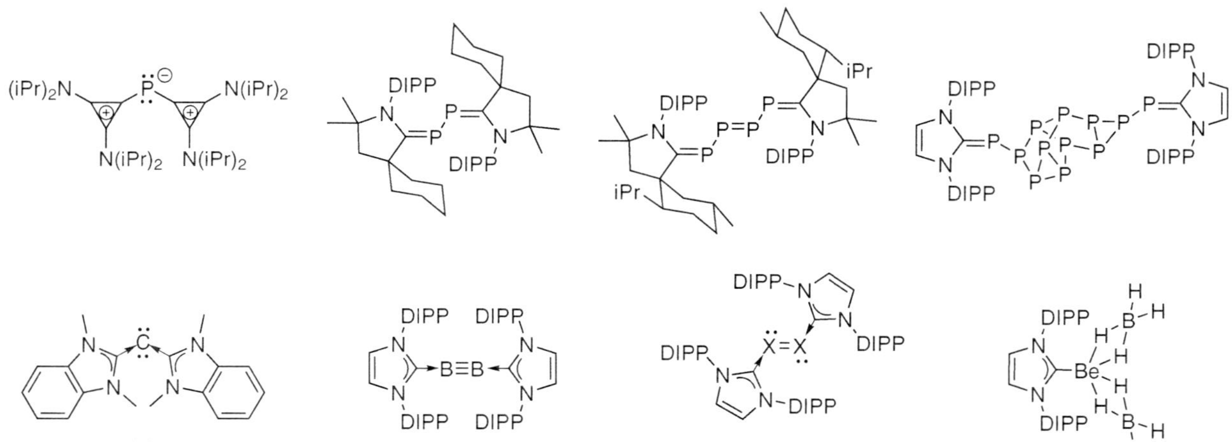

24

X = Si or Ge

Figure 2.22. Stabilization of various elemental allotropes and reactive intermediates with stable carbenes.

Considering the relatively brief history of isolable carbenes, tremendous strides have been made and the field still holds great potential. One challenge is the continued development of new scaffolds, particularly pull–pull systems, as well as novel synthetic methods for modifying existing systems to expand their chemical reactivity. Additionally, harnessing the full range of triplet carbene reactivity within an isolable carbene remains elusive. The activation of H_2, NH_3, and CO as well as the development of broadly applicable [2+1] cycloaddition chemistry have provided milestones toward achieving this goal, but intermolecular C–H insertions and the activation of more inert small molecules (e.g., methane) have not yet been realized. Perhaps one of the greatest challenges facing carbene chemistry is the ability to go beyond stabilization of reactive intermediates or stoichiometric activation of small molecules and develop catalytic processes that mimic transition metal-based catalytic methods (e.g., carbonylations, aminations, etc.). The activation of small molecules provides a foundation, but the development of new catalytic pathways may provide one of the most important advances in this field as it continues to move forward.

SUGGESTED READING

Bourissou, D., Guerret, O., Gabbaï, F. P., Bertrand, G. "Stable Carbenes," *Chem. Rev.* **2000**, *100*, 39.

Curran, D. P., Solovyev, A., Brahmi, M. M., Fensterbank, L., Malacria, M., Lacôte, E. "Synthesis and Reactions of N-Heterocyclic Carbene Boranes," *Angew. Chem. Int. Ed.* **2011**, *50*, 10294.

Dröge, T., Glorius, F. "The Measure of All Rings – N-Heterocyclic Carbenes," *Angew. Chem. Int. Ed.* **2010**, *49*, 6940–6952.

Frey, G. D., Lavallo, V., Donnadieu, B., Schoeller, W. W., Bertrand, G. "Facile Splitting of Hydrogen and Ammonia by Nucleophilic Activation at a Single Carbon Center," *Science* **2007**, *316*, 439.

Martin, D., Soleilhavoup, M., Bertrand, G. "Stable Singlet Carbenes as Mimics for Transition Metal Centers," *Chem. Sci.* **2011**, *2*, 389.

Melaimi, M., Soleilhavoup, M., Bertrand, G. "Stable Cyclic Carbenes and Related Species Beyond Diaminocarbenes," *Angew. Chem. Int. Ed.* **2010**, *49*, 8810.

Vignolle, J., Cattoën, X., Bourissou, D. "Stable Noncyclic Singlet Carbenes," *Chem. Rev.* **2009**, *109*, 3333.

Vougioukalakis, G. C., Grubbs, R. H. "Ruthenium-Based Heterocyclic Carbene-Coordinated Olefin Metathesis Catalysts," *Chem. Rev.* **2010**, *110*, 1746.

REFERENCES

1 (a) Dumas, J. B. *Ann. Chim. Phys.* **1835**, *58*, 28. (b) Regnault, H. V. *Ann. Chim. Phys.* **1839**, *71*, 427.

2 Doering, W. v. E., Knox, L. H. *J. Am. Chem. Soc.* **1956**, *78*, 4947. See footnote 9 therein.

3 Geuther, A. *Ann. Chem. Pharm.* **1862**, *123*, 121.

4 (a) Buchner, E., Curtius, T. *Ber. Dtsch. Chem. Ges.* **1885**, *8*, 2377. (b) Staudinger, H., Kupfer, O. *Ber. Dtsch. Chem. Ges.* **1911**, *44*, 2197. (c) Staudinger, H., Kupfer, O. *Ber. Dtsch. Chem. Ges.* **1912**, *45*, 501. (d) Staudinger, H., Anthes, E., Pfenninger, F. *Ber. Dtsch. Chem. Ges.* **1916**, *49*, 1928.

5 Igau, A., Grutzmacher, H., Baceiredo, A., Bertrand, G. *J. Am. Chem. Soc.* **1988**, *110*, 6463.

6 Arduengo, A. J., III, Harlow, R. L., Kline, M. *J. Am. Chem. Soc.* **1991**, *113*, 361.

7 Hirai, K., Itoh, T., Tomioka, H. *Chem. Rev.* **2009**, *109*, 3275.

8 Pauling, L. *J. Chem. Soc. Chem. Commun.* **1980**, *688*.

9 (a) Nair, V., Bindu, S., Sreekumar, V. *Angew. Chem. Int. Ed.* **2004**, *43*, 5130. (b) Dröge, T., Glorius, F. "The Measure of All Rings—*N*-Heterocyclic Carbenes," *Angew. Chem. Int. Ed.* **2010**, *49*, 6940.

10 (a) Bourissou, D., Guerret, O., Gabbaï, F. P., Bertrand, G. "Stable Carbenes," *Chem. Rev.* **2000**, *100*, 39. (b) Vignolle, J., Cattoën, X., Bourissou, D. "Stable Noncyclic Singlet Carbenes," *Chem. Rev.* **2009**, *109*, 3333. (c) Melaimi, M., Soleilhavoup, M., Bertrand, G. "Stable Cyclic Carbenes and Related Species Beyond Diaminocarbenes," *Angew. Chem. Int. Ed.* **2010**, *49*, 8810. (d) Martin, D., Soleilhavoup, M., Bertrand, G. "Stable Singlet Carbenes as Mimics for Transition Metal Centers," *Chem. Sci.* **2011**, *2*, 389.

11 (a) Hocker, J., Merten, R. *Angew. Chem. Int. Ed.* **1972**, *11*, 964. (b) Lappert, M. F. *J. Organomet. Chem.* **1988**, *358*, 185.

12 (a) Hahn, F. E., Jahnke, M. *C.* **2008**, *47*, 3122. (b) Díez-González, S., Marion, N., Nolan, S. P. *Chem. Rev.* **2009**, *109*, 3612. (c) Vougioukalakis, G. C., Grubbs, R. H. "Ruthenium-Based Heterocyclic Carbene-Coordinated Olefin Metathesis Catalysts," *Chem. Rev.* **2010**, *110*, 1746. (d) Valente, C., Çalimsiz, S., Joi, K. H., Mallik, D., Sayah, M., Organ, M. G. *Angew. Chem. Int. Ed.* **2012**, *51*, 3314.

13 (a) Marion, N., Díez-González, S., Nolan, S. P. *Angew. Chem. Int. Ed.* **2007**, *46*, 2988. (b) Enders, D., Niemeier, O., Henseler, A. *Chem. Rev.* **2007**, *107*, 5606. (c) Phillips, E. M., Chan, A., Scheidt, K. A. *Aldrichimica Acta* **2009**, *42*, 55. (d) Moore, J. L., Rovis, T. *Top. Curr. Chem.* **2010**, *291*, 77. (e) Kiesewetter, M. K., Shin, E. J., Hedrick, J. L., Waymouth, R. M. *Macromolecules* **2010**, *43*, 2093. (f) Chiang, P.-C., Bode, J. W. *Org. Lett.* **2011**, *13*, 2422. (g) Bugaut, X., Glorius, F. *Chem. Soc. Rev.* **2012**, *41*, 3511.

14 Clavier, H., Nolan, S. P. *Chem. Commun.* **2010**, *46*, 841.

15 (a) Khramov, D. M., Lynch, V. M., Bielawski, C. W. *Organometallics* **2007**, *26*, 6042. (b) Gaillard, S., Slawin, A. M. Z., Bonura, A. T., Stevens, E. D., Nolan, S. P. *Organometallics* **2010**, *29*, 394. (c) Würtz, S., Lohre, C., Fröhlich, R., Bergander, K., Glorius, F. *J. Am. Chem. Soc.* **2009**, *131*, 8344. (d) Mendoza-Espinosa, D., Donnadieu, B., Bertrand, G. *J. Am. Chem. Soc.* **2010**, *132*, 7264–7265.

16 Enders, D., Breuer, K., Raabe, G., Runsink, J., Teles, J.H., Melder, J.-P., Ebel, K., Brode, S. *Angew. Chem. Int. Ed. Engl.* **1995**, *34*, 1021.

17 (a) Krahulic, K. E., Enright, G. D., Parvez, M., Roesler, R. *J. Am. Chem. Soc.* **2005**, *127*, 4142. (b) Ishida, Y., Donnadieu, B., Bertrand, G. *Proc. Natl. Acad. Sci. USA* **2006**, *103*, 13585.

18 Despagnet-Ayoub, E., Grubbs, R. H. *J. Am. Chem. Soc.* **2004**, *126*, 10198.

19 Bantu, B., Pawar, G. M., Decker, U., Wurst, K., Schmidt, A. M., Buchmeiser, M. R. *Chem. Eur. J.* **2009**, *15*, 3103.

20 (a) Boche, G., Hilf, C., Harms, K., Marsch, M., Lohrenz, J. C. W. *Angew. Chem. Int. Ed.* **1995**, *34*, 487. (b) Alder, R. W., Blake, M. E., Bortolotti, C., Bufali, S., Butts, C. P., Linehan, E., Oliva, J. M., Orpen, A. G., Quayle, M. *J. Chem. Commun.* **1999**, 241.

21 (a) Duong, H. A., Tekavec, T. N., Arif, A. M., Louie, *J. Chem. Commun.* **2004**, 112. (b) Voutchkova, A. M., Appelhans, L. N., Chianese, A. R., Crabtree, R. H. *J. Am. Chem. Soc.* **2005**, *127*, 17624.
22 (a) Norris, B. C., Bielawski, C. W. *Macromolecules* **2010**, *43*, 3591. (b) Norris, B.C., Sheppard, D. G., Henkelman, G., Bielawski, C. W. *J. Org. Chem.* **2011**, *76*, 301.
23 Otto, M., Conejero, S., Canac, Y., Romanenko, V. D., Rudzevitch, V., Bertrand, G. *J. Am. Chem. Soc.* **2004**, *126*, 1016.
24 (a) Kuhn, N., Kratz, T. *Synthesis* **1993**, *561*. (b) Saravanakumar, S., Oprea, A. I., Kindermann, M. K., Jones, P. G., Heinicke, *J. Chem. Eur. J.* **2006**, *12*, 3143. (c) Denk, M. K., Hezarkhani, A., Zheng, F.-L. *Eur. J. Inorg. Chem.* **2007**, *3527*.
25 Alder, R. W., Allen, P. R., Murray, M., Orpen, A. G. *Angew. Chem. Int. Ed.* **1996**, *35*, 1121.
26 Snead, D. R., Inagaki, S., Abboud, K. A., Hong, S. *Organometallics*, **2010**, *29*, 1729.
27 Rosen, E. L., Sanderson, M. D., Saravanakumar, S., Bielawski, C. W. *Organometallics* **2007**, *26*, 5774.
28 Seo, H., Roberts, B. P., Abboud, K. A., Merz, K. M., Jr., Hong, S. *Org. Lett.* **2010**, *12*, 4860.
29 (a) Alder, R. W., Blake, M. E. *J. Phys. Chem. A.* **1999**, *103*, 11200. (b) Collins, M. S., Rosen, E. L., Lynch, V. M., Bielawski, C. W. *Organometallics* **2010**, *29*, 3047.
30 Denk, K., Sirsch, P., Herrmann, W. A. *J. Organomet. Chem.* **2002**, *649*, 219.
31 (a) Rosen, E. L., Sung, D. H., Chen, Z., Lynch, V. M., Bielawski, C. W. *Organometallics* **2010**, *29*, 250. (b) Seo, H., Snead, D. R., Abboud, K. A., Hong, S. *Organometallics* **2011**, *30*, 5725. (c) Wang, Y.-M., Kuzniewski, C. N., Rauniyar, V., Hoong, C., Toste, F. D. *J. Am. Chem. Soc.* **2011**, *133*, 12972. (d) Handa, S., Slaughter, L. M. *Angew. Chem. Int. Ed.* **2012**, *51*, 2912. (e) Boyarskiy, V. P., Luzyanin, K. V., Kukushkin, V. Y. *Coord. Chem. Rev.* **2012**, *256*, 2029. (f) Slaughter, L. M. ACS *Catal. 2012*, *2*, **1802**.
32 Siemeling, U., Färber, C., Bruhn, C., Leibold, M., Selent, D., Baumann, W., von Hopffgarten, M., Goedecke, C., Frenking, G. *Chem. Sci.* **2010**, *1*, 697.
33 Braun. M., Frank, W., Reiss, G. J., Ganter, C. *Organometallics* **2010**, *29*, 4418.
34 (a) Hudnall, T. W., Bielawski, C. W. *J. Am. Chem. Soc.* **2009**, *131*, 16039. (b) César, V., Lugan, N., Lavigne, G. *Eur. J. Inorg. Chem.* **2010**, 361. (c) Hudnall, T. W., Moerdyk, J. P., Bielawski, C. W. *Chem. Commun.* **2010**, *46*, 4288.
35 Hudnall, T. W., Tennyson, A. G., Bielawski, C. W. *Organometallics* **2010**, *29*, 4569.
36 Hudnall, T. W., Moorhead, E. J., Gusev, D. G., Bielawski, C. W. *J. Org. Chem.* **2010**, *75*, 2763.
37 Moerdyk, J. P., Bielawski, C. W. *Organometallics* **2011**, *30*, 2278.
38 (a) Makhloufi, A., Frank, W., Ganter, C. *Organometallics* **2012**, *31*, 2001. (b) Blake, G. A., Moerdyk, J. P., Bielawski, C. W. *Organometallics* **2012**, *31*, 3373. (c) Mushinski, R. M., Squires, B. M., Sincerbox, K. A., Hudnall, T. W. *Organometallics* **2012**, *31*, 4862.
39 Martin, D., Baceiredo, A., Gornitzka, H., Schoeller, W. W., Bertrand, G. *Angew. Chem. Int. Ed.* **2005**, *44*, 1700.
40 (a) Marrot, S., Kato, T., Gornitzka, H., Baceiredo, A. *Angew. Chem. Int. Ed.* **2006**, *45*, 2598. (b) Marrot, S., Kato, T., Cossío, F. P., Gornitzka, H., Baceiredo, A. *Angew. Chem. Int. Ed.* **2006**, *45*, 7447.
41 (a) Dixon, D. A., Arduengo, A. J., III. *J. Phys. Chem.* **1991**, *95*, 4180. (b) Schoeller, W. W., Eisner, D. *Inorg. Chem.* **2004**, *43*, 2585.
42 Merceron, N., Miqueu, K., Baceiredo, A., Bertrand, G. *J. Am. Chem. Soc.* **2002**, *124*, 6806.

43 Lavallo, V., Canac, Y., Donnadieu, B., Schoeller, W. W., Bertrand, G. *Science* **2006**, *312*, 722.
44 Kuchenbeiser, G., Donnadieu, B., Bertrand, G. *J. Organomet. Chem.* **2008**, *693*, 899.
45 Kluger, R. *Chem. Rev.* **1987**, *87*, 863.
46 Ugai, T., Tanaka, S., Dokawa, S. *J. Pharm. Soc. Jpn.* **1943**, *63*, 296.
47 Arduengo, A. J., III, Goerlich, J. R., Marshall, W. *J. Liebigs Ann.* **1997**, *365*.
48 Alder, R. W., Butts, C. P., Orpen, A. G. *J. Am. Chem. Soc.* **1998**, *120*, 11526.
49 Buron, C., Gornitzka, H., Romanenko, V., Bertrand, G. *Science* **2000**, *288*, 834.
50 Despagnet-Ayoub, E., Solé, S., Gornitzka, H., Rozhenko, A. B., Schoeller, W. W., Bourissou, D., Bertrand, G. *J. Am. Chem. Soc.* **2003**, *125*, 124.
51 Despagnet, E., Miqueu, K., Gornitzka, H., Dyer, P. W., Bourissou, D., Bertrand, G. *J. Am. Chem. Soc.* **2002**, *124*, 11834.
52 Despagnet-Ayoub, E., Gornitzka, H., Bourissou, D., Bertrand, G. *Eur. J. Org. Chem.* **2003**, 2039.
53 Solé, S., Gornitzka, H., Schoeller, W. W., Bourissou, D., Bertrand, G. *Science* **2001**, *292*, 1901.
54 Cattoën, X., Gornitzka, H., Bourissou, D., Bertrand, G. *J. Am. Chem. Soc.* **2004**, *126*, 1342.
55 (a) Lavallo, V., Mafhouz, J., Canac, Y., Donnadieu, B., Schoeller, W. W., Bertrand, G. *J. Am. Chem. Soc.* **2004**, *126*, 8670. (b) Lavallo, V., Canac, Y., Präsang, C., Donnadieu, B., Bertrand, G. *Angew. Chem. Int. Ed.* **2005**, *44*, 5705.
56 Zeng, X., Frey, G. D., Kinjo, R., Donnadieu, B., Bertrand, G. *J. Am. Chem. Soc.* **2009**, *131*, 8690.
57 Igau, A., Baceiredo, A., Trinquier, G., Bertrand, G. *Angew. Chem. Int. Ed.* **1989**, *28*, 621.
58 Merceron-Saffon, N., Baceiredo, A., Gornitzka, H., Bertrand, G. *Science* **2003**, *301*, 1223.
59 Soleilhavoup, M., Baceiredo, A., Treutler, O., Ahlrichs, R., Nieger, M., Bertrand, G. *J. Am. Chem. Soc.* **1992**, *114*, 10959.
60 Aldeco-Perez, E., Rosenthal, A. J., Donnadieu, B., Parameswaran, P., Frenking, G., Bertrand, G. *Science* **2009**, *326*, 556.
61 Gründemann, S., Kovacevic, A., Albrecht, M., Faller, J. W., Crabtree, R. H. *J. Am. Chem. Soc.* **2002**, *124*, 10473.
62 Gründemann, S., Kovacevic, A., Albrecht, M., Faller, J. W., Crabtree, R. H. *Chem. Commun.* **2001**, *2274*.
63 Matthew, P., Neels, A., Albrecht, M. *J. Am. Chem. Soc.* **2008**, *130*, 13534.
64 Mendoza-Espinosa, D., Ung, G., Donnadieu, B., Bertrand, G. *Chem. Commun.* **2011**, *47*, 10614.
65 Ung, G., Mendoza-Espinosa, D., Bertrand, G. *Chem. Commun.* **2012**, *48*, 7088.
66 (a) Ung, G., Mendoza-Espinosa, D., Bouffard, J., Bertrand, G. *Angew. Chem. Int. Ed.* **2011**, *50*, 4215. (b) Ung, G., Frey, G. D., Schoeller, W. W., Bertrand, G. *Angew. Chem. Int. Ed.* **2011**, *50*, 9923.
67 (a) Albrecht, M., Cavell, K. *J. Organomet. Chem.* **2009**, *35*, 47. (b) Ung, G., Bertrand, G. *Chem. Eur. J.* **2011**, *17*, 8269.
68 Tapu, D., Dixon, D. A., Roe, C. *Chem. Rev.* **2009**, *109*, 3385.
69 Tolman, C. A. *Chem. Rev.* **1977**, *77*, 313.
70 (a) Chianese, A. R., Li, X., Janzen, M. C., Faller, J. W., Crabtree, R. H. *Organometallics* **2003**, *22*, 1663. (b) Kelly, R. A., III, Clavier, H., Giudice, S., Scott, N. M., Stevens, E. D., Bordner, J., Samardjiev, I., Hoff, C. D., Cavallo, L., Nolan, S. P. *Organometallics* **2008**, *27*, 202.

71 (a) Wolf, S., Plenio, H. *J. Organomet. Chem.* **2009**, *694*, 1487. (b) Vorfalt, T., Leuthäußer, S., Plenio, H. *Angew. Chem. Int. Ed.* **2009**, *48*, 5191.

72 (a) Lever, A. B. P. *Inorg. Chem.* **1990**, *29*, 1271. (b) Fielder, S. S., Osborne, M. C., Lever, A. B. P., Pietro, W. J. *J. Am. Chem. Soc.* **1995**, *117*, 6990.

73 (a) Süßner, M., Plenio, H. *Chem. Commun.* **2005**, 5417. (b) Süßner, M., Plenio, H. *Angew. Chem. Int. Ed.* **2005**, *44*, 6885.

74 Arduengo, A. J., III, Dias, H. V. R., Harlow, R. L., Kline, M. *J. Am. Chem. Soc.* **1992**, *114*, 5530.

75 Cole, M. L., Jones, C., Junk, P.C. *New J. Chem.* **2002**, *26*, 1296.

76 Arduengo, A. J., III, Goerlich, J. R., Krafczyk, R., Marshall, W. J. *Angew. Chem. Int. Ed.* **1998**, *37*, 1963.

77 Ogle, J. W., Miller, S. A. *Chem. Commun.* **2009**, 5728.

78 Hahn, F. E., Wittenbecher, L., Boese, R., Bläser, D. *Chem. Eur. J.* **1999**, *5*, 1931.

79 Nonnenmacher, M., Kunz, D., Rominger, F., Oeser, T. *Chem. Commun.* **2006**, *1378*.

80 Arduengo, A. J., III, Goerlich, J. R., Marshall, W. J. *J. Am. Chem. Soc.* **1995**, *117*, 11027.

81 Iglesias, M., Beetstra, D. J., Knight, J. C., Ooi, L.-L., Stasch, A., Coles, S., Male, L., Hursthouse, M. B., Cavell, K. J., Dervisi, A., Fallis, I. A. *Organometallics* **2008**, *27*, 3279.

82 Iglesias, M., Beetstra, D. J., Kariuki, B., Cavell, K. J., Dervisi, A., Fallis, I. A. *Eur. J. Inorg. Chem.* 2009, **1913**.

83 Lu, W. Y., Cavell, K. J., Wixey, J. S., Kariuki, B. *Organometallics* **2011**, *30*, 5649.

84 Martin, D., Lassauque, N., Donnadieu, B., Bertrand, G. *Angew. Chem. Int. Ed.* **2012**, *51*, 6172.

85 Despagnet-Ayoub, E., Grubbs, R. H. *Organometallics* **2005**, *24*, 338.

86 For the close analog where the groups at C4 and C5 are methyls rather than isopropyls. See: Piel, I., Pawelczyk, M. D., Hirano, K., Fröhlich, R., Glorius, F. *Eur. J. Org. Chem.* **2011**, *5475*.

87 Lavigne, F., Maerten, E., Alcaraz, G., Saffon-Merceron, N., Acosta-Silva, C., Branchadell, V., Baceiredo, A. *J. Am. Chem. Soc.* **2010**, *132*, 8864.

88 Kato, T., Gornitzka, H., Baceiredo, A., Savin, A., Bertrand, G. *J. Am. Chem. Soc.* **2000**, *122*, 998.

89 Delaude, L. *Eur. J. Inorg. Chem.* **2009**, 1681.

90 Cheng, Y., Meth-Cohn, O. *Chem. Rev.* **2004**, *104*, 2507.

91 Lee, Y.-G., Moerdyk, J. P., Bielawski, C. W. *J. Phys. Org. Chem.* **2012**, *25*, 1027.

92 Duong, H. A., Cross, M. J., Louie, *J. Org. Lett.* **2004**, *6*, 4679.

93 (a) Winberg, H. E., Coffmann, D. D. *J. Am. Chem. Soc.* **1965**, *87*, 2776. (b) Khramov, D. M., Bielawski, C. W. *Chem. Commun.* **2005**, 4958. (c) Ono, R. J., Suzuki, Y., Khramov, D. M., Ueda, M., Sessler, J. L., Bielawski, C. W. *J. Org. Chem.* **2011**, *76*, 3239.

94 Coady, D. J., Khramov, D. M., Norris, B. C., Tennyson, A. G., Bielawski, C. W. *Angew. Chem. Int. Ed.* **2009**, *48*, 5187.

95 Tskhovrebov, A. G., Solari, E., Wodrich, M. D., Scopelliti, R., Severin, K. *Angew. Chem. Int. Ed.* **2012**, *51*, 232.

96 Leont'ev, A. V., Fomicheva, O. A., Proskurnina, M. V., Zefirov, N. S. *Russ. Chem. Rev.* **2001**, *70*, 91.

97 Tskhovrebov, A. G., Solari, E., Wodrich, M. D., Scopelliti, R., Severin, K. *J. Am. Chem. Soc.* **2012**, *134*, 1471.

98 Arduengo, A. J., III, Dias, H. V. R., Calabrese, J. C., Davidson, F. *J. Am. Chem. Soc.* **1992**, *114*, 9724.

99 (a) Alcaraz, G., Reed, R., Baceiredo, A., Bertrand, G. *J. Chem. Soc. Chem. Commun.* **1993**, *1354*. (b) Curran, D. P., Solovyev, A., Brahmi, M. M., Fensterbank, L., Malacria, M., Lacôte, E. "Synthesis and Reactions of *N*-Heterocyclic Carbene Boranes," *Angew. Chem. Int. Ed.* **2011**, *50*, 10294.
100 (a) Solovyev, A., Chu, Q., Geib, S. J., Fensterbank, L., Malacria, M., Lacôte, E., Curran, D. P. *J. Am. Chem. Soc.* **2010**, *132*, 15072. (b) McArthur, D., Butts, C. P., Lindsay, D. M. *Chem. Commun.* **2011**, *47*, 6650.
101 Lalevée, J., Telitel, S., Tehfe, M. A., Fouassier, J. P., Curran, D. P., Lacôte, E. *Angew. Chem. Int. Ed.* **2012**, *51*, 5958.
102 Pan, X., Lacôte, E., Lalevée, J., Curran, D. P. *J. Am. Chem. Soc.* **2012**, *134*, 5669.
103 Ueng, S.-H., Solovyev, A., Yuan, X., Geib, S. J., Fensterbank, L., Lacôte, E., Malacria, M., Newcomb, M., Walton, J. C., Curran, D. P. *J. Am. Chem. Soc.* **2009**, *131*, 11256.
104 Lindsay, D. M., McArthur, D. *Chem. Commun.* **2010**, *46*, 2474.
105 Nagura, K., Saito, S., Fröhlich, R., Glorius, F., Yamaguchi, S. *Angew. Chem. Int. Ed.* **2012**, *51*, 7762.
106 Frey, G. D., Masuda, J. D., Donnadieu, B., Bertrand, G. *Angew. Chem. Int. Ed.* **2010**, *49*, 9444.
107 Kinjo, R., Donnadieu, B., Celik, M. A., Frenking, G., Bertrand, G. *Science* **2011**, *333*, 610.
108 Li, H., Risko, C., Seo, J. H., Campbell, C., Wu, G., Brédas, J.-L., Bazan, G. C. *J. Am. Chem. Soc.* **2011**, *133*, 12410.
109 (a) Goumri-Magnet, S., Kato, T., Gornitzka, H., Baceiredo, A., Bertrand, *G.* **2000**, *122*, 4464. (b) Krysiak, J., Kato, T., Gornitzka, H., Baceiredo, A., Mikolajczyk, M., Bertrand, G. *J. Org. Chem.* **2001**, *66*, 8240.
110 Matsubara, S., Yamamoto, H., Oshima, K. *Angew. Chem. Int. Ed.* **2002**, *41*, 2835.
111 Moerdyk, J. P., Bielawski, C. W. *Nature Chem.* **2012**, *4*, 275.
112 Braun, M., Frank, W., Ganter, C. *Organometallics* **2012**, *31*, 1927.
113 Moss, R. A., Zhang, M., *Krog-Jespersen. Org. Lett.* **2009**, *11*, 1947.
114 (a) Keating, A. E., Merrigan, S. R., Singleton, D. A., Houk, K. N. *J. Am. Chem. Soc.* **1999**, *121*, 3933. (b) Lecea, B., Ayerbe, M., Arrieta, A., Cossio, F. P., Branchadell, V., Ortuño, R. M., Baceiredo, A. *J. Org. Chem.* **2007**, *72*, 357. (c) Xu, L., Doubleday, C. E., Houk, K. N. *J. Am. Chem. Soc.* **2011**, *133*, 17848.
115 Skell, P. S., Garner, A. Y. *J. Am. Chem. Soc.* **1956**, *78*, 3409.
116 (a) Mukai, T., Nakazawa, T., Isobe, *K. Tetrahedron Lett.* **1968**, *9*, 565. (b) Moss, R. A., Huselton, J. K. *J. Chem. Soc. Chem. Commun.* **1976**, *950*.
117 Krysiak, J., Lyon, C., Baceiredo, A., Gornitzka, H., Mikolajczyk, M., Bertrand, G. *Chem. Eur. J.* **2004**, *10*, 1982.
118 Padwa, A., Hornbuckle, S. F. *Chem. Rev.* **1991**, *91*, 263.
119 (a) Illa, O., Gornitzka, H., Baceiredo, A., Bertrand, G., Branchadell, V., Ortuño, R. M. *J. Org. Chem.* **2003**, *68*, 7707. (b) Illa, O., Gornitzka, H., Branchadell, V., Baceiredo, A., Bertrand, G., Ortuño, R. M. *Eur. J. Org. Chem.* **2003**, *3147*. (c) Illa, O., Álvarez, Á.-L., Baceiredo, A., Branchadell, V., Ortuño, R. M. *Tetrahedron: Asymmetry* **2007**, *18*, 2617.
120 Breslow, R. *J. Am. Chem. Soc.* **1958**, *80*, 3719.
121 (a) Alcaraz, G., Wecker, U., Baceiredo, A., Dahan, F., Bertrand, G. *Angew. Chem. Int. Ed. Engl.* **1995**, *34*, 1246. (b) Armhurst, R., Sanchez, M., Réau, R., Bergsträsser, U., Regitz, M., Bertrand, G. *J. Am. Chem. Soc.* **1995**, *117*, 10785. (c) Piquet, V., Baceiredo, A., Gornitzka, H., Dahan, F., Bertrand, G. *Chem. Eur. J.* **1997**, *3*, 1757.

(d) Sanchez, M., Réau, R., Gornitzka, H., Dahan, F., Regitz, M., Bertrand, G. *J. Am. Chem. Soc.* **1997**, *119*, 9720. (e) Sanchez, M., Réau, R., Marsden, C. J., Regitz, M., Bertrand, G. *Chem. Eur. J.* **1999**, *5*, 274.
122 Moerdyk, J. P., Bielawski, C. W. *J. Am. Chem. Soc.* **2012**, *134*, 6116.
123 Tonner, R., Heydenrych, G., Frenking, G. *ChemPhysChem* **2008**, *9*, 1474.
124 Magill, A. M., Cavell, K. J., Yates, B. F. *J. Am. Chem. Soc.* **2004**, *126*, 8717.
125 Arduengo, A. J., III, Calabrese, J. C., Davidson, F., Dias, H. V. R., Goerlich, J. R., Krafczyk, R., Marshall, W. J., Tamm, M., Schmutzler, R. *Helv. Chim. Acta* **1999**, *82*, 2348.
126 Lyashchuk, S. N., Skrypnik, Y. G. *Tetrahedron Lett.* **1994**, *35*, 5271.
127 Dixon, D. A., Arduengo, A. J., III, Dobbs, K. D., Khasnis, D. V. *Tetrahedron Lett.* **1995**, *36*, 645.
128 Lavallo, V., Canac, Y., Donnadieu, B., Schoeller, W. W., Bertrand, G. *Angew. Chem. Int. Ed.* **2006**, *45*, 3488.
129 Holdroyd, R. S., Page, M. J., Warren, M. R., Whittlesey, M. K. *Tetrahedron Lett.* **2010**, *51*, 557.
130 Vignolle, J., Asay, M., Miqueu, K., Bourissou, D., Bertrand, G. *Org. Lett.* **2008**, *10*, 4299.
131 Korotkikh, N. I., Rayenko, G. F., Shvaika, O. P., Pekhtereva, T. M., Cowley, A. H., Jones, J. N., Macdonald, C. L. B. *J. Org. Chem.* **2003**, *68*, 5762.
132 Lloyd-Jones, G. C., Alder, R. W., Owen-Smith, G. J. *J. Chem. Eur. J.* **2006**, *12*, 5361.
133 Schmidt, D., Berthel, J. H. J., Pietsch, S., Radius, U. *Angew. Chem. Int. Ed.* **2012**, *51*, 8881.
134 (a) Ghadwal, R. S., Roesky, H. W., Merkel, S., Henn, J., Stalke, D. *Angew. Chem. Int. Ed.* **2009**, *48*, 5683. (b) Filippou, A. C., Chernov, O., Schnakenburg, G. *Angew. Chem. Int. Ed.* **2009**, *48*, 5687.
135 Singh, A. P., Ghadwal, R. S., Roesky, H. W., Holstein, J. J., Dittrich, B., Demers, J.-P., Chevelkov, V., Lange, A. *Chem. Commun.* **2012**, *48*, 7574.
136 Arduengo, A. J., III, Davidson, F., Dias, H. V. R., Goerlich, J. R., Khasnis, D., Marshall, W. J., Prakasha, T. K. *J. Am. Chem. Soc.* **1997**, *119*, 12742.
137 Frey, G. D., Lavallo, V., Donnadieu, B., Schoeller, W. W., Bertrand, G. "Facile Splitting of Hydrogen and Ammonia by Nucleophilic Activation at a Single Carbon Center," *Science* **2007**, *316*, 439.
138 (a) Chase, P. A., Welch, G. C., Jurca, T., Stephan, D. W. *Angew. Chem. Int. Ed.* **2007**, *46*, 8050. (b) Stephan, D. W., Greenberg, S., Graham, T. W., Chase, P., Hastie, J. J., Geier, S. J., Farrell, J. M., Brown, C. C., Heiden, Z. M., Welch, G. C., Ullrich, M. *Inorg. Chem.* **2011**, *50*, 12338. (c) Axenov, K. V., Kehr, G., Fröhlich, R., Erker, G. *J. Am. Chem. Soc.* **2009**, *131*, 3454. (d) Schwendemann, S., Tumay, T. A., Axenov, K. V., Peuser, I., Kehr, G., Fröhlich, R., Erker, G. *Organometallics* **2010**, *29*, 1067.
139 Inés, B., Holle, S., Goddard, R., Alcarazo, M. *Angew. Chem. Int. Ed.* **2010**, *49*, 8389.
140 Wiggins, K. M., Moerdyk, J. P., Bielawski, C. W. *Chem. Sci.* **2012**, *3*, 2986.
141 Wang, Y., Xie, Y., Wei, P., King, R. B., Schaefer, H. F., III, Schleyer, P. v. R., Robinson, G. H. *J. Am. Chem. Soc.* **2009**, *130*, 14970.
142 Masuda, J. D., Schoeller, W. W., Donnadieu, B., Bertrand, G. *Angew. Chem. Int. Ed.* **2007**, *46*, 7052.
143 Masuda, J. D., Schoeller, W. W., Donnadieu, B., Bertrand, G. *J. Am. Chem. Soc.* **2007**, *129*, 14180.
144 Back, O., Donnadieu, B., Parameswaran, P., Frenking, G., Bertrand, G. *Nature Chem.* **2010**, *2*, 369.

145 Back, O., Kuchenbeiser, G., Donnadieu, B., Bertrand, G. *Angew. Chem. Int. Ed.* **2009**, *48*, 5530.
146 Back, O., Celik, M. A., Frenking, G., Melaimi, M., Donnadieu, B., Bertrand, G. *J. Am. Chem. Soc.* **2010**, *132*, 10262.
147 Kinjo, R., Donnadieu, B., Bertrand, G. *Angew. Chem. Int. Ed.* **2010**, *49*, 5930.
148 Weigand, J. J., Feldmann, K.-O., Henne, F. D. *J. Am. Chem. Soc.* **2010**, *132*, 16321.
149 Braunschweig, H., Dewhurst, R. D., Hammond, K., Mies, J., Radacki, K., Vargas, A. *Science* **2012**, *336*, 1420.
150 Braunschweig, H., Chiu, C.-W., Radacki, K., Kupfer, T. *Angew. Chem. Int. Ed.* **2010**, *49*, 2041.
151 Bissinger, P., Braunschweig, H., Damme, A., Dewhurst, R. D., Kupfer, T., Radacki, K., Wagner, K. *J. Am. Chem. Soc.* **2011**, *133*, 19044.
152 Gilliard, R. J., Jr., Abraham, M. Y., Wang, Y., Wei, P., Xie, Y., Quillian, B., Schaefer, H. F., III, Schleyer, P. v. R., Robinson, G. H. *J. Am. Chem. Soc.* **2012**, *134*, 9953.
153 Wang, Y., Xie, Y., Wei, P., King, B., Schaefer, H. F., III, Schleyer, P. v. R., Robinson, G. H. *Science* **2008**, *321*, 1069.
154 Sidiropoulos, A., Jones, C., Stasch, A., Klein, S., Frenking, G. *Angew. Chem. Int. Ed.* **2009**, *121*, 9881.
155 Deng, L., Holm, R. H. *J. Am. Chem. Soc.* **2008**, *130*, 9878.
156 (a) Tonner, R., Frenking, G. *Angew. Chem. Int. Ed*. **2007**, *46*, 8695. (b) Tonner, R., Frenking, G. *Chem. Eur. J.* **2008**, *14*, 3260.
157 Dyker, C. A., Lavallo, V., Donnadieu, B., Bertrand, G. *Angew. Chem. Int. Ed.* **2008**, *47*, 3206.
158 Fürstner, A., Alcarazo, M., Goddard, R., Lehmann, C. W. *Angew. Chem. Int. EdN* **2008**, *47*, 3210.
159 Dyker, C. A., Bertrand, G. *Science* **2008**, *321*, 1050.

3

ACID–BASE CHEMISTRY OF CARBENES

ANNMARIE C. O'DONOGHUE AND
RICHARD S. MASSEY
Department of Chemistry, Durham University, Durham, UK

3.1 INTRODUCTION

Singlet, spin-paired carbenes **1** may accept a proton or another electrophile, resulting in the formation of the carbocation conjugate acid **1H**$^+$ (Scheme 3.1). Thus, carbocations and carbenes may be categorized as acid–base pairs according to normal Brønsted–Lowry formalism. For many years, the evidence for this acid–base relationship was indirect; however, it is now unequivocally established that the protonation of carbenes can occur to yield carbocations.[1] It has been over 20 years since the work of several research groups clearly demonstrated that singlet diphenyl carbenes insert into O–H bonds by a proton transfer mechanism and the diphenyl carbocation intermediate was directly observed using time-resolved spectroscopy.[2–6] Singlet carbenes may also act as Lewis bases, and numerous examples of stable carbene–Lewis acid adducts have been reported; however, this chapter will focus on the Brønsted acid–base properties of carbenes.[7,8]

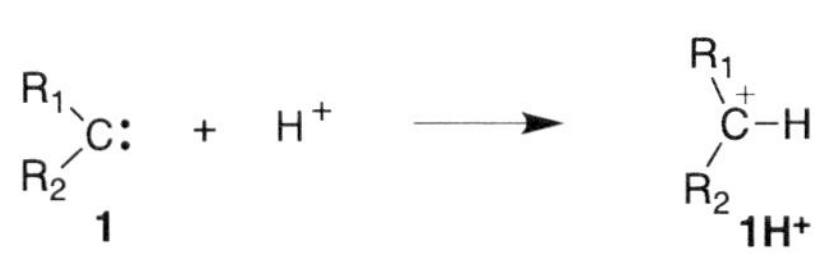

Scheme 3.1.

Contemporary Carbene Chemistry, First Edition. Edited by Robert A. Moss and Michael P. Doyle.

Prior to 2002, pK_a values of the conjugate acids of carbenes were rare. Since the first isolation and characterization of *N*-heterocyclic carbenes (NHCs),[9–12] there has been a resurgence of interest in the area of stable, predominantly nucleophilic singlet carbenes and, as a result, an increased focus on their physical properties. Typically, these carbenes are generated *in situ* by deprotonation of the conjugate acid by a strong non-nucleophilic base such as potassium hydride or hexamethyldisilazide. It has generally been assumed that the conjugate acids of such carbenes are weakly acidic. Recently, there have been a number of studies focused on the experimental and theoretical determination of pK_a values of the conjugate acids of carbenes in different solvent media. We will particularly focus on this work; however, we will also refer to relevant earlier pK_a studies where necessary. There has been more attention devoted to experimental and theoretical estimations of gas-phase basicities (GBs) and proton affinities of carbenes, and this area has been reviewed.[1] We will focus on relevant studies in this area over the past 10 years.

3.2 SOLUTION pK_as OF THE CONJUGATE ACIDS OF CARBENES

The acid ionization of a carbocation to a singlet carbene in a solvent, *S*, is shown in Scheme 3.2 where K_a is the acidity constant for this acid–base equilibrium. The conjugate acids of singlet carbenes may be classified as weak carbon acids in the large majority of cases. It is normal to report pK_a rather than K_a values (Eq. 3.1) to enable the more straightforward comparison of acidities of different carbon acids. A range of methods has been developed for the experimental determination of the pK_as of weak carbon acids,[13–17] which have been applied to the conjugate acids of carbenes.

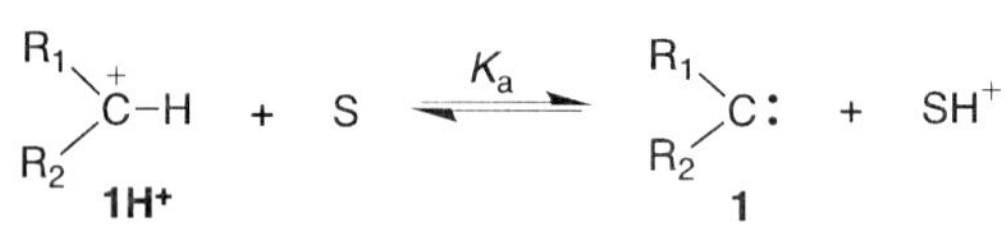

Scheme 3.2.

$$pK_a = -\log K_a \tag{3.1}$$

Experimental pK_a values have now been reported in water, dimethylsulfoxide (DMSO), and tetrahydrofuran (THF) and there have been several accounts of theoretical pK_a estimations in different solvent media. The majority of available experimental pK_a values are for the conjugate acids of NHCs. In these cases, due to the strong π-donating effects of nitrogen, the conjugate acid of the carbene is preferably written as an azolium ion with the positive charge predominantly located on nitrogen rather than carbon (Scheme 3.3). Similarly,

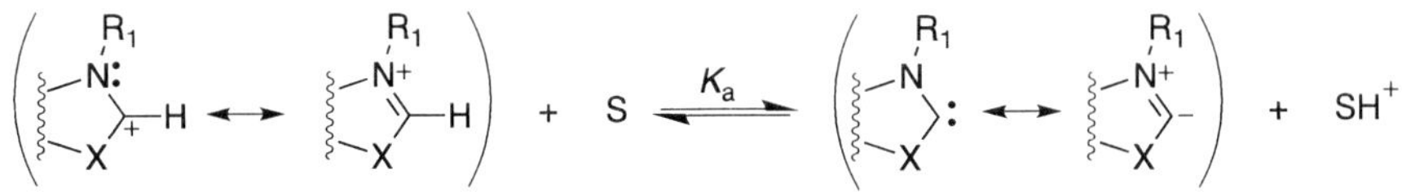

Scheme 3.3.

an ylidic resonance form of the carbene may be drawn with a formal π-bond to the carbenic carbon. Arguments have been presented, based on the magnitude of singlet–triplet energy differences, that the ylide is a better descriptor for strong π-donating substituents such as nitrogen.[18]

In Sections 3.2.1 and 3.2.2, the methodology applied to date in the determination of pK_as of the conjugate acids of carbenes in aprotic solvents and water will be discussed with reference to specific examples. Substituent and solvent effects on pK_as will be evaluated in more detail in Sections 3.2.3 and 3.2.4.

3.2.1 Experimental Methods in Aprotic Solvents

Due to the low concentrations of the conjugate bases of weak carbon acids in solution, the direct determination of equilibrium concentrations is challenging. In aprotic solvents such as DMSO, acetonitrile (MeCN), and THF, weak carbon acid pK_a values may be accessed by an "overlapping indicator method."[14–16] The position of the equilibrium (Scheme 3.4) between the acid of unknown pK_a (HA) and an indicator acid (HIn) of known pK_a is ascertained through monitoring the changes in the UV-Vis spectral properties of the indicator or unknown acid during titration against each other. Knowledge of the position of equilibrium for a range of indicators permits the "bracketing" of the unknown pK_a value to a certain range. This circumvents the need to directly determine the small equilibrium concentrations of conjugate base species, A^-, from natural ionization. Depending on experimental concentrations of indicator and acid and the solvent of choice, it is sometimes necessary to account for ion pair formation and homoconjugation.[19,20] Nuclear magnetic resonance (NMR) spectroscopy may also be utilized to quantify the equilibrium position between indicator and the weak carbon acid; however, the higher concentrations generally required can lead to increased ion pair formation.

$$\mathrm{HA} + \mathrm{In}^- \underset{}{\overset{K_{eq}}{\rightleftharpoons}} \mathrm{A}^- + \mathrm{Hn}$$

Scheme 3.4.

Early attempts towards pK_a determination in DMSO focused on the conjugate acid thiazolium ions **2H**$^+$ of thiazolyl carbenes **2** largely due to their link to the mechanism of action of the cofactor thiamin. In his early seminal studies, Breslow proposed that the thiamin conjugate base **2** (R_1=Me; R_2=5-(Pyrimidyl)-CH_2; R_3=CH_2CH_2OD), generated upon deprotonation at

C(2) of the central thiazolium ring, was the active form of catalyst.[21,22] Although the conjugate base of the thiazolium ion was generally written in ylidic form at this time, the importance of carbene character was also recognized.

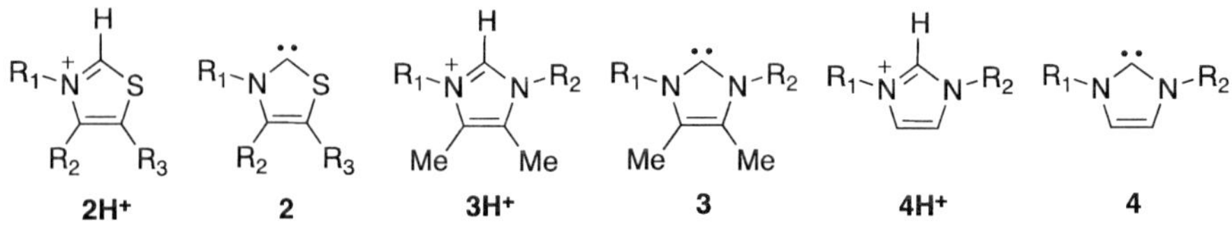

Using the overlapping indicator method, developed and applied in DMSO solution in his laboratory for a broad range of acids, Bordwell attempted the determination of a pK_a value for 3,4-dimethylthiazolium iodide **2H⁺-X⁻** ($R_1 = R_2 = Me$, $R_3 = H$; $X^- = I^-$) using UV-Vis stopped flow spectrophotometry.[23] Rapid competing carbene dimer formation was observed upon mixing the thiazolium ion with a range of indicator anions. By assuming that the rate of establishment of the proton transfer equilibrium (Scheme 3.4) was faster than dimerization and by extrapolation of indicator absorbance values to t_0 on a millisecond timescale, a pK_a value of >16.5 in DMSO could be estimated for this thiazolium salt. Several years later, Alder applied a similar method to the determination of the pK_a of the conjugate acid of imidazole-2-ylidene **3** ($R_1 = R_2 = {}^i$Pr) using ^{1}H NMR spectroscopy.[24] In this case, the problem of competing dimerization was not encountered. The ^{1}H NMR spectra of 1:1 mixtures of **3** in DMSO with fluorene (p$K_a = 22.9$) and 2,3-benzofluorene (p$K_a = 23.5$) showed separate absorptions for the indicator hydrocarbons and their anions, and integration of this data permitted an apparent pK_a of 24 to be calculated for the conjugate acid imidazolium ion **3H⁺** ($R_1 = R_2 = {}^i$Pr). These authors also made the interesting observation that only an averaged NMR spectrum was observed for **3H⁺** and **3** suggestive of fast proton transfer between these two species on the NMR timescale in contrast with the clearly slow proton transfer between the indicator hydrocarbon acid and conjugate base.

Streitwieser and coworkers have developed ion pair acidity scales for lithium and cesium salts of various hydrocarbons in THF relative to a standard taken as 22.90, the ionic pK_a of fluorene per hydrogen in DMSO.[15,16] Relative ion pair basicities of carbanions were often found to differ substantially from the ionic values in more polar solvents. In 2002, Kim and Streitwieser applied the same methodology to the determination of the basicity of 1,3-di-*tert*-butylimidazol-2-ylidene **4** ($R_1 = R_2 = {}^t$Bu) in THF using three hydrocarbon indicators that equilibrated with their anions in the presence of the pregenerated carbene.[25] The concentration of indicator anion, present as both an ion pair and free ion, at equilibrium was determined by UV-Vis spectrophotometry. The concentration data fit well to Equation (3.2), with reference to Scheme 3.5, where it is assumed that both free ions and ion pairs directly equilibrate with reactants ($K_i = K_{ip}K_d$). The results showed that concentrations of ion pairs and

$$\{R^-\} = [\text{free ion}] + [\text{ion pair}] = K_{ip}[RH][\text{carbene}] + (K_i[RH][\text{carbene}])^{1/2} \quad (3.2)$$

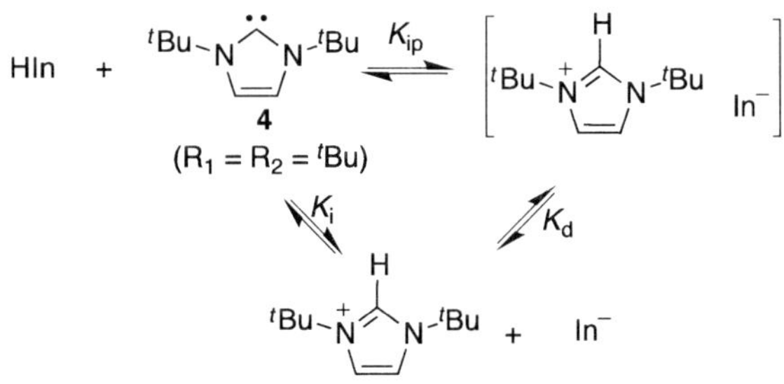

Scheme 3.5.

free ions were of similar magnitude in THF. A linear plot of the pK_{ip} obtained from the fit to Equation (3.2) against pK_{Li} for the corresponding indicator gave pK_{ip} = 0 for pK_{Li} = 20, which puts a lithium salt of this pK_{Li} and the carbene of similar basicity in THF.

Similar experiments in DMSO were accompanied by competing decomposition; however, by data analysis at early times, an ion pair pK of 22.7 was determined for the di-^{t}butylimidazolium conjugate acid of carbene **4** ($R_1 = R_2 = {}^t$Bu) by fitting to Equation (3.2).[25] The concentration data in both THF and DMSO did not fit well to equations allowing for homoconjugation in both the ion pair and free ion. Thus, C–H–C hydrogen bonding **5** between the carbene and conjugate acid imidazolium ion was shown to be negligible in THF and DMSO despite having been observed by Arduengo[26] in solid state crystal structures of a related imidazolyl carbene and corresponding imidazolium salt.

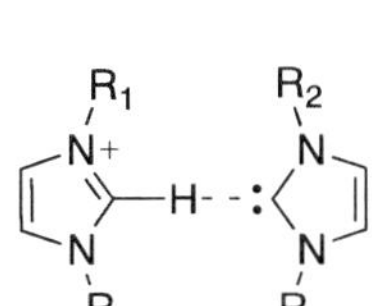

5

It was also suggested by Kim and Streitwieser that the earlier experiments by Alder using ^{1}H NMR spectroscopy could have overestimated the basicity of 1,3-diisopropylimidazol-2-ylidene **3** ($R_1 = R_2 = {}^i$Pr) in DMSO as the concentrations of ion pair and free ion species were not distinguished in the estimation of pK_a. In the case of 1,3-di-*tert*-butylimidazol-2-ylidene **4** ($R_1 = R_2 = {}^t$Bu), the assumption that indicator absorbance data reflected only the concentration of free ions yielded a higher pK of 23.07 in DMSO rather than 22.7. The higher concentrations required by ^{1}H NMR spectroscopic methods would be expected to lead to even greater contributions of ion pairing; however, Kim and Streitwieser also acknowledged that the higher pK_a of 24 observed by Alder could partly be a result of the different substitution pattern.

In 2007, Chu et al. applied the overlapping indicator method to the determination of equilibrium acidities in DMSO at 25 °C for eleven 1,3-dialkylimidazolium salts **4H$^+$-X$^-$** ($R_1 = R_2$ = alkyl; $X^- = (CF_3SO_2)_2N^-$) and one tetramethylimidazolium salt **3H$^+$-X$^-$** [$R_1 = R_2$ = Me; $X^- = (CF_3SO_2)_2N^-$] that are commonly employed as ionic liquids.[27] The effect of a range of different counter ions on pK_a in DMSO was also assessed. Two hydrocarbon indicators, estimated to have pK_a values within 2 units of the relevant imidazolium salt, were employed in each case. The indicator anion was pregenerated in DMSO following addition of a titrated solution of potassium dimsyl to the indicator acid. The imidazolium salt was subsequently titrated into this solution of indicator anion, and the position of equilibrium was determined by UV-Vis spectrophotometry. Excellent agreement was observed in each case between pK_a values determined with titration against two different indicator anions and also between the spectrophotometric pK_a and that derived from NMR titration for tetramethylimidazolium salt **3H$^+$-X$^-$** [$R_1 = R_2$ = Me; $X^- = (CF_3SO_2)_2N^-$]. The effect of a change in counteranion X^- on the observed pK_a was found to be negligible. Corrections were not made for ion pairing between indicator and imidazolyl species in these spectrophotometric titrations, and the concentrations of imidazolium salts and indicator anions employed were relatively high (50 mM). The absolute pK_a of 23.2 determined by Chu et al. for di-^{t}butylimidazolium salt **4H$^+$-X$^-$** ($R_1 = R_2 = {}^t$butyl; $X^- = (CF_3SO_2)_2N^-$) is 0.5 units higher than the value of 22.7 obtained by Kim and Streitwieser. Although ion-pairing effects were not delineated in this and another recent study[28] of imidazolium carbon acidity in DMSO, relative pK_as provide useful information regarding substituent effects on these acid–base equilibria. The true pK_a can be predicted to be about 0.5 units lower than the value uncorrected for ion pairing.

3.2.2 Experimental Methods in Aqueous Solution

To avoid the direct reaction of the strong conjugate bases of weak carbon acids with solvent and leveling effects, direct equilibrium acidity measurements may only be conducted in solvents less acidic than the acid of interest. In general, the conjugate acids of carbenes have pK_a values greater than water; thus, to avoid the leveling effect, a kinetic method can, in theory, be employed for the determination of acidities in aqueous solution. This method makes use of the relationship between the rate constants for the forward and reverse direction of the equilibrium in Scheme 3.6 and the corresponding acidity constants, K_a^{CH} and K_a^{BH}, of the conjugate acids of the carbene **1** and base, B (Eqs. 3.3 and 3.4a). In practice, this method has just been employed for the conjugate acids of NHCs.

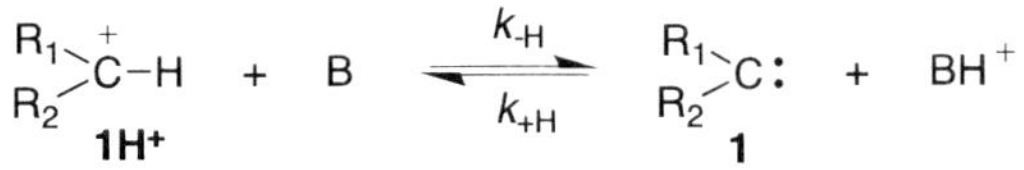

Scheme 3.6.

$$\frac{K_a^{CH}}{K_a^{BH}} = \frac{k_{-H}}{k_{+H}} \tag{3.3}$$

$$pK_a^{CH} = pK_a^{BH} + \log \frac{k_{+H}}{k_{-H}} \tag{3.4a}$$

The rate constant for deprotonation of the carbocation **1H**$^+$ to give the carbene **1** (k_{-H}), which is the uphill "slower" direction of the equilibrium in Scheme 3.6, may be accessed by isotope exchange experiments. The rate constant for deprotonation of the conjugate acid of the carbene by solvent water is often too slow for practical measurement, and a stronger base of known pK_a is employed, such as hydroxide ion. These rate constants for carbene formation (k_{-H}) can be used for the ranking of relative acidities. Direct measurement of a rate constant for the fast reverse protonation of the carbene to the carbocation (k_{+H}) in aqueous solution requires time-resolved transient spectroscopic methods. For diphenylcarbene, this reprotonation reaction has been shown to be at the physical transport limit of diffusion or solvent reorganization in hydroxylic solvent.[29]

Early aqueous acidity studies for the conjugate acids of carbenes also focused largely on thiamin **2H**$^+$ (R_1 = Me; R_2 = 5-(pyrimidyl)-CH_2; R_3 = CH_2CH_2OD). Several groups reported rate constants for base-catalyzed exchange of the C(2)–H of thiamin for deuterium.[22,30–32] There were also reports of rate constants for deuterium exchange of the C(2)–hydrogen of several imidazolium and oxazolium salts although under varying conditions of temperature and ionic strength.[33–38]

Much later, in a series of elegant experiments, Washabaugh and Jencks revisited the acidities of thiamin and related thiazolium salts.[39] p*D*-rate profiles for C(2)–hydrogen/deuterium exchange revealed contributions from p*D*-independent deprotonation by solvent D_2O at p*D* values less than zero in addition to deuteroxide-catalyzed exchange at p*D*s up to seven. Competing ring-opening of the thiazolium ions was not significant under the conditions of these experiments. Using this data, rate constants for deprotonation of the thiazolium ions by solvent and deuteroxide ion, k_{D_2O} (s^{-1}) and k_{DO} (M^{-1} s^{-1}, were determined. For three representative thiazolium salts **2H**$^+$ (R_1 = Me, Ph, CN; R_2 = Me; R_3 = H) labelled with ^{1}H, ^{2}H, and ^{3}H at C(2), Washabaugh and Jencks also determined the primary kinetic isotope effect on rate constants for deuteroxide-catalyzed C(2)–hydron exchange.[40] Primary kinetic isotope effects

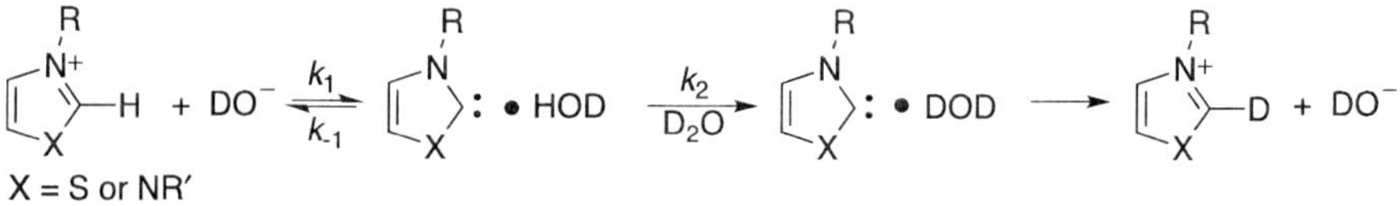

Scheme 3.7.

were observed which increase over the range $(k_{-H}/k_{-T})_{obs} = 2.9–14.7$ with increasing kinetic acidity of the thiazolium ion. The varying primary kinetic isotope effects and the observation of significant deviations of $(k_{-D}/k_{-T})_{obs}$ and $(k_{-H}/k_{-T})_{obs}$ values from the Swain–Schaad relationship, implies significant internal return (k_{-1}, Scheme 3.7) of the transferred hydron to the thiazolyl carbene from water. These observations are consistent with an Eigen mechanism[41] for proton transfer, in which both the proton transfer step (k_1) and reorganization of the thiazolyl carbene·water complex (k_2) are partially rate-limiting in the formation of the solvent-equilibrated carbene (Scheme 3.7). On this basis, by the principle of microscopic reversibility, it could be stated that the reverse protonation by solvent water was largely limited by the encounter of a molecule of H_2O and the carbene that was close to, or at, the physical transport limit for this process.

By making appropriate solvent isotope effect corrections on k_{D_2O} and k_{DO} values for deprotonation of the thiazolium ions, and by assuming the diffusion-controlled reverse protonation of the carbene by hydronium ion and water, respectively, aqueous pK_a values in the range 17–19 were calculated using Equation (3.4a). Excellent agreement was observed between pK_as calculated using k_{D_2O} or k_{DO} values. These experiments highlighted the small barriers associated with the protonation of the carbenes by hydroxylic solvent. Supporting experimental verification came in 2002 from the femtosecond transient absorption spectroscopic experiments of Kohler.[29] These authors directly measured the rate constant for protonation of diphenyl carbene by methanol and showed this process to be limited by dielectric relaxation of solvent.

In 2004 Amyes and coworkers reported second order rate constants for deuteroxide-catalyzed exchange of the C(2)–hydrogen of four imidazolium ions for deuterium in D_2O solution at 25 °C and fixed ionic strength $I = 1.0$ (KCl).[42] Substrates included the parent imidazolium salt **4H⁺-X⁻** ($R_1 = R_2 = H$; $X^- = Cl^-$), 1,3-dimethylimidazolium iodide **4H⁺-X⁻** ($R_1 = R_2 = Me$; $X^- = I^-$), and two C(4)–C(5) benzoannelated variants **6H⁺-X⁻** ($R_1 = R_2 = Me$ *or* $R_1 = R_2 = (S)$-1-Phenyl-ethyl; $X^- = I^-$). Based on the earlier observations of Washabaugh and Jencks and the direct experiments of Kohler and coworkers, Amyes et al. proposed that the reverse protonation of the corresponding imidazolyl-2-ylidenes **4** and **6** by water was limited by the physical transport step of solvent reorganization. The observed k_{DO} values for the four imidazolium salts were significantly smaller than previously observed for analogous thiazolium salts.

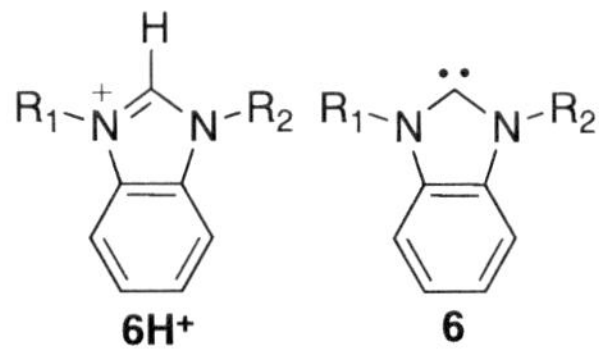

On this basis, it was argued that the reverse protonation of imidazole-2-ylidenes by water should be even more limited by solvent reorganization as these carbenes should be less stable than the thiazolylidenes relative to the azolium ion ground state. In addition, general base catalysis of deuterium exchange was not observed further supporting the claim that the solvent reorganization rather than proton transfer was limiting in the protonation of the carbenes by solvent. The data was used to obtain reliable carbon acid pK_as for the four imidazolium cations in D_2O in the range 21.2–23.8. Marcus intrinsic barriers for protonation of imidazole-2-ylidenes by water were estimated to be 11 kcal/mol lower than for protonation of enolates of monocarbonyl acids with similar carbon acid pK_as highlighting again the high intrinsic reactivities of the former towards protonation by hydroxylic solvent.

Amyes et al. also used the carbon and nitrogen acidity constants for the parent imidazolium cation, pK_a^{CH} and pK_a^{NH} to calculate the equilibrium constant for the 1,2-hydrogen shift at singlet imidazole-2-ylidene to give neutral imidazole ($K_{12} = 5 \times 10^{16}$) according to the thermodynamic cycle in Scheme 3.8 and Equation (3.4b). From analysis of substituent effects on hydrogenation, the 1,2-hydrogen shift and proton transfer reactions revealed the importance of the reference reaction when assessing the relative stabilities of NHCs. Relative to the corresponding azolium cation conjugate acid, an imidazole-2-ylidene (X=N) is substantially less stable than a thiazol-2-ylidene (X=S); however, relative to the neutral hydrogen adduct, the opposite was shown to be the case.

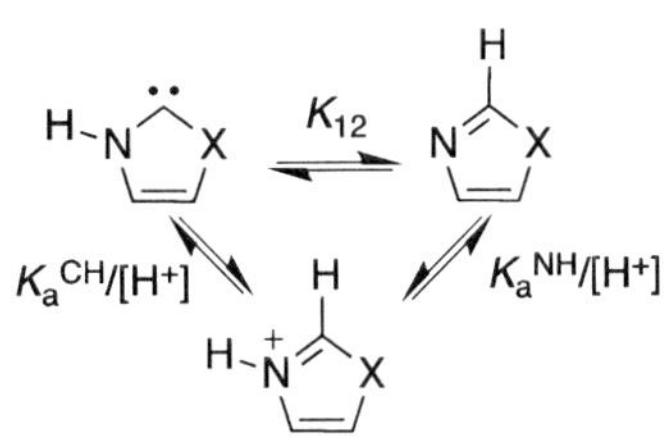

Scheme 3.8.

$$\log K_{12} = pK_a^{CH} - pK_a^{NH} \tag{3.4b}$$

Using the same kinetic method for pK_a determination, O'Donoghue and coworkers determined aqueous pK_a values for the conjugate acids of a large series of NHCs including *N*,*N*-dialkyl and diarylimidazolium ions.[43] pK_a values were also reported for the 4,5-dihydroimidazolium **7H**$^+$, bis-imidazolium **8H**$^+$, and tetrahydropyrimidinium **9H**$^+$ conjugate acids of 4,5-dihydroimidazolyl, bis-imidazolyl, and trihydropyrimidinyl carbenes **7–9**, respectively. The second order rate constants for deprotonation of this series of azolium ions by deuteroxide ion (k_{DO}) were also substantially smaller than for analogous thiazolium ions, and general base catalysis of exchange was not observed for representative members of each class of azolium ion. It was concluded that the reverse protonation of the corresponding NHCs by water was limited by solvent reorganization. The data for this large series of azolium ions enabled the evaluation of *N*-substituent, ring size, and C(4)–C(5) saturation on kinetic acidity and pK_a. As observed by Chu in DMSO,[27] the effect of the counteranion of the azolium salt on acidity was found to be negligible in water indicating that stronger interactions of both species with solvent outcompete those within the salt pair. Hydrolysis of acyclic formamidinium ion **10H**$^+$ outcompeted deuterium exchange; however, a lower limit estimate of pK_a > 29.9 was established.

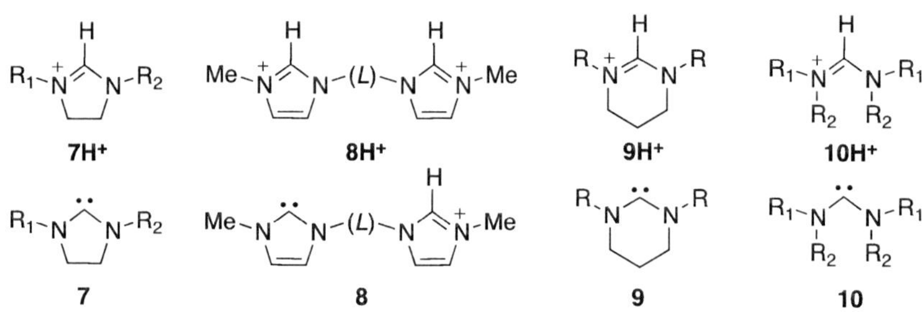

Keeffe, More O'Ferrall, and coworkers applied an alternative method for estimation of the aqueous pK_as of the conjugate acids **1H**$^+$ of a range of acyclic singlet carbenes.[44,45] For conjugate acid carbocations with a β-hydrogen, pK_a^α values could be estimated from thermodynamic cycle A in Scheme 3.9 that relates equilibrium constants for the α-deprotonation of the carbocation to the carbene (pK_a^α), the β-deprotonation to an alkene (pK_a^β), and the isomerization of the corresponding carbene and alkene (pK_{iso}). For conjugate acid carbocations without a β-hydrogen, pK_a^α values could alternatively be estimated from thermodynamic cycle B in Scheme 3.9 that relates the α-deprotonation of the carbocation to the carbene (pK_a^α) and the hydration of both the carbocation and carbene to the corresponding alcohol (pK_R and pK_{H_2O}). The combination of experimental and calculated results provided values for pK_a^β, pK_{iso}, pK_R and pK_{H_2O}, and the desired pK_a^α values could then be obtained using Equations (3.5) and (3.6). Although errors of up to 2–3 units were predicted for individual p*K*s used to calculate pK_a^α values, related compounds could still be compared because of the cancellation of systematic errors.

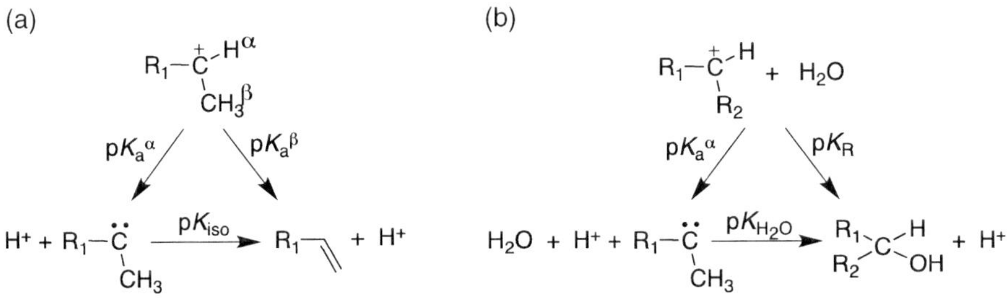

Scheme 3.9.

$$pK_a^{\alpha} = pK_a^{\beta} - pK_{iso} \tag{3.5}$$

$$pK_a^{\alpha} = pK_R - pK_{H_2O} \tag{3.6}$$

TABLE 3.1. Equilibrium Acidities for the Conjugate Acids of Thiazolyl Carbenes

Acid/Conjugate Base	X^-	pK_a[a] (DMSO)	pK_a[a] (H_2O)	pK_a[a] (MeCN)	Reference
2H⁺/2					
R_1 = Me; R_2 = R_3 = H		(14.5)[b]	(21.2)[b]	(25.6)[b]	46
R_1 = R_2 = Me; R_3 = H	I^-	>16.5			23
	Cl^-		18.8		39
R_1 = Me; R_2 = 4-NO_2Bzl; R_3 = H	Br^-		17.8		39
R_1 = Me; R_2 = 4-CH_2CN; R_3 = H	Cl^-		16.9		39
R_1 = Me; R_2 = 5-(Pyrimidyl)-CH_2; R_3 = CH_2CH_2OD	Cl^-		18.0		39
R_1 = Me; R_2 = 5-(Pyrimidinium)-CH_2; R_3 = CH_2CH_2OD	$2Cl^-$		17.6		39

[a] Experimental pK_a values.
[b] pK_a value in parentheses was calculated by computational methods.

3.2.3 Substituent Effects on pK_a

Experimental and computational pK_a values for the conjugate acids of a range of carbenes are tabulated in Table 3.1, Table 3.2, Table 3.3, and Table 3.4. Available pK_a data for thiazolium ion/thiazolylidene acid–base pairs (**2H⁺/2**) are listed in Table 3.1. The experimental aqueous pK_a values are in the narrow range 16.9–18.8; however, this data just refers to *N*-alkyl-substituted thiazolium examples and does not extend to data for *N*-aryl analogues. Alteration of the *N*-substituent from methyl to nitrobenzyl and cyanomethyl decreases the pK_a

TABLE 3.2. Equilibrium Acidities for the Conjugate Acids of Imidazolyl Carbenes

Acid/Conjugate Base	X^-	$pK_a{}^a$ (DMSO)	$pK_a{}^a$ (H_2O)	$pK_a{}^a$ (MeCN)	Reference
$4H^+/4$					
$R_1 = R_2 = H$	I^-		23.8		42
		$(21.8)^b$			47
$R_1 = R_2 = Me$	I^-		23.0		42
	$(CF_3SO_2)_2N^-$	22.0			27
		$(21.1)^b$	$(27.4)^b$	$(32.4)^b$	46
$R_1 = Et; R_2 = Me$	Cl^-		23.0		43
	$(CF_3SO_2)_2N^-$	22.1			27
$R_1 = {}^nBu; R_2 = Me$	PF_6^-	22.1			27
	PF_6^-		23.3		43
	$(CF_3SO_2)_2N^-$	22.0			27
$R_1 = {}^nHex; R_2 = Me$	PF_6^-		23.4		43
$R_1 = {}^nOct; R_2 = Me$	Br^-		23.4		43
	$(CF_3SO_2)_2N^-$	22.1			27
$R_1 = Bz; R_2 = Me$	$(CF_3SO_2)_2N^-$	21.6			27
$R_1 = Bz$ (*p*-Me); $R_2 = Me$	$(CF_3SO_2)_2N^-$	21.6			27
$R_1 = Bz$ (*p*-Br); $R_2 = Me$	$(CF_3SO_2)_2N^-$	21.2			27
$R_1 = CH_2OCH_3; R_2 = Me$	$(CF_3SO_2)_2N^-$	20.5			27
$R_1 = {}^tBu; R_2 = Me$	$(CF_3SO_2)_2N^-$	22.6			27
$R_1 = Ph; R_2 = Me$	$(CF_3SO_2)_2N^-$	19.7			27
$R_1 = R_2 = {}^iPr$		$(22.0)^b$	$(28.2)^b$	$(33.3)^b$	46
$R_1 = R_2 = {}^tBu$	Cl^-		25.2		43
	$(CF_3SO_2)_2N^-$	23.2			27
		$(22.6)^b$	$(28.3)^b$	$(33.9)^b$	46

$R_1 = R_2 =$ 1-Adamantyl	Cl^-		25.4		43
$R_1 = R_2 =$ Phenyl		(16.1)[b]	(22.0)[b]	(27.4)[b]	46
$R_1 = R_2 =$ 2,6-Dimethylphenyl		(16.8)[b]	(22.6)[b]	(28.2)[b]	46
$R_1 = R_2 =$ 2,6-Di-(^{i}Pr)-phenyl	Cl^-		21.1		43
$R_1 = R_2 =$ 2,4,6-Trimethylphenyl	Cl^-		20.8		43
$R_1 = R_2 =$ 4-Methoxyphenyl	Cl^-		20.7		43
$R_1 = R_2 =$ 4-Chlorophenyl	Cl^-		19.8		43
$3H^+/3$					
$R_1 = R_2 =$ Me	$(CF_3SO_2)_2N^-$	23.4			27
		(23.7)[b]	(29.5)[b]	(34.9)[b]	46
$R_1 = R_2 =$ ^{i}Pr	Ind^-	24.0			24
		(24.5)[b]	(30.4)[b]	(35.8)[b]	46
$R_1 = R_2 =$ ^{t}Bu	BF_4^-	24.8			28

[a] Experimental pK_a values.
[b] pK_a value in parentheses was calculated by computational methods.

TABLE 3.3. Equilibrium Acidities for the Conjugate Acids of Benzimidazolyl, 4,5-Dihydroimidazolyl, Bis-imidazolyl, Trihydropyrimidinyl, and Acyclic Diamino Carbenes

Acid/Conjugate Base	X^-	pK_a[a] (DMSO)	pK_a[a] (H_2O)	pK_a[a] (MeCN)	Reference
$6H^+/6$					
$R_1 = R_2 = Me$	I^-		21.6		42
$R_1 = R_2 = (S)$-1-Phenylethyl	I^-		21.2		42
$7H^+/7$					
$R_1 = R_2 = Me$		(22.3)[b]	(28.5)[b]	(33.6)[b]	46
$R_1 = R_2 =$ 2,6-Di-(^{i}Pr)-phenyl	Cl^-		21.5		43
$R_1 = R_2 =$ 2,4,6-Trimethylphenyl	Cl^-		21.3		43
$R_1 = R_2 =$ 4-Methoxyphenyl	Cl^-		20.7		43
$8H^+/8$					
$L = CH_2$	I^-		19.9		43
$L = CH_2CH_2$	I^-		21.6		43
$L = CH_2CH_2CH_2$	I^-		22.3		43
$L =$ 1,6-Phenyl	I^-		20.1		43
$9H^+/9$					
$R = Me$		(27.1)[b]	(33.7)[b]	(38.4)[b]	46
$R = Et$	PF_6^-		27.8		43
$R = {}^iPr$	PF_6^-		28.2		43
$10H^+/10$					
$R_1 = R_2 = Me$		(27.9)[b]	(34.0)[b]	(39.1)[b]	46
				(38.6)[b]	48
$R_1 = {}^iPr$; $R_2 = Me$				(39.3)[b]	48
$R_1 = Me$; $R_2 = {}^iPr$				(40.4)[b]	48
$R_1 = {}^tBu$; $R_2 = Me$				(39.3)[b]	48
$R_1 = R_2 = {}^iPr$	PF_6^-		>29.9		43

[a] Experimental pK_a values.
[b] pK_a value in parentheses was calculated by computational methods.

by 1.0 and 1.9 units, respectively, consistent with the greater destabilization of the cationic thiazolium ion relative to the formally neutral carbene by electron-withdrawing substituents.

The broadest range of experimental pK_a data is available for disubstituted imidazolium ions **4H**$^+$ (Table 3.2) and consists predominantly of data in DMSO and water. Imidazolium ions are less acidic by at least 4 pK units than related thiazolium ions. For example, the aqueous pK_a = 18.8 for 3,4-dimethylthiazolium chloride is 4.2 units lower than for 1,3-dimethylimidazolium chloride (pK_a = 23.0). On the basis of electronegativities, thiazolium ions would be predicted to be no more acidic, and perhaps less acidic, than imidazolium ions. One earlier explanation of the greater acidity of thiazolium ions **2H**$^+$ relative to imidazolium ions **4H**$^+$ is the presence of particular stabilization of the conjugate base by sulfur *via* *d*-orbital backbonding.[34] However, as mentioned earlier, relative to the corresponding neutral hydrogen adduct, an imidazole-2-ylidene is more stable than a thiazol-2-ylidene, thus the assessment of relative stabilities clearly depends on the reference reaction.[42]

The parent imidazolium salt **4H**$^+$**-X**$^-$ ($R_1 = R_2 = H$; $X^- = Cl^-$) has a higher aqueous pK_a value by 0.8 units than 1,3-dimethylimidazolium iodide **4H**$^+$**-X**$^-$ ($R_1 = R_2 = Me$; $X^- = I^-$) due to the increased stabilization by hydrogen bonding of the former salt in water. As this capability of hydrogen bonding to solvent is unique to the parent salt, we will use the 1,3-dimethylimidazolium ion as a reference in assessing relative substituent effects on pK_a.

The variation of one *N*-substituent from methyl, ethyl, *n*-butyl, *n*-hexyl to *n*-octyl, while keeping a methyl substituent at the other ring nitrogen, increases the experimental pK_a value by up to 0.4 units in water and 0.1 units in DMSO relative to the value for 1,3-dimethylimidazolium iodide. The small observed increases in pK_a are consistent with the greater stabilization of the azolium ion by the electron-donating *n*-alkyl substituents compared with the conjugate base imidazole-2-ylidene. These small substituent effects demonstrate that alteration of one *n*-alkyl substituent has only a small effect on the relative stabilities of imidazolium **4H**$^+$-imidazole-2-ylidene **4** acid–base pairs in both DMSO and water. The substitution of one *N*-methyl substituent for a more branched ^{t}butyl substituent increases the pK_a to the greater extent of 0.6 units in DMSO. Changing both *N*-substituents to bulky *t*-butyl or adamantyl groups increases the pK_a by 2 units in water and by 1.2 units in DMSO suggesting additivity of *N*-alkyl-substituent effects.

In contrast to the observed increases in pK_a with changes in *N*-alkyl substituents, the addition of two *N*-aryl substituents decreases the pK_a by 1.9–3.2 units relative to 1,3-dimethylimidazolium iodide. X-ray crystal structures show that the aryl and imidazole rings of *N*,*N*-diarylimidazolium salts and corresponding NHCs are not coplanar; hence, the *N*-aryl-substituent effects are mainly field/inductive and steric because of restriction of inter-ring conjugation. Electron-withdrawing substituents should stabilize the formally neutral carbene relative to the cationic azolium ion and decreases in pK_a are expected. The greater electron-withdrawing inductive effect of a 4-chlorophenyl substituent

($F = +0.18$) is consistent with the lower pK_a for **4H**$^+$**-X**$^-$ ($R_1 = R_2 = $4-chlorophenyl; $X^- = Cl^-$) relative to values for other *N,N*-diarylimidazolium ions in Table 3.2.[43] Benzoannelation at C4–C5 also results in a decrease in pK_a, although the effect is smaller than observed upon replacement of two *N*-methyl by *N*-aryl substituents. pK_a values for **6H**$^+$ (Table 3.3) are 1.4–1.8 units lower than for 1,3-dimethylimidazolium iodide.

Relative to azolium ion ground states, substituent effects on imidazolium/imidazole-2-ylidene 4H$^+$/4 and 4,5-dihydroimidazolium/4,5-dihydroimidazole-2-ylidene 7H$^+$/7 acid–base equilibria are similar. The 4,5-dihydroimidazolium salts 7H$^+$–X$^-$ have higher pKa values (Table 3.3) than analogous imidazolium salts 4H$^+$–X$^-$, however, only by up to 0.5 units. Methylene-linked bis-imidazolium ion 8H$^+$–2I$^-$ ($L = CH_2$; $X = I^-$) is significantly more acidic than 1,3-dimethylimidazolium iodide. The effect of a cationic imidazolium substituent is to lower the pK_a by 3 units likely due to the greater inductive stabilization of the NHC than the azolium conjugate acid, and the effect decreases with increasing length of the alkyl linker.

The tetrahydropyrimidinium salts **9H**$^+$**-PF**$_6^-$ are substantially less acidic than salts **1H**$^+$–**8H**$^+$**-X**$^-$, and pK_as increase by up to 8.4 units as a result of the increase in ring size by one carbon. The increased pK_as for **9H**$^+$**-PF**$_6^-$ correlate with the greater N–C–N angles in these systems compared to the five-membered ring series.[43] Diaminocarbenes prefer small N–C–N angles, for example, 102.2° in **2** ($R_1 = R_2 = {}^t$butyl)[49] compared with an N–C(H)–N angle of 109.6° in the corresponding azolium salt.[50] Enforcement of a larger angle by structural change will favor the protonated ion and increase the pK_a. In **9H**$^+$**-PF**$_6^-$ ($R_1 = R_2 = Et$), the N–C(H)–N angle is 125.3°, while the N–C–N angle in the $KN(SiMe_2)_2$ complex of **9** ($R_1 = R_2 = {}^i$Propyl; $R_3 = R_4 = H$) is 116.3°. An estimate of p$K_a > 29.9$ could be established for acyclic formamidinium salt **10H**$^+$**-PF**$_6^-$. Acyclic carbene **10** has a N–C–N angle of 121.1°, while the precursor **10H**$^+$**-OTf**$^-$ has a N–C(H)–N angle of 133.2°. This much larger N–C–N angle favors acyclic ion **10H**$^+$ resulting in a higher pK_a value.

Estimated pK_a values for the conjugate acid carbocations **1H**$^+$ of singlet methyl, ethyl, dimethyl, phenyl, phenylmethyl, diphenyl, methoxy, methoxymethyl, and dimethoxycarbenes **1** are presented in Table 3.4. These values were obtained by manipulation of the thermodynamic cycles in Scheme 3.9. Methyl substituents were found to have a slightly greater stabilizing effect on the carbocation compared to the carbene. This can be judged from the greater decrease in pK_R values for the methyl cation compared with pK_{H_2O} values for singlet methylene upon addition of a methyl substituent. Stabilization was proposed to originate in a hyperconjugative interaction between the methyl C–H bonds and an empty *p*-orbital of the carbocation or carbene. The stabilizing effect of a phenyl is greater than that of a methyl group and, again, the carbocation receives more stabilization than the carbene. By contrast, the stabilization of the carbene by a methoxy substituent was found to be substantially greater than the carbocation as judged by relative decreases in pK_R and pK_{H_2O} values. The pK_a value of 13 for the conjugate acid of dimethoxycarbene is in reason-

TABLE 3.4. Equilibrium Acidities for the Conjugate Acids of Acyclic Carbenes

Acid/Conjugate Base	pK_a (H_2O)	Reference
1H⁺/1		
$R_1 = R_2 = H$	27	44
$R_1 = Me; R_2 = H$	30	44
$R_1 = Ph; R_2 = H$	31	44
$R_1 = MeO; R_2 = H$	19	44
$R_1 = R_2 = Me$	33	44
$R_1 = Ph; R_2 = Me$	33	44
$R_1 = MeO; R_2 = Me$	23	44
$R_1 = R_2 = Ph$	29	44
$R_1 = R_2 = MeO$	13	44,45

able agreement with an earlier value of 11 calculated by Pezacki[51] by fitting Brønsted data for protonation of this carbene by a series of alcohols to theoretical Eigen curves.

3.2.4 Solvent Effects on pK_a

In all cases where comparisons are possible, experimental pK_a values in DMSO are lower by 1–2 units than in water, which is similar to the small solvent effect observed on the acidities of thiazolium and alkylamonium cations.[23,42] This could be explained by the greater stability of the cationic conjugate acid in water than in DMSO. The formally neutral carbene should be less sensitive to solvation, and, as a result, there is a higher pK_a associated with the acid ionization in water versus DMSO.[27] The data from Kim and Streitwieser additionally show the pK_a of the conjugate acid of 1,3-di-^{t}butylimidazol-2-ylidene in DMSO to be higher than in THF by almost 3 units, which the authors ascribe to the better stabilization of the conjugate acid in DMSO *via* C–H hydrogen bonding to solvent.[25]

There have been a number of computational studies focused on the calculation of pK_a values for the conjugate acids of carbenes in different solvent media and the gas phase.[46–48] In the first and most comprehensive of these studies, Yates and coworkers applied a complete basis set method CBS-QB3, in conjunction with the conductor-like polarisable continuum model (CPCM) solvation model, to predict both relative and absolute pK_as of the conjugate acids of 12 nucleophilic carbenes in DMSO, water, and MeCN.[46] The carbenes in this study included one example each of a thiazolylidene **2**, 4,5-dihydroimidazolylidene **7**, trihydropyrimidinylidene **9**, and acyclic diaminocarbene **10**, in addition to eight examples of imidazolylidenes **3** and **4** (data shown in brackets in Table 3.1, Table 3.2, and Table 3.3). The order of acidity was as follows: pK_a (DMSO) < pK_a (H_2O) < pK_a (MeCN). Calculated pK_a values in DMSO are within 0.5–0.9 units of experimental values in Table 3.1, Table 3.2, and Table 3.3.

Calculated pK_as in water are substantially higher by 3–4.5 units than experimental values; however, relative aqueous pK_as are in better agreement with calculations. The experimental pK_as in water in Table 3.1, Table 3.2, and Table 3.3 are all based on the application of Equation (3.4a) derived for Scheme 3.6, and the main uncertainty in these values is associated with the general assumption that reprotonation of the carbene is limited by solvent reorganization. If the rate constant for reprotonation were lower than these physical transport limits, this would result in lower rather than higher pK_a values. Yates and coworkers comment on the variability of $\Delta G_s^0(\mathrm{H}^+)_{\mathrm{aq}}$ and the uncertainty in determining $\Delta G_s(\mathrm{H}^+)$ for a given solvent and note that the absolute calculated pK_as may be associated with an unknown shift associated with each solvent.

3.3 GAS-PHASE BASICITIES AND PROTON AFFINITIES OF CARBENES

While pK_a provides a quantitative measure of the acid–base equilibrium between a carbene and its conjugate acid in solution, the resulting equilibrium is dependent on solvent, as described earlier. Alternative measures—the gas phase basicity (GB) and proton affinity (PA) of a species (M)—are defined in terms of the protonation reaction of the species in the gas phase (Scheme 3.10). The GB is defined as the negative of the Gibbs free energy change associated with this process (Eq. 3.7), while the PA is the negative of the corresponding enthalpy change (Eq. 3.8). Proton affinities may be obtained from GBs by means of Eq. (3.9), where TΔS_{rxn} is the entropy of protonation.

$$\mathrm{M}_{(g)} + \mathrm{H}^+_{(g)} \longrightarrow \mathrm{MH}^+_{(g)}$$

Scheme 3.10.

$$\mathrm{GB(M)} = -\Delta G_{\mathrm{rxn}} \tag{3.7}$$

$$\mathrm{PA(M)} = -\Delta H_{\mathrm{rxn}} \tag{3.8}$$

$$\mathrm{GB(M)} = \mathrm{PA(M)} + \mathrm{T}\Delta S_{\mathrm{rxn}} \tag{3.9}$$

In 2000, Wolfgang Kirmse compiled an excellent review of the proton affinities and GBs of around a dozen singlet carbenes measured experimentally and a similar number determined by computation, focusing mostly on halocarbenes.[1] This section will include carbene PA measurements since 2000, namely two novel NHC proton affinities obtained experimentally, the redetermination of PAs for several halocarbenes, and over a hundred proton affinities for a range of new carbene species obtained using computational methods. This data will be presented in order of decreasing basicity and recent experimental data will be presented alongside for comparison. First, we must briefly summarize the methods for the measurement of carbene proton affinities.

3.3.1 Experimental Methods

Experimental GB data is typically obtained by probing the proton transfer equilibrium between the species under investigation (M) and a reference base of known basicity (B) (Scheme 3.11).[52] Such measurements lead to *relative* scales of basicity that rely on the *absolute* PA of at least one species in the series. Many reference bases with well-established absolute PAs exist for use in relative basicity measurements, determined from the enthalpies of formation of both base and conjugate acid by means of Equation (3.10). Several approaches may be used to probe this equilibrium, all of which involve the detection of protonated species by mass spectrometry.

$$\mathrm{M} + \mathrm{BH^+} \rightleftharpoons \mathrm{MH^+} + \mathrm{B}$$

Scheme 3.11.

$$\mathrm{PA(B)} = \Delta H_{\mathrm{f}}(\mathrm{B}) + \Delta H_{\mathrm{f}}(\mathrm{H^+}) - \Delta H_{\mathrm{f}}(\mathrm{BH^+}) \quad (3.10)$$

Often, the equilibrium constant for the reaction can be measured directly from the ratio of ions MH^+ and BH^+ in the mass spectrum once thermodynamic equilibrium has been established. The equilibrium constant that is obtained provides the Gibbs free energy change, leading directly to the difference in GB between the two species.

For more reactive species that undergo rapid side reactions with their conjugate acids MH^+, such as carbenes, direct measurements of equilibrium constants are difficult. An alternative technique, known as the "kinetic method," involves the isolation of a proton-bound complex of the carbene and base ([M–H$^+$-B]) inside the instrument (Scheme 3.12).[53] This complex undergoes collision-induced dissociation to yield either protonated carbene (pathway A) or protonated base (pathway B). Provided the enthalpy change for each reaction is similar, the difference in PA (ΔPA) may be obtained *via* the Arrhenius relationship from the ratio of rate constants k_a and k_b, equal to the ratio of product ions MH^+ and BH^+ in the mass spectrum (Eq. 3.11).

$$\mathrm{MH^+} + \mathrm{B} \xleftarrow{\textit{Pathway A}} [\mathrm{M \cdot H^+ \cdot B}] \xrightarrow{\textit{Pathway B}} \mathrm{M} + \mathrm{BH^+}$$

Scheme 3.12.

$$\ln\left(\frac{[\mathrm{MH^+}]}{[\mathrm{BH^+}]}\right) = \ln\left(\frac{k_{\mathrm{a}}}{k_{\mathrm{b}}}\right) = \frac{\Delta \mathrm{PA}}{RT} \quad (3.11)$$

Alternatively, GBs of carbenes may be obtained by "bracketing," whereby a protonated carbene is allowed to react with a series of reference bases in the mass spectrometer. An upper and lower limit of basicity for the unknown species may be estimated from the occurrence or nonoccurrence of proton transfer with the reference compound.

Absolute proton affinities may be obtained if the enthalpies of formation of both base and conjugate acid are independently known, such as for the series of halocarbenes presented in Section 3.3.4. For these species, heats of formation for the protonated halocarbene $\Delta H_f(RR'CH^+)$ have been obtained using a combination of the enthalpy of formation for the corresponding radical and its ionization energy as shown in Scheme 3.13 and Equation (3.12). Corresponding enthalpies of formation for the halocarbenes have been determined from a combination of the heats of formation for the neutral halomethane and the halide dissociation enthalpy ($DH[RR'CX^-]$) as shown in Scheme 3.14 and Equations (3.13) and (3.14). A number of other routes towards the heats of formation of carbene and conjugate acid exist and have been reviewed in detail by Kirmse and others.[1,54,55]

$$RR'CH{-}X \xrightarrow{DH(RR'CHX)} RR'CH\cdot + X\cdot \xrightarrow{IE(RR'CH\cdot)} RR'CH^+ + X\cdot$$

Scheme 3.13.

$$\Delta H_f(RR'CH^+) = \Delta H_f(RR'CH\cdot) + IE(RR'CH\cdot) \qquad (3.12)$$

$$RR'CHX \xrightarrow{\Delta H_{acid}(RR'CHX)} RR'CX^- + H^+ \xrightarrow{DH(RR'CX^-)} RR'C\colon + X^- + H^+$$

Scheme 3.14.

$$\Delta H_f(RR'C\colon) = DH\left(RR'CX^-\right) + \Delta H_f\left(RR'CX^-\right) - \Delta H_f\left(X^-\right) \qquad (3.13)$$

$$= DH(RR'CX^-) + \Delta H_{acid}(RR'CHX) + \Delta H_f(RR'CHX) - \Delta H_{acid}(HX) - \Delta H_f(HX) \qquad (3.14)$$

3.3.2 Heteroatom-Stabilized Carbenes

In the last decade, a large number of proton affinities have been obtained by theory for carbenes stabilized by one or more heteroatoms. This broad category includes a wide variety of stabilized carbenes that have been studied by a number of groups, but we will focus first on the PAs of acyclic carbenes (Table 3.5).

The largest effect on carbene PA results from the type of heteroatom adjacent to the carbene center and stems from a carefully balanced combination of σ-withdrawal and π-donating effects. Both carbene **1** and

TABLE 3.5. Acyclic Heteroatom-Stabilized Carbene Proton Affinities Derived from Theory

Carbene	Method	PA (kcal/mol)	Reference
1 $R_1 = R_2 = SiH_3$	G3MP2	281.7	18
1 $R_1 = R_2 = NH_2$	G3MP2	249.6	18
1 $R_1 = R_2 = NHCH_3$	B3LYP/AUG-cc-pVTZ//B3LYP/6-31+G*	267.8	56
1 $R_1 = R_2 = PHCH_3$	B3LYP/AUG-cc-pVTZ//B3LYP/6-31+G*	257.2	56
1 $R_1 = R_2 = SCH_3$	B3LYP/AUG-cc-pVTZ//B3LYP/6-31+G*	248.5	56
1 $R_1 = R_2 = OCH_3$	G3MP2	235.6	18
	B3LYP/AUG-cc-pVTZ//B3LYP/6-31+G*	242.4	56
1 $R_1 = NHCH_3$; $R_2 = {}^tBu$	B3LYP/6-311++G**//B3LYP/6-31+G*	277.3	57
1 $R_1 = PHCH_3$; $R_2 = {}^tBu$	B3LYP/6-311++G**//B3LYP/6-31+G*	272.2	57
1 $R_1 = SCH_3$; $R_2 = {}^tBu$	B3LYP/6-311++G**//B3LYP/6-31+G*	260.3	57
1 $R_1 = OCH_3$; $R_2 = {}^tBu$	B3LYP/6-311++G**//B3LYP/6-31+G*	258.8	57
1 $R_1 = SiH_3$; $R_2 = H$	G3MP2	274.9	18
1 $R_1 = NH_2$; $R_2 = H$	G3MP2	241.9	18
1 $R_1 = PH_2$; $R_2 = H$	G3MP2	233.5	18
1 $R_1 = SCH_3$; $R_2 = H$	G3MP2	234.7	18
1 $R_1 = OCH_3$; $R_2 = H$	G3MP2	232.2	18
11 $R_1 = {}^tBu$; $R_2 = NH_2$; $R_3 = NH_2$	B3LYP/6-311+G**//B3LYP/6-31+G*	298.1	48
11 $R_1 = {}^tBu$; $R_2 = NH_2$; $R_3 = H$	B3LYP/6-311+G**//B3LYP/6-31+G*	294.8	48
11 $R_1 = {}^tBu$; $R_2 = H$; $R_3 = H$	B3LYP/6-311+G**//B3LYP/6-31+G*	289.5	48
11 $R_1 = {}^iPr$; $R_2 = H$; $R_3 = H$	B3LYP/6-311+G**//B3LYP/6-31+G*	284.8	48
11 $R_1 = Me$; $R_2 = H$; $R_3 = H$	B3LYP/6-311+G**//B3LYP/6-31+G*	280.2	48
10 $R_1 = {}^iPr$; $R_2 = Me$	B3LYP/6-311+G**//B3LYP/6-31+G*	275.5	48
10 $R_1 = Me$; $R_2 = {}^tBu$	B3LYP/6-311+G**//B3LYP/6-31+G*	274.4	48
10 $R_1 = Me$; $R_2 = {}^iPr$	B3LYP/6-311+G**//B3LYP/6-31+G*	272.4	48
10 $R_1 = Me$; $R_2 = Me$	B3LYP/6-311+G**//B3LYP/6-31+G*	268.6	48

(*Continued*)

TABLE 3.5. (Continued)

Carbene	Method	PA (kcal/mol)	Reference
12 R = ^{t}Bu	B3LYP/6-311+G**//B3LYP/6-31G*	274.3	58
12 R = Mes	B3LYP/6-311+G**//B3LYP/6-31G*	271.9	58
12 R = Me	B3LYP/6-311+G**//B3LYP/6-31G*	269.5	58
12 R = CF_3	B3LYP/6-311+G**//B3LYP/6-31G*	247.1	58
13 R = Mes	B3LYP/6-311+G**//B3LYP/6-31G*	270.2	58
13 R = Me	B3LYP/6-311+G**//B3LYP/6-31G*	266.8	58
13 R = CF_3	B3LYP/6-311+G**//B3LYP/6-31G*	247.8	58
13 R = $SiMe_3$	B3LYP/6-311+G**//B3LYP/6-31G*	260.7	58
13 R = SiH_3	B3LYP/6-311+G**//B3LYP/6-31G*	253.8	58

conjugate acid **1H**$^+$ are strongly stabilized by the interaction between the empty *p*-orbital and the heteroatom lone pair, although this effect has been shown to be generally more important for the carbene.[18,44] An electron-withdrawing inductive effect is also present, which although having a small stabilizing effect on the neutral carbene, strongly destabilizes the carbocation. More O'Ferrall has shown that compared to simple alkyl carbenes, the presence of an electron-withdrawing oxygen substituent will destabilize the carbocation more than the carbene, resulting in a more acidic carbocation and a lower carbene PA.[44]

Acyclic diaminocarbenes are among the most basic of all carbenes, with the highest neutral carbene PAs reported to date being those of a family of extended-ring diaminocarbenes **11** studied computationally by the Ganguly group.[48] These compounds possess an extended alkyl framework that incorporate an aryl ring directly opposite the carbene center, which may stabilize the conjugate acid through an intramolecular C–H···π nonbonding interaction. The resulting framework was seen to increase the PA by 10–25 kcal/mol compared

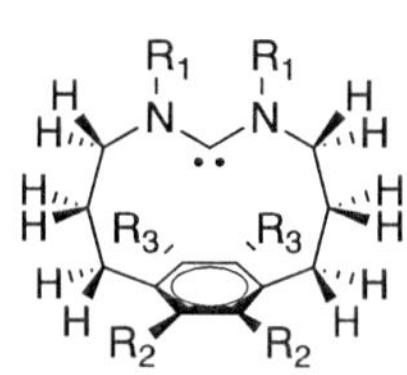

11

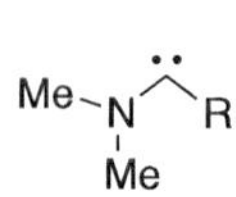

12

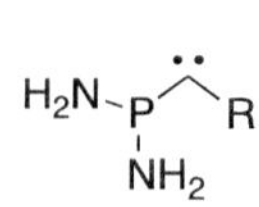

13

to analogous simple diaminocarbenes **10**, yielding PAs of up to 298 kcal/mol for more electron-rich aryl rings. These PAs are exceptionally high and even compare with those of some phosphazenes (around 274–315 kcal/mol).[59,60]

A comparative study of nitrogen- and phosphorus-containing carbenes by a number of groups place the PAs of diphosphinocarbenes **1** ($R_1 = R_2 = PHCH_3$) and phosphinoalkylcarbenes **1** ($R_1 = PHCH_3$; $R_2 = {}^tBu$) just below those of diaminocarbenes **1** ($R_1 = R_2 = NHCH_3$) and aminoalkylcarbenes **1** ($R_1 = NHCH_3$; $R_2 = {}^tBu$), with phosphinosilylcarbenes **13** (R = silyl) still lower. This small decrease may be explained in terms of a net increase in the stability of the carbene relative to the carbocation, resulting from an increase in π-donating ability to the carbene *p*-orbital upon moving from nitrogen to phosphorus. According to recent data published by Kassaee and coworkers, sulfur-stabilized carbenes are around 10 kcal/mol less basic than phosphorus containing carbenes,[56,57] reflecting the increased inductive destabilization of the carbocation from sulfur relative to phosphorus. Similarly, the highly electronegative oxygen atom results in substantial destabilization of the carbocation relative to the carbene, affording methoxycarbenes lower PAs. The effect of replacing one heteroatom for a carbon is unclear, as the available data gives conflicting results. Proton affinities obtained by Gronert suggest that this results in a small decrease in basicity, caused by a reduced π-stabilization of the cation. This is in contrast to the small increases in PA noted by Kassaee, although this could be due to the cation-stabilizing effect of the electron-pushing *t*-butyl group.

In the last decade, a considerable amount of PA data has been obtained for NHCs (Table 3.6), representing the overwhelming interest in these species as ligands and organocatalysts.[61] As with the aforementioned acyclic carbenes, the number and type of heteroatom in the azolium ring strongly influences NHC basicity and, between NHC families at least, the order of proton affinities follows the general trend in conjugate acid azolium ion pK_a discussed in Section 3.2.3 (Fig. 3.1).

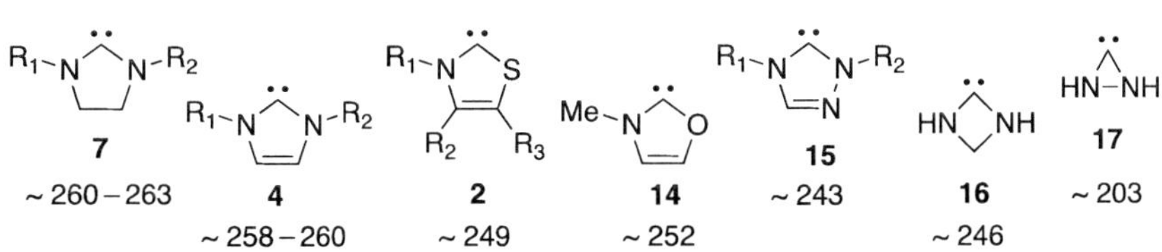

Figure 3.1. Typical PAs of parent (R = H) NHCs (kcal/mol).

Within each family, the effect of *N*-substituent on PA is smaller but still significant. Electron-donating alkyl substituents result in higher PAs, with *i*-propyl substituents typically adding 10 kcal/mol to the PA of the parent carbene. *N*-Alkyl substituents also favor the azolium ion relative to the carbene in solution and increases in pK_a are observed. Hu et al. report a difference of 18 kcal/mol between the unsubstituted and *i*-propyl *N*-substituted triazol-2-ylidene. The data from both Yates[46] and Kassaee[56] indicate an increase in basicity upon moving from alkyl to *N*-phenyl and *N*-mesityl substituents, an observation in contrast to the reported effects on solution pK_a data.

A comparison between the PAs of structurally similar imidazol-2-ylidenes **4** and 4,5-dihydroimidazol-2-ylidenes **7** show that the effect of ring saturation on basicity is small (<3 kJ/mol), as observed in solution. The degree of cyclization is considerably more important—in one study by Yates, the PA of

TABLE 3.6. NHC Proton Affinities Derived from Theory

Carbene	Method	PA (kcal/mol)	Reference
7 $R_1 = R_2 = H$	B3LYP/6-31++G**//B3LYP/6-31+G*	260.2	58
	B3LYP/AUG-cc-pVTZ// B3LYP/6-31+G*	263.0	56
7 $R_1 = R_2 = Me$	B3LYP/6-31++G**//B3LYP/6-31+G*	266.1	58
	B3LYP/6-311+G**	268.9	61
	CBS QB3	259.0	46
7 $R_1 = R_2 = Et$	B3LYP/6-31++G**//B3LYP/6-31+G*	268.8	58
7 $R_1 = R_2 = {}^iPr$	B3LYP/6-31++G**//B3LYP/6-31+G*	270.7	58
7 $R_1 = R_2 = Mes$	B3LYP/6-31++G**//B3LYP/6-31+G*	274.7	58
9 R = Me	CBS QB3	267.3	46
4 $R_1 = R_2 = H$	B3LYP/6-31++G**//B3LYP/6-31+G*	258.2	58
	B3LYP/AUG-cc-pVTZ// B3LYP/6-31+G*	260.0	56
4 $R_1 = R_2 = Me$	B3LYP/6-31++G**//B3LYP/6-31+G*	266.1	58
	B3LYP/AUG-cc-pVTZ// B3LYP/6-31+G*	268.0	57
	B3LYP/6-311+G**	267.8	61
	CBS QB3	258.3	46
4 $R_1 = R_2 = Et$	B3LYP/6-31++G**//B3LYP/6-31+G*	269.0	58
4 $R_1 = R_2 = {}^iPr$	B3LYP/6-31++G**//B3LYP/6-31+G*	271.3	58
4 $R_1 = R_2 = {}^tBu$	B3LYP/AUG-cc-pVTZ// B3LYP/6-31+G*	276.3	57
4 $R_1 = R_2 = Ph$	B3LYP/AUG-cc-pVTZ// B3LYP/6-31+G*	270.6	57
4 $R_1 = R_2 = Mes$	B3LYP/6-31++G**//B3LYP/6-31+G*	275.7	58
2 $R_1 = R_2 = R_3 = H$	B3LYP/6-31++G**//B3LYP/6-31+G*	248.6	58
2 $R_1 = R_2 = H$; $R_3 = Me$	B3LYP/6-31++G**//B3LYP/6-31+G*	253.6	58
	B3LYP/AUG-cc-pVTZ// B3LYP/6-31+G*	256.7	57
	B3LYP/6-311+G**	254.2	61
	CBS QB3	247.9	46
2 $R_1 = R_2 = H$; $R_3 = Et$	B3LYP/6-31++G**//B3LYP/6-31+G*	255.4	58
2 $R_1 = R_2 = H$; $R_3 = {}^iPr$	B3LYP/6-31++G**//B3LYP/6-31+G*	257.2	58
2 $R_1 = R_2 = H$; $R_3 = Ph$	B3LYP/AUG-cc-pVTZ// B3LYP/6-31+G*	260.0	57
2 $R_1 = R_2 = H$; $R_3 = Mes$	B3LYP/6-31++G**//B3LYP/6-31+G*	260.1	58
14	B3LYP/6-311+G**	251.9	61

TABLE 3.6. (Continued)

Carbene	Method	PA (kcal/mol)	Reference
15 $R_1 = R_2 = H$	B3LYP/6-31++G**//B3LYP/6-31+G*	242.6	58
15 $R_1 = R_2 = Me$	B3LYP/6-31++G**//B3LYP/6-31+G*	253.6	58
	B3LYP/6-311+G**	255.6	61
15 $R_1 = R_2 = Et$	B3LYP/6-31++G**//B3LYP/6-31+G*	256.9	58
15 $R_1 = R_2 = {}^iPr$	B3LYP/6-31++G**//B3LYP/6-31+G*	260.2	58
15 $R_1 = R_2 = Mes$	B3LYP/6-31++G**//B3LYP/6-31+G*	264.8	58
16	G3MP2	246.4	18
17	G3MP2	203.0	18

six-membered trihydropyrimidinyl-2-ylidene **9** ($R = Me$) was found to be 8 kcal/mol higher than that of five-membered imidazol-2-ylidene **4** ($R_1 = R_2 = Me$).[46] Similarly, More O'Ferrall and coworkers reported a PA for three-membered cyclic species **17** that was 43.4 kcal/mol lower than for four-membered **16**, implying that the carbene, rather than the azolium ion, can better accommodate the smaller bond angle at the carbene center. This sensitivity to bond angle explains why the PAs of NHCs are typically 10 kcal/mol lower than their acyclic variants, which possess relatively wide N–C–N bond angles.

Only two experimental measurements of the gas-phase proton affinities of NHCs have been reported in the literature, for imidazol-2-ylidenes **4** ($R_1 = Et$; $R_2 = Me$) and **4** ($R_1 = R_2 = Me$) (Table 3.7). The groups of Cooks and Lee independently used the kinetic method to probe the proton-transfer equilibrium between **4** ($R_1 = Et$; $R_2 = Me$) and a series of reference bases.[62,63] Both groups obtained proton affinities in excellent agreement (250–251 kcal/mol), despite the experimental PAs falling significantly lower than those predicted by *ab initio* calculations. In order to validate these measurements, proton affinities were also obtained by Lee for both species using the bracketing method.[64] Using this method, the PA of **4** ($R_1 = Et$; $R_2 = Me$) was found to lie in the range 254.0–260.0 kcal/mol, a value consistent with his own calculated PAs. Similar measurements place the PA of **4** ($R_1 = R_2 = Me$) within the range 257.4–260.6 kcal/mol, consistent with a computed value of 259.9 kcal/mol. Lee concludes that the proton affinities determined at the B3LYP/6-31G* level of theory are accurate and that proton affinities obtained by the kinetic method suffer from experimental inaccuracies.[64]

TABLE 3.7. Experimental Imidazol-2-Ylidene Proton Affinities

Carbene	PA (kcal/mol) Kinetic Method	PA (kcal/mol) Bracketing	PA (kcal/mol) Calculated[a]
4 $R_1 = Et$; $R_2 = Me$	251.3 ± 4[b] 250 ± 3[c]	254.0–260.6	261.4
4 $R_1 = R_2 = Me$	—	257.4–260.6	259.9

[a]Calculated at the B3LYP/6-31+G* level of theory.[63,64]
[b]Reported by Cooks[63].
[c]Reported by Lee[65].

TABLE 3.8. Abnormal NHC Proton Affinities Derived from Theory

Carbene	PA[a] (kcal/mol)	ΔPA(**18–4**) (kcal/mol)	Reference
18 $R_1 = R_3 =$ Dipp; $R_2 = R_4 =$ Ph	287.0		65
18 $R_1 = R_3 =$ Mes; $R_2 = R_4 =$ H	284.4	14.0	66
18 $R_1 = R_3 = {}^t$Bu; $R_2 = R_4 =$ H	283.0	12.4	66
18 $R_1 = R_3 =$ Ph; $R_2 = R_4 =$ H	279.7	15.0	66
18 $R_1 = R_3 =$ Me; $R_2 = R_4 =$ H	278.3	16.0	66
18 $R_1 = R_2 = R_3 = R_4 =$ H	270.9	16.7	66
19 $R_1 =$ Dipp; $R_2 =$ Dipp; $R_3 = {}^t$Bu	275.2		65
19 $R_1 =$ Dipp; $R_2 =$ Me; $R_3 =$ Ph	272.5		65
19 $R_1 =$ H; $R_2 =$ H; $R_3 =$ H	252.8		65

[a]Proton affinities determined at the MP2/TZVPP//BP86/SVP level of theory.

Following their initial observation in 2001[67] and recent isolation in 2009,[68] there has been considerable interest in the stability and electronic properties of "abnormal" or "mesoionic" NHCs (Table 3.8). Frenking has obtained proton affinities for abnormal imidazolylidenes **18** which are 12–17 kcal/mol higher than their "normal" counterparts **4**,[66] in agreement with experimental reports that abnormal NHCs are stronger σ-donors than normal NHCs.[69,70] Structures

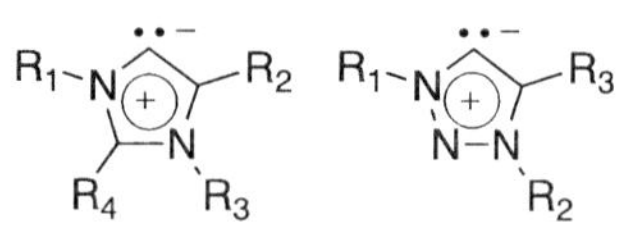

18 **19**

of protonated abnormal carbenes **18** are indistinguishable from the conjugate acids of normal carbenes **4**, thus the difference in PA, ΔPA(**18–4**), represents the energy difference between the carbene tautomers. More recently, proton affinities have been reported for abnormal triazolylidenes **19** that are comparatively less basic than their imidazolyl counterparts.[65]

3.3.3 Alkyl- and Aryl-Substituted Carbenes

In 2011, Gronert, Keeffe, and More O'Ferrall reported theoretical PA data for a large number of less stable carbenes with alkyl, alkenyl, and aryl substituents.[18] This single publication represents the largest collection of computational carbene proton affinities to date and allows for a variety of substituent effects to be probed. A representative selection of alkyl and aryl carbene PA data is presented in Table 3.9.

The stepwise replacement of hydrogen atoms for methyl groups upon moving from methylene to dimethylcarbene results in an increase in carbene

TABLE 3.9. Alkyl and Aryl Carbene Proton Affinities Derived from Theory

Carbene	Method	PA (kcal/mol)	Reference
1 $R_1 = R_2 = Me$	G3MP2	245.9	18
1 $R_1 = Me$; $R_2 = H$	G3MP2	235.6	18
1 $R_1 = R_2 = H$	G3MP2	205.2	18
1 $R_1 = Ph$; $R_2 = H$	G3MP2	259.9	18
1 $R_1 = 4\text{-}NH_2\text{-}Ph$; $R_2 = H$	G3MP2	280.9	18
1 $R_1 = 4\text{-}OH\text{-}Ph$; $R_2 = H$	G3MP2	269.6	18
1 $R_1 = 4\text{-}Me\text{-}Ph$; $R_2 = H$	G3MP2	265.6	18
1 $R_1 = 4\text{-}F\text{-}Ph$; $R_2 = H$	G3MP2	259.6	18
1 $R_1 = 4\text{-}CN\text{-}Ph$; $R_2 = H$	G3MP2	250.2	18
1 $R_1 = 4\text{-}NO_2\text{-}Ph$; $R_2 = H$	G3MP2	248.4	18
1 $R_1 = 3\text{-}NH_2\text{-}Ph$; $R_2 = H$	G3MP2	266.4	18
1 $R_1 = 3\text{-}OH\text{-}Ph$; $R_2 = H$	G3MP2	260.3	18
1 $R_1 = 3\text{-}Me\text{-}Ph$; $R_2 = H$	G3MP2	262.6	18
1 $R_1 = 3\text{-}F\text{-}Ph$; $R_2 = H$	G3MP2	255.1	18
1 $R_1 = 3\text{-}CN\text{-}Ph$; $R_2 = H$	G3MP2	249.0	18
1 $R_1 = 3\text{-}NO_2\text{-}Ph$; $R_2 = H$	G3MP2	248.6	18
1 $R_1 = R_2 = Ph$	G3MP2	268.0	18
1 $R_1 = Ph$; $R_2 = Me$	G3MP2	264.9	18

basicity by 40 kcal/mol. The authors suggest that hyperconjugative interactions between the C–H bonds of the methyl groups and empty carbon *p*-orbital are present in both carbene and conjugate acid but that the resulting stabilization effect is greater for the cation.[44] A smaller increase is observed upon addition of a second methyl group due to saturation effects. Phenyl-substituted carbenes are around 30–55 kcal/mol more basic than those of similar alkyl carbenes, which Gronert attributes to the greater carbocation-stabilizing effect of the aryl rings.[71] Proton affinities for a number of *meta*- and *para*-substituted phenylcarbenes were found to follow expected trends, with more electron-withdrawing substituents resulting in greater carbocation destabilization and less basic PAs. By correlating the proton affinities with gas-phase-substituent constants, the authors were able to quantitatively assess the type of carbocation stabilization effect occurring at the *meta* and *para* positions. Field (F) and resonance (R+) effects were found to strongly affect the stability of the carbocation for both sets, with resonance dominating at the *para* position. Carbocation stabilization by polarizability (α) was also found to have a smaller but still significant influence, particularly in the *para* position where positive charge may be delocalized.

A small number of experimental proton affinities are available for comparison with this new theoretical data, two of which were determined prior to 2000 from enthalpies of formation and are quoted by Kirmse.[1] The calculated PAs of **1** ($R_1 = CH{=}CH_2$; $R_2 = H$) and **1** ($R_1 = Ph$; $R_2 = H$) compare well with these experimental values (242 and 258 kcal/mol, respectively). The PA of the singlet methyl carbene was recently recalculated from heats of formation by Kass et al. and was found to be 206.2 kcal/mol, in good agreement with the value in Table 3.9.[54]

3.3.4 Halocarbenes

Halocarbenes are among the least basic carbenes, with proton affinities comparable with those of weakly basic amines (e.g., PA(NH_3) = 204 kcal/mol).[52] The relatively limited experimental carbene PA data in the literature is mostly for halocarbenes, and Kirmse has reviewed most of this data.[1] Here, we will briefly update the reader on a number of newly reported PAs obtained by experiment and theory.

In his review, Kirmse noted that proton affinities determined by bracketing were consistently lower than those derived from heats of formation, drawing attention to the PA of dichlorocarbene reported by Lias which differed by 15 kcal/mol between the two methods.[72] In 2007, the proton affinities of chlorocarbene and dichlorocarbene were redetermined by the Kass group using the bracketing method, generating the cationic species CH_2Cl^+ and $CHCl_2^+$ by electron ionization of chloroform or dichloromethane and reacting these species with reference bases.[54] These updated values compare more favorably with the PAs obtained from heats of formation and are shown in Table 3.10 alongside the data reported by Lias. The heats of formation reported here have been updated in accordance with changes in the thermodynamic data and differ slightly from those values reported at the time.

TABLE 3.10. Halocarbene Proton Affinities Derived from Experimental Methods

	Enthalpies of Formation			Bracketing	
Carbene	$\Delta H_{f,298}$ (:CR_1R_2)[a] (kcal/mol)	$\Delta H_{f,298}$ (:$CR_1R_2^+$)[b] (kcal/mol)	PA[c] (kcal/mol)	PA (kcal/mol)	Reference
1 $R_1 = R_2 = F$	−45.9	144.9	174.9 ± 2.4	171[d]	72
1 $R_1 = F$; $R_2 = Cl$	7.4	189	185 ± 6	185[d]	72
1 $R_1 = R_2 = Cl$	54.8	213.1	207.4 ± 1.1	205.2 ± 1.9	54
				191 ± 1[d]	72
1 $R_1 = F$; $R_2 = H$	35.8	201	201 ± 3	191[d]	72
1 $R_1 = Cl$; $R_2 = H$	76.3	229.8	212.2 ± 1.2	209.7 ± 2.2	54
				209[d]	72

[a] Recommended carbene enthalpies of formation from Reference 74.
[b] Carbocation enthalpies of formation calculated using Equation (3.12). Recommended radical enthalpies of formation from Reference 74 and recommended radical ionization energies quoted in Reference 55.
[c] PAs determined using Equation (3.4a and b) and $\Delta H_f(H^+)$ = 365.7 kcal/mol, from Reference 74. PAs differ slightly from those reported in Reference 54 due to changes in the most recent recommended thermochemical data.
[d] From the experimental GB data reported in Reference 74. As calculated in Reference 1 using ΔS_p from Reference 54.

TABLE 3.11. Halocarbene Proton Affinities Derived from Theory

Carbene	Method	PA (kcal/mol)	Reference
1 $R_1 = R_2 = F$	G3MP2	175.7	18
	G2MP2	177.4	73
1 $R_1 = F$; $R_2 = H$	G3MP2	199.1	18
1 $R_1 = R_2 = Cl$	G3MP2	207.1	18
	G2MP2	209.6	73
1 $R_1 = Cl$; $R_2 = H$	G3MP2	212.9	18

Halocarbene proton affinities calculated at the G3MP2 level of theory were recently reported by Gronert[18] and update the existing values determined by De Almeida in 1997 (Table 3.11).[73] Although there is reasonable agreement between the two data sets, the new data more closely matches those values obtained experimentally which may be due to the higher level of theory used.

3.4 CONCLUSION AND OUTLOOK

Since the first isolation of imidazolylidenes in 1991,[10] stable carbenes have become the focus of intensive research.[75, 76] Isolable carbenes at room temperature typically have N or P-atoms directly attached to the carbene carbon. The majority of carbenes have extremely short lifetimes and may be classed as "Reactive Intermediates."[77] The study of the acid–base equilibria of carbenes in solution has to date been mostly restricted to stable isolable aminocarbenes, although More O'Ferrall and Keeffe[44] have used a combination of calculation and experiment to estimate aqueous pK_as for a range of less stable alkyl and aryl carbenes. Even within the stable aminocarbene series, there remains a range of structural classes for which no experimental pK_a data is available, for example, the cycloalkylamino carbenes. PA and GB data is available for a much broader range of carbene structural types including diphosphino, phosphinoalkyl, and phosphinosilyl carbenes. Where comparisons can be made, the effects of substituents on carbene acidity–basicity in the gas phase mirror those observed in solution. One notable exception is the effect of *N*-aryl substituents of NHCs which increase basicity in the gas phase yet decrease pK_as in both DMSO and water. The challenge that remains is the determination of solution pK_as for a broader range of carbenes. Within the NHC class, theoretical calculations have provided access to predicted pK_a values in DMSO for carbenes with boron, nitrogen, and phosphorus backbones;[47] however, there are no available experimental values for comparison. PA data is available for a limited number of "mesoionic" carbenes; however, there have been no reports of solution pK_a data for this increasingly important carbene class. Given that protonation is implicated in the mechanism of many fundamental reactions of carbenes including dimerization,[78] further understanding of their acid–base chemistry is essential.

SUGGESTED READING

Amyes, T. L., Diver, S. T., Richard, J. P., Rivas, F. M., Toth, K. *J. Am. Chem. Soc.* **2004**, *126*, 4366.

Gronert, S., Keeffe, J. R., More O'Ferrall, R. A. *J. Am. Chem. Soc.* **2011**, *133*, 3381.

Higgins, E. M., Sherwood, J. A., Lindsay, A. G., Armstrong, J., Massey, R. S., Alder, R. W., O'Donoghue, A. C. *Chem. Commun.* **2011**, *47*, 1559.

Kim, Y. J., Streitwieser, A. *J. Am. Chem. Soc.* **2002**, *124*, 5757.

Kirmse, W. "Carbene Protonation." In *Advances in Carbene Chemistry*. Brinker, U. H. Ed. JAI Press: London, **2000**, pp. 1–52.

Magill, A. M., Cavell, K. J., Yates, B. F. *J. Am. Chem. Soc.* **2004**, *126*, 8717.

REFERENCES

1 Kirmse, W. "Carbene Protonation." In *Advances in Carbene Chemistry*. Brinker, U. H., Ed. JAI Press: London, **2000**, pp. 1–52.
2 Belt, S. T., Bohne, C., Charette, G., Sugamori, S. E., Scaiano, J. C. *J. Am. Chem. Soc.* **1993**, *115*, 2200.
3 Chateauneuf, J. E. *J. Chem. Soc. Chem. Commun.* **1991**, 1437.
4 Dix, E. J., Goodman, J. L. *J. Phys. Chem.* **1994**, *98*, 12609.
5 Kirmse, W., Krzossa, B., Steenken, S. *J. Am. Chem. Soc.* **1996**, *118*, 7473.
6 Kirmse, W., Kilian, J., Steenken, S. *J. Am. Chem. Soc.* **1990**, *112*, 6399.
7 Li, H. P., Risko, C., Seo, J. H., Campbell, C., Wu, G., Bredas, J. L., Bazan, G. C. *J. Am. Chem. Soc.* **2011**, *133*, 12410.
8 Kirmse, W. *Eur. J. Org. Chem.* **2005**, 237.
9 Arduengo, A. J. *Acc. Chem. Res.* **1999**, *32*, 913.
10 Arduengo, A. J., Harlow, R. L., Kline, M. *J. Am. Chem. Soc.* **1991**, *113*, 361.
11 Bourissou, D., Guerret, O., Gabbai, F. P., Bertrand, G. *Chem. Rev.* **2000**, *100*, 39.
12 Igau, A., Baceiredo, A., Trinquier, G., Bertrand, G. *Angew. Chem. Int. Ed.* **1989**, *28*, 621.
13 Bordwell, F. G. *Acc. Chem. Res.* **1988**, *21*, 456.
14 Matthews, W. S., Bares, J. E., Bartmess, J. E., Bordwell, F. G., Cornforth, F. J., Drucker, G. E., Margolin, Z., McCallum, R. J., McCollum, G. J., Vanier, N. R. *J. Am. Chem. Soc.* **1975**, *97*, 7006.
15 Streitwieser, A., Ciula, J. C., Krom, J. A., Thiele, G. *J. Org. Chem.* **1991**, *56*, 1074.
16 Streitwieser, A., Wang, D. Z., Stratakis, M., Facchetti, A., Gareyev, R., Abbotto, A., Krom, J. A., Kilway, K. V. *Can. J. Chem.* **1998**, *76*, 765.
17 Gronert, S. In *Reactive Intermediate Chemistry*. Moss, R. A., Platz, M. S., Jones, M., Jr., Eds. Wiley-Interscience: Hoboken, NJ, **2004**, p. 69.
18 Gronert, S., Keeffe, J. R., More O'Ferrall, R. A. *J. Am. Chem. Soc.* **2011**, *133*, 3381.
19 Bordwell, F. G., Branca, J. C., Hughes, D. L., Olmstead, W. N. *J. Org. Chem.* **1980**, *45*, 3305.
20 Olmstead, W. N., Bordwell, F. G. *J. Org. Chem.* **1980**, *45*, 3299.
21 Breslow, R. *J. Am. Chem. Soc.* **1958**, *80*, 3719.
22 Breslow, R. *Ann. N. Y. Acad. Sci.* **1962**, *98*, 445.
23 Bordwell, F. G., Satish, A. V. *J. Am. Chem. Soc.* **1991**, *113*, 985.
24 Alder, R. W., Allen, P. R., Williams, S. J. *Chem. Commun.* **1995**, 1267.

25 Kim, Y. J., Streitwieser, A. *J. Am. Chem. Soc.* **2002**, *124*, 5757.
26 Arduengo, A. J., Gamper, S. F., Tamm, M., Calabrese, J. C., Davidson, F., Craig, H. A. *J. Am. Chem. Soc.* **1995**, *117*, 572.
27 Chu, Y., Deng, H., Cheng, J. P. *J. Org. Chem.* **2007**, *72*, 7790.
28 Grishina, A. A., Polyakova, S. M., Kunetskiy, R. A., Cisarova, I., Lyapkalo, I. M. *Chem. Eur. J.* **2011**, *17*, 96.
29 Peon, J., Polshakov, D., Kohler, B. *J. Am. Chem. Soc.* **2002**, *124*, 6428.
30 Crosby, J., Lienhard, G. E. *J. Am. Chem. Soc.* **1970**, *92*, 5707.
31 Kemp, D. S., Obrien, J. T. *J. Am. Chem. Soc.* **1970**, *92*, 2554.
32 Kluger, R. *Chem. Rev.* **1987**, *87*, 863.
33 Olofson, R. A., Michelmay, J. S., Thompson, W. R. *J. Am. Chem. Soc.* **1964**, *86*, 1865.
34 Haake, P., Bausher, L. P., Miller, W. B. *J. Am. Chem. Soc.* **1969**, *91*, 1113.
35 Vaughan, J. D., Mughrabi, Z., Wu, E. C. *J. Org. Chem.* **1970**, *35*, 1141.
36 Bradbury, J. H., Chapman, B. E., Pellegri.Fa *J. Am. Chem. Soc.* **1973**, *95*, 6139.
37 Wong, J. L., Keck, J. H. *J. Org. Chem.* **1974**, *39*, 2398.
38 Takeuchi, Y., Yeh, H. J. C., Kirk, K. L., Cohen, L. A. *J. Org. Chem.* **1978**, *43*, 3565.
39 Washabaugh, M. W., Jencks, W. P. *Biochemistry* **1988**, *27*, 5044.
40 Washabaugh, M. W., Jencks, W. P. *J. Am. Chem. Soc.* **1989**, *111*, 683.
41 Eigen, M. *Angew. Chem. Int. Ed.* **1964**, *3*, 1.
42 Amyes, T. L., Diver, S. T., Richard, J. P., Rivas, F. M., Toth, K. *J. Am. Chem. Soc.* **2004**, *126*, 4366.
43 Higgins, E. M., Sherwood, J. A., Lindsay, A. G., Armstrong, J., Massey, R. S., Alder, R. W., O'Donoghue, A. C. *Chem. Commun.* **2011**, *47*, 1559.
44 Keeffe, J. R., More O'Ferrall, R. A. *Arkivoc* **2008**, 183.
45 Guthrie, J. P., More O'Ferrall, R. A., O'Donoghue, A. C., Waghorne, W. E., Zrinski, I. *J. Phys. Org. Chem.* **2003**, *16*, 582.
46 Magill, A. M., Cavell, K. J., Yates, B. F. *J. Am. Chem. Soc.* **2004**, *126*, 8717.
47 Kausamo, A., Tuononen, H. M., Krahulic, K. E., Roesler, R. *Inorg. Chem.* **2008**, *47*, 1145.
48 Lo, R., Ganguly, B. *Chem. Commun.* **2011**, *47*, 7395.
49 Arduengo, A. J., Bock, H., Chen, H., Denk, M., Dixon, D. A., Green, J. C., Herrmann, W. A., Jones, N. L., Wagner, M., West, R. *J. Am. Chem. Soc.* **1994**, *116*, 6641.
50 Chase, P. A., Stephan, D. W. *Angew. Chem. Int. Ed.* **2008**, *47*, 7433.
51 Pezacki, J. P. *Can. J. Chem.* **1999**, *77*, 1230.
52 (a) Lias, S. G., Liebman, J. F., Levin, R. D. *J. Phys. Chem. Ref. Data* **1984**, *13*, 695, (b) Hunter, E. P. L., Lias, S. G. *J. Phys. Chem. Ref. Data* **1998**, *27*, 413.
53 McLuckey, S. A., Cameron, D., Cooks, R. G. *J. Am. Chem. Soc.* **1981**, *103*, 1313.
54 Tian, Z. X., Kass, S. R. *Int. J. Mass Spectrom.* **2007**, *267*, 288.
55 Poutsma, J. C., Paulino, J. A., Squires, R. R. *J. Phys. Chem. A* **1997**, *101*, 5327.
56 Kassaee, M. Z., Ghambarian, M., Shakib, F. A., Momeni, M. R. *J. Phys. Org. Chem.* **2011**, *24*, 351.
57 Kassaee, M. Z., Shakib, F. A., Momeni, M. R., Ghambarian, M., Musavi, S. M. *J. Org. Chem.* **2010**, *75*, 2539.
58 Lai, C. L., Guo, W. H., Lee, M. T., Hu, C. H. J. *Organomet. Chem.* **2005**, *690*, 5867.
59 Despotovic, I., Kovacevic, B., Maksic, Z. B. *Org. Lett.* **2007**, *9*, 4709.
60 Margetic, D., Ishikawa, T., Kumamoto, T. *Eur. J. Org. Chem.* **2010**, 6563.
61 Holloczki, O., Kelemen, Z., Nyulaszi, L. *J. Org. Chem.* **2012**, *77*, 6014.
62 Chen, H., Justes, D. R., Cooks, R. G. *Org. Lett.* **2005**, *7*, 3949.
63 Liu, M., Yang, I., Buckley, B., Lee, J. K. *Org. Lett.* **2010**, *12*, 4764.

64 Liu, M., Chen, M., Zhang, S. S., Yang, I., Buckley, B., Lee, J. K. *J. Phys. Org. Chem.* **2011**, *24*, 929.
65 Bouffard, J., Keitz, B. K., Tonner, R., Guisado-Barrios, G., Frenking, G., Grubbs, R. H., Bertrand, G. *Organometallics* **2011**, *30*, 2617.
66 Tonner, R., Heydenrych, G., Frenking, G. *ChemPhysChem* **2008**, *9*, 1474.
67 Grundemann, S., Kovacevic, A., Albrecht, M., Faller, J. W., Crabtree, R. H. *Chem. Commun.* **2001**, 2274.
68 Aldeco-Perez, E., Rosenthal, A. J., Donnadieu, B., Parameswaran, P., Frenking, G., Bertrand, G. *Science* **2009**, *326*, 556.
69 Crabtree, R. H. J. *Organomet. Chem.* **2005**, *690*, 5451.
70 Ung, G., Bertrand, G. *Chem. Eur. J.* **2011**, *17*, 8269.
71 Gronert, S., Keeffe, J. R., More O'Ferrall, R. A. *J. Org. Chem.* **2009**, *74*, 5250.
72 Lias, S. G., Karpas, Z., Liebman, J. F. *J. Am. Chem. Soc.* **1985**, *107*, 6089.
73 Pliego, J. R., De Almeida, W. B. J. *Chem. Soc., Faraday Trans.* **1997**, 93, 1881.
74 Sander, S. P., Friedl, R. R., Barker, J. R., Golden, D. M., Kurylo, M. J., Wine, P. H., Abbatt, J. P. D., Burkholder, J. B., Kolb, C. E., Moortgat, G. K., Huie, R. E., Orkin, V. L. *Chemical Kinetics and Photochemical Data for Use in Atmospheric Studies Number 17.*
75 Kirmse, W. *Angew. Chem. Int. Ed.* **2004**, *43*, 1767.
76 Kirmse, W. *Angew. Chem. Int. Ed.* **2010**, *49*, 8798.
77 Jones, M., Jr., Moss, R. A. In *Reactive Intermediate Chemistry*. Moss, R. A., Platz, M. S., Jones, M., Jr., Eds. Wiley-Interscience: Hoboken, NJ, **2004**, p. 273.
78 Alder, R. W., Blake, M. E., Chaker, L., Harvey, J. N., Paolini, F., Schutz, J. *Angew. Chem. Int. Ed.* **2004**, *43*, 5896.

4

COMPUTATIONAL METHODS FOR THE STUDY OF CARBENES AND THEIR EXCITED STATES

HOI LING LUK, SHUBHAM VYAS, AND CHRISTOPHER M. HADAD
Department of Chemistry and Biochemistry, The Ohio State University, Columbus, OH, USA

4.1 INTRODUCTION

Organic chemists would like to understand the details of every bond-making and bond-breaking step in a chemical reaction. There are many practical reasons for this level of understanding, including the targeted synthesis of new materials or pharmaceuticals. These reactions often proceed in several steps involving reactive intermediates. As a class of reactive intermediates, carbenes are among the most important, as evidenced by extensive studies since the 1950s. Moreover, the chemistry and rearrangements of carbenes contribute to high-value industries, including photolithography for the manufacture of silicon microprocessors.

A wide variety of experimental studies have been performed to investigate carbenes. In support of these experiments, computational chemistry established itself in carbene chemistry upon a reinvestigation of the structures and energies of the various states of methylene (CH_2), a molecule proposed as an example for computational quantum chemistry by Schaefer.[1] In the 1960s, triplet methylene was generally believed to be linear.[2] Even though the bent structure was suggested by some theoretical calculations,[3] chemists were initially hesitant to believe the theoretical results until they were confirmed by additional

Contemporary Carbene Chemistry, First Edition. Edited by Robert A. Moss and Michael P. Doyle.

experimental analysis. The bent structure of triplet methylene was proposed by theory, and this prediction attracted chemists to reinterpret the original spectroscopic results in the 1970s, thereby confirming the bent geometry. Computational chemistry expanded its reputation through success in predicting the correct geometry as well as the singlet–triplet energy gap of methylene in the late 1970s.[4]

Carbenes are often short-lived, high-energy, and highly reactive molecules that are crucial in determining the outcome of chemical reactions. The lifetimes of transient carbenes can range from picoseconds to minutes. In order to observe these transient species, different spectroscopic tools have been used over the years, varying from matrix isolation spectroscopic methods[5–7] in the 1960s, to nanosecond laser flash photolysis[8,9] in the 1980s, and to ultrafast spectroscopic methods[10–12] in the late 1990s, and into the present. In order to understand these spectroscopic results, different computational tools have been applied to provide an accurate prediction of the corresponding potential energy surfaces as well as other chemical properties. Combining theory with experiment, chemists can understand the observed transient spectra and formulate viable mechanisms for the diverse photochemical reactions—many of which are valuable for practical applications, including photolithography and photo-cross-linking.[13]

With the development of ultrafast spectroscopic methods[10–12] in the late 1990s, researchers could observe phenomena on the femtosecond timescale, including internal conversion and the spectroscopic signatures of the singlet excited states of some carbenes and their precursors. These experimental developments challenged computational chemists to provide accurate information for these excited-state species. Over the past decade, some carbene studies have evolved from simply computing the properties of the lowest-energy singlet and triplet states of the carbene itself to exploring instead the excited state(s) of the precursor used to generate the carbene.

This chapter focuses on the roles that theory and computation have played in this collaborative effort to unravel the complex chemistry and properties of carbenes and their excited-state precursors. The first part of the chapter describes the structure and electronic states of carbenes and includes a brief overview of the fate of the carbenes. The second part describes different examples of Rearrangement in Excited States (RIES), which challenge the computational chemist to understand how these excited-state precursors form the carbene as a reactive intermediate. Then, various computational methods used in studying carbenes, as well as their excited-state precursors, are described. Finally, the chapter concludes with a review of an exciting new area of study in the combined investigation of carbenes and their primary precursors, diazirines and diazo compounds. Computational studies of the excited-state precursors of carbenes yield insights into the dynamics of these species, and how the excited-state potential energy surfaces of the photo-precursors connect with the corresponding surfaces of the nascent carbenes.

4.2 CARBENES

Carbenes are defined as a class of organic molecules containing a divalent carbon atom which has six electrons at the carbon center. Carbenes are equivalent to nitrenes which also have six valence electrons at the nitrogen center. The simplest carbene is methylene (CH_2), a species that is isoelectronic with the simplest nitrene, imidogen (NH).

The first question investigated by chemists was the geometry and electronic structure of methylene. Methylene exists in both singlet and triplet states, each with a bent structure. The triplet state for methylene is the ground state in which one electron occupies the in-plane sp^2 orbital and the other electron is in the out-of-plane *p* orbital on the carbene center. The lowest singlet state of methylene has a closed-shell description in which both electrons occupy the sp^2 orbital, along with an empty *p* orbital (Fig. 4.1).

The relative energies of the singlet and triplet electronic states of carbenes can be explained qualitatively by the energy gap between the carbene's frontier orbitals and the electron–electron repulsion of electrons with the opposite spins. Whenever the energy gap between the carbene's frontier orbitals is small, the ground state of the carbene is the triplet state in order to minimize electron–electron repulsion. When the energy separation between the in-plane sp^2 orbital and the *p* orbital is large, the carbene will have a closed-shell singlet ground state so that both electrons can reside in the lower-energy sp^2 orbital. The singlet–triplet energy gap for many carbenes (excepting those substituted with heteroatomic substituents) is relatively small (less than 10 kcal/mol). Moreover, the intersystem crossing rate between the singlet and triplet states of carbenes is relatively fast (within 100 ps), as compared to nitrenes (*vide infra*).[14]

Carbenes are commonly generated by the thermal or photochemical decomposition of diazirines or diazo compounds. Upon decomposition, these

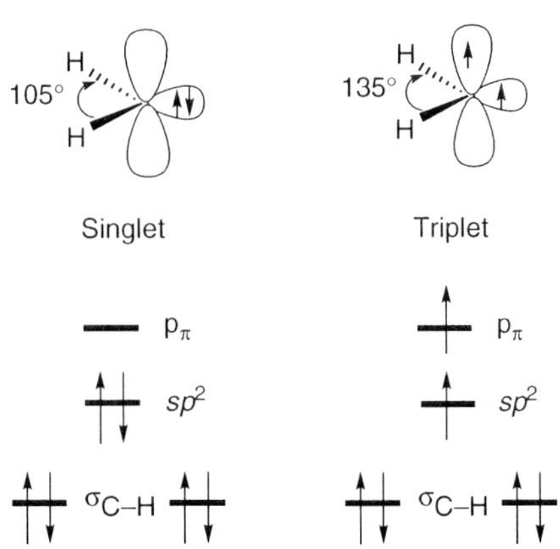

Figure 4.1. Closed-shell singlet state (left) and the open-shell triplet state (right) of methylene.

compounds extrude molecular nitrogen (N_2) to form the corresponding carbene. Once closed-shell singlet carbenes are generated, these species can undergo several reactions, both inter- or intramolecularly.

Intermolecularly, closed-shell singlet carbenes can react with alkenes to form cyclopropyl products stereospecifically. This is one of the most studied organic reactions because the preparation of cyclopropanes is synthetically useful. This reaction is also mechanistically interesting because the degree of stereospecificity of the addition reaction serves as a critical test to distinguish by chemical methods the intermediacy of a singlet or triplet state of the corresponding carbene;[15,16] see Scheme 4.1. These reactivity patterns are traditionally used as criteria to distinguish the spin states of carbenes, and are commonly known as *the Skell-Woodworth rules*.[17]

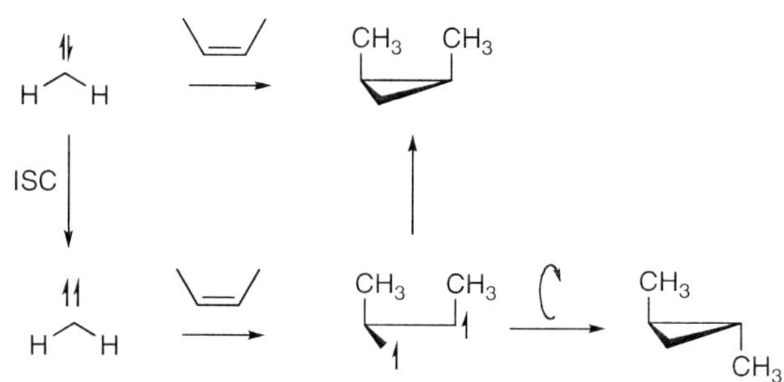

Scheme 4.1.

Furthermore, a closed-shell singlet state carbene can react with nucleophiles to form ylides; see Scheme 4.2. A well-known example is the reaction with pyridine to generate the corresponding carbene–pyridine ylide; this reaction is particularly valuable as it can be used to indirectly observe the formation of a carbene in nanosecond laser flash photolysis experiments using UV-vis or infrared (IR) detection.[18] Other examples include alcohol-ylides[19] and acetonitrile-ylides.[20]

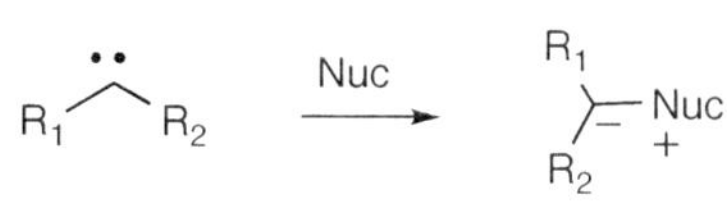

Scheme 4.2.

Moreover, closed-shell singlet carbenes can undergo insertion reactions with X–H bonds including O–H, N–H, and Si–H, as well as C–H bonds.[21] Indeed, intramolecular C–H bond insertion reactions are one of most important reactions of alkylcarbenes, whereby the different intramolecular C–H insertion reactions lead to various alkenes or cyclic products. Scheme 4.3 shows examples of the intramolecular carbon–hydrogen insertion reactions of ethylmethylcarbene (2-butylidene).

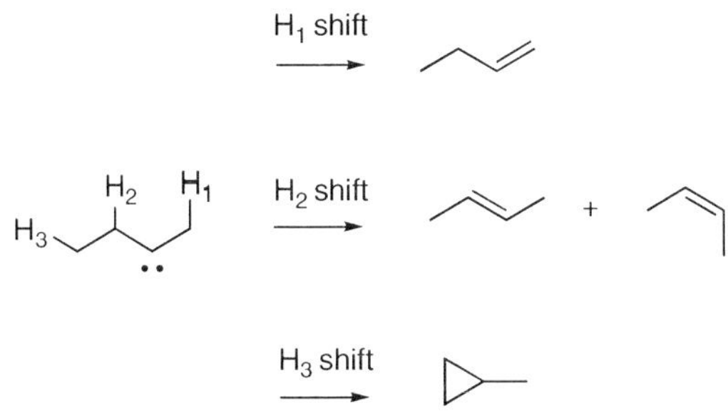

Scheme 4.3.

4.3 REARRANGEMENT IN EXCITED STATES (RIES)

It is economical to think that photolysis is equivalent to thermolysis; in both cases, energy is supplied to precursors to generate the common carbene intermediates. However, many experiments have demonstrated that this viewpoint is overly simplistic. In order to explain observed differences, it has been suggested that carbene rearrangement products can be formed via two pathways: (i) from a free carbene intermediate or (ii) directly from an excited state of the carbene precursor ("Rearrangement in the Excited State (RIES)," as designated by Liu[22]).

With the recent development of ultrafast time-resolved transient absorption spectroscopy, there have been numerous indirect reports supporting the RIES mechanism. For studies of diazirine and diazo compounds, there are observations of the formation of rearranged products from a "nontrappable carbene" route. In ultrafast, time-resolved UV-vis transient absorption studies of *p*-biphenylyldiazoethane ($BpCN_2CH_3$) and *p*-biphenylyldiazomethane ($BpCN_2H$), the observed quantum yield of formation of carbene $^1BpCCH_3$ is ~40% lower than that of ^1BpCH.[23] This was rationalized as an example of an RIES process, which leads directly from the diazo excited state of $BpCN_2CH_3$ to *p*-vinylbiphenyl, and consequently decreases the carbene yield. Product studies using a deuterated alcohol in cyclohexane (R–OD in cyclohexane) indicate that there is an ~25% yield of *p*-vinylbiphenyl formed via RIES (see Scheme 4.4).

Another example is Wolff rearrangement in concert with nitrogen extrusion, proceeding after photoexcitation of diazo carbonyl compounds. Back in 1966, Kaplan and Meloy[24] proposed that direct ketene formation is favored upon decomposition of the *syn* conformer of a carbonyl diazo compound, while the *anti* conformer preferentially decomposes to form trappable carbenes (see Scheme 4.5). Later, ultrafast time-resolved studies indicated that both the concerted and stepwise mechanism occurred. The corresponding ketenes can be formed from the diazo carbonyl excited state on a fast timescale (<0.3 ps),[25] while the stepwise process proceeds with a time constant of 1–1000 ps depending on the carbene's structure. These studies provide evidence for the RIES mechanism, even though further experiments are needed to

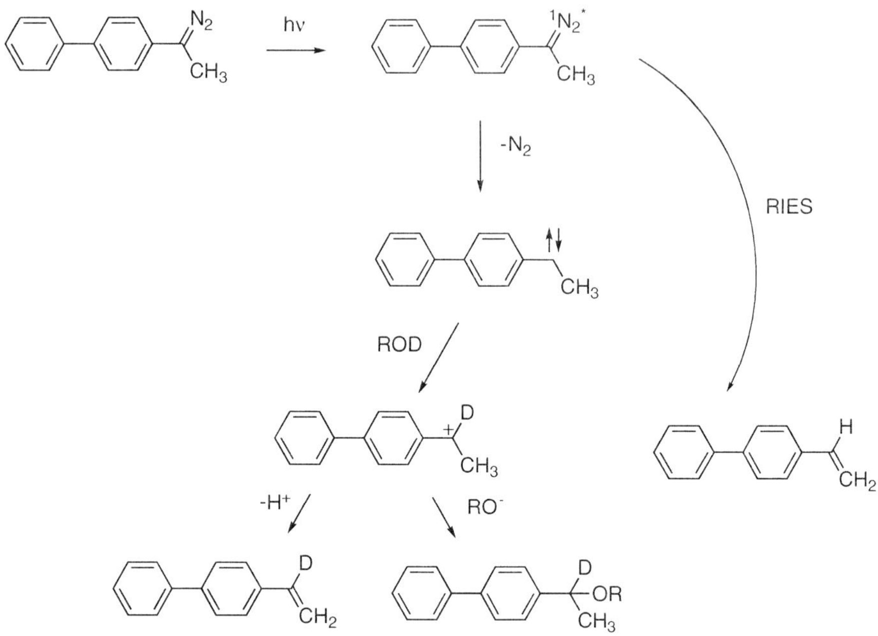

Scheme 4.4.

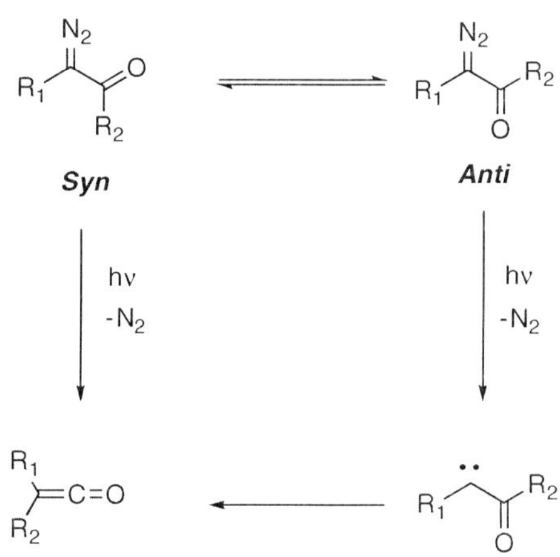

Scheme 4.5.

determine how the photochemistry of diazo carbonyl compounds depends on the structure, solvent, and excitation wavelength.

Ultrafast time-resolved IR studies of diazo ester compounds also favor the RIES mechanism (see Scheme 4.6).[26] It was observed that the decay of the diazo excited states accompany rearrangements to give the alkene products in less than 1 ps. The RIES mechanism was confirmed by the instantaneous rise of the IR transient absorption bands assigned to the alkene products.[26]

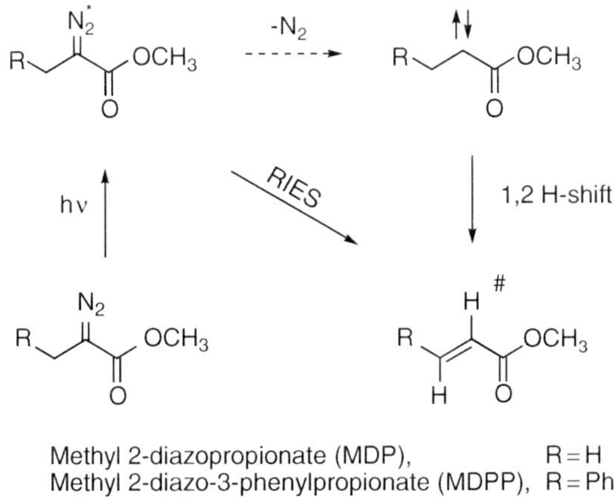

Scheme 4.6.

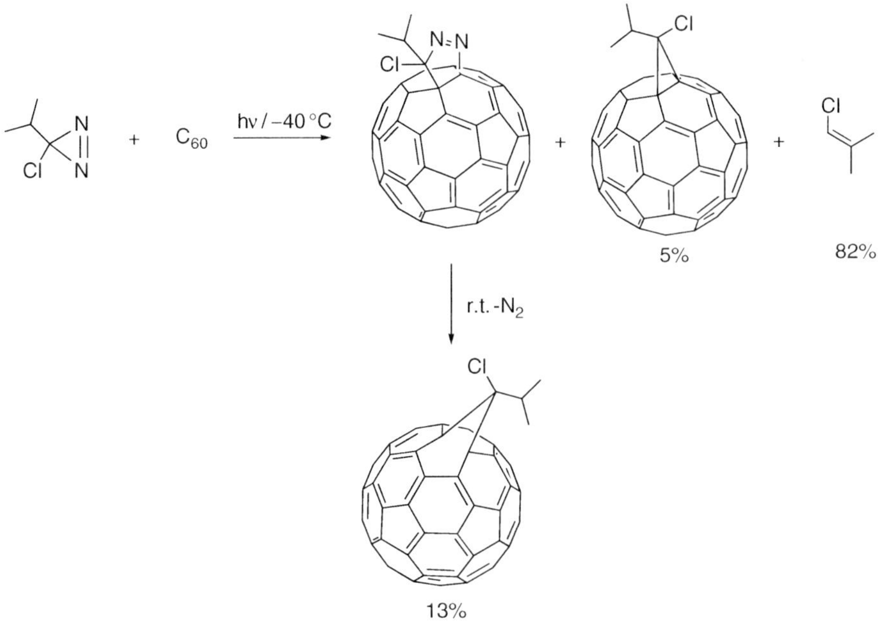

Scheme 4.7.

Studies of diazirines have also provided evidence for RIES, even though there are additional complications due to the photoisomerization of diazirines to diazo compounds. However, Nagase et al.[27] successfully used C_{60} as a mechanistic probe for the formation of carbene (5%), diazo compound (13%), and rearranged product (82%) via RIES in the photolysis of 3-chloro-3-isopropyldiazirine (Scheme 4.7).

Another classic example is the photolysis and thermolysis of 3-ethyl-3-methyldiazirine (EMD), which give very different product distributions

TABLE 4.1. Experimental Product Distributions (%) for the Decomposition of Nitrogenous Precursors of Ethylmethylcarbene (2-Butylidene)[a].

Product	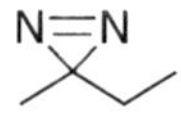 hυ (313 nm)	 Δ
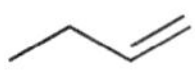	23.2	3.3
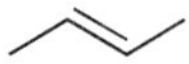	38.0	66.6
	34.7	29.5
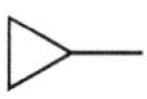	3.7	0.5
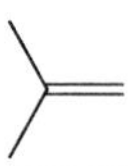	0.3	0.0

[a] From Reference 28.

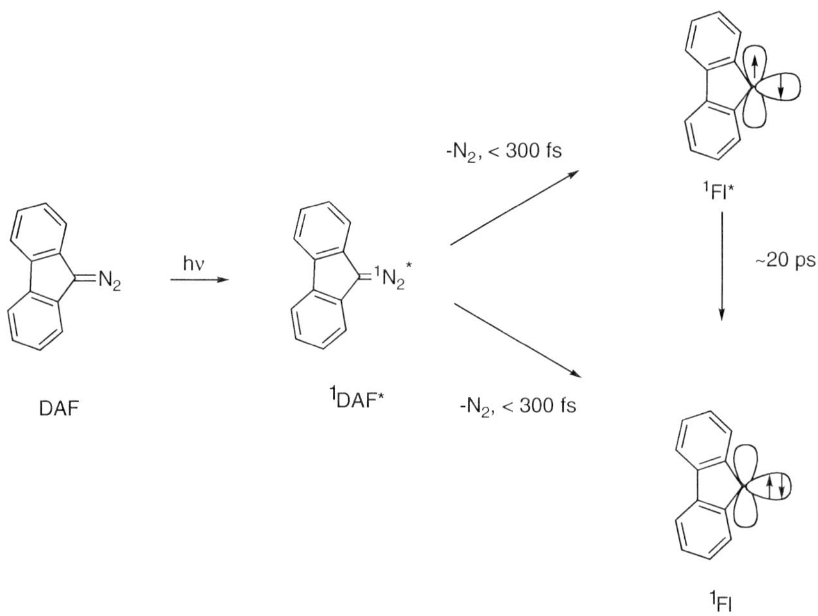

Scheme 4.8.

upon thermal or photochemical activation[28] (Table 4.1). In order to explain the differences, it has been suggested that the photo-generated, product-forming species is (at least in part) a diazirine (or carbene) excited state.

A recent ultrafast time-resolved study[29] of 9-diazofluorene (DAF) revealed that the excited state of the diazo compound fragments to form fluorenylidene in both its lowest-energy singlet state (^{1}Fl) and in a higher-energy open-shell singlet state (^{1}Fl*), as shown in Scheme 4.8. This motivates the computational chemist to search for means to describe the potential energy surface of the carbene's open-shell excited singlet state so as to explain the differences in product distributions upon thermal and photochemical activation of the carbene's precursor.

For all of these experimental results, it is important to have the complementary computational tools that can accurately predict the excited state of the precursors, as well as the intermediates and conical intersections. Spectroscopic properties must be accurately predicted by these computational tools in order to assist in the interpretation of the ultrafast time-resolved transient spectra. In the next section, we will give an overview of the theoretical methodologies available to study the potential energy surface of an excited state.

4.4 ADVANCES IN COMPUTATIONAL INVESTIGATIONS OF CARBENES

Unlike thermal reactions that only involve a single ground-state potential energy surface, photochemical reactions may involve multiple potential energy surfaces, along with different spin multiplicities. In order to accurately calculate the properties of electronically excited states, computational chemistry approaches must incorporate efficient means to deal with the mixture of closed-shell and open-shell descriptions of these different excited states. A good quantum chemical method for the calculation of excited states should have the following properties:[30]

1. The method should have a balanced treatment of the electronic ground state and all excited states relevant to the photochemical process.
2. The accuracy of the method should allow direct comparison with experimental data.
3. The method should be a black-box method that can be used without prior knowledge of the system, while also giving reasonable results.
4. The method should be computationally inexpensive with respect to memory and CPU resources.

To study excited-state reactivity, many aspects of the excitation process must be understood—specifically, one needs to compute the vertical excitation energies in the Franck–Condon[31] region, the various geometric and energetic minima for the different excited states above the ground state, transition states along those different potential energy surfaces, and conical intersections that connect one energy surface to another. The computational method, therefore, must be able to describe both ground and excited states in a balanced way for both vertical and relaxed geometries, as well as in the region where two electronic states are degenerate.

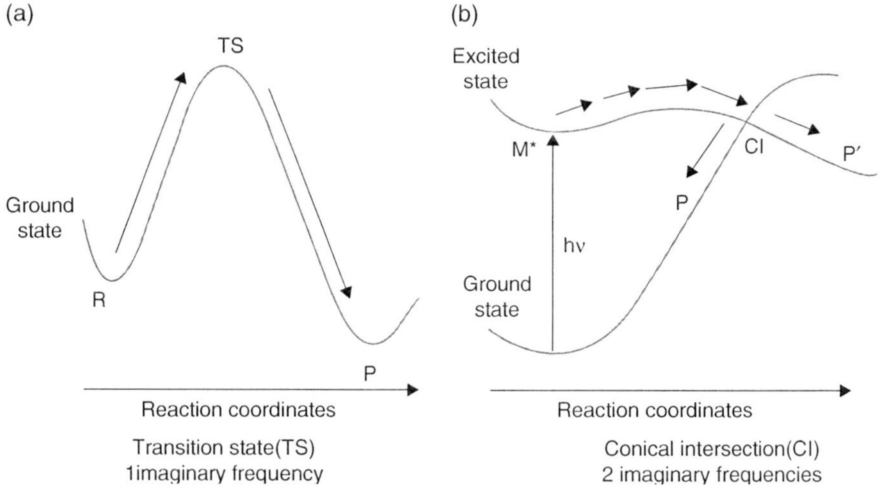

Figure 4.2. Comparison between (a) transition state (TS, 1 imaginary frequency) and (b) conical intersection (CI, 2 imaginary frequencies).

Another challenge for excited-state calculations is that the multiple potential energy surfaces introduce new critical points called conical intersections. Conical intersections are sets of molecular geometry points where two potential energy surfaces are degenerate, thereby providing a relaxation channel for an excited-state species to decay back to the ground state, or perhaps for one excited state to relax to a lower-energy excited state. Figure 4.2 compares a transition state (TS) and a conical intersection (CI) involved in a reaction profile. Panel A in Figure 4.2 shows a TS connecting the reactant R and product P, in a thermal reaction. The TS is the highest-energy point along this reaction coordinate, and such a stationary point is an energy maximum in one dimension (the reaction coordinate), while being an energy minimum in all other dimensions. A TS has one imaginary vibrational frequency, along with a normal mode for this vibration—the normal mode at the TS corresponds to the reaction coordinate that connects the reactant(s) to the product(s). On the other hand, Panel B in Figure 4.2 shows a CI, which is a point where the excited-state and ground-state potential energy surfaces intersect. In photochemical reactions, the ground-state species is vertically excited to M*. This M* state can relax back to the ground state nonadiabatically through the CI to give multiple products (P or P′).

Conical intersections (CI) are typically denoted as a double cone (see Fig. 4.3a). Unlike a TS, the optimized geometry of a CI yields two imaginary vibrational modes upon computation of the second derivatives of the energy with regard to the internal coordinates of the molecular system. The corresponding vibrational modes (x and y in Fig. 4.3a) are directions that would decrease the degeneracy of the CI. These two vectors would form a

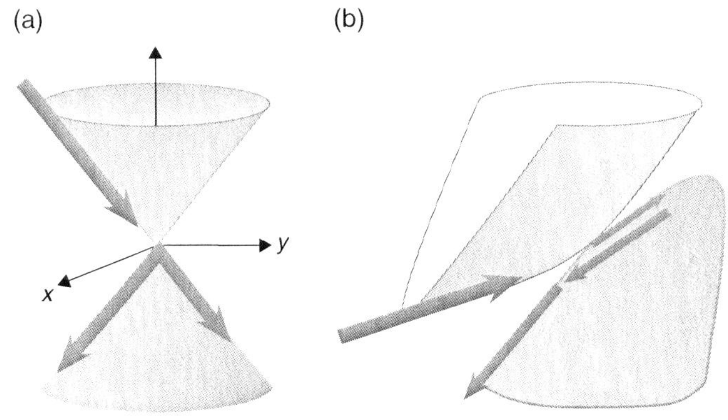

Figure 4.3. Two types of conical intersections: (a) Peaked and (b) Sloped.[32]

two-dimensional branching space that separates the two states. Depending on the topology in the lower-energy state, the decay through such a CI will normally lead to one or more new chemical species through photodegradation, or proceed in another direction to regenerate the reactant.

Besides the typical funnel type CI, which is known as *peaked*, as shown in Figure 4.3a, Figure 4.3b shows the other type of CI that is referred to as *sloped*.[32] As illustrated by the arrows, relaxation from the excited-state surface through this CI would primarily occur in one direction due to the sloped topology of the lower-energy state. This type of CI would preferably generate a single product after photoexcitation.

4.4.1 Configuration Interaction

Configuration interaction is the most accurate quantum chemical method. It is based on the use of the Hartree–Fock (HF) ground-state wavefunction ψ_0; the electronic many-body wavefunction can be constructed as a linear combination of the ground-state Slater determinant and so-called excited determinants as shown in Equation (4.1).

$$\Psi_{CI} = (1 + \hat{c}_1 + \hat{c}_2 + \hat{c}_3 + \ldots)\Psi_0 \qquad (4.1)$$

The symbols $\hat{c}_1, \hat{c}_2, \hat{c}_3$ and so on are the single, double, and triple excitation operators, respectively. Thus, if one considers the singlet ground state as having a set of doubly occupied molecular orbitals, then the $\hat{c}_1$ operator would generate all configurations in which only one electron is excited from an occupied orbital and into a previously unoccupied orbital. Similarly, other high-order operators would generate double, triple, etc., configurations. For each of these configurations, the molecular orbital coefficients for the HF wavefunction are unchanged—thus, only the weighting factors for the admixture of the single, double, triple, etc., excited configurations are optimized as an application of the variational principle. If all possible "excited" determinants or configurations were included, this total wavefunction would be the full configuration interaction wavefunction, corresponding to the exact numerical solution of the Schrödinger equation (assuming an infinitely flexible basis set). Without much effort, the number of these configurations increases to billions, trillions, and larger, even for a small molecule. Thus, such calculations are virtually impossible with the exception of very small molecules and, even then, usually limited to small basis sets. The computational cost increases exponentially as the number of electrons and states increase because the number of Slater determinants would also increase exponentially.

The simplest and crudest approximation is to truncate Equation (4.1) to include only single excitations, and this method is known as configuration interaction with single excitations, CIS. In this approach, the CIS method would be used to compute the properties of the excited state, comparing them to the analogous properties of the HF solution for the ground state. The CIS approach for excited states is variational and size-consistent, but excitation energies computed with the CIS method are usually overestimated.[33] This large error is due to the poor first-order estimates for the true excited-state energies, as well as the poor treatment of the ground state at the HF level (i.e., no inclusion of electron correlation in the ground state). Hence, CIS has been mostly supplanted by time-dependent density functional theory (TD-DFT)[34,35] and other more accurate and less expensive methods. All configuration interaction methods are variational, but some truncation methods, such as only single and double excitations, eliminate size-consistency.

Configuration interaction methods have been long applied to carbenes, including a seminal study by Foster and Boys in 1960 for the properties of methylene (CH_2).[3] This is one of the earliest calculations performed, and it yielded a good prediction for the geometry of singlet methylene as bent rather than linear.

4.4.2 Complete Active Space Self-Consistent Field (CASSCF) Theory

Because full configuration interaction is limited to very small systems, another type of approximation can be applied so that the approach, in a modified form, can be used in larger systems.[36] Instead of limiting the kind of excitations (single vs. single and double, etc.), one can restrict the number of molecular orbitals and the number of electrons for which the electronic excitations take place. As a result, the number of configurations for a large chemical system is restricted. In other words, one can define an active space of occupied and virtual orbitals in which all possible electronic combinations are constructed (Fig. 4.4). This wavefunction-based method is most widely used for photochemical reactions and is known as the complete active space self-consistent field (CASSCF) method. There is one significant difference between CASSCF

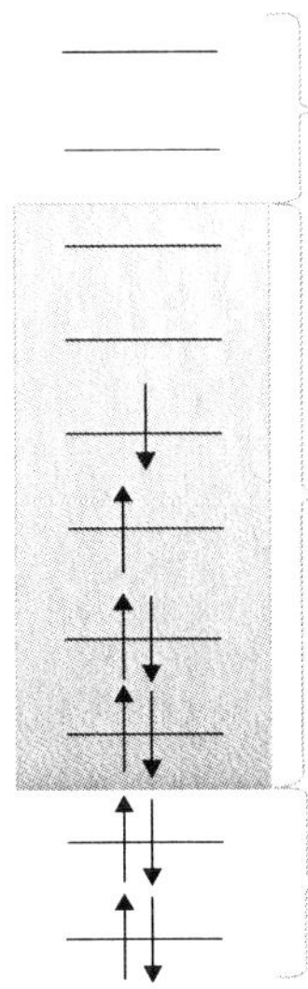

Figure 4.4. Definition of active space for CASSCF calculations with inactive orbitals restricted to being doubly occupied, and virtual orbitals restricted to being unoccupied.

and configuration interaction methods. For the CASSCF wavefunction, the weighting factors for each of the excited configurations, as well as the molecular orbital coefficients for each unique configuration, are optimized in a variational manner leading to flexibility in the wavefunction, but with significant computational cost.

Similar to full configuration interaction (i.e., by including all excitations), CASSCF methods are variational and size-consistent. However, the size consistency of CASSCF is highly dependent on how the active space is chosen. Selection of the correct active space is critical because an intuitive failure to include certain orbitals might result in the inaccurate computation of certain types of excited states. Also, using flexible basis sets, especially those with diffuse functions, can lead to poor convergence of the CASSCF equations.

One of the major limitations of CASSCF calculations is the high computational cost for large active spaces. At present, active spaces with 12 electrons in 12 orbitals (12e, 12o) constitute the feasible limit for such calculations;[30] although, new theoretical developments, termed restricted active space self-consistent field (RASSCF), hold promise to remove this limitation.[37,38] However, other problems arise from choosing which orbitals are doubly occupied, except for a maximum number of holes allowed in this orbital subspace, and which orbitals are unoccupied, except for a maximum number of electrons allowed in this orbital subspace. If one makes use of symmetry, the practical limit for CASSCF can increase up to 16 active electrons.[39] The high computational cost of such calculations is due to the factorial dependence of computing time on both the number of electrons and the number of orbitals involved in generating the many-electron configurations. Equation (4.2) shows the number of Slater determinants in full configuration interaction methods, where M is the number of active orbitals, and where $N\alpha$ and $N\beta$ are the numbers of active electrons with α and β spins.

$$n_{\text{Slater}} = \binom{M}{N_\alpha}\binom{M}{N_\beta} \tag{4.2}$$

Equation (4.3) shows the number of configurations as given by the Weyl-Robinson formula:[40]

$$n_{\text{config}} = \frac{2S+1}{M+1}\binom{M+1}{M-(N/2)-S}\binom{M+1}{(N/2)-S} \tag{4.3}$$

where S is the total spin and N is the total number of active electrons. Both equations illustrate that increasing the number of active electrons and active orbitals will significantly increase the computational cost.

Another problem that arises for CASSCF calculations is which state(s) of the orbitals are to be optimized in order to compute the properties of the desired state. (Moreover, this feature is not unique to CASSCF calculations, but unfortunately plagues all methods used to compute electronically excited states.) Figure 4.5 shows a CI between State 1 and State 2 of an arbitrary system. One starts at the geometric position of A and selects State 2 for the optimization process. As the optimization proceeds, the energy of State 2 may go below that of State 1, for example, geometric position B. As the optimization crosses near to the conical interaction, the optimization process often will fail because the computation may suddenly switch from optimizing the orbitals and configurations for State 2 to optimizing them instead for State 1.

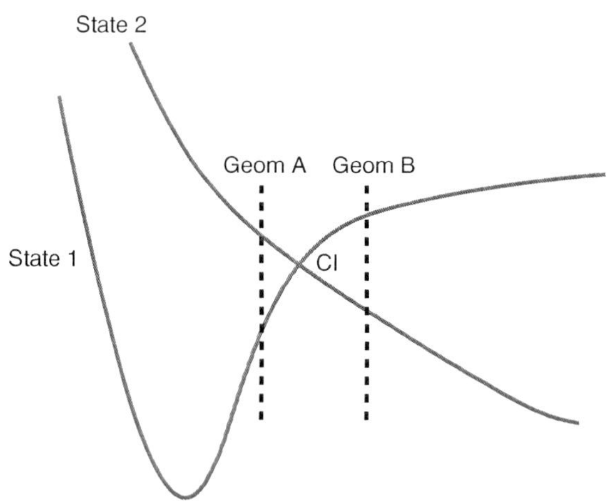

Figure 4.5. Two electronic states of an arbitrary system with a CI between State 1 and State 2 (see discussion in text).

To overcome this problem, one can optimize the wavefunction with state-averaged orbitals, in which the orbitals are optimized with respect to a weighted mean of all states of interest. This approach is commonly used when optimizing conical intersections, and it is often valuable whenever the energy difference between two states is very small. The drawback, however, is that one can introduce some bias in the optimization toward a certain state.

A third problem that arises in a CASSCF calculation is that the method neglects dynamical electron correlation. In order to overcome this, second-order perturbation theory based on the CASSCF wavefunction (e.g., CASPT2[41,42]) can be used. Such calculations have been successfully employed to study various photochemical processes.[43]

Complete active space self-consistent field methods have been used widely in studying the excited states of carbene precursors such as diazirines and diazo compounds. These investigations will be discussed in Section 4.4.5.

4.4.3 Coupled-Cluster Theory with Resolution-of-the-Identity Approximation (RI-CC2)

The aforementioned computational methodologies are multireference methods that allow accurate descriptions of both the ground-state and excited-state properties of chemical systems.[44] However, the major disadvantage of these multireference methods is that they are extremely computationally expensive for large systems and there is an inherent challenge of choosing the active space for the system at hand. In order to overcome this problem, one can use a single-reference coupled-cluster theory method, such as the second-order approximate coupled cluster model (CC2).[45] The excitation energies are obtained by standard coupled-cluster linear response theory.

Even though CC2 methods are based on a single reference wavefunction, the computational cost for the CC2 is still high. To overcome this, the resolution-of-the-identity (RI) approximation[46–48] can be used to reduce the computational cost. This approximation makes use of an auxiliary basis set that reduces the four-center, two-electron integrals to a number of three-center, two-electron integrals. This can reduce the computational cost without significant loss of accuracy.[49]

One major disadvantage, however, of single-reference methods is that conical intersections cannot be located with coupled cluster methods. Conical intersections are often described as biradical-like; thus, the application of single-reference methods is not appropriate.

4.4.4 Time-Dependent Density Functional Theory (TD-DFT)

Differing from the aforementioned theories, which are based on an electronic wavefunction Ψ, density functional theory (DFT)[50,51] describes the electronic energy as a function of the electronic density. In order to obtain a particular electronically excited state, one can introduce certain constraints, like a different spin multiplicity than the ground state or a different irreducible representation of a molecular point group (e.g., A_1 vs. A_2 for a C_{2v} symmetric molecule). The excitation energies are given as the difference of the total energies of the electronic ground state and the excited state, as obtained in two independent calculations. This method is also commonly known as the ΔSCF method.[52] However, this method fails when one is interested in excited states of the same spin multiplicity and irreducible representation.

To overcome this, one can employ unrestricted DFT with an equal number of α and β electrons and use broken-symmetry solutions.[53] This approach is particularly useful for computing the properties of open-shell singlet states, as found in nitrenes. The resulting broken-symmetry solutions should be regarded as describing a 50:50 mixture of singlet and triplet spin states. By the ΔSCF method, one can determine the lowest singlet excited state. This method has been shown to successfully give results comparable with CASPT2 for a variety of nitrenes.[54] However, the limitation of this method is that it is only applicable to the lowest singlet excited states. Moreover, these computations suffer from spin contamination and tend to be less desirable.

As an implementation of linear response theory with Kohn–Sham (DFT) orbitals, TD-DFT[34,35] can be used to calculate excitation energies. Major advantages of TD-DFT are that this method can compute electronic spectra without computing all of the excited-state wavefunctions, and some electron correlation is included for the ground (reference) state and the excited states. A limitation of this approach, however, is that the computed properties of the excited states depend on the ground-state wavefunction. Thus, bond-cleavage reactions in the excited state can be difficult to compute because a restricted (closed-shell) ground-state reference wavefunction may be inappropriate. On the positive side, TD-DFT calculations are relatively inexpensive compared to other methods.

Nevertheless, TD-DFT is still a single-reference method; hence, it also suffers from the same deficiency as other single determinantal reference methods, rendering TD-DFT methods incapable of computing conical intersections. This limitation can be overcome by a modification of TD-DFT that permits a formally multideterminantal target state to be described in spin-flip excitations from a single determinantal reference state of different spin (SF-TD-DFT).[55] However, the SF-TD-DFT method is still in its early stages of development, especially for optimizing conical intersections.[56]

Another disadvantage of the TD-DFT approach is that typical DFT functionals substantially underestimate the relative energies between charge-transfer excited states and the ground state.[57] Long-range correction can be applied to improve these deficiencies, but, at present, such long-range corrected functionals lack an analytical first derivative and therefore do not allow an optimization of the excited-state's geometry.[58]

4.4.5 MOLECULAR DYNAMICS (MD)

The theoretical levels described earlier allow only the establishment of the minimum energy pathways for excited states, which give only a static picture of the potential energy surface. In order to understand the dynamical picture of the bond-breaking or bond-formation processes, especially the particular timing for these individual processes, molecular dynamics simulations can be performed. A few examples for excited-state simulations are given in the next section of this chapter and, in greater detail, in Chapter 5 of this volume.

4.5 THEORETICAL STUDIES OF THE PHOTOCHEMISTRY OF CARBENE PRECURSORS

Diazirines and diazo compounds are well-known photo-precursors of carbenes. However, there have been relatively few theoretical investigations of the photochemistry of diazirines in comparison to the many experimental studies. The first *ab initio* calculations on the diazirine-diazomethane system were performed by Devaquet and coworkers in 1978.[59] Then, Muller-Rammers and Jug[60] reported a semiempirical study in 1985. After almost a decade, Olivucci and Robb[61] performed a detailed CASSCF study on the reaction mechanism for photolysis of 3H-diazirine (CH_2N_2) using an active space of (6e, 6o) along with the 6-31G* basis set. They mapped out the potential energy surfaces corresponding to the ground state, the first and second singlet excited states, and the first triplet state. Detailed results are summarized in Figure 4.6.

Their results show that photolysis promotes the ground-state diazirine (S_0 DZ) to the first excited singlet state. A TS connects (S_1 DZ) to the S_1/S_0 conical intersection CI1, which leads to the ground-state diazomethane (S_0 DM). Another S_1/S_0 conical intersection CI2 connects the first excited-state diazomethane (S_1 DM) with the closed-shell singlet carbene and molecular nitrogen. Finally, there is a S_2/S_1 conical intersection (CI3) that connects the second excited state of diazirine to the S_1 surface. Three pathways from S_1 DZ were proposed, although no TS connecting the CI2 could be located. The first pathway to CI2 requires vibrational leakage from CI1. The second pathway requires the population of the second excited state of the diazirine, potentially by passing through the CI3. The third route goes directly from S_1 DZ to CI2, through a barrier which is determined to be less than 4 kcal/mol (and might be dynamically preferred over passage through CI1) to form the closed-shell singlet carbene and molecular nitrogen. Moreover, it is concluded that the triplet state manifold does not seem to be involved in the formation of the carbene.

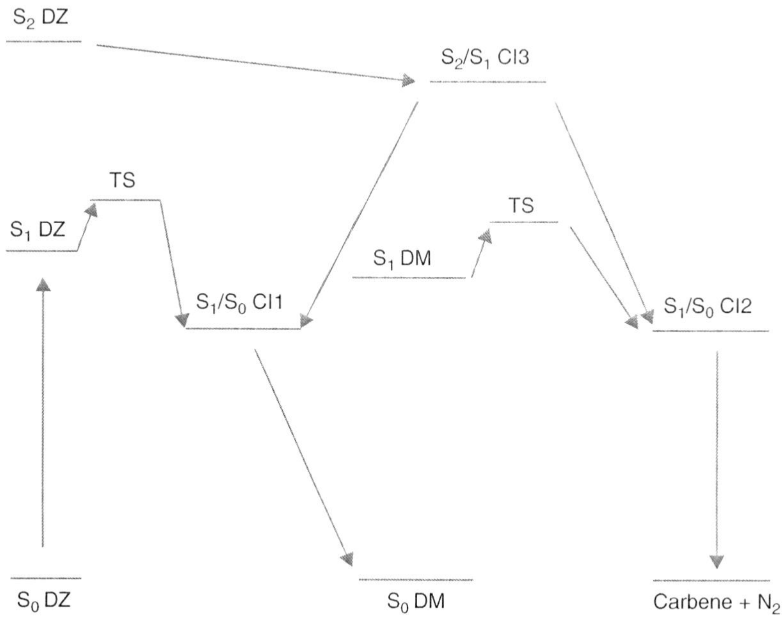

Figure 4.6. Computed scheme for the ground state and the first and second excited singlet surfaces of 3H-diazirine (CH_2N_2) at the CASSCF(6e, 6o)/6–31G* level of theory (DZ = diazirine, DM = diazomethane).

Later, Soto[62] reported calculations on 3,3-dimethyldiazirine [$(CH_3)_2CN_2$] at the CASSCF and CASMP2 levels of theory using 6-31G* and cc-pVDZ basis sets. They were unable to find any evidence for extrusion of nitrogen concerted with rearrangement on either the ground- or excited-state surfaces. The only reaction path they located started on the S_1 excited state surface, passed over a 1.0 kcal/mol barrier, and then through a S_1/S_0 CI leading to 2-diazopropane on the ground state.

More recently, Arenas and coworkers[63] reinvestigated the parent 3H-diazirine system at the CASPT2//CASSCF level of theory using a cc-pVTZ basis set. They made use of a larger active space (12e, 10o) and they also performed limited direct dynamics trajectory calculations on the S_1 and S_2 surfaces. Detailed results are summarized in Figure 4.7.

Two TSs (TS1 and TS2) were found on the ground-state surface connecting the diazirine to either the ground-state carbene or S_0 of diazomethane. It was found that the diazirine would prefer to dissociate into the singlet state of the carbene (and N_2), rather than diazomethane. On the first excited-state surface, the S_1 DZ is a first-order saddle point instead of a local minimum, as was reported when a small active space CASSCF(6e, 6o) and a small 6-31G* basis set was used by Olivucci and Robb.[61] Arenas et al. found that both S_1 DM and S_1 DZ would lead directly to the S_1/S_0 CI. They

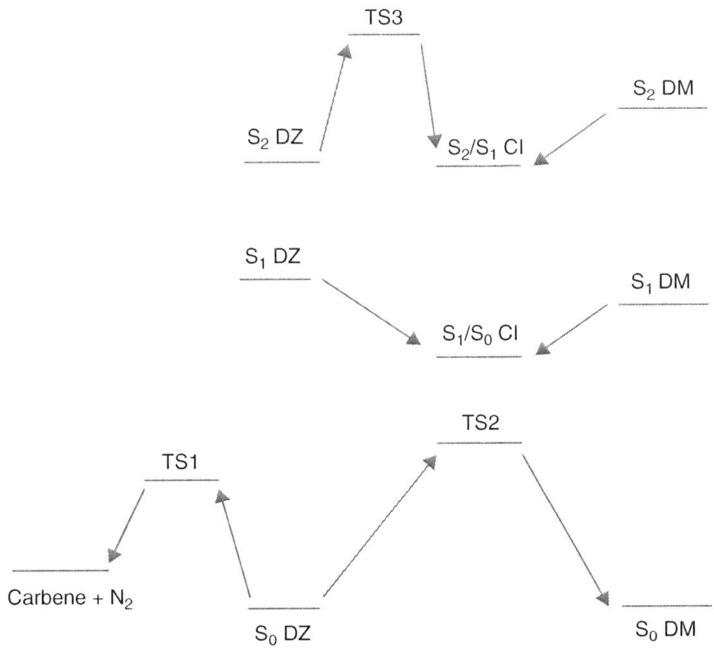

Figure 4.7. Scheme for the ground state and the first and second excited singlet surfaces of 3H-diazirine at the CASSCF(12e, 10o)/cc-pVTZ level of theory (DZ = diazirine, DM = diazomethane).

concluded that a molecule which transited this CI would have enough energy to pass over the barrier to form the closed-shell singlet carbene and molecular nitrogen.

On the second excited-state surface, two stationary points for diazirine (S_2 DZ) and diazomethane (S_2 DM) were found to be first-order saddle points. The S_2 DZ connected to a transition state (TS3) on S_2 via a S_2/S_1 CI. Regardless of the precursors, the observed products were always singlet carbene and molecular nitrogen. The differences in results from the earlier work were suggested to be due to the orbitals chosen for the active space.

Because of limitations of the active space for CASSCF calculations, different computational methods were employed for some larger photochemical precursors. An *ab initio* study of the ground- and excited-state chemistry of phenyldiazirine and phenyldiazomethane was performed by Hadad and Platz.[64] Both RI-CC2 and TD-DFT calculations predicted that formation of phenylcarbene is favored on the ground-state surface of the phenyldiazirine, whereas the S_1 excited state of the diazirine favors isomerization to the first excited state of phenyldiazomethane, which would extrude nitrogen to form the carbene. These results were similar to the CASSCF results obtained for 3,3-dimethyldiazirine and noted earlier.[62]

Later, a CASSCF study was performed by Fang and Cui[65] on phenyldiazomethane. They predict that upon excitation to the S_1 state, direct dissociation through a TS with an activation barrier of 14 kcal/mol occurs to form the (open-shell) singlet excited carbene, or the excited phenyldiazomethane passes through an S_1/S_0 conical intersections to regenerate the ground-state diazo compound. The excited singlet carbene can also decay to the closed-shell ground-state carbene.

Recently, Blancafort and coworkers[66] performed quantum dynamics calculations on a model surface based on CASPT2//CASSCF calculations for the light-induced Wolff rearrangement of 2-diazo-1-naphthoquinone (DNQ), as shown in Scheme 4.9.

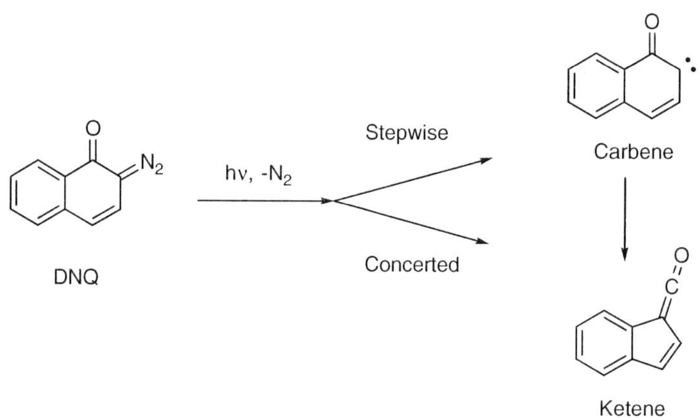

Scheme 4.9.

They found that conical intersections are not isolated points, but exist as an extended seam. By going through an extended seam of conical intersections (Fig. 4.8), photoexcitation leads to a competition between concerted ketene formation and a stepwise mechanism involving a carbene. They obtained results that could not be deduced from the minimum-energy pathway, which predicts the regeneration of the diazo compound. Instead, they showed that the inertia of the excited DNQ would take it in the direction that allows complete dissociation to form the carbene. And by going through different points on the extended seam of conical intersection, the excited state of the DNQ could evolve to the ketene. These results demonstrate the importance of a detailed picture of the potential energy surface and the photodynamics through the CI seams. Moreover, they emphasize that conical intersections are a multidimensional seam of intersections leading to various products.

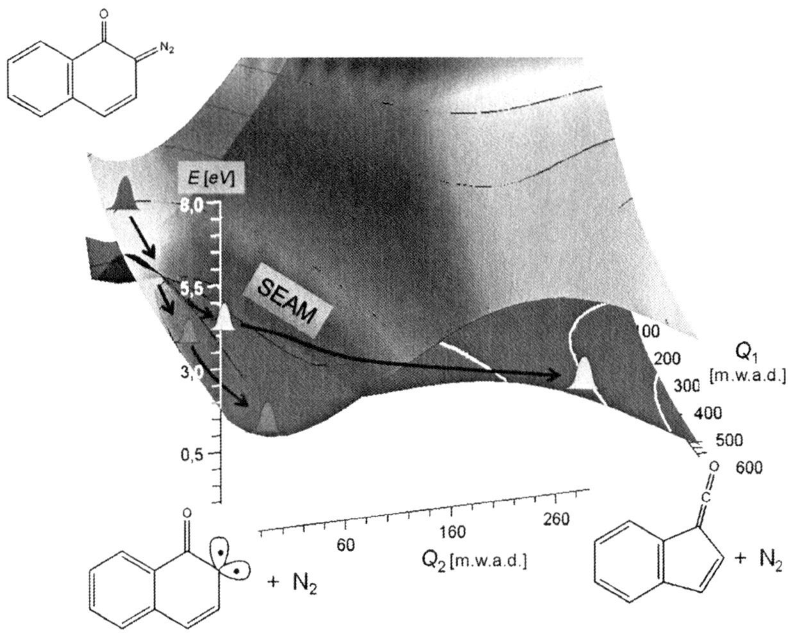

Figure 4.8. Two-dimensional plot of the model surface for the dynamics along Q_1 (C–N dissociation coordinate) and Q_2 (ring rearrangement coordinate) (projection for $Q_3 = 0$; m.w.a.d.: mass-weighted atomic displacements). The purple and blue–green surfaces correspond to the closed- and open-shell diabatic states, correlating with the ketene and carbene products, respectively. The minima for the two photoproducts are found at large values of $Q1$ on either side of the plot. Snapshots of the wave packet evolution at different time steps are also shown on the surface. The wave packet at the beginning of the dynamics is shown in blue. The red wave packet corresponds to the fraction that decays at the region of the seam leading to the carbene minimum, and the yellow one corresponds to the fraction that decays at the region leading to the ketene. (Reprinted with permission from Li, Q., Migani, A., Blancafort, L. *J. Phys. Chem. Lett.* **2012**, *3*, 1056–1061. Copyright 2012 American Chemical Society.) For color details, please see color plate section.

4.6 CONCLUSION AND OUTLOOK

Over the past decade, the development of computational methods has allowed one to perform calculations for unique excited states and to characterize the conical intersections that interconnect the various ground and excited states. These conical intersections, similar to TSs for thermal reactions of the ground state, control how multiple products may be formed after photoexcitation of the precursor. Conical intersections are well-known to be important in the mechanism for radiationless decay in a photochemical reaction; however, they may also control product formation after photoexcitation. This advance will stimulate computational chemists to perform calculations to understand how carbenes are formed from the excited states of their corresponding precursors. Thus, excited states have become "new" reactive intermediates and, as chemists, we are only beginning to learn how to control their reactivity and how to influence the corresponding product formation. In this regard, computational studies will be critical for understanding the potential energy surfaces and the interplay of geometry and energy. In the future, dynamics calculations will reveal the importance of extended seams of conical intersections, and allow computational chemists to quantitatively predict the yields of carbenes or other rearrangement products from RIES mechanisms.

SUGGESTED READING

Blancafort, L., Ogliaro, F., Olivucci, M., Robb, M. A., Bearpark, M. J., Sinicropi, A. "Computational Investigation of Photochemical Reaction Mechanisms." In *Computational Methods in Photochemistry*. Kutateladze, A. Ed. CRC Press: Boca Raton, FL, **2005**, pp. 31–110.

Klessinger, M., Michl, J. *Excited States and Photochemistry of Organic Molecules*. VCH Publishers: New York, **1994**.

Olivucci, M., Ed. *Computational Photochemistry*. Elsevier B. V.: Amsterdam, **2005**.

Robb, M. A. "Conical Intersections in Organic Photochemistry." In *Conical Intersections: Theory, Computation and Experiment*. Domcke, W., Yarkony, D., Köppel, H., Eds. Advanced Series in Physical Chemistry, Vol. *17*. Singapore: World Scientific, **2011**, 3–50.

Yarkony, D. "Nonadiabatic Quantum Chemistry-Past, Present, and Future." *Chem. Rev.* **2012**, *112*, 481–498.

REFERENCES

1 Schaefer, H. F., III. *Science* **1986**, *231*, 1100–1107.

2 Paulsen, S. R. *Angew. Chem.* **1960**, *72*, 781–782.

3 Foster, J. M., Boys, S. F. *Rev. Mod. Phys.* **1960**, *32*, 305–307.

4 (a) Harding, L. B., Goddard, W. A., III. *Chem. Phys. Lett.* **1978**, *55*, 217–220. (b) Zittel, P. F., Ellison, G. B., ONeil, S. V., Herbst, E., Lineberger, W. C., Reinhardt, W. P. *J. Am. Chem. Soc.* **1976**, *98*, 3731–3732.

5 Whittle, E., Dows, D. A., Pimentel, G. C. *J. Chem. Phys.* **1954**, *22*, 1943.

6 Becker, E. D., Pimentel, G. C. *J. Chem. Phys.* **1956**, *25*, 224–228.

7 Norman, I., Porter, G. *Nature* **1954**, *174*, 508–509.

8 Nobel Foundation Website (**1967**) http://www.nobelprize.org/nobel_prizes/chemistry/laureates/1967/ (accessed July 13, 2012).

9 Small, R. D., Jr., Scaiano, J. C. *J. Am. Chem. Soc.* **1978**, *100*, 296–298.

10 Zewail, A. H. *Angew. Chem. Int. Ed.* **2000**, *39*, 2586–2631.

11 Zewail, A. H. *J. Phys. Chem. A* **2000**, *104*, 5660–5694.

12 Nobel Foundation Website (**1999**) http://www.nobelprize.org/nobel_prizes/chemistry/laureates/1999/ (accessed July 13, 2012).

13 Breslow, D. In *Azides and Nitrenes*. E. F. V. Scriven, Ed. Academic Press: New York, **1984**, p. 491.

14 Borden, W. T., Gritsan, N. P., Hadad, C. M., Karney, W. L., Kemnitz, C. R., Platz, M. S. *Acc. Chem. Res.* **2000**, *33*, 765–771. See also: Borden, W. T. In *Reactive Intermediate Chemistry*. Moss, R. A., Platz, M. S., Jones, M., Jr., Eds. Wiley-Interscience: New York, **2004**, pp. 961–1004.
15 Jones, M., Jr., Ando, W., Hendrick, M. E., Kulczycki, A., Jr., Howley, P. M., Hummel, K. R., Malament, D. S. *J. Am. Chem. Soc.* **1972**, *94*, 7469–7479.
16 Turro, N. J., Cha, Y., Gould, I. R. *J. Am. Chem. Soc.* **1987**, *109*, 2101–2107.
17 Woodworth, R. C., Skell, P. S. *J. Am. Chem. Soc.* **1959**, *81*, 3383–3386.
18 (a) Jackson, J. E., Soundararajan, N., Platz, M. S., Liu, M. T. H. *J. Am. Chem. Soc.* **1988**, *110*, 5595–5596. (b) Moss, R. A., Xue, S., Liu W. *J. Am. Chem. Soc.* **1994**, *116*, 1583–1584.
19 Xue, J., Luk, H. L., Platz, M. S. *J. Am. Chem. Soc.* **2011**, *133*, 1763–1765.
20 Scaiano, J. C., McGimpsey, W. G., Casal, H. L. *J. Am. Chem. Soc.* **1985**, *107*, 7204–7206.
21 (a) Brinker, U. H., Ed. *Advances in Carbene Chemistry*, Vol. *1*. JAI Press: Greenwich, CT, **1994**. (b) Brinker, U. H., Ed. *Advances in Carbene Chemistry*, Vol. *3*. Elsevier: Amsterdam, **2001**.
22 Bonneau, R., Liu, M. T. H., Kim, K. C., Goodman, J. L. *J. Am. Chem. Soc.* **1996**, *118*, 3829–3837.
23 Wang, J., Burdzinski, G., Gustafson, T. L., Platz, M. S. *J. Am. Chem. Soc.* **2007**, *129*, 2597–2606.
24 Kaplan, F., Meloy, G. K. *J. Am. Chem. Soc.* **1966**, *88*, 950–956.
25 Burdzinski, G., Wang, J., Gustafson, T. L., Platz, M. S. *J. Am. Chem. Soc.* **2008**, *130*, 3746–3747.
26 Burdzinski G., Zhang, Y., Selvaraj, P., Sliwa, M., Platz, M. S. *J. Am. Chem. Soc.* **2010**, *132*, 2126–2127.
27 Wakahara, T., Niino, Y., Kato, T., Maeda, Y., Akasaka, T., Liu, M. T. H., Kobayashi, K., Nagase, S. *J. Am. Chem. Soc.* **2002**, *124*, 9465–9468.
28 Mansoor, A. M., Stevens, I. D. R. *Tetrahedron Lett.* **1966**, *16*, 1733–1737.
29 Wang, J., Kubicki, J., Hilinski, E. F., Mecklenburg, S. L., Gustafson, T. L., Platz, M. S. *J. Am. Chem. Soc.* **2007**, *129*, 13683–13690.
30 Grimme, S. *Rev. Comput. Chem.* **2004**, *20*, 153–218.
31 Franck, J., Dymond, E. G. *Trans. Faraday Soc.* **1926**, *21*, 536–542.
32 Atchity, G. J., Xantheas, S. S., Ruedenberg, K. *Chem. Phys.* **1991**, *95*, 1862–1876.
33 (a) Foresman, J. B., Head-Gordon, M., Pople, J. A., Frisch, M. J. *J. Phys. Chem.* **1992**, *96*, 135–149. (b) Stanton, J. F., Gauss, J., Ishikawa, N., Head-Gordon, M. *J. Chem. Phys.* **1995**, *103*, 4160–4174. (c) Hirata, S., Head-Gordon, M., Bartlett, R. J. *J. Chem. Phys.* **1999**, *111*, 10774–10786.
34 Furche, F., Ahlrichs, R. *J. Chem. Phys.* **2002**, *117*, 7433–7447.
35 Furche, F., Ahlrichs, R. *J. Chem. Phys.* **2004**, *121*, 12772–12773.
36 Bearpark, M. J., Robb, M. A., Schlegel, H. B. *Chem. Phys. Lett.* **1994**, *223*, 269–274.
37 Olsen, J., Roos, B. O., Jørgensen, P., Jensen, H. J. A. *J. Chem. Phys.* **1988**, *89*, 2185–2192.
38 Klene, M., Robb, M. A., Blancafort, L., Frisch, M. J. *J. Chem. Phys.* **2003**, *119*, 713–728.
39 Bearpark, M. J., Ogliaro, F., Vreven, T., Boggio-Pasqua, M., Frisch, M. J., Larkin, S. M., Morrison, M., Robb, M. A. *J. Photochem. Photobiol. A: Chem.* **2007**, *190*, 207–227.
40 Pauncz, R. *The Symmetric Group in Quantum Chemistry*. CRC Press: Boca Raton, FL, **2005**.

41 Andersson, K., Malmqvist, P.-Å., Roos, B. O., Sadlej, A. J., Wolinski, K. *J. Phys. Chem.* **1990**, *94*, 5483–5488.
42 Andersson, K., Malmqvist, P.-Å., Roos, B. O. *J. Chem. Phys.* **1992**, *96*, 1218–1226.
43 (a) Roos, B. O. *Acc. Chem. Res.* **1999**, *32*, 137–144. (b) Pulay, P. *Int. J. Quant. Chem.* **2011**, *111*, 3273–3279.
44 Hättig, C., Weigend, F. *J. Chem. Phys.* **2000**, *113*, 5154–5161.
45 Christiansen, O., Koch, H., Jørgensen, P. *Chem. Phys. Lett.* **1995**, *243*, 409–418.
46 Whitten, J. L. *J. Chem. Phys.* **1973**, *58*, 4496–4501.
47 Dunllap, B. I., Connolly, J. W. D., Sabin, J. R. *J. Chem. Phys.* **1979**, *71*, 3396–3402.
48 Vahtras, O., Almlof, J. E., Feyereisen, M. W. *Chem. Phys. Lett.* **1993**, *213*, 514–518.
49 (a) Weigend, F., Köhn, A., Hättig, C. *J. Chem. Phys.* **2001**, *116*, 3175–3183. (b) Hättig, C., Köhn. A. *J. Chem. Phys.* **2002**, *117*, 6939–6951.
50 Becke, A. D. *J. Chem. Phys.* **1993**, *98*, 5648–5652.
51 Lee, C., Yang, W., Parr, R. G. *Phys. Rev. B* **1988**, *37*, 785–789.
52 Ziegler, T., Rauk, A., Baerends, E. J. *Theor. Chim. Acta* **1977**, *43*, 261–271.
53 Cramer, C. J., Dulles, F. J., Giesen, D. J., Almlöf, J. *Chem. Phys. Lett.* **1995**, *245*, 165–170.
54 Johnson, W. T. G., Sullivan, M. B., Cramer, C. J. *Int. J. Quant. Chem.* **2001**, *85*, 492–508.
55 Shao, Y., Head-Gordon, M., Krylov, A. I. *J. Chem. Phys.* **2003**, *118*, 4807–4818.
56 Minezawa, N., Gordon, M. S. *J. Phys. Chem. A* **2009**, *113*, 12749–12753.
57 (a) Dreuw, A., Weisman, J. L., Head-Gordon, M. *J. Chem. Phys.* **2003**, *119*, 2943–2946. (b) Tozer, D. J. *J. Chem. Phys.* **2003**, *119*, 12697–12699. (c) Sobolewski, A. L., Domcke, W. *Chem. Phys.* **2003**, *294*, 73–83. (d) Dreuw, A., Head-Gordon, M. *J. Am. Chem. Soc.* **2004**, *126*, 4007–4016.
58 Tawada, Y., Tsuneda, T., Yanagisawa, S., Yanai, T., Hirao, K. *J. Chem. Phys.* **2004**, *120*, 8425–8433.
59 Bigot, B., Ponec, R., Sevin, A., Devaquet, A. *J. Am. Chem. Soc.* **1978**, *100*, 6575–6580.
60 Muller-Rammers, P. L., Jug, K. *J. Am. Chem. Soc.* **1985**, *107*, 7275–7284.
61 Yamamoto, N., Bernardi, F., Bottoni, A., Olivucci, M., Robb, M. A., Wilsey, S. *J. Am. Chem. Soc.* **1994**, *116*, 2064–2074.
62 Bernardi, F., Olivucci, M., Robb, M. A., Vreven, T., Soto, J. *J. Org. Chem.* **2000**, *65*, 7847–7857.
63 Arenas, J. F., López-Tocón, I., Otero, J. C., Soto, J. *J. Am. Chem. Soc.* **2002**, *124*, 1728–1735.
64 Zhang, Y., Vyas, S., Hadad, C. M., Platz, M. S. *J. Phys. Chem. A* **2010**, *114*, 5902–5912.
65 Fang, W., Cui, G. *Chem. Phys. Chem.* **2011**, *12*, 1689–1696.
66 Li, Q., Migani, A., Blancafort, L. *J. Phys. Chem. Lett.* **2012**, *3*, 1056–1061.

5

DYNAMICS IN CARBENE REACTIONS

DINA C. MERRER

Department of Chemistry, Barnard College, New York, NY, USA

K. N. HOUK

Department of Chemistry and Biochemistry, University of California, Los Angeles, CA, USA

LAI XU

Department of Chemistry and Biochemistry, Texas Tech University, Lubbock, TX, USA

5.1 INTRODUCTION

5.1.1 Carbenes

Since Hine discovered that the hydrolysis of chloroform produced dichlorocarbene (CCl_2),[1] and Doering and Hoffmann subsequently found that the reaction of CCl_2 with alkenes yielded cyclopropanes,[2] carbenes have inspired a great deal of investigation.[3–16] The mechanisms of the reactions of these divalent carbon intermediates, both intramolecular and intermolecular, have been explored mechanistically and their relative and absolute kinetics measured via classical and modern techniques of product studies and fast and ultrafast laser methods, respectively.[17–24]

Over the past 20 years, singlet carbenes moved from exclusively transient species at ambient temperature to stable species in the absence of air and water.[25–31] The discovery of the first crystalline *N*-heterocyclic carbene by Arduengo in 1991[32] opened the floodgates to the synthesis of related nitrogen-, phosphorus-,

Contemporary Carbene Chemistry, First Edition. Edited by Robert A. Moss and Michael P. Doyle.

and sulfur-substituted stable singlet carbenes.[25–31] Stable singlet carbenes have been used as transition metal ligands in the development of organometallic catalysts.[30,33–55]

In synthesis, carbenes have been utilized for the installation of cyclopropyl moieties by building upon Doering and Hoffmann's initial work.[2] The cycloadditions of carbenes to π bonds were first proposed to proceed in a concerted, synchronous manner by Skell and Garner.[56] Later, Roald Hoffmann showed a C_{2v}-symmetric, cyclic, four-electron transition state to be symmetry-forbidden.[57] Therefore, in order for carbene cycloadditions to occur, they must proceed nonlinearly, involving electrophilic and nucleophilic phases.[58]

Singleton, Houk, and coworkers used kinetic isotope effects and quantum mechanical calculations to show that the transition state of CCl_2 + 1-butene is asymmetric, where CCl_2 adds in an asynchronous yet concerted manner (Fig. 5.1).[59] Analogous work by Merrer and Rablen on CCl_2 + cyclopropene showed that the mechanism of carbene addition to this strained π system proceeded similarly.[60] Hence, for ground-state singlet electrophilic carbenes such as CCl_2 and other halocarbenes, concerted asynchronous cycloadditions to π bonds are the norm. An exception is difluorocarbene (CF_2) addition to alkenes, where the mechanism proceeds with considerable biradical character.[61,62]

More recently, the studies of carbene cycloadditions to unstrained and strained π systems have benefited from the use of quasi-classical direct dynamics trajectory calculations.[60,64] Trajectory calculations have provided greater insight into the concerted-versus-stepwise nature of CCl_2 and CF_2 additions to unstrained alkenes.[62] In the case of halocarbene additions to strained C–C π bonds, dynamic factors have been found to play a substantial role in observed reactivity.[60,64] These results have opened the door to the likelihood of such nonstatistical behavior in related carbene reactions with strained substrates[65–68] and in other carbene-mediated reactions.[69,70]

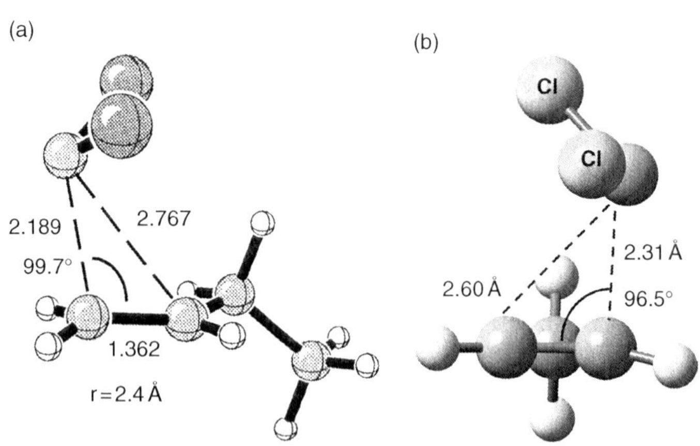

Figure 5.1. Transition structures of CCl_2 addition to (a) 1-butene.[59,63] (Copyright © 1999 American Chemical Society.) and (b) cyclopropene.[60]

5.1.2 Nonstatistical Reaction Dynamics

In order to understand the influence of nonstatistical dynamics on reactions involving carbenes, we will briefly review transition state theory (TST). An organic chemist's first introduction to the intersection of thermodynamics and kinetics in an organic reaction is often pictorial, in the form of a reaction coordinate diagram. Energy (Gibbs free energy or enthalpy) is plotted against the reaction coordinate, with the reaction proceeding from left to right. Between the starting materials and products lie reactive intermediates and the transition states connecting these energy minima. The connections form a one-dimensional potential energy surface (PES) of a particular reaction (Fig. 5.2a).

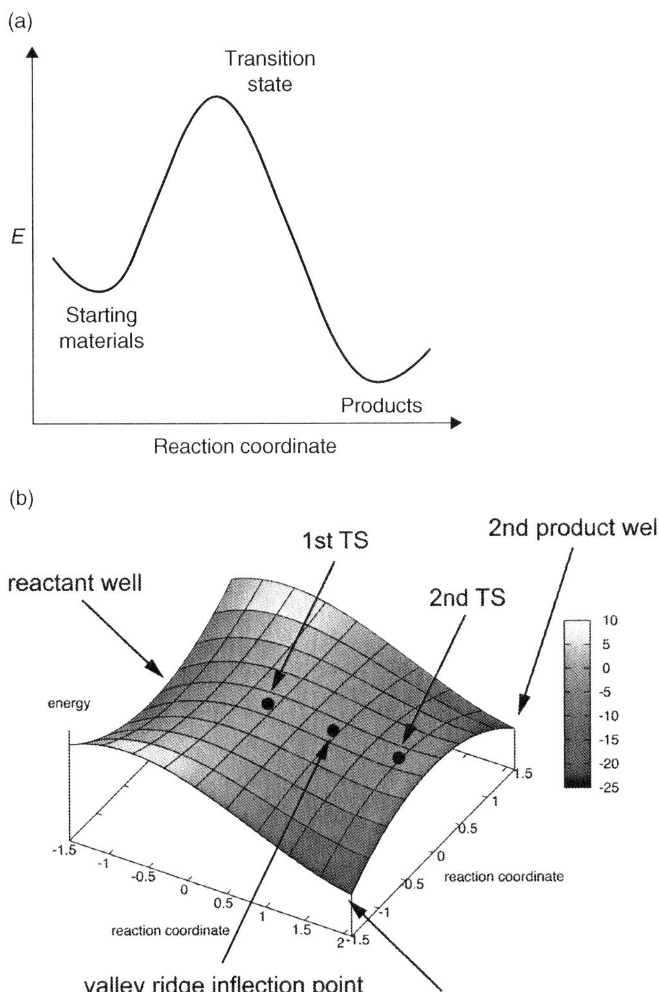

Figure 5.2. (a) A one-dimensional PES; (b) a two-dimensional PES (hypersurface).[71] (Copyright © 2012 American Chemical Society.) For color details, please see color plate section.

The PES is even more informative when an additional dimension is included, thus generating a two-dimensional PES or reaction hypersurface. Introductory organic students learn that a reaction proceeds over the PES via the path of least resistance, akin to a hiker choosing the least strenuous route through a mountain pass (Fig. 5.2b).[71] Thus, the topology of the surface determines the formation of a particular product(s) and also the quantities of that product(s). For an organic reaction with a sizable activation energy and clear rate-determining step, each product is formed from its own unique transition state, and the product distributions are determined statistically by the heights of the energy barriers leading to the products. These are the tenets of TST.

Transition state theory holds well for reactions with unambiguous lowest-energy paths. However, a reaction's PES may contain topological features that induce results not predicted well or accurately by TST. These features may include negligible or nonexistent energy barriers, a plateau-like surface, significant exothermicity (i.e., ΔH_{rxn} or ΔH from the rate-determining transition state to product(s) on the order of −100 kcal/mol), and/or a single transition state leading to multiple products or reaction routes. In these cases, the energies of the stationary-point minima and maxima do not dictate the progress of a reaction and TST is inadequate for predicting a reaction's outcome. Rather, for reactions possessing complex potential energy hypersurfaces, molecular motions may play a much larger role, where the preferred reaction paths are controlled by factors such as kinetic energy, momentum, and vibrations of the molecules. In other words, these types of reactions are controlled by nonstatistical dynamic effects.[72]

Nonstatistical reaction dynamics have been implicated in a variety of organic reactions (Fig. 5.3).[72–82] Reactions proceeding over plateau-type regions of their PESs, such as the vinylcyclopropene-cyclopentene rearrangement,[83–88] are one class. Another class of dynamically controlled reactions includes a PES in which an intermediate in a reasonably deep potential energy well (i.e., 10–20 kcal/mol) is avoided, as in the [3,3]-sigmatropic rearrangement of 1,2,6-heptatriene[89–91] and in the deazetization of 2,3-diazabicyclo[2.2.1]hept-1-ene.[92–96] Reaction dynamics has also been found to influence reactions that contain bifurcating PESs or one transition state leading to multiple products.[73,78,97–111] Several reactions in this class have been discovered recently,[73] including some Diels–Alder,[103,104,111] Beckmann,[106] and Schmidt[101] reactions; carbocation rearrangements involved in natural product biosyntheses;[80,98,99,107,110] and even transition-metal catalyzed reactions.[108,109]

We will review here the use of direct dynamics trajectories as a valuable tool to determine the synchronicity of dihalocarbene cycloadditions to unstrained alkenes.[62] Furthermore, we will see how nonstatistical dynamics have influenced the mechanisms of dihalocarbene additions to the strained C–C π systems of cyclopropenes (**1**),[60,62,64] benzocyclopropenes (**2**),[65] and cyclooctyne (**3**),[66] and the strained C–C σ system of bicyclo[1.1.0]butanes (**4**).[67]

The PESs of all of these reactions are characterized by post-transition state branching, low activation barriers, and exothermicities on the order of 100 kcal/mol. Finally, we will discuss the role of molecular dynamics in the Wolff rearrangement of acylcarbenes[70] and in carbene photochemical generation.[69]

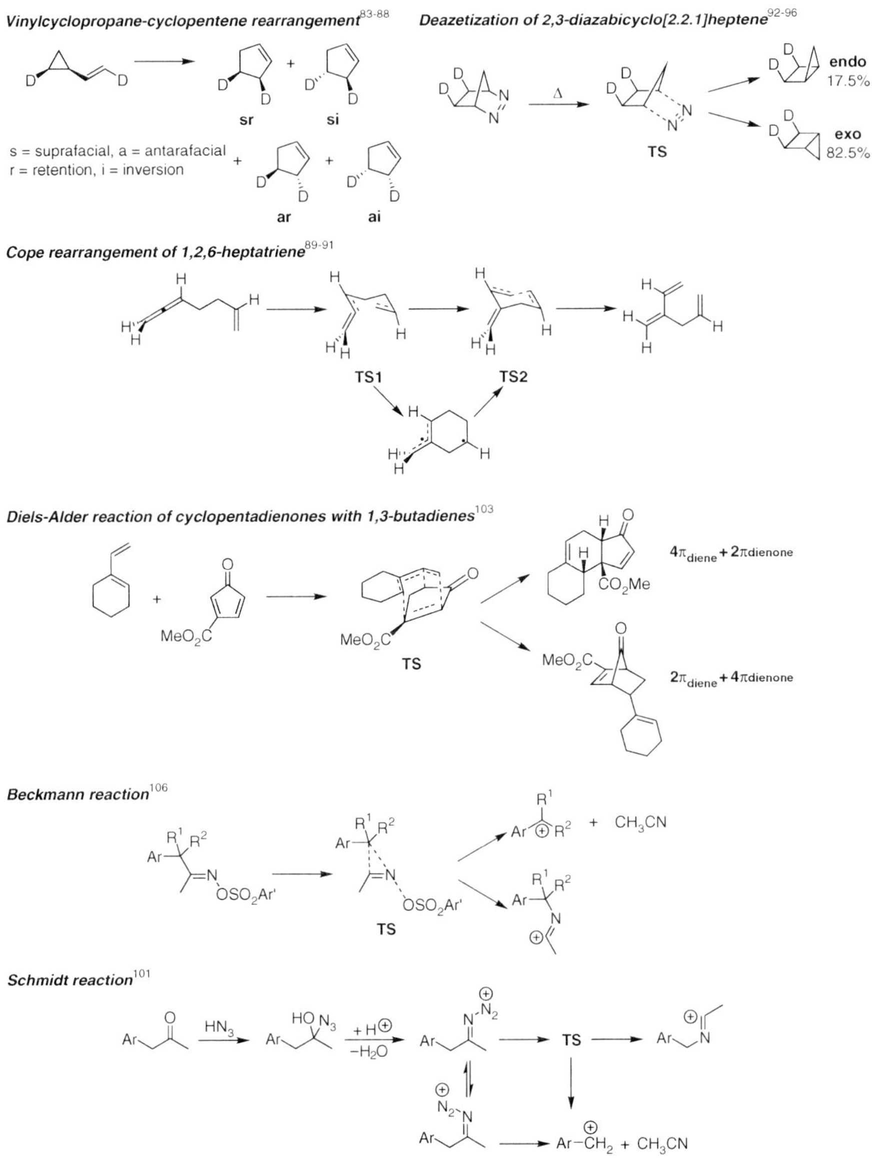

Figure 5.3. Examples of organic reactions controlled by reaction dynamics.

Carbocation rearrangements in abetadiene biosynthesis[107]

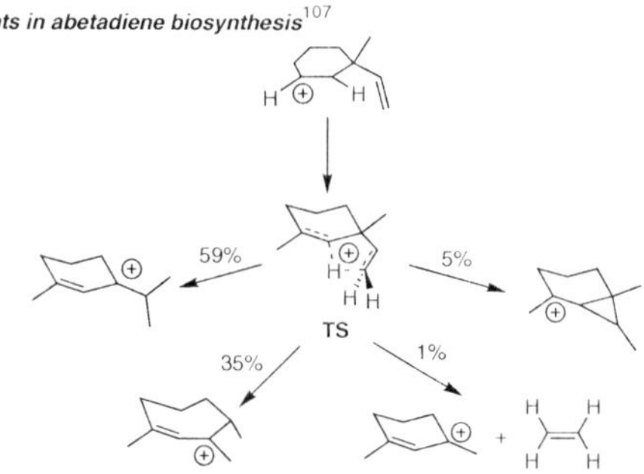

Transition-metal catalyzed process in tetracycline indole formation[108]

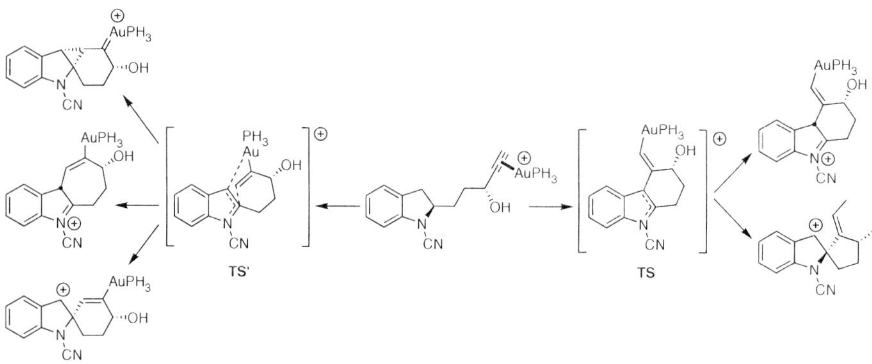

Figure 5.3. (Continued).

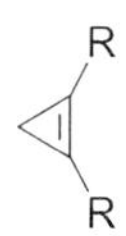

1

a, R = H
b, R = CH_3
c, R = $CH{=}CH_2$
d, R = Ph
e, R = (p-CH_3O)C_6H_4
f, R = (p-CF_3)C_6H_4

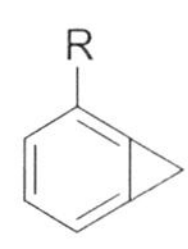

2

a, R = H
b, R = CH_3
c, R = Br
d, R = CHO

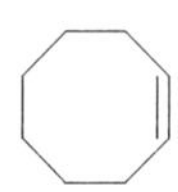

3

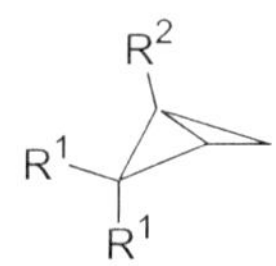

4

a, $R^1 = R^2 = H$
b, $R^1 = R^2 = CH_3$

5.2 DYNAMICS OF CARBENE CYCLOADDITIONS TO ALKENES AND ALKYNES

5.2.1 Carbene Additions to Unstrained Alkenes

The use of reaction dynamics to investigate carbene reactions dates to the pioneering studies of the reaction of singlet CH_2 with H_2 by Wang and Karplus in 1973.[112] Although they used only semiempirical quantum mechanical methods, the revolutionary aspect of their communication was reflected in the title: "Dynamics of Organic Reactions"!

5.2.1.1 Ethylene and 2-Butene. Xu, Doubleday, and Houk reported the first quasi-classical direct dynamics studies of cycloadditions of carbenes, the reactions of singlet CCl_2 and CF_2 with ethylene, using density functional theory for the quantum mechanical computations.[62] Figure 5.4 shows the overlay of individual starting geometries for 453 reactive trajectories using UB3LYP/6-31G* for CCl_2 + ethylene (left), and 298 reactive trajectories on the mUB3LYP/6-31G* surface for CF_2 + ethylene (right). The use of a modified (by Doubleday) B3LYP functional in the latter case was necessitated by the spurious minimum on the CF_2 plus ethylene surface for the 1,1-difluorotrimethylene diradical as compared to Nagase's MR-MP2 results that show a flat platform but no minimum for this diradical.[113] In the mB3LYP functional, the fraction of HF exchange is reduced from 20% to 12%, with a compensatory adjustment in the contribution of local exchange to 88%.[114]

The initial structures for the reaction of CCl_2 with ethylene are widely scattered, and this is consistent with the very low barrier and early transition state. The overlay of structures for the reaction of CF_2 with ethylene is tighter than that for CCl_2 due to the more advanced, less flexible transition state.

Figure 5.5 shows snapshots of representative trajectories for the reactions of CCl_2 and CF_2 with ethylene.[115] For the CCl_2 + ethylene trajectory in Figure 5.5a, the time gap between formation of the two bonds of the cyclopropane product (defined as a C–C distance of 1.59 Å)[116] is 47 fs. In Figure 5.5b, the time gap between bond formations is 118 fs. Trajectory animations have been placed on the website (http://www.chem.ucla.edu/~lxu01pku/carbene_dynamics). The time

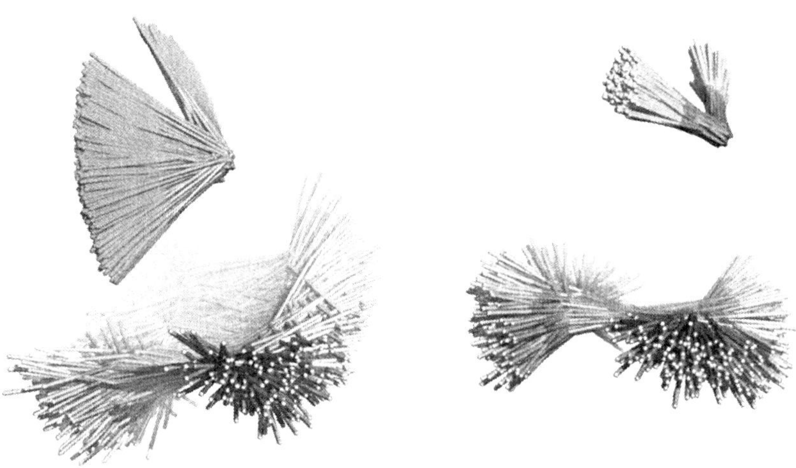

Figure 5.4. Left: Overlay of starting geometries of 453 reactive trajectories for the reaction of CCl_2 with ethylene. Right: Overlay of starting geometries of 298 reactive trajectories for CF_2 + ethylene.[115] (Copyright © 2011 American Chemical Society.) For color details, please see color plate section.

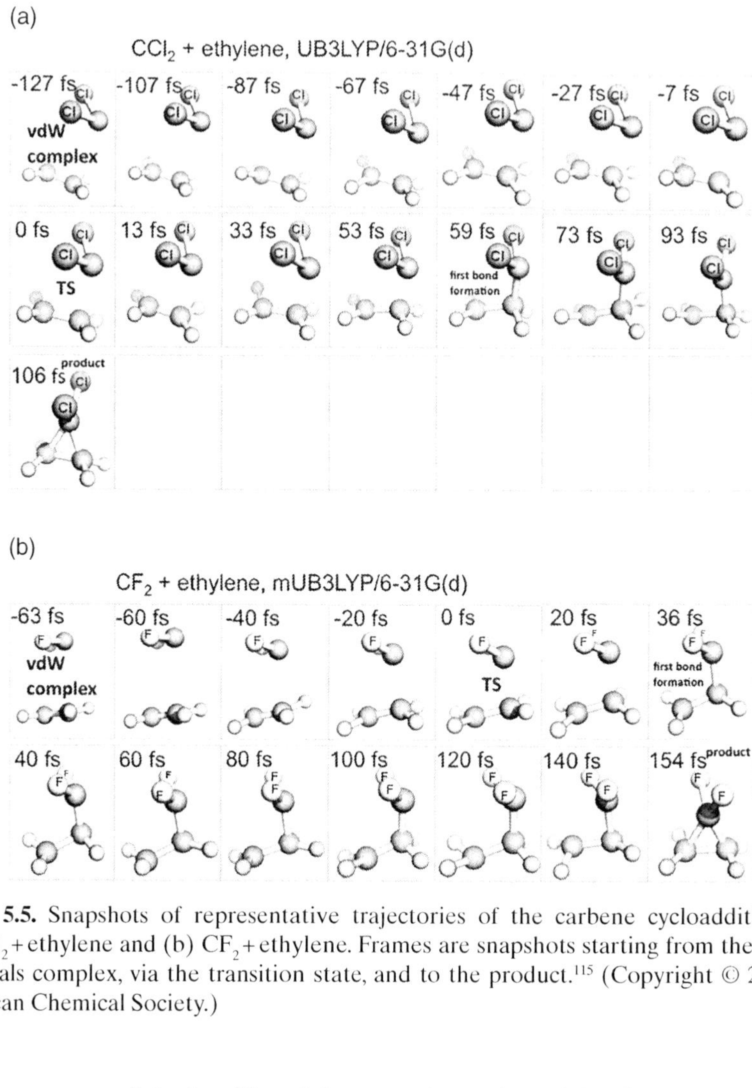

Figure 5.5. Snapshots of representative trajectories of the carbene cycloadditions. (a) CCl_2+ethylene and (b) CF_2+ethylene. Frames are snapshots starting from the van der Waals complex, via the transition state, and to the product.[115] (Copyright © 2011 American Chemical Society.)

gap is a measure of the lag (if any) between formation of the first and second bond. The bond formation is defined as a C–C distance less than 1.59 Å. The formation of the first bond defines the appearance of the diradical, and formation of the second bond indicates its disappearance.

Figure 5.6 shows the lifetimes of the diradicals computed by the time gap of bond formation. There is a large difference in the lifetimes of the diradicals derived from CCl_2 and CF_2 additions to ethylene at the UB3LYP level. However, the CF_2 diradical lifetimes at the mUB3LYP level are much shorter.

The UB3LYP description of CF_2 addition differs from that of mUB3LYP; there is no diradical minimum with mUB3LYP by design to match the MRMP2 results, and this contributes to the difference. As found from this study, the flat region of the IRC path at the mUB3LYP surface was also severely truncated

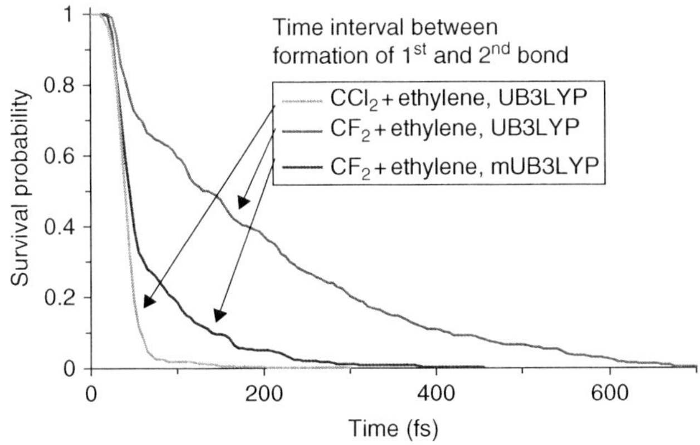

Figure 5.6. Lifetimes of diradicals formed in three sets of trajectories, computed as the time gap between formation of the first and second bond (defined as C–C < 1.59 Å). Survival probability at time t is the fraction of diradicals that have not yet cyclized at time t. Time zero is the formation of the first bond.[115] (Copyright © 2011 American Chemical Society.)

relative to the UB3LYP and MRMP2(10,10) surfaces. One measure of the flat region is the value of the CCC angle at the point along the minimum energy path at which the energy begins to decrease rapidly. The point where the gradient reaches –5 kcal/mol per bohr-amu$^{1/2}$ was taken as the point where the energy plunges to the cyclopropane product. For UB3LYP and MRMP2(10,10), this point is reached when CCC = 92°, while it is 97° for mUB3LYP. This gives the diradical at the mUB3LYP level a narrower boundary within which to move while avoiding cyclization. Dynamics simulations of other diradicals have shown that trajectories with an extensive flat region might be trapped and behave as an intermediate.[86,117,118] On the other hand, a reduction in the length of the flat region is expected to shorten the lifetime of the trajectories. The UB3LYP and mUB3LYP descriptions of CF_2 addition are actually likely to bracket the true situation. UB3LYP overestimates lifetimes because it predicts a potential energy minimum, while mUB3LYP gives the diradical a rather small range of motion within which to avoid cyclization.

Figure 5.6 illustrates the decay patterns, which show a range of complexity. There is no decay in the first 20–25 fs. This is the minimum time gap in the bond forming steps. That is the time needed for the diradical termini to move directly to within 1.59 Å, after formation of the first bond. Each set of trajectories shows at least a fraction of very fast decay, followed by a variable amount of slower decay, after 25 fs. The transition state for CCl_2 addition decays sharply, with 95% decay in 65 fs after the first bond is formed. As shown in Figure 5.6, decays of the CF_2 plus ethylene diradicals have both a long and short component. These were fitted to biexponential decays. Time constants of 15 and 190 fs were obtained for the UB3LYP level calculations (amplitudes 0.55 and 0.45), and of 15 and 70 fs for mUB3LYP level (amplitudes 0.15 and 0.85). This indicates a combination of direct cyclization and more complex motion leading to slower decay.

The molecular dynamics study gave diradical lifetimes ranging from well under 100 to over 600 fs with biexponential decay for CF_2, and approximately single exponential decay in the CCl_2 case. These reactions are well-known to be stereospecific, and their classification as concerted reactions is not controversial. However, an intermediate might be identified spectroscopically at a femtosecond time resolution, whether or not it can be trapped or its stereochemistry affected. Indeed, Zewail has generated trimethylene and measured a lifetime of only 120 fs.[119] It has been pointed out by Carpenter[76,120] that the concerted/stepwise classification encounters a problem if the definition of an intermediate is not clear. Here, direct product formation occurs within 60–70 fs after the first bond is formed in these reactions. This is an appropriate time frame for direct (concerted) motion in these reactions, since 60–70 fs is the timescale of CCC angle bending and C–C stretching, which are associated with the reaction coordinate for cyclization.

The reactions of CF_2 with *trans*-2-butene and *cis*-2-butene are stereospecific experimentally.[121] The dynamics calculations of the reaction of CF_2 with *trans*-2-butene and *cis*-2-butene are being investigated (Xu, L., Doubleday, C. E., Houk, K. N., unpublished results). For the trajectories obtained so far, Xu, Doubleday, and Houk observed that there is no rotation along the double bond during the cycloaddition, as is found in experiment. This is consistent with a mechanism that Firestone proposed long ago for 1,3-dipolar and Diels–Alder cycloadditions.[122] However, Houk et al.'s simulations of 1,3-dipolar cycloadditions[123,124] and Diels–Alder reactions[82] do not show this behavior.

Table 5.1 shows the time gaps of bond formation among carbene cycloadditions, 1,3-dipolar cycloadditions, and Diels–Alder cycloadditions. As expected, the largest time gaps in bond formation occur for carbene cycloadditions, which include a large contribution from complex dynamics in the difluorodiradicals. 1,3-Dipolar cycloadditions are among the clearest examples of concerted bimolecular reactions.

The quasi-classical trajectories provide a detailed dynamical picture of how carbenes react with alkenes, especially the timing of bond formation and a detailed description of dihalodiradical cyclization. The reaction of CCl_2 with alkenes is called a dynamically concerted process with an average time gap of bond formation of around 50 fs. The reaction of CF_2 with ethylene has biexponential decays of short- and longtime constants. The short component is a concerted

TABLE 5.1. Comparison of Time Gap of Bond Formation Among Carbene Cycloadditions, 1,3-Dipolar Cycloadditions, and Diels–Alder Cycloadditions

	Time Gap of Bond Formation (fs)[a]
Carbene cycloadditions	50 ~ 250
1,3-Dipolar cycloadditions	10 ~ 30[b]
Diels–Alder cycloadditions	5 ~ 60[c]

[a] Average values.
[b] Data adapted from previous papers on 1,3-dipolar cycloadditions (References 123 and 124).
[c] Reference 82.

process. The long component is dynamically complex, in the sense of a temporarily trapped diradical trajectory in a region with no potential energy minimum. This can also be termed dynamically stepwise, where the second step involves decay of an intermediate that could, in principle, be spectroscopically observed.

5.2.1.2 Acetylene. In 2005, Yu and Muckerman performed direct dynamics calculations of the reaction of singlet methylene with acetylene (Scheme 5.1).[125] The energetics of the reaction of $^1CH_2 + C_2H_2 \rightarrow H + C_3H_3$ were accurately calculated by the CBS method in combination with the cc-pVDZ, cc-pVTZ, and cc-pVQZ basis sets. The energy plot for the $^1CH_2 + C_2H_2 \rightarrow H + C_3H_3$ reaction is shown in Figure 5.7. The reaction proceeds from the starting materials essentially barrierlessly to cyclopropene. The intermediate cyclopropene then rearranges via either of transition state TS1 to form allene or transition state TS3 to form propyne (Fig. 5.8). The enthalpies of reaction to form these two products are −109 to −110 kcal/mol. The ΔH of the $^1CH_2 + C_2H_2 \rightarrow H + C_3H_3$ reaction was also calculated as −20.3 kcal/mol at 0 K. This value is close to the experimental result[126–129] of −19 ± 3 kcal/mol.

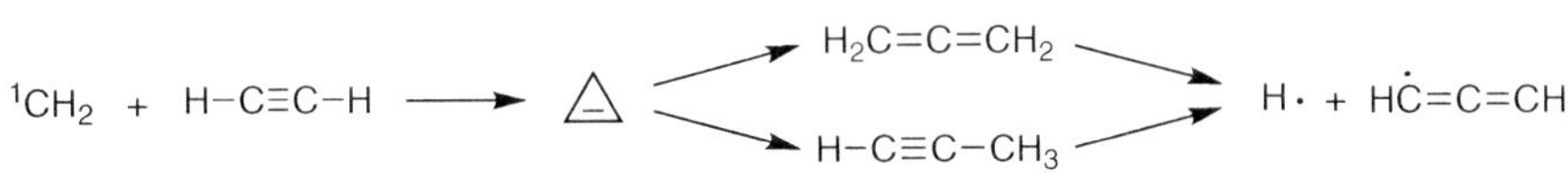

Scheme 5.1.

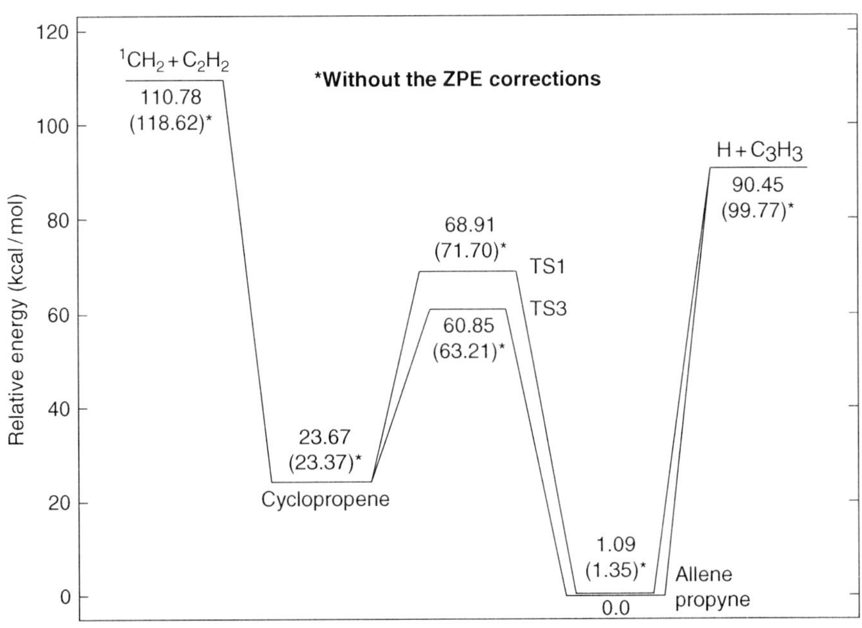

Figure 5.7. Potential energy surface for $^1CH_2$ + acetylene $\rightarrow H + C_3H_3$ calculated using the CBS-2 method.[130] (Copyright © 2005 American Chemical Society.)

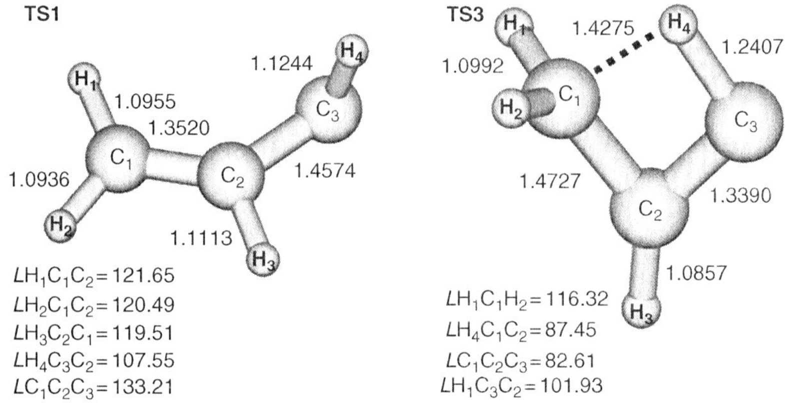

Figure 5.8. UCCSD/cc-pVDZ-optimized geometries of TS1 and TS3 for the rearrangement of cyclopropene to allene and propyne, respectively. Lengths are in Ångstroms; angles are in degrees.[130] (Copyright © 2005 American Chemical Society.)

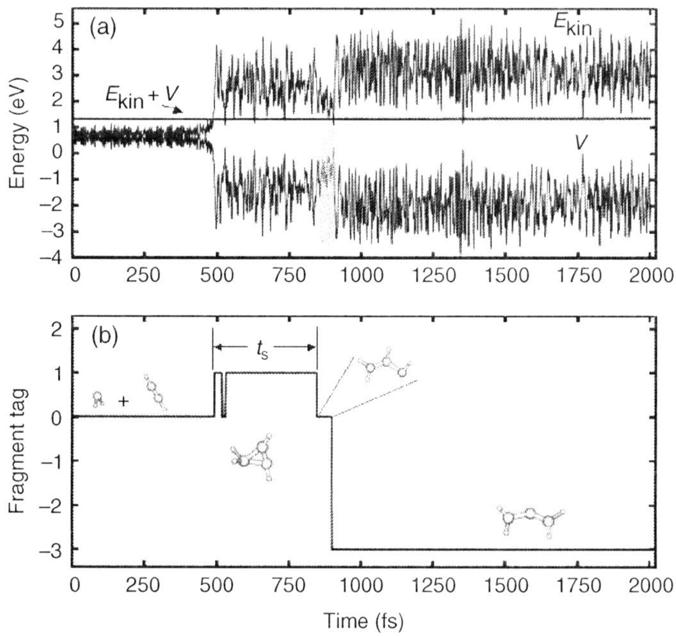

Figure 5.9. (a) Energy variation and conservation and (b) structure evolution for a typical $^1CH_2$ + acetylene → cyclopropene → allene reaction.[130] (Copyright © 2005 American Chemical Society.)

The PES of this system was also computed at UCCSD(SAC)/cc-pVDZ. Because the UCCSD(SAC)/cc-pVDZ energies were in excellent agreement with the experimental values and the CBS results, the authors used this level of theory for dynamics simulations. Figure 5.9 shows a typical trajectory for the

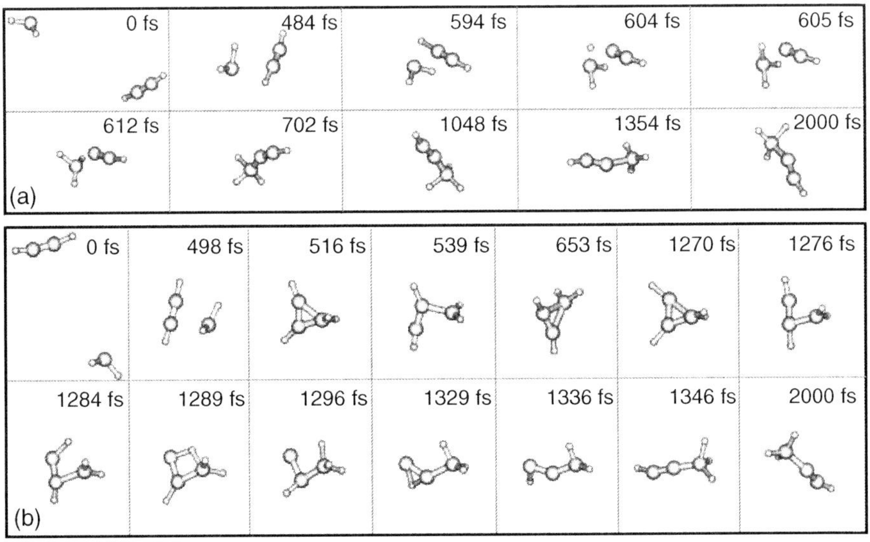

Figure 5.10. Two typical trajectories for the formation of propyne from $^1CH_2$ + cyclopropene: (a) a direct abstraction/addition mechanism and (b) a normal addition/ring-opening mechanism.[130] (Copyright © 2005 American Chemical Society.)

formation of allene. Figure 5.9a shows clearly that the total energy—kinetic energy (E_{kin}) + potential energy (V)—is well-conserved throughout the 2-ps duration of the trajectory. Figure 5.9b shows that $^1CH_2$ adds to acetylene to give the expected cyclopropene intermediate. Although cyclopropene resides in a deep potential energy well (Fig. 5.7), this intermediate is short-lived, with $\tau = 3$ fs. From there, cyclopropene rearranges, via TS1, to allene.

The dynamics simulations also help to explore the formation of propyne. It turned out that two reaction mechanisms produce propyne, as shown via representative trajectories in Figure 5.10. Figure 5.10a shows an abstraction–inversion–dissociation mechanism. After $^1CH_2$ and acetylene approached each other over 600 fs, one hydrogen atom in acetylene began to transfer to $^1CH_2$. The duration of this abstraction process was a few femtoseconds. A quick inversion of the methyl radical umbrella mode then occurred, followed by the formation of propyne. Finally, propyne vibrated, rotated, and dissociated into products H + C_3H_3. The rotational period was estimated to be 1.6 ps for this trajectory.

The expected addition/ring-opening mechanism is shown in Figure 5.10b. First, a cyclopropene intermediate was formed when methylene approached acetylene. Large-amplitude vibrations occurred during a complex rotation process to give a vinylidene CH_3CHC species. At 1329 fs, a 1,2 H-shift in the CH_3CHC species produced propyne.

These results illustrate the importance of dynamics in showing how a seemingly simple cycloaddition can produce subsequent products, and how geometrically very different trajectories may give the same overall reaction.

5.2.2 Carbene Additions to Strained π Systems

As seen earlier, additions of carbenes to unstrained alkenes have been prevalent in experimental and computational chemistry. In contrast, experimental studies of carbene additions to small-ring π systems have been considerably fewer in number, with many conducted in the early 1970s.[131–135] Trost and Atkins[135] and Dehmlow[132] found that CCl_2 added to 1,1,2-trimethylcyclopropene and a substituted spirocyclopropene, respectively, to yield expected bicyclobutane products which then ring-expanded to the corresponding cyclobutenes (Scheme 5.2a). Mahler's addition of CF_2 to hexafluoro-2-butyne first produced octafluorodimethylcyclopropene. Addition of a second equivalent of CF_2 to this cyclopropene gave the expected bicyclobutane, but also 2,3-bis(trifluoromethyl)-1,1,4,4-tetrafluorobutadiene (Scheme 5.2b).[134] Hart and Nitta[133] and Dehmlow[131,132] reacted CCl_2 with hexamethyldewarbenzene and cyclopentadiene (Scheme 5.2c). In these cases, initial CCl_2 addition was followed by subsequent ring expansion to a variety of products.

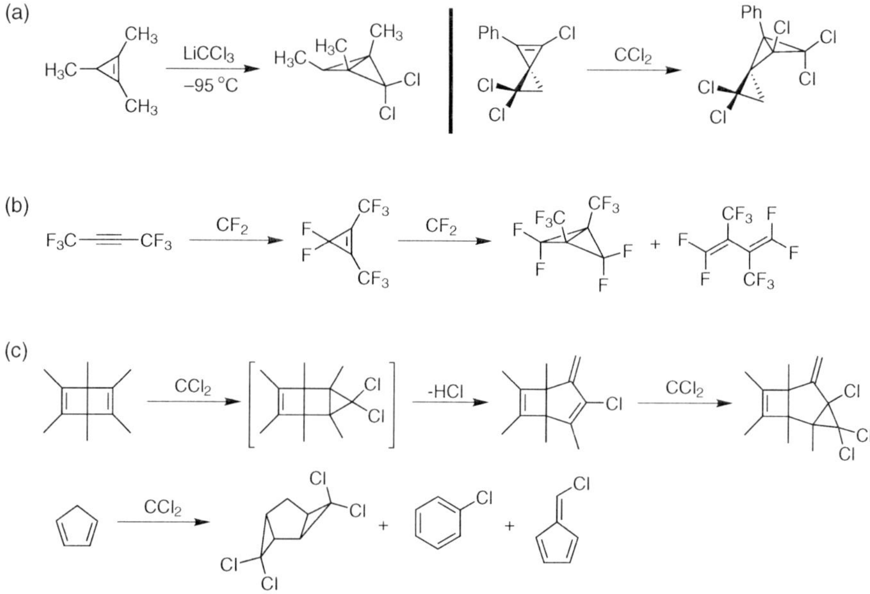

Scheme 5.2. Addition of dihalocarbenes to small cycloalkenes.[131–135]

5.2.2.1 *Cyclopropene.*

For the ensuing 20 years, experimental studies of halocarbene additions to small cycloalkenes remained dormant. In the late 1990s, Brinker and coworkers investigated CCl_2 and CF_2 additions to 1,2-disubstituted cyclopropenes **1d–f** (Scheme 5.3).[136,137] They found the major products (80%) of this reaction to be cyclobutenes **7d–f**, which resulted from a cyclopropylallyl rearrangement of bicyclobutanes **5d–f**. Surprisingly, they also observed the formation of 20% of butadienes **8d–f**, which did not form via rearrangement of

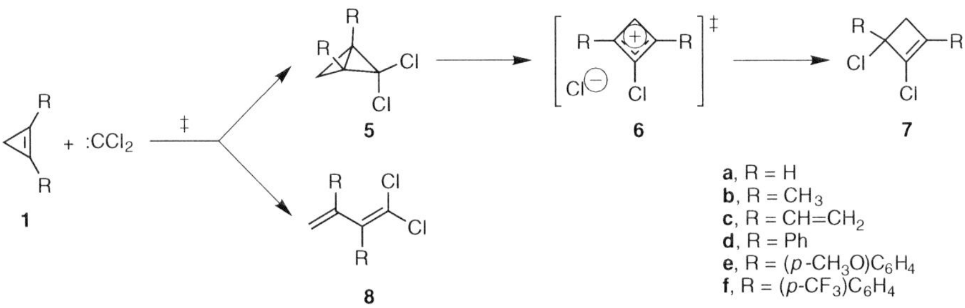

Scheme 5.3. Dihalocarbene addition to cyclopropene.[142]

5d–f. Subsequent experiments with *para*-substituted aryl substituents (p-CF_3 vs. p-CH_3O), along with an increased amount of butadiene **8d–f** formation in a more polar solvent (acetonitrile), led Brinker et al. to propose that **8d–f** formed via a stepwise, polar carbene addition to the starting cyclopropene via zwitterions **6d–f** or a polarized species.[136] Although nucleophilic carbenes (e.g., dimethoxycarbene) can add in a stepwise, dipolar manner to electron-poor alkenes,[138–141] this marked the first time that an electrophilic carbene addition mechanism, mediated by a zwitterion or a polarized species, was proposed.

The possibility of a stepwise mechanism proceeding through a zwitterion or zwitterion-like species led Merrer and Rablen to computationally investigate CCl_2 addition to cyclopropenes **1a–d** at B3LYP/6-31G*.[60] They found that a zwitterion (i.e., **6a–d**) was not a stationary point on the PES of this reaction, regardless of how well the cyclopropenyl substituents might stabilize a putative positive charge (e.g., 1-aminocyclopropene), or whether the reaction was modeled in the gas phase or in a polar solvent continuum such as acetonitrile.

Furthermore, they found that only one transition state existed for CCl_2 + cyclopropene (Fig. 5.11): the formation of both bicyclobutane and butadiene products occurred in a concerted manner from one transition state (TS).[60] The activation barrier was only 0–2 kcal/mol. The transition state then proceeded dramatically exothermically (−75 to −110 kcal/mol) with post-transition state branching to produce both Brinker's major experimental bicyclobutane product (yellow path) and minor butadiene product (red path). The route to bicyclobutane must overcome a secondary energy barrier of ~2 kcal/mol after the branching point, whereas the path to butadiene proceeds completely energetically downhill after the bifurcation of the reaction path.[60]

Merrer and Rablen proposed that bicyclobutane formation was favored experimentally due to dynamic control of the reaction.[60] The secondary barrier for bicyclobutane formation was suggested to be unproblematic because its path is direct and the momentum accumulated as the molecules tumble down the PES provided enough energy to overcome the slight 2 kcal/mol barrier. In contrast, the route to butadiene required a dramatic change in momentum with an almost 90° turn on the PES. This is to say: better dynamic matching[76,95,144,145] favors the formation of bicyclobutane over butadiene.

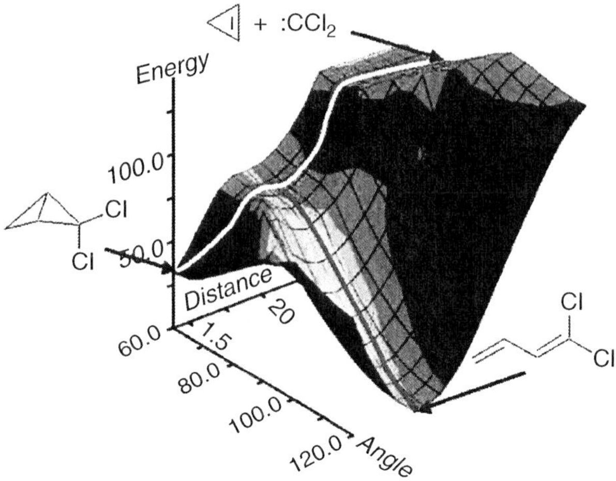

Figure 5.11. PES of CCl_2 + cyclopropene **1a** at varying carbene-alkene distances, *r*, and cyclopropenyl bond angles, θ (CCSD(T)/6-31G*//B3LYP/6-31G*).[143] (Copyright © 2005 American Chemical Society.) For color details, please see color plate section.

Following the proposal that CCl_2 addition to cyclopropene (**1a**) may be controlled by reaction dynamics, Merrer and Doubleday computed quasi-classical direct dynamics trajectories for this system.[64] The 392 trajectories were initiated at the transition state of carbene addition and proceeded to form bicyclobutane **5a** and butadiene **8a**. Figure 5.12 shows representative trajectories for the formation of **5a** and **8a**.

The product ratio obtained from these trajectories was in a ratio of bicyclobutane:butadiene of 4.7:1, which corresponds very well to Brinker et al.'s experimental 4:1 ratio for products **5**:**8**.[136,137] Furthermore, both products were formed in a concerted manner within 100 fs of the start of the trajectories, with butadiene formation occurring somewhat more quickly at 30–70 fs versus 50–100 fs for **5a** (Fig. 5.13).[64]

The trajectories also showed bicyclobutane formation to either be terminal, or (in eight cases) to rearrange to cyclobutene **7a**, as was observed experimentally by Brinker.[136,137] The formation of **7a** occurred over a longer trajectory (reaction) time, typically over 200 fs, as shown for a representative trajectory in Figure 5.14.

Notable also is that in none of the trajectories does **5a** rearrange to butadiene, also in accord with the experimental studies. Because of the excellent correlation between the trajectory product distribution and the experimental product ratio, the direct dynamics study lends strong support to the probability that the addition of dichlorocarbene to cyclopropene proceeds via dynamic control.

5.2.2.2 Benzocyclopropene. In the case of carbene additions to annelated cyclopropenes, Kagabu and Saito found that dichloro and dibromocarbene added

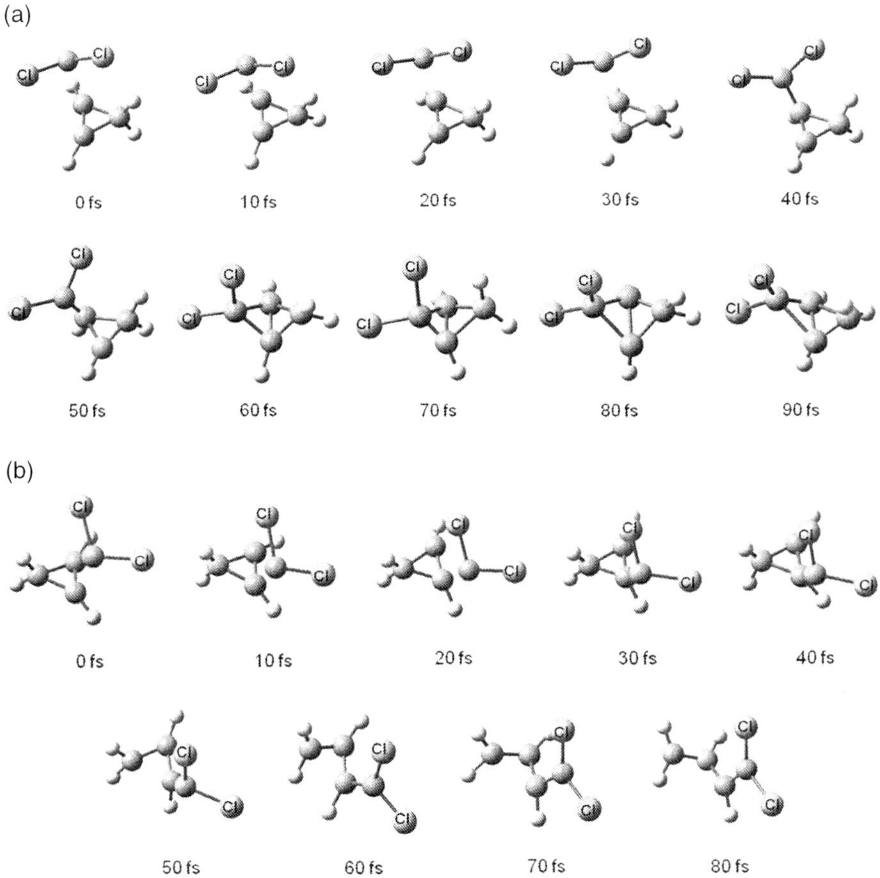

Figure 5.12. Representative trajectories for the reaction of CCl_2 + cyclopropene to produce (a) bicyclobutane **5a** and (b) butadiene **8a**.[142] (Copyright © 2011 John Wiley and Sons.)
(a) CCl_2 + cyclopropene → bicyclobutane **5a**
(b) CCl_2 + cyclopropene → butadiene **8a**

to benzocyclopropene (**2a**), yielding benzocyclobutene (**10a**);[146] Scheme 5.4. They proposed a mechanism for this addition that proceeded through a propellane intermediate (**9a**), but did not provide any evidence, spectroscopic or otherwise, for the propellane. In other work, the Billups group determined that a methylene ruthenium carbenoid added to **2a** via a xylylene intermediate.[147]

Merrer and coworkers took these past studies into consideration when they investigated the mechanism of CCl_2 and CBr_2 additions to benzocyclopropene **2**.[65] They devised mechanistic paths that accounted for Kagabu and Saito's proposed propellane intermediate **9**,[146] as well as Billups et al.'s xylylene intermediate[147] (Scheme 5.4). They also considered the possibility of stepwise electrophilic carbene addition to **2** via zwitterions **12** and **13**. These mechanisms were examined experimentally and computationally by utilizing 2-aryl substitution on **2**, where

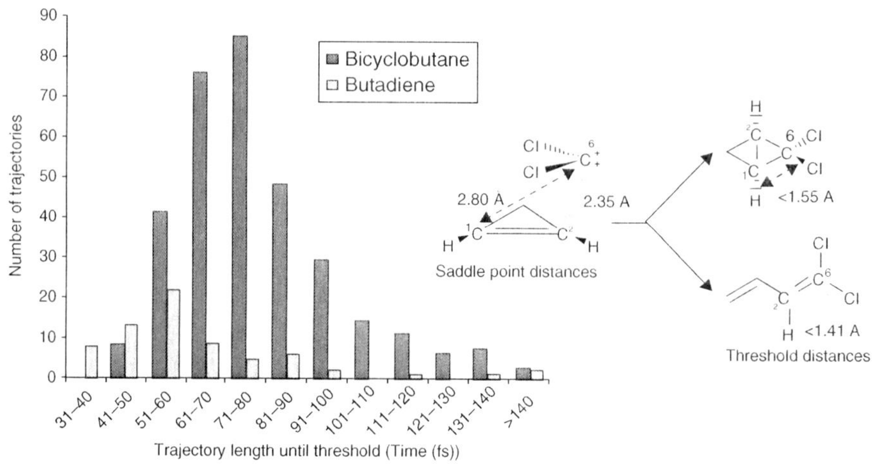

Figure 5.13. Histogram of numbers of trajectories versus time to reach threshold interatom distances for trajectories that form or proceed through bicyclobutane **5a** and those that form butadiene **8a**. The threshold distance for **5a** is $C^1–C^6 < 1.55$ Å; the threshold for **8a** is $C^2–C^6 < 1.41$ Å.[142] (Copyright © 2011 John Wiley and Sons.)

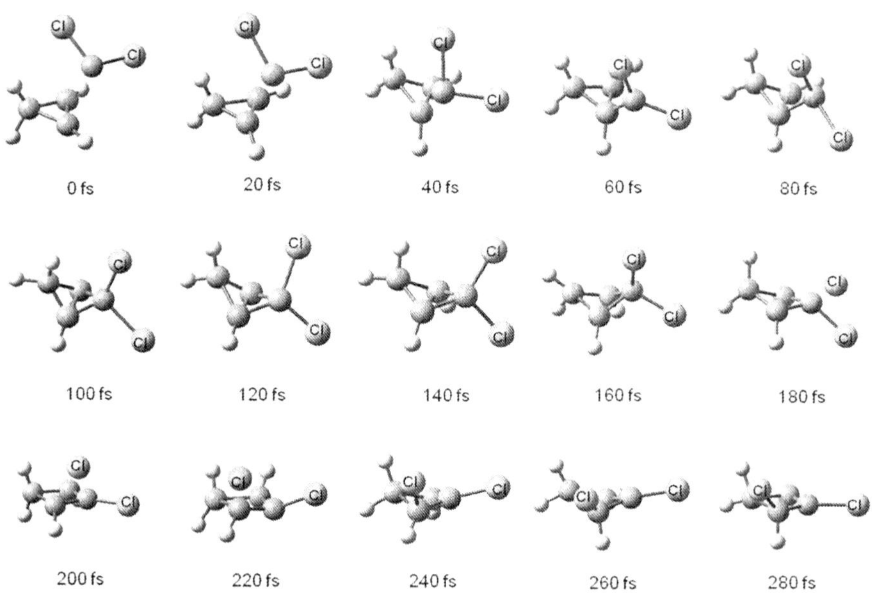

Figure 5.14. Representative trajectory for the reaction of CCl_2 + cyclopropene to produce cyclobutene **7a**.[142] (Copyright © 2011 John Wiley and Sons.)

the electronic nature of R was varied. If the regioisomeric product ratio **10**:**11** varied with R, a symmetric intermediate such as propellane **9** would be disfavored in deference to the paths proceeding through asymmetric intermediates (i.e., **12** and **13** and/or **14** and **15**).

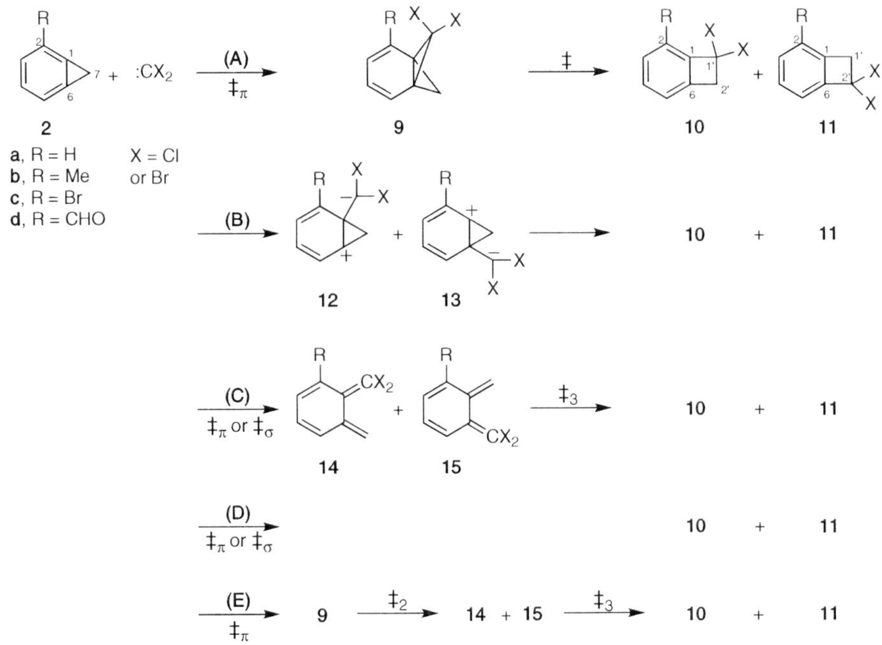

Scheme 5.4. Dihalocarbene additions to benzocyclopropene **2**.[148] (Copyright © 2008 American Chemical Society.)

Addition of CX_2 (X = Cl, Br) was found to be regioselective with respect to aryl substitution.[65] When R was an electron donor, the reaction showed preference for carbene addition proximal to R: when R = CH_3, the ratio of **10b**:**11b** was 2:1 for both CCl_2 and CBr_2 addition. When R was an electron-withdrawing group, the regioselectivity reversed and carbene addition distal to R was favored: for example, when R = Br, the ratio of **10c**:**11c** was 1:5 or 1:3 for CCl_2 or CBr_2 additions, respectively.[65] Computationally, the stationary-point PES for CCl_2 addition to **2** showed the possibility of propellane **9** as an intermediate in this reaction (Fig. 5.15);[65] however, as stated earlier, the experimental regioselectivity of CX_2 addition to **2** indicates that **9** is an unlikely intermediate.

In terms of the other proposed mechanistic paths for CX_2 + **2**, the computed PES (B3LYP/6-31G*) of this reaction also shows the possible existence of xylylenes **14** and **15**, consistent with the experimental findings.[65] In contrast, a stepwise dipolar carbene addition proceeding through zwitterions **12** and **13** is believed unlikely. These species were not found to be energy minima, in accord with the calculations on the proposed zwitterion for CCl_2 addition to cyclopropene (Scheme 5.3).

Other similarities between the addition of CCl_2 to benzocyclopropenes **2** and cyclopropene include the negligible barrier for addition (0–2 kcal/mol), and the extreme exothermicity of the reaction ($\Delta H_{rxn} \approx -110$ kcal/mol). The most interesting result, however, is that the transition state for CCl_2 addition connects the reactants, CCl_2 plus **2**, to each of the propellane **9** and xylylene intermediates **14** and **15**, and also directly to benzocyclobutenes **10** and **11**

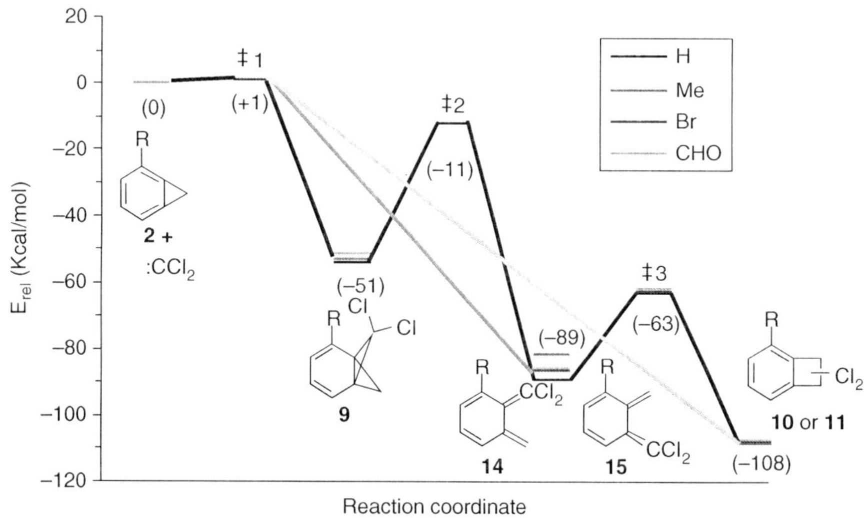

Figure 5.15. Potential energy surface for CCl_2 + benzocyclopropene **2**. Relative energies are in kcal/mol at B3LYP/6-31G* and are ZPE-corrected (ZPE's are scaled by 0.9806).[148] (Copyright © 2011 American Chemical Society.) For color details, please see color plate section.

(Fig. 5.15).[65] Therefore, as is the case for CCl_2 addition to cyclopropene, CCl_2 plus benzocyclopropene exhibits a PES that appears to have one transition state leading to a number of products.

In order to further narrow the possible mechanistic pathways, Merrer et al. attempted to trap the putative xylylene intermediates **14** and **15** via Diels–Alder chemistry in the reaction of CBr_2 + **2b**. The trapping of **14** and **15** was thought to be likely, as these intermediates reside in a deep potential energy well. However, they were unable to detect the formation of any Diels–Alder adducts; only benzocyclobutene products **10** and **11** were formed.[65] The xylylenes' untrappability was attributed to the highly energized states in which they were expected to form, due to the very exothermic pathway by which they were generated in this system ($\Delta H = -90$ kcal/mol from starting materials to xylylene). This untrappability is similar to Brinker et al.'s inability to isolate bicyclobutane intermediates **5** in their solution-phase reactions of CX_2 + 1,2-diarylcyclopropenes **1d–f**.[136,137] Additionally, Yu and Muckerman's direct dynamics study of $^1CH_2$ addition to acetylene found that, similar to xylylenes **14** and **15**, an intermediate cyclopropene was situated in a deep potential well.[125] However, the cyclopropene generated in this manner was able to escape the well to form propyne and allene products because of the cyclopropene's short lifetime ($\tau = 3$ fs), and the excess energy with which it was formed.[68] Hence, just because xylylenes **14** and **15** could not be trapped does not mean they were not formed. The xylylenes' untrappability may be due to the energetic manner in which they were generated, thus giving them brief lifetimes, lifetimes perhaps too short to allow chemical trapping.[65]

Dichloro- and dibromocarbene additions to benzocyclopropene (**2**) thus exhibit many characteristics similar to carbene additions to cyclopropene. These similarities include a negligible activation barrier, reaction exothermicity in excess of 100 kcal/mol, and a single transition state leading to possible multiple paths. Because of these similarities, the reaction of CX_2 with **2** was also suspected to proceed under dynamic control.[65] Calculations of direct dynamics trajectories for CCl_2 addition to benzocyclopropene are underway by the Merrer group.

5.2.2.3 Cyclooctyne. The strain energies of cyclopropene and benzocyclopropene are quite high: 54 and 68 kcal/mol, respectively.[149] It is believed that the strain in these C–C π substrates contributes significantly to the role of reaction dynamics in the mechanisms of carbene additions to these compounds. The smallest isolable cycloalkyne at room temperature is cyclooctyne (**3**). Its strain energy is notable, at 16 kcal/mol,[150–152] yet is significantly less than that of cyclopropene or benzocyclopropene. Thus, whether reactions of dihalocarbenes with **3** will adhere to TST or be subject to dynamic effects is questionable. A recent computational study of CCl_2 and phenylchlorocarbene (PhCCl) additions to **3** shed some light on this matter.[66]

In their investigation, the Merrer group found the addition of both CCl_2 and PhCCl to cyclooctyne to proceed over a very flat surface with no barrier to addition (at MP2/6-311+G*, B3LYP/6-311+G**, or B3LYP/6-31G*) (Fig. 5.16).[66] The clear potential energy sink is the expected carbene cycloaddition product, cyclopropene **16**; however, the formation of vinylcarbene **17** may also be possible. Because carbene **17** lies in a very shallow energy well, if it is indeed formed, it is

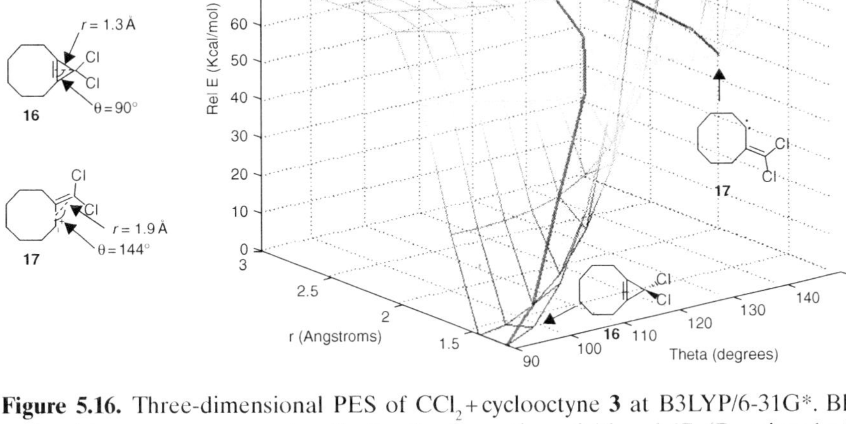

Figure 5.16. Three-dimensional PES of CCl_2 + cyclooctyne **3** at B3LYP/6-31G*. Blue and red traces are proposed paths for the formation of **16** and **17**. (Reprinted with permission from Reference 66; Copyright 2008 by the American Chemical Society.) For color details, please see color plate section.

expected to have a transient lifetime, likely on the order of femtoseconds. Future experiments and planned direct dynamics trajectory calculations are necessary to help elucidate whether these additions proceed with dynamic control or not.

5.2.3 Carbene Additions to Strained σ Systems

Experimental investigations of carbene additions to strained C–C π systems are few in number, but even fewer are those of strained C–C σ substrates. Notable examples, however, are CF_2 and CCl_2 additions to 1,2,2-trimethylbicyclo[1.1.0]butane (**4b**).[153–156] Several years ago, Jones, Jackson, and coworkers found that CCl_2 to added to **4b** to produce a single diene product, **20b-Cl$_2$** (Scheme 5.5).[153–156] To account for the formation of this lone regioisomeric product, they proposed the "pluck" mechanism, in which two C–C sigma bonds reacted in a concerted, regiospecific manner (via **18b-Cl$_2$**) to give **20b-Cl$_2$**. Specifically, they proposed that the carbene approached from the less hindered side of **4b** with the chlorines positioned away from the geminal dimethyls.[153–156]

Given the implications of reaction dynamics for the mechanism of the addition of CCl_2 to cyclopropene,[60,64] Rablen et al. undertook a thorough computational exploration of the two-dimensional PES for CCl_2 plus trimethylbicyclobutane **4b** to investigate the possibility of dynamic control of this system.[67] This system proved to be quite challenging to compute accurately. Consequently, Rablen and Jones used a variety of computational methods for geometry optimization and single-point energy calculations. For geometry optimizations, they used HF/3-21G*, HF/6-31G*, QCISD/6-31G*, CCSD/6-31G* and, where possible, B3LYP/6-31G*. Single-point energies were computed with QCISD/6-31G*, QCISD(T)/6-31G*, CCSD/6-31G*, and CCSD(T)/6-31G* (on the HF/6-31G* optimized geometries) and CCSD(T)/6-31G* and CCSD(T)/6-311 + G** (on the CCSD/6-31G* optimized geometries). Furthermore, even more accurate energies were computed with the CBS-4 procedure and G3-CCSD.[67] They investigated a variety of approach angles of CCl_2 to **4** and also the possibility of forming a variety of products, **20–24;** see Scheme 5.5.

Rablen, Jones, and coworkers found CCl_2 to add to **4b** and parent bicyclobutane **4a** through the pluck mechanism (via **18**) *or* via a stepwise mechanism through a zwitterionic intermediate (**19**); the distinction depends upon the

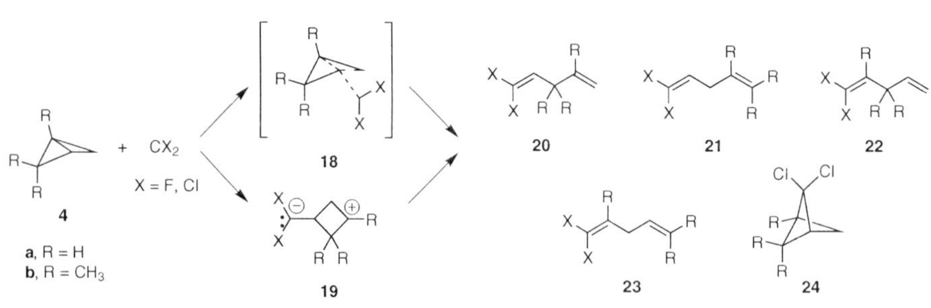

Scheme 5.5. Dihalocarbene addition to bicyclo[1.1.0]butane **4**.

degree of bicyclobutane substitution.[67] The zwitterionic mechanism was not competitive for CCl_2 addition to the parent bicyclobutane **4a**; thus, this reaction appeared to proceed in a concerted manner. In the trimethyl-substituted system—CCl_2 plus **4b**—the zwitterionic pathway appeared competitive with two regioisomeric concerted processes, as indicated by comparable barriers for each of these paths. Therefore, the computational results are inconsistent with Jones and Jackson's experimental regiospecificity. Rablen and Jones thus concluded that the experimental behavior of CCl_2 + **4b** may be nonstatistical and controlled by dynamic factors.[67] Direct dynamics trajectory calculations await.

5.3 DYNAMICS OF OTHER CARBENE-MEDIATED REACTIONS

5.3.1 Wolff Rearrangement

To this point, we have discussed the use of dynamics trajectories and the impact of nonstatistical dynamic effects on bimolecular reactions involving carbenes. Litovitz, Keresztes, and Carpenter explored whether unimolecular carbene chemistry is subject to dynamical effects.[70] They used a combination of experiment and theory to examine the Wolff rearrangement of diacetylcarbene **27** (Scheme 5.6).

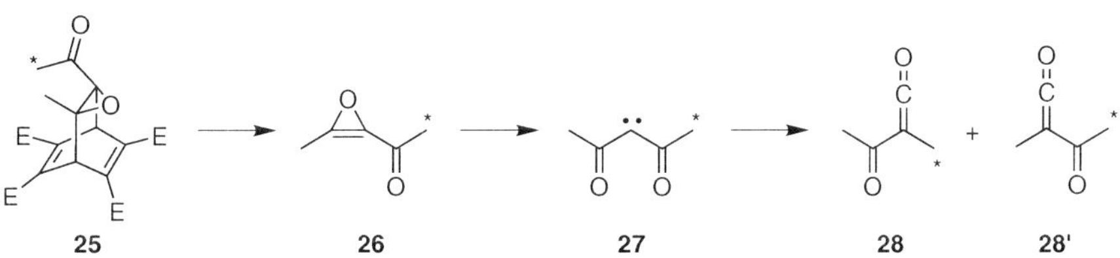

Scheme 5.6. The generation of diacetylcarbene **27** from retro-Diels-Alder precursor **25**. (Copyright © 2008 American Chemical Society.)

A potential energy profile for the reaction **25a** → **26** → **27** → **28** (+**28′**) was computed at MPWB1K/6-31+G(*d*,*p*) (Fig. 5.17). Several features of this one-dimensional PES led Carpenter to suspect that this reaction was subject to nonstatistical dynamics.[157] First of all, the rate-determining retro-Diels–Alder step was followed by a 16-kcal/mol drop in potential energy. Also, the barrier heights for the ensuing unimolecular rearrangements were low or nonexistent. For example, no transition structure for the rearrangement of oxirene **26** to carbene **27** was found. Both of these characteristics indicated that the reactive intermediates could be generated with substantial selective excitation of specific vibrational modes. In addition, formation of the ketene products might occur before intramolecular vibrational redistribution (IVR) was complete.

The experimental results also suggested the possibility of dynamic intervention in this system. The C_2 symmetry of diacetylcarbene (**27**) would lead to a 1:1 isotopomeric ratio of products derived from ketenes **28**:**28′**. However, flash vacuum pyrolysis of **25b** gave ketene-derived products in a ratio of 1 to 4.5

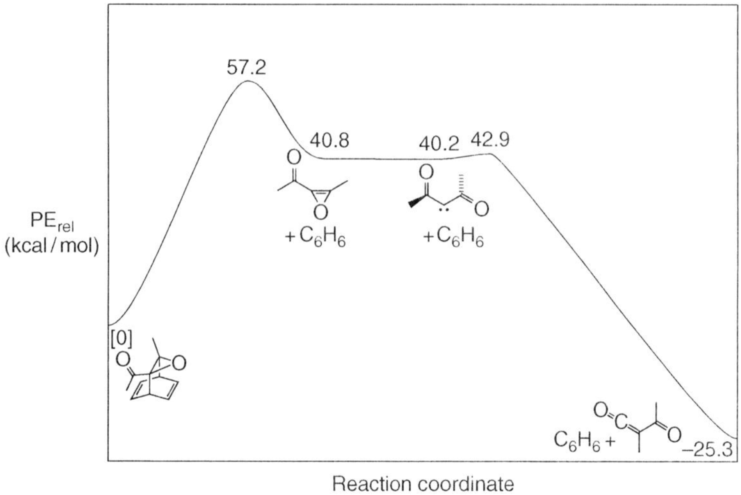

Figure 5.17. The MPWB1K/6-31+G(d,p) PE profile for retro-Diels–Alder fragmentation of **25a**.[158] (Copyright © 2008 American Chemical Society.)

from **28** and **28′**.[70] This isotopomer ratio may be due to nonstatistical effects stemming from branching from carbene **27**.

The answer proved to be more complex. Additional ^{13}C-labeling experiments revealed that most of the ketene-derived products were generated from methylpyruvoyl carbene **29**, even though **29** was 8.7 kcal/mol higher in enthalpy than diacetylcarbene (**27**).[70] Rather than undergoing retro-Diels–Alder cycloaddition, starting material **25b** underwent an unexpected unimolecular rearrangement followed by fragmentation to **29**. It was determined computationally that carbene **29** then either underwent acetyl group migration to produce ketene **28′** directly, or converted to (an isotopomer of) oxirene **26**. Oxirene **26** ring-opened to diacetylcarbene **27**, which subsequently underwent Wolff rearrangement to give **28** and **28′**. The acetyl-migration path of **29** to **28′** was faster than the conversion of **29** to **26**. These paths are summarized in Scheme 5.7.

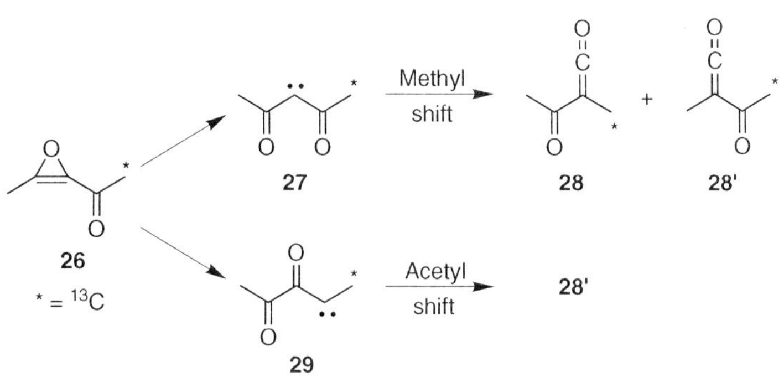

Scheme 5.7. Formation of ketene **28′** by two different routes. (Copyright © 2008 American Chemical Society.)

Even taking into account the preferential formation of carbene **29** and products formed therefrom, the ratio of ketene-derived products that could be traced to diacetylcarbene was still nonstatistical, 1:2.3 ± 0.4. In order to determine the cause of the still-unexpected unequal product ratio, direct dynamics calculations were performed with quasi-classical normal-mode sampling from a canonical ensemble at 923 K (the experimental reaction temperature). One hundred 500-fs trajectories were computed at each of three different DFT levels—B3LYP,[159,160] O3LYP,[161] and MPWB1K,[162] using a 3-21G basis set for each. The small basis set for these initial calculations was used to balance computational time with a reasonable number of trajectories.

The starting point for the trajectories was chosen in the vicinity of the transition state for the rearrangement of carbene **29** to carbene **27** via oxirene **26**. Four different types of trajectories were found. The first two types resulted in Wolff rearrangements. Type 1 proceeded via carbene **27** followed by a shift of the C1 methyl group to form the final ketene product (**28**+**28′**), whereas type 2 formed carbene **27** followed by a shift of the C5 methyl group. The third type of trajectory produced carbene **27** which did not undergo Wolff rearrangement in the allotted 500-fs time span. The authors assumed these trajectories would eventually lead to ketene with an equal probability of C1 or C5 methyl migrations. Lastly, some of the trajectories recrossed their barriers, returning to carbene **29** and, in most cases, underwent an acetyl migration. The recrossing trajectories were ignored in the calculation of the fractionation factor *f*. The results of the dynamics calculations are summarized in Table 5.2.[70]

As can be seen in Table 5.2, although there is difference among the fractionation factors, all three DFT models predicted the preferred migration of the C5 methyl group. These results lend strong support to the role of nonstatistical dynamic effects in the Wolff rearrangement of diacetylcarbene.

TABLE 5.2. Results of Dynamics Simulations on the Rearrangement of Carbene 29 to Carbene 27 Followed by Wolff Rearrangement[a]

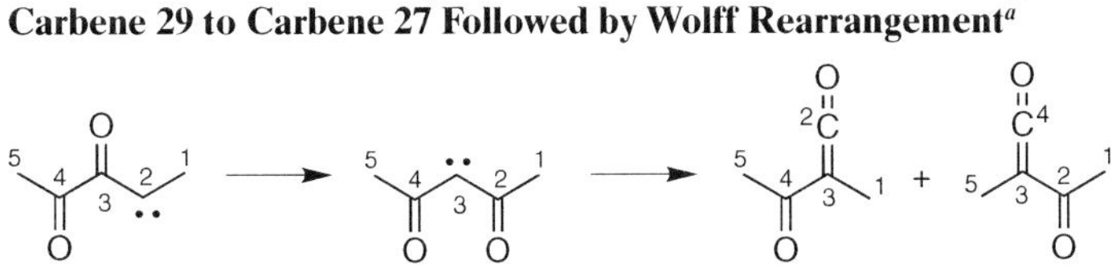

29 27 28 28'

	DFT Model		
	B3LYP	O3LYP	MPWB1K
Number of C1 shifts	13	27	7
Number of C5 shifts	69	54	79
Number failing to form product in 500 fs	8	12	8
Number of recrossings	10	7	6
f	4.3	1.8	7.6

[a]All calculations use the 3-21G basis set.

5.3.2 Photochemical Carbene Formation

There are clear nonstatistical dynamic effects in reactions of carbenes. Can dynamics also impact the generation of carbenes? Arenas and coworkers investigated this phenomenon in a direct dynamics computational study of the thermal and photochemical generation of methylene from 3*H*-diazirine and diazomethane (Scheme 5.8).[69] Their most prominent finding was that methylene is produced in its lowest singlet state from radiationless electronic decay of its precursor, regardless of whether that precursor is diazirine or diazomethane.

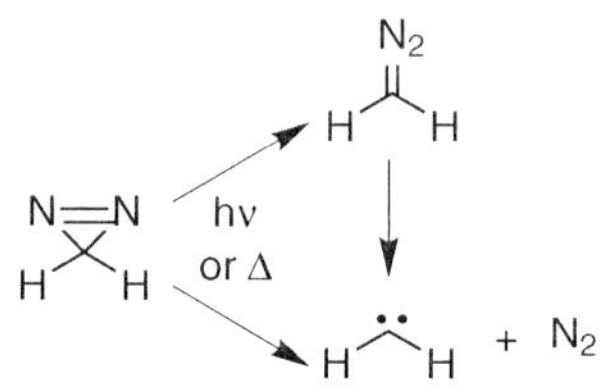

Scheme 5.8. Photolysis or thermolysis of 3*H*-diazirine or diazomethane.

To reach this conclusion, Arenas et al. started by calculating PESs for the photochemical and thermal reactions of 3*H*-diazirine and diazomethane. Energies and geometries of the stationary points were computed at complete active space self-consistent field (CASSCF) (12,10)/cc-pVTZ; stationary point energies were further calculated at CASPT2(12,10) using the CASSCF geometries. The ground-state surface calculations show singlet methylene formation (via N_2 extrusion) to be favored over diazirine-to-diazomethane isomerization.

On the first- and second-excited-state PESs (S_1 and S_2, respectively), geometry-optimized 3*H*-diazirine and diazomethane are first-order saddle points. Thus, because these PESs consist of two sequential first-order saddle points, the authors expected a valley-ridge inflection point or bifurcation to exist along the minimum energy path between the excited-state diazirine and excited state diazomethane.[163] Furthermore, on these excited-state surfaces, IRC calculations showed well-defined paths terminating in conical intersections: an S_1/S_0 conical intersection for the S_1 surface and an S_2/S_1 conical intersection for the S_2 PES.[69]

To better understand the topologies of the excited S_1 and S_2 PESs, ab initio direct[164,165] semiclassical trajectories with surface hopping[166,167] were propagated on the fly. The trajectories were computed at CASSCF(8,7)/6-31G*.[168–170] For the first-excited-state PES, trajectories were initiated at the S_1 diazomethane transition state, to which only translational (and not rotational) energy was added along some selected directions corresponding to excited normal modes of diazomethane.

Figure 5.18 illustrates the formation of methylene in its lowest singlet state after radiationless deactivation of the precursor, S_1 diazomethane. The energy profiles of the S_1 and S_0 states along this trajectory are represented in Figure 5.18a. The distortions of the molecular geometry as a function of time are shown in Figures 5.18b and c. The surface hop occurs quite early along the potential surface, and then the S_0 diazomethane minimum is reached. The excess energy from electronic decay funnels into stretching of the C–N bond,

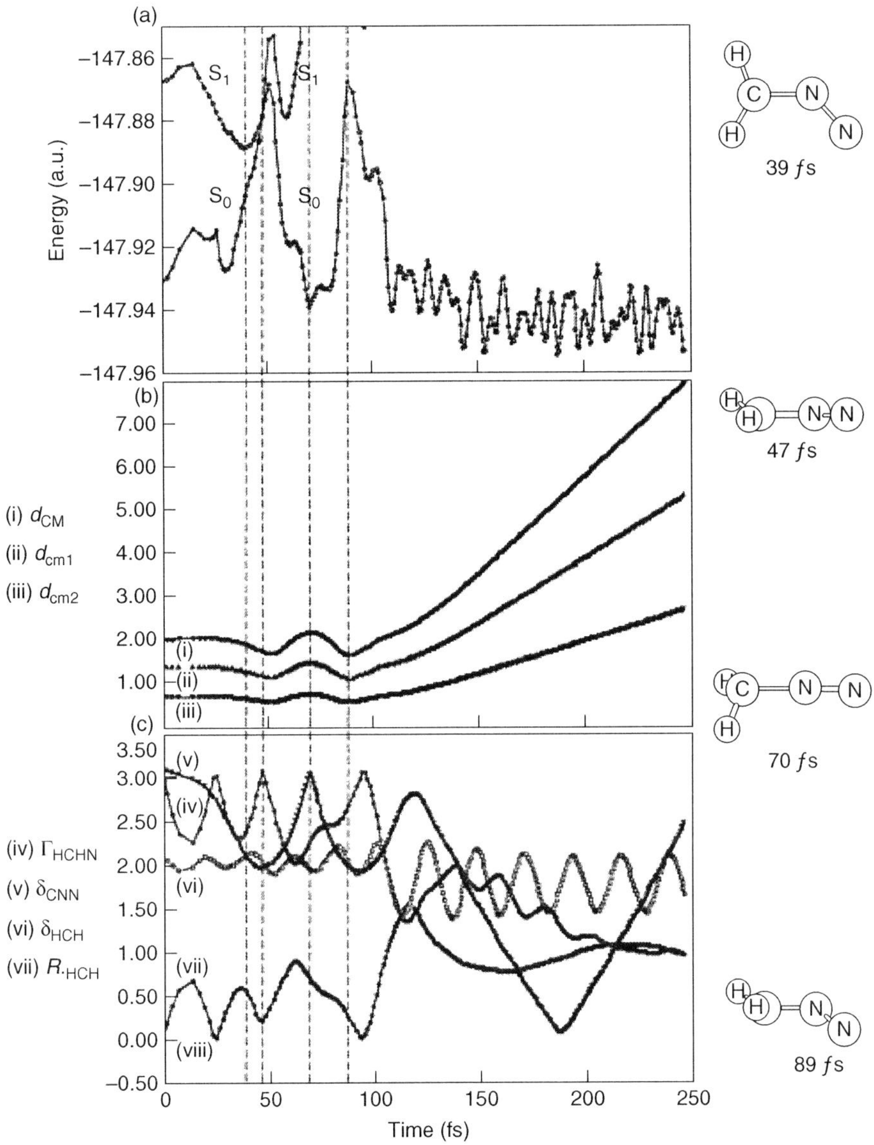

Figure 5.18. Direct trajectories of diazomethane fragmenting to singlet CH_2 and N_2 on the S_1 surface. (a) (top third) Energy profiles of S_1 and S_0 states versus time. (b) (middle third) Distance between the centers of mass of (i) CH_2 and N_2 fragments, (ii) CH_2 and the whole molecule, and (iii) N_2 and the whole molecule. (c) (bottom third) Time evolution of vibrational and rotational modes.[171] (Copyright © 2002 American Chemical Society.)

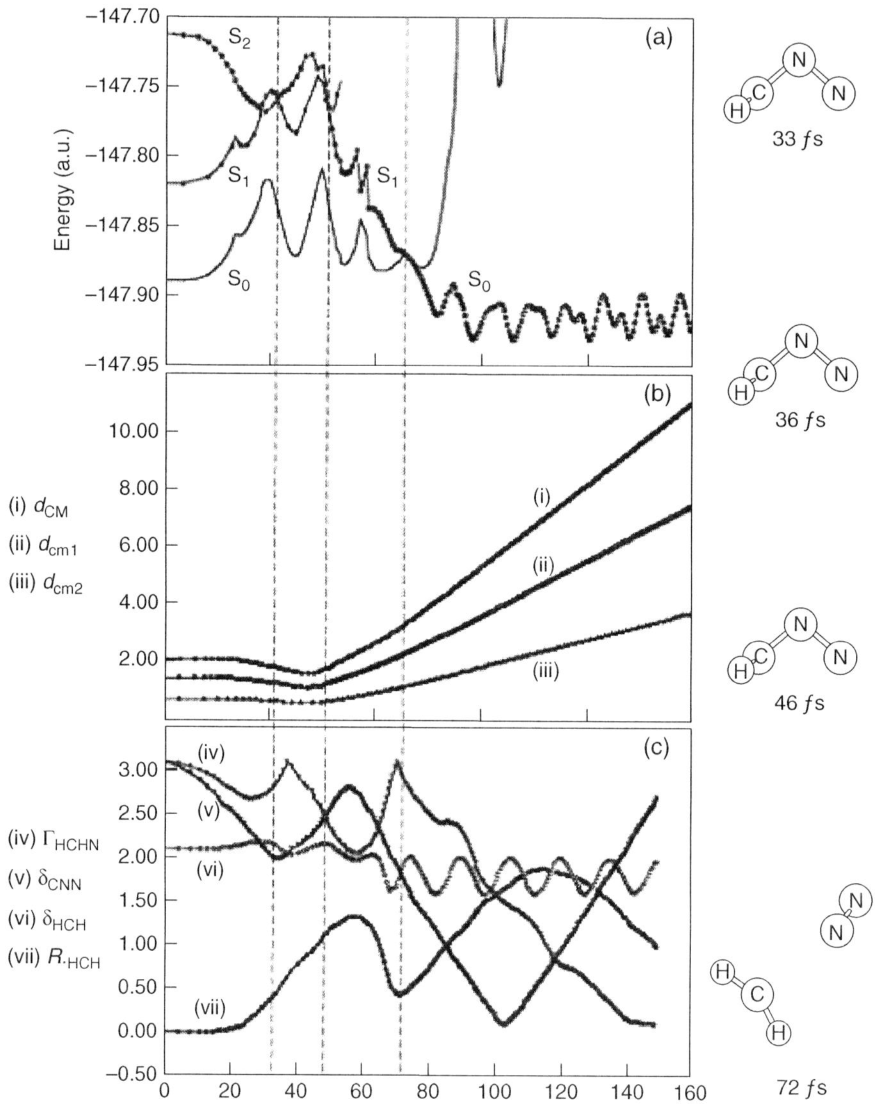

Figure 5.19. Direct trajectories of diazomethane fragmenting to singlet CH_2 and N_2 on the S_2 surface. (a) (top third) Energy profiles of S_2, S_1, and S_0 states versus time. (b) (middle third) Distance between the centers of mass of (i) CH_2 and N_2 fragments, (ii) CH_2 and the whole molecule, and (iii) N_2 and the whole molecule. (c) (bottom third) Time evolution of vibrational and rotational modes.[171] (Copyright © 2002 American Chemical Society.)

and this elongation leads to dissociation of the molecule after passing near the transition structure TS1 on the S_1 surface.

Thus, trajectories begun at either S_1 diazirine or S_1 diazomethane will flow to the S_1/S_0 conical intersection, which subsequently undergoes nonradiative decay. The system retains enough energy to eclipse the barrier to fragmentation and then produce the S_0 CH_2 and N_2 products.

The trajectory computed on the S_2 surface was started at the Franck – Condon region of S_0 diazomethane (Fig. 5.19). At the third time the S_2 and S_1 surfaces touch each other, the molecule hops. After the hop, the C – N internuclear distance starts to increase, leading to dissociation of the molecule into $CH_2 + N_2$. Hence, as was the case on the S_1 surface, a conical intersection (i.e., S_2/S_1 here) serves as a funnel for the reaction, and ultimately the generation of CH_2.

This study by Arenas and coworkers supports a mechanism for the photochemical generation of singlet methylene differing from the one found by Olivucci and Robb et al.[172] Olivucci and Robb et al. proposed that CH_2 is formed via a second S_1/S_0 conical intersection. In contrast, the dynamics simulations by Arenas and coworkers showed no methylene generation in the excited state.[69] Rather, an excited state (S_1 or S_2) of the precursor (3*H*-diazirine or diazomethane) leads to a conical intersection. This conical intersection then funnels to the S_0 surface, where CH_2 is ultimately formed. Furthermore, it does not matter which particular excited-state surface is populated: the same final product is formed.

5.4 CONCLUSION AND OUTLOOK

Dynamics simulations have been performed to study reaction mechanisms and the nonstatistical behavior of carbene cycloadditions and other types of carbene rearrangements, as well as carbene generation. The dynamical picture of reactions provides details about mechanisms and intimate detail about how reactions occur. We are approaching a more complete understanding of reactivity of carbenes through these types of molecular dynamics simulations.

SUGGESTED READING

Carpenter, B. K. "Dynamic Behavior of Organic Reactive Intermediates," *Angew. Chem. Int. Ed.* **1998**, *37*, 3340–3350.

Carpenter, B. K. "Potential Energy Surfaces and Reaction Dynamics." In *Reactive Intermediate Chemistry*. Moss, R. A., Platz, M. S., Jones, M., Jr., Eds. Wiley: Hoboken, NJ, **2004**, pp. 925–960.

Merrer, D. C., Doubleday, C. "Dynamic Control of Dichlorocarbene Addition to Cyclopropene," *J. Phys. Org. Chem.* **2011**, *24*, 947–951.

Merrer, D. C., Rablen, P. R. "Dichlorocarbene Addition to Cyclopropenes: A Computational Study," *J. Org. Chem.* **2005**, *70*, 1630–1635.

Wang, I. S. Y., Karplus, M. "Dynamics of Organic Reactions," *J. Am. Chem. Soc.* **1973**, *95*, 8160–8164.

Xu, L., Doubleday, C. E., Houk, K. N. "Dynamics of 1,3-Dipolar Cycloaddition Reactions of Diazonium Betaines to Acetylene and Ethylene: Bending Vibrations Facilitate Reaction," *Angew. Chem. Int. Ed.* **2009**, *48*, 2746–2748.

Xu, L., Doubleday, C. E., Houk, K. N. "Dynamics of 1,3-Dipolar Cycloadditions: Energy Partitioning of Reactants and Quantitation of Synchronicity," *J. Am. Chem. Soc.* **2010**, *132*, 3029–3037.

Xu, L., Doubleday, C. E., Houk, K. N. "Dynamics of Carbene Cycloadditions," *J. Am. Chem. Soc.* **2011**, *133*, 17848–17854.

REFERENCES

1 Hine, J. *J. Am. Chem. Soc.* **1950**, *72*, 2438–2445.

2 Doering, W. v. E., Hoffmann, A. K. *J. Am. Chem. Soc.* **1954**, *76*, 6162–6165.

3 Bertrand, G., Ed. *Carbene Chemistry: From Fleeting Intermediates to Powerful Reagents*. Marcel Dekker: New York, **2002**.

4 Bertrand, G. In *Reactive Intermediate Chemistry*. Moss, R. A., Platz, M. S., Jones, M., Jr., Eds. John Wiley and Sons: Hoboken, NJ, **2004**, pp. 329–374.

5 Bethell, D. In *Organic Reactive Intermediates*. McManus, S. P., Ed. Academic Press: New York, **1973**, pp. 61–126.

6 Brinker, U. H., Ed. *Advances in Carbene Chemistry*, Vol. 1. JAI Press: Stamford, CT, **1994**.

7 Brinker, U. H., Ed. *Advances in Carbene Chemistry*, Vol. 2. JAI Press: Stamford, CT, **1998**.

8 Brinker, U. H., Ed. *Advances in Carbene Chemistry*, Vol. 3. Elsevier: Amsterdam, **2001**.

9 Doyle, M. P. In *Reactive Intermediate Chemistry*. Moss, R. A., Platz, M. S., Jones, M., Jr., Eds. John Wiley and Sons: Hoboken, NJ, **2004**, pp. 561–592.

10 Jones, M., Jr., Moss, R. A. In *Reactive Intermediate Chemistry*. Moss, R. A., Platz, M. S., Jones, M., Jr., Eds. John Wiley and Sons: Hoboken, NJ, **2004**, pp. 273–328.

11 Kirmse, W., Ed. *Carbene Chemistry*, 2nd ed. Academic Press: New York, **1971**.

12 Jones, M., Jr., Moss, R. A., Eds. *Carbenes*, Vol. 1. John Wiley and Sons: New York, **1973**.

13 Moss, R. A., Jones, M., Jr., Eds. *Carbenes*, Vol. 2. John Wiley and Sons: New York, **1975**.

14 Platz, M. S., Ed. *Kinetics and Spectroscopy of Carbenes and Biradicals*. Plenum Press: New York, **1990**.

15 Shevlin, P. B. In *Reactive Intermediate Chemistry*. Moss, R. A., Platz, M. S., Jones, M., Jr., Eds. John Wiley and Sons: Hoboken, NJ, **2004**, pp. 463–500.

16 Tomioka, H. In *Reactive Intermediate Chemistry*. Moss, R. A., Platz, M. S., Jones, M., Jr., Eds. John Wiley and Sons: Hoboken, NJ, **2004**, pp. 375–462.

17 Scaiano, J. C. In *Reactive Intermediate Chemistry*. Moss, R. A., Platz, M. S., Jones, M., Jr., Eds. John Wiley and Sons: Hoboken, NJ, **2004**, pp. 847–871.

18 Merrer, D. C., Moss, R. A. In *Advances in Carbene Chemistry*, Vol. *3*. Brinker, U. H., Ed. Amsterdam: Elsevier, **2001**, pp. 53–113.

19 Burdzinski, G., Rehault, J., Wang, J., Platz, M. S. *J. Phys. Chem. A* **2008**, *112*, 10108–10112.

20 Wang, J., Burdzinski, G., Gustafson, T. L., Platz, M. S. *J. Am. Chem. Soc.* **2007**, *129*, 2597–2606.
21 Zhang, Y., Burdzinski, G., Kubicki, J., Platz, M. S. *J. Am. Chem. Soc.* **2009**, *131*, 9646–9647.
22 Jackson, J. E., Platz, M. S. In *Advances in Carbene Chemistry*, Vol. 1. Brinker, U. H., Ed. JAI Press: Stamford, CT, **1994**, pp. 89–160.
23 Moss, R. A. In *Advances in Carbene Chemistry*, Vol. 1. Brinker, U. H., Ed. JAI Press: Stamford, CT, **1994**, pp. 59–88.
24 Toscano, J. P. In *Reviews of Reactive Intermediate Chemistry*. Platz, M. S., Moss, R. A., Jones, M., Jr., Eds. John Wiley and Sons: Hoboken, NJ, **2007**, pp. 183–205.
25 Bourissou, D., Guerret, O., Gabbaie, F. P., Bertrand, G. *Chem. Rev.* **2000**, *100*, 39–91.
26 Canac, Y., Soleilhavoup, M., Conejero, S., Bertrand, G. *J. Organomet. Chem.* **2004**, *689*, 3857–3865.
27 Kato, T., Maerten, E., Baceiredo, A. *Top. Organomet. Chem.* **2010**, *30*, 131–147.
28 Kirmse, W. *Angew. Chem. Int. Ed.* **2004**, *43*, 1767–1769.
29 Martin, D., Soleilhavoup, M., Bertrand, G. *Chem. Sci.* **2011**, *2*, 389–399.
30 Melaimi, M., Soleilhavoup, M., Bertrand, G. *Angew. Chem. Int. Ed.* **2010**, *49*, 8810–8849.
31 Vignolle, J., Cattoen, X., Bourissou, D. *Chem. Rev.* **2009**, *109*, 3333–3384.
32 Arduengo, A. J., III, Harlow, R. L., Kline, M. *J. Am. Chem. Soc.* **1991**, *113*, 361–363.
33 Trnka, T. M., Grubbs, R. H. *Acc. Chem. Res.* **2001**, *34*, 18–29.
34 Arduengo, A. J., III, Iconaru, L. I. *Dalton Trans.* **2009**, 6903–6914.
35 Beligny, S., Blechert, S. In *N-Heterocyclic Carbenes in Synthesis*; Nolan, S. P., Ed.; Wiley-VCH: Weinheim, Germany, **2006**, pp. 1–25.
36 Burling, S., Paine, B. M., Whittlesey, M. K. In *N-Heterocyclic Carbenes in Synthesis*; Nolan, S. P., Ed.; Wiley-VCH: Weinheim, Germany, **2006**, pp. 27–53.
37 Cesar, V., Bellemin-Laponnaz, S. *Actual. Chim.* **2009**, *326*, 8–14.
38 Clavier, H., Nolan, S. P. *Annu. Rep. Prog. Chem., Sect. B Org. Chem.* **2007**, *103*, 193–222.
39 Correa, A., Nolan, S. P., Cavallo, L. *Top. Curr. Chem.* **2011**, *302*, 131–155.
40 de Fremont, P., Marion, N., Nolan, S. P. *Coord. Chem. Rev.* **2009**, *253*, 862–892.
41 Despagnet-Ayoub, E., Ritter, T. *Top. Organomet. Chem.* **2007**, *21*, 193–218.
42 Diez-Gonzalez, S., Marion, N., Nolan, S. P. *Chem. Rev.* **2009**, *109*, 3612–3676.
43 Diez-Gonzalez, S., Nolan, S. P. *Annu. Rep. Prog. Chem., Sect. B Org. Chem.* **2005**, *101*, 171–191.
44 Fuerstner, A. *Chem. Commun.* **2011**, *47*, 6505–6511.
45 Hahn, F. E., Jahnke, M. C. *Angew. Chem. Int. Ed.* **2008**, *47*, 3122–3172.
46 Haibach, M. C., Kundu, S., Brookhart, M., Goldman, A. S. *Acc. Chem. Res.* **2012**, *45*, 947–958.
47 Kotha, S., Dipak, M. K. *Tetrahedron* **2012**, *68*, 397–421.
48 Martin, D., Melaimi, M., Soleilhavoup, M., Bertrand, G. *Organometallics* **2011**, *30*, 5304–5313.
49 Peris, E., Crabtree, R. H. In *Iridium Complexes in Organic Synthesis*; Oro, L. A., Claver, C., Eds., Wiley-VCH: Weinheim, Germany, **2009**, pp. 39–54.
50 Szadkowska, A., Samojlowicz, C., Grela, K. *Pure Appl. Chem.* **2011**, *83*, 553–563.
51 Wang, F., Liu, L.-j., Wang, W., Li, S., Shi, M. *Coord. Chem. Rev.* **2012**, *256*, 804–853.
52 Whited, M. T., Grubbs, R. H. *Acc. Chem. Res.* **2009**, *42*, 1607–1616.
53 Zhang, W., Moore, J. S. *Adv. Synth. Catal.* **2007**, *349*, 93–120.
54 Zhong, C., Shi, X. *Eur. J. Org. Chem.* **2010**, 2999–3025.
55 Zinn, F. K., Viciu, M. S., Nolan, S. P. *Annu. Rep. Prog. Chem., Sect. B Org. Chem.* **2004**, *100*, 231–249.

56 Skell, P. S., Garner, A. Y. *J. Am. Chem. Soc.* **1956**, *78*, 5430–5433.
57 Hoffmann, R. *J. Am. Chem. Soc.* **1968**, *90*, 1475–1485.
58 Kraka, E., Cremer, D. *Acc. Chem. Res.* **2010**, *43*, 591–601.
59 Keating, A. E., Merrigan, S. R., Singleton, D. A., Houk, K. N. *J. Am. Chem. Soc.* **1999**, *121*, 3933–3938.
60 Merrer, D. C., Rablen, P. R. *J. Org. Chem.* **2005**, *70*, 1630–1635.
61 Bernardi, F., Bottoni, A., Canepa, C., Olivucci, M., Robb, M. A., Tonachini, G. *J. Org. Chem.* **1997**, *62*, 2018–2025.
62 Xu, L., Doubleday, C. E., Houk, K. N. *J. Am. Chem. Soc.* **2011**, *133*, 17848–17854.
63 Reprinted with permission from Keating, A. E., Merrigan, S. R., Singleton, D. A., Houk, K. N. *J. Am. Chem. Soc.* **1999**, *121*, 3933–3938. Copyright 1999 American Chemical Society.
64 Merrer, D. C., Doubleday, C. *J. Phys. Org. Chem.* **2011**, *24*, 947–951.
65 Khrapunovich, M., Zelenova, E., Seu, L., Sabo, A. N., Flaherty, A., Merrer, D. C. *J. Org. Chem.* **2007**, *72*, 7574–7580.
66 Mo, X. Y., Bernard, S. E., Khrapunovich, M., Merrer, D. C. *J. Org. Chem.* **2008**, *73*, 8537–8544.
67 Rablen, P. R., Paiz, A. A., Thuronyi, B. W., Jones, M, Jr. *J. Org. Chem.* **2009**, *74*, 4252–4261.
68 Yu, H.-G., Francisco, J. S., Muckerman, J. T. *J. Chem. Phys.* **2008**, *129*, 064301.
69 Arenas, J. F., Lopez-Tocon, I., Otero, J. C., Soto, J. *J. Am. Chem. Soc.* **2002**, *124*, 1728–1735.
70 Litovitz, A. E., Keresztes, I., Carpenter, B. K. *J. Am. Chem. Soc.* **2008**, *130*, 12085–12094.
71 Reprinted with permission from Siebert, M. R., Manikandan, P., Sun, R., Tantillo, D. J., Hase, W. L. *J. Chem. Theory Comput.* **2012**, *8*, 1212–1222. Copyright 2012 American Chemical Society.
72 Carpenter, B. K. In *Reactive Intermediate Chemistry*, Moss, R. A., Platz, M. S., Jones, M., Jr., Eds., Wiley: Hoboken, NJ, **2004**, pp. 925–960.
73 Rehbein, J., Carpenter, B. K. *Phys. Chem. Chem. Phys.* **2011**, *13*, 20906–20922.
74 Bachrach, S. M. *Annu. Rep. Prog. Chem., Sect. B Org. Chem.* **2010**, *106*, 407–427.
75 Birney, D. M. *Curr. Org. Chem.* **2010**, *14*, 1658–1668.
76 Carpenter, B. K. *Angew. Chem. Int. Ed.* **1998**, *37*, 3341–3350.
77 Carpenter, B. K. *Annu. Rev. Phys. Chem.* **2005**, *56*, 57–89.
78 Ess, D. H., Wheeler, S. E., Iafe, R. G., Xu, L., Celebi-Olcum, N., Houk, K. N. *Angew. Chem. Int. Ed.* **2008**, *47*, 7592–7601.
79 Yamataka, H. *Adv. Phys. Org. Chem.* **2010**, *44*, 173–222.
80 Tantillo, D. J. *Nat. Prod. Rep.* **2011**, *28*, 1035–1053.
81 Quijano, L. M. M., Singleton, D. A. *J. Am. Chem. Soc.* **2011**, *133*, 13824–13827.
82 Black, K., Liu, P., Xu, L., Doubleday, C., Houk, K. N. *Proc. Natl. Acad. Sci. USA.* **2012**, *109*, 12860–12865.
83 Baldwin, J. E. *Chem. Rev.* **2003**, *103*, 1197–1212.
84 Baldwin, J. E., Leber, P. A. *Org. Biomol. Chem.* **2008**, *6*, 36–47, and references therein.
85 Hudlicky, T., Kutchan, T. M., Naqvi, S. M. *Org. React.* **1985**, *33*, 247–335.
86 Doubleday, C. *J. Phys. Chem. A* **2001**, *105*, 6333–6341.
87 Doubleday, C., Li, G., Hase, W. L. *Phys. Chem. Chem. Phys.* **2002**, *4*, 304–312.
88 Doubleday, C., Nendel, M., Houk, K. N., Thweatt, D., Page, M. *J. Am. Chem. Soc.* **1999**, *121*, 4720–4721.

89 Debbert, S. L., Carpenter, B. K., Hrovat, D. A., Borden, W. T. *J. Am. Chem. Soc.* **2002**, *124*, 7896–7897.
90 Hrovat, D. A., Duncan, J. A., Borden, W. T. *J. Am. Chem. Soc.* **1999**, *121*, 169–175.
91 Roth, W. R., Wollweber, D., Offerhaus, R., Rekowski, V., Lennartz, H. W., Sustmann, R., Mueller, W. *Chem. Ber.* **1993**, *126*, 2701–2715.
92 Khuong, K. S., Houk, K. N. *J. Am. Chem. Soc.* **2003**, *125*, 14867–14883.
93 Lyons, B. A., Pfeifer, J., Peterson, T. H., Carpenter, B. K. *J. Am. Chem. Soc.* **1993**, *115*, 2427–2437.
94 Reyes, M. B., Carpenter, B. K. *J. Am. Chem. Soc.* **1998**, *120*, 1641–1642.
95 Reyes, M. B., Carpenter, B. K. *J. Am. Chem. Soc.* **2000**, *122*, 10163–10176.
96 Yamamoto, N., Olivucci, M., Celani, P., Bernardi, F., Robb, M. A. *J. Am. Chem. Soc.* **1998**, *120*, 2391–2407.
97 Hansen, J. H., Gregg, T. M., Ovalles, S. R., Lian, Y., Autschbach, J., Davies, H. M. L. *J. Am. Chem. Soc.* **2011**, *133*, 5076–5085.
98 Hong, Y. J., Tantillo, D. J. *Nat. Chem.* **2009**, *1*, 384–389.
99 Hong, Y. J., Tantillo, D. J. *Org. Biomol. Chem.* **2010**, *8*, 4589–4600.
100 Itoh, S., Yamataka, H. *Chem.—Eur. J.* **2011**, *17*, 1230–1237.
101 Katori, T., Itoh, S., Sato, M., Yamataka, H. *J. Am. Chem. Soc.* **2010**, *132*, 3413–3422.
102 Kelly, K. K., Hirschi, J. S., Singleton, D. A. *J. Am. Chem. Soc.* **2009**, *131*, 8382–8383.
103 Thomas, J. B., Waas, J. R., Harmata, M., Singleton, D. A. *J. Am. Chem. Soc.* **2008**, *130*, 14544–14555.
104 Wang, Z., Hirschi, J. S., Singleton, D. A. *Angew. Chem. Int. Ed.* **2009**, *48*, 9156–9159.
105 Xu, H., Wade, P. A., Sohlberg, K. *Tetrahedron* **2010**, *66*, 845–851.
106 Yamamoto, Y., Hasegawa, H., Yamataka, H. *J. Org. Chem.* **2011**, *76*, 4652–4660.
107 Siebert, M. R., Zhang, J., Addepalli, S. V., Tantillo, D. J., Hase, W. L. *J. Am. Chem. Soc.* **2011**, *133*, 8335–8343.
108 Noey, E. L., Wang, X., Houk, K. N. *J. Org. Chem.* **2011**, *76*, 3477–3483.
109 Ye, L., Wang, Y., Aue, D. H., Zhang, L. *J. Am. Chem. Soc.* **2012**, *134*, 31–34.
110 Siebert, M. R., Manikandan, P., Sun, R., Tantillo, D. J., Hase, W. L. *J. Chem. Theory Comput.* **2012**, *8*, 1212–1222.
111 Gonzalez-James, O. M., Kwan, E. E., Singleton, D. A. *J. Am. Chem. Soc.* **2012**, *134*, 1914–1917.
112 Wang, I. S. Y., Karplus, M. *J. Am. Chem. Soc.* **1973**, *95*, 8160–8164.
113 Gao, X., Ohtsuka, Y., Ishimura, K., Nagase, S. *J. Phys. Chem. A* **2009**, *113*, 9852–9860.
114 In Gaussian 09, mUB3LYP is specified by UBLYP with IOp(3/76=1000001200), IOp(3/77=0720008800), IOp(3/78=0810010000).
115 Reprinted with permission from Xu, L., Doubleday, C. E., Houk, K. N. *J. Am. Chem. Soc.* **2011**, *133*, 17848–17854. Copyright 2011 American Chemical Society.
116 We have previously used the arbitrary values of 1.6 and 2.0 Å in *J. Am. Chem. Soc.* **2010**, *132*, 3029. The latter is closer to the TS and gives a time gap of 52 fs for the trajectory.
117 Doubleday, C., Bolton, K., Peslherbe, G. H., Hase, W. L. *J. Am. Chem. Soc.* **1996**, *118*, 9922–9931.
118 Doubleday, C., Suhrada, C. P., Houk, K. N. *J. Am. Chem. Soc.* **2006**, *128*, 90–94.
119 Pedersen, S., Herek, J. L., Zewail, A. H. *Science* **1994**, *266*, 1359–1364.
120 Carpenter, B. K. *Acc. Chem. Res.* **1992**, *25*, 520–528.

121 Mitsch, R. A. *J. Am. Chem. Soc.* **1965**, *87*, 758–761.
122 Firestone, R. A. *J. Org. Chem.* **1968**, *33*, 2285–2290.
123 Xu, L., Doubleday, C. E., Houk, K. N. *Angew. Chem. Int. Ed.* **2009**, *48*, 2746–2748.
124 Xu, L., Doubleday, C. E., Houk, K. N. *J. Am. Chem. Soc.* **2010**, *132*, 3029–3037.
125 Yu, H.-G., Muckerman, J. T. *J. Phys. Chem. A* **2005**, *109*, 1890–1896.
126 Pedley, J. B., Naylor, R. D., Kirby, S. P. *Thermochemical Data of Organic Compounds*, 2nd ed. Chapman and Hall: London, **1986**.
127 Robinson, M. S., Polak, M. L., Bierbaum, V. M., Depuy, C. H., Lineberger, W. C. *J. Am. Chem. Soc.* **1995**, *117*, 6766–6778.
128 National Institute of Standards and Technology, *NIST Standard Reference Database*, NIST Chemistry WebBook, http://webbook.nist.gov.
129 Janaf Thermochemical Tables. *Nat. Stand. Ref. Data Ser.* Nat. Bur. Stand. (US): Washington, DC, **1985**.
130 Reprinted with permission from Yu, H.-G., Muckerman, J. T. *J. Phys. Chem. A* **2005**, *109*, 1890–1896. Copyright 2005 American Chemical Society.
131 Dehmlow, E. V. *Tetrahedron* **1972**, *28*, 175–179.
132 Dehmlow, E. V. *Tetrahedron Lett.* **1975**, 203–206.
133 Hart, H., Nitta, M. *Tetrahedron Lett.* **1974**, 2109–2112.
134 Mahler, W. *J. Am. Chem. Soc.* **1962**, *84*, 4600–4601.
135 Trost, B. M., Atkins, R. C. *J. Chem. Soc. D.* **1971**, 1254–1255.
136 Weber, J., Brinker, U. H. *Angew. Chem. Int. Ed. Engl.* **1997**, *36*, 1623–1626.
137 Weber, J., Xu, L., Brinker, U. H. *Tetrahedron Lett.* **1992**, *33*, 4537–4540.
138 Warkentin, J. In *Advances in Carbene Chemistry*, Vol. 2. Brinker, U. H., Ed. JAI Press: Stamford, CT, **1998**, pp. 245–295.
139 Warkentin, J. *Perkin Trans. 1* **2000**, 2161–2169.
140 Warkentin, J. *Acc. Chem. Res.* **2009**, *42*, 205–212.
141 Zhou, H., Mloston, G., Warkentin, J. *Org. Lett.* **2005**, *7*, 487–489.
142 Reprinted with permission from Merrer, D. C., Doubleday, C. *J. Phys. Org. Chem.* **2011**, *24*, 947–951. Copyright 2011 John Wiley and Sons.
143 Reprinted with permission from Merrer, D. C., Rablen, P. R. *J. Org. Chem.* **2005**, *70*, 1630–1635. Copyright 2005 American Chemical Society.
144 Carpenter, B. K. *J. Am. Chem. Soc.* **1995**, *117*, 6336–6344.
145 Suhrada, C. P., Houk, K. N. *J. Am. Chem. Soc.* **2002**, *124*, 8796–8797.
146 Kagabu, S., Saito, K. *Tetrahedron Lett.* **1988**, *29*, 675–676.
147 Litosh, V. A., Saini, R. K., Guzman-Jimenez, I. Y., Whitmire, K. H., Billups, W. E. *Org. Lett.* **2001**, *3*, 65–66.
148 Reprinted with permission from Mo, X. Y., Bernard, S. E., Khrapunavich, M., Merrer, D. C. *J. Org. Chem.* **2008**, *73*, 8537–8544. Copyright 2008 American Chemical Society.
149 Anslyn, E. V., Dougherty, D. A. *Modern Physical Organic Chemistry*, University Science: Sausalito, CA, **2006**, p.111.
150 Bach, R. D. *J. Am. Chem. Soc.* **2009**, *131*, 5233–5243.
151 Roberts, J. D., Caserio, M. C. In *Basic Principles of Organic Chemistry*, 2nd ed. W. A. Benjamin: Menlo Park, CA, **1977**, p. 475*f*.
152 Turner, R., Jarrett, A. D., Goebel, P., Mallon, B. J. *J. Am. Chem. Soc.* **1973**, *95*, 790–792.
153 Jackson, J. E., Misslitz, U., Jones, M., Jr., De Meijere, A. *Tetrahedron* **1987**, *43*, 653–662.

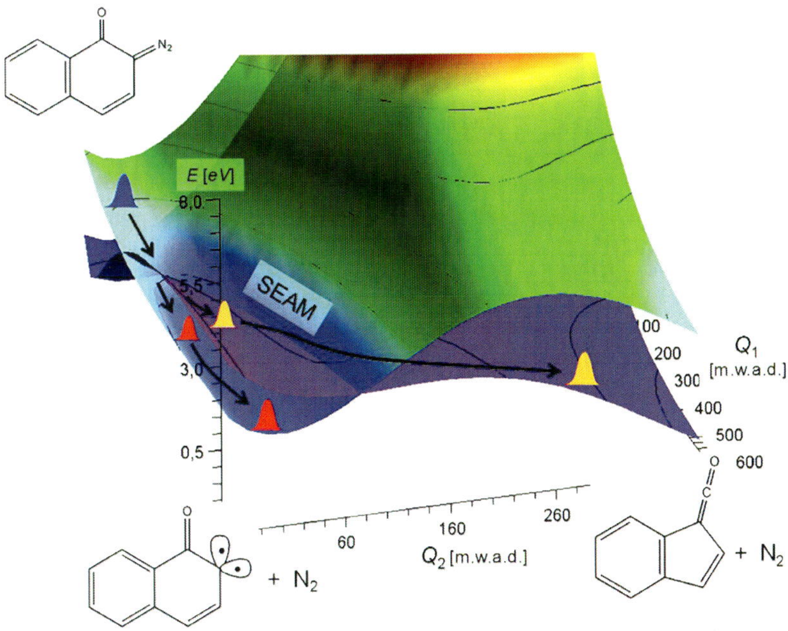

Plate 4.8. Two-dimensional plot of the model surface for the dynamics along Q_1 (C–N dissociation coordinate) and Q_2 (ring rearrangement coordinate) (projection for $Q_3=0$; m.w.a.d.: mass-weighted atomic displacements). The purple and blue–green surfaces correspond to the closed- and open-shell diabatic states, correlating with the ketene and carbene products, respectively. The minima for the two photoproducts are found at large values of *Q*1 on either side of the plot. Snapshots of the wave packet evolution at different time steps are also shown on the surface. The wave packet at the beginning of the dynamics is shown in blue. The red wave packet corresponds to the fraction that decays at the region of the seam leading to the carbene minimum, and the yellow one corresponds to the fraction that decays at the region leading to the ketene. (Reprinted with permission from Li, Q., Migani, A., Blancafort, L. *J. Phys. Chem. Lett.* **2012**, *3*, 1056–1061. Copyright 2012 American Chemical Society.)

Contemporary Carbene Chemistry, First Edition. Edited by Robert A. Moss and Michael P. Doyle.

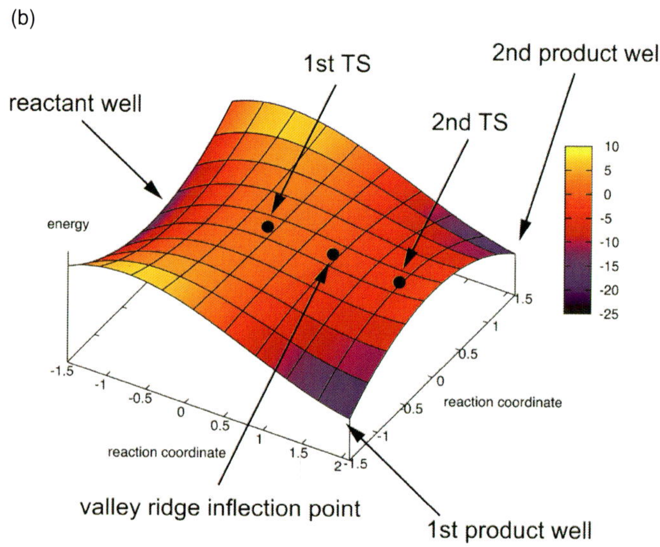

Plate 5.2. (b) a two-dimensional PES (hypersurface). (Copyright © 2012 American Chemical Society.)[71]

Plate 5.4. Left: Overlay of starting geometries of 453 reactive trajectories for the reaction of CCl_2 with ethylene. Right: Overlay of starting geometries of 298 reactive trajectories for CF_2 + ethylene.[115] (Copyright © 2011 American Chemical Society.)

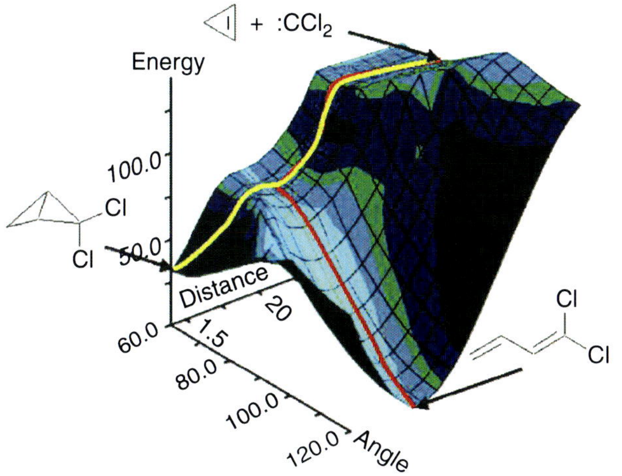

Plate 5.11. PES of CCl_2 + cyclopropene **1a** at varying carbene-alkene distances, *r*, and cyclopropenyl bond angles, θ (CCSD(T)/6-31G*//B3LYP/6-31G*). (Copyright © 2012 American Chemical Society.)[143]

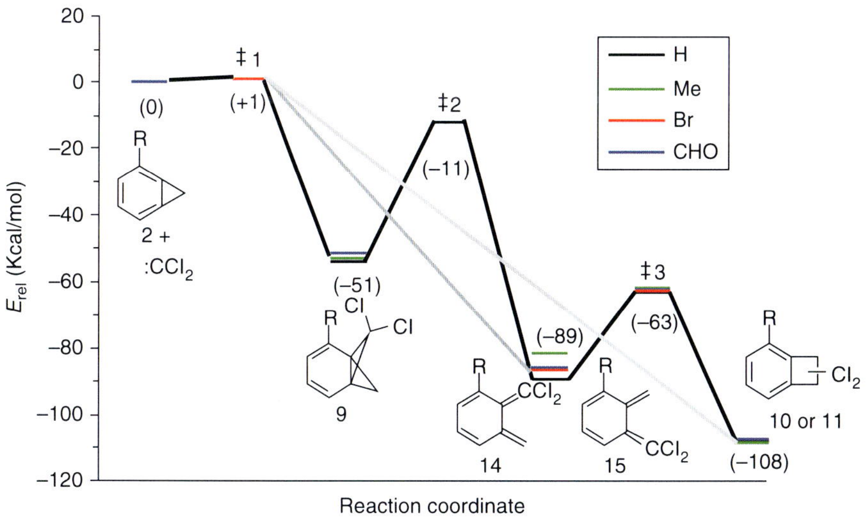

Plate 5.15. Potential energy surface for CCl_2 + benzocyclopropene **2**. Relative energies are in kcal/mol at B3LYP/6-31G* and are ZPE-corrected (ZPE's are scaled by 0.9806). (Copyright © 2011 American Chemical Society.)[148]

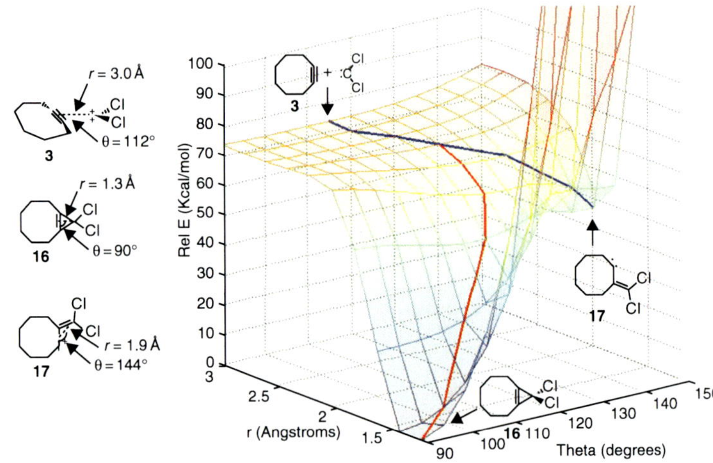

Plate 5.16. Three-dimensional PES of CCl_2 + cyclooctyne **3** at B3LYP/6-31G*. Blue and red traces are proposed paths for the formation of **16** and **17**. (Reprinted with permission from Reference 66; Copyright 2008 by the American Chemical Society.)

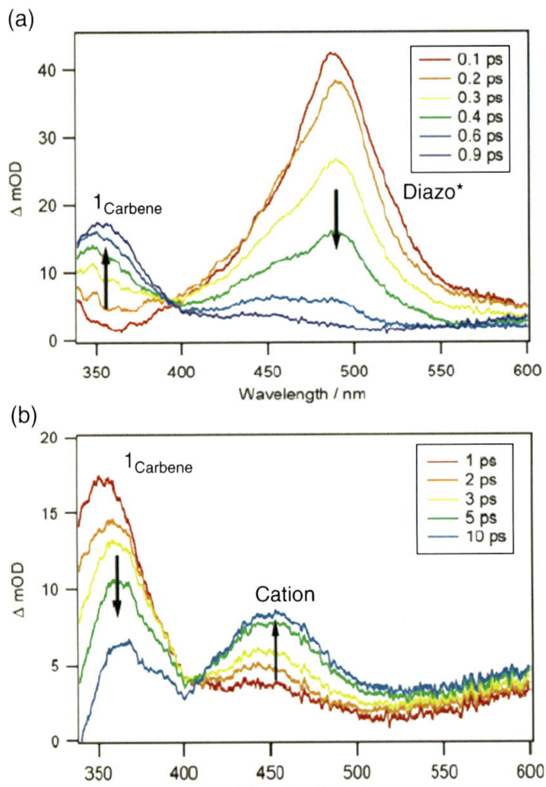

Plate 6.1. (a) Transient absorption UV-Vis spectra depicting the decay of the diazo excited state and the rise of the singlet carbene. (b) In methanol, the carbene forms the cation (λ_{exc} = 311 nm). (Reprinted with permission from Reference 5; Copyright 2006 by the American Chemical Society.)

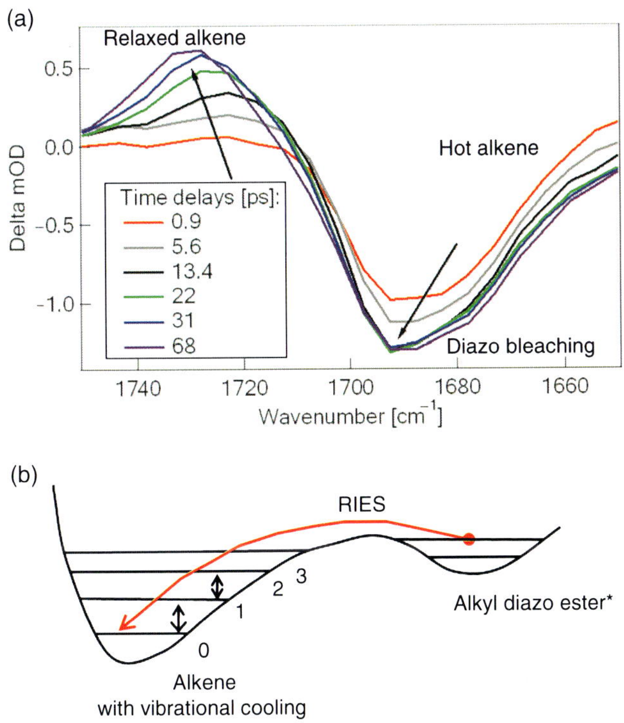

Plate 6.2. Ultrafast rise of a vinyl ester species (<0.4 ps) due to RIES after photoexcitation of methyl 2-diazopropionate at 266 nm (a) and alkene vibrational cooling (b). (Reprinted with permission from Reference 21; Copyright 2010 by the American Chemical Society.)

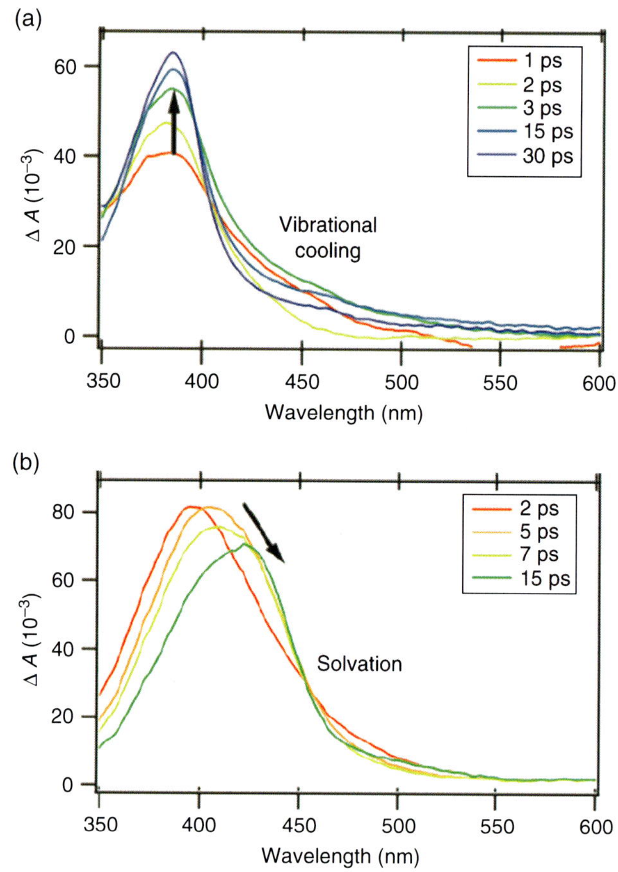

Plate 6.3. Carbene absorption spectra in (a) nonpolar cyclohexane and in (b) polar and coordinating methanol solvent produced by ultrafast photolysis ($\lambda_{exc} = 308\,nm$) of *p*-biphenylyltrifluoromethyl diazomethane. (Reprinted with permission from reference 8; Copyright 2008 by the American Chemical Society.)

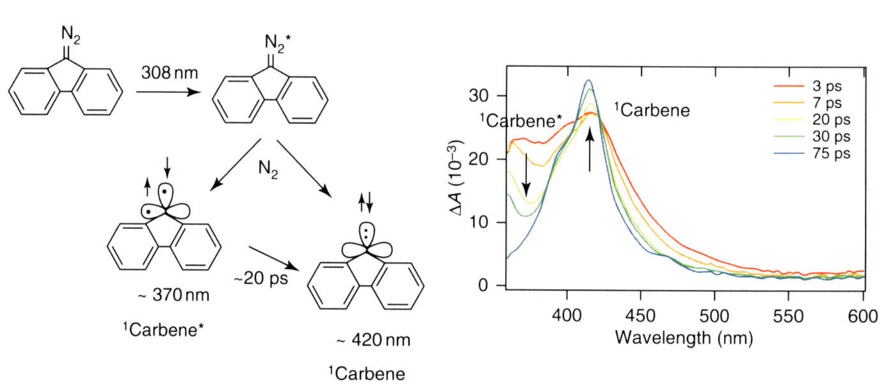

Plate 6.6. Postulated mechanism for the decay of photoexcited 9-diazofluorene* (the asterisk denotes the singlet excited state). The time-resolved UV-Vis transient absorption spectra were recorded over a 3–75 ps time window. (Reprinted with permission from Reference 11; Copyright 2007 by the American Chemical Society.)

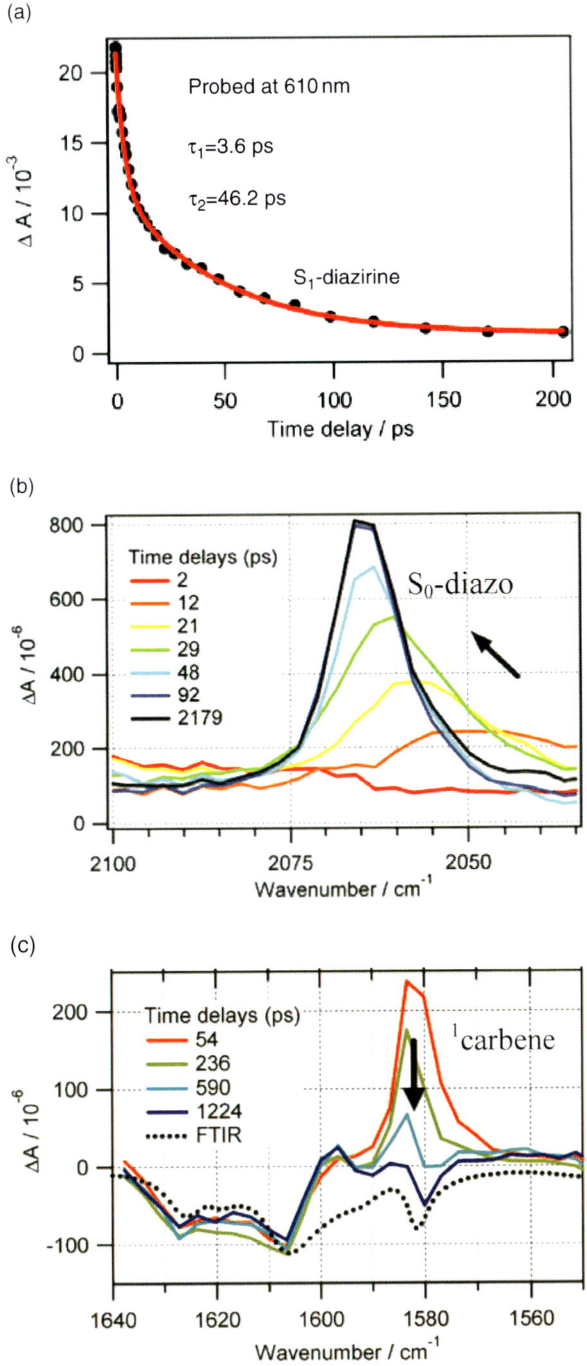

Plate 6.7. Ultrafast results recorded for phenyldiazirine using UV-Vis and IR spectroscopies: (a) the decay of the S_1-diazirine in chloroform ($\lambda_{exc}=375$ nm), (b) the rise of the S_0-diazo from S_1-diazirine isomerization in acetonitrile, and (c) the ^{1}carbene decay in chloroform ($\lambda_{exc}=270$ nm). (Reprinted with permission from References 24 and 40; Copyright 2009 by the American Chemical Society.)

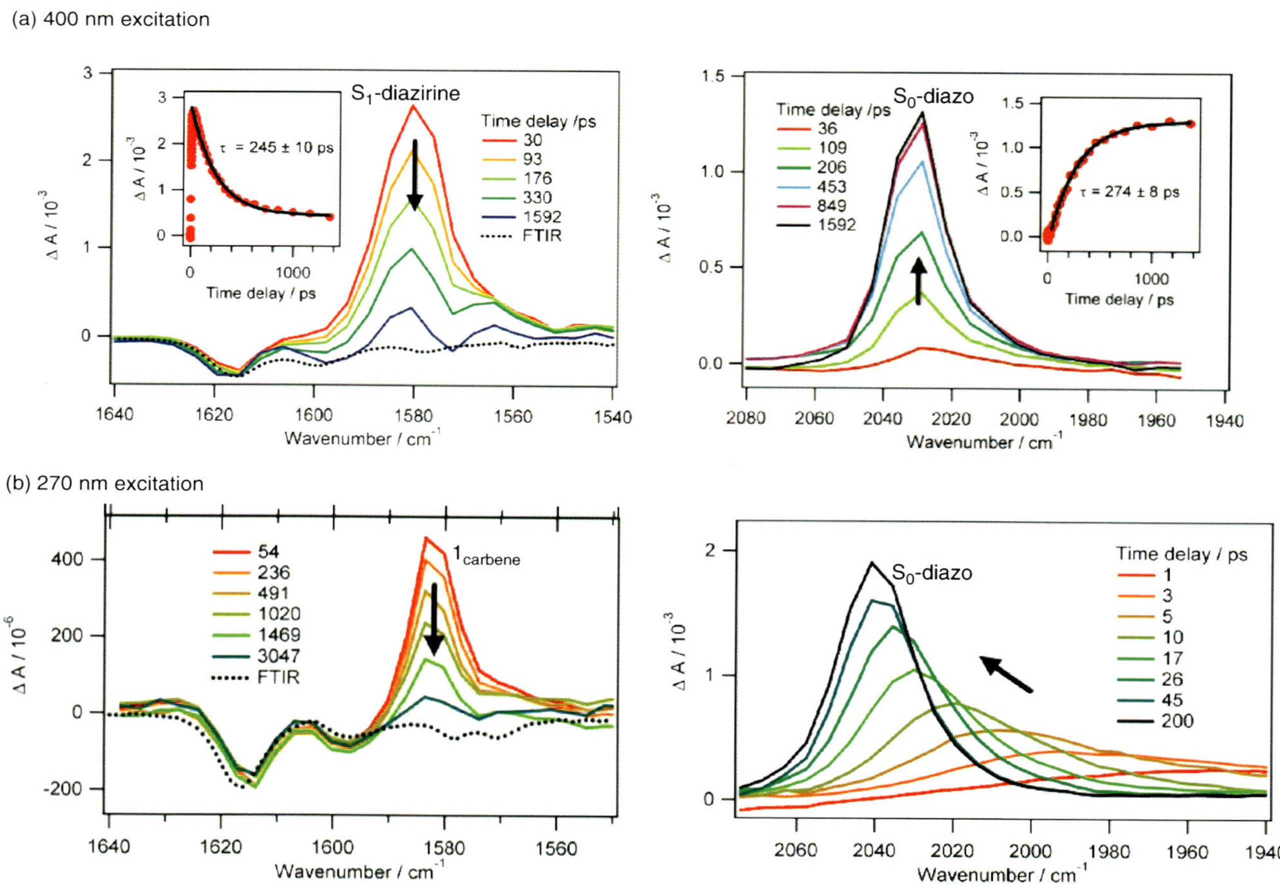

Plate 6.11. Time-resolved IR spectra recorded upon excitation of *p*-methoxy-3-phenyl-3-methyldiazirine in chloroform with light of (a) 400 nm and (b) 270 nm. (Reprinted with permission from Reference 1; Copyright 2009 by the American Chemical Society.)

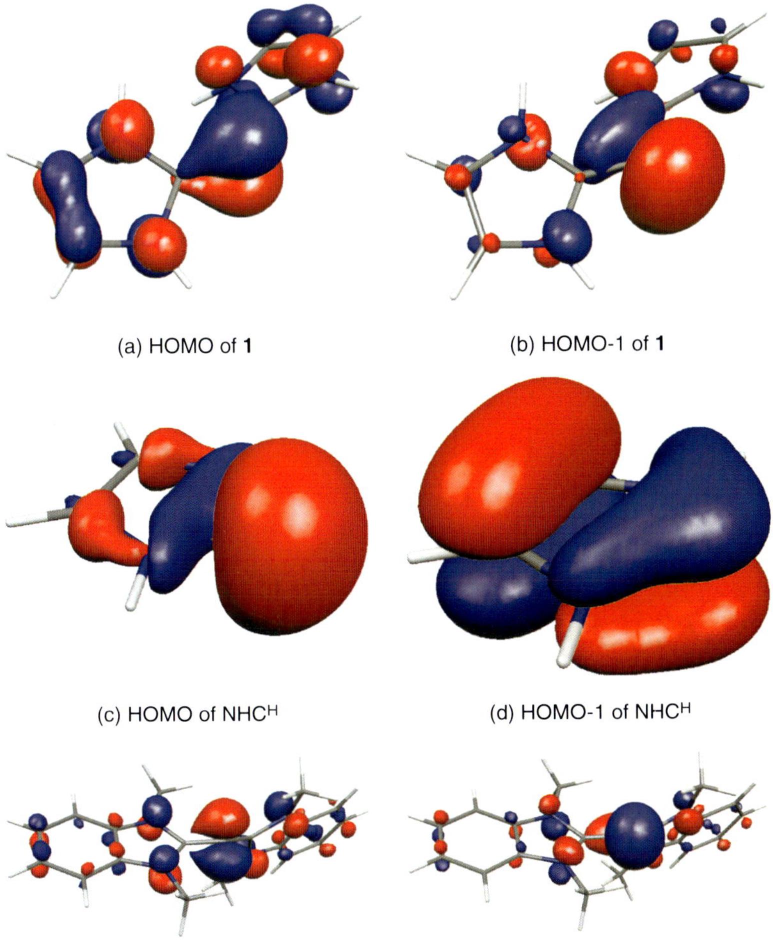

(a) HOMO of **1** (b) HOMO-1 of **1**

(c) HOMO of NHC^H (d) HOMO-1 of NHC^H

(e) HOMO of **3** (f) HOMO-1 of **3**

Plate 8.2. Plot of the two highest lying occupied orbitals HOMO and HOMO-1 of the carbodicarbenes **1** and **3** and the carbene NHC^H; see References 4b and 5b.

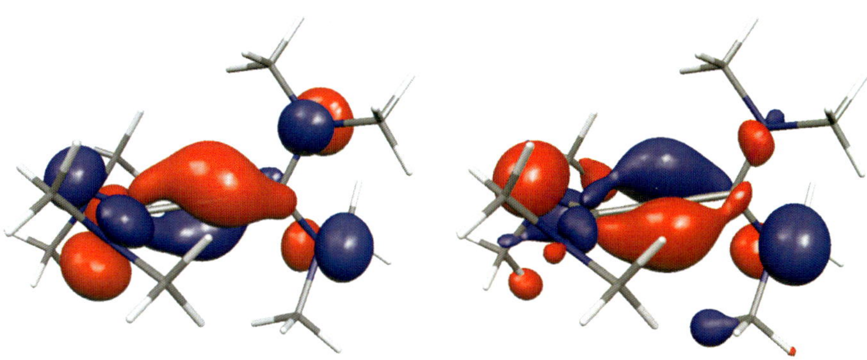

(a) HOMO of **5** (b) HOMO-1 of **5**

Plate 8.6. Plot of the two highest lying occupied orbitals HOMO and HOMO-1 of tetraaminoallene **5**; see Reference 4b.

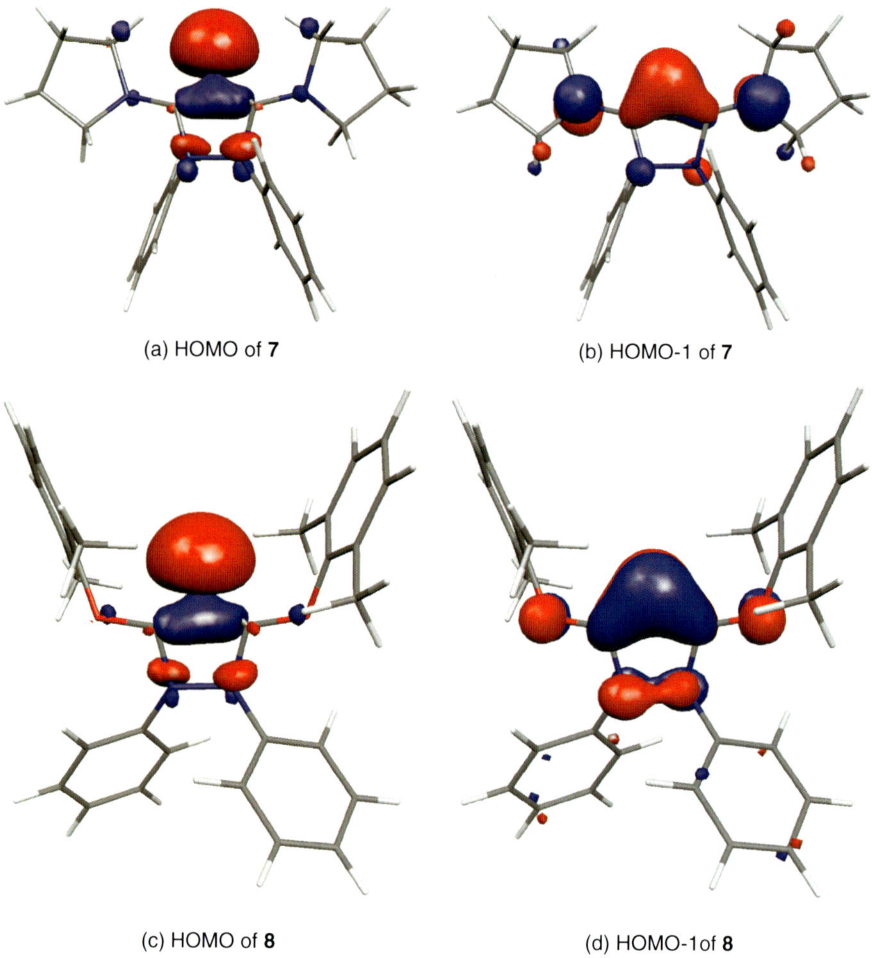

Plate 8.8. Plot of the two highest lying occupied orbitals HOMO and HOMO-1 of the "bent allenes" **7** and **8**; see Reference 5b.

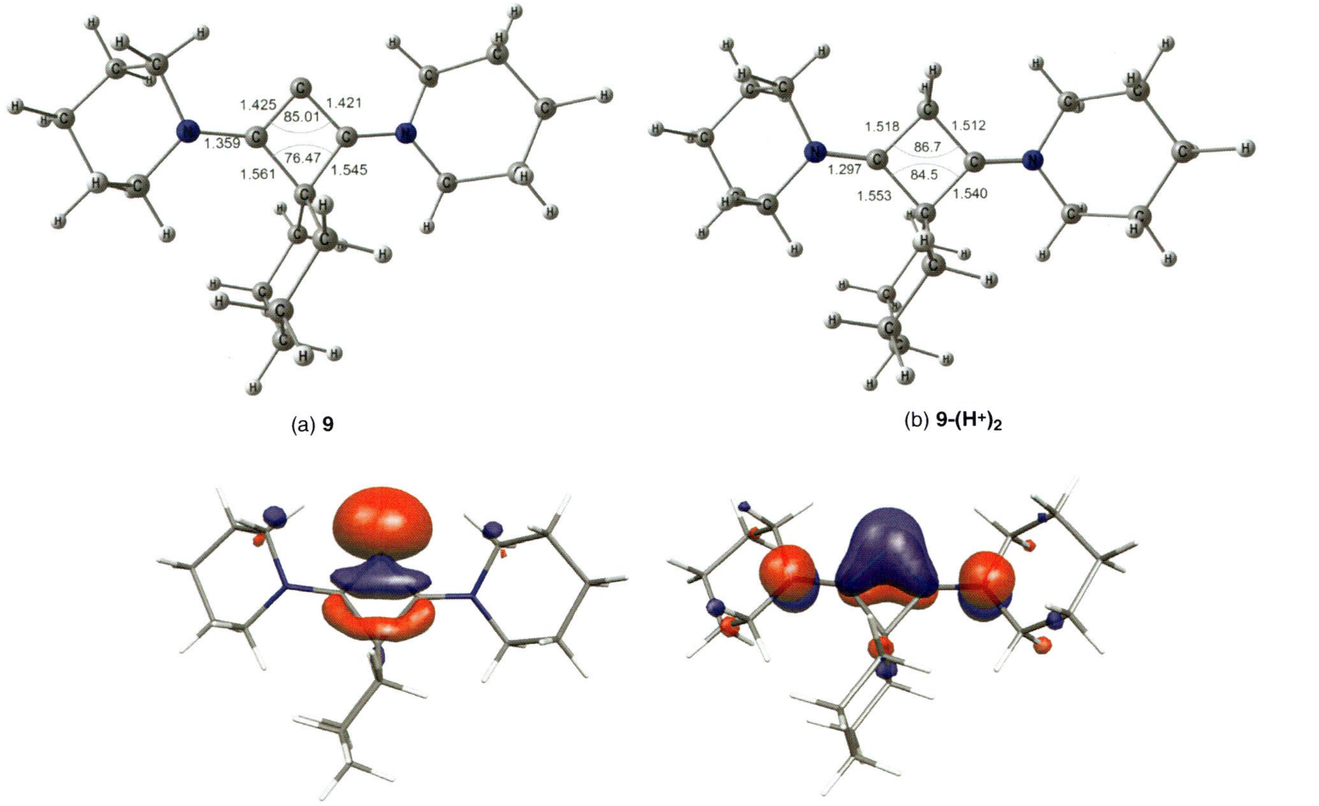

(a) **9**

(b) **9-(H+)₂**

(c) HOMO of **9**

(d) HOMO-1 of **9**

Plate 8.9. Structure of an all-carbon four-membered ring allene (**9**) together with the doubly protonated species (**9-(H$^+$)$_2$**) and (c) HOMO and (d) HOMO-1 of **9**; see Reference 13c.

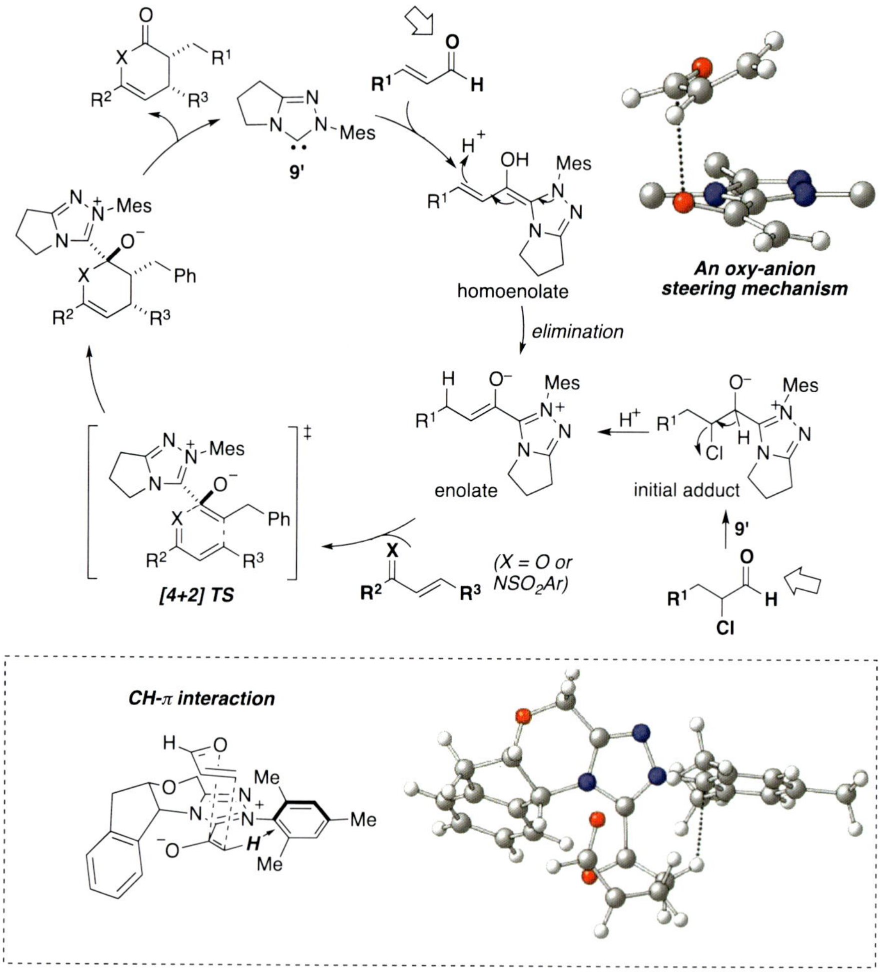

Plate 9.10. The mechansim of NHC-catalyzed Diels–Alder reactions.

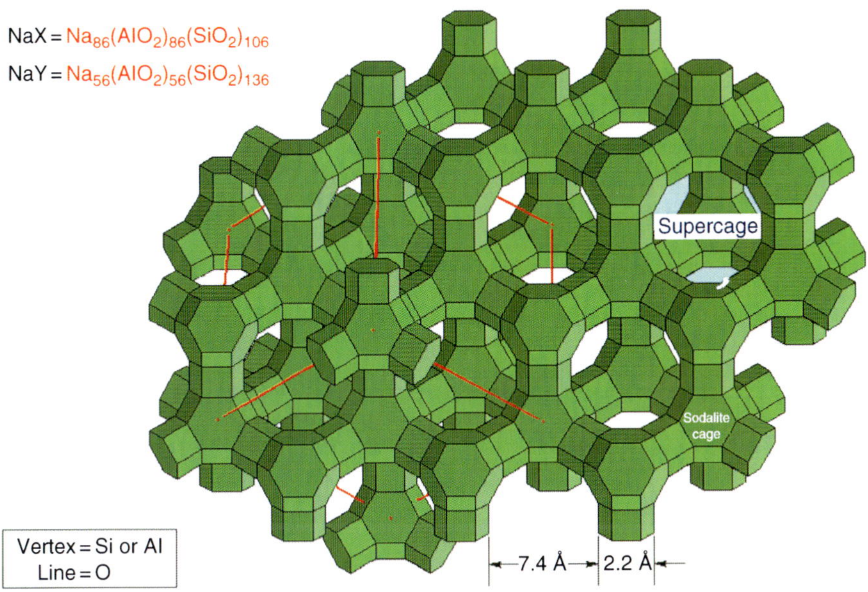

Plate 10.6. Structure of faujasite zeolites X and Y with the diamond-like unit cell outlined. Carbenes are generated within 13-Å diameter supercages that are interconnected by a tetrahedral array of 7.4-Å apertures. The guest-to-supercage ratio is known as the loading factor (‹s›) and can be used to control inter- versus intramolecularity (e.g., when ‹s› = 1/4, each carbene-filled supercage is (on average) surrounded by four vacant ones).

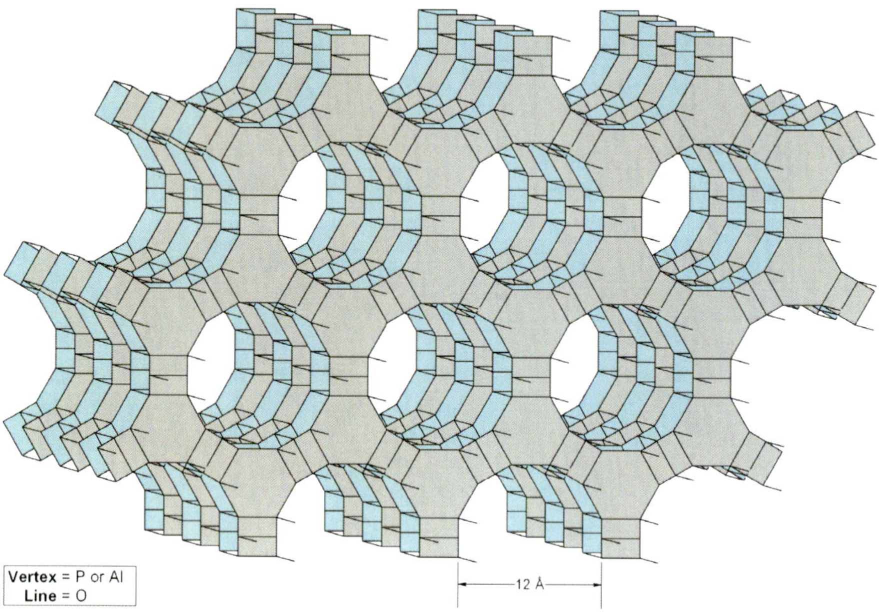

Plate 10.7. The aluminophosphate zeolite VPI-5 has channels that are 12 Å wide. It is more hospitable to acid-sensitive diazo compounds due to the exact balance of charge between the $(AlO_2)^-$ and $(PO_2)^+$ moieties (i.e., Al to P ratio = 1:1).

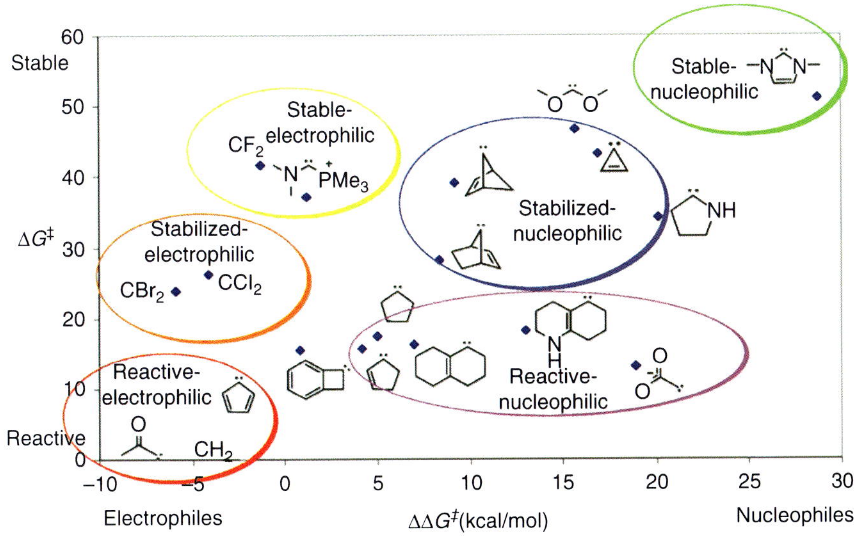

Plate 10.8. Carbenes may be categorized in six ways according to a 2-D surface that accounts for their reactivity and philicity. (Reprinted from Reference 44.)

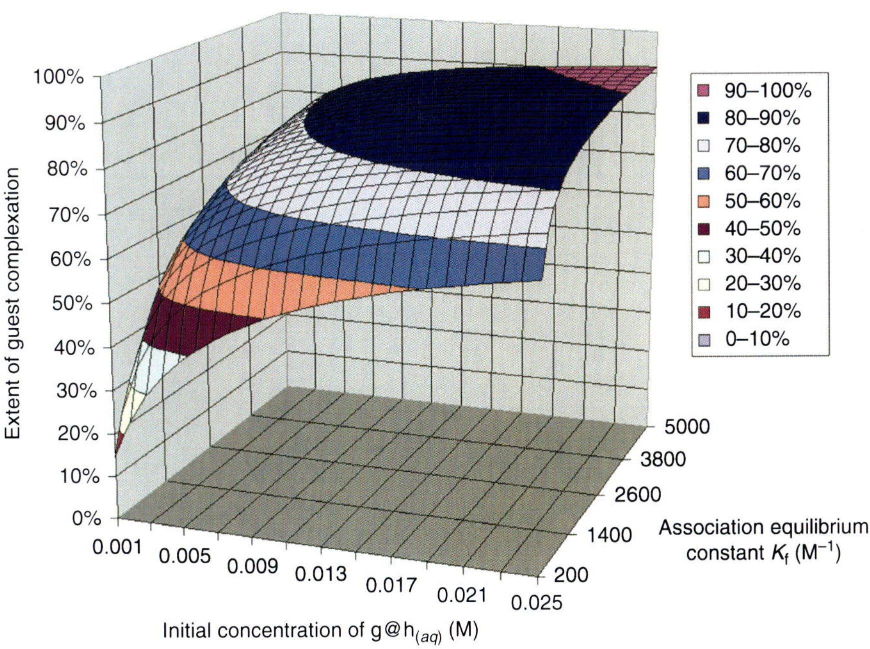

Plate 10.10. The extent of guest complexation reaches a plateau after a certain concentration or association complex equilibrium constant (K_f) of formation.

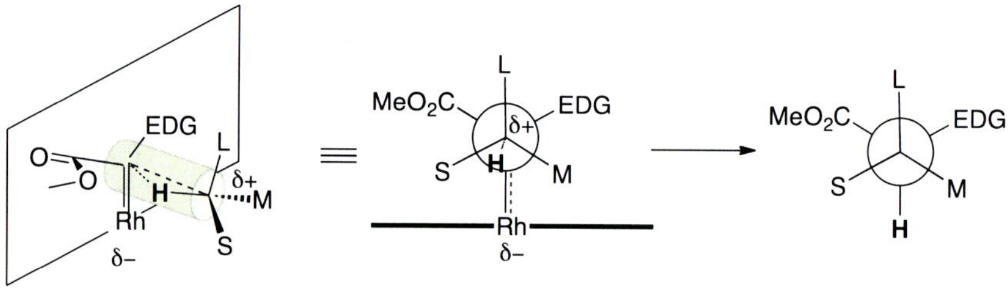

Plate 12.3. Transition state for C–H insertion.

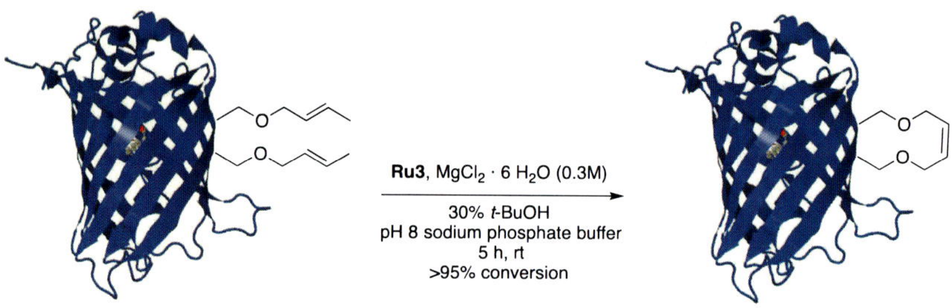

Equation 13.65.

154 Jackson, J. E., Mock, G. B., Tetef, M. L., Zheng, G., Jones, M., Jr. *Tetrahedron* **1985**, *41*, 1453–1464.
155 Mock, G. B., Jones, M., Jr. *Tetrahedron Lett.* **1981**, *22*, 3819–3822.
156 Shiue, G. H., Misslitz, U., Ding, X. t., Jones, M., Jr., De Meijere, A. *Tetrahedron Lett.* **1985**, *26*, 5399–5402.
157 Carpenter, B. K. *J. Phys. Org. Chem.* **2003**, *16*, 858–868.
158 Reprinted with permission from Litovitz, A. E., Keresztes, I., Carpenter, B. K. *J. Am. Chem. Soc.* **2008**, *130*, 12085–12094. Copyright 2008 American Chemical Society.
159 Becke, A. D. *Phys. Rev. A* **1988**, *38*, 3098–3100.
160 Lee, C., Yang, W., Parr, R. G. *Phys. Rev. B* **1988**, *37*, 785–789.
161 Cohen, A. J.: Handy, N. C. *Mol. Phys.* **2001**, *99*, 607–615.
162 Zhao, Y., Truhlar, D. G. *J. Phys. Chem. A* **2004**, *108*, 6908–6918.
163 Castaño, O., Frutos, L.-M., Palmeiro, R., Notario, R., Andrés, J. L., Gomperts, R., Blancafort, L., Robb, M. A. *Angew. Chem. Int. Ed.* **2000**, *39*, 2095–2097.
164 Helgaker, T., Uggerud, E., Jensen, H. J. A. *Chem. Phys. Lett.* **1990**, *173*, 145–150.
165 Helgaker, T., Uggerud, E. *J. Am. Chem. Soc.* **1992**, *114*, 4265–4268.
166 Tully, J. C. *J. Chem. Phys.* **1990**, *93*, 1061–1071.
167 Tully, J. C., Preston, R. K. *J. Chem. Phys.* **1971**, *55*, 562–572.
168 Ditchfield, R., Hehre, W. J., Pople, J. A. *J. Chem. Phys.* **1971**, *54*, 724–728.
169 Hariharan, P. C., Pople, J. A. *Theor. Chim. Acta* **1973**, *28*, 213–219.
170 Hehre, W. J., Ditchfield, R., Pople, J. A. *J. Chem. Phys.* **1972**, *56*, 2257–2261.
171 Reprinted with permission from Arenas, J. F., Lopez-Tocon, I., Otero, J. C., Soto, J. *J. Am. Chem. Soc.* **2002**, *124*, 1728–1735. Copyright 2002 American Chemical Society.
172 Yamamoto, N., Bernardi, F., Bottoni, A., Olivucci, M., Robb, M. A., Wilsey, S. *J. Am. Chem. Soc.* **1994**, *116*, 2064–2074.

6

ULTRAFAST KINETICS OF CARBENE REACTIONS

GOTARD BURDZINSKI

Quantum Electronics Laboratory, Faculty of Physics, Adam Mickiewicz University, Poznan, Poland

MATHEW S. PLATZ

Department of Chemistry, The Ohio State University, Columbus, OH, USA

6.1 INTRODUCTION

Diazirines and diazo compounds are the most common photochemical precursors of carbenes. Activation of these precursors produces singlet carbenes with relatively large but not unit quantum yields.[1–3] Details of photofragmentation and of the competing photochemical pathways in precursor excited states will be discussed in this chapter. The quantum yield of singlet carbene formation obviously decreases when the carbene precursor excited state undergoes decay through other pathways (photochemical and/or photophysical as internal conversion).

Singlet carbenes are reactive intermediates that undergo intermolecular reactions with solvent molecules or intramolecular processes such as isomerization or intersystem crossing (ISC) to form a triplet carbene when the triplet is the ground state. Often the triplet is the ground state of an aryl, alkyl, and/or many carbonyl substituted carbenes, while dialkyl carbenes and halo-substituted carbenes typically have closed-shell singlet ground states. In the latter examples, ISC does not contribute to singlet carbene decay.

Contemporary Carbene Chemistry, First Edition. Edited by Robert A. Moss and Michael P. Doyle.

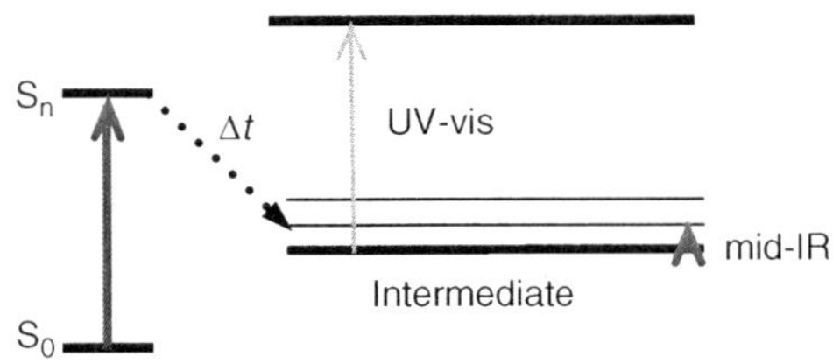

Scheme 6.1.

Direct observation of the early events in the photochemistry and photophysics of diazirines and diazo compounds can be achieved using ultrafast time-resolved spectroscopy. The experiment is triggered by the absorption of a pulse of UV radiation by the precursor, while the presence of intermediates can be probed by a delayed, second pulse of light (UV-Vis or infrared (IR), Scheme 6.1). On the basis of the recorded UV-Vis or IR transient absorption spectra, mechanistic, and kinetic aspects of the reaction can be deduced.

To complement ultrafast time-resolved measurements, the persistent photoproducts are characterized by stationary spectroscopy (NMR, IR, MS), and mechanistic interpretations are increasingly supported by quantum mechanical calculations.

This chapter will focus on recent, and in our subjective opinion, the most interesting results characterizing the carbene species formed by ultrafast photolysis of diazirine and diazo compounds obtained in our laboratories. Since keto/ester carbenes were recently reviewed elsewhere,[4] we will not discuss the ultrafast time-resolved photochemistry of diazocarbonyl compounds in this chapter. In recent years, many mechanistic aspects of photochemical reactions have been clarified, but new and important questions have also appeared, and they will stimulate future experimental and theoretical investigations.

6.2 ULTRAFAST UV-VIS STUDIES OF THE INTERMOLECULAR REACTIVITY OF *p*-BIPHENYLYLCARBENE (BpCH)

Arylcarbenes can be conveniently formed upon photolysis of diazo compounds (Scheme 6.2).[5] The nascent singlet carbene can undergo intermolecular reactions to form products (P_S) or relax by ISC to the lower energy triplet carbene. The latter species undergoes distinctly different chemical reactions than the singlet carbene and these reactions lead to a different spectrum of persistent products (P_T). Aryl carbene chemistry is controlled by the competition between carbene ISC rates k_{ST} and k_{TS}, and chemical reaction rates (k_s and k_T); see Scheme 6.2.

Our studies in this area commenced with *p*-biphenyl carbene systems[4–8] because their favorable spectroscopic properties facilitate direct observation with UV-Vis detection. The longer wavelength absorption spectra (compared to the phenyl derivative) of these diazo compounds and carbenes provide

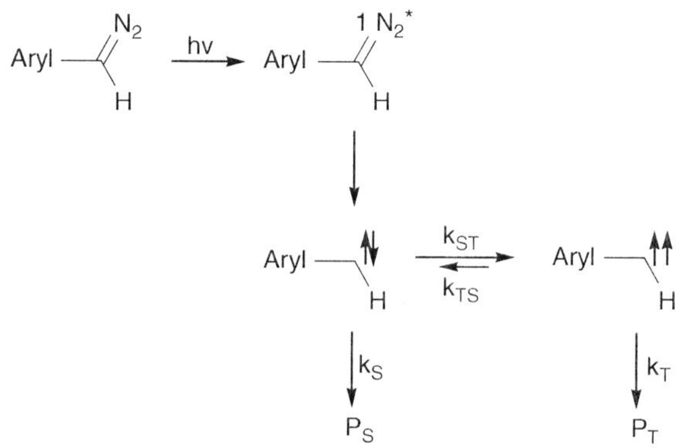

Scheme 6.2.

more detailed information than the phenyl analogues. In particular, a singlet *p*-biphenylylcarbene absorption band was detected at 360 nm in acetonitrile, in good agreement with the predictions of theory. The carbene precursor, an excited state of *p*-biphenyldiazomethane was observed as a short-lived species (less than 300 fs) with an absorption band at 470 nm (Fig. 6.1a).

As the diazo excited state decays, the growth of the singlet carbene absorption band at 360 nm is observed. The short (sub-ps) lifetimes $\tau_{diazo}*$ of excited state diazo compounds is typical of this functional group as revealed by ultrafast time-resolved studies. This property is actually an advantage for direct studies of singlet carbenes, because if $\tau_{diazo}* > \tau_{carbene}$, then the instantaneous concentration of carbene might well be too low to allow its facile detection.

The reactivity of singlet *p*-biphenylylcarbene in acetonitrile, cyclohexane, and methanol solvents has been studied (Scheme 6.3). The nature of the solvent strongly influences the *p*-biphenylylcarbene lifetime: 200 ps (in acetonitrile), 77 ps (in cyclohexane), and 7.9 ps (in methanol).[5] Scheme 6.3 shows the reaction products formed in these solvents.

In acetonitrile, a carbene-nitrile ylide is formed, which was detected by an absorption band peaking at 370 nm. The decay of such ylides has been previously observed on the microsecond timescale as a result of cyclization to form the corresponding azirine. In cyclohexane solvent, the formation of the triplet carbene is the dominant decay pathway. Intersystem crossing of singlet aryl carbenes will be discussed in detail subsequently. The C–H insertion reaction of the singlet carbene with cyclohexane proceeds efficiently, as deduced from proton NMR studies. This reaction mode transpires from a singlet–triplet equilibrium mixture of carbene in cyclohexane at ambient temperature.

In neat methanol, the carbene abstracts a proton from the solvent and the fast decay of the singlet carbene (7.9 ps) is accompanied by the rise of the cation (Fig. 6.1b) at 450 nm. The cation lifetime is only 6.7 ps, because the ion pair collapses to form a persistent ether product. However, cation formation, while

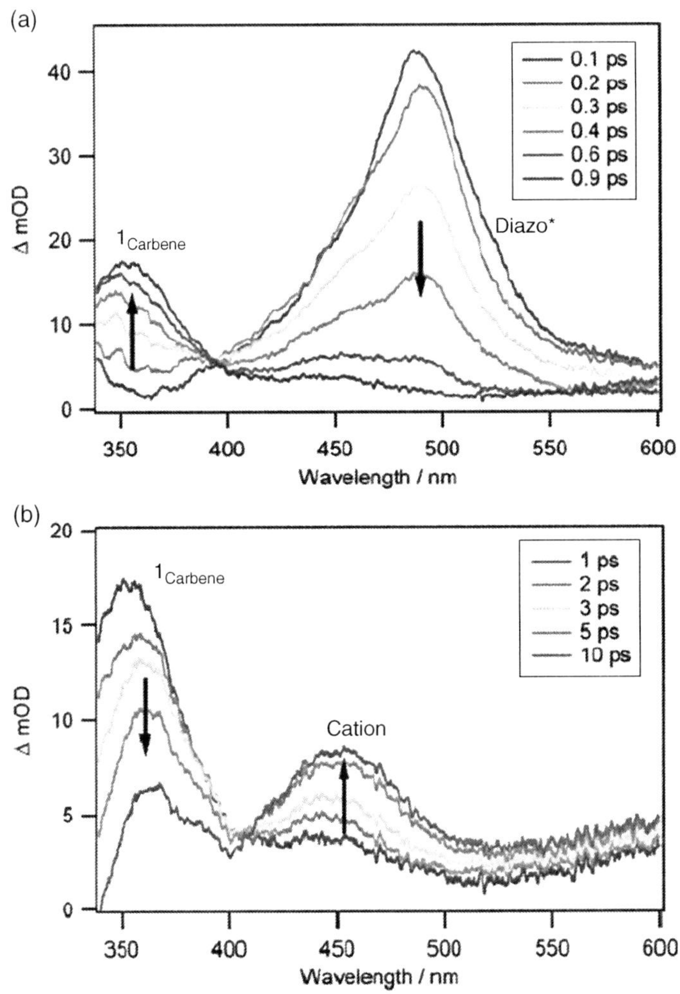

Figure 6.1. (a) Transient absorption UV-Vis spectra depicting the decay of the diazo excited state and the rise of the singlet carbene. (b) In methanol, the carbene forms the cation (λ_{exc} = 311 nm). (Reprinted with permission from Reference 5; Copyright 2006 by the American Chemical Society.) For color details, please see color plate section.

spectroscopically prominent, is likely a minor reaction path for this carbene, based on prior studies. Kohler's studies,[9] which considered a variety of mechanisms for the carbene–alcohol reaction, concluded that the direct but invisible carbene insertion into the OH bond is the major pathway in aryl carbene decay in methanol (Scheme 6.3). The stable methoxy ether product is a dark species in the 340–600 nm spectral window, in contrast to the UV-Vis active cation, which explains why the major pathway is not observed by time-resolved spectroscopy.

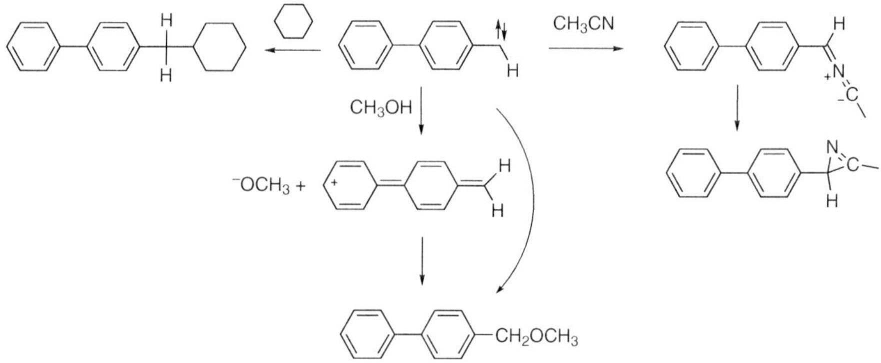

Scheme 6.3.

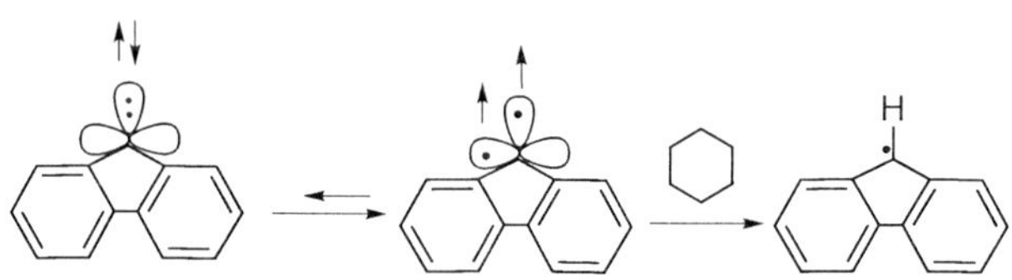

Scheme 6.4.

Scheme 6.3 shows a typical example of singlet carbene specific products. When ISC is a prominent decay channel of the singlet carbene, either the triplet carbene alone or an equilibrium mixture of singlet and triplet carbenes react with solvent. Picosecond time-resolved spectroscopy can then be used to monitor the formation of triplet-derived (free radical) product. For example, fluorenylidene has a small singlet–triplet energy separation in cyclohexane (calculations predict ~2.36 kcal/mol). The singlet to triplet ISC time constant is ~100 ps in this solvent, thus singlet–triplet equilibrium is realized 500 ps after the laser pulse. On these timescales, the equilibrium populations of the singlet and triplet states react with the cyclohexane solvent over 2–3 ns (Scheme 6.4).[10,11] The equilibrium population of singlet fluorenylidene reacts by spectroscopically invisible CH insertion, but a hydrogen abstraction reaction of the triplet carbene proceeds to form the 9-fluorenyl radical with a sharp absorption maximum at 500 nm.[11] The small solvent (cylcohexane versus cyclohexane-d_{12}) isotope effect on the lifetime of the spin-equilibrated singlet–triplet carbene mixture argues that C–H insertion reactions are the dominant but not exclusive decay mode of certain spin-equilibrated carbenes in hydrocarbon solvents.[11–13] Diaryl carbenes have larger singlet–triplet separations because of their wide bond angles due to steric factors.[14,15] Lack of a solvent isotope effect indicates that these spin-equilibrated carbenes decay by triplet carbene hydrogen atom abstraction.[7]

6.3 REARRANGEMENTS IN THE EXCITED STATE OF THE CARBENE PRECURSOR

On the basis of extensive photochemical studies of diazo compounds and diazirines, numerous chemists have proposed that formal carbene rearrangement products can be formed by at least two pathways:[16,17] (i) directly from the excited state of the precursor ("Rearrangement in the Excited State," termed RIES by Liu[18]) or (ii) from a free carbene species.[19] Historically, the dual mechanisms were deduced from studies of persistent photolysis products. Support for the RIES mechanism was provided by the observation of a "non trappable carbene" route to rearranged products.[16,17,19]

Recently, three different ultrafast time-resolved studies have been reported that are consistent with the RIES mechanism (concerted molecular nitrogen extrusion and rearrangement): (i) the cyclization of a diazoamide,[20] (ii) a Wolff rearrangement of *syn* diazo ketones,[4] and (iii) 1,2 hydrogen shifts in alkyl diazo[6,21] and diazirine compounds[22] (Scheme 6.5).

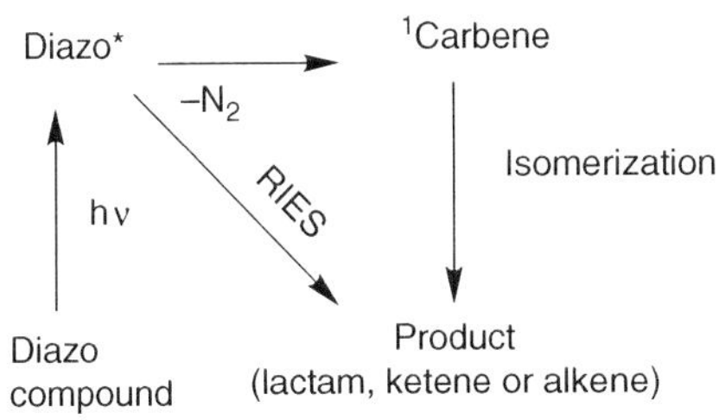

Scheme 6.5.

The latter experiment will be considered in some detail as the diazo carbonyl processes have been reviewed elsewhere. Generally, these studies indicate that an ultrafast formation rate (<0.4 ps) of product (lactam, ketene, or alkene) is too fast to attribute to an electronically or vibrationally excited carbene process and is most consistent with an RIES pathway. A stepwise process involving carbene, if operative, should result in a slower rise of the product. In such a case, the factor limiting the dynamics is the (hot or relaxed) carbene lifetime, and the shortest lifetime measured to date is 2.3 ps.[23] This is too slow to be associated with a diazo excited state (RIES) process as diazo excited states have sub-ps lifetimes.

A carbene with an intrinsically short lifetime (~2 ps) is difficult to intercept bimolecularly (e.g., trapping by alcohol); in such cases the origin of stable reaction products is ambiguous without ultrafast (sub-ps) detection of the formation of rearranged product.[23]

Initial ultrafast time-resolved UV-Vis studies involving 1,2 hydrogen shift reactions were performed with *para*-biphenylyldiazoethane ($BpCN_2CH_3$) and *para*-biphenylyldiazomethane ($BpCN_2H$) under identical conditions to allow meaningful comparison of the data.[6] The experimentally observed absorption bands at 360 nm were assigned to the corresponding singlet carbenes with lifetimes ranging from a few hundred picoseconds in acetonitrile to ~10 ps in a carbene-scavenging solvent such as methanol. In three solvents (acetonitrile, cyclohexane, and methanol) the observed quantum yield of formation of the carbene $^1BpCCH_3$ was 30–40% lower than that of ^1BpCH, assuming that the two carbenes have similar extinction coefficients, as indicated by Time-dependent density functional theory (TD DFT) calculations. This result was taken as an example of an RIES process; a process which leads directly from the diazo excited state to *para*-vinylbiphenyl (Scheme 6.6), and as a result decreases the yield of carbene.

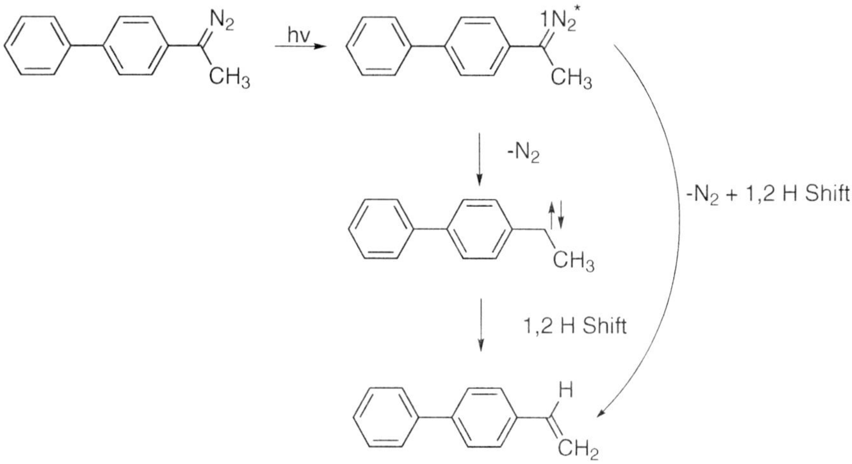

Scheme 6.6.

Subsequent studies were performed with alkyl diazo esters to detect the rise of the alkene product using ultrafast time-resolved IR spectroscopy by monitoring the C=O vibration of the vinyl ester product.[21] Note, that the ester group is less prone to Wolff rearrangement and ketene formation (due to the loss of the ester resonance energy) relative to the corresponding diazo ketone. In the case of the photochemistry of methyl 2-diazopropionate ($CH_3CN_2CO_2CH_3$) two distinctive spectral features are observed (see Fig. 6.2a): (i) a negative band at 1693 cm^{-1} corresponding to the diazo bleaching signal without any signal evolution due to lack of a diazo ground state recovery process, thus $\Phi_{IC}=0$, and (ii) a positive wide vinyl ester band (C=O stretching), initially observed at 1660 cm^{-1} with subsequent frequency upshift to 1733 cm^{-1} due to vibrational cooling over a 70 ps time window (Fig. 6.2b).

The presence of substantial alkene absorption band intensity at short delay times (~1 ps) indicates that the RIES process is the major pathway of alkene formation. The intermediacy of a singlet carbene was not observed in this system by spectroscopic methods. Photolysis (270 nm) of this diazo compound in neat methanol failed to produce the ether expected from carbene trapping, further arguing against photochemical fragmentation of the diazo precursor to

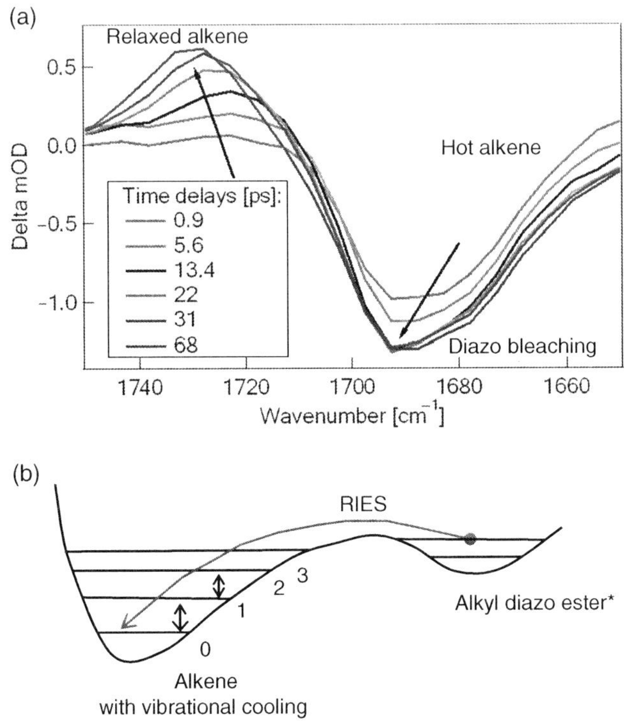

Figure 6.2. Ultrafast rise of a vinyl ester species (<0.4 ps) due to RIES after photoexcitation of methyl 2-diazopropionate at 266 nm (a) and alkene vibrational cooling (b). (Reprinted with permission from Reference 21; Copyright 2010 by the American Chemical Society.) For color details, please see color plate section.

form free carbene. Future work should attempt to observe alkenes formed by both RIES and relaxed carbenic processes in order to understand how structure and wavelength control the relative proportions of these two pathways.

6.4 DYNAMICS OF CARBENE VIBRATIONAL COOLING AND SOLVATION

When diazo or diazirine compounds are employed as photochemical precursors, one commonly observes that carbenes are born with an excess of vibrational energy and undergo subsequent cooling by collisions with surrounding solvent molecules. Vibrational cooling proceeds with a time constant of a few (5–30 ps) picoseconds.[8] This phenomenon is clearly indicated by the observation of carbene absorption band narrowing (in both the UV-Vis and IR transient absorption experiments) and is sometimes accompanied by a blue shift of the absorption maximum.[8,16,24,25]

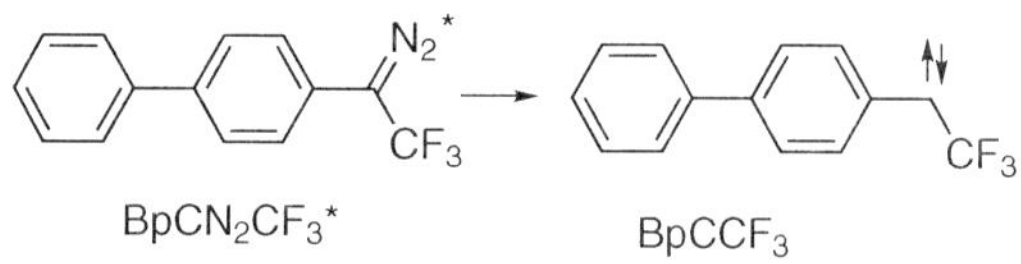

Scheme 6.7.

We prepared and studied the photochemistry of *para*-biphenylyltrifluoromethyl diazomethane $BpCN_2CF_3$ (Scheme 6.7), which is not expected to undergo 1,2 shift reactions. We expected to compare its photochemistry with $BpCN_2CH_3$, but found a more interesting effect.

Vibrational cooling is clearly visible in nonpolar solvents (such as cyclohexane, Fig. 6.3a), while in the polar solvent methanol (Fig. 6.3b) this process is accompanied by solvation.[8] Solvation in polar solvents is evidenced by a red shift of the carbene absorption band, typically over a 1–15 ps time window in UV–vis transient absorption experiments. An analogous red shift has also been observed for the carbene carbonyl stretch in 2-naphthyl(carbomethoxy)carbene using ps time-resolved IR spectroscopy.[2]

The dynamics of carbene solvation were studied in detail using ultrafast time-resolved UV-Vis spectroscopy with *p*-biphenylyltrifluoromethylcarbene ($BpCCF_3$, Scheme 6.7) in solvents of different polarity, including solvents that form hydrogen bonds.[8] In a polar solvent (acetonitrile), the initial carbene absorption band undergoes a red shift from 400 to 410 nm with a time constant of 0.4 ps. In a more interacting and more polar solvent (methanol), the red shift is even more dramatic (from 394 to 427 nm). Solvation proceeds more slowly in the hydrogen bonding solvent, with a time constant ~10 ps (Fig. 6.3b). These observations are thought to be a consequence of solvent–carbene interactions. The singlet carbene $BpCCF_3$ has a closed-shell electronic structure with a filled and an empty nonbonding orbital on the carbene carbon. The presence of the trifluoromethyl group makes the carbene center electron-deficient, which enhances the strength of the interaction between carbene's empty *p*-orbital and the nonbonding electrons of a coordinating solvent such as acetonitrile. In methanol, the red shift is more dramatic than in acetonitrile, because of the greater polarity in the former solvent, and because the alcohol solvent can interact with both the filled and the empty nonbonding orbitals on the carbene carbon (Scheme 6.8).

The oxygen atom of the alcohol will interact with the empty orbital and the acidic alcohol proton (of the same or yet another alcohol molecule) can coordinate with the filled in-plane nonbonding orbital of the carbene. There is a kinetic O–H/O-D isotope effect on the red shift in alcohol solvents revealing that solvation is associated with the formation of a new hydrogen bond. Additionally, experiments performed with a homologous series of alcohols show a linear dependence between solvent viscosity and the time constant of the red shift (from ten to hundreds of ps, Fig. 6.4).

This observation is in agreement with intuition that solvent viscosity plays an important role in the rate of solvation. A similar pattern was realized for

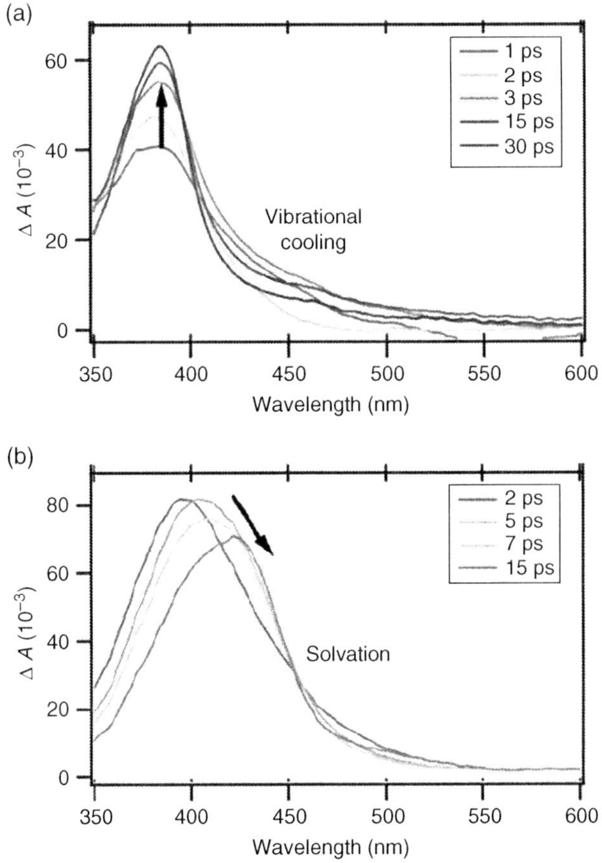

Figure 6.3. Carbene absorption spectra in (a) nonpolar cyclohexane and in (b) polar and coordinating methanol solvent produced by ultrafast photolysis ($\lambda_{exc} = 308\,nm$) of *p*-biphenylyltrifluoromethyl diazomethane. (Reprinted with permission from reference 8; Copyright 2008 by the American Chemical Society.) For color details, please see color plate section.

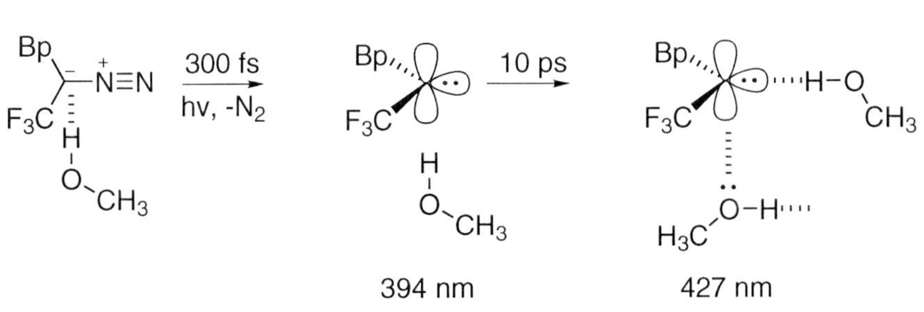

Scheme 6.8.

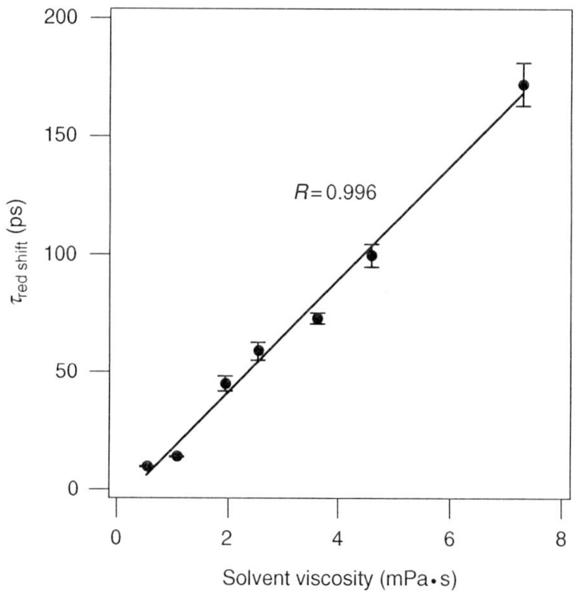

Figure 6.4. Dependence of *p*-biphenylyltrifluoromethylcarbene red shift time constant on alcohol solvent viscosity. (Reprinted with permission from Reference 8; Copyright 2008 by the American Chemical Society.)

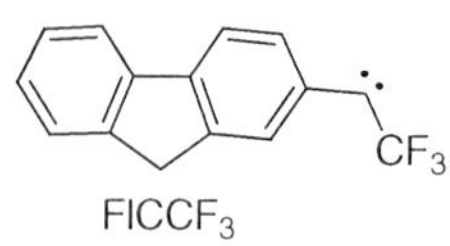

Scheme 6.9.

2-flurorenyltrifluoromethylcarbene ($FlCCF_3$, Scheme 6.9), a rigid analogue of $BpCCF_3$, so the observations assigned to solvation and vibrational cooling cannot be explained by the torsional motions of the latter's two phenyl rings.[8]

6.5 INFLUENCE OF SOLVENT ON CARBENE INTERSYSTEM CROSSING RATES

The first systematic study of the influence of solvent on carbene singlet to triplet ISC rates was reported by Eisenthal and coworkers with diphenylcarbene.[26–29] In this pioneering work, a linear dependence was found between the log of the first-order rate constant of ISC versus the Dimroth $E_T(30)$ parameter. This study utilized a variety of nonhalogenated solvents. The fastest ISC rates were observed in nonpolar solvents with correspondingly small $E_T(30)$ values. In nonpolar solvents, the S–T energy gap is relatively large, because the zwitterionic closed-shell singlet carbene is poorly stabilized relative to the diradical-like triplet spin isomer. Thus, the fast ISC rate in nonpolar solvents apparently contradicts the "Golden Rule"[30–32] of radiationless transitions, which states that smaller energy separations promote faster radiationless transitions. Eisenthal and coworkers accepted that polar solvents preferentially stabilize the singlet state relative to the triplet state of the arylcarbene, as confirmed later by theoretical[33,34] and experimental studies.[5–8,11] To explain the slower than expected ISC rates in polar solvents, the Eisenthal group concluded that the diphenylcarbene system fortuitously suffered from poor S–T vibronic coupling.

Recent ultrafast time-resolved studies demonstrate that ISC rates do not solely depend on bulk solvent polarity, but respond to specific interactions between the carbene and solvent molecules. Two alternative scenarios were proposed[7] to explain the retardation of the rate of ISC for fluorenylidene in its lowest, closed-shell singlet configuration, in coordinating solvents (Fig. 6.5).

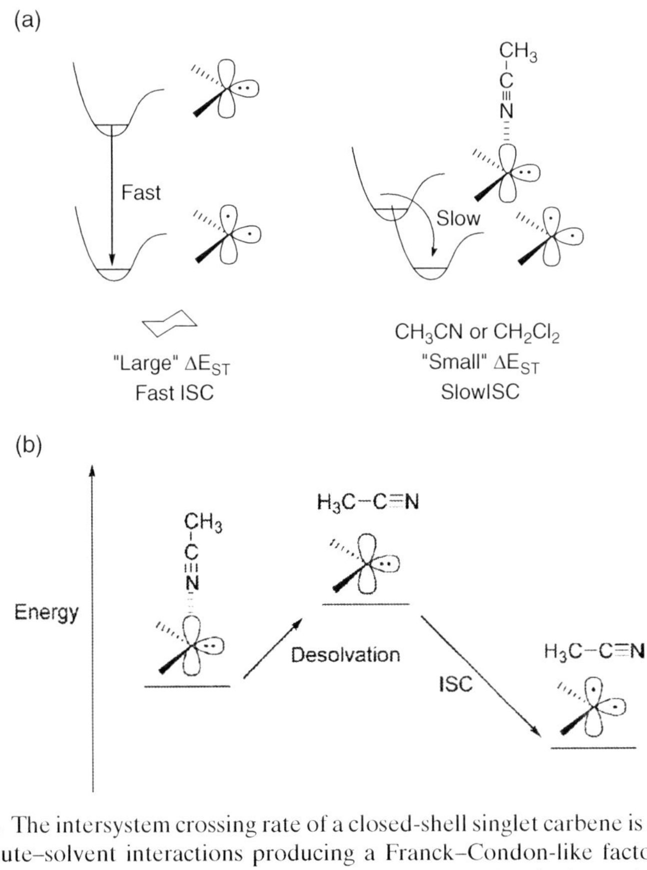

Figure 6.5. The intersystem crossing rate of a closed-shell singlet carbene is decelerated by: (a) solute–solvent interactions producing a Franck–Condon-like factor or (b) an energy barrier to desolvation to be surmounted to release the singlet carbene prior to ISC. (Reprinted with permission from Reference 7; Copyright 2008 by the American Chemical Society.)

In the case of a polar solvent such as acetonitrile, the significantly different solvation of the singlet and triplet carbenes may produce a Franck–Condon-like factor (Fig. 6.5a), because the carbene–solvent complex will be structurally different in the two spin states. Alternatively, (Fig. 6.5b), the rate limiting step might be the surmounting of a small energy barrier to release "free" carbene prior to the ISC process. The latter explanation is reminiscent of the well-known ability of protic solvents to depress S_N2 reaction rates relative to dipolar aprotic solvents.[35]

Table 6.1 reveals that the ISC rates in (polar) acetonitrile are slower than the rates in either cyclohexane or benzene, two nonpolar, zero dipole moment solvents. However, a slow ISC rate, comparable to acetonitrile, was found[7,11] in a nonpolar, halogenated solvent (hexafluorobenzene) with zero dipole moment. These data were explained by the formation of a pseudo ylide complex by interaction between the empty p orbital of the singlet carbene with a nonbonding pair of electrons of the halogen. This type of solvation was first proposed by Turro for the simplest singlet carbene, methylene.[36,37]

In an aromatic solvent, another possible type of interaction is the formation of a π complex stabilizing the singlet carbene.[38] Structural differences between the carbene spin isomers, singlet versus triplet, may also play a role in retarding the ISC rate, as observed in case of *para*-biphenyl carbene ester.[16]

It should be emphasized, that with singlet diphenylcarbene and fluorenylidene in these solvents, ISC to the ground triplet state is the main deactivation pathway.[7,11]

We note that the carriers of the transient spectra attributed to the lowest energy singlet and triplet carbenes can be assigned with considerable confidence in a manner not possible 20 years ago. Time-dependent density functional theory calculations can be used to support the assignment of transient absorption spectra to singlet or triplet carbenes with reliability. The assignment of spectra to triplet carbenes is easily confirmed in cryogenic matrix isolation experiments or by their sensitivity to oxygen in nanosecond laser flash photolysis experiments. In a similar fashion, the sensitivity of singlet carbene lifetimes to neat alcohols confirms the assignment of transient absorption bands to these species.

TABLE 6.1. Singlet Carbene Lifetimes (Intersystem Crossing Rates)[7–11] in Various Solvents and Associated $E_T(30)$ Parameters[39]

Carbene	Solvent	Lifetime (ps)	$E_T(30)$ (kcal/mol)
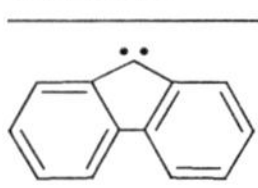	Acetonitrile	440 ± 40	45.6
	Cyclohexane	88 ± 11	30.9
	Hexafluorobenzene	490 ± 90	34.2
	Benzene	100 ± 12	34.3
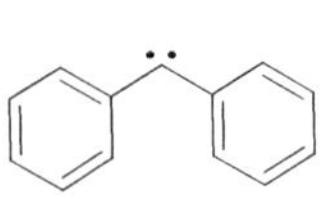	Acetonitrile	340 ± 20	45.6
	Cyclohexane	120 ± 10	30.9
	Hexafluorobenzene	260 ± 20	34.2
	Benzene	200 ± 7	34.3

6.6 ELECTRONICALLY EXCITED (OPEN SHELL) SINGLET CARBENES

In ultrafast time resolved experiments, we commonly observe that singlet carbenes are born in a vibrationally excited state (#) when diazo compounds or diazirines are used as photochemical precursors. An intriguing question is whether the carbene is initially formed in the lowest (closed shell) zwitterionic-like singlet electronic state, or in a higher (open shell) diradical-like excited state (*). Typically, the C=N double bond energy in a diazo compound is low (e.g., 30.6 kcal/mol in diazomethane).[6] Thus, when UV excitation is used (~310 nm corresponds to 92 kcal/mol) the formation of a ^{1}carbene* species is energetically feasible and must be considered.[6,9] We have postulated that ^{1}fluorenylidene* (^{1}Fl*) is produced upon UV photolysis of 9-diazofluorene (Fig. 6.6).

The carrier of the decaying band at 370 nm was tentatively assigned to ^{1}Fl* (see Fig. 6.6), and the rising band at 420 nm to the ^{1}carbene in the lowest (closed shell) singlet state.[11] The estimated lifetime of the open-shell singlet carbene was 20 ps.

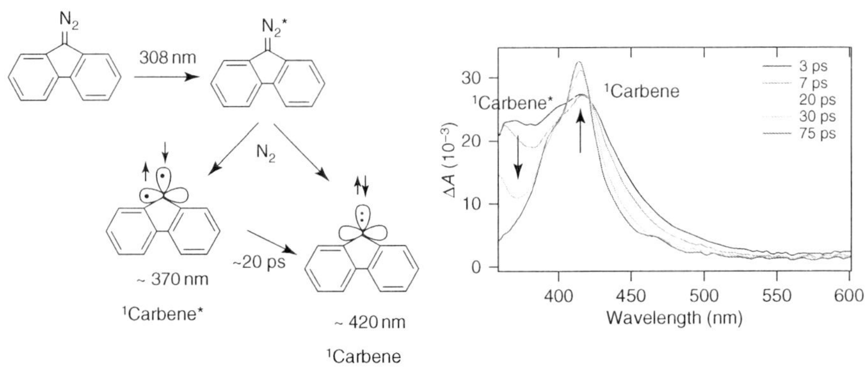

Figure 6.6. Postulated mechanism for the decay of photoexcited 9-diazofluorene* (the asterisk denotes the singlet excited state). The time-resolved UV-Vis transient absorption spectra were recorded over a 3–75 ps time window. (Reprinted with permission from Reference 11; Copyright 2007 by the American Chemical Society.) For color details, please see color plate section.

6.7 PARENT PHENYLDIAZIRINE—MECHANISTIC ASPECTS OF SINGLET CARBENE FORMATION

Early events in photochemistry of simple aryldiazirines leading to nitrogen extrusion and the formation of singlet carbenes have been studied by ultrafast time-resolved spectroscopy.[24,40] Mechanistic aspects of carbene formation

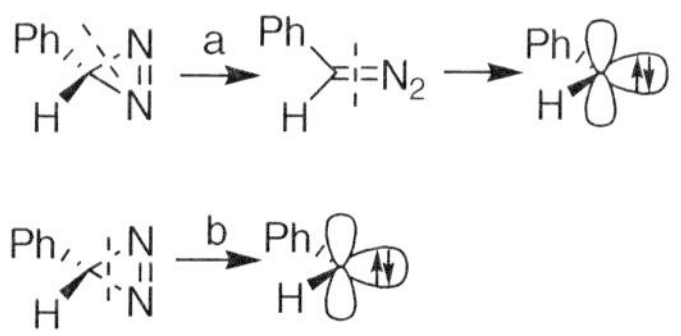

Scheme 6.10.

from phenyldiazirine have also been recently investigated by theory.[41] Two alternative decomposition pathways leading to carbene formation (Scheme 6.10) were considered. Pathway (a) is suggested by intuition, as single C–N bond cleavage should be more facile than simultaneous cleavage of two C–N bonds. This simple analysis was recently confirmed by advanced calculations.[41] These recent theoretical findings will be used in the following to interpret previously published experimental results.[24,40]

On the basis of calculated UV-Vis transition energies (TD-B3LYP and RI-CC2) one can assume that 375 and 270 nm excitation promotes ground state phenyldiazirine to the S_1 and S_2 states, respectively. Ultrafast photolysis at 375 nm of phenyldiazirine in solution results in a broad transient band peaking at about 625 nm, which can be assigned to the absorption of the diazirine in the first excited singlet state (S_1). The signal decay is biexponential, with a fast component of 1–4 ps corresponding to the decay of hot S_1-diazirine, and a second component (13.3, 46.2, and 61.6 ps in cyclohexane, chloroform, and acetonitrile, respectively) assigned to the lifetime of thermalized S_1-diazirine (Fig. 6.7a).[40]

These results clearly indicate that the S_1 state lifetime of phenyldiazirine increases with increasing solvent polarity. This effect is not surprising, since the calculated dipole moment for the S_1 state of the diazirine is large (4.75 D) relative to the ground state (1.82 D).[41] The initial broad, red-located absorption band assigned to the S_1 state of phenyldiazirine was also observed for halophenyl diazirines[40,42] and *p*-methoxy-3-phenyl-3-methyldiazirine.[1] In the case of 270 nm excitation, the S_2 state of phenyldiazirine is formed initially and it undergoes ultrafast internal conversion from $S_2 \rightarrow S_1$ (Scheme 6.11).

The RI-CC2/TZVP calculation predicts that the S_2 state is largely $\pi \rightarrow \pi^*$ in nature, and is localized primarily on the benzene ring. The S_2-diazirine reactivity, either diazo or carbene formation, is lower than the S_1 state, since the calculated C–N bonds are stronger and shorter by 0.6 Å in the higher energy state. When phenyldiazirine is excited with a short-wavelength laser pulse (270 vs. 375 nm), the observation of a larger contribution of the short component (1–4 ps) in the kinetic trace, characteristic of a hot S_1-diazirine, is consistent with the presence of the $S_2 \rightarrow S_1$ internal conversion channel.[42]

Even though vibrational relaxation can be invoked to rationalize the fast decay of the S_1 state of phenyldiazirine another process, a *hot diazirine → diazo* isomerization mechanism in the vibrationally (and electronically) excited S_1 state manifold, must also be considered. The isomerization process is in agreement with the predicted $\sigma_{C-N} \rightarrow \pi^*_{N=N}$ transition to the S_1 state which is

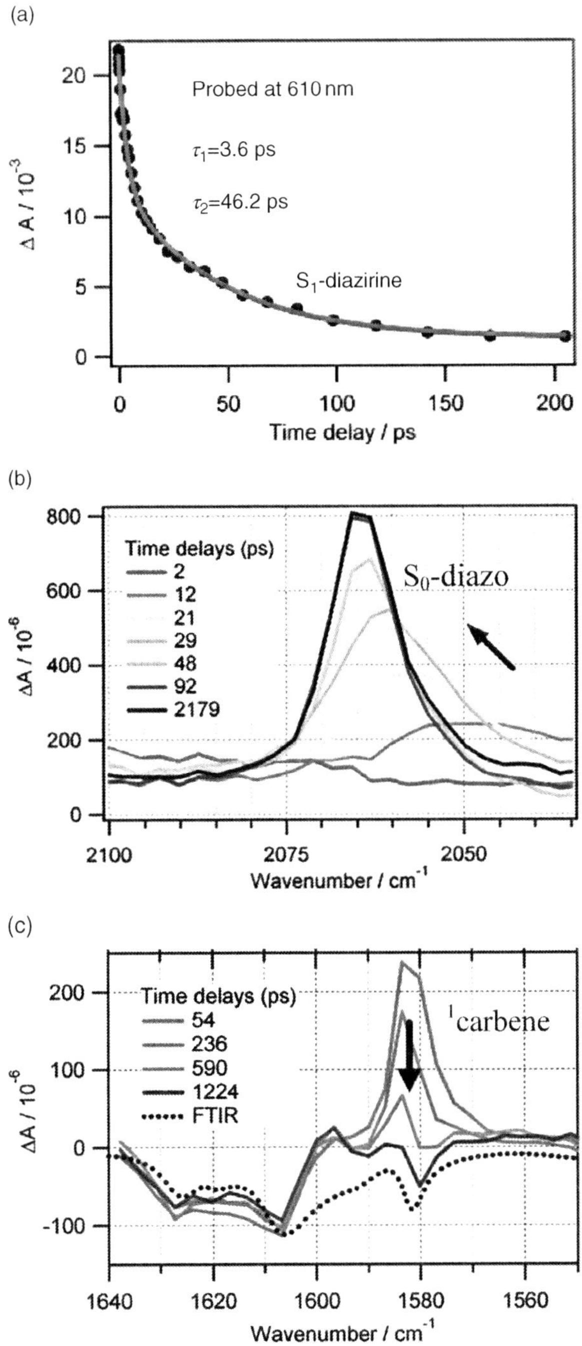

Figure 6.7. Ultrafast results recorded for phenyldiazirine using UV-Vis and IR spectroscopies: (a) the decay of the S_1-diazirine in chloroform (λ_{exc}=375 nm), (b) the rise of the S_0-diazo from S_1-diazirine isomerization in acetonitrile, and (c) the ^{1}carbene decay in chloroform (λ_{exc}=270 nm). (Reprinted with permission from References 24 and 40; Copyright 2009 by the American Chemical Society.) For color details, please see color plate section.

localized on the diazirine ring. As a result of the excitation energy resident on the three-membered ring, the predicted activation energy barrier to ring opening is very small (2.5 kcal/mol) in the first excited state.[41]

Excess vibrational energy in a weakly stabilized S_1 state of a diazirine can accelerate the isomerization process, as reported for other classes of molecules.[43,44] The isomeric diazo compound, in its first electronically excited singlet state, is formed rapidly from the vibrationally hot S_1 excited state of the diazirine, and/or more slowly from the thermalized S_1 state of phenyldiazirine (Scheme 6.11). The S_1 state of the diazo intermediate has not been observed in ultrafast UV-vis and IR studies with diazirine precursors, likely due to a low instantaneous concentration (assuming a short lifetime <300 fs, a typical value for other excited diazo compounds, see Section 6.2).

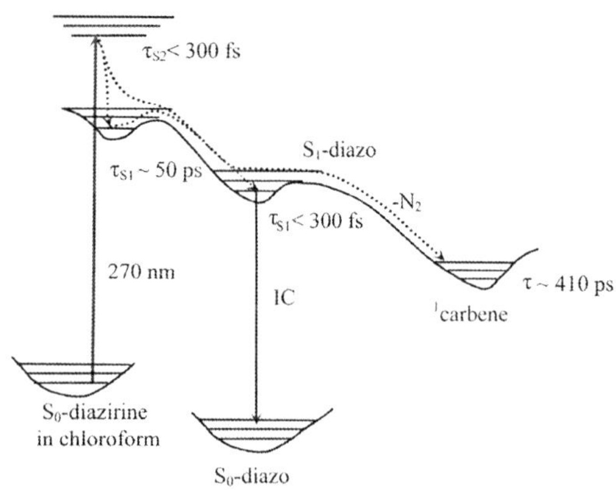

Scheme 6.11.

The decay of the S_1 state of the diazo compound can proceed by at least three pathways: (i) by $S_1 \rightarrow S_0$ internal conversion; (ii) by fragmentation leading to a closed-shell singlet carbene and N_2; and (iii) by fragmentation to an open-shell singlet carbene and nitrogen, as described for fluorenylidene in Section 6.6. The formation of two products, S_0 (ground state) diazo and closed-shell singlet carbene, was observed in the phenyldiazirine experiments. Open-shell singlet phenylcarbene was not detected. The isomerization to form ground state (S_0) diazo compound was monitored using time-resolved IR experiments (N=N stretching at 2064 cm^{-1}, Fig. 6.7b). The formation of closed-shell singlet phenylcarbene absorption was shown by a characteristic vibrational band (C=C at 1582 cm^{-1}) after photolysis of phenyldiazirine with 270 nm excitation. The fragmentation of the S_1 state of the diazo compound is supported by calculations[41] which predict that the S_1 state ($\pi_{C=N} \rightarrow \pi^*_{N=N}$) is the dissociative state, and its denitrogenation process requires only 0.3–2.7 kcal/mol.

Theory also predicts there will be a large excess of vibrational energy (over 50 kcal/mol) in nascent S_1-diazo compound as a result of highly exothermic S_1-diazirine isomerization[41] and this might facilitate carbene formation. An intense singlet carbene IR band area is formed within a few picoseconds of the laser pulse,[24] so that the precursors of singlet phenylcarbene are themselves short-lived reactive intermediates (270 nm excitation).

Singlet phenylcarbene decays with a lifetime of 414 ps in chloroform (Fig. 6.7c), as a result of ISC to the triplet ground state and/or reaction with solvent by Cl atom abstraction.[24] Neither triplet phenylcarbene nor α-chlorobenzyl radical were experimentally observed,[24] presumably due to their low IR intensities.

6.8 INFLUENCE OF HALO-SUBSTITUENT ELECTRON-DONATING CAPACITY ON DIAZIRINE DECAY IN THE FIRST EXCITED SINGLET STATE

Prior to the ultrafast time-resolved studies in our laboratory, very little was known about the photophysics and photochemistry of arylhalodiazirine precursors.[45] The first experiments using ultrafast time-resolved UV-Vis spectroscopy[42] with halophenyldiazirines detected transients with $\lambda_{max} \approx$ 600–700 nm with picosecond lifetimes (an example is shown in Fig. 6.8a), tentatively assigned to a ring-opened diradical/zwitterion (Scheme 6.12). Simple intuition based on the dependence of transient lifetimes on the electron-donating capacity of halo-substituent X (Table 6.2), and the solvent polarity (Tables 6.2 and 6.3), led to this interpretation.

Further experimental[1,40] and later theoretical studies[40,41] modified this assignment. Presently, the S_1 excited state of the diazirine is predicted to have two equivalent C–N bonds, shown in Figure 6.9, and is thought to be the carrier of the red-absorbing transient. Clearly, polar solvents such as acetonitrile, should kinetically stabilize the S_1 state of the diazirine. This is consistent with RI-CCR/TZVP predictions of an increase in the dipole moment by 2.5 D upon $S_0 \rightarrow S_1$ promotion in chlorophenyldiazirine.[40]

In the case of the fluoro derivative in acetonitrile, the decay of the S_1-diazirine is monoexponential, in contrast to the biexponential behavior observed in cyclohexane and for other diazirine derivatives in both solvents (Table 6.2). The monoexponential kinetic trace (Fig. 6.8b) can be explained by the absence of a fast channel in the hot S_1-diazirine decay (Scheme 6.11). This might be a consequence of the larger energy barrier for diazirine–diazo isomerization in the S_1 state manifold deduced from the Arrhenius equation, as the S_1 lifetime in acetonitrile is exceptionally long (~350 ps).

RI-CC2/TZVP calculations performed for chlorophenyldiazirine in its S_1 state indicate that it has a "quinoidal" structure where the diazirine C–N bonds are lengthened by 0.8 Å compared to the S_0 state (Fig. 6.9). It is clear, therefore, that the S_1 state is disposed toward dissociation.[40] Calculations of Mulliken charge distribution predict that substantial negative charge is localized on the sterically unencumbered nitrogens (−0.12*e*, −0.13*e*) in the S_1 state. Thus, it is

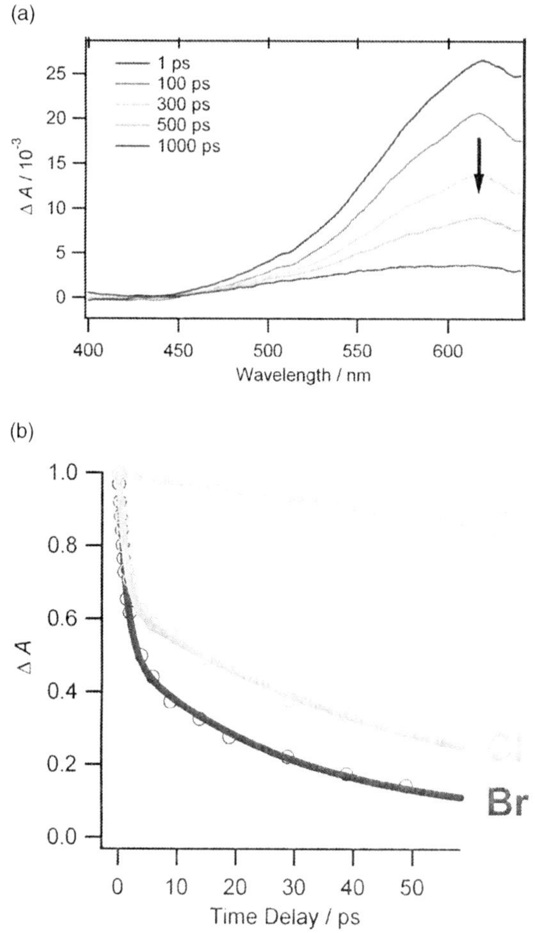

Figure 6.8. (a) Decay of the S_1-fluorophenyldiazirine absorption band in acetonitrile, generated by ultrafast photolysis at 360 nm. (b) Normalized kinetic traces of fluorophenyldiazirine, chlorophenyldiazirine, and bromophenyldiazirine in acetonitrile. (Reprinted with permission from Reference 42; Copyright 2006 by the American Chemical Society.)

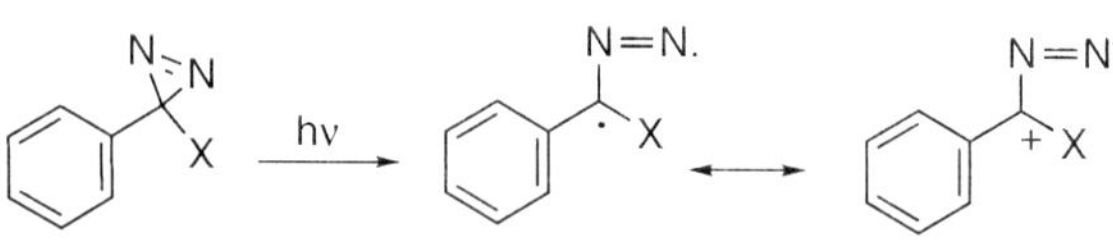

Scheme 6.12.

likely that these atoms play an important role in interaction with solvent dipoles (solvation).

The observed dependence of the relaxed diazirine S_1 state lifetimes on the halo substituent can be explained by resonance donation from X to the positively charged diazirine carbon ($+0.09e$).[40]

TABLE 6.2. Decay of Halophenyldiazirine in the First Excited State Measured at 600 nm After Photolysis at 360 nm

Halogen	Solvent	A_1	Hot S_1 Diazirine Lifetime τ_1 (ps)	A_2	Relaxed S_1 Diazirine Lifetime τ_2 (ps)	A_∞
F	Acetonitrile			0.89	352 ± 27	0.11
Cl	Acetonitrile	0.41	0.92 ± 0.1	0.52	50 ± 4	0.07
Br	Acetonitrile	0.53	0.88 ± 0.1	0.42	28 ± 2	0.05
F	Cyclohexane	0.34	2.4 ± 0.4	0.57	117 ± 13	0.10
Cl	Cyclohexane	0.58	0.45 ± 0.1	0.37	22 ± 3	0.06
Br	Cyclohexane	0.79	0.32 ± 0.2	0.16	10 ± 3	0.04

TABLE 6.3. Vibrationally Relaxed S_1 Lifetimes of *p*-Y-Substituted Chlorophenyldiazirines

		Relaxed S_1 Lifetime (ps)		
Y	σ_p^+	Acetonitrile	Chloroform	Cyclohexane
p-MeO	−0.78	760 ± 22	644 ± 78	132 ± 12
p-CH_3	−0.31	150 ± 8.1	114 ± 8.3	25.8 ± 1.6
p-H	0.00	61.6 ± 7.4	46.2 ± 5.9	13.3 ± 3.1
p-Cl	0.11	50.0 ± 5.9	40.6 ± 8.1	8.0 ± 0.8
p-CF_3	0.53	10.6 ± 0.7	6.9 ± 0.8	2.5 ± 0.2

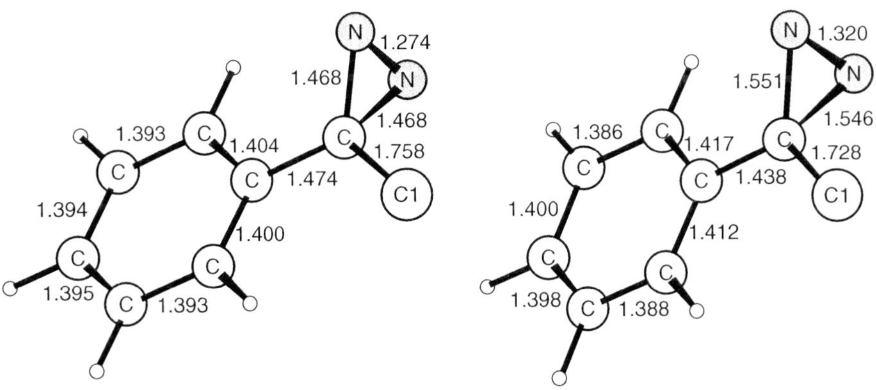

Figure 6.9. Optimized geometries of S_0 (left) and S_1 (right) states of chlorophenyldiazirine; the bond lengths are in Ångstroms. (Reprinted with permission from Reference 40; Copyright 2009 by the American Chemical Society.)

With various *para*-Y substituents (Scheme 6.13), a direct interaction of the Y substituent with the positively charged *para* carbon in the S_1 state accounts for the observed Hammett correlations between logarithm of the diazirine S_1 state lifetime and the σ_p^+ value (Table 6.3 and Fig. 6.10). Such Hammett correlations for the excited state lifetimes are quite rare.[40]

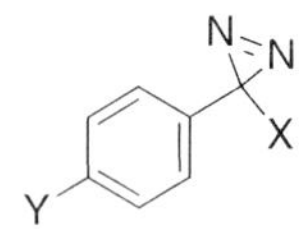

Scheme 6.13.

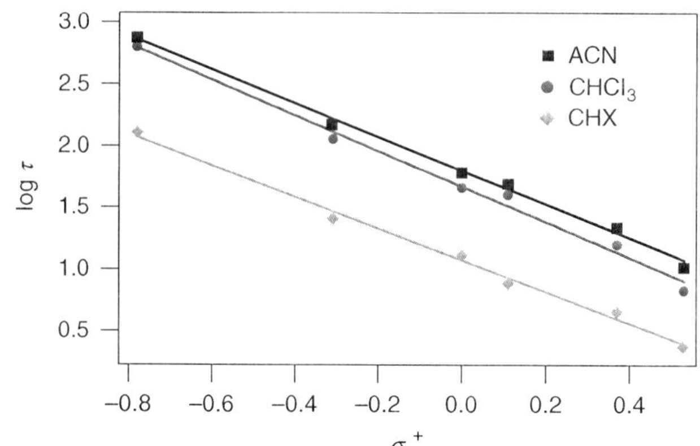

Figure 6.10. Linear Hammett correlations with determined slopes, ρ = −1.37 in acetonitrile, −1.43 in chloroform, and −1.27 in cyclohexane. (Reprinted with permission from Reference 40; Copyright 2009 by the American Chemical Society.)

6.9 THE INFLUENCE OF EXCITATION WAVELENGTH ON THE PHOTOCHEMISTRY OF DIAZIRINES

Derivatives of phenyl diazirine have a distinct absorption band located between 350 and 400 nm corresponding to an $S_0 \rightarrow S_1$ transition. The transition to the higher energy S_2 state is located around 270 nm. This absorption spectrum, with two distinct bands, provides photochemists a unique possibility to study the influence of excitation wavelength on diazirine photochemistry by pumping the precursor to a well-defined, initial photoexcited state. In the case of diazo compounds, the situation usually is less favorable because the $S_0 \rightarrow S_1$ transition typically has a low absorption coefficient. This is problematical in transient absorption spectroscopy because it requires a prohibitively large concentration of precursor, and UV photoexcitation leads to a mixture of highly excited states of the diazo precursor.

The influence of excitation wavelength on diazirine photochemistry was studied in detail with *p*-methoxy-3-phenyl-3-methyldiazirine using ultrafast time-resolved spectroscopy.[1] Upon 400 nm excitation, a 1580 cm^{-1} vibrational band, assigned to the S_1-state of the diazirine, is present immediately after the laser pulse (300 fs), and decays with a ~245 ps time constant in chloroform (Fig. 6.11a). The IR band correlates with the formation and decay of a band at

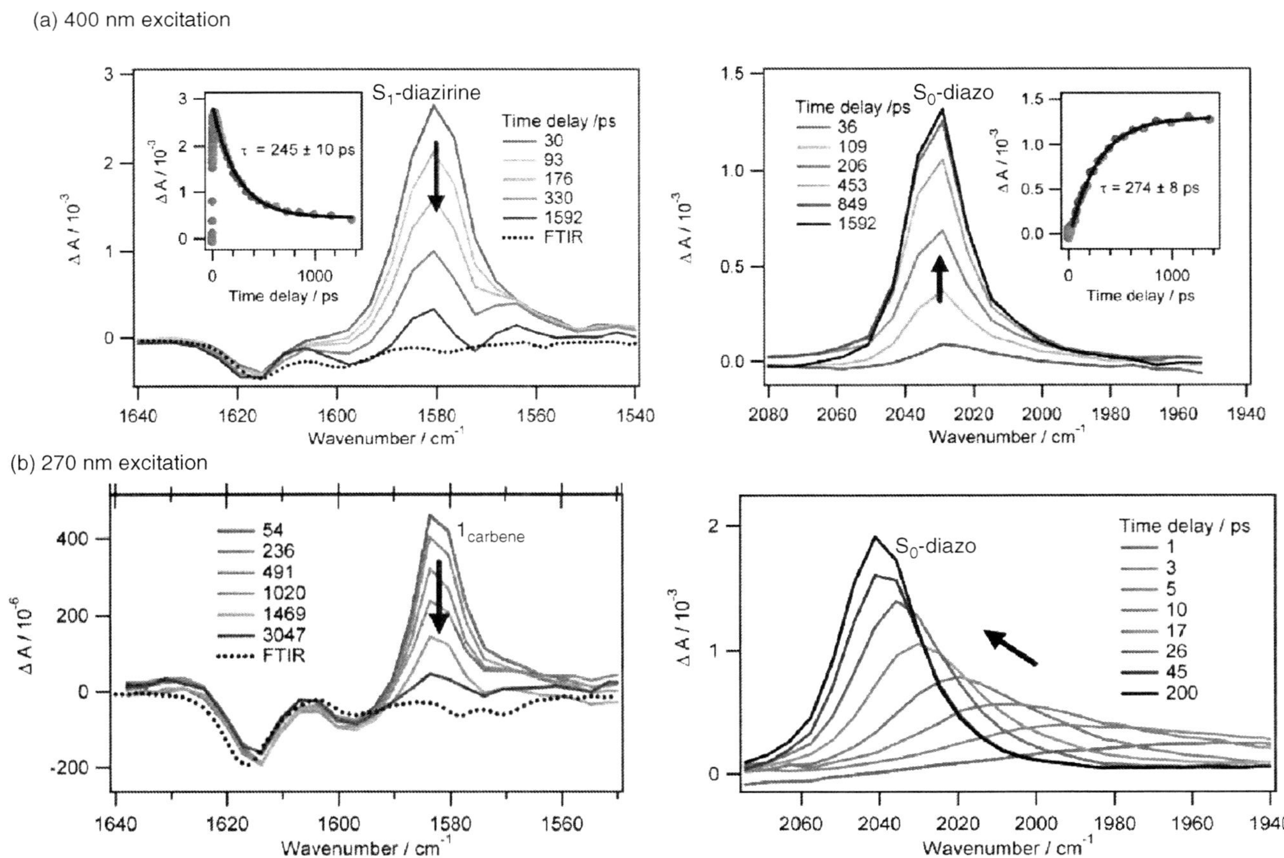

Figure 6.11. Time-resolved IR spectra recorded upon excitation of *p*-methoxy-3-phenyl-3-methyldiazirine in chloroform with light of (a) 400 nm and (b) 270 nm. (Reprinted with permission from Reference 1; Copyright 2009 by the American Chemical Society.) For color details, please see color plate section.

650 nm, observed in time-resolved UV-Vis experiments (~270 ps), indicating that the same species is responsible for the UV-Vis and IR transient spectra. The carrier of the transient absorbing at 650 nm is assigned to the diazirine in the S_1 state. It cannot be assigned to the carbene because the lifetime of the transient absorption lengthens in neat methanol, a carbene-scavenging solvent.

The assignment of the vibrational 1580 cm^{-1} band to the S_1 state was also strongly supported by RI-CC2/TZVP calculations which predict a C=C stretching vibration in this vicinity.[1] The predicted dipole moment of the S_1 state is about 3D higher than that of S_0 and S_2, consistent with the observation of solvation, which explains the observed shifts in the UV-Vis and IR transient spectra of the S_1 state of the diazirine in polar solvents (chloroform or methanol) relative to cyclohexane.[1]

A dramatic change is observed upon ultrafast photolysis with 270 nm, relative to 400 nm, excitation light (Fig. 6.11b). With each excitation wavelength a transient is observed in chloroform at 1585 cm^{-1}. We propose, however, that the carrier of this band with 400 nm light is the S_1 state of the diazirine, but is largely the singlet carbene with 270 nm excitation. The carbene assignment is supported by two pieces of evidence. First, with 270 nm light, the transient lifetime is 1.3 ns, which seems rather long for a diazirine excited state and is certainly longer than that observed with 400 nm excitation (~245 ps). Second, and more significantly, with 270 nm excitation there is a significant shortening of the lifetime in the carbene-scavenging solvent methanol (~35 ps). Calculations predict that it is reasonable that both the S_1 state of the diazirine and the carbene have intense C=C stretching bands around 1585 cm^{-1} because both species (*vide infra*) have quinoidal structures.

The S_0 ground state isomeric diazo compound is formed in substantial yield within a few ps upon 270 nm photolysis of phenyldiazirine. Vibrational relaxation of the diazo photoproduct is indicated by frequency blue shift and band-narrowing over a 100 ps window. Interestingly, integration of the IR band of the diazo isomer reveals the presence of a minor, rising component with a 240 ps time constant. This result indicates the presence of a slow isomerization channel from the relaxed S_1 state of the diazirine to the ground state of the diazo compound. In turn, this requires that at least some of the initially formed S_2 state can relax to the S_1 state by internal conversion. Unfortunately, it is not yet possible, to estimate the efficiency of this decay channel. The proposed photoreaction mechanism is shown in Scheme 6.14.

With 270 nm excitation, the fast singlet carbene formation channel can be attributed to either of two precursor candidates: the initially formed S_2 state of the diazirine and/or the vibrationally hot S_1 state of the diazirine. Either or both of these two precursors may be responsible for the fast S_0 state diazo formation pathway through the S_1 diazo intermediate.

As mentioned previously, there is also a minor rise of the S_0 state diazo population proceeding on long timescales (hundreds of ps), resulting from relaxed S_1 state diazirine isomerization. This indicates that at least some population of the S_2 state diazirine decays by internal conversion to the S_1

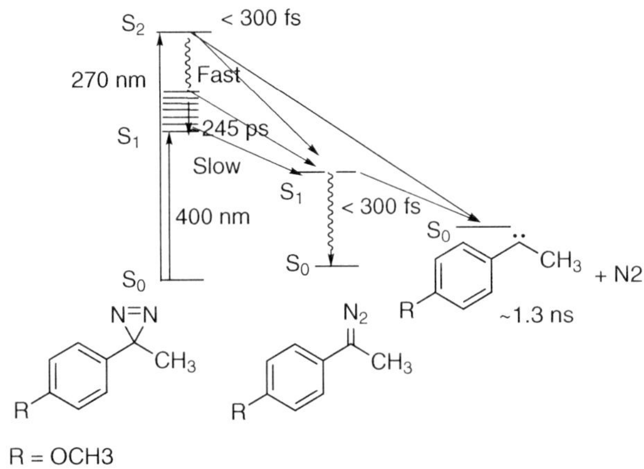

Scheme 6.14.

state. In the case of 400 nm excitation, the S_1 state of the diazirine is formed directly and its decay is accompanied by the slow rise (~270 ps) of the S_0 state of the diazo compound. A weak contribution of the singlet carbene is present at 1580 cm^{-1} in Figure 6.11a.

6.10 CONCLUSION AND OUTLOOK

This chapter has discussed some aspects of the nature and the dynamics of chemical transformations that proceed during the first 0.3–2000 ps after a diazirine or diazo compound absorbs a 300 fs pulse of light. Many old and new questions relevant to carbene chemistry can now be answered using a combination of ultrafast time-resolved spectroscopy and advanced theoretical calculations. Particularly intriguing is the process of rearrangement on the diazo/diazirine/carbene excited state surface proceeding on an ultrafast fast timescale (<0.3 ps).

In some experiments, we failed to directly or indirectly detect a carbene species. This is not necessarily proof that carbenes are not involved in the photochemistry. In these cases, advanced theoretical calculations are of critical importance to help us understand the reaction mechanism.

Mechanistic photochemistry on ultrafast timescales seeks to understand (i) which diazo or diazirine excited state is initially populated, (ii) which excited state decomposes, (iii) the pathway from the Franck–Condon state to the dissociative state, (iv) the competition between internal conversion and fragmentation in diazo and diazirine excited states, (v) whether or not excited states undergo carbene "mimetic" rearrangements, (vi) how the excess of vibrational energy influences the carbene's reactivity, and finally (vii) how the variations of structure, excitation wavelength, and solvent influence the answers to the previous

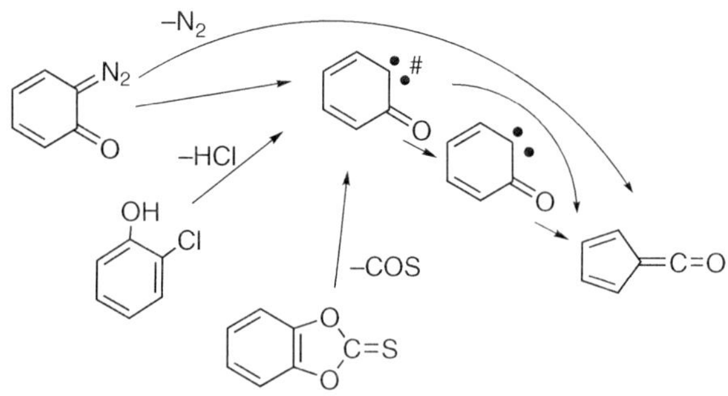

Scheme 6.15.

questions. Most ultrafast studies to date have been performed with nitrogenous precursors, but one can safely predict that analogous studies with other precursors will soon appear. A promising new direction is photogeneration of the "same" carbene from different precursors (Scheme 6.15 shows an example).

Another new direction will be the study of the conical intersection leading from the lowest singlet electronic excited state (S_1) of diazo precursors to open-shell, diradical-like singlet carbenes. Recall that this pathway was predicted by theory and observed with fluorenylidene. Can this pathway be detected for other carbenes? How do structure, solvent, and exciting wavelength influence the yield of this process? Can we begin to study and understand the intra and intermolecular chemistry of these excited, open-shell singlet carbenes? Perhaps future research will resolve many-decades-old puzzles in the product analysis literature.

One can also imagine future studies of carbenes in very highly excited singlet (S_n) and triplet states (T_n, $n > 1$) using ultrafast spectroscopy. Such excited carbenes may be generated by double laser pulse excitation (the first pulse generates the carbene in the singlet or triplet state, the second pulse, appropriately delayed, excites the carbene to the highly excited state). The goal would be characterization of these putative transients (spectral signature, lifetime, and expected high reactivity).

SUGGESTED READING

Bally, T. "Matrix Isolation." In *Reactive Intermediate Chemistry*. Moss, R. A., Platz, M. S., Jones, M., Jr., Eds. John Wiley & Sons, Inc.: Hoboken, NJ, **2004**, p. 797*f*.

Baldwin, J. E. "Reactions on the Femtosecond Time Scale." In *Reactive Intermediate Chemistry*. Moss, R. A., Platz, M. S., Jones, M., Jr., Eds. John Wiley & Sons, Inc.: Hoboken, NJ, **2004**, p. 899*f*.

Bearpark, M. J., Robb, M. A. "Conical Intersection Species as Reactive Intermediates." In *Reviews of Reactive Intermediate Chemistry*. Platz, M. S., Moss, R. A., Jones, M., Jr., Eds. John Wiley & Sons, Inc.: Hoboken, NJ, **2004**, p. 379*f*.

Borden, W. T. "The Partnership between Electronic Structure Calculations and Experiments in the Study of Reactive Intermediates." In *Reactive Intermediate Chemistry*. Moss, R. A., Platz, M. S., Jones, M., Jr., Eds. John Wiley & Sons, Inc.: Hoboken, NJ, **2004**, p. 961*f*.

Carpenter, B.K. "Potential Energy Surfaces and Reaction Dynamics." In *Reactive Intermediate Chemistry*. Moss, R. A., Platz, M. S., Jones, M., Jr., Eds. John Wiley & Sons, Inc.: Hoboken, NJ, **2004**, p. 925*f*.

Hilinski, E "The Picosecond Realm." In *Reactive Intermediate Chemistry*. Moss, R. A., Platz, M. S., Jones, M., Jr., Eds. John Wiley & Sons, Inc.: Hoboken, NJ, **2007**, p. 873*f*.

Jones, M., Jr., Moss, R. A. "Singlet Carbenes." In *Reactive Intermediate Chemistry*. Moss, R. A., Platz, M. S., Jones, M., Jr., Eds. John Wiley & Sons, Inc.: Hoboken, NJ, **2004**, p. 273*f*.

Phillips, D. L., Kwok, W. M., Ma, C. "An Introduction to Time-Resolved Resonance Raman Spectroscopy and its Application to Reactive Intermediates." In *Reviews of Reactive Intermediate Chemistry*. Platz, M. S., Moss, R. A., Jones. M., Jr., Eds. John Wiley & Sons, Inc.: Hoboken, NJ, **2007**, p. 123*f*.

Scaiano, J. C. "Nanosecond Laser Flash Photolysis: A Tool for Physical Organic Chemistry." In *Reactive Intermediate Chemistry*. Moss, R. A., Platz, M. S., Jones, M., Jr., Eds. John Wiley & Sons, Inc.: Hoboken, NJ, **2004**, p. 847*f*.

Sheridan, R. S. "Quantum Mechanical Tunneling in Organic Reactive Intermediates." In *Reviews of Reactive Intermediate Chemistry*. Moss, R. A., Platz, M. S., Jones, M., Jr., Eds. John Wiley & Sons, Inc.: Hoboken, NJ, **2004**, p. 415*f*.

REFERENCES

1 Zhang, Y., Burdzinski, G., Kubicki, J., Vyas, S., Hadad, C. M., Sliwa, M., Poizat, O., Buntinx, G., Platz, M. S. *J. Am. Chem. Soc.* **2009**, *131*, 13784.
2 Zhang, Y., Kubicki, J., Wang, J., Platz, M. S. *J. Phys. Chem. A* **2008**, *112*, 11093.
3 Burdzinski, G., Zhang, Y., Wang, J., Platz, M. S. *J. Phys. Chem. A* **2010**, *114*, 13065.
4 Burdzinski, G., Platz, M. S. *J. Phys. Org. Chem.* **2010**, *23*, 308.
5 Wang, J., Burdzinski, G., Gustafson, T. L., Platz, M. S. *J. Org. Chem.* **2006**, *71*, 6221.
6 Wang, J., Burdzinski, G., Gustafson, T. L., Platz, M. S. *J. Am. Chem. Soc.* **2007**, *129*, 2597.
7 Wang, J., Kubicki, J., Peng, H., Platz, M. S. *J. Am. Chem. Soc.* **2008**, *130*, 6604.
8 Wang, J., Kubicki, J., Gustafson, T., Platz, M. S. *J. Am. Chem. Soc.* **2008**, *130*, 2304.
9 Peon, J., Polshakov, D., Kohler, B. *J. Am. Chem. Soc.* **2002**, *124*, 6428.
10 Grasse, P. B., Brauer, B. E., Zupancic, J. J., Kaufmann, K. J., Schuster, G. B. *J. Am. Chem. Soc.* **1983**, *105*, 6833.
11 Wang, J., Kubicki, J., Hilinski, E. F., Mecklenburg, S. L., Gustafson, T. L., Platz, M. S. *J. Am. Chem. Soc.* **2007**, *129*, 13683.
12 Griller, D., Hadel, L., Nazran, A. S., Platz, M. S., Wong, P. C., Savino, T. G., Scaiano, J. C. *J. Am. Chem. Soc.* **1984**, *106*, 2227.
13 Barcus, R. L., Hadel, L. M., Johnston, L. J., Platz, M. S., Savino, T. G., Scaiano, J. C. *J. Am. Chem. Soc.* **1986**, *108*, 3928.
14 Hadel, L. M., Platz, M. S., Scaiano, J. C. *J. Am. Chem. Soc.* **1984**, *106*, 283.
15 Hadel, L. M., Platz, M. S., Wright, B. B., Sacaiano, J. C. *Chem. Phys. Lett.* **1984**, *105*, 539.

16 Wang, J., Burdzinski, G., Kubicki, J., Platz, M. S. *J. Am. Chem. Soc.* **2008**, *130*, 11195.
17 Burdzinski, G., Wang, J., Gustafson, T. L., Platz, M. S. *J. Am. Chem. Soc.* **2008**, *130*, 3746.
18 Bonneau, R., Liu, M. T. H., Kim, K. C., Goodman, J. L. *J. Am. Chem. Soc.* **1996**, *118*, 3829.
19 Platz, M. S. In *Advances in Carbene Chemistry*. Brinker, U. H., Ed., Vol. *2*, p. 133. JAI Press: Stamford, CT, **1998**.
20 Zhang, Y., Burdzinski, G., Kubicki, J., Platz, M. S. *J. Am. Chem. Soc.* **2009**, *131*, 9646.
21 Burdzinski, G., Zhang, Y., Selvaraj, P., Sliwa, M., Platz, M. S. *J. Am. Chem. Soc.* **2010**, *132*, 2126.
22 Zhang, Y., Kubicki, J., Platz, M. S., *Org. Lett.* **2010**, *12*, 3182.
23 Burdzinski, G., Rehault, J., Wang, J., Platz, M. S. *J. Phys. Chem. A* **2008**, *112*, 10108.
24 Zhang, Y., Burdzinski, G., Kubicki, J., Platz, M. S. *J. Am. Chem. Soc.* **2008**, *130*, 16134.
25 Wang, J., Yunlong, Z., Kubicki, J., Platz, M. *Photochem. Photobiol. Sci.* **2008**, *7*, 552.
26 Langan, J. G., Sitzmann, E. V., Eisenthal, K. B. *Chem. Phys. Lett.* **1984**, *110*, 521.
27 Langan, J. G., Sitzmann, E. V., Eisenthal, K. B. *Chem. Phys. Lett.* **1986**, *124*, 59.
28 Sitzmann, E. V., Langan, J. G., Griller, D., Eisenthal, K. B. *Chem. Phys. Lett.* **1989**, *161*, 353.
29 Sitzmann, E. V., Langan, J. G., Eisenthal, K. B. *J. Am. Chem. Soc.* **1984**, *106*, 1868.
30 Jortner, J., Rice, S. A., Hochstrasser, R. M. *Adv. Photochem.* **1969**, *7*, 149.
31 Robinson, G. W., Frosch, R. P. *J. Chem. Phys.* **1962**, *37*, 1962.
32 Robinson, G. W., Frosch, R. P. *J.Chem. Phys.* **1963**, *38*, 1187.
33 Geise, C. M., Wang, Y., Mykhaylova, O., Frink, B. T., Toscano, J. P., Hadad, C. M. *J. Org. Chem.* **2002**, *67*, 3079.
34 Wang, Y., Hadad, C. M., Toscano, J. P. *J. Am. Chem. Soc.* **2002**, *124*, 1761.
35 Smith, M., March, *J. Advanced Organic Chemistry*. Wiley Interscience: New York, **2007**.
36 Turro, N. J., Cha, Y., Gould, I. R. *J. Am. Chem. Soc.* **1987**, *109*, 2101.
37 Eisenthal, K. B., Moss, R. A., Turro, N. J. *Science* **1984**, *225*, 1439.
38 Moss, R. A., Yan, S., Krogh-Jespersen, K. *J. Am. Chem. Soc.* **1998**, *120*, 1088.
39 Reichardt, C. *Chem. Rev.* **1994**, *94*, 2319.
40 Zhang, Y., Wang, L., Moss, R. A., Platz, M. S. *J. Am. Chem. Soc.* **2009**, *131*, 16652.
41 Zhang, Y., Vyas, S., Hadad, C. M., Platz, M. S. *J. Phys. Chem. A* **2010**, *114*, 5902.
42 Wang, J., Burdzinski, G., Kubicki, J., Platz, M. S., Moss, R. A., Fu, X., Piotrowiak, P., Myahkostupov, M. *J. Am. Chem. Soc.* **2006**, *128*, 16446.
43 Burdzinski, G., Middleton, C., Gustafson, T., Platz, M. S. *J. Am. Chem. Soc.* **2006**, *128*, 14804.
44 Schmierer, T., Shreier, W. J., Koller, F. O., Schrader, T. E., Gilch, P. *Phys. Chem. Chem. Phys.* **2009**, *11*, 11596.
45 Moss, R. A., Turro, N. J. In *Kinetics and Spectroscopy of Carbenes and Biradicals*. Platz, M. S., Ed. Plenum: New York, **1990**, p. 213f.

7

TUNNELING IN THE REACTIONS OF CARBENES AND OXACARBENES

DENNIS GERBIG AND PETER R. SCHREINER
Institute of Organic Chemistry, Justus-Liebig University, Giessen, Germany

7.1 INTRODUCTION: LIGHT- AND HEAVY-ATOM TUNNELING

Tunneling is a common phenomenon in the physics of fundamental particles, for example, for electrons in (semi)conducting materials. Yet, examples of comparable importance of tunneling in chemical transformations are still limited. However, this picture is changing rapidly. In chemical kinetics, tunneling of entire atoms is generally evaluated only as a quantum mechanical correction to the rate constants derived from various classical or semiclassical approaches. A reaction mechanism that only becomes feasible due to the existence of quantum mechanical tunneling was not known till the advent of hydroxycarbenes, which occupy a considerable part of this chapter. Reactions that largely depend on tunneling are, for example, the inversion of ammonia (in which tunneling leads to an observable splitting of vibrational signals) (**1**);[1,2] the rapid interconversion between the Jahn–Teller distorted rectangular forms of cyclobutadiene, a process also referred to as automerization (**2**);[3–7] the rearrangement of the carbon frameworks in well-known fluxional molecules, for example, in barbaralane (**3**); semibullvalene (**4**); as well as the quick scrambling of hydrogen atoms in the methonium ion CH_5^+ (**5**).[8–19]

Tunneling in chemical reactions is often classified in terms of atomic mass. Hence, one distinguishes between light- and heavy-atom tunneling. Yet, it is important to mention that the transition between the two is not sharp and

Contemporary Carbene Chemistry, First Edition. Edited by Robert A. Moss and Michael P. Doyle.

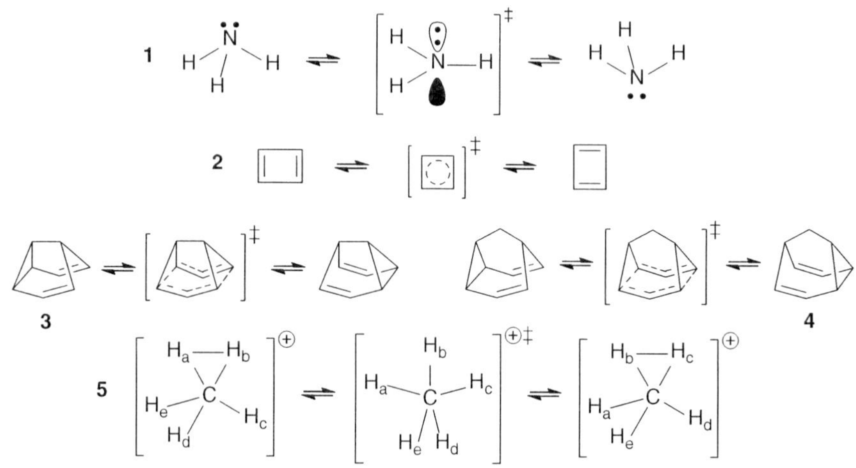

Scheme 7.1.

tunneling of heavy atoms is often accompanied by the concurrent movement of lighter atoms as well. As the probability of tunneling depends on the square root of the moving particles' mass, reactions involving the tunneling of heavy atoms (e.g., carbon in the automerization of **2**) are generally less common than those including lighter ones (e.g., hydrogen in the inversion of **1**). However, this does not imply that light-atom tunneling must naturally be faster than heavy-atom tunneling. Another important factor that governs the probability of tunneling is the height of the barrier impeding a certain molecular transformation. In a pictorial explanation of the tunneling process, the particle moves through the barrier rather than traversing it. Reactions featuring a high barrier, that is, a large enthalpy of activation $\Delta H^{\ddagger}$, tend to be less prone to tunneling. Of similar importance is the length of the arc that particles travel during a reaction, that is, the extent of the overall atomic movement inside a molecule (encompassing both the actual reacting moiety as well as the molecular periphery). The larger the arc length of the moving particles, the less likely it is for tunneling to be involved in the transformation. In a typical reaction profile, the intrinsic reaction coordinate (IRC), as defined by Fukui, is usually chosen to exemplify the progress of a reaction.[20,21] By definition, the IRC in mass-weighted coordinates (in units of $amu^{1/2}\cdot bohr$) encompasses the reduced masses and the traveled arc lengths of all atoms in the molecule and can hence represent the advancement of the reaction in a *single* dimension. The derivative of the IRC $\mathbf{x}$ with respect to the arc length s is mathematically related to the negative normalized gradient vector $\mathbf{g}$ of molecular energy, that is, the gradient vector divided by its own length (Eq. 7.1). This allows the formulation of a basic IRC-following algorithm that generates the actual one-dimensional reaction profile by computing the molecular energy at every step $s+\Delta s$ along the IRC (Eq. 7.2). Such algorithms are routinely employed and are available in most modern electronic structure codes. The step size Δs, which represents the arc length interval after which the energy is recomputed, can usually be selected by the user.

$$\frac{d\mathbf{x}(s)}{ds} = -\frac{\mathbf{g}[\mathbf{x}(s)]}{|\mathbf{g}[\mathbf{x}(s)]|} \tag{7.1}$$

$$\mathbf{x}(s+\Delta s) = \mathbf{x}(s) - \Delta s\frac{\mathbf{g}[\mathbf{x}(s)]}{|\mathbf{g}[\mathbf{x}(s)]|} \tag{7.2}$$

As a result, the mass and arc length dependence of tunneling is well captured within the IRC so that following the reaction path can give viable information about the likelihood of the involvement of tunneling in a reaction. The tunneling probability depends both on the barrier height and the width of the IRC path, the latter capturing both the influence of mass and arc length. As it turns out, light- and heavy-atom tunneling show different dependencies on barrier heights and widths, respectively, so that different characteristic profiles can be found for either category of tunneling. For the same arc length, the IRC path for lighter atoms is significantly smaller than that for heavy atoms. Hence, reactions involving heavy atoms are only prone to tunneling if the traveled arc length of all participating atoms is very small, which is often the case in carbon skeletal rearrangements. Generally speaking, light atoms may cover more ground and still undergo tunneling. Both forms of tunneling profit from small barriers. Yet, due to typically broader IRC paths, barrier height is more limiting for heavy-atom tunneling. On the other hand, light particles may cross narrow barriers that are remarkably high.

The different tunneling behavior of light and heavy atoms has a most profound impact on the (primary) kinetic isotope effect (KIE) of a reaction. Heavier isotopes result in broader IRC paths and hence a decreased tunneling probability. This effect is most pronounced for the substitution of protium H with deuterium D due to a twofold increase in mass. Hence, tunneling of hydrogen can be significantly slowed or even shut down entirely by deuteration. Rate constants involving tunneling can be evaluated using multiple formalisms, some of which are briefly described here. The simplest semiclassical tunneling methodology is the one-dimensional Wentzel–Kramers–Brillouin (WKB) formalism,[22–24] which determines the tunneling probability $P(\varepsilon)$ depending on the molecular energy ε by computing action integrals $\theta(\varepsilon)$ along the length s of the minimum energy path (MEP). The MEP is constructed from the IRC path by adding zero-point vibrational energy corrections obtained by the evaluation of projected frequencies along the path and thus defines a potential $V(s)$. The action integral $\theta(\varepsilon; s_1, s_2)$ corresponds to the area under the potential curve limited by the classical turning points s_1 and s_2, which are defined as the points where the zero-point energy level of the active frequency ν_{IRC} bisects the curve, that is, where $s = \varepsilon$. The active (starting material) frequency ν_{IRC} corresponds to the frequency that propagates the reaction, that is, the vibration that leads toward the barrier. In the course of the reaction, this specific degree of freedom translates directly into the IRC, so that only 3*N*-7 normal modes remain along the MEP for a nonlinear molecule. The tunneling probability, and hence the tunneling rate constant, may then be determined (Eqs. 7.3 and 7.4) by assuming that every attempt of nonclassical barrier penetration has the same probability of success.

$$k = \nu_{\mathrm{IRC}} \cdot P(\varepsilon) = \frac{\nu_{\mathrm{IRC}}}{1+\exp[2\theta(\varepsilon)]} \tag{7.3}$$

$$\theta(\varepsilon) = \int_{s_1}^{s_2} \sqrt{2[V(s)-\varepsilon]}\,\mathrm{d}s \tag{7.4}$$

In stark contrast to conventional Transition State Theory (TST),[25,26] which only refers to barrier height for the explanation of chemical reactivity, it becomes evident from Equation (7.4) that barrier width is of even greater importance for the occurrence of tunneling. Whereas the barrier height enters the equation in the form of the square root of the potential $V(s)$, the barrier width enters linearly.

A multidimensional approach for the description of tunneling in molecular systems is Canonical Variational Transition State Theory (CVT or Canonical VTST), which is an extension of the classical formulation of TST and includes corrections for quantum mechanical effects on rate constants.[27–31] In classical TST, starting material and product regions along the reaction path are defined by a dividing surface that runs through the classical transition state (TS) and is perpendicular to the molecular energy gradient along the path. The no-crossing rule states that a molecular trajectory may not return to its region of origin after crossing the dividing surface. The MEP is exclusively chosen as the reaction path. In the VTST extension, recrossing is allowed and the dividing surface does not necessarily contain the classical TS. Moreover, the point of maximum potential (the generalized TS) along the reaction path does not need to correspond to a saddle point, that is, the reaction may proceed on a path other than the MEP. The now unconstrained position of the dividing surface is variationally optimized to minimize recrossing and to minimize the molecular flux from starting material to product region. Tunneling corrections along this generalized path are then added by the evaluation of action integrals similar to the WKB method.

Rate constants obtained in this fashion encompass contributions from both thermal reactivity and tunneling and are referred to as zero-curvature tunneling (ZCT) results. If the reaction path is allowed to deviate largely from the MEP, so that it is shifted to the concave side of the latter, it results in small-curvature tunneling (SCT). The SCT methodology allows for a more strongly curved path, and hence shorter reaction path, that may go through a higher energy structure for the benefit of an enhanced tunneling rate, leading to a net increase of the overall reaction rate. This has been referred to as "corner cutting" of a chemical reaction, an effect that cannot be properly captured by limiting the system to travel only along the MEP. As a result, CVT/SCT rate constants are usually larger than CVT/ZCT or WKB rate constants.

7.2 ALKYL- AND HALOCARBENES

7.2.1 Cyclopropylcarbenes

The aforementioned differences in tunneling behavior are particularly evident for the reactions of cyclopropylcarbenes. Tunneling in cyclopropyl-derived systems was recently discovered in the ring-opening reaction of the cyclopropylmethyl radical (**6**) to the linear but-1-en-4-yl radical (**7**; Scheme 7.2).[32,33]

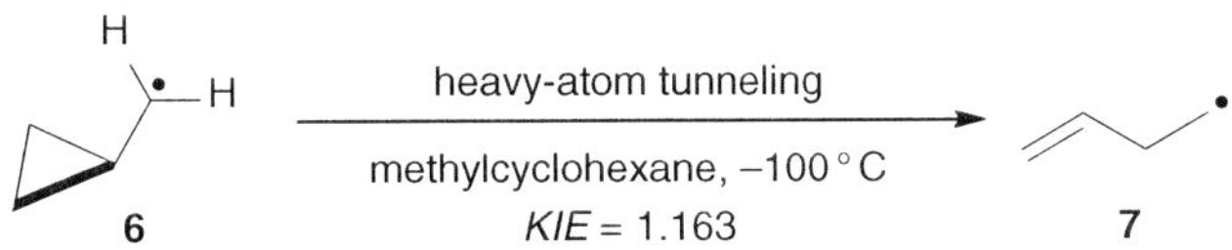

Scheme 7.2.

Motivated by these findings, the same reaction was studied computationally for cyclopropylmethylene (**8**), the related carbene species. Employing CVT with ZCT and SCT approximations using the B3PW91/cc-pVTZ method as well as the semiclassical WKB approach at the AE-CCSD(T)/cc-pCVTZ and CCSD(T)/cc-pVTZ levels of theory, tunneling was found to play a major role in the corresponding ring expansion process (Scheme 7.3) to cyclobutene (**9**).[34] Extending the investigations to the higher homologue, cyclopropylmethylcarbene (**10**), showed that the latter is prone to undergo a tunneling-assisted [1,2] H-migration to yield vinylcyclopropane (**11**) (Scheme 7.3).

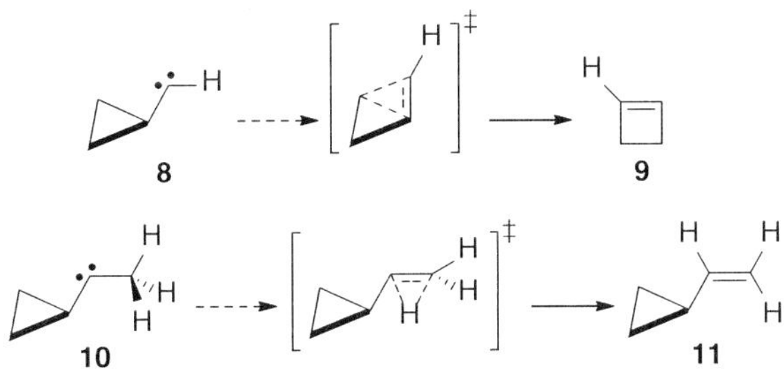

Scheme 7.3.

From comparison of the reaction profiles of carbon and hydrogen tunneling (Fig. 7.1), it becomes evident that carbon tunneling (i.e., ring expansion of **8**) occurs along a much broader IRC path. Yet, the barrier that impedes the rearrangement is sufficiently small for tunneling to contribute to the rate constant of the reaction. In contrast, hydrogen tunneling (i.e., [1,2]H-migration in **10**) takes place along a smaller IRC path and is capable of crossing a higher enthalpic barrier.

The primary KIE associated with an exchange of hydrogen for deuterium in the [1,2]H-migration of **10** is evident in the IRC profile of the reaction. Although the reaction has the exact same electronic barrier (i.e., omitting ZPVE corrections), the reaction path is much broader, thus resulting in a lower rate of

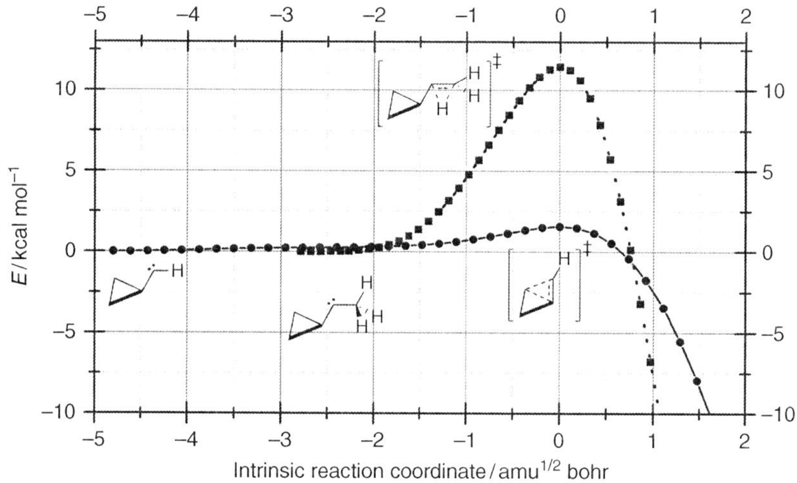

Figure 7.1. Intrinsic reaction coordinate paths (B3PW91/cc-pVTZ) for the reactions of cyclopropylmethylene (**8**; circles) and cyclopropylmethylcarbene (**10**; squares) to cyclobutene (**9**) and vinylcyclopropane (**11**), respectively.

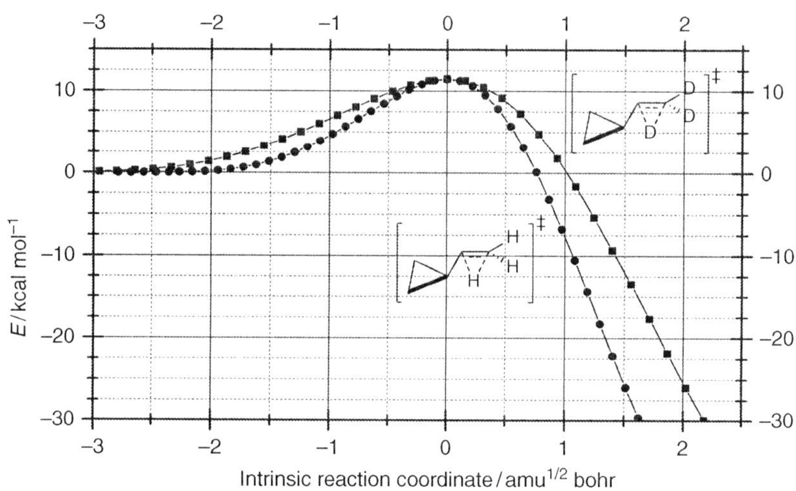

Figure 7.2. Intrinsic reaction coordinate paths (B3PW91/cc-pVTZ) for the [1,2]-migration of hydrogen (circles) and deuterium (squares) in cyclopropylmethylcarbene (**10**).

tunneling (Fig. 7.2). Still, deuteration does not yet impede tunneling completely in the case of **10** despite a KIE as large as 10^5. Substitution of ^{12}C for ^{13}C in **8** only has a subtle influence on the rate of ring expansion. Due to the relatively small increase in mass, a much smaller KIE of 1.22 was determined computationally.

The rates of both light- and heavy-atom tunneling are temperature independent at very low temperatures (Fig. 7.3). However, at temperatures significantly above the zero point, tunneling may benefit from thermal activation: A slight population

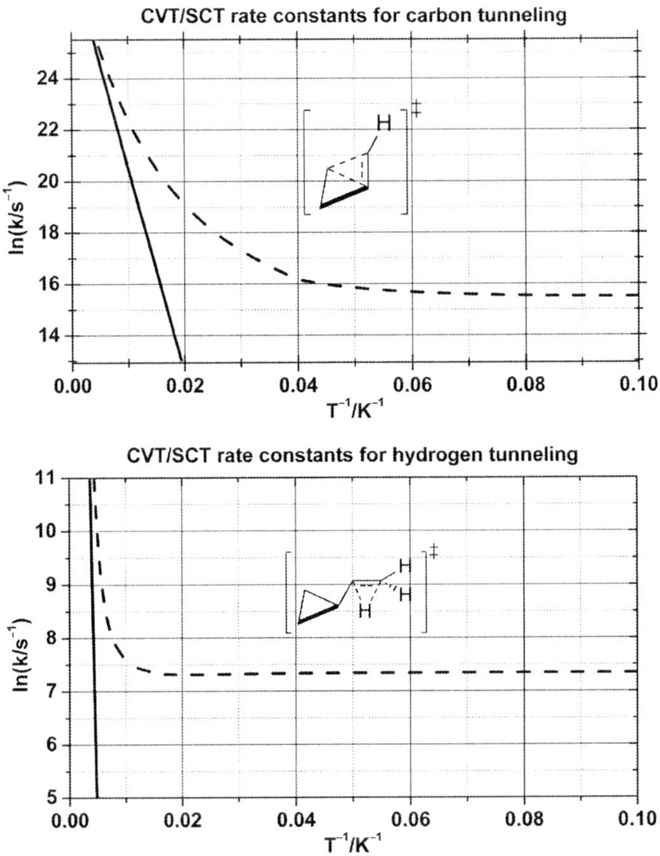

Figure 7.3. Arrhenius profiles for carbon tunneling (top) and hydrogen tunneling (bottom) computed at the B3PW91/cc-pVTZ level of theory; solid lines show the classical Arrhenius behavior (CVT), dashed lines show rates including CVT/SCT tunneling corrections. Whereas carbon tunneling is roughly temperature independent only up to 20 K, hydrogen tunneling is not affected by temperature up to almost 100 K.

of molecular vibrational levels above the ground state may lead to enhanced rates due to a decrease of the effective reaction barrier. The effect of this so-called vibrationally assisted tunneling (VAT)[35,36] is most pronounced for reactions with small enthalpic barriers because the separation of vibrational energy levels is large compared to the barrier height. Consequently, heavy-atom tunneling is more prone to thermal activation already at lower temperatures. In contrast, light-atom tunneling through higher barriers is often temperature independent over a much larger temperature interval.

In summary, cyclopropylcarbenes are promising species for studying the influence of tunneling on reactions involving both light and heavy atoms on a theoretical basis. Due to the high rate constants at which their reactions proceed, experimental proof of tunneling is rather difficult to obtain.

7.2.2 Ethylidene and Phenylmethylcarbene

Ethylidene (i.e., methylcarbene, **12**) is the simplest carbene for which a [1,2] H-shift can be envisioned. However, despite several efforts to generate this congener of methylene, which likewise features a triplet ground state, **12** has evaded direct spectroscopic characterization until today.[37–46] Yet, the ethylidene analogues silaethylidene (**13**) and 2,2,2-trifluoroethylidene (**14**) could be successfully generated and characterized in low-temperature noble gas matrices.[47,48] However, neither undergoes hydrogen or fluorine tunneling, respectively.

The elusiveness of **12** has attracted much attention among the community of theoretically oriented chemists as well, and even during the early days of computational chemistry, several *ab initio* studies appeared.[49–56] It was established that **12** has a small singlet–triplet energy separation (ΔE_{S-T}) of approximately 5–7 kcal/mol and that the [1,2]H-shift, operative only in singlet **12**, features a very small thermal barrier of less than 2 kcal/mol.[51,54–56] Hence, both the singlet–triplet gap and the barrier may be overcome even at low temperatures (Scheme 7.4). For some time, it even seemed unclear if the structure of **12** corresponds at all to a local minimum on the C_2H_4 potential energy hypersurface or if it rather represents a TS for the scrambling of hydrogens in ethylene (**15**).[54] Whereas it is now generally accepted that **12** is a minimum energy structure, albeit within a very shallow potential well, the role of quantum mechanical tunneling in the [1,2]H-shift to **15** remains unclear. The barrier that impedes hydrogen migration is very small, so that a definitive distinction between over or through the barrier reactivity cannot be made.

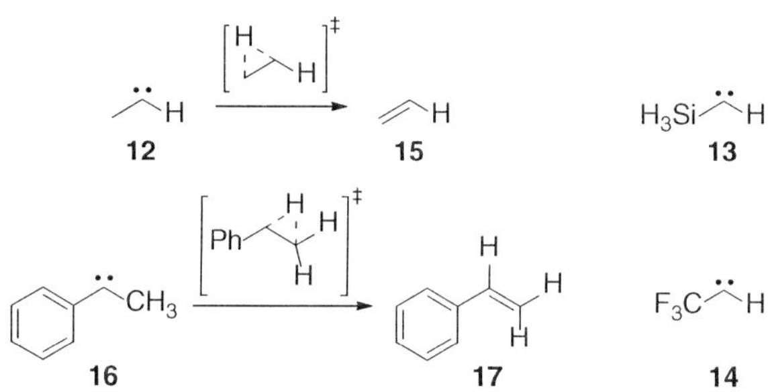

Scheme 7.4.

However, substitution of the hydrogen atom with a phenyl moiety, as in **16**, kinetically stabilizes the carbene and increases the barrier height of the [1,2] H-shift on the singlet potential energy surface to approximately 5 kcal/mol, whereas the singlet–triplet gap of phenylmethylcarbene (i.e., 1-phenylethylidene; **16**; Scheme 7.4) remains comparable to that of **12** (approximately 4–7 kcal/mol).[57–60] Thus, **16** could be characterized in various experiments.[58,61–71] In addition, numerous theoretical investigations have been conducted.[59,60,67,72–77] Because the [1,2]H-shift to styrene (**17**) is impeded by a larger barrier on the T_0 surface, it is presumed that hydrogen migration rather occurs from the S_1 state in a surface crossing mechanism similar to that expected to be at work in **12**.[60,64]

The experimental activation energy for the rearrangement of **16** on the singlet surface at 65 K was determined to be slightly less than 5 kcal/mol and is hence in excellent agreement with the computed barrier value.[64] Consequently, the effect of tunneling on the kinetics of [1,2]H-migration, if any, might be rather small. Moreover, the finding of experimental evidence for hydrogen tunneling, for instance in the course of matrix isolation experiments under cryogenic conditions,[64,69] is further complicated by a plethora of possible concurrent rearrangements involving the phenyl moiety. In summary, it can be stated that hydrogen tunneling in both **12** and **16** is likely, but it remains difficult to envision a suitable experimental setup to provide unambiguous proof. This situation improves upon consideration of alkyl- and arylhalocarbenes, which will now be discussed.

7.2.3 Methylchlorocarbene and Benzylchlorocarbene

Chlorine substitution of alkyl- and arylcarbenes drastically changes the characteristics of the [1,2]H-migration in these molecules. First and foremost, chlorine strongly stabilizes the singlet state so that halocarbenes usually possess a singlet ground state. Whereas it could not be unequivocally determined to what extent tunneling plays a role in the reactions of **12** and **16**, H-tunneling could be shown to be of notable importance in the rearrangements of their halo-substituted congeners, methylchlorocarbene (**18**) and benzylchlorocarbene (**19**) to vinyl chloride (**20**) and β-chlorostyrene (**21**), respectively (Scheme 7.5).[78–90]

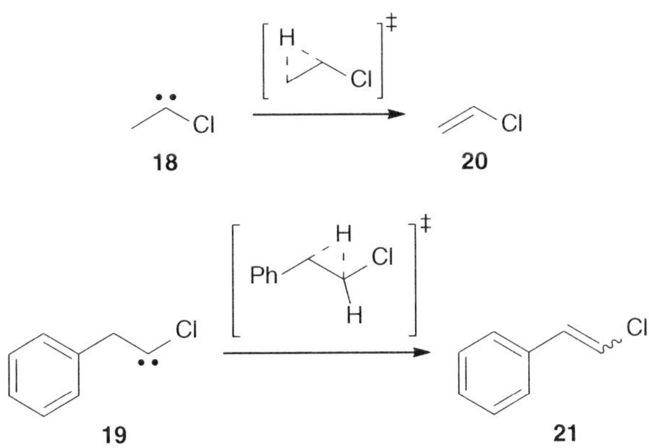

Scheme 7.5.

The enthalpic barrier of the [1,2]H-shift in **18** was evaluated theoretically to be around 11 kcal/mol.[52,90] Yet, the experimentally observed kinetics result in an activation energy of roughly 4–5 kcal/mol,[82] a dichotomy that suggests tunneling.[87] Carbene **19** features an experimental activation energy of approximately 3–6 kcal/mol for the [1,2]H-migration.[85,91,92] Computationally, a value of ~5 kcal/mol was determined for the thermal barrier. Hence, as the measured effective activation energy is in good agreement with the computed barrier height, it seems that tunneling is not overly important for **19** at ambient temperatures.[92] However, tunneling in the [1,2]H-migration in **19** could be observed directly in a cryogenic matrix at 10–30 K.[88]

Alkylchlorocarbenes also show tunneling effects in reactions other than [1,2]H-migrations. Rearrangements in halocarbenes that occur concomitantly with the relocation of heavy atoms are discussed next.

7.2.4 *tert*-Butylchlorocarbene

Singlet *tert*-butylchlorocarbene (**22**) has been generated both in solution and in cryogenic matrices.[79,93,94] When isolated in solid nitrogen at a temperature of 11 K, irradiation at short wavelengths promotes insertion into a neighboring C–H bond of the *tert*-butyl moiety yielding 2-chloro-1,1-dimethylcyclopropane (**23**; Scheme 7.6). Yet, even in the dark, **22** slowly undergoes insertion, a process which was determined to be insensitive to temperature in the range of 11–30 K.[94] The kinetically derived activation energy was estimated to be as low as 0.5 kcal/mol.

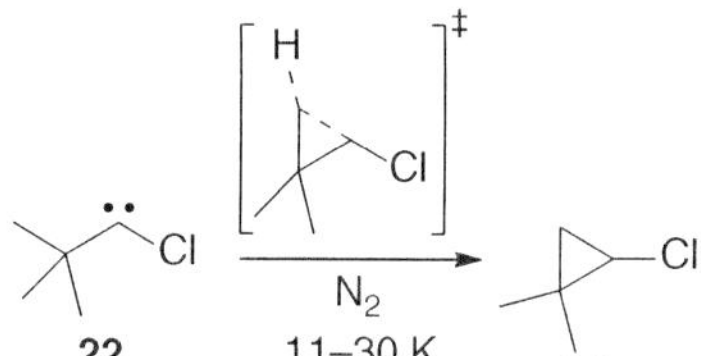

22 23

Scheme 7.6.

In contrast, perdeutero-*tert*-butylchlorocarbene is stable towards C–D insertion up to temperatures of at least 40 K. The pronounced KIE as well as the observed temperature independence strongly suggest the involvement of tunneling. Remarkably, the reaction not only encompasses a hydrogen migration but also comprises the formation of a new carbon–carbon bond. Hence, the elongation of the methyl C–H bond must be coupled to heavy-atom movement. This example nicely demonstrates that the formal distinction between light- and heavy-atom tunneling is not strictly applicable.

7.2.5 1-Methylcyclobutylfluorocarbene

Carbene species incorporating small rings in direct vicinity to the carbene center are prone to ring expansion reactions. Tunneling was found to play a key role in the ring expansion reaction of singlet 1-methylcyclobutylfluorocarbene (**24**), which was photochemically generated in solid nitrogen and argon matrices at temperatures as low as 8 K (Scheme 7.7).[36] Insertion into the cyclobutyl ring, yielding 1-fluoro-2-methylcyclopentene (**25**), even proceeds in the dark due to heavy-atom tunneling. The rate of insertion increased significantly when the matrix was warmed up to 16 or even 25 K.

Hence, the involvement of VAT was suggested in agreement with the expected tunneling profile for heavy atoms. The enthalpic barrier for the rearrangement of **24** was computed to be around 6 kcal/mol. Yet, the corresponding barrier in 1-methylcyclobutylchlorocarbene (**26**) was computationally determined as 3 kcal/mol, suggesting an even faster insertion process that yields

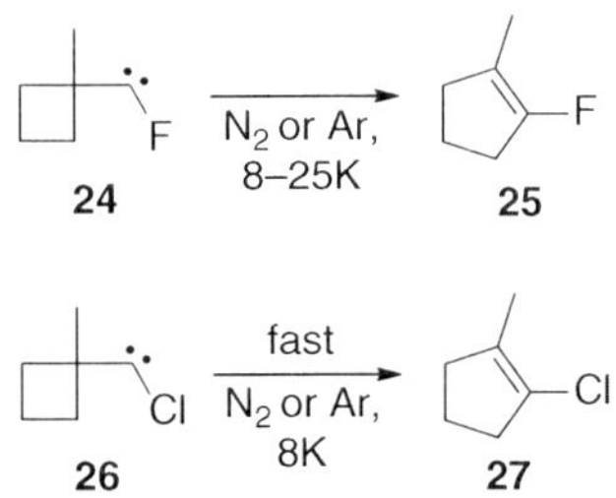

Scheme 7.7.

1-chloro-2-methylcyclopentene (**27**). Indeed, it was found experimentally that **26** could not be stabilized in a matrix even at 8 K, most probably due to an extremely fast tunneling process.[36]

7.2.6 Noradamantylchlorocarbene

An example of heavy-atom tunneling in carbene cage compounds is provided by the ring expansion rearrangement of noradamantylchlorocarbene (**28**) to 2-chloroadamantene (**29**), which was observed in a cryogenic nitrogen matrix at temperatures ranging from 9 to 23 K within a time frame of several days (Scheme 7.8). The barrier of insertion was computed to be around 5 kcal/mol. Carbene **28** is the first chlorocarbene in which heavy-atom tunneling could unequivocally be characterized experimentally.[95]

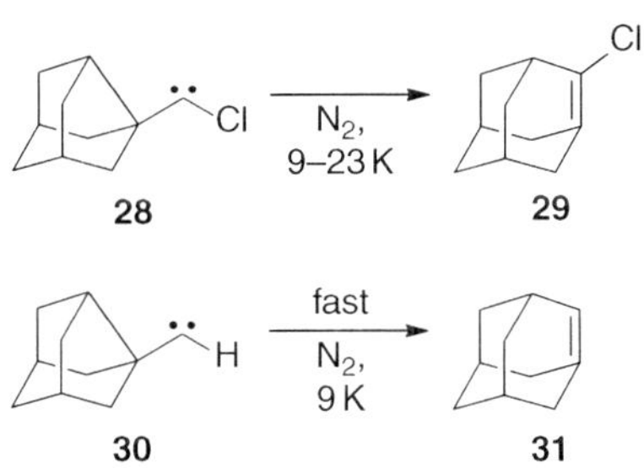

Scheme 7.8.

As in the case of **26**, unsubstituted noradamantylcarbene (**30**) is likely to undergo tunneling to 1-adamantene (**31**) too rapidly to be observable in the matrix. In accordance, the respective thermal barrier was computationally determined to be less than 1 kcal/mol.

7.3 THE FORMOSE REACTION AND HYDROXYCARBENES

7.3.1 Hydroxymethylene

In 1861, Butlerow discovered that upon treatment of aqueous formaldehyde (**32**) with the aqueous bases of alkali earth metals, several simple aldose sugars formed, up to a chain length of four. This reaction is known as the *formose* or

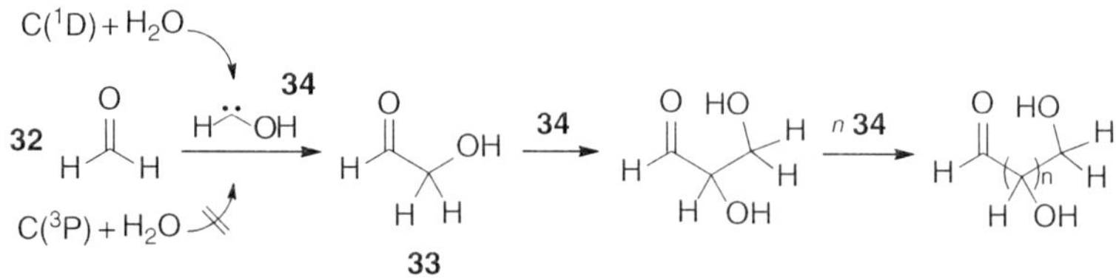

Scheme 7.9.

Butlerow reaction.[96,97] The active species, which was intermittently referred to as activated **32**, remained elusive and evaded experimental detection for more than a century.[98,99] In 1983, Shevlin and coworkers reacted atomic carbon, generated in an electrical arc in its excited ^{1}D singlet state, with gaseous water and found considerable amounts of glycolaldehyde (**33**) alongside **32** as the reaction products (Scheme 7.9). The researchers had discovered a variant of the formose reaction with an active species produced from carbon and water, which they proposed to be hydroxymethylene (**34**) on the basis of theoretical considerations.[100,101] In the light of a seemingly new pathway of sugar formation, it was speculated that there might be a spaceborne variant of the formose reaction, for example, in dense molecular clouds or within the primitive atmosphere of prebiotic Earth.

However, the probability of such a reaction pathway was shown to be exceptionally low in a matrix isolation study of the reaction between water and carbon in its ^{3}P triplet ground state, which was generated by laser ablation from graphite (Scheme 7.9). No formation of **34** was observed at cryogenic temperatures and very low pressures, that is, at conditions closely resembling those in molecular clouds;[102] hence, without a source of singlet carbon, the generation of sugars starting from **34** seems extremely unlikely.

Yet, in the course of a recent matrix isolation experiment, **34** could be generated and characterized by means of infrared (IR) and ultraviolet/visible (UV/Vis) spectroscopy.[22,103] High-vacuum flash pyrolysis (HVFP) of glyoxylic acid (**35**) afforded the carbene, which was immediately trapped in solid argon at 10 K. However, **34** was observed to undergo a fast transformation to **32** by hydrogen migration with a half-life τ of around 2 h (Scheme 7.10). The enthalpic barrier of the thermal [1,2]H-shift was determined to be as large as 30 kcal/mol by employing a focal point analysis (FPA) at the AE-CCSD(T) level in conjunction with correlation consistent basis sets including core correlation.[104] The singlet–triplet gap was deduced to be 28.0 kcal/mol. Given this formidable barrier and the cryogenic conditions, a classical over-the-barrier process can most certainly be excluded. Hence, [1,2]H-tunneling must be responsible for the observed reactivity, which marks the transformation of **34** to **32**, the first reaction to exhibit tunneling through a barrier of more than 30 kcal/mol. The half-life of **34** was estimated as 122 min by a WKB treatment featuring an intrinsic reaction path evaluated at the coupled cluster level. Upon *O*-deuteration, tunneling in *O*-*d*-**34** is effectively suppressed due to the twofold mass of deuterium compared to protium. As a result, *O*-*d*-**34** is stable under matrix conditions. Transformation to d_1-**32** can only be enforced by visible-light irradiation.

Tunneling in **34** has profound consequences for a possible spaceborne formose reaction: Given a limited half-life of approximately 2 h, a reaction with interstellar **32** seems improbable even in the densest molecular clouds.

Upon HVFP of oxalic acid (**36**) and its monomethyl ester, dihydroxy- (**37**) and methoxyhydroxycarbene (**38**), respectively, were similarly generated and characterized inside an argon matrix at 10 K (Scheme 7.10).[105] In stark contrast to **34**, neither **37** nor **38** undergo [1,2]H-tunneling to formic acid (**39**) or methylformate (**40**), respectively. The stability towards tunneling can be attributed to the twofold π-donating effect of the oxygen substituents, which compensates for the carbene's intrinsic electron deficiency. Electron donation thus raises the enthalpic barrier for [1,2]H-migration, whereas the width of the reaction path barely changes. As a result, the probability of H-tunneling is greatly reduced.

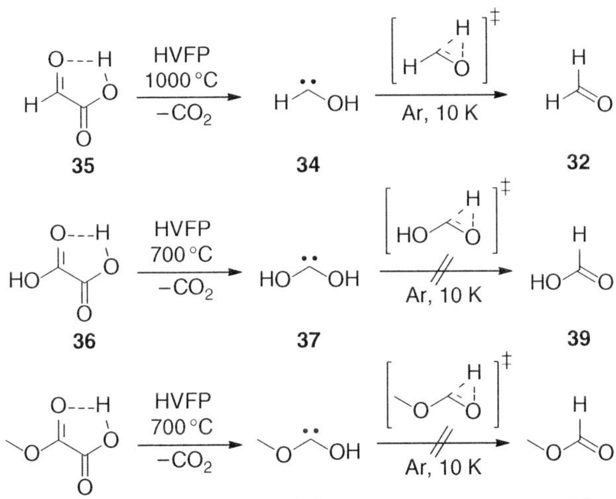

Scheme 7.10.

7.3.2 Methylhydroxycarbene and Tunneling Control

In order to probe for the effect of a methyl group on the half-life of **34**, methylhydroxycarbene (**41**) was generated by HVFP of pyruvic acid (**42**) and subsequent matrix isolation (Scheme 7.11).[106,107] Remarkably, **41** showed a shorter half-life ($\tau = 1.1$ h) than **34** in Ar. A WKB analysis using a similar approach as described for **34** (*vide supra*) resulted in a half-life of 66 min for **41**. When isolated in other noble gas matrices, **41** showed different lifetimes: In krypton and xenon, the half-life was determined to be $\tau = 3.3$ and 4.2 h, respectively. A comparable situation was previously found for several aliphatic and aromatic carboxylic acids.[108–110] The main product of HVFP was acetaldehyde (**43**), which was also the tunneling product of **41**. In addition, vinyl alcohol (**44**) was detected in the matrix after pyrolysis (Scheme 7.11). The occurrence of the latter can be explained as the result of a thermal [1,2]H-shift from the methyl

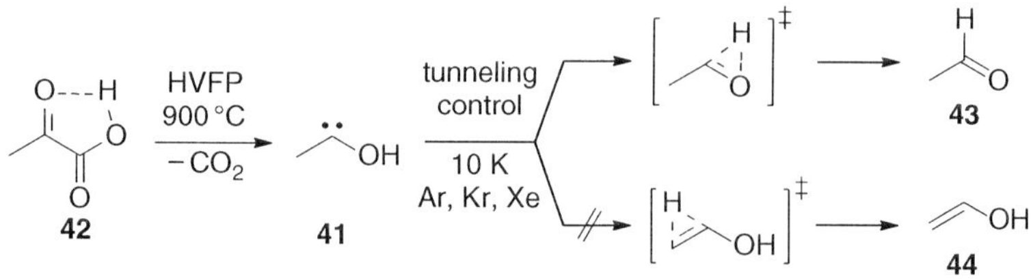

Scheme 7.11.

group to the carbene carbon, which occurs during the passage of the hot pyrolysis tube. Yet, once matrix isolated, an increase of concentration could only be detected for **43** but not for **44**.

When HVFP was repeated with *O*-deuterated **42**, neither d_1-**43** nor *O*-*d*-**44** could be observed as products of a tunneling mechanism. Hence, it can be concluded that under cryogenic conditions, **41** exclusively rearranges to **43** by [1,2] H-tunneling despite the existence of a second reaction pathway leading to **44**. Thus, only one of two conceivable [1,2]H-migrations in **41** selectively occurs by tunneling (Scheme 7.11). It is only upon irradiation of the matrix that both **43** and **44** are generated alongside each other.

The potential energy surface surrounding **41** was mapped out using the FPA extrapolation scheme with molecular geometries obtained at the AE-CCSD(T)/cc-pCVQZ level. Intriguingly, the thermal barrier for the rearrangement to **44** ($\Delta H^{\ddagger}$ = 22.6 kcal/mol) was found to be lower than the barrier for the formation of **43** ($\Delta H^{\ddagger}$ = 28.0 kcal/mol). Consequently, tunneling clearly favors the reaction with the higher barrier, a situation that is not explicable in terms of classical TST.[111,112] The concept of kinetic control implies that at low temperatures, such as during matrix isolation, chemical reactivity selectively favors the reaction with the lowest activation barrier. The reason for the apparent dichotomy is due to the neglect of barrier width in the original formulation of TST. A WKB analysis yields a broader barrier for the rearrangement of **41** to **44**, which results in a lower tunneling probability compared to [1,2]H-migration to **43** (Fig. 7.4). Due to the shortcomings of TST to account for this phenomenon within the framework of thermodynamic and kinetic control alone, the nonclassical principle of *tunneling control* needs to be introduced. Furthermore, the example of **41** unambiguously shows that tunneling must not only be considered as a mere correction of classical reaction rates but that it can, in fact, completely *control* the outcome of a reaction.

7.3.3 Arylhydroxycarbenes

In order to elucidate the effect of a π-system in direct vicinity of the carbene carbon on the half-life of [1,2]H-migration, phenylhydroxycarbene (**45**) was generated by HVFP of phenylglyoxylic acid (**46**) and immediate matrix isolation in solid Ar (Scheme 7.12). The tunneling half-life of **45** was spectroscopically

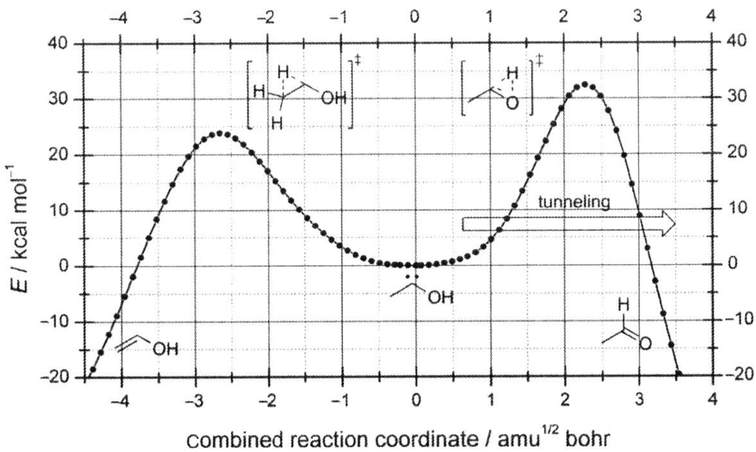

Figure 7.4. Reaction profile for the formation of acetaldehyde (**43**) and vinyl alcohol (**44**) starting from methylhydroxycarbene (**41**) along a combined reaction coordinate. H-tunneling only occurs towards **43** through the higher yet narrower barrier.

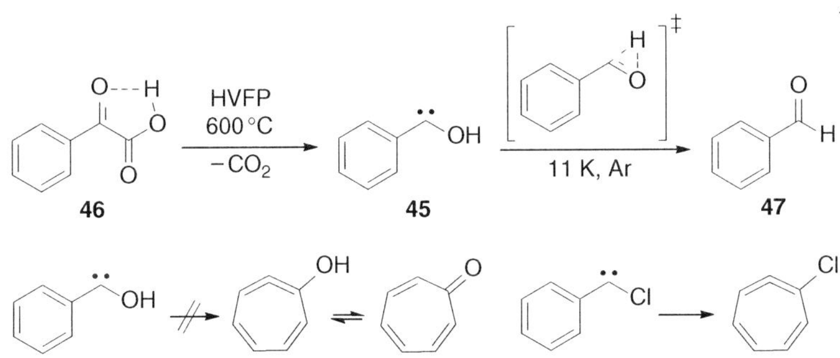

Scheme 7.12.

determined to be approximately 2.5 h in Ar, which is comparable to that of parent **34**. As anticipated, **45** produces benzaldehyde (**47**) upon H-migration. A WKB analysis encompassing high level CCSD(T)/cc-pVTZ energies on geometries generated at the MP2/aug-cc-pVDZ level gave a theoretical half-life of τ=3.3 h, which is in good agreement with experiment. *O*-deuteration renders **45** stable towards tunneling under cryogenic conditions. The small increase in half-life can be attributed to the slight π-donating ability of the phenyl moiety, an effect that is even more noticeable in the case of cyclopropylhydroxycarbene with a half-life of about 18 h. During the matrix experiment, no ring expansion product of **45**, that is, 1-hydroxycyclohepta-1,2,4,6-tetraene (**48**) or tautomer tropone (**49**) could be detected (Scheme 7.12). This behavior is in contrast to other phenylcarbenes, for example, phenylchlorocarbene (**50**), which readily rearranges to 1-chlorocyclohepta-1,2,4,6-tetraene (**51**).[71,113] Irradiation or heating of the matrix did not lead to a ring expansion product. In accordance, computations with the M06–2X functional revealed a sizeable thermal barrier for ring expansion.[114,115]

The first C–H bond insertion of a hydroxycarbene could be shown to occur in (*o*-methoxyphenyl)hydroxycarbene (**52**), which was generated by HVFP of *o*-methoxyphenyl-glyoxylic acid (**53**) and subsequent matrix isolation in solid Ar (Scheme 7.13). No traces of **52** could be detected inside the matrix and only minimal quantities of *o*-methoxybenzaldehyde (**54**) were present. The much more abundant insertion product was identified as 2,3-dihydrobenzofuran-3-ol (**55**), which was detected alongside its dehydration product benzo[*b*]furan (**56**) by means of IR spectroscopy and electron- impact mass spectrometry (EI–MS).[116] Hence, insertion of the carbene carbon into a C–H bond of the methoxy group is clearly kinetically favored compared to [1,2]H-tunneling. *O*-deuteration of the precursor acid did not have any impact on the outcome of the reaction. A computational investigation of the reaction resulted in a C–H insertion barrier of 21.0 kcal/mol, which suggests that insertion may easily occur within the pyrolysis zone. As the reaction path is broad due to extensive reorganization of carbon atoms, a tunneling contribution to the insertion process can safely be ruled out.

The experiment was subsequently repeated with (*o*-trifluoromethoxyphenyl)glyoxylic acid. Yet, (*o*-trifluoromethoxyphenyl)hydroxycarbene (**57**) could not be identified experimentally, most probably due to its thermal instability. However, almost quantitative amounts of the corresponding aldehyde were found, which is most likely formed by a thermal [1,2]H-shift within the pyrolysis tube. As expected, no product of C–F insertion was detected (Scheme 7.13).

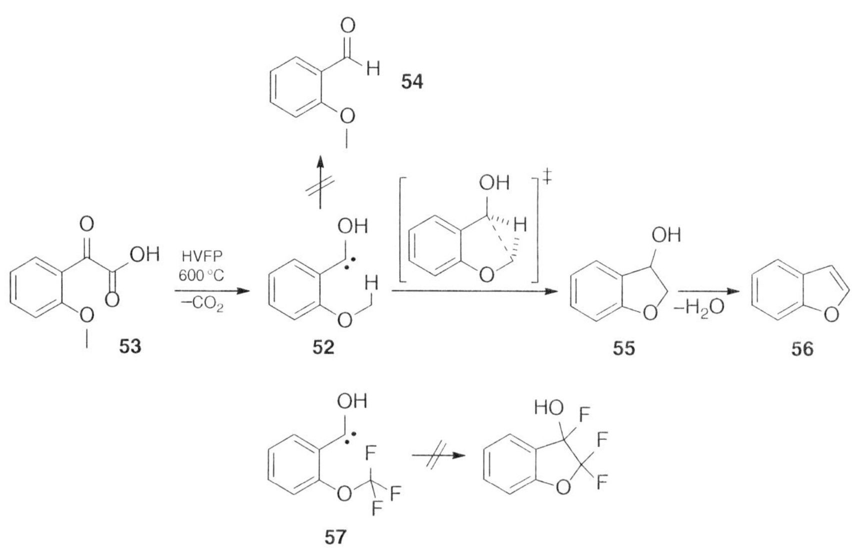

Scheme 7.13.

7.3.4 Cyclopropylhydroxycarbene

The cyclopropyl group is an excellent π-donor, but less strong than a second hetero-atom donor (O, S, N), which would shut down [1,2]H-tunneling completely as observed in dihydroxy- and methoxyhydroxycarbene. Still, the cyclopropyl group can be expected to display a stronger π-interaction with the carbene center than the phenyl moiety in phenylhydroxycarbene, resulting in a considerably larger rotational barrier due to the strong interaction of the cyclopropyl moiety's Walsh orbitals with the empty *p*-orbital located on the carbene center.

Upon HVFP of cyclopropylglyoxylic acid (**58**) and subsequent matrix isolation in Ar, cyclopropylhydroxycarbene (**59**) could be generated and characterized (Scheme 7.14).[117] Initially after pyrolysis, only small amounts of cyclopropylcarboxaldehyde (**60**) were detected in the matrix, but (**60**) noticeably increased in concentration in the course of hours, along with an observable decrease of the carbene signals. This process was again attributed to [1,2]H-tunneling, in accord with the previously examined hydroxycarbenes. The tunneling half-life was determined to be $\tau = 17.8$ h in good agreement with the WKB result of $\tau = 16.6$ h as evaluated by using the economical M06–2X functional. To date, this is the largest half-life of a hydroxycarbene that undergoes H-tunneling.

Unexpectedly, the main product of HVFP was identified as methylvinyl ketone (**61**), a product that is not directly apparent. The generation of the ketone was rationalized by a series of consecutive rearrangements within the pyrolysis zone (Scheme 7.14): Ring expansion of **59** initially leads to novel cyclobuten-1-ol (**62**), which instantly undergoes electrocyclic ring opening to buta-1,3-dien-2-ol (**63**). Finally, the dienol undergoes a thermal [1,3]H-shift to **61**. Remarkably, irradiation of the matrix did not generate **60** but rather **62**. Hence, the products of

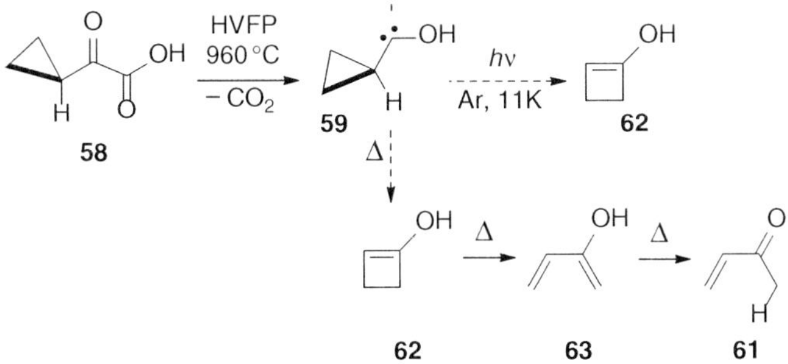

Scheme 7.14.

[1,2]H-tunneling and irradiation were *not* identical. Matrix-isolated **62** did not undergo further rearrangement. Computations with the M06-2X/6-311++G (d,p) functional revealed that the activation barrier for ring expansion of cyclopropylhydroxycarbene ($\Delta H^{\ddagger}=21.9$ kcal/mol) is smaller than that for the thermal [1,2]H-shift ($\Delta H^{\ddagger}=30.4$ kcal/mol). Hence, [1,2]H-tunneling under matrix conditions constitutes another example of *tunneling control*. *O*-deuteration stabilizes *O-d*-**59** and impedes tunneling.

By comparing the relative π-donor strength of all hydroxycarbenes bearing an α-carbon atom, a qualitative correlation was found between the amount of π-donation and half-life. This relationship also becomes apparent when the reaction enthalpies of the following isodesmic equations are considered (Scheme 7.15):

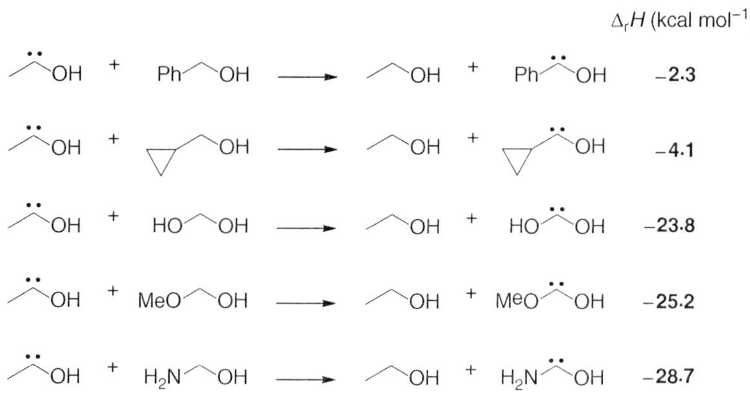

Scheme 7.15.

Hence, hydroxycarbenes bearing substituent groups with strong π-donating properties have significantly lengthened tunneling lifetimes. The π-effects subtly influence the barrier height of the [1,2]H-migration at work in hydroxycarbenes, whereas their effect on the barrier width is negligible.

7.4 CONCLUSION AND OUTLOOK

The distinctive reactivity of carbenes clearly demonstrates the chemical importance of tunneling, both for light and heavy atoms. It could be demonstrated that the half-life of [1,2]H-tunneling in hydroxycarbenes depends on the substituent and that tunneling can hence be influenced by straightforward chemical means. Yet, in order to fully comprehend the electronic nature of the tunneling mechanism, more experimental examples need to be uncovered. Studies on the cyclopropyl framework lead to the conclusion that tunneling in hydrocarbon rearrangements is probably more common than currently appreciated, especially when only little atomic movement is involved. Most importantly, however, it has been confirmed that barrier width usually trumps barrier height in tunneling processes. The insights about tunneling presented herein will most certainly be applicable to other systems as well, for example, in the rearrangements of protonated carbenes (i.e., carbocations), hydrogenation, and many others.

SUGGESTED READING

Carpenter, B. K. "Heavy-Atom Tunneling as the Dominant Pathway in a Solution-Phase Reaction? Bond Shift in Antiaromatic Annulenes." *J. Am. Chem. Soc.* **1983**, *105*, 1700–1701.

Ma, B., Schaefer III, H. F. "Singlet Methylcarbene: Equilibrium Geometry or Transition State?" *J. Am. Chem. Soc.* **1994**, *116*, 3539–3542.

Moss, R. A., Sauers, R. R., Sheridan, R. S., Tian, J., Zuev, P. S. "Carbon Tunneling in the Ring Expansion of Noradamantylchlorocarbene." *J. Am. Chem. Soc.* **2004**, *126*, 10196–10197.

Schreiner, P. R., Reisenauer, H. P., Pickard, F. C., Simmonett, A. C., Allen, W. D., Mátyus, E., Császár, A. G. "Capture of hydroxymethylene and its fast disappearance through tunnelling." *Nature* **2008**, *453*, 906–909.

Schreiner, P. R., Reisenauer, H. P., Ley, D., Gerbig, D., Wu, C.-H., Allen, W. D. "Methylhydroxycarbene: Tunneling Control of a Chemical Reaction." *Science* **2011**, *332*, 1300–1303.

Truhlar, D. G., Gordon, M. S. "From Force Fields to Dynamics: Classical and Quantal Paths." *Science* **1990**, *249*, 491–498.

Wierlacher, S., Sander, W., Liu, M. T. H. "Photolysis of Alkylhalodiazirines and Direct Observation of Benzylchlorocarbene in Cryogenic Matrices." *J. Am. Chem. Soc.* **1993**, *115*, 8943–8953.

Zhang, X., Hrovat, D. A., Borden, W. T. "Calculations Predict That Carbon Tunneling Allows the Degenerate Cope Rearrangement of Semibullvalene to Occur Rapidly at Cryogenic Temperatures." *Org. Lett.* **2010**, *12*, 2798–2801.

Zuev, P., Sheridan, R. S. "Tunneling in the CH Insertion of a Singlet Carbene: *tert*-Butylchlorocarbene." *J. Am. Chem. Soc.* **1994**, *116*, 4123–4124.

Zuev, P. S., Sheridan, R. S., Albu, T. V., Truhlar, D. G., Hrovat, D. A., Borden, W. T. "Carbon Tunneling from a Single Quantum State." *Science* **2003**, *299*, 867–870.

REFERENCES

1 Dennison, D. M., Uhlenbeck, G. E. *Phys. Rev.* **1932**, *41*, 313–321.
2 Gordon, J. P., Zeiger, H. J., Townes, C. H. *Phys. Rev.* **1954**, *95*, 282–284.
3 Whitman, D. W., Carpenter, B. K. *J. Am. Chem. Soc.* **1982**, *104*, 6473–6474.
4 Carpenter, B. K. *J. Am. Chem. Soc.* **1983**, *105*, 1700–1701.
5 Huang, M. J., Wolfsberg, M. *J. Am. Chem. Soc.* **1984**, *106*, 4039–4040.
6 Dewar, M. J. S., Merz, K. M., Stewart, J. J. P. *J. Am. Chem. Soc.* **1984**, *106*, 4040–4041.
7 Orendt, A. M., Arnold, B. R., Radziszewski, J. G., Facelli, J. C., Malsch, K. D., Strub, H., Grant, D. M., Michl, J. *J. Am. Chem. Soc.* **1988**, *110*, 2648–2650.
8 Zhang, X., Hrovat, D. A., Borden, W. T. *Org. Lett.* **2010**, *12*, 2798–2801.

9 Tal'roze, V. L., Lyubimova, A. K. *Dokl. Akad. Nauk. SSSR* **1952**, *86*, 909–912.
10 Sefcik, M. D., Henis, J. M. S., Gaspar, P. P. *J. Chem. Phys.* **1974**, *61*, 4321–4328.
11 Scuseria, G. E. *Nature* **1993**, *366*, 512–513.
12 Schreiner, P. R., Kim, S.-J., Schaefer III, H. F., Schleyer, P. v. R. *J. Chem. Phys.* **1993**, *99*, 3716–3720.
13 White, E. T., Tang, J., Oka, T. *Science* **1999**, *284*, 135–137.
14 Kramer, G. M. *Science* **1999**, *286*, 1051.
15 Marx, D., Parrinello, M. *Science* **1999**, *284*, 59–61.
16 Asvany, O., Kumar P., P., Redlich, B., Hegemann, I., Schlemmer, S., Marx, D. *Science* **2005**, *309*, 1219–1222.
17 Kumar P., P., Marx, D. *Phys. Chem. Chem. Phys.* **2006**, *8*, 573–586.
18 Ivanov, S. D., Asvany, O., Witt, A., Hugo, E., Mathias, G., Redlich, B., Marx, D., Schlemmer, S. *Nature Chem.* **2010**, *2*, 298–302.
19 Witt, A., Ivanov, S. D., Mathias, G., Marx, D. *J. Phys. Chem. Lett.* **2011**, *2*, 1377–1381.
20 Fukui, K. *J. Phys. Chem.* **1970**, *74*, 4161–4163.
21 Fukui, K. *Acc. Chem. Res.* **1981**, *14*, 363–368.
22 Schreiner, P. R., Reisenauer, H. P., Pickard, F. C., Simmonett, A. C., Allen, W. D., Mátyus, E., Császár, A. G. *Nature* **2008**, *453*, 906–909.
23 Rosa, C., Brandão, *J. Chem. Phys. Lett.* **2008**, *461*, 150–154.
24 Kiselev, V. G., Swinnen, S., Nguyen, V. S., Gritsan, N. P., Nguyen, M. T. *J. Phys. Chem. A* **2010**, *114*, 5573–5579.
25 Eyring, H. *J. Chem. Phys.* **1935**, *3*, 107–115.
26 Evans, M. G., Polanyi, M. *Trans. Faraday. Soc.* **1935**, *31*, 875–894.
27 Truhlar, D. G., Isaacson, A. D., Skodje, R. T., Garrett, B. C. *J. Chem. Phys.* **1982**, *86*, 2252–2261.
28 Truhlar, D. G., Gordon, M. S. *Science* **1990**, *249*, 491–498.
29 Fernandez-Ramos, A., Ellingson, B. A., Garrett, B. C., Truhlar, D. G. In *Rev. Comput. Chem.* Lipkowitz, K. B., Cundari, T. R., Eds. Wiley-VCH: Hoboken, **2007**, Vol. *23*, pp. 125–232.
30 Isaacson, A. D., Truhlar, D. G., Rai, S. N., Steckler, R., Hancock, G. C., Garrett, B. C., Redmon, M. J. *Comput. Phys. Commun.* **1987**, *47*, 91–102.
31 Steckler, R., Hu, W.-P., Liu, Y.-P., Lynch, G. C., Garrett, B. C., Isaacson, A. D., Melissas, V. S., Lu, D., Truong, T. N., Rai, S. N., Hancock, G. C., Lauderdale, J. G., Joseph, T., Truhlar, D. G. *Comput. Phys. Commun.* **1995**, *88*, 341–343.
32 Zhang, X., Datta, A., Hrovat, D. A., Borden, W. T. *J. Am. Chem. Soc.* **2009**, *131*, 16002–16003.
33 Gonzalez-James, O. M., Zhang, X., Datta, A., Hrovat, D. A., Borden, W. T., Singleton, D. A. *J. Am. Chem. Soc.* **2010**, *132*, 12548–12549.
34 Gerbig, D., Ley, D., Schreiner, P. R. *Org. Lett.* **2011**, *13*, 3526–3529.
35 Dewar, M. J. S., Merz, K. M., Stewart, J. J. P. *J. Chem. Soc., Chem. Commun. 1985*, 166–168.
36 Zuev, P. S., Sheridan, R. S., Albu, T. V., Truhlar, D. G., Hrovat, D. A., Borden, W. T. *Science* **2003**, *299*, 867–870.
37 Rice, F. O., Glasebrook, A. L. *J. Am. Chem. Soc.* **1934**, *56*, 741–743.
38 Bawn, C. E. H., Milsted, *J. Trans. Faraday Soc.* **1939**, *35*, 889–896.
39 Volman, D. H., Leighton, P. A., Blacet, F. E., Brinton, R. K. *J. Chem. Phys.* **1950**, *18*, 203–206.
40 Brinton, R. K., Volman, D. H. *J. Chem. Phys.* **1951**, *19*, 1394–1395.

41 Frey, H. M. *J. Chem. Soc.* **1962**, 2293–2297.
42 Kramer, K. A. W., Wright, A. N. *Tetrahedron Lett.* **1962**, *3*, 1095–1096.
43 Tschuikow-Roux, E., McNesby, J. R., Jackson, W. M., Faris, J. L. *J. Phys. Chem.* **1967**, *71*, 1531–1533.
44 Seburg, R. A., McMahon, R. J. *J. Am. Chem. Soc.* **1992**, *114*, 7183–7189.
45 Modarelli, D. A., Platz, M. S. *J. Am. Chem. Soc.* **1993**, *115*, 470–475.
46 Frey, H. M., Stevens, I. D. R. *J. Chem. Soc.* **1965**, 1700–1706.
47 Schreiner, P. R., Reisenauer, H. P., Sattelmeyer, K. W., Allen, W. D. *J. Am. Chem. Soc.* **2005**, *127*, 12156–12157.
48 O'Gara, J. E., Dailey, W. P. *J. Am. Chem. Soc.* **1994**, *116*, 12016–12021.
49 Altmann, J. A., Csizmadia, I. G., Yates, K. *Chem. Phys. Lett.* **1976**, *41*, 500–502.
50 Harding, L. B. *J. Am. Chem. Soc.* **1981**, *103*, 7469–7475.
51 Ha, T.-K., Nguyen, M. T., Vanquickenborne, L. G. *Chem. Phys. Lett.* **1982**, *92*, 459–461.
52 Evanseck, J. D., Houk, K. N. *J. Phys. Chem.* **1990**, *94*, 5518–5523.
53 Gallo, M. M., Schaefer III, H. F. *J. Phys. Chem.* **1992**, *96*, 1515–1517.
54 Ma, B., Schaefer III, H. F. *J. Am. Chem. Soc.* **1994**, *116*, 3539–3542.
55 Miller, D. M., Schreiner, P. R., Schaefer III, H. F. *J. Am. Chem. Soc.* **1995**, *117*, 4137–4143.
56 Nguyen, M. T., Matus, M. H., Lester, W. A., Dixon, D. A. *J. Phys. Chem. A* **2007**, *112*, 2082–2087.
57 Gerbig, D., Reisenauer, H. P., Wu, C.-H., Ley, D., Allen, W. D., Schreiner, P. R. *J. Am. Chem. Soc.* **2010**, *132*, 7273–7275.
58 Sugiyama, M. H., Celebi, S., Platz, M. S. *J. Am. Chem. Soc.* **1992**, *114*, 966–973.
59 Schreiner, P. R., Karney, W. L., Schleyer, P. v. R., Borden, W. T., Hamilton, T. P., Schaefer III, H. F. *J. Org. Chem.* **1996**, *61*, 7030–7039.
60 Cramer, C. J., Truhlar, D. G., Falvey, D. E. *J. Am. Chem. Soc.* **1997**, *119*, 12338–12342.
61 Trozzolo, A. M., Murray, R. W., Wasserman, E. *J. Am. Chem. Soc.* **1962**, *84*, 4990–4991.
62 Gaspar, P. P., Hsu, J.-P., Chari, S., Jones Jr., M. *Tetrahedron* **1985**, *41*, 1479–1507.
63 Sander, W. *Angew. Chem. Int. Ed.* **1985**, *24*, 988–989.
64 McMahon, R. J., Chapman, O. L. *J. Am. Chem. Soc.* **1987**, *109*, 683–692.
65 Sander, W., Bucher, G., Wierlacher, S. *Chem. Rev.* **1993**, *93*, 1583–1621.
66 Platz, M. S. *Acc. Chem. Res.* **1995**, *28*, 487–492.
67 Matzinger, S., Bally, T., Patterson, E. V., McMahon, R. J. *J. Am. Chem. Soc.* **1996**, *118*, 1535–1542.
68 Admasu, A., Platz, M. S., Marjcinek, A., Michalak, J., Gudmundsdóttir, A. D., Gebicki, J. *J. Phys. Org. Chem.* **1997**, *10*, 207–220.
69 Matzinger, S., Bally, T. *J. Phys. Chem. A* **2000**, *104*, 3544–3552.
70 Zhang, Y., Burdzinski, G., Kubicki, J., Platz, M. S. *J. Am. Chem. Soc.* **2008**, *130*, 16134–16135.
71 Noller, B., Poisson, L., Maksimenka, R., Gobert, O., Fischer, I., Mestdagh, J. M. *J. Phys. Chem. A* **2009**, *113*, 3041–3050.
72 Cramer, C. J., Dulles, F. J., Falvey, D. E. *J. Am. Chem. Soc.* **1994**, *116*, 9787–9788.
73 Wong, M. W., Wentrup, C. *J. Org. Chem.* **1996**, *61*, 7022–7029.
74 Karney, W. L., Borden, W. T. *J. Am. Chem. Soc.* **1997**, *119*, 1378–1387.
75 Geise, C. M., Hadad, C. M. *J. Org. Chem.* **2000**, *65*, 8348–8356.
76 Geise, C. M., Hadad, C. M. *J. Org. Chem.* **2002**, *67*, 2532–2540.

77 Cui, G., Fang, W. *ChemPhysChem* **2011**, *12*, 1689–1696.
78 Moss, R. A., Mamantov, A. *J. Am. Chem. Soc.* **1970**, *92*, 6951–6956.
79 Moss, R. A., Munjal, R. C. *J. Chem. Soc., Chem. Commun.* **1978**, 775–776.
80 Liu, M. T. H., Subramanian, R. *J. Chem. Soc., Chem. Commun.* **1984**, 1062–1064.
81 Jackson, J. E., Soundararajan, N., White, W., Liu, M. T. H., Bonneau, R., Platz, M. S. *J. Am. Chem. Soc.* **1989**, *111*, 6874–6875.
82 LaVilla, J. A., Goodman, J. L. *J. Am. Chem. Soc.* **1989**, *111*, 6877–6878.
83 Liu, M. T. H., Bonneau, R. *J. Am. Chem. Soc.* **1989**, *111*, 6873–6874.
84 Liu, M. T. H., Suresh, R. V. *J. Org. Chem.* **1989**, *54*, 486–488.
85 Liu, M. T. H., Bonneau, R. *J. Am. Chem. Soc.* **1990**, *112*, 3915–3919.
86 Moss, R. A., Ho, G. J., Liu, W. *J. Am. Chem. Soc.* **1992**, *114*, 959–963.
87 Dix, E. J., Herman, M. S., Goodman, J. L. *J. Am. Chem. Soc.* **1993**, *115*, 10424–10425.
88 Wierlacher, S., Sander, W., Liu, M. T. H. *J. Am. Chem. Soc.* **1993**, *115*, 8943–8953.
89 Liu, M. T. H., Bonneau, R., Wierlacher, S., Sander, W. *J. Photochem. Photobiol. A: Chem.* **1994**, *84*, 133–137.
90 Kraka, E., Cremer, D. *J. Phys. Org. Chem.* **2002**, *15*, 431–447.
91 Liu, M. T. H. *J. Chem. Soc., Chem. Commun.* **1985**, 982–984.
92 Merrer, D. C., Moss, R. A., Liu, M. T. H., Banks, J. T., Ingold, K. U. *J. Org. Chem.* **1998**, *63*, 3010–3016.
93 Jackson, J. E., Soundararajan, N., Platz, M. S., Liu, M. T. H. *J. Am. Chem. Soc.* **1988**, *110*, 5595–5596.
94 Zuev, P., Sheridan, R. S. *J. Am. Chem. Soc.* **1994**, *116*, 4123–4124.
95 Moss, R. A., Sauers, R. R., Sheridan, R. S., Tian, J., Zuev, P. S. *J. Am. Chem. Soc.* **2004**, *126*, 10196–10197.
96 Butlerow, A. *Ann. Chem. Pharm.* **1861**, *120*, 295–298.
97 Breslow, R. *Tetrahedron Lett.* **1959**, *1*, 22–26.
98 Baly, E. C. C., Heilbron, I. M., Barker, W. F. *J. Chem. Soc. Trans.* **1921**, *119*, 1025–1035.
99 Baly, E. C. C., Heilbron, I. M., Hudson, D. P. *J. Chem. Soc. Trans.* **1922**, *121*, 1078–1088.
100 Ahmed, S. N., McKee, M. L., Shevlin, P. B. *J. Am. Chem. Soc.* **1985**, *107*, 1320–1324.
101 Ahmed, S. N., McKee, M. L., Shevlin, P. B. *J. Am. Chem. Soc.* **1983**, *105*, 3942–3947.
102 Schreiner, P. R., Reisenauer, H. P. *ChemPhysChem* **2006**, *7*, 880–885.
103 Bucher, G. *Angew. Chem. Int. Ed.* **2008**, *47*, 6957–6958.
104 Li, B., Ren, Y., Bian, W. *ChemPhysChem* **2011**, *12*, 2419–2422.
105 Schreiner, P. R., Reisenauer, H. P. *Angew. Chem. Int. Ed.* **2008**, *47*, 7071–7074.
106 Schreiner, P. R., Reisenauer, H. P., Ley, D., Gerbig, D., Wu, C.-H., Allen, W. D. *Science* **2011**, *332*, 1300–1303.
107 Carpenter, B. K. *Science* **2011**, *332*, 1269–1270.
108 Pettersson, M., Maçôas, E. M. S., Khriachtchev, L., Lundell, J., Fausto, R., Räsänen, M. *J. Chem. Phys.* **2002**, *117*, 9095–9098.
109 Maçôas, E. M. S., Khriachtchev, L., Pettersson, M., Fausto, R., Räsänen, M. *J. Chem. Phys.* **2004**, *121*, 1331–1338.
110 Amiri, S., Reisenauer, H. P., Schreiner, P. R. *J. Am. Chem. Soc.* **2010**, *132*, 15902–15904.
111 Korneev, S. M. *Eur. J. Org. Chem.* **2011**, *2011*, 6153–6175.

112 Martinů, T., Böhm, S., Hanzlová, E. *Eur. J. Org. Chem.* **2011**, *2011*, 6254–6260.
113 Sander, W. *Spectrochim. Acta* **1987**, *43A*, 637–646.
114 Zhao, Y., Truhlar, D. G. *Acc. Chem. Res.* **2008**, *41*, 157–167.
115 Zhao, Y., Truhlar, D. G. *Theor. Chem. Acc.* **2008**, *120*, 215–241.
116 Gerbig, D., Ley, D., Reisenauer, H. P., Schreiner, P. R. *Beilstein J. Org. Chem.* **2010**, *6*, 1061–1069.
117 Ley, D., Gerbig, D., Wagner, J. P., Reisenauer, H. P., Schreiner, P. R. *J. Am. Chem. Soc.* **2011**, *133*, 13614–13621.

8

CARBODICARBENES

GERNOT FRENKING AND RALF TONNER
Fachbereich Chemie, Philipps-Universität Marburg, Marburg, Germany

8.1 INTRODUCTION

This chapter deals with four different classes of two-coordinated carbon compounds which are depicted in Scheme 8.1. Carbenes (**A**) are divalent carbon (II) compounds CR_2 where the two-coordinated carbon atom has two electron-sharing bonds to the substituents R and the remaining nonbonding electrons have either the same spin (triplet carbene) or opposite spin (singlet carbene). Carbenes were introduced into contemporary synthetic chemistry with the syntheses of the first stable carbenes by Bertrand[1] and by Arduengo,[2] who isolated the first stable *N*-heterocyclic carbene (NHC). Because singlet carbenes have a formally empty p(π) orbital at carbon, they are strong σ donors and weak to moderate π acceptors.[3]

Carbones (**B**) are divalent carbon (0) compounds CL_2 which have two donor–acceptor bonds $L\rightarrow C\leftarrow L$ to a carbon atom in the excited ^1D singlet state and which possess *two* lone pair orbitals with σ and π symmetry.[4–6] Carbones are thus σ donors and π donors. The ligands L in typical carbones are strong σ donors and weak π acceptors. The first carbone which was synthesized already in 1961 by Ramirez is the carbodiphosphorane $C(PPh_3)_2$.[7] The synthetic chemistry of carbodiphosphoranes was extensively studied in experimental work in the following years,[8] but the peculiar donor–acceptor bonding situation $Ph_3P\rightarrow C\leftarrow PPh_3$ was only recently recognized in quantum chemical studies of complexes where $C(PPh_3)_2$ serves as mono- and bidendate Lewis bases.[4,9] A particular class of carbones CL_2 are *carbodicarbenes* $C(CR_2)_2$ where the

Contemporary Carbene Chemistry, First Edition. Edited by Robert A. Moss and Michael P. Doyle.

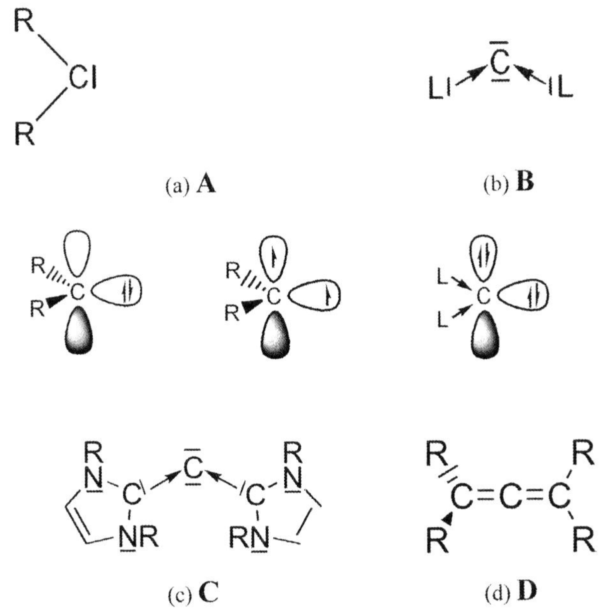

Scheme 8.1.

phosphane ligands of carbodiphosphoranes are replaced by singlet carbenes. Quantum chemical calculations predicted that carbodicarbenes which have NHC ligands $C(NHC)_2$ (**C**) should be stable compounds that might be isolable in a condensed phase.[5] Phosphanes PR_3 and NHCs are both considered to be ligands that are strong σ donors and weak to moderate π acceptors, depending on the nature of the substituent R.[10] Carbodicarbenes exhibit the unusual feature that there are carbon–carbon donor–acceptor bonds (NHC)→C←(NHC). This prediction was quickly verified by the synthesis of substituted $C(NHC)_2$ compounds by Bertrand[11] and by Fürstner.[12]

The Lewis structure for the carbodicarbene **C** is valid when the π back-donation of the carbon p(π) lone pair to the carbene donor atom is negligible. If the π back-donation becomes stronger, the bonding situation converges toward the situation which is found in allenes (**D**), where the two-coordinated carbon(IV) atom has four electron-sharing bonds with the terminal CR_2 groups, that is, two σ and two π bonds. It becomes obvious that the bonding situations **C** and **D** are strongly influenced by the nature of R. The donor–acceptor bonding **C** sketches a situation where the carbene ligands bind in their electronic singlet state, whereas the carbene ligands in **D** bind in the triplet state. There may be cases where the real situation is between the two extreme cases and the formal oxidation state of the two-coordinated carbon atoms is between zero and four. This is the reason that some species were termed "bent allenes."[11,13] There are also molecules whose bonding situation lies between **A** and **B**. Bonding models in chemistry are qualitative sketches of extreme situations which serve as orientation for classifying real molecules whose electronic structure is usually between two extremes. This is often done by using mesomeric forms of Lewis structures for a molecule. In the past, carbones **B** were sometimes sketched as diylides with formal charges $L^+–C^{2-}–L^+$.[14] Systematic studies showed that carbones are best described as donor–acceptor complexes where arrows are introduced in Lewis structures for specifying the bonding situation $L \rightarrow C \leftarrow L$.[4,5]

Our topic is the electronic structure and reactivity of carbodicarbenes **C** and related dicoordinated carbon compounds whose bonding situation belongs to the spectrum which is shown in Scheme 8.1. Carbodicarbenes have been known only since 2007[5,11,12] but the theoretical and experimental work carried out so far indicates that they are a promising class of compounds in an infant area that is largely unexplored. Unless otherwise noted, the following results stem from quantum chemical calculations using the BP86/def2-SVP approximation for structural optimization and subsequent calculation of improved energies with the def2-TZVPP basis set. Further details are given in the literature which is cited next.

8.2 CARBODICARBENES WITH *N*-HETEROCYCLIC LIGANDS $C(NHC)_2$

Quantum chemical calculations predicted that the carbodicarbenes $C(NHC^{H})_2$ (**1**) and $C(NHC^{Me})_2$ (**2**), which have hydrogen or methyl attached to nitrogen, have geometries with strongly bent C–C–C angles α of 126.6° for **1** and 132.3° for **2**.[4b] The X-ray structure analysis[11] of the benzoannelated carbodicarbene $C(NHC^{Bz})_2$ (**3**) gave a bending angle of $\alpha = 134.8(2)°$ which is in reasonable agreement with the calculated value of 142.5° for **3**.[5b] The bending potential for the C–C–C angle of carbodicarbenes is rather shallow and the difference between the theoretical and experimental value may be due to solid state effects. The calculated bending angle for the saturated carbodicarbene $C(NHC_{sat}^{Me})_2$ (**4**) which has a C–C single bond in the NHC ring is slightly larger (141.6°) than in **2**.[5b] The optimized geometries of **1–4** are shown in Figure 8.1.

A characteristic feature of carbones is the occurrence of two lone pair orbitals. Figure 8.2a and b show plots of the HOMO and HOMO-1 of the parent carbodicarbene **1**. The HOMO-1 is clearly a σ-type lone pair orbital which is localized at the central carbon atom C1. The HOMO is a π-type orbital which also has the largest coefficient at C1. The energy ordering of HOMO and HOMO-1 is opposite to what is found for the NHC^{H} (Figs. 8.2c and d) where the HOMO is a σ-type lone pair orbital, whereas the HOMO-1 is a delocalized π-type orbital. The shape of the occupied frontier orbitals of the isolated carbodicarbene $C(NHC^{Bz})_2$ (**3**), which are shown in Figure 8.2e and f, closely resemble the HOMO and HOMO-1 of **1**.

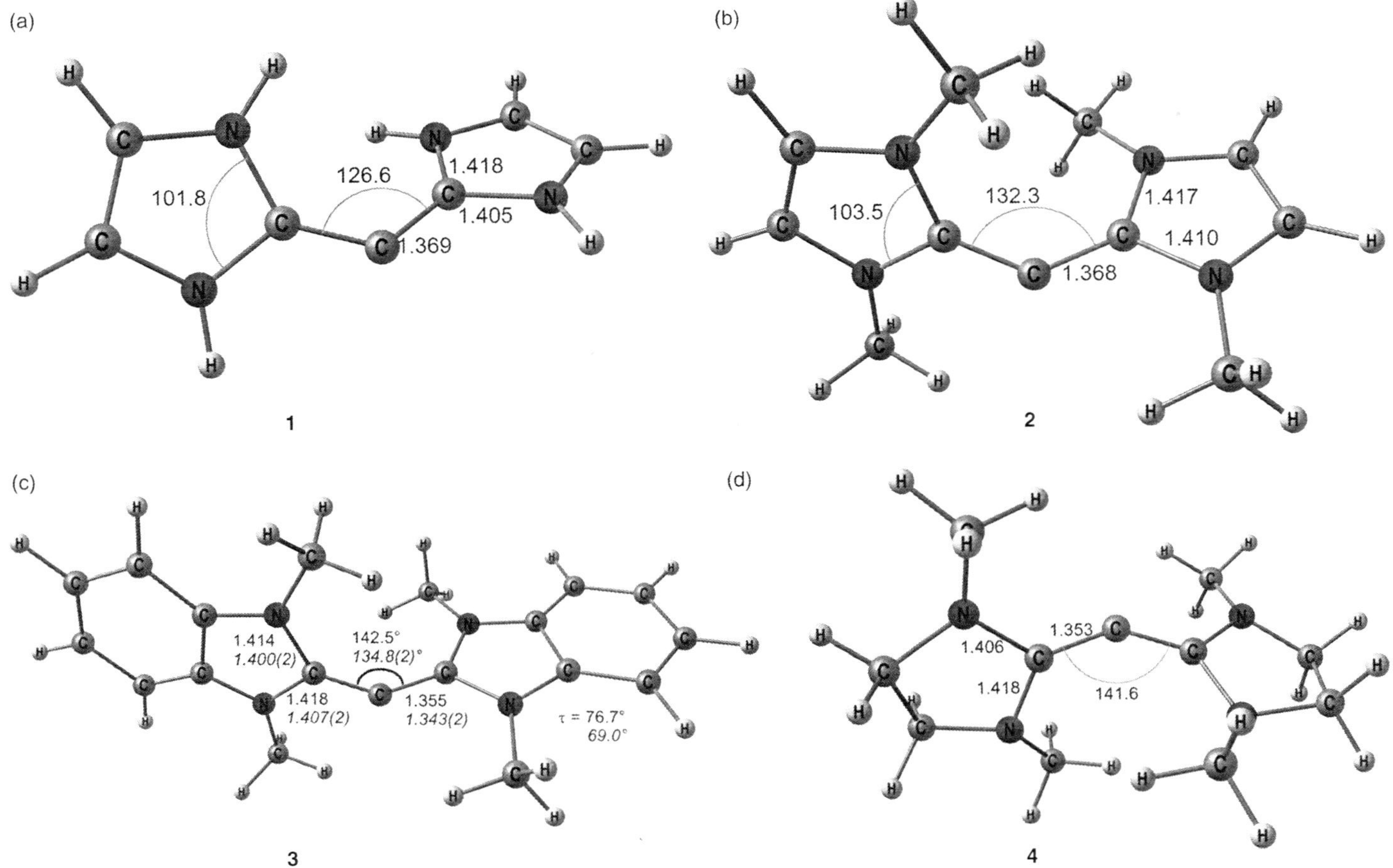

Figure 8.1. Calculated equilibrium geometries of the carbodicarbenes **1–4**. Bond lengths in angstrom; angles in degrees. Experimental values for **3** are given in italics. The values are taken from References 4b and 5b.

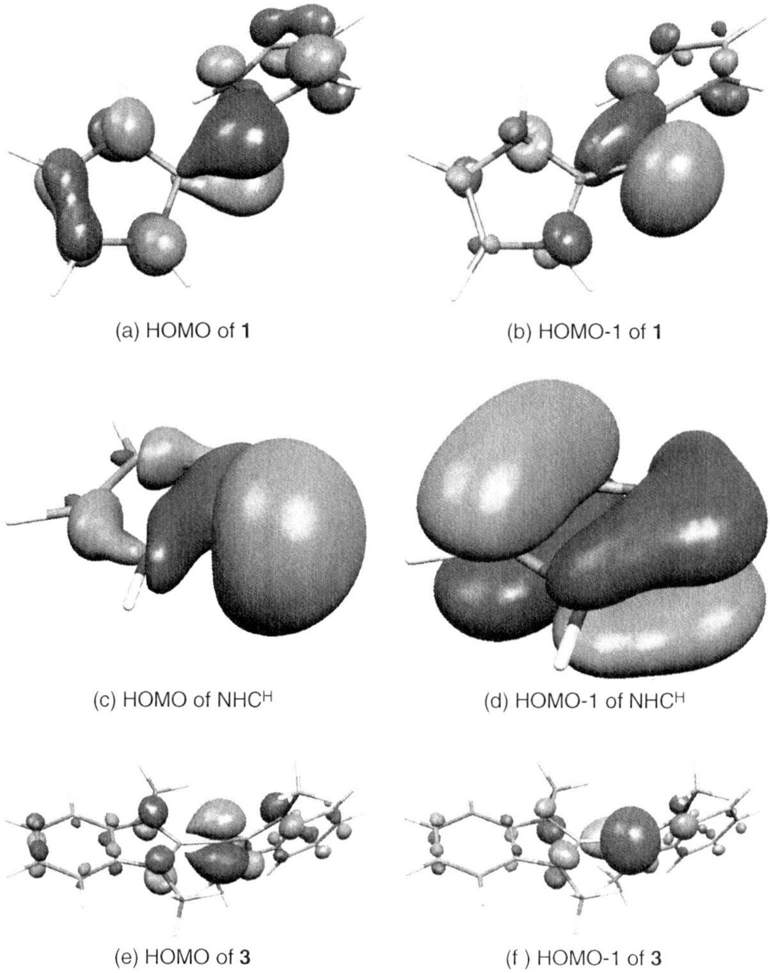

Figure 8.2. Plot of the two highest lying occupied orbitals HOMO and HOMO-1 of the carbodicarbenes **1** and **3** and the carbene NHC^H; see References 4b and 5b. For color details, please see color plate section.

The difference between the carbene NHC^{Me} and the carbones **1–4** becomes manifest by comparison of the calculated second proton affinities (PAs) which are given in Table 8.1. Whereas the first PAs of all compounds have a comparable magnitude, the second PA of NHC^{Me} (71.8 kcal/mol) is much lower than the second PAs of the carbodicarbenes **1–4** which are between 155.3 and 168.4 kcal/mol. A systematic investigation of the first and second PAs of carbon bases showed that the second PA of NHCs is in the range 40–100 kcal/mol, whereas carbones CL_2 exhibit values between 110 and 200 kcal/mol.[15] The very large value for the second PAs of the carbodicarbenes suggests that doubly protonated species

TABLE 8.1. First and Second Proton Affinities (PA) of Compounds 1–8 and NHC^{Me} in Kilocalories per Mole at the MP2/TZVPP//BP86/SVP Level of Theory

	N	N–(H^+)		N–$(H^+)_2$	
Compound	q(C1)	1. PA	q(C1)	2. PA	q(C1)
1	−0.51	292.3	−0.46	155.3	−0.52
2	−0.50	294.3	−0.52	168.4	−0.53
3	−0.43	284.7	−0.51	167.8	−0.53
4	−0.40	282.2	−0.48	157.5	−0.54
5	−0.21	282.5	−0.47	151.6	−0.51
6	−0.22	268.2	−0.47	175.8	−0.57
7	−0.48	296.5	−0.42	158.7	−0.52
8	−0.48	284.3	−0.44	133.3	−0.61
9	−0.46	306.7[a]	−0.37	152.3[a]	−0.57
NHC^{Me}	0.04	262.3	0.21	71.8	−0.12

Atomic partial charges at the central carbon atom q(C1) in electrons (BP86/TZVPP//BP86/SVP). Data are taken from References 5b, 13c, and 15.
[a]Calculated at BP86/TZVPP//BP86/SVP.

should be stable species. This is in agreement with experimental results. The dication **3-$(H^+)_2$** was easily prepared as a salt, which actually served as a precursor for the isolation of **3** through bis-deprotonation.[11]

Table 8.1 also shows the atomic partial charges at the dicoordinated carbon atom. The donor–acceptor bonding model L→C←L suggests that the carbon atom carries a negative partial charge which could become smaller in cases of π back-donation. The calculations give negative charges for the carbodicarbenes **1–4**, whereas NHC^{Me} is essentially neutral. We point out that the negative charge at carbon becomes only slightly smaller upon protonation in **1-(H^+)**, whereas it even becomes more negative in **2-(H^+)–4-(H^+)**. Even more surprisingly, the partial charge q(C1) becomes even more negative upon second protonation in all four systems **N-$(H^+)_2$**. The loss of electronic charge from the lone pair electrons to the protons makes the carbon atom become a stronger Lewis acid than neutral C(0).

The occurrence of a second lone pair orbital makes carbodicarbenes double Lewis donors not only for protons but also for other Lewis acids. Figure 8.3a–d shows the optimized geometries of complexes of carbodicarbenes **1–4** with two Lewis acids BH_3 and the theoretically predicted bond dissociation energies D_e for loss of the first and the second Lewis acid. The calculated data indicate that the carbodicarbenes bind the second Lewis acid still rather strongly. In contrast, calculations predict that NHC^H binds only one BH_3 ligand at carbon.[4c] However, the number of monodentate Lewis acids that bind at carbon may not always be a faithful probe for the donor multiplicity of the Lewis base. Figure 8.3e shows that the carbene NHC^{Me} binds two AuCl moieties at the divalent carbon atom and that the bond association energy (BDE) of the second AuCl is still rather large (D_e=36.4 kcal/mol).[4f] However, the latter value is significantly smaller than the BDE of the second AuCl of **2-$(AuCl)_2$** which is displayed in Figure 8.3f (D_e=51.1 kcal/mol), while the BDE of the first AuCl

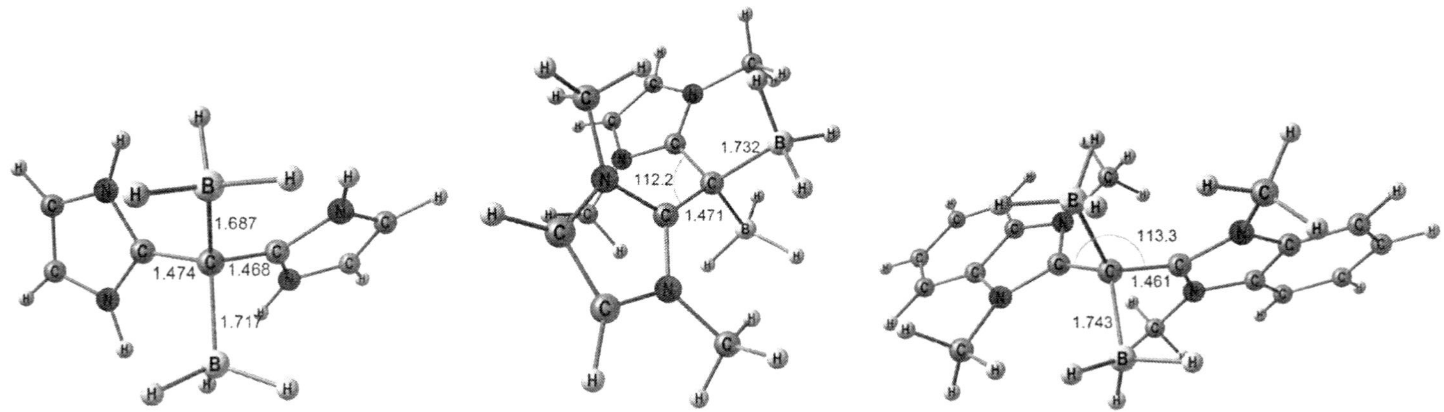

(a) **1-$(BH_3)_2$**

D_e (1. BH_3) = 23.5 kcal/mol
D_e (2. BH_3) = 64.4 kcal/mol

(b) **2-$(BH_3)_2$**

D_e (1. BH_3) = 24.3 kcal/mol
D_e (2. BH_3) = 46.1 kcal/mol

(c) **3-$(BH_3)_2$**

D_e (1. BH_3) = 18.6 kcal/mol
D_e (2. BH_3) = 39.7 kcal/mol

Figure 8.3. Optimized geometries of carbodicarbene complexes with two BH_3 ligands **1-$(BH_3)_2$–4-$(BH_3)_2$** and complexes with two AuCl ligands **NHC^{Me}-$(AuCl)_2$** and **2-$(AuCl)_2$**. Bond lengths in angstrom; angles in degrees. Calculated bond dissociation energies D_e for the first and the second ligand in kcal/mol. The values are taken from References 4c,f, and 5b.

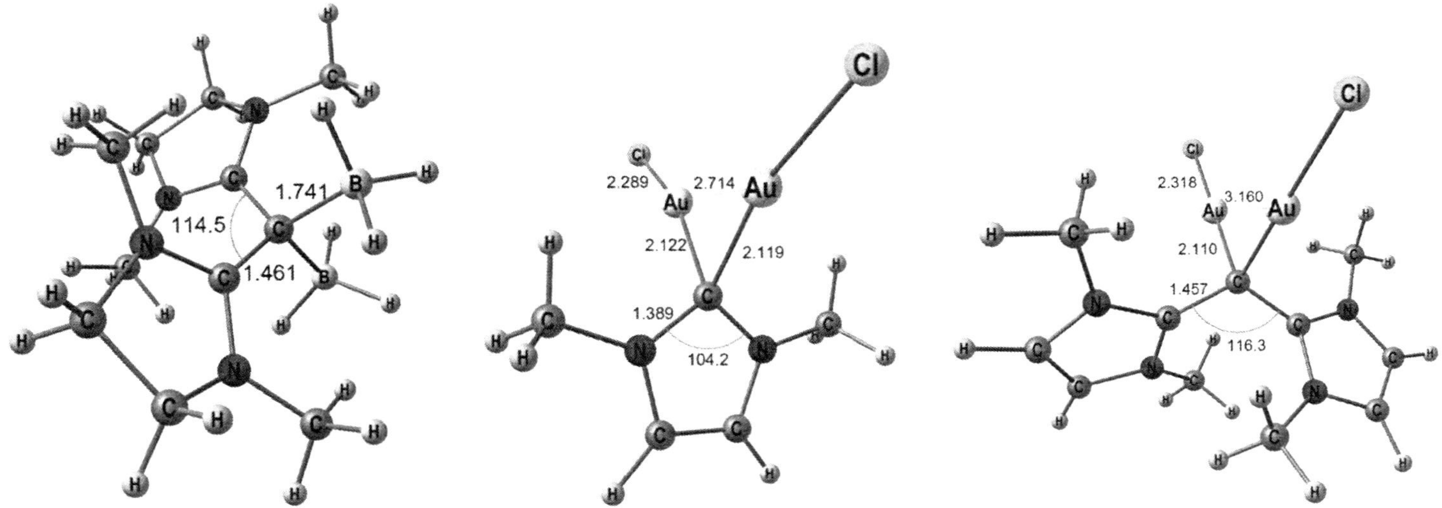

(d) **4-(BH_3)$_2$**

D_e (1. BH_3) = 22.2 kcal/mol
D_e (2. BH_3) = 37.9 kcal/mol

(e) **NHC^{Me}-(AuCl)$_2$**

D_e (1. AuCl) = 36.4 kcal/mol
D_e (2. AuCl) = 78.3 kcal/mol

(f) **2-(AuCl)$_2$**

D_e (1. AuCl) = 51.1 kcal/mol
D_e (2. AuCl) = 74.1 kcal/mol

Figure 8.3 (Continued).

ligand in **NHC^{Me}(AuCl)$_2$** (D_e = 78.3 kcal/mol) is even larger than the value for **2-(AuCl)$_2$** (D_e = 74.1 kcal/mol). *The second BDE of a carbon base is a rather sensitive probe for the carbone character of a dicoordinated carbon atom.*

Inspection of the equilibrium geometry of **NHC^{Me}(AuCl)$_2$** (Fig. 8.3e) reveals a rather short Au–Au distance which is much shorter (2.714 Å) than in the carbodicarbene complex **2-(AuCl)$_2$** which is shown in Figure 8.3f (3.160 Å). The former value suggests that the second AuCl ligand is partly bonded through aurophilic attraction to the first AuCl ligand rather than genuine C→Au bonding. It is well known that Au–Au attraction, which is due to relativistic effects, can be quite strong.[16]

The subtle differences between the carbon–gold bonding in the gold–carbene complex **NHC^{Me}(AuCl)$_2$** and the gold–carbone adduct **2-(AuCl)$_2$** become visible when the Laplacian distribution of the electronic charge $\nabla^2\rho(r)$ for the two compounds are examined. Figure 8.4 shows that there are two areas of charge concentration ($\nabla^2\rho(r) < 0$, solid lines) at the carbon donor atom of **2-(AuCl)$_2$** which point along the C–Au bond paths into the direction of the AuCl ligands. In contrast, the Laplacian distribution at the carbon donor atom of **NHC^{Me}(AuCl)$_2$** exhibits a single broad area of charge concentration towards both ligands. Also, the C–Au–Cl bond angle in **2-(AuCl)$_2$** is essentially linear, while in **NHC^{Me}(AuCl)$_2$** it is clearly bent with an angle of 160.5°.[4f]

Few experimental studies have been carried out to study the behavior of **3** as a ligand in transition metal complexes. Bertrand and coworkers isolated the

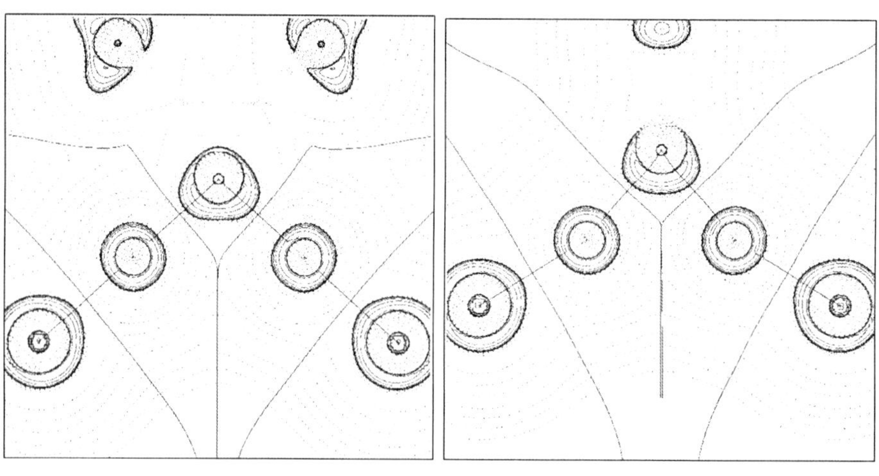

Figure 8.4. Contour line diagrams of the Laplacian $\nabla^2\rho(r)$ of the AuCl complexes **2-(AuCl)$_2$** and **NHC^{Me}-(AuCl)$_2$** in the Au–C–Au planes. Solid lines indicate areas of charge concentration ($\nabla^2\rho(r) < 0$), while dashed lines show areas of charge depletion ($\nabla^2\rho(r) > 0$). The solid lines connecting the atomic nuclei are the bond paths. The solid lines separating the atomic basins indicate the zero-flux surfaces crossing the molecular plane; see Reference 4f.

adduct $Cl(CO)_2Rh$-**3** and they reported the X-ray structure analysis and spectroscopic properties of the compound.[11] They noted that the average C–O stretching frequency (2014 cm^{-1}) is significantly lower than typically observed values for analogous NHC complexes which are in the range 2036–2058 cm^{-1}. The difference was explained by weaker π acceptance of the carbodicarbene ligand compared with NHC. Complexes where **3** binds to two metal fragments have thus far not been isolated.

8.3 TETRAAMINOALLENES AND "HIDDEN" CARBODICARBENES

Calculations of two-coordinated carbon compounds with acyclic diaminocarbene ligands $C[C(NR_2)_2]_2$ showed that the equilibrium geometry for R = Me has a linear arrangement $(Me_2N)_2C{=}C{=}C(NMe_2)_2$ (**5**) where the planes of the diaminocarbene ligands are orthogonal to each other (Fig. 8.5a).[5a] The optimized geometry and the analysis of the bonding situation of **5** suggest that the molecule should be considered as tetraaminoallene (TAA) where the two-coordinated carbon atom has the formal oxidation state four. Applying the def2-SVP basis set, both tetraaminoallenes (TAAs) **5** and **6** are computed to be virtually linear (Fig. 8.5). If a basis set of higher quality is used (BP86/TZ2P), the optimized geometry of the ethyl homologue $(Et_2N)_2C{=}C{=}C(NEt_2)_2$ (**6**) has a bent structure where the C–C–C angle is 169.5°.[5a] This result indicates the shallow bending potential of carbodicarbenes and analogues which has been found in numerous studies.[4,5] Compounds **5** and **6**, which were already synthesized in 1973 by Viehe et al., possess a chemical reactivity which is atypical for allenes.[17] For example, compound **5** readily reacts with the weak Lewis acids CO_2 and CS_2 yielding complexes where the central carbon atom binds via a donor–acceptor bond in $[(NMe_2)_2C]_2C \rightarrow CX_2$ (X = O, S).[17a] The latter complexes exhibit the same bonding motif as in the carbone complexes $(Ph_3P)_2C$–CX_2.[9a] There are further similarities between the reactivity of TAAs and carbones. For example, TAAs can bind two protons at the central carbon atom yielding the crystallographically characterized dication $[\{(NHR)_2C\}_2CH_2]^{2+}$ (R = *tert*-butyl).[17d] The protonation of the dicoordinated C(IV) atom rather than the nitrogen atoms of the amino substituents, which carry lone pair electrons, is surprising at first sight. There is a striking analogy between the protonation of TAAs and carbodiphosphorane compounds. The dication $[(Ph_3P)_2CH_2]^{2+}$ has also been synthesized and characterized by X-ray analysis.[18]

Calculations were carried out for TAAs **5** and **6** where the central C–C–C angle was kept frozen at 125.8° which is the equilibrium angle of the carbodicarbene **1**.[5a] The energies of the bent species were only 8.7 kcal/mol (**5**) and 8.8 kcal/mol (**6**) higher than the energy minima (BP86/TZ2P). This shows that the bending potential of the **5** and **6** is very shallow. It appears that the reactivity of TAAs is determined by the bonding situation that occurs in the distorted geometry with a bent C–C–C angle which is not obvious when

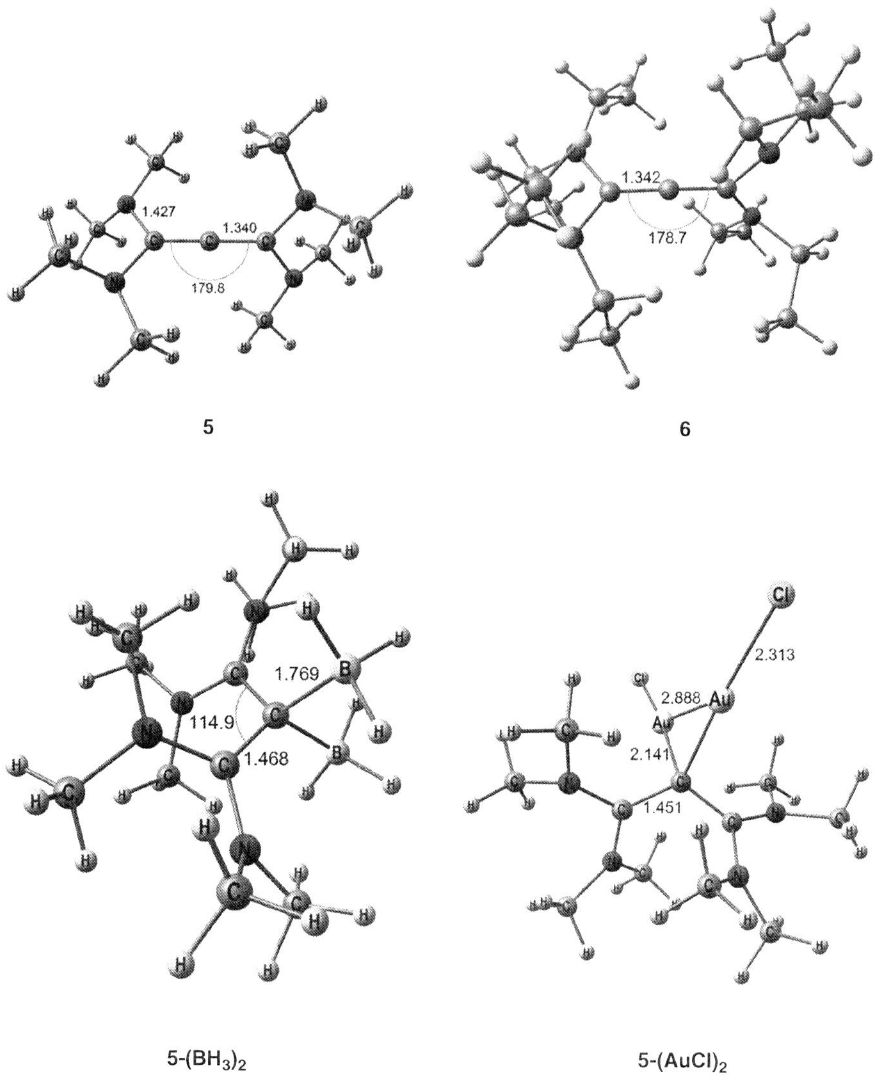

D_e (1. BH_3) = 9.5 kcal/mol
D_e (2. BH_3) = 37.2 kcal/mol

D_e (1. AuCl) = 44.1 kcal/mol
D_e (2. AuCl) = 63.9 kcal/mol

Figure 8.5. Optimized geometries of tetraaminoallenes **5** and **6** and complexes of **5** with two BH_3 ligands **5-(BH_3)$_2$** and two AuCl ligands **5-(AuCl)$_2$**. Bond lengths in angstrom; angles in degrees. Calculated bond dissociation energies D_e for the first and the second ligand in kilocalories per mole. The values are taken from References 4c, 5a, and 4f.

the equilibrium geometry is considered. Tetraaminoallenes have therefore been termed "hidden carbones."[4c] While the equilibrium geometry of TAAs suggests that the compounds are allenes, the compounds chemically behave like acyclic carbodicarbenes. The bonding situation which is depicted in Scheme 8.1

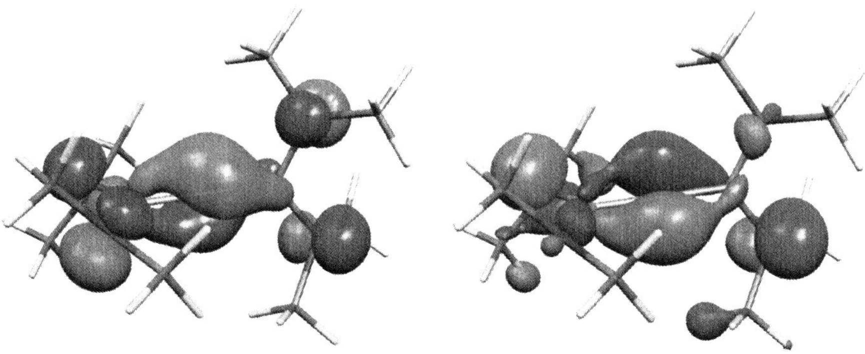

Figure 8.6. Plot of the two highest lying occupied orbitals HOMO and HOMO-1 of tetraaminoallene **5**; see Reference 4b. For color details, please see color plate section.

for **C** explains the reactivity of compounds $C[C(NR_2)_2]_2$ at room temperature where the bent form is sufficiently populated to become the reactive species. Figure 8.6 shows the HOMO and HOMO-1 of **5**. The largest coefficient of the orbitals is at the dicoordinated carbon atom, but the shape of the two highest lying MOs suggests that they are mainly C–C π-bonding orbitals.

Further evidence for the carbodicarbene character of TAAs comes from the calculated second PAs and from the Lewis basicity with respect to two Lewis acids. Table 8.1 shows that the second PA of **6** (175.8 kcal/mol) is even higher than the second PA of **1**, while the second PA of **5** (151.6 kcal/mol) is only slightly less than that of **1**. The very large PAs are not obvious when the calculated partial charges are considered which were calculated for the linear equilibrium structures. The values for q(C1) in **5** and **6** are clearly smaller than for the carbodicarbenes **1–4** (Table 8.1). Calculations have been carried for complexes of **5** with one and two Lewis acids BH_3 and AuCl.[4c,f] Figure 8.5 shows the equilibrium structures of **5-(BH$_3$)$_2$** and **5-(AuCl)$_2$** and the theoretically predicted BDEs for the ligands. The calculated data show that the BDEs of the second ligands BH_3 (9.5 kcal/mol) and AuCl (44.1 kcal/mol) in the TAA complexes are smaller than in the adducts of **1–4** (Fig. 8.3). This suggests that the carbone character of acyclic carbodicarbenes is somewhat weaker than that of the cyclic homologues $C(NHC)_2$. Note that the Au–Au distance in **5-(AuCl)$_2$** (2.888 Å) signifies aurophilic interaction which contributes to the BDE of the second AuCl ligand.

The ligand behavior of the TAA **5** has experimentally been studied by Fürstner and coworkers.[12b] The authors also addressed the question whether mixed carbones $L^1 \rightarrow C \leftarrow L^2$, which carry one carbene ligand L^1 while the second ligand L^2 is a phosphine PPh_3, might be prone to binding two transition metals at carbon. They did not succeed in isolating a complex where **5** binds to two metals. However, they synthesized the cationic complex **5(H$^+$)-AuPPh$_3$** where one lone pair orbital binds to $AuPPh_3$ while the second lone pair orbital is protonated. It was concluded that steric congestion prevents the formation of doubly aurated complexes of **5**. This was achieved, however, when the carbone $Ph_3P \rightarrow C \leftarrow C(OEt)_2$, which has one carbene ligand bonded to C(0), was reacted with an excess of AuCl(SMe). The authors obtained in good yield the complex L^1L^2C-$(AuCl)_2$, where $L^1 = C(OEt)_2$ and $L^2 = PPh_3$, which could be structurally characterized by X-ray analysis.[12b] Doubly aurated complexes L_2C-$(AuR)_2$ were previously reported.[19]

8.4 BENT ALLENES

The foregoing theoretical and experimental results reveal that π-donor substituents may influence the chemical reactivity of allenes such that they behave like carbodicarbenes. A similar effect could be achieved when the allene becomes part of a cyclic structure which enforces bending of the C–C–C moiety. A combination of electronic effects and ring constraint has been employed in the synthesis of "bent allenes" by Bertrand and coworkers.[13] Figure 8.7 shows the optimized geometries of compounds **7** and **8** where the allene fragment is part of a five-membered ring that has two nitrogen atoms in the 3,4 positions. Compound **7** may be considered as a cyclic TAA, whereas **8** is a cyclic diaminodioxoallene. Both compounds have been synthesized and structurally characterized by X-ray analysis. Figure 8.7 also gives the most important calculated and experimental bond lengths and angles, which are in good agreement.

Should **7** and **8** be considered as bent allenes or as cyclic carbodicarbenes? Figure 8.8 displays the HOMO and HOMO-1 of the molecules. Both orbitals have the largest coefficient at the two-coordinated carbon atom. The HOMO can be identified as a σ lone pair MO, whereas the HOMO-1 is a π-type orbital. The energy ordering is thus inverted compared to the HOMO and HOMO-1 of the carbodicarbene **1** and is rather in line with the highest lying frontier orbitals of the NHC (Fig. 8.2). However, the calculated second PAs for **7** and **8** suggest that the compounds behave chemically like carbodicarbenes. Table 8.1 shows that the second PA of **7** (158.7 kcal/mol) and **8** (133.3 kcal/mol) are much higher than for NHC^{Me} (71.8 kcal/mol). The atomic partial charges q(C1) for **7** and **8** have values similar to the carbodicarbenes **1–4**. The carbone character of the "bent allenes" is also revealed by the complexation of **7** and **8** with two BH_3 ligands. The calculated structures of **7-(BH$_3$)$_2$** and **8-(BH$_3$)$_2$** are shown in Figure 8.7. The BDEs of the second BH_3 ligand are not very large, but the finding that the latter species are minima on the potential energy surface (PES) suggests that **7** and **8** should be considered as divalent C(0) compounds, that is, carbodicarbenes rather than divalent C(IV) species (bent allenes).

The steric strain on the "bent allene" fragment increases when the ring becomes smaller. Bertrand and coworkers synthesized the spiro compound **9** which has a four-membered ring with a $C(CR_2)_2$ moiety. The molecule **9** could not be isolated and structurally characterized by X-ray analysis, but the ^{13}C NMR signals

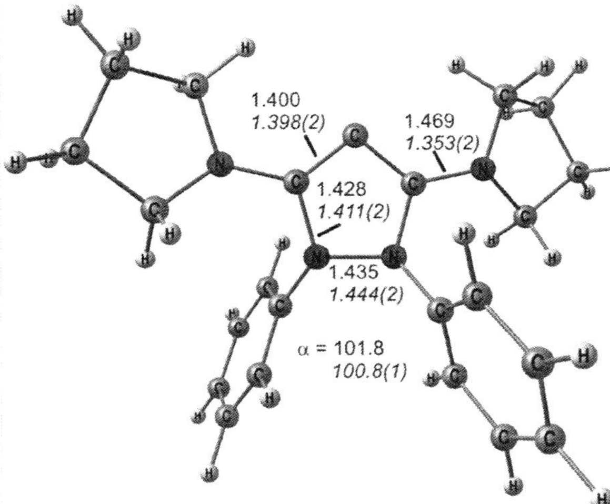

7

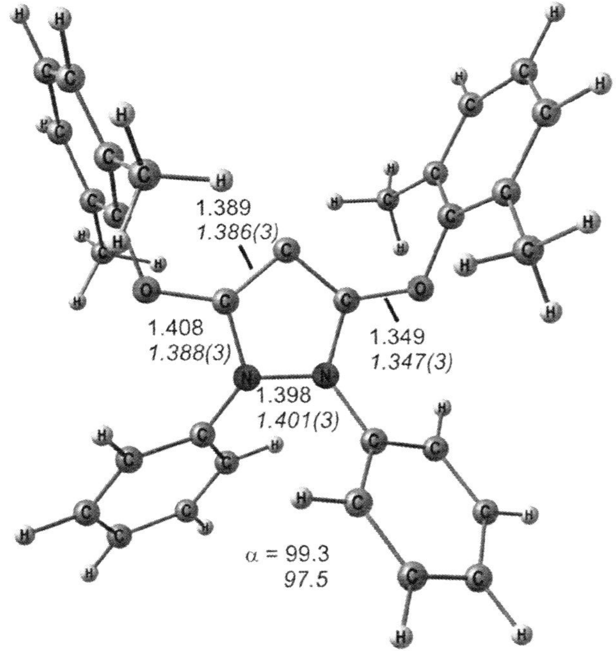

8

Figure 8.7. Optimized geometries of the "bent allenes" **7** and **8** and complexes with two BH_3 ligands **7-(BH$_3$)$_2$** and **8-(BH$_3$)$_2$**. Bond lengths in angstrom; angles in degrees. Experimental values for **7** and **8** are given in italics. Calculated bond dissociation energies D_e for the first and the second ligand in kilocalories per mole. The values are taken from Reference 5b.

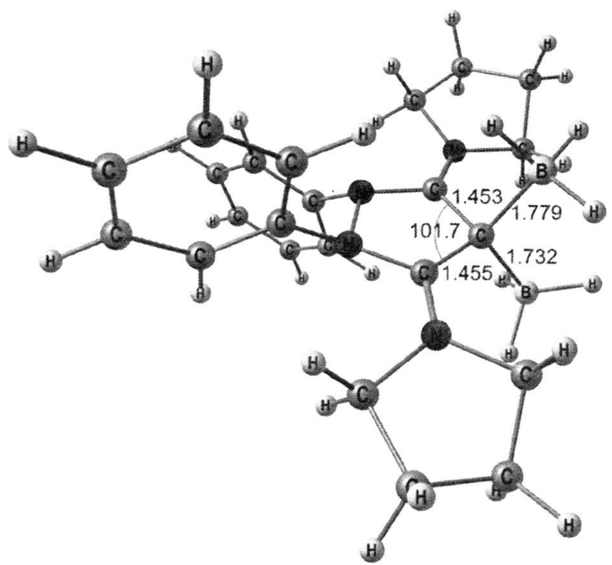

7-$(BH_3)_2$

D_e (1. BH_3) = 14.2 kcal/mol
D_e (2. BH_3) = 55.9 kcal/mol

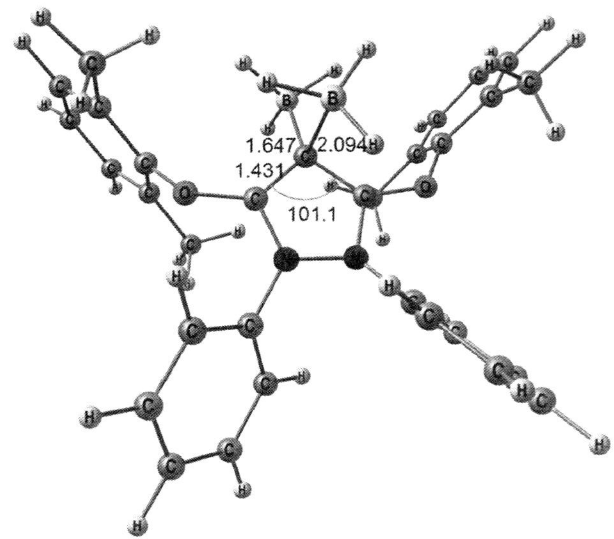

8-$(BH_3)_2$

D_e (1. BH_3) = 2.6 kcal/mol
D_e (2. BH_3) = 48.2 kcal/mol

Figure 8.7. (Continued).

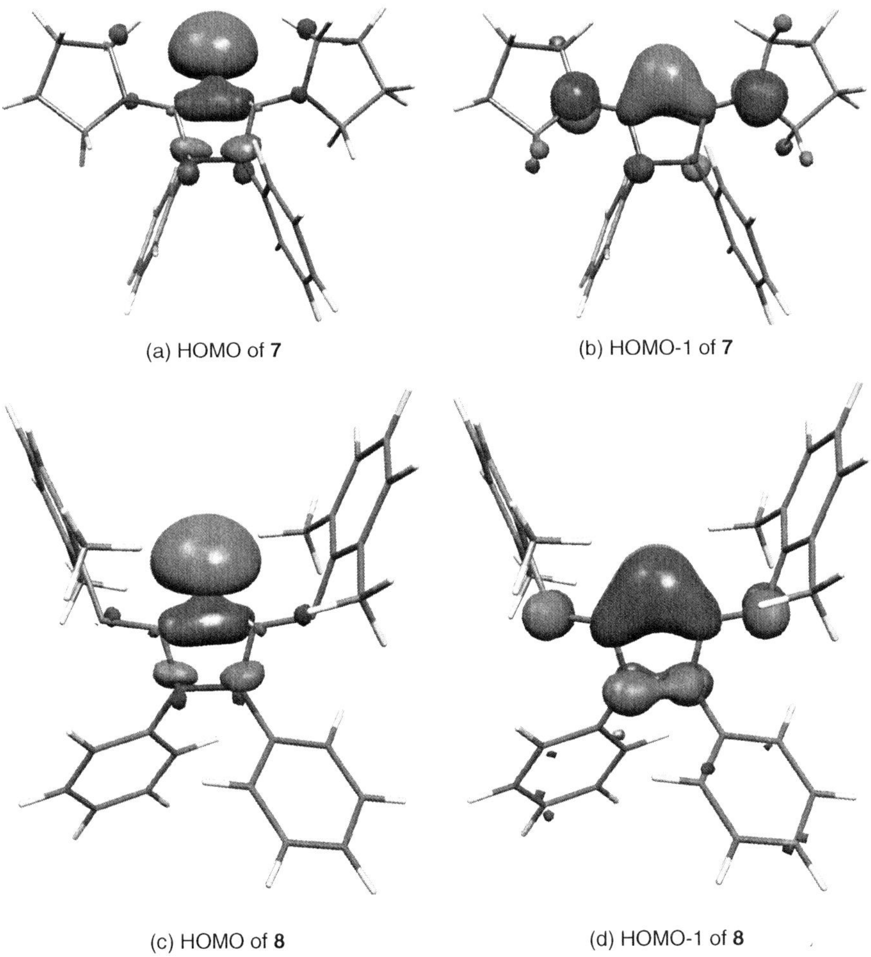

Figure 8.8. Plot of the two highest lying occupied orbitals HOMO and HOMO-1 of the "bent allenes" **7** and **8**; see Reference 5b. For color details, please see color plate section.

made it possible to identify the compound. Figure 8.9 shows the calculated equilibrium geometry of **9** which has a very acute bond angle of 85.0° for C1–C2–C3. The C–C distances of the dicoordinated carbon atoms are somewhat longer (1.421 and 1.425 Å) than in the five-membered "bent allenes" **7** (1.400 Å) and **8** (1.389 Å) and clearly longer than in the carbodicarbenes **1–4** (1.353–1.369 Å). Analysis of the electronic structure shows that the HOMO is a σ orbital while the HOMO-1 is a π orbital (Figs. 8.9c and d), where both have lone pair character. The theoretically predicted very large values for the first and second PAs and the calculated partial charges q(C1), clearly characterize **9** as a carbodicarbene (Table 8.1). While the neutral species could not be isolated, the singly and doubly protonated cations **9-(H$^+$)** and **9-(H$^+$)$_2$** have been obtained

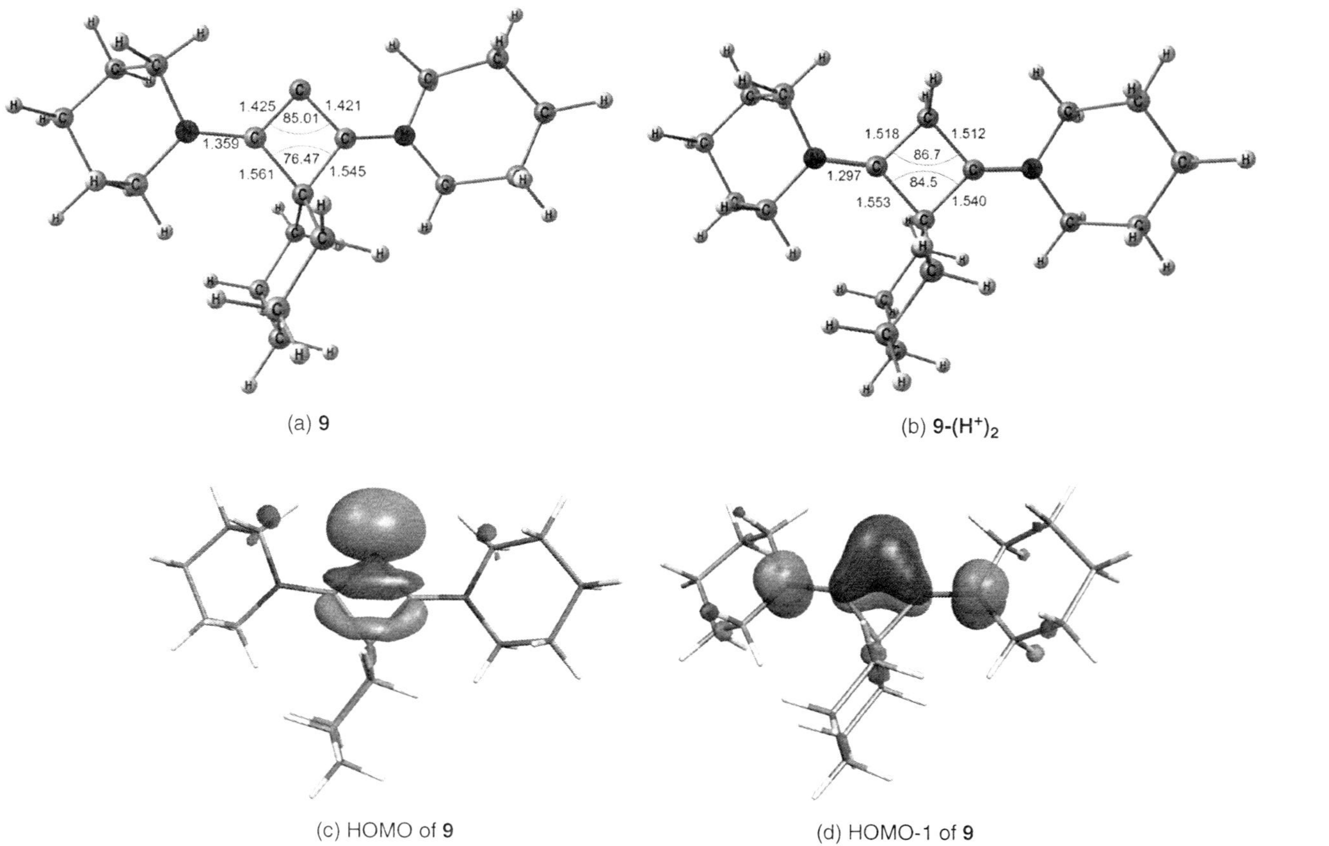

Figure 8.9. Structure of an all-carbon four-membered ring allene (**9**) together with the doubly protonated species (**9-(H$^+$)$_2$**) and (c) HOMO and (d) HOMO-1 of **9**; see Reference 13c. For color details, please see color plate section.

as salts that could be structurally characterized by X-ray analysis. Figure 8.9b shows the optimized geometry of **9-(H$^+$)$_2$** and the most important theoretical bond lengths and angles, which are in good agreement with the experimental values (not shown).

8.5 RELATED COMPOUNDS

There are further compounds which are not genuine carbodicarbenes but which are related to them. Fürstner and coworkers reported theoretical and experimental work on "mixed carbones" $L^1 \rightarrow C \leftarrow L^2$ which carry one carbene ligand L^1 while the second ligand L^2 is PPh_3.[12b] Variation of the nature of the carbene ligand makes it possible to achieve fine-tuning of the donor strength of the "carbomonocarbene" CL^1L^2, which was monitored by synthesizing the adducts $Cl(CO)_2Rh\text{-}CL^1L^2$ and measuring the C–O stretching frequencies of the metal complexes.

Another class of related compounds is heavy-atom homologues which are valence isoelectronic with carbodicarbenes. Theoretical studies have been reported for the group-14 compounds **10E–14E** (E = C–Pb) which are shown in Scheme 8.2.[20] The systems **10E**, **12E**, and **13E** are group-14 homologues of the parent carbodicarbene **1**. The compounds **11E** do not have a nitrogen atom in the α position of the five-membered ligands, and thus they might be considered classical allenes. The carbon species **11C** is indeed calculated with a linear C–C–C arrangement at the central dicoordinated carbon atom and was identified as an allene.[20a] However, the heavier homologues **11E** (E = Si–Pb) exhibit very acute bending angles at the dicoordinated atom E, where the angle is between 76.4° (E = Sn) and 79.4° (E = Ge).[20b] This is in agreement with experimental studies by Kira and coworkers of substituted analogues of **11Si** and **11Ge** which possess strongly bent C–E–C angles of 136.5° (**11Si'**) and 122.6° (**11Ge'**).[21] The latter compounds were termed by the authors "trisilaallene" and "trigermaallene." The analysis of the bonding situation in the model compounds **10E–14E** (E = Si–Pb) and in the real systems **11Si'** and **11Ge'** suggests that the molecules should be considered as tetrylones EL_2 where the divalent E(0) atom has the formal oxidation state zero. The larger bending angle E–E–E in **11Si'** and **11Ge'**, which was also found in the calculations, was explained by steric repulsion between the bulky ligands. The first tristannaallene $[(t\text{-but})_3Si]_2Sn{=}Sn{=}Sn[(t\text{-but})_3Si]_2$, which has a Sn–Sn–Sn bending angle of 156.0°, had already been reported prior to trisilaallene and trigermaallene.[22]

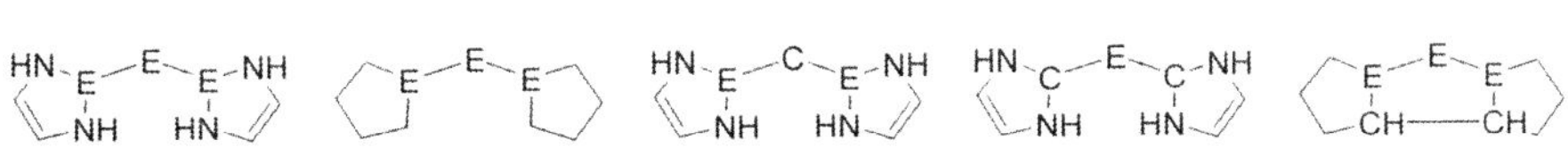

10E **11E** **12E** **13E** **14E**

Scheme 8.2.

8.6 CONCLUSION AND OUTLOOK

Carbodicarbenes are a very young class of chemical compounds which have been known only since 2007. Theoretical and experimental research on carbodicarbenes is still in an infant stage, but the results which have been reported thus far let us expect that the field will greatly expand in the coming years. The unusual carbon→carbon donor–acceptor bonding in carbodicarbenes offers a potentially wide variety of modifications through modulating the donor strength and the steric bulk of the carbene donor ligands. This holds for both free carbodicarbenes and for complexes with carbodicarbenes as ligands. The advent of this new class of compounds, which was first theoretically predicted and then experimentally verified, is a challenge for inventive chemists.

SUGGESTED READING

Alcarazo, M., Lehmann, W., Anoop, A., Thiel, W., Fürstner, A. "Coordination Chemistry at Carbon." *Nat. Chem.* **2009**, *1*, 295.

Dyker, C. A., Lavallo, V., Donnadieu, B., Bertrand, G. "Synthesis of an Extremely Bent Acyclic Allene (A "Carbodicarbene"): A Strong Donor Ligand." *Angew. Chem. Int. Ed.* **2008**, *47*, 3206.

Esterhuysen, C., Frenking, G. "Distinguishing Carbones from Allenes by Complexation to AuCl." *Chem. Eur. J.* **2011**, *17*, 9944.

Frenking, G., Tonner, R. "Carbodicarbenes—Divalent Carbon(0) Compounds Exhibiting Carbon–Carbon Donor–Acceptor Bonds." *WIREs Comput. Mol. Sci.* **2011**, *1*, 869.

Klein, S., Tonner, R., Frenking, G. "Carbodicarbenes and Related Divalent Carbon(0) Compounds." *Chem. Eur. J.* **2010**, *16*, 10160.

Tonner, R., Frenking, G. "$C(NHC)_2$: Divalent Carbon(0) Compounds with *N*-Heterocyclic Carbene Ligands—Theoretical Evidence for a Class of Molecules with Promising Chemical Properties." *Angew. Chem. Int. Ed.* **2007**, *46*, 8695.

REFERENCES

1 (a) Igau, A., Grützmacher, H., Baceiredo, A., Bertrand, G. *J. Am. Chem. Soc.* **1988**, *110*, 6463. (b) Igau, A., Baceiredo, A., Trinquier, G., Bertrand, G. *Angew. Chem. Int. Ed.* **1989**, *28*, 621.

2 Arduengo, III, A. J., Harlow, R. L., Kline, M. *J. Am. Chem. Soc.* **1991**, *113*, 2801.

3 In carbene–transition metal complexes, the strength of the transition metal → carbene π back-donation strongly depends on the nature of the substituents R of the carbene CR_2. Recent theoretical studies suggest that the π back-donation to NHC ligands may be stronger than generally assumed, but that in some cases there might even be π donation from the NHC ligand to the metal: (a) Tulloch, A. A. D., Danopoulos, A. A., Kleinhenz, S., Light, M. E., Hursthouse, M. B., Eastham, G. *Organometallics* **2001**, *20*, 2027. (b) Hu, X., Tang, Y., Gantzel, P., Meyer, K. *Organometallics* **2003**, *22*, 612. (c) Hu, X., Castro-Rodriguez, I., Olsen, K., Meyer, K. *Organometallics* **2004**, *23*, 755. (d) Nemcsok, D., Wichmann, K., Frenking, G. *Organometallics* **2004**, *23*, 3640.

(e) Scott, N. M., Dorta, R., Stevens, E. D., Correa, A., Cavallo, L., Nolan, S. P. *J. Am. Chem. Soc.* **2005**, *127*, 3516. (f) Jacobsen, H. J. *Organomet. Chem.* **2005**, *690*, 6068. (g) Jacobsen, H., Correa, A., Costabile, C., Cavallo, L. *J. Organomet. Chem.* **2006**, *691*, 4350.

4 (a) Tonner, R., Öxler, F., Neumüller, B., Petz, W., Frenking, G. *Angew. Chem. Int. Ed.* **2006**, *45*, 8038. (b) Tonner, R., Frenking, G. *Chem. Eur. J.* **2008**, *14*, 3260. (c) Tonner, R., Frenking, G. *Chem. Eur. J.* **2008**, *14*, 3273. (d) Frenking, G., Tonner, R. *Pure Appl. Chem.* **2009**, *81*, 597. (e) Klein, S., Frenking, G., *Angew. Chem. Int. Ed.* **2010**, *49*, 7106. (f) Esterhuysen C., Frenking, G. *Chem. Eur. J. 17*, 9944 (2011).

5 (a) Tonner, R., Frenking, G. *Angew. Chem. Int. Ed.* **2007**, *46*, 8695. (b) Klein, S., Tonner, R., Frenking, G. *Chem. Eur. J.* **2010**, *16*, 10160. (c) Frenking, G., Tonner, R. *WIREs Comput. Mol. Sci.* **2011**, *1*, 869.

6 (a) Jablonski, M., Palusiak, M. *Phys. Chem. Chem. Phys.* **2009**, *11*, 5711. (b) Hanninen, M. M., Peuronen, A., Tuononen, H. M. *Chem. Eur. J.* **2009**, *15*, 7287.

7 Ramirez, F., Desai, N. B., Hansen, B., McKelvie, N. *J. Am. Chem. Soc.* **1961**, *83*, 3539.

8 Recent reviews: (a) Petz, W., Frenking, G. *Top. Organomet. Chem.* **2010**, *30*, 49. (b) Jones, N. D., Cavell, R. G. *J. Organomet. Chem.* **2005**, *690*, 5485.

9 (a) Petz, W., Kutschera, C., Heitbaum, M., Frenking, G., Tonner, R., Neumüller, B. *Inorg. Chem.* **2005**, *44*, 1263. (b) Petz, W., Öxler, F., Neumüller, B., Tonner, R., Frenking, G. *Eur. J. Inorg. Chem.* **2009**, 4507.

10 (a) Tonner, R., Heydenrych, G., Frenking, G. *Chem. Asian J.* **2007**, *2*, 1555. (b) Frenking, G., Wichmann, K., Fröhlich, N., Grobe, J., Golla, W., Le Van, D., Krebs, B., Läge, M. *Organometallics* **2002**, *21*, 2921.

11 Dyker, C. A., Lavallo, V., Donnadieu, B., Bertrand, G. *Angew. Chem. Int. Ed.* **2008**, *47*, 3206.

12 (a) Fürstner, A., Alcarazo, M., Goddard, R., Lehmann, C. W. *Angew. Chem. Int. Ed.* **2008**, *47*, 3210. (b) Alcarazo, M., Lehmann, W., Anoop, A., Thiel, W., Fürstner, A. *Nat. Chem.* **2009**, *1*, 295.

13 (a) Lavallo, V., Dyker, C. A., Donnadieu, B., Bertrand, G. *Angew. Chem. Int. Ed.* **2008**, *47*, 5411. (b) Fernandez, I., Dyker, C., De Hope, A., Donnadieu, B., Frenking, G., Bertrand, G. *J. Am. Chem. Soc.* **2009**, *131*, 11875. (c) Melaimi, M., Parameswaran, P., Donnadieu, B., Frenking, G., Bertrand, G. *Angew. Chem. Int. Ed.* **2009**, *48*, 4792.

14 Kaska, W. C., Mitchell, D. K., Reichelderfer, R. F. *J. Organomet. Chem.* **1973**, *47*, 391.

15 Tonner, R., Heydenrych, G., Frenking, G. *ChemPhysChem* **2008**, *9*, 1474.

16 Schmidbaur, H., Cronje, S., Djordjevic, B., Schuster, O. *Chem. Phys.* **2005**, *311*, 151.

17 (a) Viehe, H. G., Janousek, Z., Gompper, R., Lach, D. *Angew. Chem. Int. Ed. Engl.* **1973**, *12*, 566. (b) Oeser, E. *Chem. Ber.* **1974**, *107*, 627. (c) Gompper, R., Schelble, J., Schneider, C. S. *Tetrahedron Lett.* **1978**, *41*, 3897. (d) Taylor, M. J., Surman, P. W. J., Clark, G. R. *J. Chem. Soc., Chem. Commun.* **1994**, 2517.

18 (a) Walker, J. D., Poli, R. *Polyhedron* **1989**, *8*, 1293. (b) Jensen, W. P., Gehrke, H., Jones, D. R., Suh, I.-H., Jacobson, R. A. *Z. Kristallogr.* **1996**, *211*, 829.

19 (a) Schmidbaur, H., Gasser, O. *Angew. Chem. Int. Ed. Engl.* **1976**, *15*, 502. (b) Vicente, J., Singhal, A. R., Jones, P. G. *Organometallics* **2002**, *21*, 5887. (c) Marrot, S., Kato, T., Gornitzka, H., Baceiredo, A. *Angew. Chem. Int. Ed.* **2006**, *45*, 2598.

20 (a) Takagi, N., Shimizu, T., Frenking, G. *Chem. Eur. J.* **2009**, *15*, 3448. (b) Takagi, N., Shimizu, T., Frenking, G. *Chem. Eur. J.* **2009**, *15*, 8593. (c) Takagi, N., Frenking, G. *Theoret. Chem. Acc.* **2011**, *129*, 615.
21 Iwamoto, T., Masuda, H., Kabuto, C., Kira, M. *Organometallics* **2005**, *24*, 197.
22 Wiberg, N., Lerner, H.-W., Vasisht, S.-K., Wagner, S., Karaghiosoff, K., Nöth, H., Ponikwar, W. *Eur. J. Inorg. Chem.* **1999**, 1211.

9

CATALYTIC REACTIONS WITH *N*-MESITYL-SUBSTITUTED *N*-HETEROCYCLIC CARBENES

JESSADA MAHATTHANANCHAI AND JEFFREY W. BODE
Laboratorium für Organische Chemie, Department of Chemistry and Applied Biosciences, Swiss Federal Institute of Technology, ETH Zürich, Zürich, Switzerland

9.1 INTRODUCTION

Reactions catalyzed by *N*-heterocyclic carbenes (NHC) have been known longer than the NHCs themselves. The thiamine-catalyzed benzoin reaction engendered great excitement over its unusual—and for a while incomprehensible—mechanism, but the connection between the thiazolium ylide identified as the key nucleophilic catalyst and its characterization as an NHC grew only slowly. Nowadays, the idea of nucleophilic carbenes acting as highly versatile catalysts is widely accepted and new NHC-catalyzed transformations appear almost weekly in the primary literature. In addition to a renaissance in benzoin and Stetter chemistry, NHC-catalyzed reactions are now most closely associated with the remarkable collection of transformation of α,β-unsaturated and α-functionalized aldehydes. This field has exploded since 2004 and now constitutes one of the most dynamic areas of homogeneous catalysis.

Even a brief survey of the literature reveals one recurring catalyst motif of *N*-mesityl-substituted NHCs, with the triazolium derivatives capturing the majority of the catalysts used. In this chapter, we seek to give an overview of the development of these catalysts, a survey of the diverse and mechanistically

Contemporary Carbene Chemistry, First Edition. Edited by Robert A. Moss and Michael P. Doyle.

intricate reactions they catalyzed, and an explanation for the origin of the superiority of these catalysts on new generation of NHC-catalyzed reactions.

9.1.1 Historical Background

von Liebig reported the first benzoin dimerization of aldehyde catalyzed by cyanide ion in 1832,[1] but it was not until 1903 that Lapworth elucidated the mechanism of this remarkable reaction.[2] These seminal papers were important not only for the nascent field but also constituted the first full mechanistic elucidation of an organic reaction. In 1943, almost 100 years after Liebig's report of the cyanide-catalyzed benzoin reaction, Ugai[3] disclosed the use of thiamine (vitamin B1) as the catalyst for the same transformation (Fig. 9.1). This unexpected finding emerged from biological studies of thiamine-dependent enzymes involved in many biological pathways.[4] Breslow[5] extensively studied the mechanism of the thiamine-catalyzed benzoin reaction and proposed the structure of the key enaminol intermediate, now commonly referred to as "the Breslow intermediate." While these efforts were thorough, reexaminations of the mechanism have continued and revealed complicated nuances in the rate-determining step[6] and the nature of the reactive intermediates.[7]

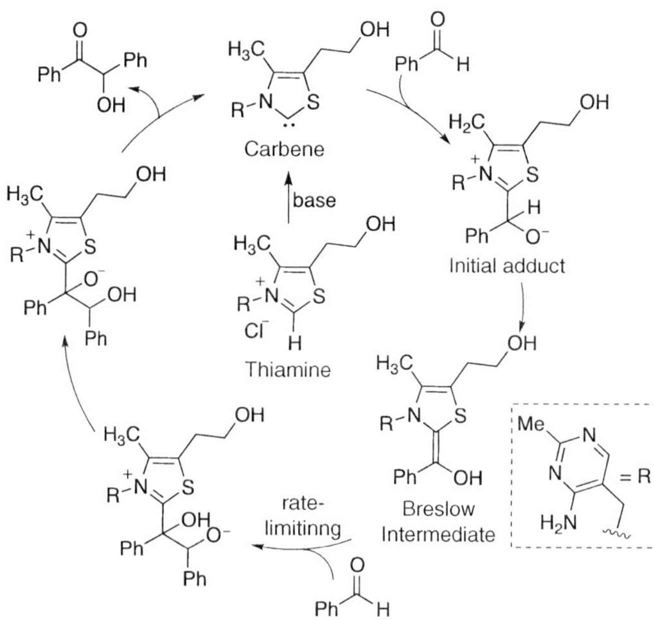

Figure 9.1. The mechanism of thiamine-catalyzed benzion reaction.

9.1.2 State of the Art Prior to 2004

Ugai's report encouraged improvements in the efficiency of the benzoin reaction and attempts to control the stereochemistry. The first major innovation

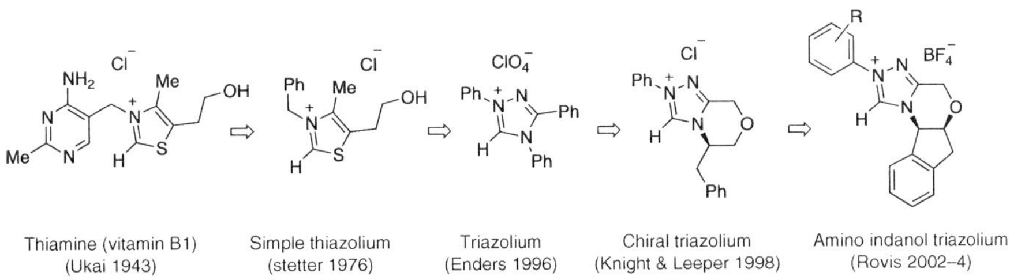

Figure 9.2. The progression of NHC precatalysts.

arrived when Stetter simplified thiamine by replacing the heterocycle with an aryl or alkyl groups in 1976.[8] Another 20 years progressed before Enders and Teles disclosed a new catalyst design and detailed studies of a triazolium salt (Fig. 9.2).[9] Contemporaneously, Miyashita introduced imidazolium NHC catalyst in his studies.[10] The introduction of both the imidazole and triazole cores for NHC sparked the surge in finding more reaction and essentially led to the development of chiral NHC precatalysts for both the benzoin and the related Stetter reactions.[11] Enders and Teles were able to improve the enantioselectivity of both processes in comparison to the first attempt by Sheehan with a chiral thiazolium salt,[12] but it was still many years before an effective catalyst was to be found. Another breakthrough in designing chiral triazolium salt for carbene catalysis arose from the use of chiral amino alcohol, reported by Knight and Leeper in 1998. Chiral triazolium salts (such as **1** in Scheme 9.1a) were found to be effective for dimerization of aldehyde, affording the products up to ~80% enantiomeric excess (%ee).[13] A significant improvement was reported by Rovis, who employed chiral aminoindanol for the preparation of a series of catalyst of type **2**,[14] which were competent for highly enantioselective intramolecular Stetter[15] and benzoin[16] reactions (Scheme 9.1b). Modifying the protocol from Knight and Leeper, Rovis was able to introduce the electron-withdrawing *N*-C_6F_5 moiety into the chiral triazolium salts (i.e., **3**) and used it to catalyze a highly enantioselective intermolecular Stetter reaction (Scheme 9.1c).[17] The modular nature of this new procedure[18] allowed the generation and identification of diverse, new NHC catalysts for challenging transformations.

9.1.3 *N*-Mesityl Catalysts as the Key Innovation

At the outset of our studies in 2003, we postulated that the combination of an NHC catalyst and α-functionalized aldehyde could unlock catalytically generated reactive species beyond that of the traditional "acyl anion equivalent" formed in the benzoin and Stetter reactions. One of our primary goals was to develop esterification and amidation reactions without the need for coupling reagents, which we believed could be achieved by the catalytic generation of "acyl azoliums," which would serve as equivalents to activated esters. One of our initial test reactions for this process was the conversion of cinnamaldehyde to dihydrocinnamic esters. Initially, we believed that the success of this reaction would require us to suppress competing benzoin and Stetter reactions. Our

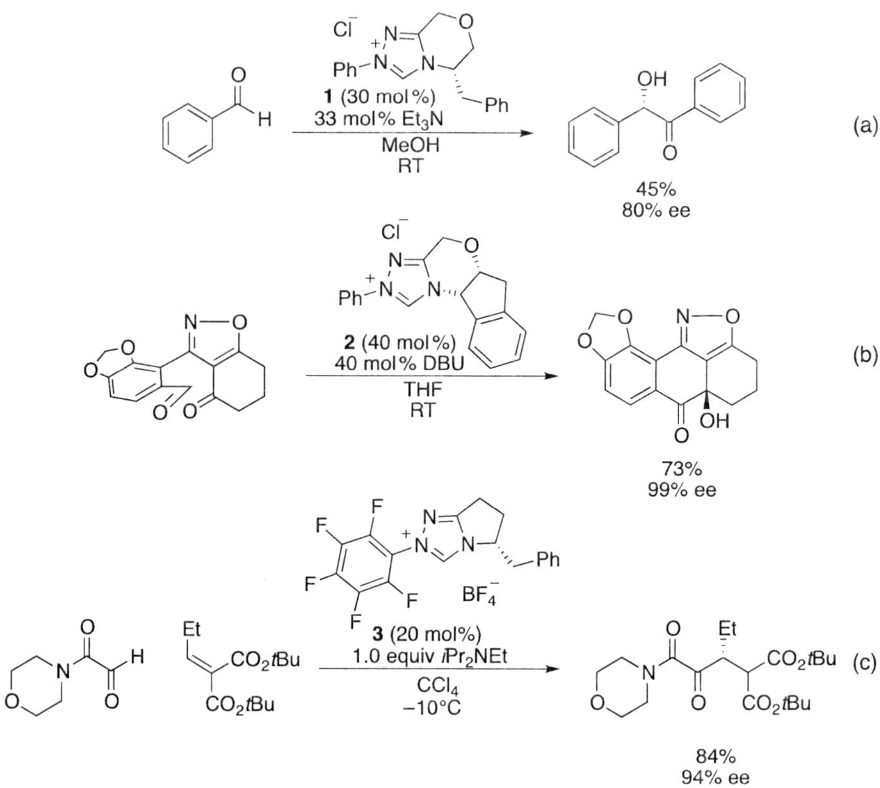

Scheme 9.1.

thinking was that bulky groups on the NHC precatalyst would lead to the formation of a Breslow intermediate too sterically hindered to undergo reactions via the acyl anion equivalent. We reasoned that the bulky mesityl moiety would shepherd the reactions of the conjugated Breslow intermediate to the distal position. Indeed, NHC precatalysts lacking *N*-mesityl groups gave poor results for the desired reaction (Fig. 9.3). The bis *N*-mesityl imidazolium **8** was a good catalyst for cinnamaldehyde, but we found that **8** was poorly effective for the redox esterification of other α,β-unsaturated aldehydes. By combining the *N*-mesityl substituent and the triazolium core of **7**, Bode and coworkers synthesized the novel catalyst **9**, which has emerged as the prototypical catalyst for a remarkable range of new NHC-catalyzed reactions.[19] As this chapter will show, nearly all of these reactions require the *N*-mesityl group or a similar substituent for effective reactions. (In this chapter, the authors will consider other ortho, ortho-dialkyl-substituted *N*-aryl catalysts to be of essentially the same effect as the *N*-mesityl counterparts.)

Further synthetic effort from our group along with mechanistic investigations have led to the development of the chiral *N*-mesityl triazolium salt **10**[20] and a chiral *N*-mesityl imidazolium salt **11** (Fig. 9.4).[21] After our original report in 2005,

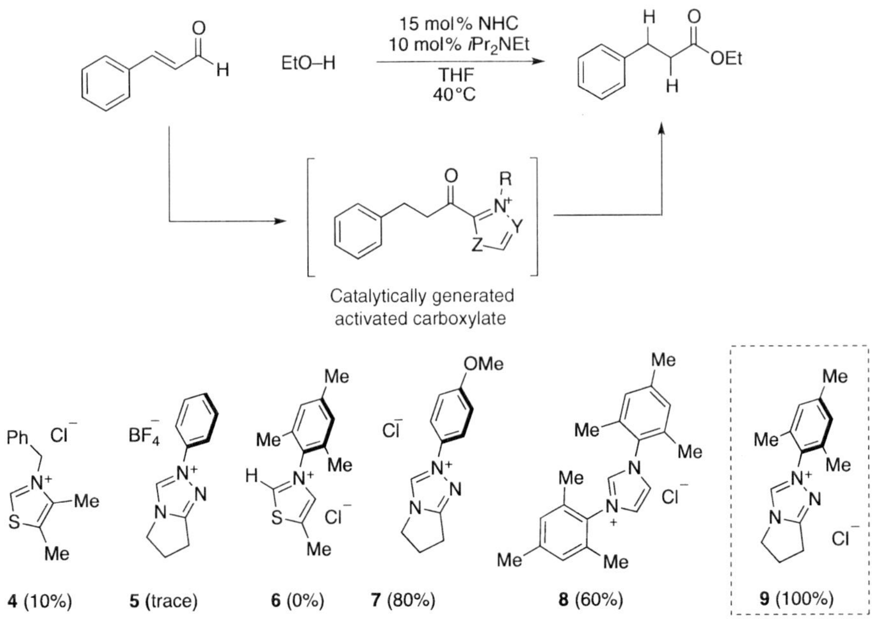

Figure 9.3. Our impetus for developing *N*-mesityl NHC catalysts.

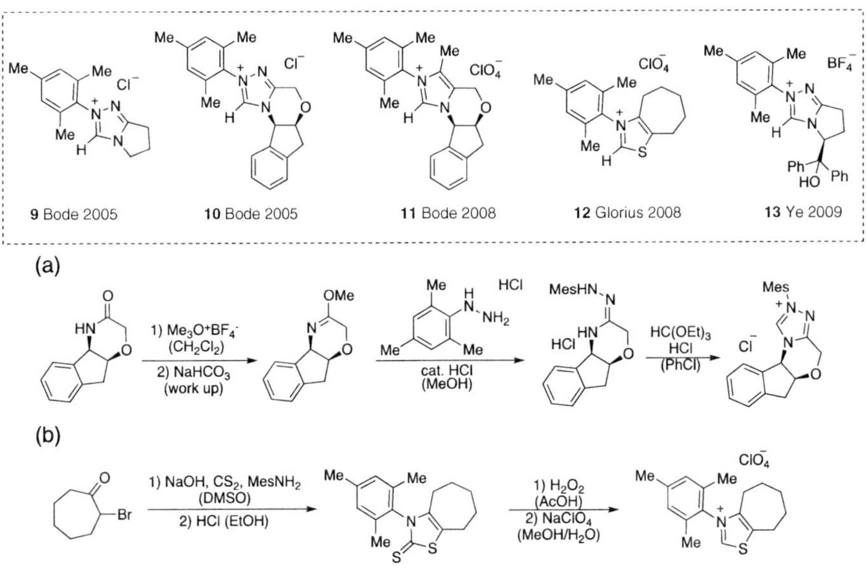

Figure 9.4. *N*-mesityl NHC catalysts and their syntheses. (a) Synthesis of *N*-mesityl chiral triazolium salt **10** (Bode) and (b) Synthesis of *N*-mesityl achiral thiazolium salt **12** (Glorius).

other researchers have extended the use of the mesityl moiety to furnish new catalyst types with interesting reactivity (i.e., **12**[22] or **13**[23]). Arguably, the *N*-mesityl catalysts, especially triazolium salts **9** and **10**, are the field's key innovation and have emerged as the most versatile and widely used NHC catalysts. Figure 9.4 depicts the syntheses of azolium salts bearing the *N*-mesityl moiety.

9.2 THE *N*-MESITYL GROUP: A MECHANISTIC ASPECT

9.2.1 Catalytic Generation of Reactive Species via *N*-Mesityl NHCs

The exceptional nature of NHCs as small organic molecule catalysts lies in their ability to access multiple catalytically generated reactive species from the simple combination of commercial substrate and catalyst. For example, five discrete reactive intermediates can be accessed from an *N*-mesityl catalyst precursor and an α,β-unsaturated aldehyde (Fig. 9.5). The acyl anion, homoenolate, and enolate equivalents serve as nucleophilic species, and the acyl azolium and unsaturated acyl azolium are excellent electrophiles with remarkable chemistry. Most importantly, all five intermediates can be accessed in catalytic, enantioselective reactions by employing a chiral NHC precursor. Previously, our group reviewed the chemistry of these catalytically generated species[24] but we were not able to explain the mechanistic rationale for the exceptional catalytic activity of triazolium NHCs bearing the *N*-mesityl moiety. In the subsequent sections, we will elucidate this phenomenon and highlight recent, exciting chemistry that these catalysts offer.

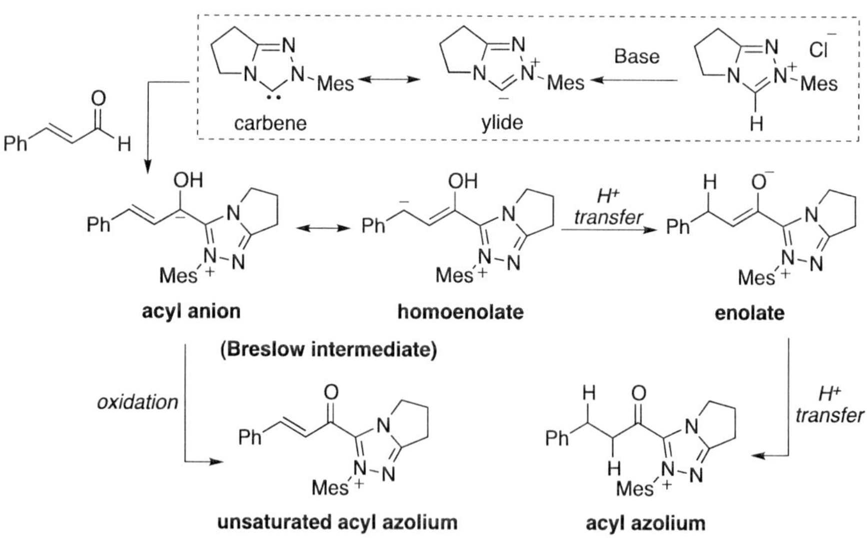

Figure 9.5. Catalytic generation of five discreet intermediates via NHC catalysis.

9.2.2 The Kinetic Effect of the *N*-Mesityl Group

A brief survey of the literature, along with accumulated experience, confirms that essentially all reactions with α,β-unsaturated aldehydes require the *N*-mesityl catalysts (Fig. 9.6). While other NHC catalysts such as those bearing *N*-C_6F_5 groups have proven to be useful for many transformations (*vide supra*), they are less reactive or in certain cases displayed no reactivity for enals.

In 2008, Rovis reviewed the effect of the various *N*-aryl substitution on triazolium-derived NHCs in terms of reactivity, acidity, and enantioselectivity for various Stetter reactions.[25] Triazolium salts bearing highly electron-donating group such as the *N*-mesityl group are less acidic than the *N*-C_6F_5 variants. Indeed, titration experiments revealed complete deprotonation of **14** by DBU in CD_2Cl_2, while the extent of deprotonation for **9** was negligible (Fig. 9.7).[26] We were therefore very surprised to find that the *N*-mesityl salts were able to catalyze many transformations even in the absence of an external base additive, while other classes of azoliums salts could not do this![27] These findings raised the question of how a triazolium salt that is much less acidic generates enough active carbene for a given catalytic cycle. Taking together this knowledge and the reactions in Figure 9.6 as mechanistic probes, we undertook careful comparison studies by catalyst analog synthesis, labeling experiments, and consideration of all the elementary steps. Contemporaneous

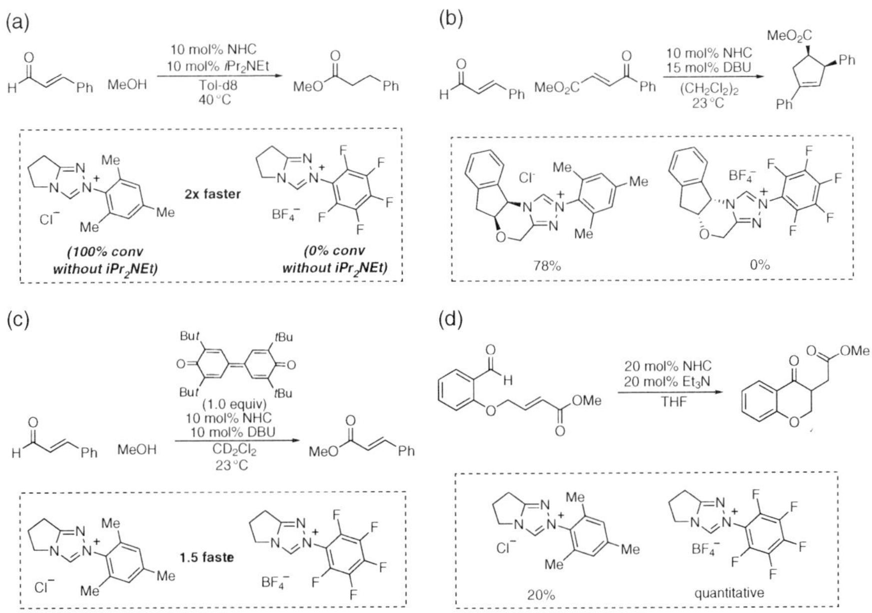

Figure 9.6. Representative reactions and activities of *N*-mesityl versus *N*-C_6F_5 NHCs. (a) Redox esterification via acyl azolium intermediate; (b) Homoenolate annulation via Breslow intermediate; (c) Oxidative esterification via acyl azolium intermediate; and (d) An intramolecular Stetter reaction via acyl anion intermediate.

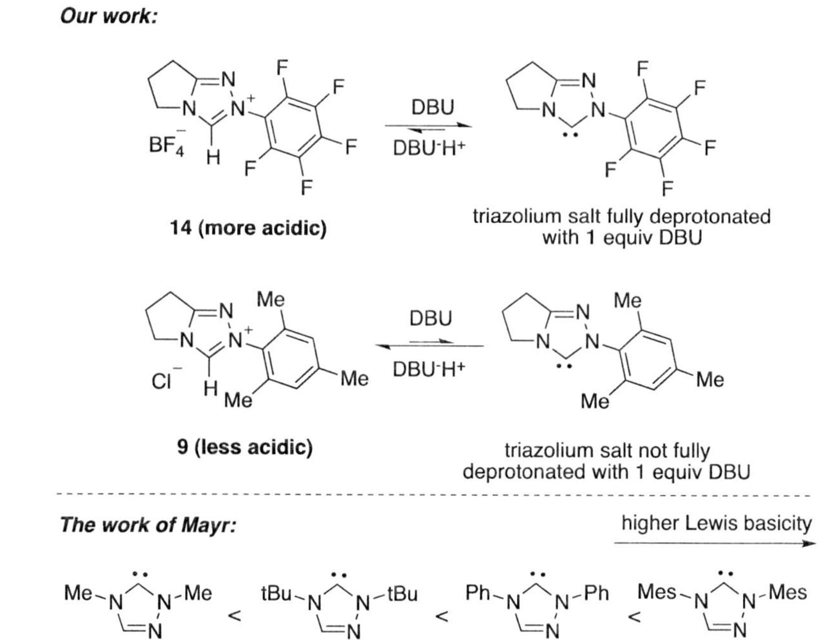

Figure 9.7. Relative basicity and acidity of various NHCs.

with our investigation, Mayr reported that the *N*-mesityl carbenes are not very nucleophilic but are highly Lewis basic (Fig. 9.7).[28]

Our investigations revealed that the *N*-mesityl group alters the kinetic profile of the catalytically generated reactive intermediates rather than dramatically influencing their thermodynamic stability. This effect has its origin in both the steric and the electronic nature of the catalyst (Fig. 9.8). First, the formation of the initial aldehyde–NHC adduct is essentially irreversible with the *N*-mesityl catalyst but is reversible for less sterically demanding catalysts (i.e., **14**). The steric bulk of the mesityl moiety appears to prolong the lifetime of the intermediate and makes this NHC a slower leaving group, which renders the barrier of the back reaction higher in energy ($\boldsymbol{k}_{-a}$; Fig. 9.8). Second, the formation ($\boldsymbol{k}_b$) of the Breslow intermediate (i.e., homoenolate or acyl anion equivalents) is accelerated and appears to also be irreversible. The electron-rich nature of the mesityl moiety results in a highly reactive Breslow intermediate towards electrophile or oxidant. This step ($\boldsymbol{k}_c$) for the *N*-mesityl catalyst is faster than the other NHCs. The two energy diagrams in Figure 9.8 summarize our finding and compare the reaction profile for the *N*-C_6F_5 and *N*-mesityl catalysts.

It is worth noting here that other triazolium salts do show good catalytic reactivity for substrates other than enals. α-Epoxyaldehyde,[29] α-haloaldehyde,[30] ketene,[31] acyl silane,[32] and ester[33] have all been shown to be competent-reaction partners in various NHC-catalyzed acylation and annulation[34] reaction with simpler NHC precatalysts (Scheme 9.2). We consider the fact that these substrates

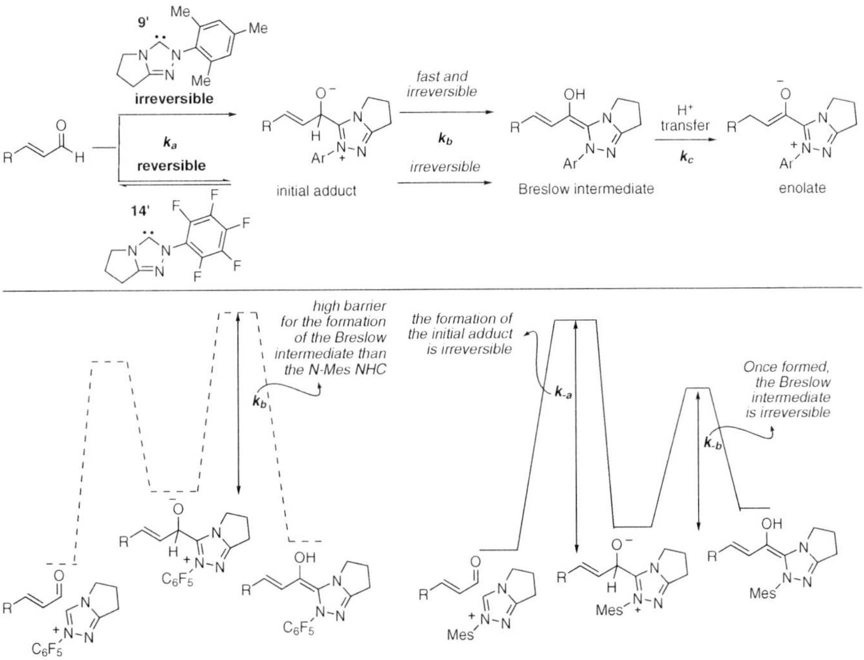

Figure 9.8. The kinetic effects of the *N*-mesityl group.

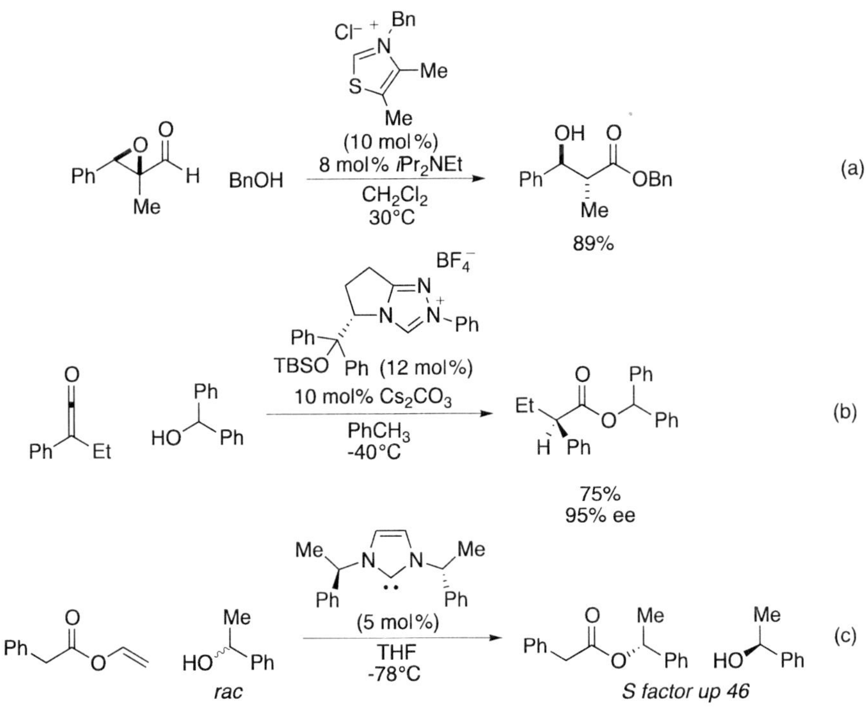

Scheme 9.2.

do not necessarily require the kinetic enhancement by the mesityl moiety to be "the exceptions that prove the rule."

9.3 NHC CATALYSIS BY CLASS OF REACTIVE INTERMEDIATES

9.3.1 Acyl Anion Equivalent

The *N*-mesityl catalysts are not always required for simple benzoin and Stetter reactions, but still are common among the best catalysts for enantioselective or product-selective variants. In certain cases, we[26] and others[35] have noted a decrease in catalytic activity of the *N*-mesityl catalysts, such as during an intramolecular cyclization event where the steric bulk of the mesityl group impedes the reaction rate. Despite this, recent literature has shown that exciting new reactivities can be garnered from the *N*-mesityl triazolium catalysts for challenging transformations. Glorius has reported two impressive examples of chemoselective crossed benzoin reactions[36] between two aromatic aldehydes (Scheme 9.3a).[37] The sterically hindered catalyst **12** prefers to add at the least hindered aldehyde, resulting in a highly nucleophilic Breslow intermediate/acyl anion that adds quickly to the electron-deficient aldehyde. Glorius also reported the addition of an aromatic aldehyde to paraformaldehyde (Scheme 9.3b).[38] While the formation of the Breslow intermediate derived from simple aldehydes is reversible,[26] the lower effective concentration of the needed formaldehyde in equilibrium led to the NHC addition first to benzaldehyde and the resulting chemoselectivity.

Apart for the benzoin reaction, the *N*-mesityl catalysts have recently gained attention for being able to catalyze challenging Stetter reactions—the conjugate addition reaction between acyl anion and α,β-unsaturated acceptors. While a large majority of reports on Stetter reactions shows the *N*-C_6F_5 catalysts developed by Rovis[14,17,18,25,39] as most reactive and selective, the use of enals in

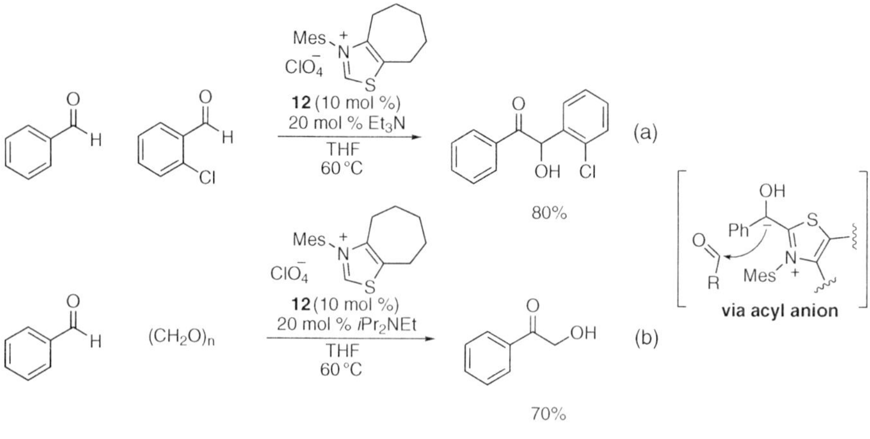

Scheme 9.3.

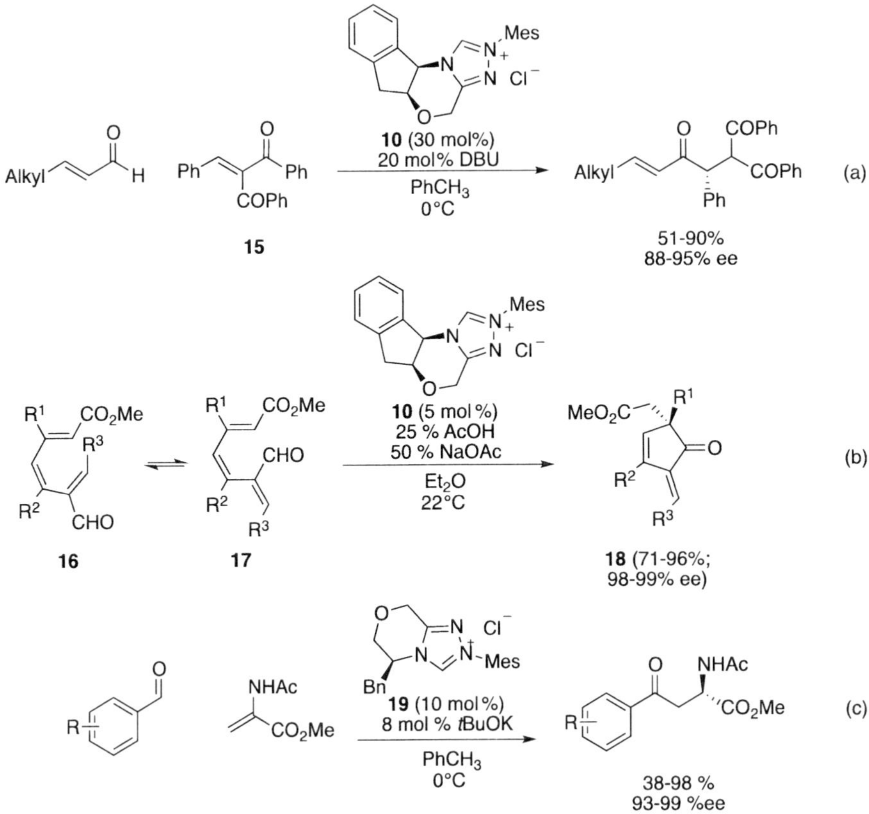

Scheme 9.4.

combination with the *N*-mesityl catalyst offers a new mode of reactivity for the Stetter reaction. To demonstrate this, Chi[40] disclosed an intermolecular enantioselective Stetter reaction of various aliphatic α,β-unsaturated aldehydes with chalcones (i.e., **15** in Scheme 9.4a). Equally impressive is the enantioselective intramolecular Stetter reaction of trienes such as **16** to afford **18** reported by Xu[41] (Scheme 9.4b). The use of acetic acid as the cocatalyst was found to be advantageous in preventing catalyst decomposition. From a synthetic utility point of view, Glorius[42] developed a highly enantioselective synthesis of α-amino acid derivatives by means of the NHC-catalyzed enantioselective intermolecular Stetter reaction (Scheme 9.4c). Mechanistic studies revealed a diastereoselective protonation as the key stereocontrol element.

By extending the concepts of the Stetter reaction, Glorius was able to develop various hydroacylation reactions[43] by addition of the acyl anion to unactivated alkene[44], benzyne[45], and cyclopropene[46] (Scheme 9.5). Computational studies showed that these reactions occur via a concerted Conia-ene partway rather than a stepwise mechanism. Similarly, Rovis had suggested a reverse Cope elimination as a possible mechanism.[47] While the two *N*-mesityl catalysts shown

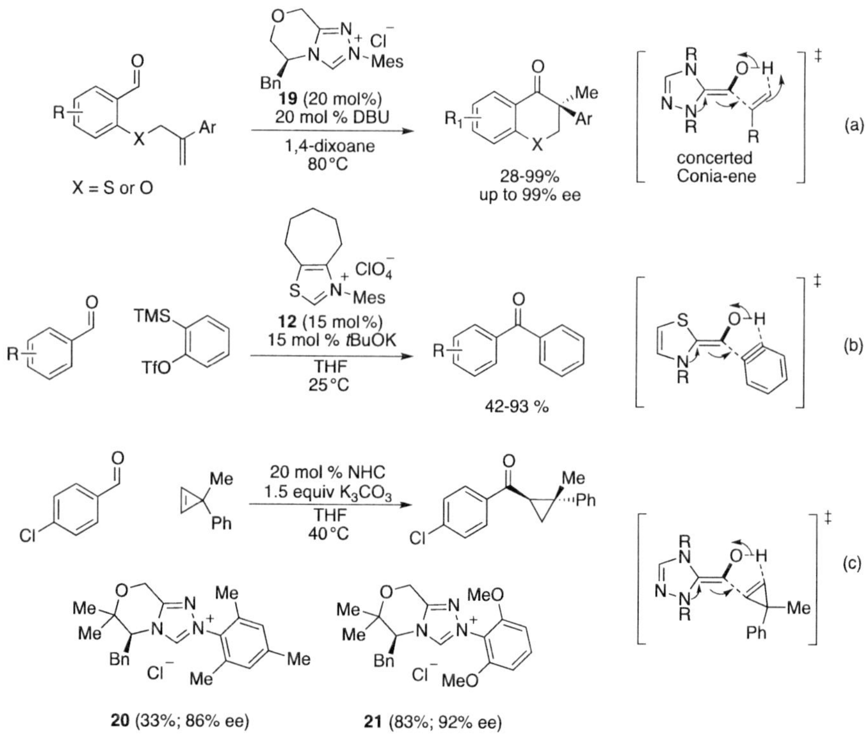

Scheme 9.5.

in Scheme 9.5 perform well for these reactions, a more electron-rich catalyst such as **21** (Scheme 9.5c)[48] was necessary for more challenging olefin partners in hydroacylation reactions involving the Breslow intermediate. The irreversible nature of the formation of this intermediate species nicely rationalizes the need for electron-rich, sterically hindered NHCs in these transformations by precluding the reverse reactions that might otherwise occur prior to the favored pathways.

9.3.2 Homoenolate Equivalent

9.3.2.1 1,2-Addition Reactions of the Homoenolate Equivalent. As part of our initial thinking about the generation of new reactive species by the combination of NHC and α,β-unsaturated aldehyde, we recognized the potential for the catalytic generation of "homoenolate equivalent." At the outset of this research in 2003, the chemical community was captivated by emerging results on the generation of enolate equivalents from aldehydes and ketones using proline-derived catalysts. The synthetically valuable homoenolate equivalent, in contrast, could be effectively formed only under harsh or operationally difficult conditions. Furthermore, no enantioselective homoenolate additions were known. Our successful conversion of cinnamaldehyde to dihydrocinnammic esters (Fig. 9.3) implied the generation of a homoenolate equivalent and invited attempts to trap it with carbon electrophiles rather than allowing it to undergo protonation. With this idea formulated, its execution proved to be surprisingly straightforward: simply combining cinnamaldehyde and an electrophilic aldehyde in the presence of IMesCl (**8**) and base led to the formation of a γ-lactone product resulting from electrophilic trapping of a catalytically generated homoenolate equivalent in which both our group[49] and Glorius'[50] group simultaneously reported (Scheme 9.6a). Our work also extended to the syntheses of γ-lactam via homoenolate addition to

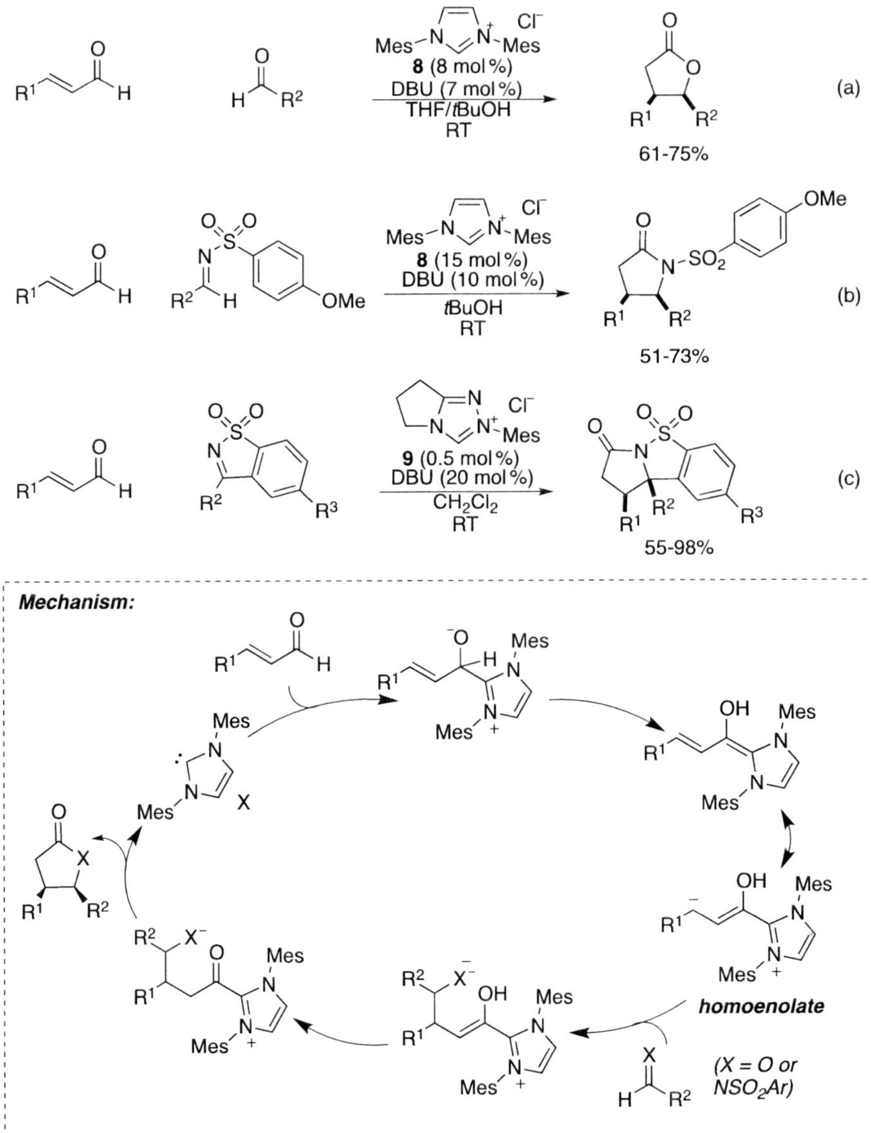

Scheme 9.6.

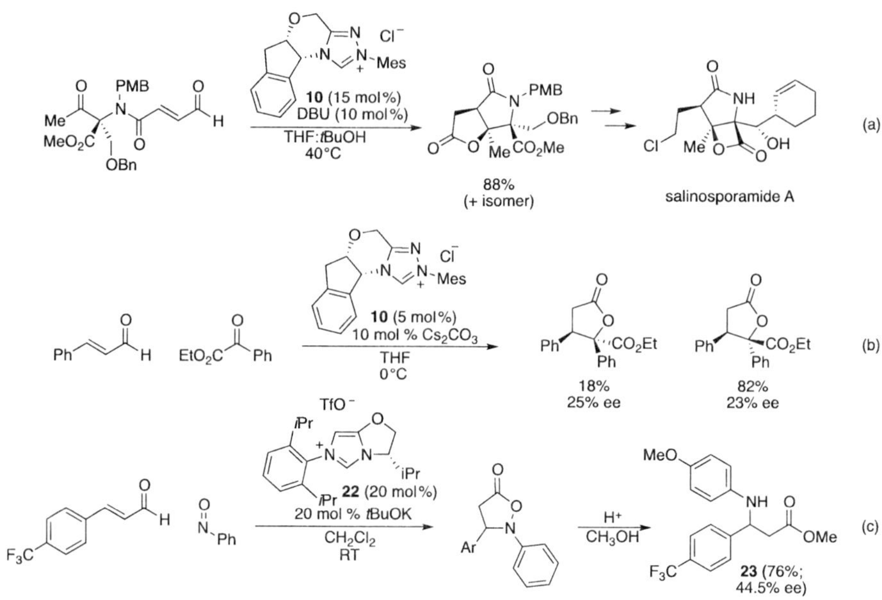

Scheme 9.7.

N-sulfonyl imines[51] (Scheme 9.6b) and saccharine-derived imines (Scheme 9.6c).[52] A general catalytic cycle of these reactions is shown next.

Although these reactions are often diastereoselective, rendering these processes enantioselective using the existing chiral *N*-mesityl catalysts is still challenging.[53] For example, low enantioselectivity was observed during the formal synthesis of salinosporamide[54] (Scheme 9.7a). A similar level of enantioinduction was also documented by You[55] during their attempt to render γ-lactone synthesis enantioselective (Scheme 9.7b) and by Zhang and Ying[56] during their synthesis of protected β-amino ester **23** (Scheme 9.7c). While a solution by new catalyst design has not truly emerged, Scheidt focuses on the use of nitrone substrates[57] or preorganization by Lewis acids[58] with either an uncommon protecting group[59] or a more activated substrate[60] (Scheme 9.8).

9.3.2.2 1,4-Addition Reactions of the Homoenolate Equivalent. The conjugated Breslow intermediate can also act as a nucleophile for conjugate addition with various unsaturated electrophiles.[61] One commonly used catalyst for this transformation is again the bismesityl imidazolium salt **8** (IMesCl). Nair reported the first example of what appeared to be the trapping of a catalytically generated homoenolate by a Michael-type addition. Surprisingly, the products observed were not the expected cyclopentanones but rather cyclopentenes, which are formed by an intramolecular aldol cyclization, lactonization, and decarboxylation.[62] Later they were able to further optimize the reaction to employ chalcones bearing heterocycles (Scheme 9.9a).[63] Using the same protocol, certain dienones such as **27** were found to be excellent coupling partners, this time forming the originally expected cyclopentenones (Scheme 9.9b).[64]

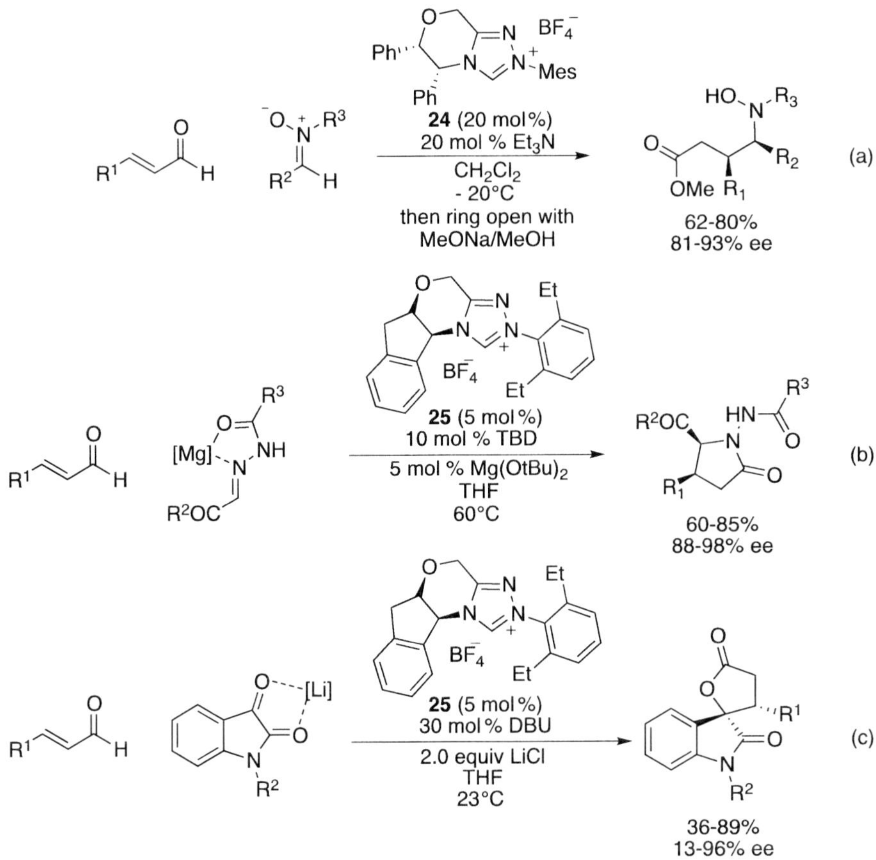

Scheme 9.8.

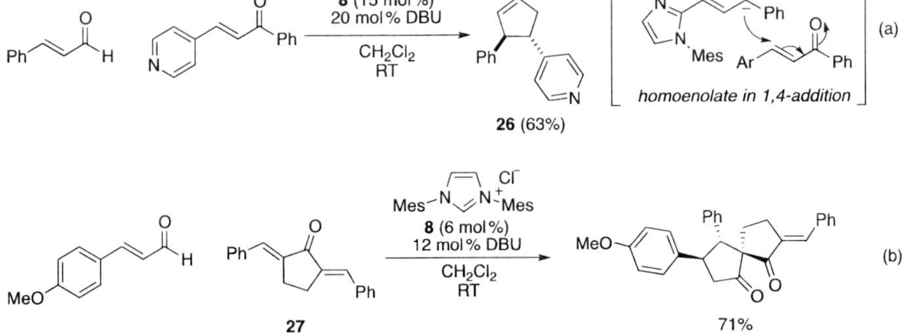

Scheme 9.9.

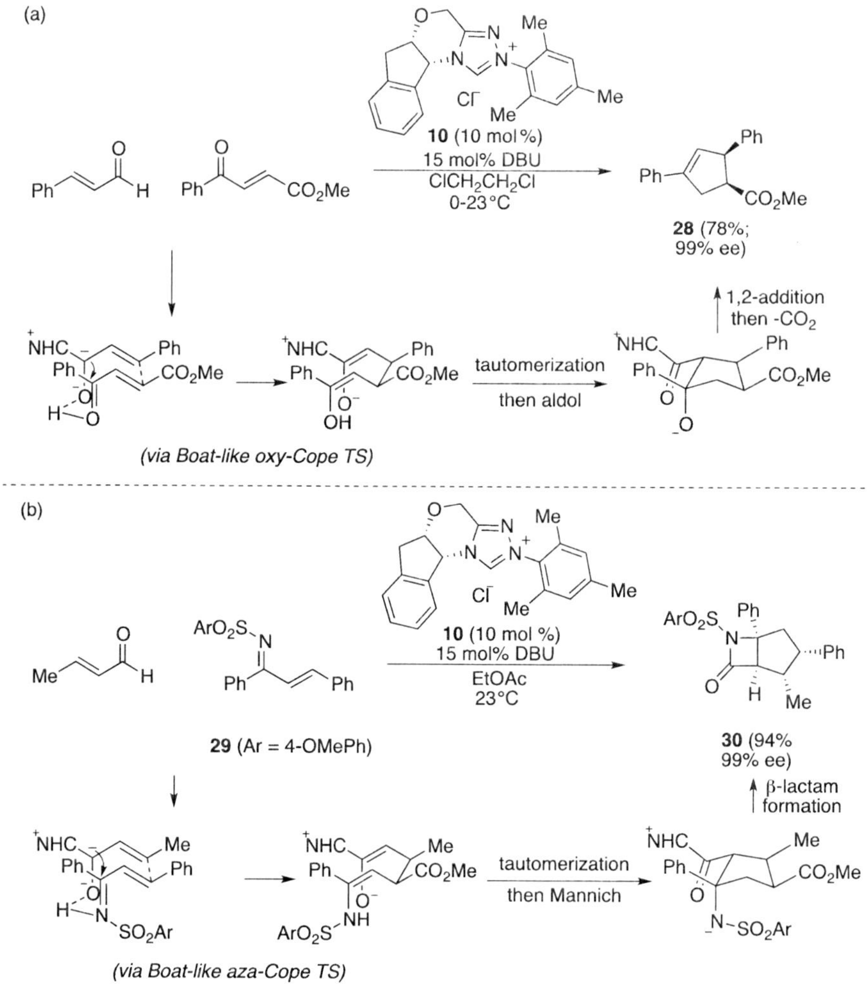

Scheme 9.10.

Chiral *N*-mesityl-substituted triazolium precatalysts have proven to be exceptional catalysts for highly enantioselective variants of the cyclopentene-forming annulation. Unexpectedly, *cis*-substituted cyclopentenes (rather than the trans-products observed in Nair's work earlier) were formed. We currently attribute this to a highly organized transition state in the addition of the homoenolate equivalent to the enone, one that can be thought of as involving a stepwise or possibly concerted crossed benzoin–oxy-Cope reaction (Scheme 9.10a).[65] The key decarboxylative formation of cyclopentenes **26** and **28** results in the loss of a stereocenter and a synthetic handle. This stereochemical information can be preserved in other variations of this annulation. For example, unsaturated imines (i.e., **29**) undergo a remarkable annulation leading to the formation of bicyclic β-lactams with outstanding yields and

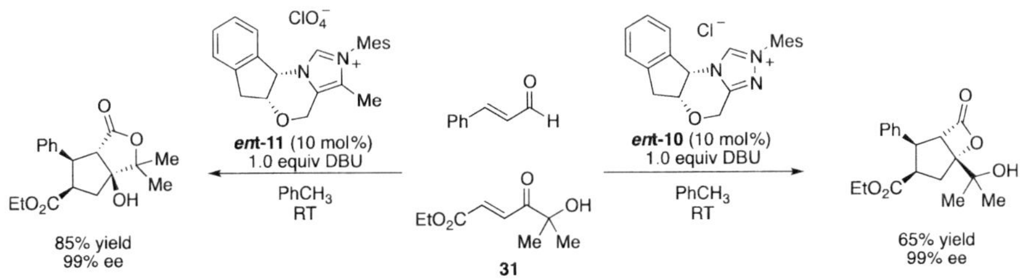

Figure 9.9. Stereodivergence from chiral *N*-mesityl triazolium versus imidazolium catalysts.

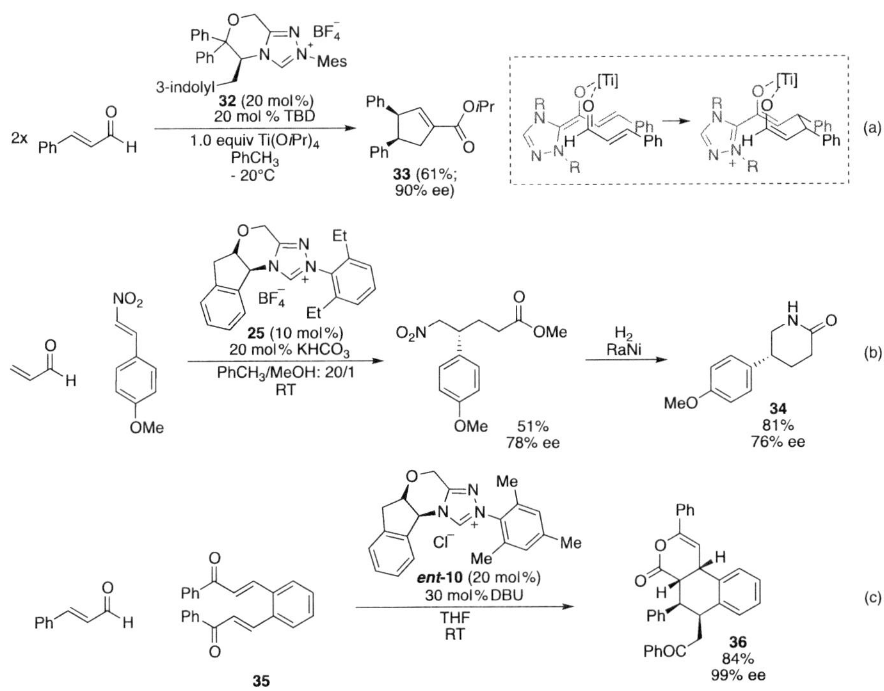

Scheme 9.11.

enantioselectivities (Scheme 9.10b).[66] Alternatively, enone **31** contains a pendant hydroxyl group, which can intercept the activated carboxylate formed during the cascade annulation reaction and prevent the cyclopentene-forming decarboxylation reaction (Fig. 9.9).[67] Remarkably, changing the catalyst from *N*-mesityl-substituted triazolium **10** to the nearly identical *N*-mesityl-substituted imidazolium **11** resulted in different products.

As seen in the previous section with 1,2-addition of homoenolate generated via NHC catalysis, conjugate addition of the same species can also be rendered enantioselective using chiral *N*-mesityl precatalysts, in this case, usually with high selectivity. Scheidt was able to report a dimerization reaction of enals in the absence of other electrophiles via the homoenolate pathway (Scheme 9.11a), in which the additive $Ti(OiPr)_4$ serves the role of Lewis acid for chelation and an external nucleophile ($iPrO^-$) for catalyst turnover.[68] Besides enones or enals as conjugate acceptors, β-nitrostyrene derivatives[69] can also be used as demonstrated by Liu, who showcases the versatility of this method in the synthesis of δ-lactam **34** (Scheme 9.11b).[70] Chi shows that the use of dienones such as **35** can furnish exquisite homoenolate–enolate cyclization cascade, yielding enantiopure product with high level of molecular complexity (**36** in Scheme 9.11c).[71]

9.3.3 Enolate Equivalent

Inspection of the reaction pathways shown in Figure 9.5 reveals the potential formation of yet another catalytically generated reactive enolate species by the protonation of the homoenolate/Breslow intermediate.[20a] The existence of this intermediate can also be inferred in the redox esterification of α,β-unsaturated and α-haloaldehydes. At the outset of our studies, the enantioselective trapping of this catalytically generated ester enolate equivalent with carbon electrophiles seemed like a far-fetched idea. We were, however, pleasantly surprised to find that both the generation and utility of this species is one of the most facile and versatile enantioselective C–C bond-forming processes to have been disclosed in NHC catalysis.

In 2006, we reported the first successful generation and enantioselective C–C bond formation featuring the catalytically generated ester enolate equivalents. This work, which also debuted *N*-mesityl-substituted aminoindanol-derived triazolium salt **10**, revealed the azadiene Diels–Alder reaction of unsaturated imines and the catalytically generated enolates (Scheme 9.12a).[20a] Already, we recognized that the generation of the NHC-bound, chiral ester enolate equivalents was not limited to α,β-unsaturated aldehydes as substrates. Our group pioneered highly enantioselective NHC-catalyzed oxodiene Diels–Alder reactions from chloroaldehydes (Scheme 9.12b)[72] and their bisulfite salts adduct surrogates (Scheme 9.12c).[73] With improved underlying mechanistic understanding garnered over the years, our group demonstrated that protonation of homoenolate derived from simple aldehyde (i.e., cinnamaldehyde instead of **37**) requires a weak base—which generates a strong conjugate acid—and we employed this knowledge for oxodiene Diels–Alder reactions of simple enals[74] (Scheme 9.12d). Computational studies in collaboration with the group of Marisa Kozlowski at the University of Pennsylvania indicated that the reaction proceeds via the deprotonated enolate, in which the oxygen of the enolate preassociates with the carbonyl of the diene (so called "oxyanion steering mechanism" in Fig. 9.10).[75] The calculated transition state follows a concerted, yet highly asynchronous [4+2] cycloaddition rather than an alternative stepwise Michael addition–cyclization pathway.

We have noted that many NHC-catalyzed reactions of substrates other than enals do not require the *N*-mesityl substituent catalysts; however, these NHCs

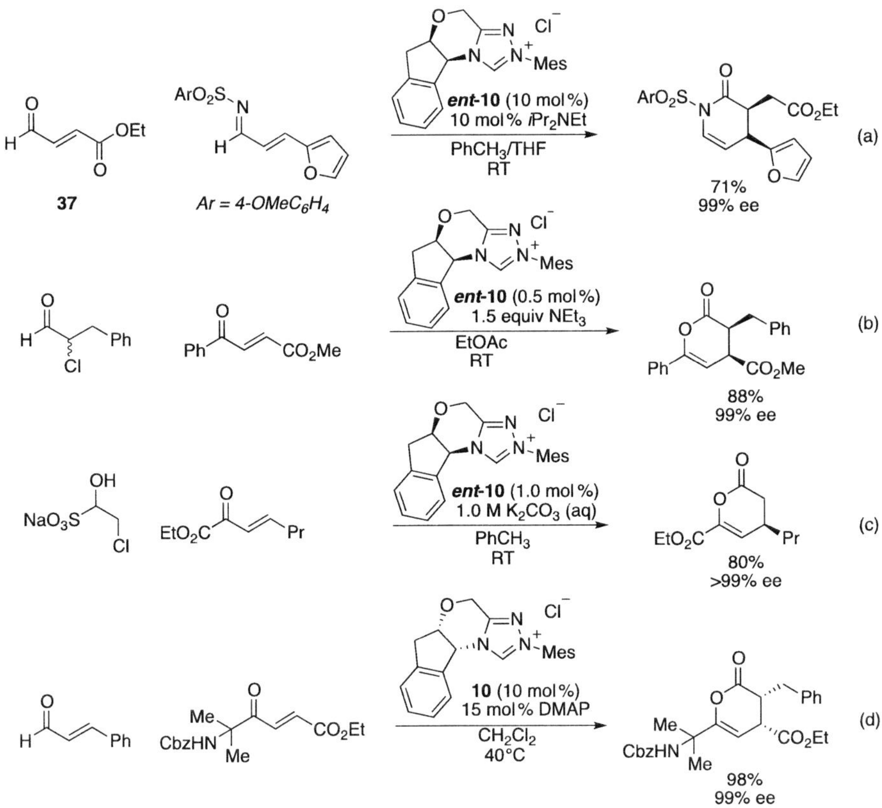

Scheme 9.12.

are required for high conversion and improved enantioselectivity in C–C bond-forming reactions involving the enolate species.[26] In order to better understand the role of the mesityl group in these reactions, ours and Kozlowski's group performed extensive computational and experimental studies on the NHC-catalyzed hetero-Diels–Alder reaction, in which the *N*-substitution of the triazolium salts influences the stereochemical outcome. The use of other *N*-aryl NHCs is possible here due to the fact that the redox reaction of α-chloroaldehyde proceeds via an elimination mechanism proposed by Nowak[76] and not through the Breslow intermediate (the rate acceleration effect from the mesityl moiety is not essential). The yields, however, were far lower than with the *N*-mesityl-substituted catalysts. The computational model suggests a C–H π interaction between the substrate and the *N*-aryl moiety of the catalyst as the key enantiocontrol element (Fig. 9.10). Electron-withdrawing groups decrease this interaction and cause erosion in the enantioselectivity of the product (i.e., 78% ee for the C_6F_5 group) in comparison to the mesityl group (>99% ee), which maximizes this favorable interaction.

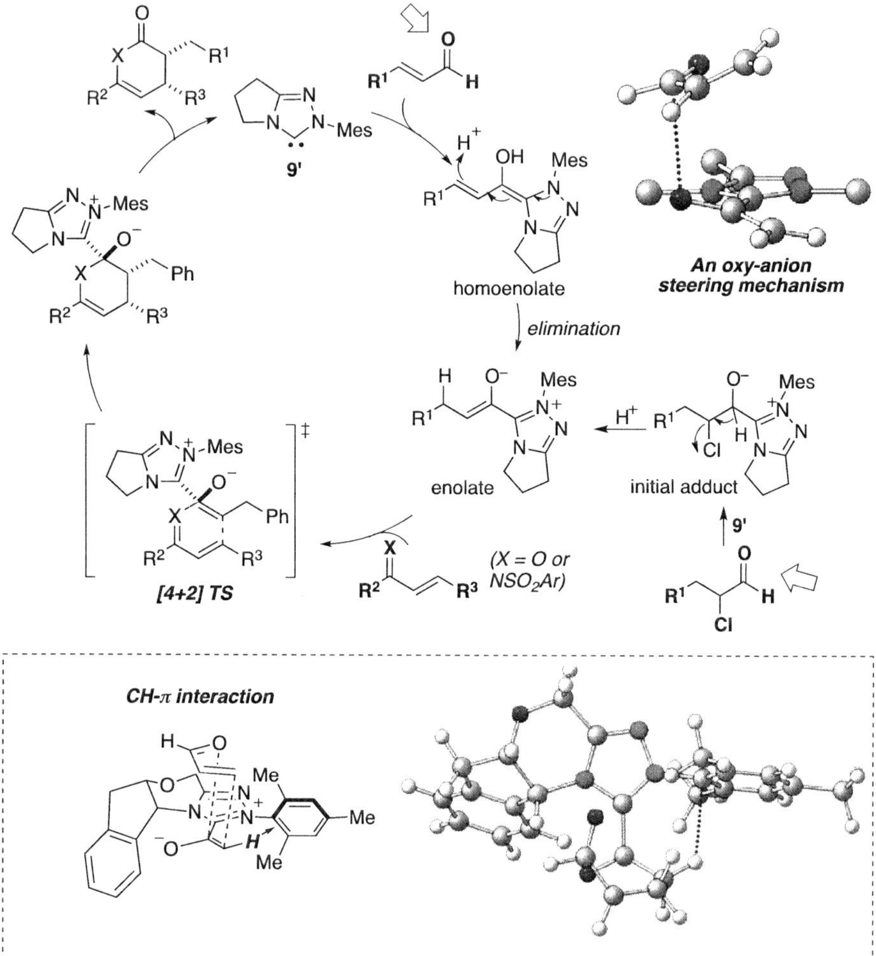

Figure 9.10. The mechansim of NHC-catalyzed Diels–Alder reactions. For color details, please see color plate section.

Beyond the work from our group, important contributions from Ye (Scheme 9.13a)[23] and Chi (Scheme 9.13b)[77] have extended the suitable precursors to stable ketenes, other α-functionalized aldehydes, and even saturated esters.[78] Subsequent works have expanded the substrate scope[79] of NHC-catalyzed [4+2] cycloadditions to include, for example, those of enals with vinyl ketones[80] or modified chalcones,[81] and Scheidt[82] also demonstrated an intramolecular variant of these reactions (Scheme 9.13c).

Besides the Diels–Alder manifold, other uses of NHC-bound chiral ester enolate equivalent as nucleophiles for aldol[83] or Mannich[84] reactions have been reported by the group of Scheidt (Scheme 9.14a,b); the latter comprises an enantioselective synthesis of protected β-amino acid derivatives. More interesting mechanistically is the catalytic generation of a conjugated enolate as a

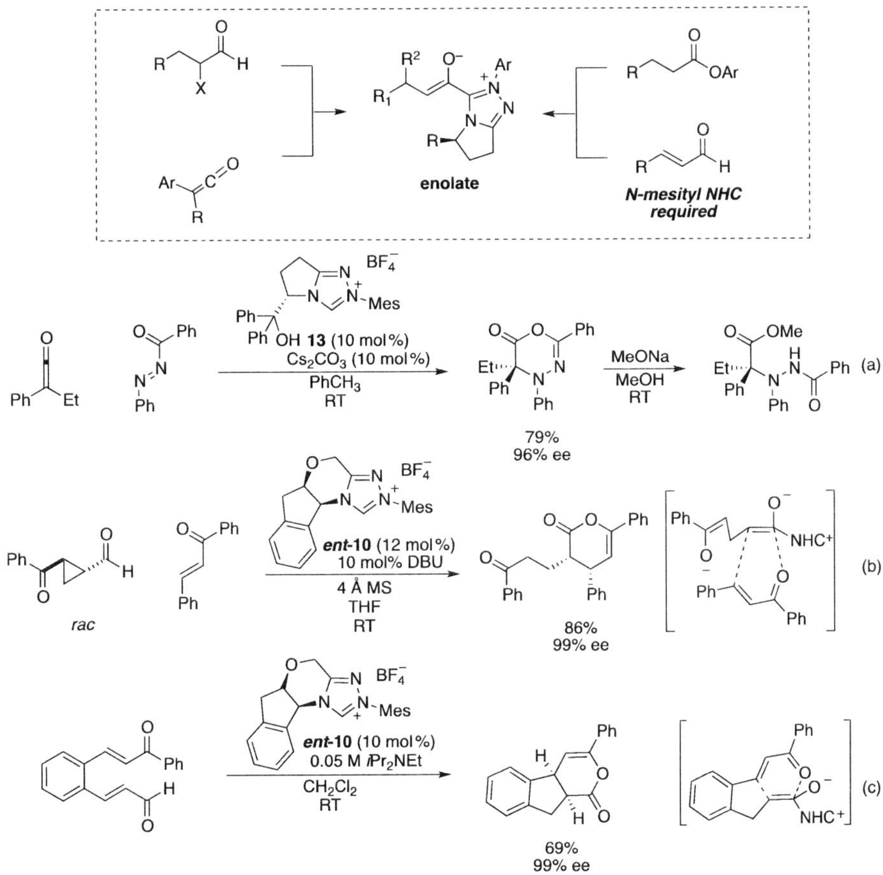

Scheme 9.13.

novel nucleophilic species in NHC chemistry. The group of Lin and Sun has reported an enantioselective fluorination by mean of internal redox reaction of enals bearing a leaving group at the γ position, using NFSI as the electrophilic fluorinating agent (Scheme 9.14c).[85] Using trisubstituted enals together with an external oxidant **38**, Chi reported an enantioselective aldol-cyclization cascade from unsubstituted conjugated enolate to trifluoromethyl ketones (Scheme 9.14d).[86] We anticipate that this mode of reactivity via conjugated enolate will find new discoveries and more applications in the future.

9.3.4 Activated Carboxylate Equivalent

Our original entry into NHC catalysis was motivated by the desire to develop catalytic methods for amide and ester formation. The typical synthesis of carboxylic esters, amides, and thioesters utilizes superstoichiometric amounts of expensive coupling reagents, bases, and additives to affect the union of a carboxylic acid and a nucleophile (Fig. 9.11a). Such reactions routinely produce

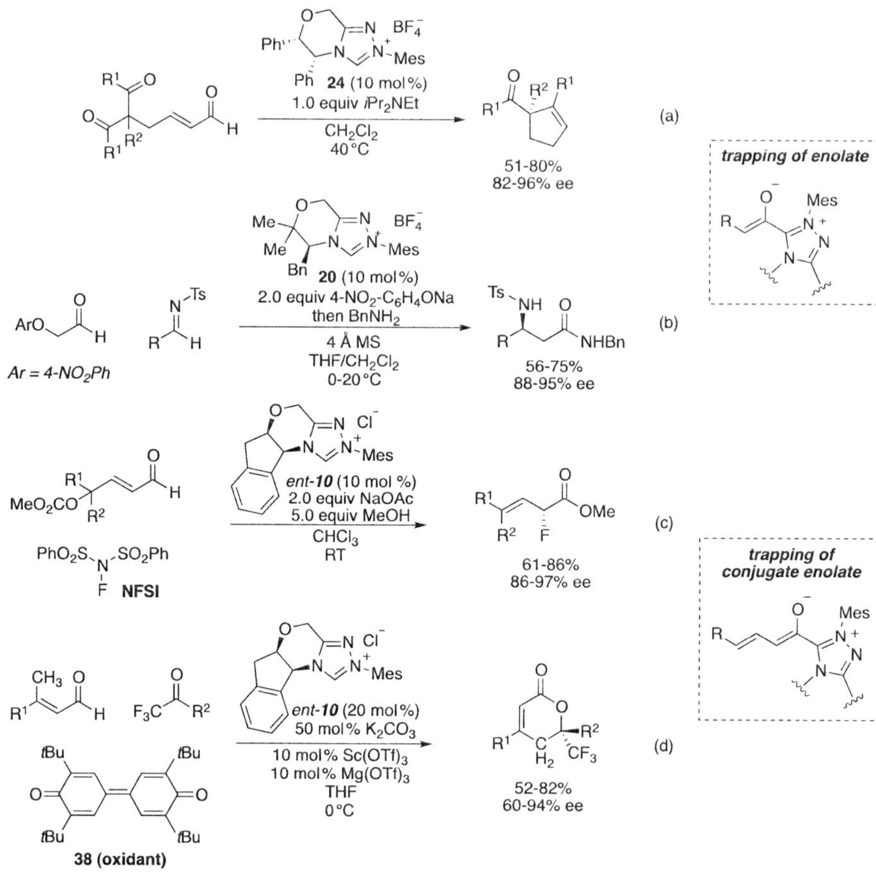

Scheme 9.14.

several equivalents or more of byproducts, necessitating complicated workups or other strategies, such as solid supports, for isolating the desired products. Furthermore, despite the ubiquity of carboxylic acid derivatives containing neighboring stereocenters, that is, amino acid residues, there are few catalytic methodologies for the direct synthesis of carboxylic acid derivatives from constituent fragments with concomitant introduction of stereochemical complexity.[87] The synthesis of carboxylic acid derivatives typically progresses through the intermediacy of "activated carboxylates" usually formed by the combination of a carboxylic acid and a coupling reagent. Given that these intermediates are rarely isolated, we believed that ester and amide forming processes featuring catalytically generated activated carboxylates should be feasible. This idea, coupled with a report from Townsend[88] on the thiamine-mediated biosynthesis of clavulanic acid, led us to postulate that NHCs could affect the catalytic generation of activated carboxylates from α-functionalized aldehydes. The advantage of using NHC is the ability to catalytically generate

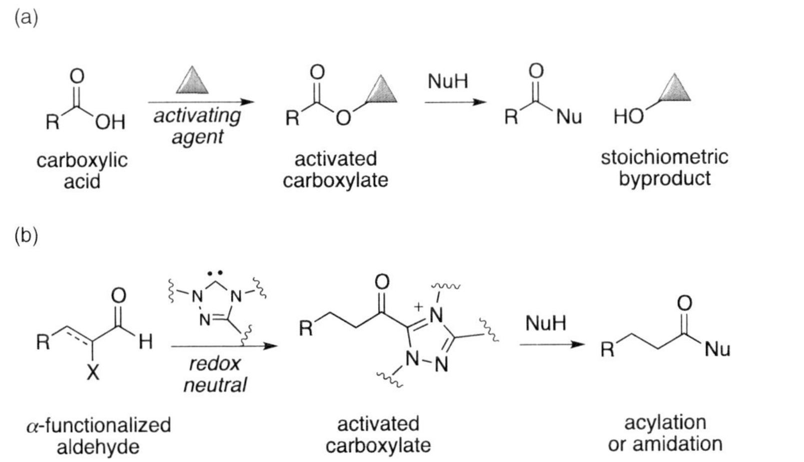

Figure 9.11. Generation of activated carboxylate for acyltion and amidation chemistry. (a) Traditional approach to genenration of activated carboxylate and (b) Catalytic generation of activated carboxylate via NHC catalysis.

activated carboxylate in a redox neutral fashion and without any superstoichiometric byproducts (Fig. 9.11b).

Since the first reports of such reactions on α-epoxyaldehydes (Scheme 9.2a)[29] from our group and on α-haloaldehydes from the Rovis' group,[30] the substrate scope has been extended to just about any aldehyde with a reducible functionality. With many of these aldehydes, the *N*-mesityl-substituted catalysts are not essential; the *N*-C_6F_5 ones are often superior. But in redox reactions of unsaturated aldehydes, in particular, the use of *N*-mesityl catalysts is preferred. Our group has also reported a redox esterification reaction with achiral *N*-mesityl salt **9** and formylcyclopropanes bearing an electron-withdrawing group (Scheme 9.15a).[89] The mechanism of this reaction follows the course of protonation of enolate **39** to form acyl azolium **40**, which serves the role of the activated carboxylate for the esterification step (Fig. 9.12). Similar to this approach, Smith has reported another redox esterification of α-aroyloxyaldehydes using the same catalyst (Scheme 9.15b),[90] while the intramolecular variants of this particular reaction with ribose and protected carbohydrate derivatives were reported by the group of Wendeborn at Syngenta (Scheme 9.15c).[91] This example is synthetically useful for rapid diversification of sugar commodities.

This redox neutral strategy using sterically hindered NHCs can be applied to ring expansion reactions of α-functionalized aldehydes with intramolecular acylation as the catalyst turnover strategy. Using the same mechanistic rational as in Figure 9.12, You has shown that dihydropyranone **41** can be synthesized from the same formylcyclopropane in the absence of an external nucleophile and a proton source (Scheme 9.16a).[92] Likewise, She combined the redox processes of enal and epoxyaldehyde in a relay redox-/ring-expansion cascade leading to another dihydropyranone (**43** in Scheme 9.16b).[93] Lastly, Gravel

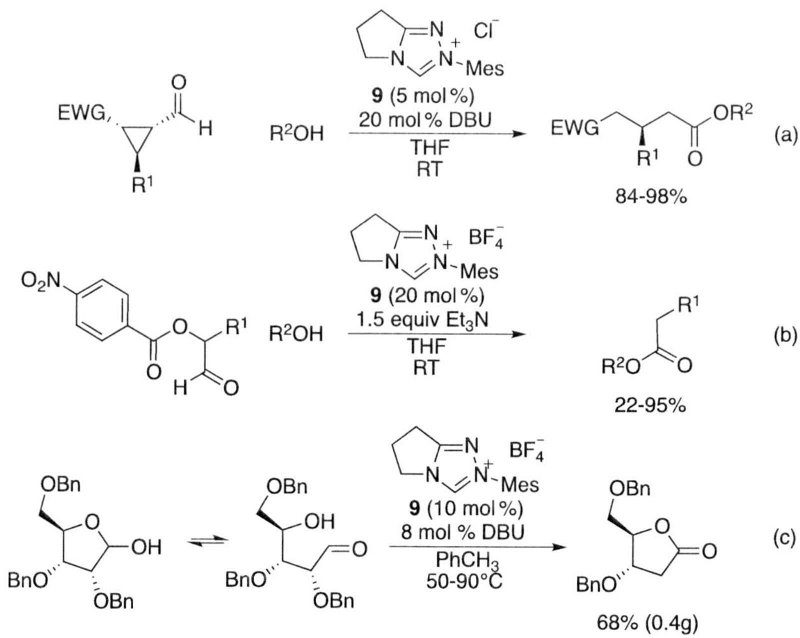

Scheme 9.15.

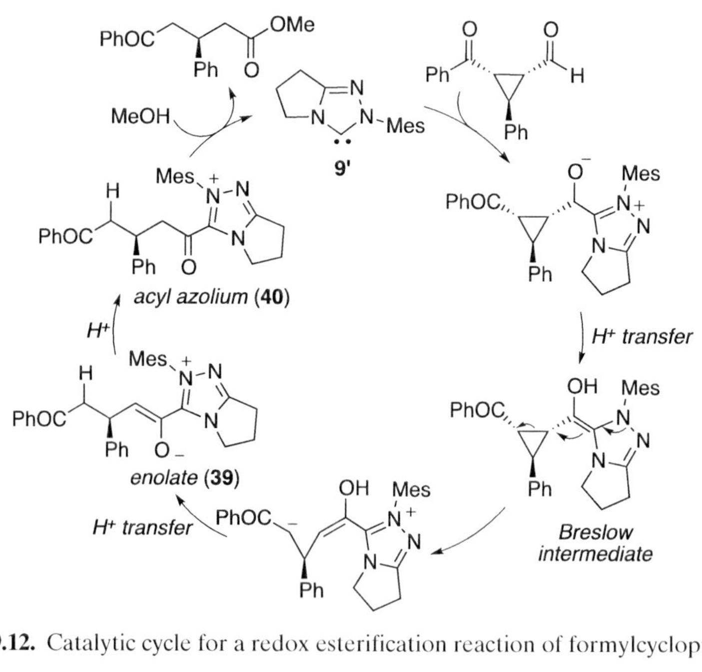

Figure 9.12. Catalytic cycle for a redox esterification reaction of formylcyclopropane.

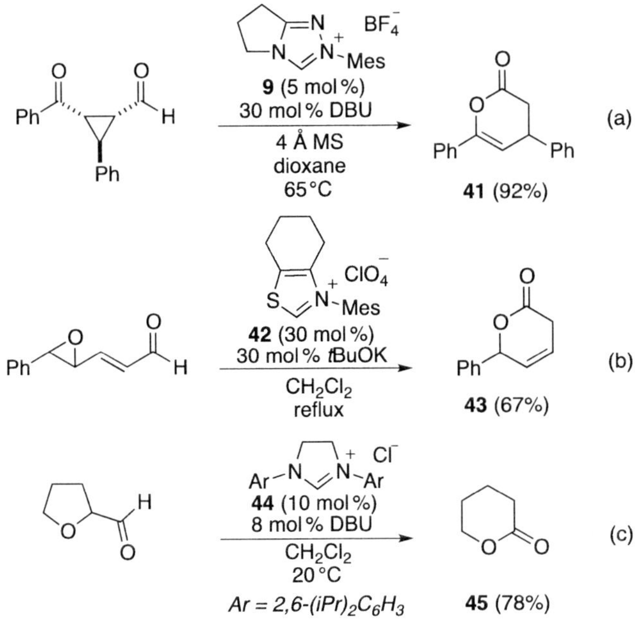

Scheme 9.16.

has shown that tetrahydrofurfural can be opened and lactonized intramolecularly by NHC **44** (Scheme 9.16c).[94]

The first transesterification catalyzed by *N*-mesityl azolium salts was the IMesCl catalyzed reaction pioneered by Nolan, Waymouth, and Hedrick in 2002.[95] Based on this and biological studies of thiamine phosphate–dependent enzymes,[5b] Movassaghi observed that acyl azolium does not react with amine[96]—an intriguing property that was used for amidation of amino alcohols by mean of acyl transfer via the ester (Scheme 9.17a).[97] While this is impressive, a direct catalytic amidation is still highly desirable and potentially synthetically powerful. In 2007, our group and Rovis' simultaneously reported the use of cocatalytic reagent such as imidazole[98] or HOAt[99] as a general solution to NHC-catalyzed amidation reactions of α-functionalized aldehydes (Scheme 9.17b). To avoid the formation of imine from aldehyde and amine, our group has introduced α-hydroxyenone[100] as aldehyde surrogate for amidation[101] and others[102] (Scheme 9.17c).

With this knowledge in hand, our group was also able to engineer a general solution to catalytic kinetic resolution of cyclic secondary amines[103]—compounds of contemporaneous interests for which other catalytic methods of resolution have not proven successful. With the help of the chiral hydroxamic acid cocatalyst **49**, this reaction has a broad substrate scope; substituted piperidines, morpholines, piperazines, and azepane all reacted with high selectivity. Figure 9.13 summarizes the synergistic catalytic cycle of this kinetic resolution. It is worth noting here that all approaches to catalytic amidation rely on the same strategy of transferring the acyl azolium (**46**; unreactive towards amine) to acyl imidazole **47**, acyl triazole **47**, or acyl hydroxamate **50** (Scheme 9.17).

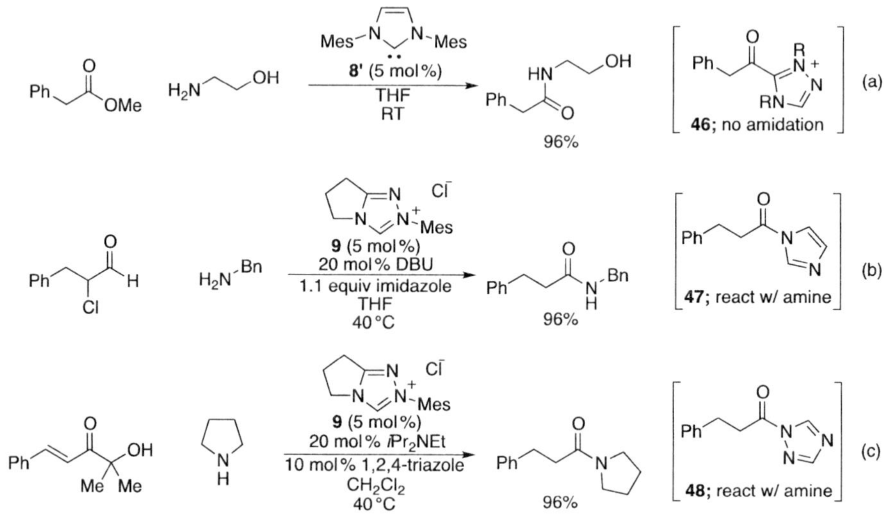

Scheme 9.17.

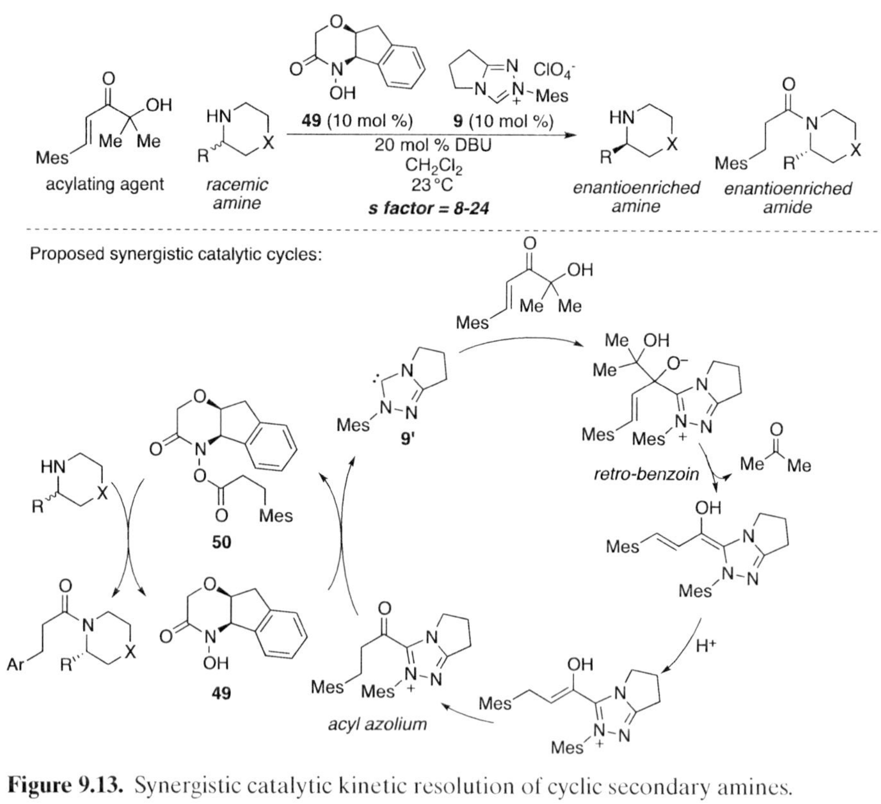

Figure 9.13. Synergistic catalytic kinetic resolution of cyclic secondary amines.

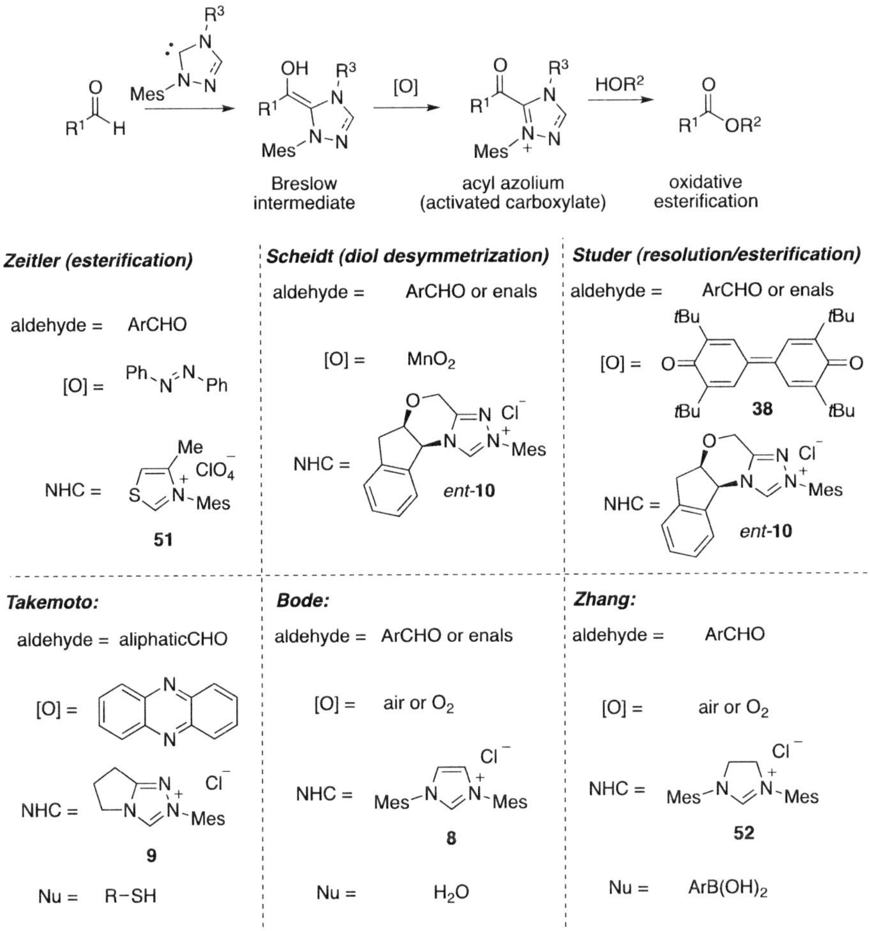

Figure 9.14. A brief summary of NHC-catalyzed oxidative acylation reactions.

While the previous redox approach to esters and amides is desirable in terms of efficiency and sustainability, oxidative esterification, in which the aldehyde stating materials can be oxidized to generate the activated carboxylate in mild condition, is also synthetically useful. This reaction operates by oxidation of the Breslow intermediate directly to the acyl azolium[104] without the protonation sequence (Fig. 9.14). This concept has gained much attention from researchers in the field. Many oxidants such as azobenzene,[105] MnO_2,[106] or diquinone 38[107] have been used together with a number of *N*-mesityl NHCs for oxidative esterification of simple aromatic aldehydes and enals. Contemporaneously, the works from our group[108] and Zhang[109] have shown that air (O_2) can be used as the stoichiometric oxidant instead of small molecule organic oxidants. The last remaining challenge in this field is the NHC-catalyzed oxidation of aliphatic aldehydes, in which Takemoto has reported the only solution up to date.[110] Figure 9.14 summarizes these approaches.

9.3.5 α,β-Unsaturated Activated Carboxylate Equivalent

The catalytically generated acyl azolium species are the key intermediates in NHC-catalyzed esterification and amidation reactions. The same principles also make possible the catalytic generation of α,β-unsaturated acyl azolium (Fig. 9.15), which undergoes a diverse range of enantioselective C–C bond formation. In 2006, Zeitler reported a protocol for the stereoselective generation of unsaturated acyl azolium with IMesCl (**8**) and ynals as acylating agents (Scheme 9.18a).[111] A conceptually related redox esterification from α-bromo-enals has been recently reported by Du (Scheme 9.18b).[112] Zeitler's mechanistic proposal involves the intermediacy of conjugated Breslow intermediate such as **53** and its allenoate tautomer, whose existence has been hinted in another study by Sun (Scheme 9.18c).[113]

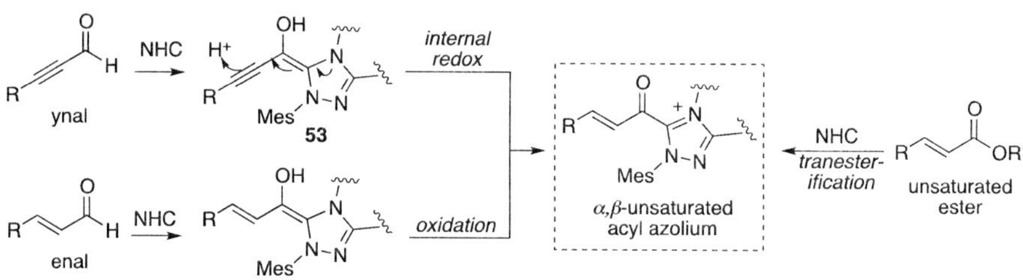

Figure 9.15. Catalytic generation of α,β-unsaturated acyl azolium for NHC catalysis.

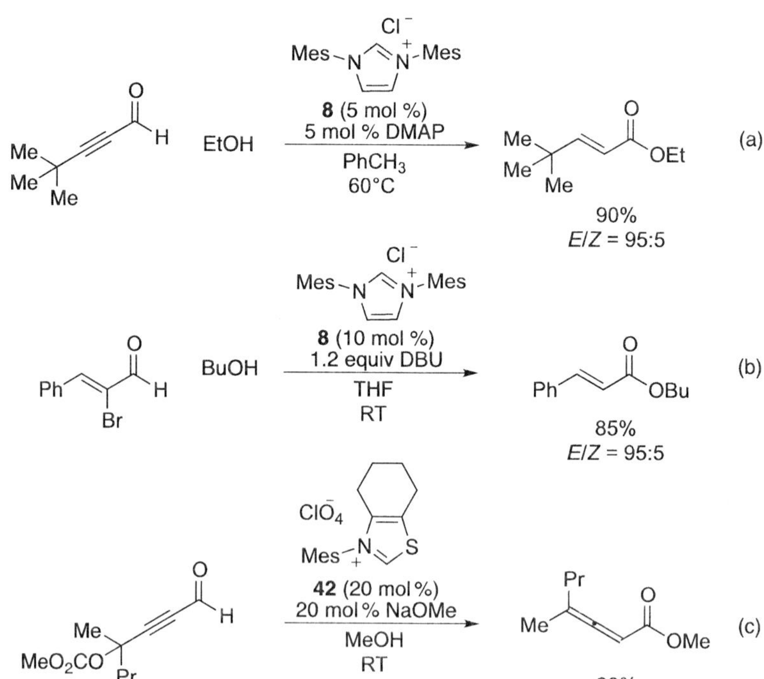

Scheme 9.18.

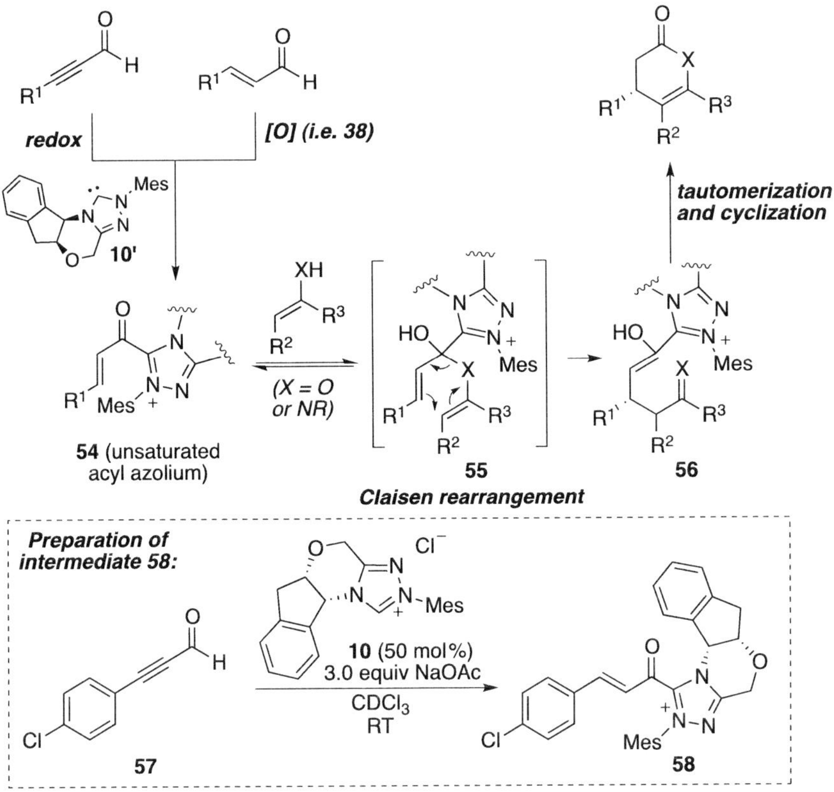

Figure 9.16. A general mechanism of NHC-catalyzed Claisen rearrangements.

The electrophilic nature of this intermediate suggested that it could be used for new reactions. Given the long-known fact that acyl azolium species have a metastable hydrate form in equilibrium,[5b,96] we began our investigations by trapping α,β-unsaturated acyl azolium (**54**) generated from ynal with stable enols. This combination leads to a hemiacetal (i.e., **55**), which is rendered long-lived enough by the *N*-mesityl catalyst (slow turnover) to undergo a Claisen rearrangement to give intermediate **56** (Fig. 9.16, top). Upon tautomerization, lactone formation finishes the catalytic cycle and affords the resulting dihydropyranones in excellent yield and enantioselectivity. The synthetic utility of this methodology is demonstrated in the functionalizations of ethyl pyruvate **59** and kojic acid derivative **60** (Scheme 9.19a,b).[27a] We were able to extend this methodology to the coupling of stable, unprotected enamines (i.e., **61**) with enals in the presence of oxidant **38** to afford enantioenriched dihydropyridinones, which are of pharmaceutical interest but had no previous method for catalytic asymmetric synthesis (Scheme 9.19c).[114] More recently, we showed that the enamine tautomer of various saccharine-derived *N*-sulfonylimines such as **62** may be intercepted in a highly stereoselective manner with trisubstituted enals[115]—a class of substrates whose reaction is a long-standing challenge in organocatalysis (Scheme 9.19d).

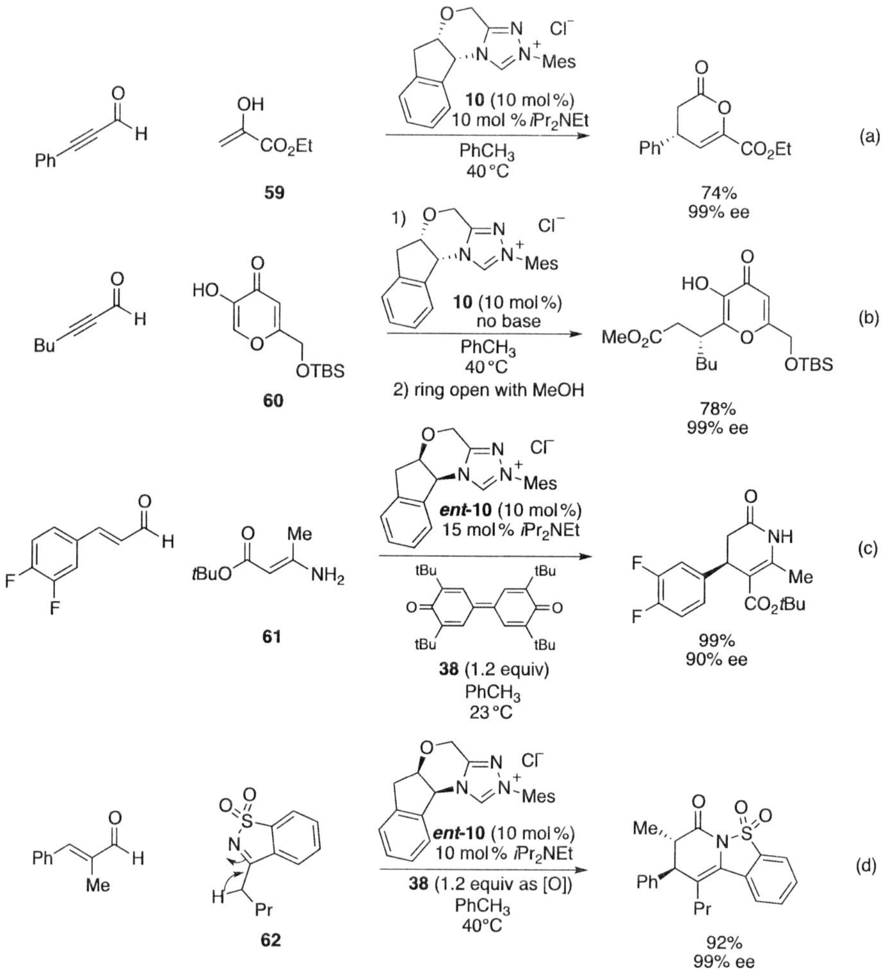

Scheme 9.19.

As a part of our mechanistic research program, we were able to prepare α,β-unsaturated acyl azolium **58** by a simple combination of ynal **57** and *N*-mesityl chiral triazolium salt **10** (Fig. 9.16).[116] The mesityl moiety provided kinetic stability for this highly reactive intermediate and allowed for a sufficient lifetime during characterization (UV-Vis, 2D NMR, and high-resolution mass spectrometry). Our detailed kinetic studies regarding the reactions of acyl azolium **58** revealed that catalyst turnover is the rate-limiting step and that it had similar physical properties as a biosynthetic intermediate previously studied by Townsend[117] in 2007 and then later by Mück-Lichtenfeld, Mayr, and Studer in 2012.[118] Both these studies argued for the possibility of thiazolium- and imidazolium-derived unsaturated acyl azolium as Michael acceptors—a situation we ruled unlikely for the Claisen reactions based on the related works

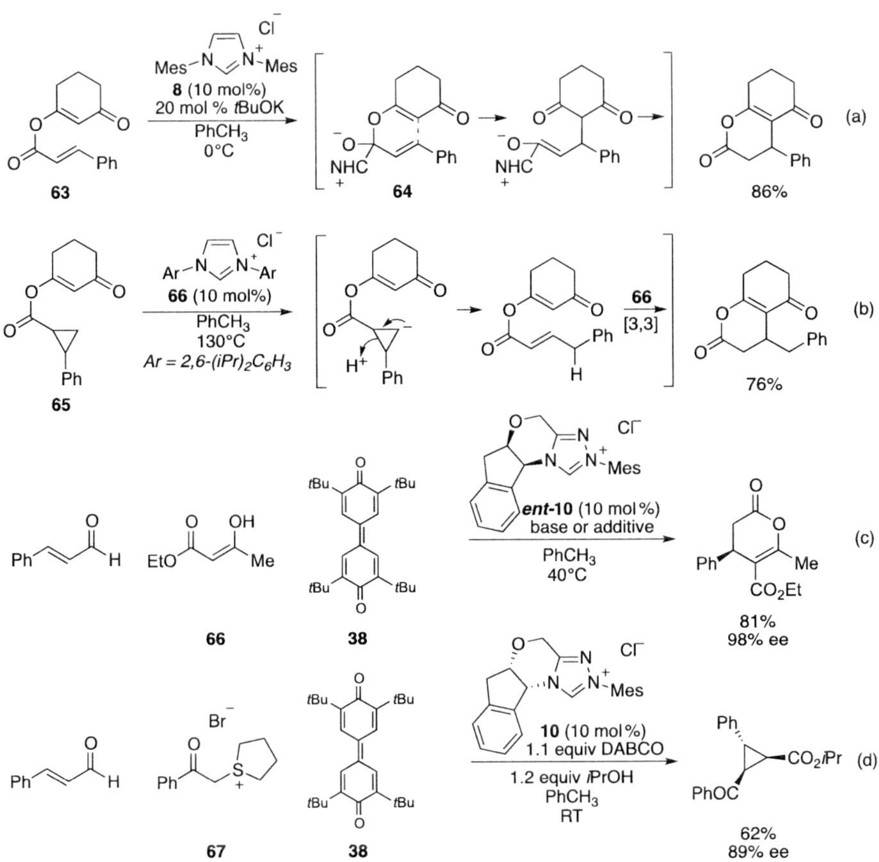

Scheme 9.20.

of Coates and Curran,[119] our extensive kinetic data, and the observation that good Michael donors we tested (thiol, hydroxamic acid, coumarin, etc.) all failed to react.[120] Computational studies by Schoenebeck rationalized the origin of stereocontrol and confirmed once again the proposed [3,3]-sigmatropic rearrangement mechanism.[121]

Just a few months prior to our first publication,[27a] Lupton reported a similar reaction involving α,β-unsaturated acyl azoliums generated via transesterification of enol esters (i.e., **63**) and IMesCl (**8**) to afford dihydropyranone products (Scheme 9.20a).[122] The original proposed mechanism also postulated a direct conjugate addition from the breakdown of hemiacetal **64**, but was later revised to implicate the Claisen reaction as the key C–C bond-forming step based on a crossover experiment[123] and observation of byproduct, at least for some substrate classes.[124] Most impressively, Lupton has utilized this rearrangement strategy in the total synthesis of deoxyloganin[125] and the cyclization of cyclopropyl enol ester **65** (Scheme 9.20b).[126] Following our work and that of Lupton, Xiao[127] and Studer[128] employed 1,3-dicarbonyl compounds such as **66** as the enol component in a related cascade reaction via the intermediacy of unsaturated acyl azolium in good yield and enantioselectivity (Scheme 9.20c). This mode of reactivity has received much attention between 2011 and 2012.[129] Recently, Studer has further increased the scope of activated ketones to include sulfur ylide **67** in his oxidative cyclopropanation reaction of enals with *i*PrOH as the terminal nucleophile (Scheme 9.20d).[130]

9.4 CONCLUSION AND OUTLOOK

Although NHC catalysis is by no means limited to catalysts bearing *N*-mesityl or related groups, the role of *N*-mesityl-substituted triazolium precatalysts in unlocking many new reaction manifolds cannot be underestimated. Few would have anticipated the diversity of the reactions and products or the outstanding stereocontrol offered by essentially one class of catalysts. As this field progresses, opportunities for new catalyst designs open further pathways and expand the class of substrates amenable to NHC catalysis abound. Most importantly, we anticipate that future innovations in catalyst design will make possible reactions from even simpler starting materials so that NHC catalysis can move beyond its current niche of fine chemical products to impacting the industrial production of organic molecules—a field currently dominated by heterogeneous metal catalysis.

SUGGESTED READING

Chiang, P.-C., Bode, J. W. *TCI MAIL.* **2011**, *149*, 2–17.

Chiang, P.-C., Bode, J. W. "N-heterocyclic carbenes as organic catalysts." In *N-Heterocyclic Carbenes*. The Royal Society of Chemistry: Cambridge, UK, **2011**, pp. 399–435.

Douglas, J., Churchill, G., Smith, A. D. *Synthesis* **2012**, *44*, 2295–2309.

Kaeobamrung, J., Kozlowski, M. C., Bode, J. W. *Proc. Natl. Acad. Sci.* **2010**, *107*, 20661–20665.

Mahatthananchai, J., Bode, J. W. *Chem. Sci.* **2012**, *3*, 192–197.

Nair, V., Menon, R. S., Biju, A. T., Sinu, C. R., Paul, R. R., Jose, A., Sreekumar, V. *Chem. Soc. Rev.* **2011**, *40*, 5336–5346.

Sohn, S. S., Rosen, E. L., Bode, J. W. *J. Am. Chem. Soc.* **2004**, *126*, 14370–14371.

Sohn, S. S., Bode, J. W. *Org. Lett.* **2005**, *7*, 3873–3876.

Struble, J. R., Kaeobamrung, J., Bode, J. W. *Org. Lett.* **2008**, *10*, 957–960.

Vora, H. U., Wheeler, P., Rovis, T. *Adv. Synth. Catal.* **2012**, *354*, 1617–1639.

REFERENCES

1 Wöhler, F., von Liebig, J. *Ann. der Pharm.* **1832**, *3*, 249–282.

2 (a) Lapworth, A. *J. Chem. Soc., Trans.* **1903**, *83*, 995–1005.(b) Lapworth, A. J. *Chem. Soc., Trans.* **1904**, *85*, 1206–1214.

3 Ugai, T., Tanaka, S., Dokawa, S. *J. Pharm. Soc. Jpn.* **1943**, *63*, 296–300.

4 (a) Frank, R., Leeper, F., Luisi, B. *Cell. Mol. Life Sci.* **2007**, *64*, 892–905. (b) Kluger, R., Tittmann, K. *Chem. Rev.* **2008**, *108*, 1797–1833.
5 (a) Breslow, R. *J. Am. Chem. Soc.* **1958**, *80*, 3719–3726. (b) Breslow, R., McNelis, E. *J. Am. Chem. Soc.* **1960**, *82*, 2394–2395. (c) Breslow, R., Kim, R. *Tetrahedron Lett.* **1994**, *35*, 699–702.
6 White, M. J., Leeper, F. J. *J. Org. Chem.* **2001**, *66*, 5124–5131.
7 (a) Teles, J. H., Melder, J.-P., Ebel, K., Schneider, R., Gehrer, E., Harder, W., Brode, S., Enders, D., Breuer, K., Raabe, G. *Helv. Chim. Acta* **1996**, *79*, 61–83. (b) Berkessel, A., Elfert, S., Etzenbach-Effers, K., Teles, J. H. *Angew. Chem.* **2010**, *122*, 7063–7063.(c) Berkessel, A., Elfert, S., Yatham, V. R., Neudörfl, J.-M., Schlörer, N. E., Teles, J. H. *Angew. Chem. Int. Ed.* **2012**, *51*, 12370–12374.
8 Stetter, H. *Angew. Chem. Int. Ed.* **1976**, *15*, 639–647.
9 (a) Enders, D., Breuer, K., Raabe, G., Runsink, J., Teles, J. H., Melder, J.-P., Ebel, K., Brode, S. *Angew. Chem. Int. Ed.* **1995**, *34*, 1021–1023.(b) Enders, D., Breuer, K., Runsink, J., Henrique Teles, J. *Liebig Ann.* **1996**, *1996*, 2019–2028.
10 (a) Miyashita, A., Matsuoka, Y., Suzuki, Y., Iwamoto, K., Higashino, T., *Chem. Pharm. Bull.* **1997**, *45*, 1235–1242. (b) Miyashita, A., Suzuki, Y., Nagasaki, I., Ishiguro, C., Iwamoto, K., Higashino, T., *Chem. Pharm. Bull.* **1997**, *45*, 1254–1258.
11 (a) Enders, D., Breuer, K., Teles, J. H. *Helv. Chim. Acta* **1996**, *79*, 1217–1221.(b) Enders, D., Breuer, K., Runsink, J., Teles, J. H. *Helv. Chim. Acta* **1996**, *79*, 1899–1902.
12 Sheehan, J. C., Hunneman, D. H. *J. Am. Chem. Soc.* **1966**, *88*, 3666–3667.
13 Knight, R. L., Leeper, F. J. *J. Chem. Soc., Perkin Trans. 1* **1998**, 1891–1893.
14 Kerr, M. S., Read de Alaniz, J., Rovis, T. *J. Org. Chem.* **2005**, *70*, 5725–5728.
15 Kerr, M. S., Read de Alaniz, J., Rovis, T. *J. Am. Chem. Soc.* **2002**, *124*, 10298–10299.
16 Takikawa, H., Hachisu, Y., Bode, J. W., Suzuki, K. *Angew. Chem. Int. Ed.* **2006**, *45*, 3492–3494.
17 Liu, Q., Perreault, S., Rovis, T. *J. Am. Chem. Soc.* **2008**, *130*, 14066–14067.
18 Vora, H. U., Lathrop, S. P., Reynolds, N. T., Kerr, M. S., Read de Alaniz, J., Rovis, T. *Org. Synth.* **2010**, *87*, 350–361.
19 Sohn, S. S., Bode, J. W. *Org. Lett.* **2005**, *7*, 3873–3876.
20 (a) He, M., Struble, J. R., Bode, J. W. *J. Am. Chem. Soc.* **2006**, *128*, 8418–8420. (b) Struble, J. R., Bode, J. W. *Org. Synth.* **2010**, *87*, 362–376.
21 (a) Struble, J. R., Kaeobamrung, J., Bode, J. W. *Org. Lett.* **2008**, *10*, 957–960. (b) Struble, J. R., Bode, J. W. *Tetrahedron* **2008**, *64*, 6961–6972.
22 Lebeuf, R., Hirano, K., Glorius, F. *Org. Lett.* **2008**, *10*, 4243–4246.
23 Huang, X.-L., He, L., Shao, P.-L., Ye, S. *Angew. Chem. Int. Ed.* **2009**, *48*, 192–195.
24 (a) Chiang, P.-C., Bode, J. W. *TCI MAIL*. **2011**, *149*, 2–17. (b) Mahatthananchai, J., Bode, J. W. In *Asymmetric Synthesis: The Essentials II*. Wiley-VCH: Weinheim, *2012*, pp. 67–77.
25 (a) Rovis, T. *Chem. Lett.* **2008**, *37*, 2–7. (b) For a recent study about the pKa values of triazolium NHCs, see Massey, R. S., Collett, C. J., Lindsay, A. G., Smith, A. D., O'Donoghue, A. C. *J. Am. Chem. Soc.* **2012**, *134*, 20421–20432.
26 Mahatthananchai, J., Bode, J. W. *Chem. Sci.* **2012**, *3*, 192–197.
27 (a) Kaeobamrung, J., Mahatthananchai, J., Zheng, P., Bode, J. W. *J. Am. Chem. Soc.* **2010**, *132*, 8810–8812. (b) Bugaut, X., Liu, F., Glorius, F. *J. Am. Chem. Soc.* **2011**, *133*, 8130–8133.
28 (a)Maji, B., Breugst, M., Mayr, H. *Angew. Chem. Int. Ed.* **2011**, *50*, 6915–6919.(b) See also Collett, C. J., Massey, R. S., Maguire, O. R., Batsanov, A. S., O'Donoghue, A. C., Smith, A. D. *Chem. Sci.* **2013**, *4*, 1514–1522.

29 Chow, K. Y.-K., Bode, J. W. *J. Am. Chem. Soc.* **2004**, *126*, 8126–8127.
30 Reynolds, N. T., Read de Alaniz, J., Rovis, T. *J. Am. Chem. Soc.* **2004**, *126*, 9518–9519.
31 (a) Concellón, C., Duguet, N., Smith, A. D. *Adv. Synth. Catal.* **2009**, *351*, 3001–3009. (b) Wang, X.-N., Lv, H., Huang, X.-L., Ye, S. *Org. Biomol. Chem.* **2009**, *7*, 346–350.
32 Johnson, J. S., *Angew. Chem. Int. Ed.* **2004**, *43*, 1326–1328.
33 (a) Kano, T., Sasaki, K., Maruoka, K. *Org. Lett.* **2005**, *7*, 1347–1349. (b) Suzuki, Y., Muramatsu, K., Yamauchi, K., Morie, Y., Sato, M. *Tetrahedron* **2006**, *62*, 302–310.
34 (a) Duguet, N., Campbell, C. D., Slawin, A. M. Z., Smith, A. D. *Org. Biomol. Chem.* **2008**, *6*, 1108–1113. (b) He, L., Lv, H., Zhang, Y.-R., Ye, S. *J. Org. Chem.* **2008**, *73*, 8101–8103.(c) Zhang, Y.-R., Lv, H., Zhou, D., Ye, S. *Chem. Eur. J.* **2008**, *14*, 8473–8476.(d) Shao, P.-L., Chen, X.-Y., Ye, S. *Angew. Chem. Int. Ed.* **2010**, *49*, 8412–8416.(e) Jian, T.-Y., Shao, P.-L., Ye, S. *Chem. Commun.* **2011**, *47*, 2381–2383.
35 (a) Rong, Z.-Q., Li, Y., Yang, G.-Q., You, S.-L. *Synlett* **2011**, *7*, 1033–1037. (b) Kim, S. M., Jin, M. Y., Kim, M. J., Cui, Y., Kim, Y. S., Zhang, L., Song, C. E., Ryu, D. H., Yang, J. W. *Org. Biomol. Chem.* **2011**, *9*, 2069–2071.(c) Mao, H., An, S. L., Kim, S.-M., Yang, J.-W., *Bull. Korean Chem. Soc.* **2011**, *32*, 4408–4410.
36 Other examples include (a) Jin, M. Y., Kim, S. M., Han, H., Ryu, D. H., Yang, J. W., *Org. Lett.* **2011**, *13*, 880–883.(b) O'Toole, S. E., Rose, C. A., Gundala, S., Zeitler, K., Connon, S. J. *J. Org. Chem.* **2010**, *76*, 347–357.
37 Piel, I., Pawelczyk, M. D., Hirano, K., Fröhlich, R., Glorius, F. *Eur. J. Org. Chem.* **2011**, *2011*, 5475–5484.
38 Kuhl, N., Glorius, F. *Chem. Commun.* **2011**, *47*, 573–575.
39 Read de Alaniz, J., Rovis, T. *Synlett* **2009**, *8*, 1189–1207.
40 Fang, X., Chen, X., Lv, H., Chi, Y. R., *Angew. Chem. Int. Ed.* **2011**, *50*, 11782–11785.
41 Liu, G., Wilkerson, P. D., Toth, C. A., Xu, H. *Org. Lett.* **2012**, *14*, 858–861.
42 Jousseaume, T., Wurz, N. E., Glorius, F. *Angew. Chem. Int. Ed.* **2011**, *50*, 1410–1414.
43 Biju, A. T., Kuhl, N., Glorius, F. *Acc. Chem. Res.* **2011**, *44*, 1182–1195.
44 Piel, I., Steinmetz, M., Hirano, K., Fröhlich, R., Grimme, S., Glorius, F. *Angew. Chem. Int. Ed.* **2011**, *50*, 4983–4987.
45 Biju, A. T., Glorius, F. *Angew. Chem. Int. Ed.* **2010**, *49*, 9761–9764.
46 Liu, F., Bugaut, X., Schedler, M., Fröhlich, R., Glorius, F. *Angew. Chem. Int. Ed.* **2011**, *50*, 12626–12630.
47 DiRocco, D. A., Rovis, T. *Angew. Chem. Int. Ed.* **2011**, *50*, 7982–7983.
48 Schedler, M., Fröhlich, R., Daniliuc, C.-G., Glorius, F., *Eur. J. Org. Chem.* **2012**, *2012*, 4164–4171.
49 Sohn, S. S., Rosen, E. L., Bode, J. W. *J. Am. Chem. Soc.* **2004**, *126*, 14370–14371.
50 Burstein, C., Glorius, F. *Angew. Chem. Int. Ed.* **2004**, *43*, 6205–6208.
51 He, M., Bode, J. W. *Org. Lett.* **2005**, *7*, 3131–3134.
52 Rommel, M., Fukuzumi, T., Bode, J. W. *J. Am. Chem. Soc.* **2008**, *130*, 17266–17267.
53 Zheng, P., Gondo, C. A., Bode, J. W. *Chem. Asian J.* **2011**, *6*, 614–620.
54 Struble, J. R., Bode, J. W. *Tetrahedron* **2009**, *65*, 4957–4967.
55 Li, Y., Zhao, Z.-A., He, H., You, S.-L. *Adv. Synth. Catal.* **2008**, *350*, 1885–1890.
56 Seayad, J., Patra, P. K., Zhang, Y., Ying, J. Y. *Org. Lett.* **2008**, *10*, 953–956.
57 Phillips, E. M., Reynolds, T. E., Scheidt, K. A. *J. Am. Chem. Soc.* **2008**, *130*, 2416–2417.
58 Cohen, D. T., Scheidt, K. A. *Chem. Sci.* **2012**, *3*, 53–57.
59 Raup, D. E. A., Cardinal-David, B., Holte, D., Scheidt, K. A. *Nat. Chem.* **2010**, *2*, 766–771.
60 Dugal-Tessier, J., O'Bryan, E. A., Schroeder, T. B. H., Cohen, D. T., Scheidt, K. A. *Angew. Chem. Int. Ed.* **2012**, *51*, 4963–4967.

61 Nair, V., Vellalath, S., Babu, B. P. *Chem. Soc. Rev.* **2008**, *37*, 2691–2698.
62 Nair, V., Vellalath, S., Poonoth, M., Suresh, E., *J. Am. Chem. Soc.* **2006**, *128*, 8736–8737.
63 Nair, V., Paul, R. R., Padmaja, D. V. M., Aiswarya, N., Sinu, C. R., Jose, A., *Tetrahedron* **2011**, *67*, 9885–9889.
64 Nair, V., Babu, B. P., Vellalath, S., Suresh, E., *Chem. Commun.* **2008**, 747–749.
65 Chiang, P.-C., Kaeobamrung, J., Bode, J. W., *J. Am. Chem. Soc.* **2007**, *129*, 3520–3521.
66 He, M., Bode, J. W., *J. Am. Chem. Soc.* **2008**, *130*, 418–419.
67 Kaeobamrung, J., Bode, J. W., *Org. Lett.* **2009**, *11*, 677–680.
68 Cohen, D. T., Cardinal-David, B., Roberts, J. M., Sarjeant, A. A., Scheidt, K. A., *Org. Lett.* **2011**, *13*, 1068–1071.
69 Nair, V., Sinu, C. R., Babu, B. P., Varghese, V., Jose, A., Suresh, E., *Org. Lett.* **2009**, *11*, 5570–5573.
70 Maji, B., Ji, L., Wang, S., Vedachalam, S., Ganguly, R., Liu, X.-W., *Angew. Chem. Int. Ed.* **2012**, *51*, 8276–8280.
71 Fang, X., Jiang, K., Xing, C., Hao, L., Chi, Y. R., *Angew. Chem. Int. Ed.* **2011**, *50*, 1910–1913.
72 He, M., Uc, G. J., Bode, J. W. *J. Am. Chem. Soc.* **2006**, *128*, 15088–15089.
73 He, M., Beahm, B. J., Bode, J. W. *Org. Lett.* **2008**, *10*, 3817–3820.
74 Kaeobamrung, J., Kozlowski, M. C., Bode, J. W. *Proc. Natl. Acad. Sci.* **2010**, *107*, 20661–20665.
75 (a) Allen, S. E., Mahatthananchai, J., Bode, J. W., Kozlowski, M. C. *J. Am. Chem. Soc.* **2012**, *134*, 12098–12103. (b) The azolium-enolate species has been studied since 1970: Regitz, M., Hocker, J., Weber, B. *Angew. Chem. Int. Ed.* **1970**, *9*, 375.
76 Nowak, R. M. *J. Org. Chem.* **1963**, *28*, 1182–1187.
77 Fang, X., Chen, X., Chi, Y. R. *Org. Lett.* **2011**, *13*, 4708–4711.
78 Hao, L., Du, Y., Lv, H., Chen, X., Jiang, H., Shao, Y., Chi, Y. R. *Org. Lett.* **2012**, *14*, 2154–2157.
79 (a) Kobayashi, S., Kinoshita, T., Uehara, H., Sudo, T., Ryu, I., *Org. Lett.* **2009**, *11*, 3934–3937. (b) Rong, Z.-Q., Jia, M.-Q., You, S.-L., *Tetrahedron* **2011**, *67*, 9329–9333.(c) Jian, T.-Y., Sun, L.-H., Ye, S., *Chem. Commun.* **2012**, *48*, 10907–10909.
80 Nair, V., Paul, R. R., Seetha Lakshmi, K. C., Menon, R. S., Jose, A., Sinu, C. R. *Tetrahedron Lett.* **2011**, *52*, 5992–5994.
81 Fang, X., Chen, X., Chi, Y. R. *Org. Lett.* **2011**, *13*, 4708–4711.
82 Phillips, E. M., Wadamoto, M., Chan, A., Scheidt, K. A. *Angew. Chem. Int. Ed.* **2007**, *46*, 3107–3110.
83 Wadamoto, M., Phillips, E. M., Reynolds, T. E., Scheidt, K. A. *J. Am. Chem. Soc.* **2007**, *129*, 10098–10099.
84 Kawanaka, Y., Phillips, E. M., Scheidt, K. A. *J. Am. Chem. Soc.* **2009**, *131*, 18028–18029.
85 Zhao, Y.-M., Cheung, M. S., Lin, Z., Sun, J. *Angew. Chem. Int. Ed.* **2012**, *51*, 10359–10363.
86 Mo, J., Chen, X., Chi, Y. R. *J. Am. Chem. Soc.* **2012**, *134*, 8810–8813.
87 Pattabiraman, V., Bode, J. W. *Nature* **2011**, *480*, 471–479.
88 Phaleeli, N., Li, R., Townsend, C. A., *J. Am. Chem. Soc.* **1999**, *121*, 9223–9224.
89 Sohn, S. S., Bode, J. W. *Angew. Chem. Int. Ed.* **2006**, *45*, 6021–6024.
90 Ling, K. B., Smith, A. D. *Chem. Commun.* **2011**, *47*, 373–375.
91 Wendeborn, S., Mondière, R., Keller, I., Nussbaumer, H. *Synlett* **2012**, *4*, 541–544.
92 Li, G.-Q., Dai, L.-X., You, S.-L. *Org. Lett.* **2009**, *11*, 1623–1625.

93 Qi, J., Xie, X., He, J., Zhang, L., Ma, D., She, X. *Org. Biomol. Chem.* **2011**, *9*, 5948–5950.

94 Wang, L., Thai, K., Gravel, M. *Org. Lett.* **2009**, *11*, 891–893.

95 (a) Grasa, G. A., Kissling, R. M., Nolan, S. P. *Org. Lett.* **2002**, *4*, 3583–3586. (b) Nyce, G. W., Lamboy, J. A., Connor, E. F., Waymouth, R. M., Hedrick, J. L. *Org. Lett.* **2002**, *4*, 3587–3590.

96 (a) Daigo, K., Reed, L. J. J. *Am. Chem. Soc.* **1962**, *84*, 659–662. (b) White, F. G., Ingraham, L. L. J. *Am. Chem. Soc.* **1962**, *84*, 3109–3111.(c) Bruice, T. C., Kundu, N. G. *J. Am. Chem. Soc.* **1966**, *88*, 4097–4098.(d) Lienhard, G. E. *J. Am. Chem. Soc.* **1966**, *88*, 5642–5649.(e) Owen, T. C., Richards, A. *J. Am. Chem. Soc.* **1987**, *109*, 2520–2521.(f) Owen, T. C., Harris, J. N. *J. Am. Chem. Soc.* **1990**, *112*, 6136–6137.(g) Ferreira, L. M., Lobo, A. M., Prabhakar, S., Marcelo-Curto, M. J., Rzepa, H. S., Yi, M. Y. *J. Chem. Soc., Chem. Commun. 1991*, 1127–1128.

97 (a) Movassaghi, M., Schmidt, M. A. *Org. Lett.* **2005**, *7*, 2453–2456. (b) Schmidt, M. A., Miiller, P., Movassaghi, M. *Tetrahedron Lett.* **2008**, *49*, 4316–4318.

98 Bode, J. W., Sohn, S. S. *J. Am. Chem. Soc.* **2007**, *129*, 13798–13799.

99 Vora, H. U., Rovis, T. *J. Am. Chem. Soc.* **2007**, *129*, 13796–13797.

100 Unlike enals, this substrate can be synthesized from an aldol reaction in one step from inexpensive materials: (a) Palomo, C., Oiarbide, M., García, J. M., González, A., Arceo, E., *J. Am. Chem. Soc.* **2003**, *125*, 13942–13943. (b) Monge, D., Martín-Zamora, E., Vázquez, J., Alcarazo, M., Álvarez, E., Fernández, R., Lassaletta, J. M. *Org. Lett.* **2007**, *9*, 2867–2870.(c) Lee, S., Lim, C. J., Kim, S., Subramaniam, R., Zimmerman, J., Sibi, M. P. *Org. Lett.* **2006**, *8*, 4311–4313. (d) Sakai, T., Yamawaki, A., Ito, H., Utaka, M., Takeda, A. *J. Heterocycl. Chem.* **1986**, *23*, 1199–1201.

101 Chiang, P. C., Kim, Y., Bode, J. W. *Chem. Commun.* **2009**, 4566–4568.

102 Chiang, P.-C., Rommel, M., Bode, J. W., *J. Am. Chem. Soc.* **2009**, *131*, 8714–8718.

103 (a) Binanzer, M., Hsieh S.-Y., Bode, J. W. *J. Am. Chem. Soc.* **2011**, *133*, 19698–19701. (b) Hsieh, S.-Y., Binanzer, M., Kreituss, I., Bode, J. W. *Chem. Commun.* **2012**, *48*, 8892–8894.

104 (a) Inoue, H., Higashiura, K. *J. Chem. Soc. Chem. Commun.* **1980**, 549– 550. (b) Tam, S.-W., Jimenez, L., Diederich, F. *J. Am. Chem. Soc.* **1992**, *114*, 1503–1505. For an alternative pathway, see (c) Castells, J., Llitjos, H., Moreno-Manas, M. *Tetrahedron Let.* **1977**, *18*, 205–206. (d) Delany, E. G., Fagan, C.-L., Gundala, S., Zeitler, K., Connon, S. J. *Chem. Commun.* **2013**, *49*, 6513–6515.

105 (a) Rose, C. A., Zeitler, K. *Org. Lett.* **2010**, *12*, 4552–4555. (b) For a related report, see Noonan, C., Baragwanath, L., Connon, S. J. *Tetrahedron Lett.* **2008**, *49*, 4003–4006.

106 Maki, B. E., Chan, A., Phillips, E. M., Scheidt, K. A. *Org. Lett.* **2007**, *9*, 371–374.

107 (a) De Sarkar, S., Biswas, A., Song, C. H., Studer, A. *Synthesis* **2011**, *12*, 1974–1983. (b) For the preparation of the oxidant, Kharasch, M. S., Joshi, B. S. *J. Org. Chem.* **1957**, *22*, 1439–1443.

108 (a) Chiang, P.-C., Bode, J. W. *Org. Lett.* **2011**, *13*, 2422–2425. (b) For a related report, see Maji, B., Vedachalan, S., Ge, X., Cai, S., Liu, X.-W. *J. Org. Chem.* **2011**, *76*, 3016–3023.

109 (a) Meng, J.-J., Gao, M., Wei, Y.-P., Zhang, W.-Q. *Chem. Asian J.* **2012**, *7*, 872–875. (b) For a related report, see Arde, P., Ramanjaneyulu, B. T., Reddy, V., Saxena, A., Anand, R. V. *Org. Biomol. Chem.* **2012**, *10*, 848–851.

110 Uno, T., Inokuma, T., Takemoto, Y. *Chem. Commun.* **2012**, *48*, 1901–1903.

111 (a) Zeitler, K., *Org. Lett.* **2006**, *8*, 637–640. (b) The cyanide catalyzed variant was reported much earlier: Walia, J. S., Vishwakarma, L. C., J. *Chem. Soc., Chem. Commun.* **1969**, 396.
112 (a) Wang, X.-B., Zou, X.-L., Du, G.-F., Liu, Z.-Y., Dai, B. *Tetrahedron* **2012**, *68*, 6498–6503. (b) The cyanide catalyzed variant was also reported much earlier: Roedig, A., Hagedorn, F., *Liebigs Ann.* **1965**, *683*, 30–41.
113 Zhao, Y.-M., Tam, Y., Wang, Y.-J., Li, Z., Sun, J., *Org. Lett.* **2012**, *14*, 1398–1401.
114 Wanner, B., Mahatthananchai, J., Bode, J. W., *Org. Lett.* **2011**, *13*, 5378–5381.
115 Kravina, A. G., Mahatthananchai, J., Bode, J. W., *Angew. Chem. Int. Ed.* **2012**, *51*, 9433–9436.
116 Mahatthananchai, J., Zheng, P., Bode, J. W., *Angew. Chem. Int. Ed.* **2011**, *50*, 1673–1677.
117 Merski, M., Townsend, C. A., *J. Am. Chem. Soc.* **2007**, *129*, 15750–15751.
118 Samanta, R. C., Maji, B., De Sarkar, S., Bergander, K., Fröhlich, R., Mück-Lichtenfeld, C., Mayr, H., Studer, A., *Angew. Chem. Int. Ed.* **2012**, *51*, 5234–5238.
119 (a) Coates, R. M., Rogers, B. D., Hobbs, S. J., Curran, D. P., Peck, D. R., *J. Am. Chem. Soc.* **1987**, *109*, 1160–1170. (b) Coates, R. M., Hobbes, S. J. *J. Org. Chem.* **1984**, *49*, 140–152.
120 Mahatthananchai, J., Kaeobamrung, J., Bode, J. W., *ACS Catal.* **2012**, *2*, 494–503.
121 Lyngvi, E., Bode, J. W., Schoenebeck, F. *Chem. Sci.* **2012**, *3*, 2346–2350.
122 Ryan, S. J., Candish, L., Lupton, D. W. *J. Am. Chem. Soc.* **2009**, *131*, 14176–14177.
123 Candish, L., Lupton, D. W. *Org. Biomol. Chem.* **2011**, *9*, 8182–8189.
124 (a) Ryan, S. J., Candish, L., Lupton, D. W. *J. Am. Chem. Soc.* **2011**, *133*, 4694–4697. (b) Ryan, S. J., Stasch, A., Paddon-Row, M. N., Lupton, D. W. *J. Org. Chem.* **2012**, *77*, 1113–1124.
125 Candish, L., Lupton, D. W. *Org. Lett.* **2010**, *12*, 4836–4839.
126 Candish, L., Lupton, D. W. *Chem. Sci.* **2012**, *3*, 380–383.
127 (a) Zhu, Z. Q., Xiao, J. C. *Adv. Syn. Catal.* **2010**, *352*, 2455–2458. (b) Zhu, Z.-Q., Zheng, X.-L., Jiang, N.-F., Wan, X., Xiao, J.-C. *Chem. Commun.* **2011**, *47*, 8670–8672.
128 (a) De Sarkar, S., Studer, A. *Angew. Chem. Int. Ed.* **2010**, *49*, 9266–9269. (b) Biswas, A., Sarkar, S. D., Frohlich, R., Studer, A. *Org. Lett.* **2011**, *13*, 4966–4969.
129 (a) Rong, Z.-Q., Jia, M.-Q., You, S.-L. *Org. Lett.* **2011**, *13*, 4080–4083. (b) Sun, F.-G., Sun, L.-H., Ye, S. *Adv. Synth. Catal.* **2011**, *353*, 3134–3138.(c) Yao, C., Wang, D., Lu, J., Li, T., Jiao, W., Yu, C. *Chem. Eur. J.* **2012**, *18*, 1914–1917.(d) Du, D., Hu, Z., Jin, J., Lu, Y., Tang, W., Wang, B., Lu, T. *Org. Lett.* **2012**, *14*, 1274–1277.(e) Romanov-Michailidis, F., Besnard, C. L., Alexakis, A. *Org. Lett.* **2012**, *14*, 4906–4909.
130 Biswas, A., De Sarkar, S., Tebben, L., Studer, A. *Chem. Commun.* **2012**, *48*, 5190–5192.

10

SUPRAMOLECULAR CARBENE CHEMISTRY

UDO H. BRINKER,[†] JEAN-LUC MIEUSSET
Department of Chemistry, Institute of Organic Chemistry, University of Vienna, Vienna, Austria
MURRAY G. ROSENBERG
[†] *Department of Chemistry, The State University of New York at Binghamton, Binghamton, NY, USA*

10.1 INTRODUCTION

The world of carbene chemistry has been extensively explored.[1] Truly, they are intriguing reaction intermediates. However, classical, unmediated carbene reactions have had limited impact on synthetic organic chemistry beyond simple transformations, such as the cyclopropanation of alkenes and the methylation of alcohols with diazomethane.[2] Carbenes readily insert into the C–H bonds of hydrocarbons as well. But, even when this is synthetically desired, the intermolecular reactions are hardly selective. For example, the relative preference of (methoxycarbonyl)carbene for representative 1°, 2°, and 3° C–H bonds is scant at 1, 1.62, and 1.84, respectively.[3] This is due to the indiscriminate nature of most transient carbenes. The lack of carbene selectivity largely stems from their high potential energy and kinetic instability (cf. Chapter 1).

Laser flash photolysis (LFP) is a modern technique that provides the absolute rate constant (k) for carbene decay.[4,5] The effects of substituents, reaction partners, solvent, and temperature can thus be gauged. Fleeting carbenes have also been immobilized within cryogenic matrices (e.g., liquid-helium-cooled

Contemporary Carbene Chemistry, First Edition. Edited by Robert A. Moss and Michael P. Doyle.

argon at $T = -269\,°C$ or liquid-nitrogen-cooled glasses at $T = -196\,°C$).[6–12] This too has provided chemists with much useful information about these unusual reaction intermediates. The frigid temperature impedes the nuclear movements associated with classical elementary steps—even those with small activation enthalpies—thereby extending a carbene's lifetime (τ) almost indefinitely. When present, however, thermodynamically independent processes, such as nonclassical quantum mechanical (QM) tunneling (cf. Chapter 7),[10,13,14] can dominate the chemistry of carbenes at very low temperatures.

Thermodynamically dependent processes, such as skeletal rearrangement or intersystem crossing (isc), are impeded for carbenes in cryogenic matrices. For practical purposes, one needs to conduct carbene reactions at ambient temperatures. Yet, the formation of unwanted solvent insertion products, alone, calls for an improved way of conducting condensed-phase carbene reactions at room temperature. The reactions of carbenes generated *inside* host compounds, or supramolecular carbene chemistry,[15–22] have shown great promise in this regard because they are generally performed under specifically tailored conditions at ambient temperatures. The outcome of reactions with kinetically significant carbene intermediates is often altered by the host when thermodynamically dependent processes, such as skeletal rearrangement or isc, are allowed to proceed at $T = 25\,°C$.

The primary goal of carbene@host chemistry is simple—to give chemists full control over carbene lifetimes and reaction selectivities. Constraining microenvironments can affect the mechanisms of decay that are available to the usually short-lived carbenes—if they are depleted at all. In some cases (*vide infra*),[19h] supramolecular carbenes persist for so long that terms such as "transient intermediate" lose their customary meaning. Molecular constraint can also favor one conformational isomer (i.e., conformer) among many due to topological restrictions. The confinement of carbenes within larger entities limits which substrates are available for intermolecular reaction, if such is even wanted. Eliminating undesired dimeric and polymeric material produced from intermolecular reactions is a necessary first step toward focussing efforts on modifying the intramolecular behavior of carbene guests.

Of course, achieving this aim will not always be possible. First, carbenes that undergo intramolecular rearrangement cannot be isolated in a host. Indeed, the lifetime of a carbene is substantially diminished whenever a rapid rearrangement is possible. For example, carbene rearrangements by 1,2-H shifts can be so fast, due to QM tunneling, that modification is impossible. In such cases, only the reaction product(s) can be detected. Thus, it is necessary to focus on carbenes with inherently inaccessible intramolecular reaction pathways if lifetime enhancement is desired or on carbenes with truly competitive intramolecular reaction pathways whenever product selectivity is sought. Second, reactive carbenes can decay by reacting irreversibly with the host. These so-called innermolecular reactions are undesirable side effects. Thus, even carbenes that decay only by intermolecular reaction can remain transient. Nevertheless, the carefully planned choice of carbene guests and their hosts will increase the overall utility of carbenes.

10.2 TYPES OF HOSTS USED IN SUPRAMOLECULAR CARBENE CHEMISTRY

The following is a general overview of the host compounds that have been used to modify carbene reactions normally performed under more traditional conditions. The most significant results in supramolecular carbene chemistry have been obtained using cyclodextrins (CyD), zeolites, and hemicarcerands (HCs) as microreactors (*vide infra*). Within a host's confining spaces, guest molecules have restricted mobility and are deprived of reaction partners. This often leads to a dramatic increase in the selectivity of the carbenes.

10.2.1 Cyclodextrins

The first supramolecular carbene reactions were performed within the natural macrocyclic oligosaccharides α-, β-, and γ-cyclodextrin (CyD).[16,17,23,24] They comprise α-D-(1$\rightarrow$4)-glucopyranose (α-D-Glc*p*) monomers. The structure of the toroidal cavitand cycloheptaamylose is shown in Figure 10.1. A guest@CyD inclusion complex (IC) is readily formed when a compatibly sized organic compound is introduced into an aqueous solution of CyD because CyD ICs are only slightly soluble in water. There are several factors that can contribute to complex formation: (i) electron-dispersive van der Waals forces, (ii) hydrophobic interactions, (iii) electrostatic dipole–dipole interactions, (iv) hydrogen bonding (O–H$\cdots$X; X = N, O, or F), (v) release of CyD distortional energy by the merging guest, and (vi) expulsion of "high-energy" water from the CyD cavity upon complexation.[25]

Cyclodextrin cavities are not very deep and the openings are broad. Also, the walls of γ-CyD are relatively flexible.[24d] Thus, many guests are not fully protected from surrounding water, and CyD complexes of carbenes tend to form the corresponding alcohols resulting from ostensible insertion into H_2O molecules. This tendency is reinforced by the kinetic lability of most CyD complexes. The carbene guests remain quite mobile within the cavity and may be pushed out of the host into the bulk solution before undergoing intra- or intermolecular reactions. Therefore, supramolecular carbene chemistry with CyDs is mainly performed in the solid state.

10.2.2 Hemicarcerands

Hemicarcerand vessels may be the best hosts for stabilizing carbenes because they fully enclose the guest and do not allow its release at room temperature (Fig. 10.2). The encapsulated guest remains indefinitely, even in solution. The carbene is photochemically generated from an entrapped precursor. In the restricted space available, the possibilities for the carbene to react are limited. Clearly, carbene dimerization is precluded and other intermolecular reactions are restricted. Moreover, reduced mobility may hinder the reactive intermediate from adopting the necessary orientation for it to undergo a potentially available

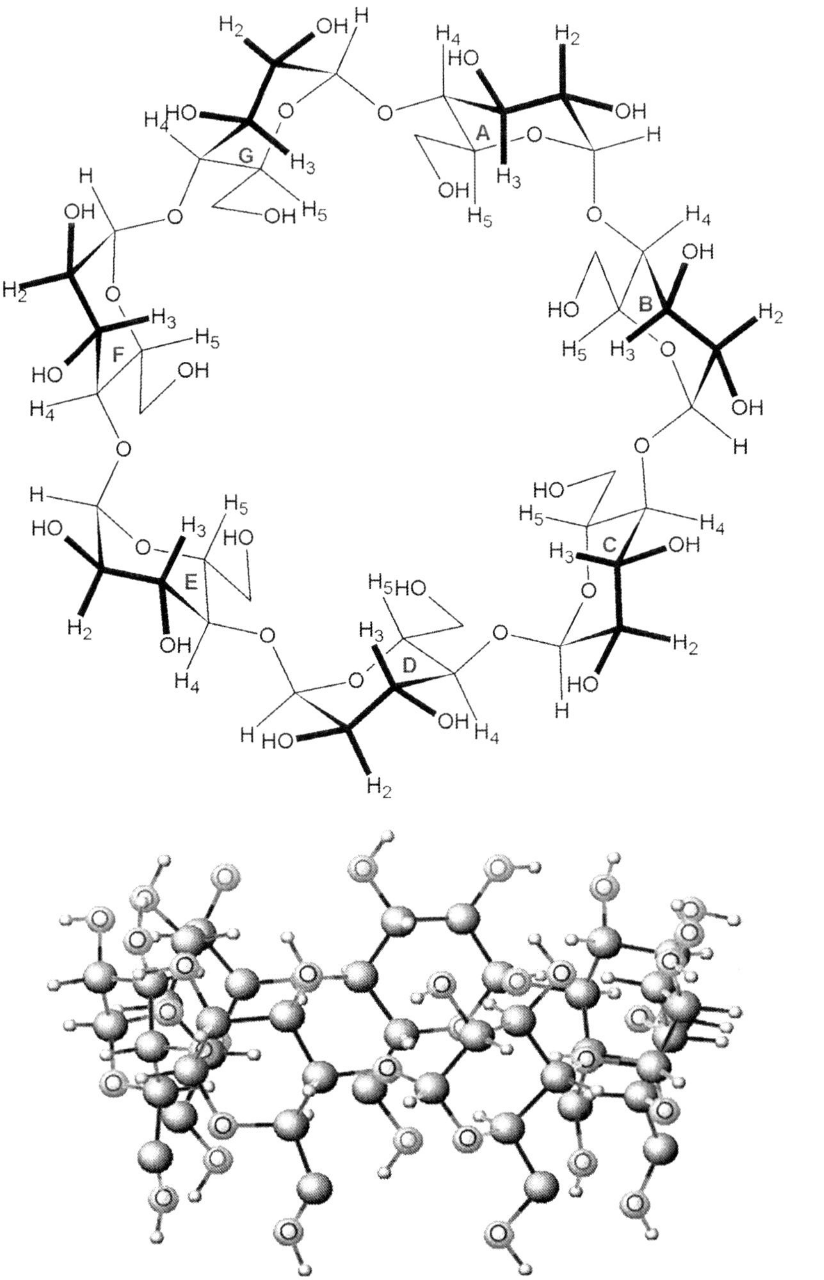

Figure 10.1. Axial and equatorial views of the basketball-hoop-shaped heptamer β-cyclodextrin. The nuclear Overhauser effect between guest protons and the inward-pointing H-3 and H-5 atoms indicates where the guest is situated.

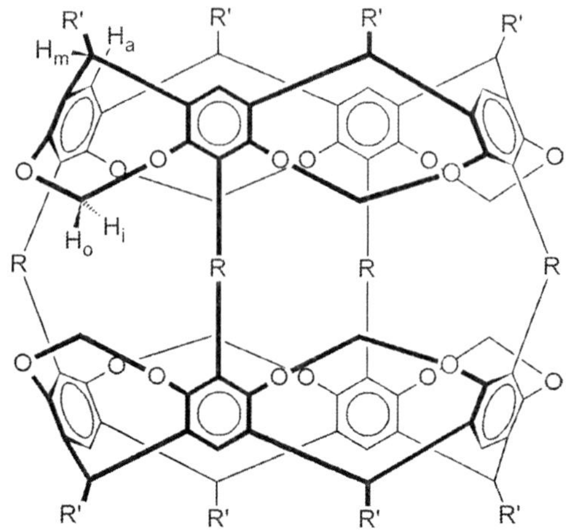

Figure 10.2. General hemicarcerand (HC) structure (e.g., $R = O(CH_2)_nO$, $R' = (CH_2)_4CH_3$).

intramolecular reaction. However, the carbene might react innermolecularly with its host, although this is not typical. Therefore, the carbene's lifetime can be enhanced significantly (*vide infra*).[19h]

10.2.3 Resorcin[4]arene-Based Cavitands

Resorcin[4]arene-based cavitand hosts are cyclic tetramers founded on 1, 3-benzenediol (Fig. 10.3). Their cavity size and solubility characteristics make them suitable for guest encapsulation in organic solvents.

10.2.4 Octa Acid

An octa acid (OA) is another rigid resorcin[4]arene-based cavitand that bears eight carboxylic acid functional groups (Fig. 10.4). It is a water-soluble host because it is ionized in alkaline buffer solutions (e.g., sodium tetraborate). Therefore, supramolecular carbene chemistry within this host can be conducted in aqueous solution. Striking results, however, were also observed when powdered carbene@OA complexes were adsorbed on silica (*vide infra*).[21]

10.2.5 Cucurbit[n]urils

Cucurbit[n]urils are named for their pumpkin-like shape (Fig. 10.5).[26–28] The macrocyclic (n = 5, 6, 7, 8, and 10) cavitand hosts are composed of *cis*-2,5-dioxo-1,4-diylomethyl-octahydroimidazo[4,5-*d*]imidazolidin-3,6-diyl units. They are well suited hosts for slightly acidic aqueous media and have a high affinity for positively charged organic molecules, such as ammonia compounds. Their rigid structures make them good at molecular recognition.[28]

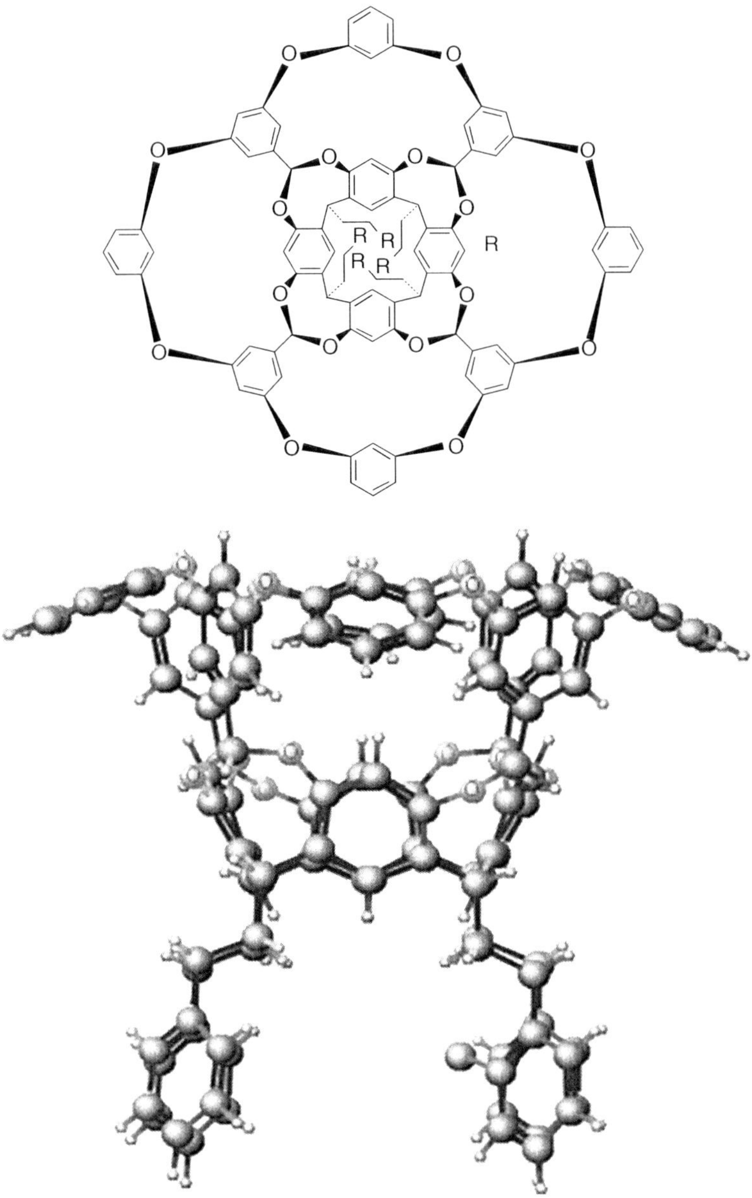

Figure 10.3. Axial and equatorial views of a resorcin[4]arene-based cavitand ($R = C_6H_5$).

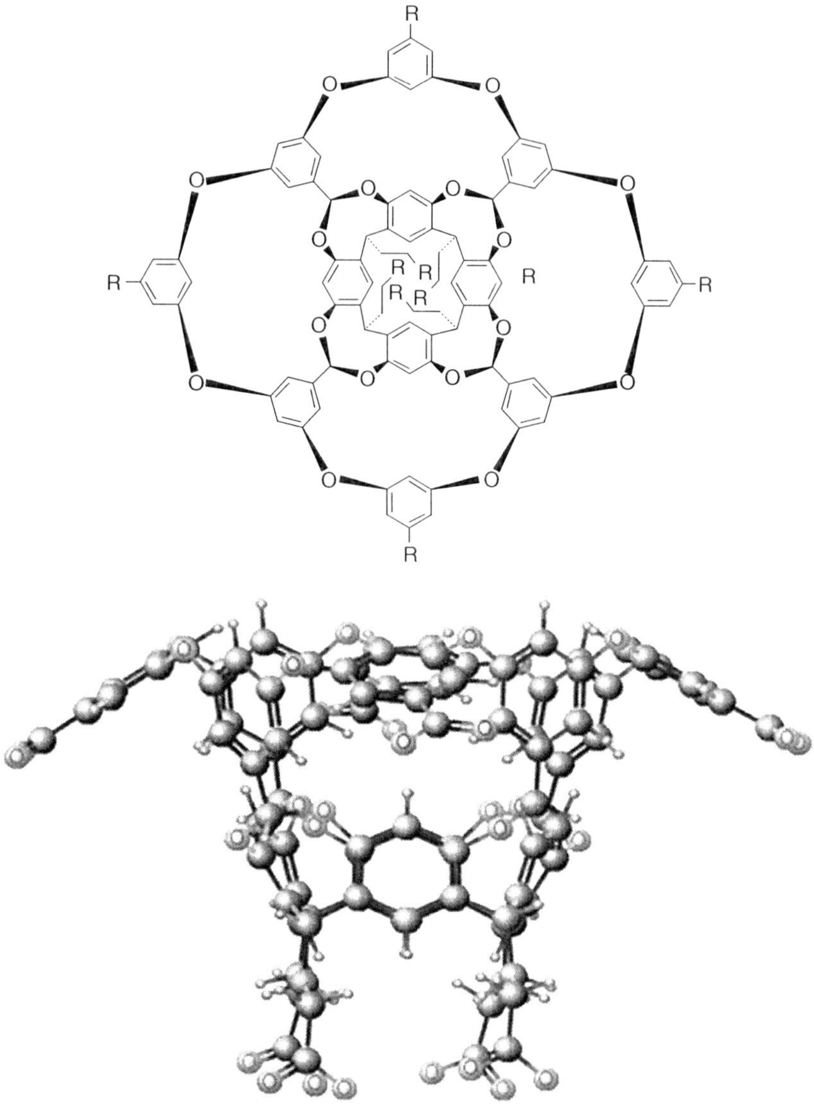

Figure 10.4. Axial and equatorial views of OA ($R = CO_2H$).

10.2.6 Zeolites

The name zeolite is derived from the Greek words *zein* (to boil) and *lithos* (a stone) because stilbite zeolites appeared to "boil" when heated by their discoverer.[29] But zeolites have come a long way since then.[30] Now, they are widely used for the stabilization of various reactive intermediates.[31] Typically, these

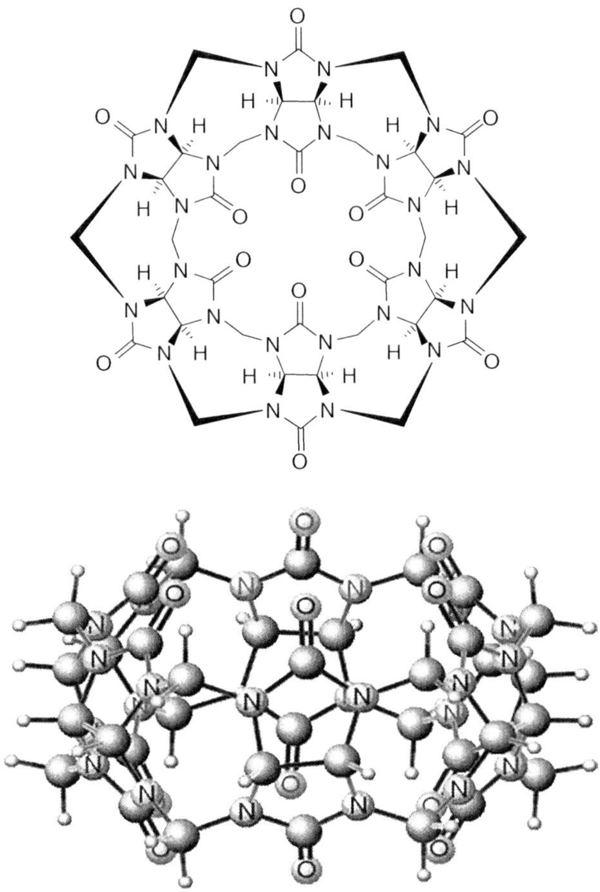

Figure 10.5. Axial and equatorial views of the pumpkin-shaped hexamer cucurbit[6]uril (CB[6]).

microporous inorganic hosts are infused with a solution of the carbene precursor and then photolyzed in the solid state. There is a broad range of natural and synthetic zeolites with different cavity sizes and properties, so that the main challenge lies with the choice of zeolite host.

10.2.6.1 Aluminosilicate. Faujasite (FAU) zeolites are crystalline, highly polar, multicameral aluminosilicates wherein organic molecules may be entrapped. Two forms, denoted Linde Types X and Y, are characterized by a tetrahedrally arranged network of interconnected sodalite cages (Fig. 10.6).[32] This creates relatively large 13-Å supercages that are tetrahedrally interconnected by 7.4-Å apertures. Their size makes it possible for several guest molecules to be present in the same supercage. Inspection of the unit cell reveals that there are 8 sodalite cages with an equal number of supercages (i.e., 1 whole supercage in the

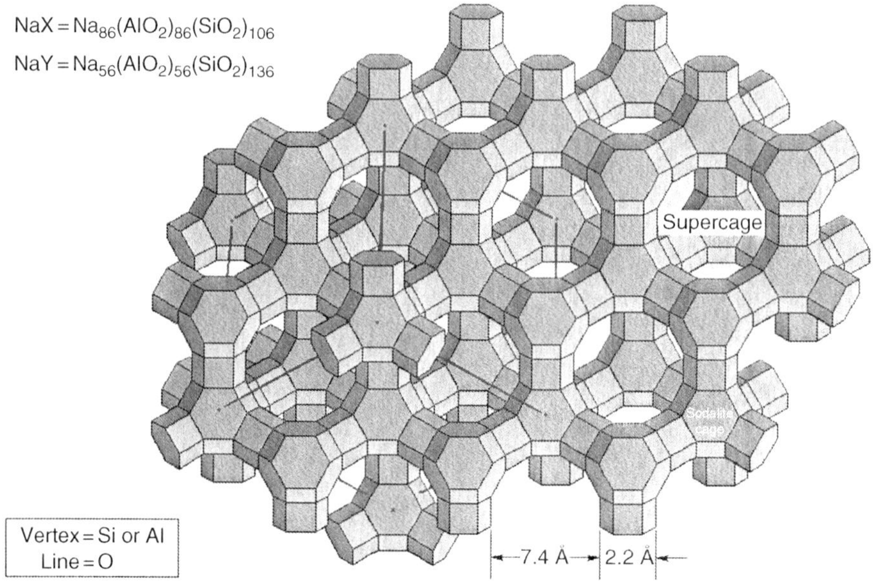

Figure 10.6. Structure of faujasite zeolites X and Y with the diamond-like unit cell outlined in red. Carbenes are generated within 13-Å diameter supercages that are interconnected by a tetrahedral array of 7.4-Å apertures. The guest-to-supercage ratio is known as the loading factor (‹s›) and can be used to control inter- versus intramolecularity (e.g., when ‹s› = 1/4, each carbene-filled supercage is (on average) surrounded by four vacant ones). For color details, please see color plate section.

center, 4 whole supercages occupying the other 4 tetrahedral sites, and 12 quarter supercages occupying the edges). In this environment, the guest is mobile but its motion is restricted. The negatively charged FAU framework is typically compensated by Na^+ counterions, but other alkali metal cations may be exchanged. This is the concept behind softening hard water.[33] Larger cations are useful because they reduce the free volume within the inclusion compound and they also exhibit lower Lewis acidities.[32,34–36] The proportion of Si within aluminosilicates is always higher than that of Al,[37] but the Y-type FAUs have an even higher Si to Al ratio (2.43) than do X-type FAUs (1.23). Thus, the former should exhibit less Lewis acidity than the latter. Lewis acidic sites within the zeolite lattice are unwelcome because they could interfere with carbenes or their precursors.

10.2.6.2 Aluminophosphate. The aluminophosphate VPI-5 is a neutral molecular sieve with the chemical formula $AlPO_4$.[38] It does not require charge-compensating cations and is therefore free of Brønsted–Lowry acidic sites. It comprises large unidimensional channels with 12-Å diameters (Fig. 10.7), so that its voids are roomy enough to accommodate relatively large guest molecules.

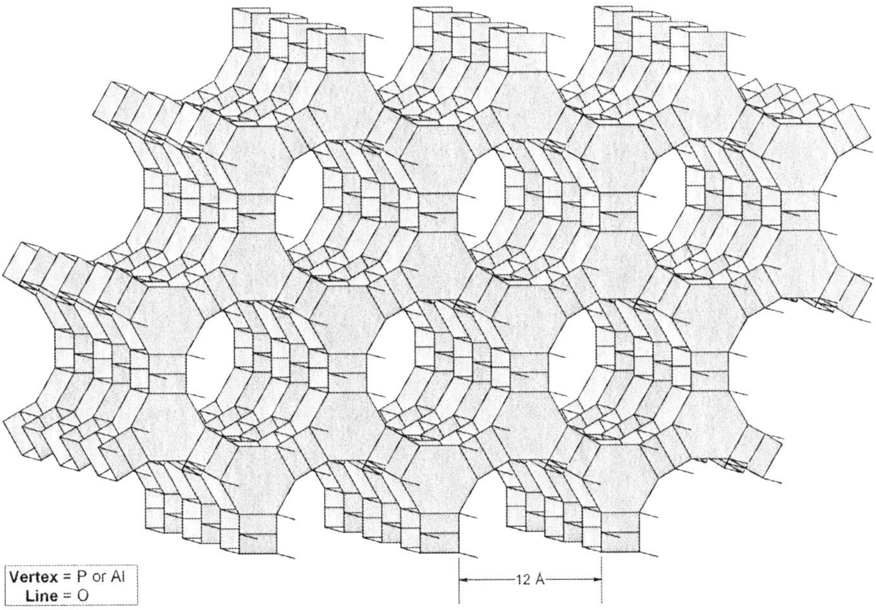

Figure 10.7. The aluminophosphate zeolite VPI-5 has channels that are 12 Å wide. It is more hospitable to acid-sensitive diazo compounds due to the exact balance of charge between the $(AlO_2)^-$ and $(PO_2)^+$ moieties (i.e., Al to P ratio = 1:1). For color details, please see color plate section.

10.3 CHOOSING THE RIGHT CARBENE GUEST

Carbenes were initially regarded as molecular curiosities because their most basic feature is the presence of a divalent carbon atom.[1] They encompass a subclass of neutral reaction intermediates and exhibit diverse reactivities. Traditionally, carbenes are highly reactive, strongly electrophilic, and ephemeral molecules. But their stability and philicity can vary greatly depending on their spin multiplicity and the functional groups that flank the divalent carbon (cf. Chapters 1 and 2). Their most intrinsic feature is their electron deficiency (i.e., they possess a carbon atom with an incomplete valence shell). Nevertheless, singlet carbenes are formally ambiphilic. They can accept an electron pair but also donate their own. Indeed, some singlet carbenes act as electron-pair donors. Such carbenes will even form their corresponding carbenium ion conjugate acid by serving as a Brønsted–Lowry base (cf. Chapter 3).[39–41]

Hence, it is necessary to gauge the intrinsic chemical properties of carbenes in order to understand their affinities for supramolecular host compounds, which possess a variety of chemical microenvironments that could unduly interfere with the carbene reaction of interest. There are different classification schemes for assessing carbene reactivity. A popular method has been to use a 1-D philicity scale (i.e., electrophilic, nucleophilic, or ambiphilic) whose values (m) are based

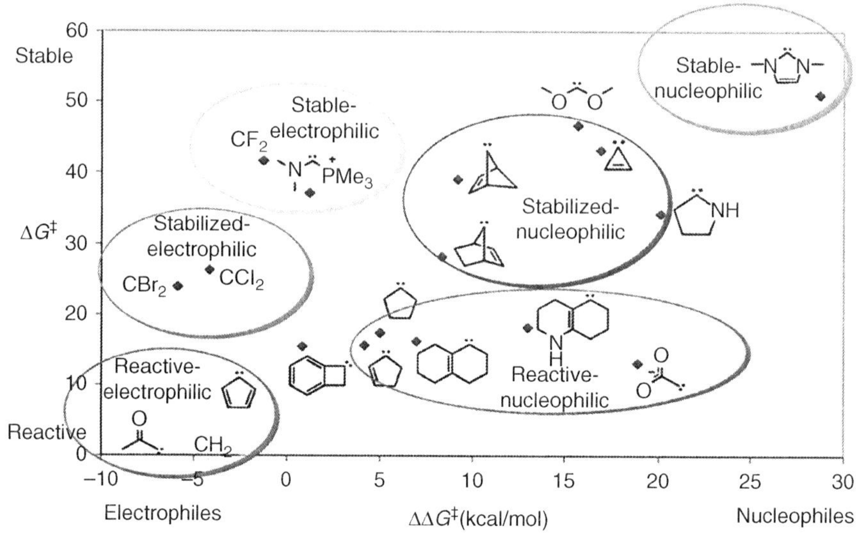

Figure 10.8. Carbenes may be categorized in six ways according to a 2-D surface that accounts for their reactivity and philicity. (Reprinted from Reference 44.) For color details, please see color plate section.

on competitive cyclopropanations of a set of electron-rich alkenes by various carbenes at $T = 25\,°C$ versus those for a standard carbene—dichlorocarbene.[42] This approach is limited, however, because the standard carbene is not overtly ambiphilic and electron-poor alkenes are not included. Refinements using HOMO–LUMO energy differences and absolute rate constants help to resolve these issues.[1g,43] Carbenes have also been categorized by their ability to undergo intermolecular C–H insertion.[44] Here, carbenes are sorted according to their position on a 2-D reactivity surface whereupon two criteria are considered: (i) propensity to react (i.e., stable to reactive) and (ii) philicity (i.e., electrophilic to nucleophilic). Six categories of carbenes may be defined in this way (Fig. 10.8).[44]

Singlet carbenes insert into C–H bonds in one elementary step, approaching them perpendicularly.[45,46] The partial charges developed within the activated complex are stabilized by polar solvents or within polar host cavities. The reduction in energy at the transition state promotes C–H insertion. Of course, the bond dissociation energy of the specific C–H bond is a major contributing factor toward selectivity. Triplet carbenes also insert into C–H bonds but in two elementary steps due to their *gem*-diradical character. First, the open-shell carbene approaches the hydrogen atom coaxially and abstracts it from the carbon atom.[47] Next, the resulting intimate radical pair usually combines to yield the formal insertion product. Attempts to stabilize triplet carbenes within host molecules suffer from this mechanism because a C–H bond is more sterically approachable end-on than sideways. Also, effective cavitands enclose their guests well, but this necessarily brings a triplet carbene closer to the hydrogen ends of the surrounding C–H bonds. One could use a deuterated host to obviate this potential difficulty, because C–D bonds are stronger than C–H bonds,[19] or one could dispense with C–H bonds altogether and generate a carbene within an inorganic host, such as a zeolite.

Carbenes can also react with pnictogens and chalcogens to form ylides. For example, *N*-ylides are used for measuring a carbene's lifetime during LFP experiments.[4] They also have their own characteristic chemistry (cf. Section 10.6.2). The ylidic zwitterion is formed reversibly or irreversibly depending on the heteroatom's propensity for electron donation (cf. Section 10.6.9.2). This can be fine-tuned by choosing the appropriate Lewis base. Those that have weaker donation ability, such as ethers or arenes, have the potential to form electron donor–acceptor (EDA) complexes. This is not the case for *reactive electrophilic* carbenes (Fig. 10.8) because their reactivity is not attenuated by intramolecular electron donation. Therefore, they are extremely unstable and indiscriminate in their free state. They form *O*-ylides rather than ICs even with relatively weak electron donors, such as ethers. Thus, these carbenes are unsuitable for supramolecular complexation.

10.4 DIAZIRINES AS SUITABLE SUPRAMOLECULAR CARBENE PRECURSORS

Supramolecular carbenes are normally generated by the UV photolysis of diazirines.[48] These carbene precursors are not decomposed by water and thus survive aqueous complexation procedures, unlike moisture-sensitive Bamford–Stevens reagents.[49] They are less basic than ammonia because their nitrogen atoms are fairly resistant to protonation.[50] For example, the azi functional group is unaffected during the acid-catalyzed hydrolysis of 17-azi-3-(1,3-dioxolan-2-yl)-Δ^5-estrene (**1**) to 17-azi-3-oxo-Δ^4-estrene (**2**) (Scheme 10.1).[51] The strained azi functional group is often more thermally stable than others, such as the linear diazo function.[52] Nonetheless, diphenyldiazomethane is known[53,54] and 3,3-diphenyldiazirine is not.[55] The isomeric diazo compounds are easily protonated, however, even by weak hydroxylic acids, such as H_2O. Indeed, diazomethane is typically used to protect carboxylic acids by converting them to their methyl esters.[2] Thus, diazo compounds are incompatible with aqueous complexation processes. Oddly, diazirines are considerably less polar than the three-membered heterocyclic ring would suggest.[50a] They are, therefore, optimal carbene precursors for host pockets that are apolar.

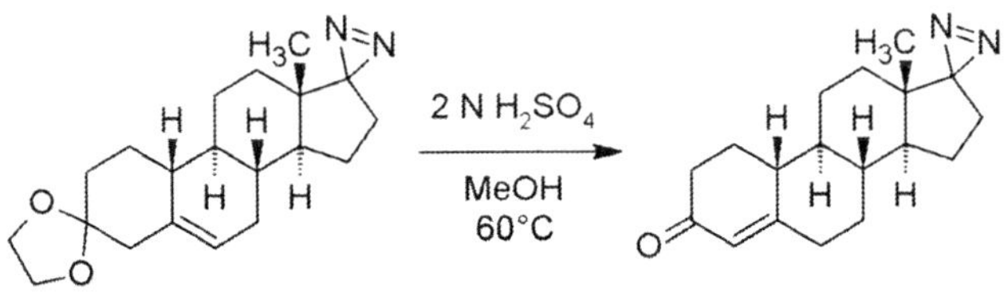

1 2

Scheme 10.1.

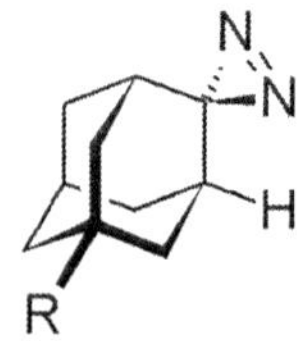

3a: R = H

3b: R = OH

3c: R = Br

Figure 10.9. Aziadamantane and γ-substituted aziadamantanes.

The supramolecular complexation of a diazirine can noticeably improve the labile guest's thermal stability. This was observed during differential scanning calorimetry analysis of 4-azi-1-bromoadamantane (**3c**) included within a resorcin[4]arene-based cavitand (Fig. 10.3, Fig. 10.9). The expulsion of N_2 from neat diazirine **3c** starts at $T = 118\,°C$, but it commences at 131 °C for the guest@host complex.[20] However, stabilization of diazirines toward fragmentation is not universal. In fact, denitrogenation usually tends to be slightly accelerated as was observed with aryldiazirines encapsulated in HCs.[56] That is because the reductions in enthalpy ($\Delta\Delta H$) of the transition state and the ensuing carbene are both greater than that experienced by the diazirine reactant (e.g., $\Delta\Delta H = 2$–3 kcal/mol). This stabilization effect might be especially pronounced when a host possesses highly polarizable groups capable of dispersion interactions.[56b]

Irradiation of diazirine@CyD ICs readily generates carbene reaction intermediates, even in the solid state, because CyDs are fully transparent in the absorption region of the azi chromophore ($\lambda = \sim 370$ nm). In addition, induced circular dichroism (ICD) is an advantageous method for determining the association equilibrium constant for diazirine@CyD complexes for two reasons: (i) the azi chromophore is easily observable in the near-UV region and (ii) CyDs naturally occur as enantiomerically pure host molecules.[57,58]

The most severe disadvantage of using nitrogenous carbene precursors is that they are labile compounds that have the potential to explode. Precautions should be taken to limit unnecessary exposure to heat and light. Another problem, with regard to reaction productivity, is that the precursors easily trap the carbenes as *N*-ylides (cf. pyridine ylides)[5] whenever possible.[59,60] These intermediates subsequently form aldazines or ketazines (henceforth "azines"). This unproductive interaction can be reduced in liquid-phase solutions by working with dilute samples.[17c,f]

Furthermore, photochemically generated carbene intermediates are sometimes circumvented to some degree by diazirine rearrangement in the excited state (RIES).[61–63] Thus, N_2 expulsion and 1,2-H shifts can sometimes occur simultaneously rather than in tandem. Alkyl- and dialkyldiazirines that have relatively weak α-C–H bonds are especially susceptible to RIES during their photolysis reactions. When a carbene is formed, the presence of an α-C–H bond will often lead to the swift formation of an alkene. But rapid 1,2-H shifts by classical reaction, by nonclassical QM tunneling, or by both processes within alkyl- and dialkylcarbenes can severely diminish their lifetimes. Furthermore, this facile isomerization can overwhelm other possible rearrangements within the carbene. For example, isomerizations that transpire more slowly by several orders of magnitude will not be observed. Therefore, a distinguishing reaction product only attributable to a carbene helps to confirm its involvement. Nevertheless, diazirines have proven to be an excellent class of carbene precursors.[48,64]

10.5 ARCHITECTURE OF THE GUEST@HOST COMPLEX

Supramolecular complexation can stabilize a reactive guest by the combined effects of weak noncovalent interactions. The spontaneous formation (i.e., net free energy $\Delta G < 0$) of an IC depends on the sometimes related balance between changes in enthalpy (ΔH) and entropy (ΔS) (Eq. 10.1). At first glance, the reduction in the number of particles (i.e., guest + host → guest@host) is expected to lower the system's entropy. Increased order due to the loss in the guest's degrees of freedom (i.e., its flexibility and motion are curtailed) upon complexation will further reduce ΔS.[65] But sometimes ΔS actually increases when a guest forms a complex, for example, with a CyD. This is because the inclusion of one organic guest molecule ejects several "guest" molecules of water from the cavity, increasing the number of particles in the system. Desolvation also increases ΔS because particles of water are released from the solvated guest as it enters the cavity as well as from the cavitand's hydroxy group–lined peripheries. Changes in the CyD host's hydrogen bond network make it more flexible, which further increases ΔS. The net effect of these entropic factors is hard to predict. For example, some adamantane derivatives afford stronger complexes with heptakis(2,6-di-*O*-methyl)-β-CyD than with natural β-CyD but weaker ones with heptakis(2,3,6-tri-*O*-methyl)-β-CyD (TRIMEB)[66,67] due to very unfavorable entropy changes.[68] However, the order imposed by hosts upon flexible and mobile guests usually outweighs other aspects, resulting in a net negative value for ΔS. Thus, there must often be a marked decrease in the system's enthalpy of complexation to sufficiently offset this adverse entropic condition. This will minimize ΔG and concomitantly raise the equilibrium constant (K_f) of complex formation (Eq. 10.1).

$$\Delta G = \Delta H - T\Delta S = -RT\ln(K_f) \tag{10.1}$$

The association constant K_f reflects the strength of the noncovalent guest@host interaction and thus the extent of guest complexation (Fig. 10.10). Strong binding entails large K_f values, which can be determined in various ways.[69]

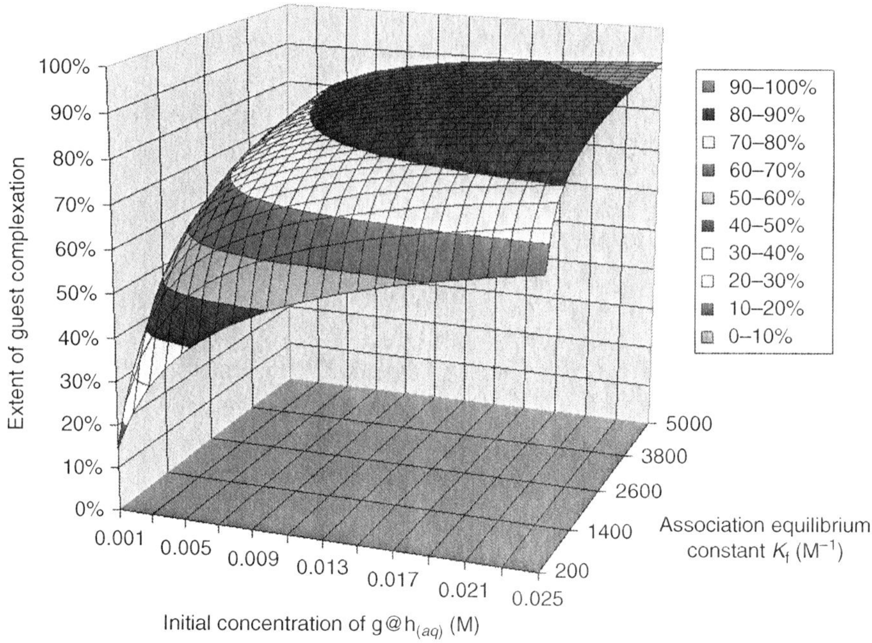

Figure 10.10. The extent of guest complexation reaches a plateau after a certain concentration or association complex equilibrium constant (K_f) of formation. For color details, please see color plate section.

For instance, spectrofluorometric monitoring of competition experiments has been used.[70] Also, ICD has been used successfully to assess K_f values for diazirine@CyD ICs.[17d,57,58]

The structure of a supramolecular aggregate is important because it affects the chemical behavior of a reactive intermediate generated therein. Product selectivity often correlates with guest@host stoichiometry and the guest's orientation within the host(s) (*vide infra*). Therefore, a detailed knowledge of the arrangement and structure of each supramolecular IC is warranted. In this regard, aziadamantane (**3a**) has proven to be an ideal compound (Fig. 10.9).[71] It forms a very strong 1:1 complex with β-CyD in water ($K_f = \sim 4 \times 10^4\, M^{-1}$). With the smaller α-CyD, **3a** is encapsulated by two α-CyDs to give the dimeric **3a@(α-CyD)$_2$** IC (Fig. 10.11). Such an enclosure isolates the embedded guest from the surrounding polar environment. This assertion is supported by the ICD and UV/vis spectra of the IC in aqueous solution because the azi chromophore exhibits the phenotype and fine structure otherwise observed only with nonpolar solvents (e.g., hexane).[57]

Substituted aziadamantanes were also examined for three reasons: (i) to determine how substituents affect the formation of the CyD ICs, (ii) to analyze CyD IC structures, and (iii) to investigate the reactivity of the corresponding adamantylidenes.[72] Not only was the crystal structure of **(3b@β-CyD)$_2$** obtained (Fig. 10.12a)[73] but structural information for its reaction in solution also was

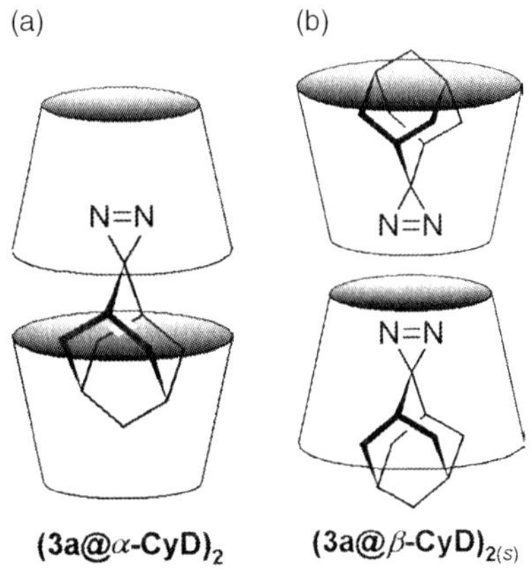

Figure 10.11. The aziadamantane@cyclodextrin inclusion complex: (a) α-CyD forms a 2:1 ensemble that isolates the carbene precursor from other molecules, whereas that with (b) β-CyD forms a crystal with a 2:2 stoichiometry that is conducive to azine formation.

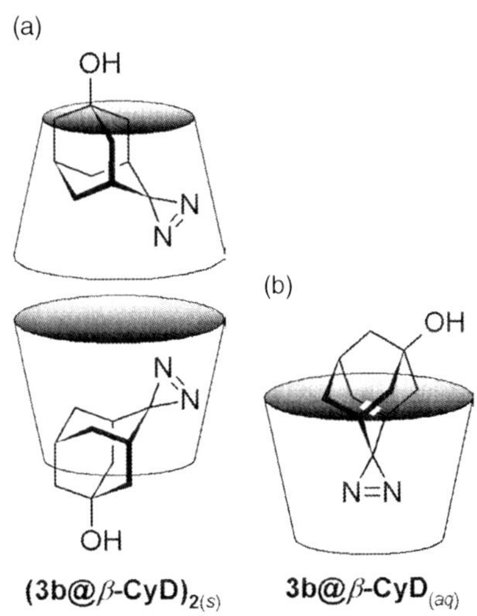

Figure 10.12. The structure of the inclusion complex 4-aziadamantanol@β-CyD is (a) dimeric in the solid phase and (b) monomeric in water, where the carbene precursor's orientation is also reversed.

gained through two-dimensional rotating-frame Overhauser effect spectroscopy (2-D ROESY NMR) experiments.

The most direct method to determine the structure of an IC is by X-ray diffraction analysis. As noted, the crystal structure of 4-aziadamantanol@β-CyD was determined in this way (Fig. 10.12a).[73] However, **3b** is an exceptional case because suitable crystals of other ICs are lacking. Thus, structural information must be deduced from ICD and 2-D ROESY NMR experiments.

For example, 2-D ROESY can be used to assess the proximity of the guest to the inward-pointing CyD protons H-3 and H-5, thereby indicating how deep within the cavity the guest is submerged.[74] This analysis was performed on the 4-aziadamantanol@β-CyD IC (Fig. 10.12b).[73] Once the dimeric **(3b@β-CyD)$_2$** units are cleaved in water-d_2, the orientation of the carbene precursor is reversed. This is due to solvation of its polar hydroxy group (Fig. 10.12).[73] Such repositionings and full reversals of orientation are uncommon but known.[75]

In addition, the molecular dynamics of the aqueous β-CyD ICs with 4-aziadamantanol,[76] 4-aziadamantanamine,[77] and unsubstituted aziadamantane were simulated to corroborate experimental findings.[78] These complexes were analyzed with respect to relative orientation and interaction energies, with aqueous solvent molecules explicitly considered. The calculations revealed two stable orientations for each guest within the host. They had different interaction energies. The enthalpy of complex formation for simple aziadamantane was calculated to be -9.3 ± 1.7 kcal/mol. This value is in close agreement with that obtained from experimental ICD measurements ($\Delta H = -8.6 \pm 1.0$ kcal/mol).[78]

Although these sophisticated methods address the structure of supramolecular complexes in solution, compiling various structural data for the precursor complex helps one to consider the solid-state reactivity of carbene intermediates. A supramolecular carbene will be transient whenever it can be consumed innermolecularly. In these cases, one must assume that topological differences between the diazirine@host complex and the carbene@host complex are negligible due to the small size of the N_2 molecule that is released. This hypothesis should be most valid for solid-phase reactions. Of course, the expulsion of N_2 from the complex affects the carbene molecule generated therein: (i) it frees up extra volume within the cavitand for the carbene to undergo skeletal rearrangement, ring expansion, or fragmentation; (ii) it may reorient a carbene due to the change in polarity (e.g., carbenes with *gem*-zwitterionic character); and (iii) it could alter the axis or the rate of carbene tumbling.[79]

Finally, an interesting characteristic of diazirine@CyD ICs was uncovered using D_{2d}-symmetric 2,6-diaziadamantane (**4**).[50a,80] The guest@host arrangements of the aqueous phase **4@(α-CyD)$_2$** (Fig. 10.13a) and **4@β-CyD** (Fig. 10.13b)[81] complexes were examined by ICD using the guest as a bilateral probe.[82] The spectrum of **4@(α-CyD)$_2$** was nearly identical in phenotype and of doubled intensity to that of **3a@(α-CyD)$_2$** as one would expect. But the ICD spectrum of **4@β-CyD** unexpectedly showed positive and negative contributions by the twin chromophores, which are arranged perfectly perpendicularly to each other according to a single-crystal X-ray diffraction analysis of **4**.[50a] It was concluded that a 1:1 complex was being observed, wherein one azi function was implanted within the apolar β-CyD cavity and the other one was exposed to the aqueous solution.[82] This reconciled the unusual observation with established rules predicting the ICD of chiral supramolecular systems.[83–85]

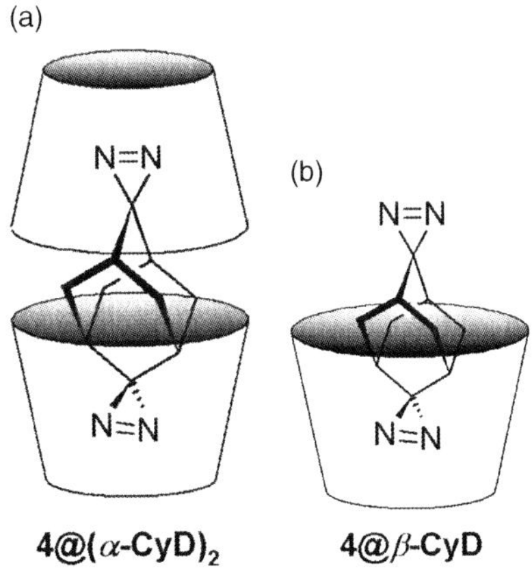

Figure 10.13. In solution, 2,6-diaziadamantane forms (a) a 1:2 complex with α-CyD but (b) a 1:1 complex with β-CyD.

10.6 CASE STUDIES

10.6.1 Carbenes with Available 1,2-H Shifts

10.6.1.1 Methyl(phenyl)carbene. Nitrogenous carbene precursors have been enlisted to selectively derivatize CyDs,[16] which are notoriously difficult to monofunctionalize.[17d] For example, the photolysis of 3-methyl-3-phenyldiazirine (**5**), complexed within β-CyD, produced innermolecular insertion products, such as *O*-(1-phenylethyl)-β-CyD, in 30% yield (Scheme 10.2).[16a] It is possible that the derivatization of β-CyD by diazo compound **6** occurred because diazo compounds decompose during the formation of CyD ICs.[16d] Therefore, the intermediacy of methyl(phenyl)carbene (**7**) is not certain in this derivatization. Protonation of diazo compound **6** would lead to the corresponding diazonium ion[52b] which would readily expel N_2 when a hydroxy group of β-CyD attacks it.

The intramolecular rearrangement product styrene (**8**) was also formed during the photolysis of diazirine **5**. It may stem from RIES of **5***,[61–63] classical 1,2-H shift within carbene **7**, QM tunneling within carbene **7**, or by a combination of these processes. Alkene **8** underwent a secondary reaction giving *cis*- and *trans*-1-methyl-1,2-diphenylcyclopropanes (**10**). However, these products are not necessarily formed by addition of carbene **7** to **8**. The formation of pyrazoline **9**, by the facile [3+2] dipolar addition of diazo compound **6** to alkene **8**, followed by homolytic scission of **9** would also give cyclopropane **10** (Scheme 10.2).

10.6.1.2 Cyclooctylidene. The effect of molecular confinement on reactant conformation can be astounding. If a reactive intermediate is deprived of its normal decay route, then new reaction alternatives may flourish. This is the case with cyclooctylidene (**15**) (Scheme 10.3).[17b,86,87] Upon photolysis of azicyclooctane (**11**),

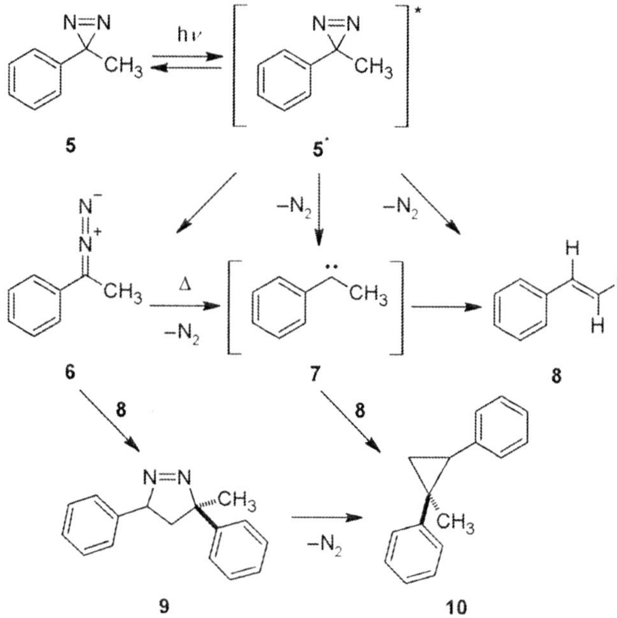

Scheme 10.2.

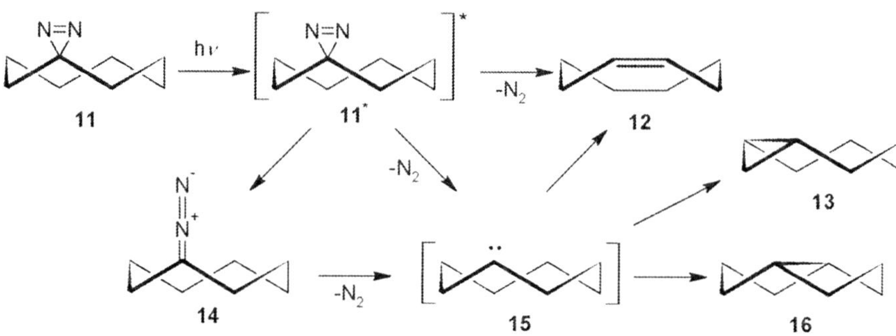

Scheme 10.3.

cis-cyclooctene (*cis*-**12**) is readily formed, but so too are products derived from intramolecular C–H insertions of carbene **15**. Of special interest is *cis*-octahydropentalene (*cis*-**16**), which is created from a transannular 1,5-CH insertion within carbene **15**.[88] The free volume of a CyD depends on how many monomer units it comprises, and *cis*-fused **16** needs space to form. Therefore, it is not surprising that these factors conspire to limit the formation of (*cis*-**16**) when diazirine **11** is photolyzed within α-CyD.[17b] The occurrence of transannular C–H insertion becomes less probable as the cavity within the molecular reaction vessel gets smaller: 42% *cis*-**16** in γ-CyD, 37% *cis*-**16** in β-CyD, and 29% *cis*-**16** in α-CyD.

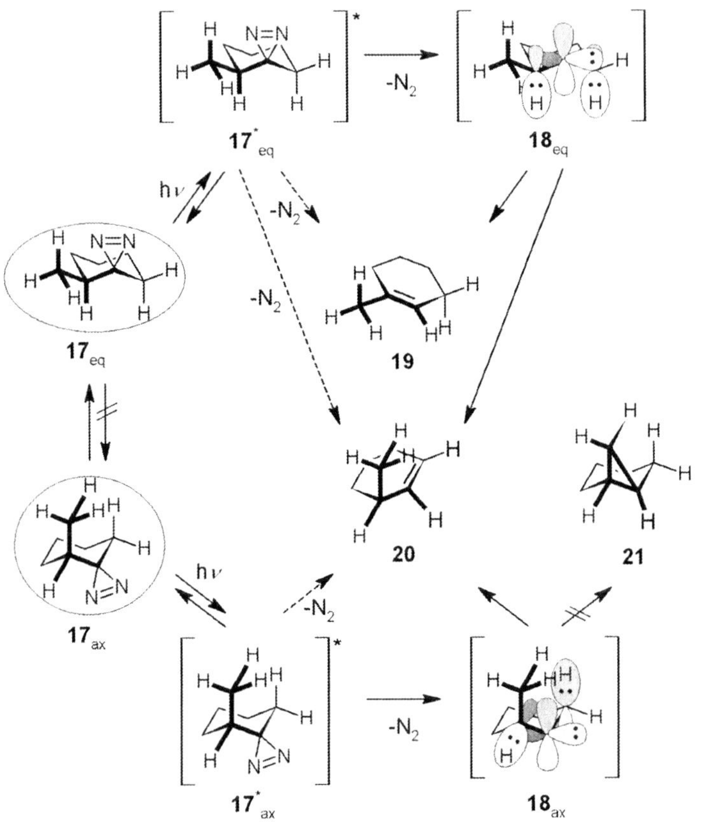

Scheme 10.4.

10.6.1.3 2-Methylcyclohexylidene. 2-Methylcyclohexylidene (**18**) was examined to determine whether it adopts its higher energy conformation within confined spaces.[17c] The spheroidally shaped axial form of the precursor $\mathbf{17}_{ax}$ is more compact than the ellipsoidally shaped equatorial conformer $\mathbf{17}_{eq}$ (Scheme 10.4). Thus, the shape-selective aspect of CyD cavitands was used in an effort to shift the conformational equilibrium in favor of the tighter fitting $\mathbf{17}_{ax}$. This situation was hypothesized to have two simultaneous consequences: (i) an increase in the production of norcarane (**21**) from intramolecular 1,3-CH insertion reaction due to the decrease in the carbene$\cdots CH_3$ distance and (2) the suppression of 1-methylcyclohexene (**19**) in favor of 3-methylcyclohexene (**20**) due to the imposition of orbital misalignment of the C–H σ-bond (HOMO) that must migrate to the divalent carbon atom's vacant p-orbital (LUMO) during 1,2-H shift (Scheme 10.4). The statistical 1:2 ratio (i.e., 0.5) between cycloalkenes **19** and **20** from carbene **18** was altered due to the unequal inherent migratory aptitudes of their hydrogen atoms.[89] Thus, the ratio was found to be 1.8–2.3 in solution and 1.9–2.6 for the CyD ICs.[17c] Indeed, this is close to the 2.2 ratio of

19 to **20** found for the microwave-induced dehydration of 2-methylcyclohexanol@NaY FAU IC. One must therefore conclude that the equilibrium between $\mathbf{17}_{eq}$ and $\mathbf{17}_{ax}$ was unaffected because no norcarane was formed. The elusive **21** is only produced in trace amounts from the thermal Bamford–Stevens reaction.[90]

10.6.1.4 (1-Phenylethyl)(4-phenylphenyl)carbene. The conformation that a newly generated carbene adopts can sometimes factor into the outcome of its reaction. Thus, limiting this variable could enhance one product relative to the others. This is not feasible in liquid-phase solution due to structural fluctuation, but sometimes only one conformer dominates in the crystalline solid phase. In such cases, a higher or perhaps unusual selectivity can be obtained. This experimental approach has met with great success.[91,92] For example, the photolysis of 1-diazo-2-phenyl-1-(4-phenylphenyl)propane (**22**) in solution generates the corresponding carbene **23**, which yields a mixture of products deriving from 1,2-H shift, 1,2-aryl migration, and reaction with the solvent.[93] In the crystalline state, however, stereoselective 1,2-H shift was observed giving (*Z*)-2-phenyl-1-(4-phenylphenyl)propene *Z*-**24** with the near exclusion of the *E*-isomer (Scheme 10.5).[93a,b,94]

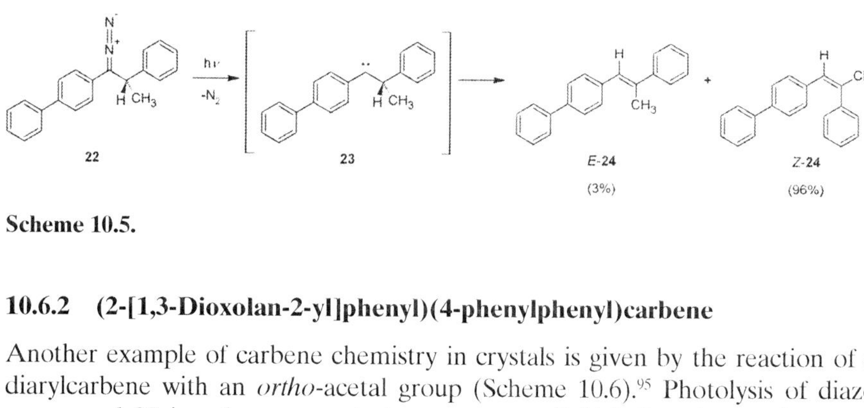

Scheme 10.5.

10.6.2 (2-[1,3-Dioxolan-2-yl]phenyl)(4-phenylphenyl)carbene

Another example of carbene chemistry in crystals is given by the reaction of a diarylcarbene with an *ortho*-acetal group (Scheme 10.6).[95] Photolysis of diazo compound **25** in a benzene solution generates (2-[1,3-dioxolan-2-yl]phenyl)(4-phenylphenyl)carbene (**28**), which is poised to undergo two different intramolecular reactions: (i) H-abstraction by the triplet carbene followed by radical recombination to give the C–H insertion product 2-(4-phenylphenylbenzocyclobutanone [1,3]-dioxalone (**29**) and (ii) oxonium ylide formation leading to other assorted isomers (*vide infra*). In the liquid phase, only products stemming from *O*-ylide **31** were observed. It rearranged to the enol ether 1-(4-phenylphenyl)-3,4-dihydro-2,5-benzodioxocin (**30**) and to the separate Stevens-rearrangement products 4a-(4-phenylphenyl)-3,8b-dihydro-2*H*-1,4-dioxabiphenylene (**32**) and 1-(4-phenylphenyl)-9,12-dioxatricyclo[6.3.1.0^{2,7}]dodeca-2,4,6-triene (**33**). In contrast, photolysis of the crystalline diazo compound **25** greatly favored the formation of carbene C–H insertion product **29** over the oxonium ylide–derived products at low conversion. The latter increased, however, as the reaction progressed. This was attributed to the buildup of defect sites within the triclinic pinacoidal crystal.

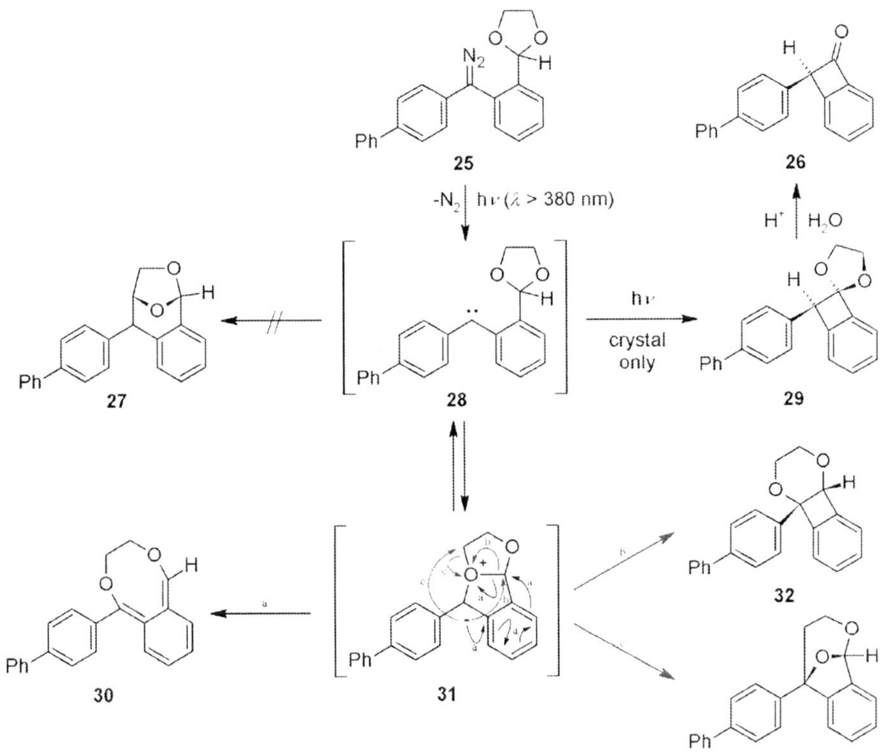

Scheme 10.6.

However, this solid-state approach has some limitations and is sometimes unsuccessful. The main disadvantage is that the crystal structure often enforces a single conformation that may not lead to the desired product. Another drawback is that the reactants are not separated from each other. Therefore, they can easily combine intermolecularly to give polymeric material. For example, carbene occupants may react with proximate nitrogenous precursors to produce mostly azines. Finally, the reaction usually causes significant changes in topology that can radically alter the conditions for optimal crystal structure. An increase in crystal defects may foster side effects prior to full conversion. Depending on the desired result, inclusion of carbenes within individual pockets might address these shortcomings.

10.6.3 4-Oxocyclohexa-2,5-dienylidene

Photolysis of 4-chlorophenol (**34**) generates the 4-hydroxyphenyl cation (**35**), which is in equilibrium with 4-oxocyclohexa-2,5-dienylidene (**36**) (Scheme 10.7). Both intermediates are relatively long lived and β-CyD was found to trap the carbene intermediate much more efficiently.[96]

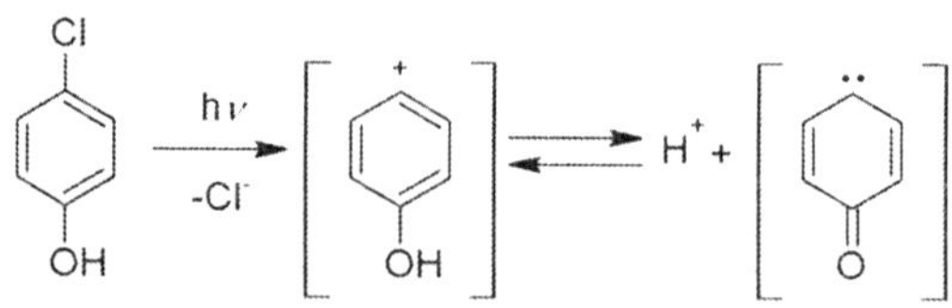

Scheme 10.7.

10.6.4 Dichlorocarbene

Enhanced regioselectivity was observed when the Reimer–Tiemann reaction was conducted in the presence of CyD (Scheme 10.8).[97,98] Dehydrochlorination of $CHCl_3$ by hydroxide ion was retarded due to the formation of $CHCl_3$@CyD ICs. Thus, the overall yield of *p*-hydroxybenzaldehyde (*p*-**38**) was even lower than usual. Indeed, at pH = 13, the CyD hosts were likely deprotonated and behaved much like anionic micelles, which cannot transport the electrically repulsive HO^- base needed for $CHCl_3$ dehydrochlorination. This was a further barrier toward CCl_2 formation.[99] However, α-elimination of $CHCl_3$ by the deprotonated CyD host itself was inexplicably deemed to be "less important."[98b] Encapsulation of CCl_2 within the hydrophobic CyD cavity is expected to impede its hydrolysis to carbon monoxide. Although possible,[100] no innermolecular chlorination by CCl_2 of the CyD hydroxy groups occurred.[98b] Instead, the aryl end of the phenolate anions dipped into the nonpolar cavity of the CyD and reacted with the ordinarily short-lived CCl_2 intermediates. Considering this and the likelihood that CCl_2 access to the phenolate's *ortho* carbon atoms was restricted by steric hindrance, increased *para* selectivity resulted (i.e., *p*-**38**/*o*-**38** = 4.65). In the absence of α-CyD, *p*-**38**/*o*-**38** = 0.71,[98b] close to the theoretical limit of 0.50.

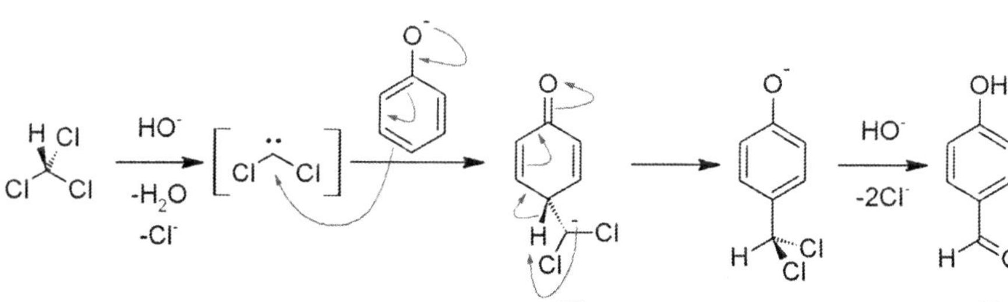

Scheme 10.8.

10.6.5 Adamantylidene

Adamantylidene (**39a**) is conveniently generated from aziadamantane (**3a**).[71] The presence of the carbene's tertiary α-C–H bond might lead one to think that a facile 1,2-H shift dominates the chemistry of carbene **39a**. However,

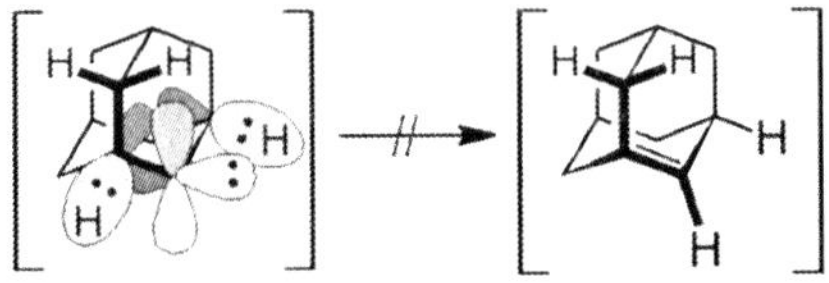

39a 40

Scheme 10.9.

adamantene (**40**) is an anti-Bredt alkene with a twisted double bond positioned next to a bridgehead (Scheme 10.9).[101] This strained compound is highly reactive and is computed to be only 17.2 kcal/mol lower in energy than carbene **39a**.[59b] Such a low thermodynamic driving force is not representative of carbene reactions. In addition, the 1,2-H shift reaction within carbene **39a** giving **40** is kinetically stymied; the computed energy barrier is very high ($\Delta G^{\ddagger} = 52.7$ kcal/mol).[59b] The resistance to 1,2-H shift within carbene **39a** can be explained by FMO theory. A straightforward 1,2-H shift within carbene **39a** is precluded due to orthogonal misalignment of the α-C–H σ-bond (HOMO) that must migrate to the divalent carbon atom's vacant p-orbital (LUMO) (Scheme 10.9).

Adamantylidene does not last forever, despite the absence of 1,2-H shift. Even in the absence of intermolecular reaction, carbene **39a** decays by a 1,3-CH insertion reaction, at a relatively slow rate ($k = 1.4$–$1.8 \times 10^6\,s^{-1}$; $\Delta G^{\ddagger} = 11.8$ kcal/mol),[59b,104a,b] to form the three-membered ring in 2,4-didehydroadamantane (**46a**) (Scheme 10.10).[102–104] With the γ-substituted adamantylidenes, symmetric (*s*) and asymmetric (*as*) isomers of **46** can be formed. Moreover, the 1,3-CH insertion reactions of carbene **39** leading to *s*- and *as*-**46** occur unequally. That is because the divalent carbon atom bridge in **39** leans over preferentially to one side of the molecule when $R \neq H$ (e.g., $\omega = 17°$).[103b] Indeed, the divalent carbon atom bridge in carbene **39** experiences conformational oscillation resulting in an equilibrium between distal and proximal conformers *dist*-**39** and *prox*-**39**, respectively (Scheme 10.10). *Ab initio* computations indicate that the preference for *dist*-**39** or *prox*-**39** depends on the σ-bond inductive effect of the adamantylidene's γ-substituent R.

As shown in Figure 10.11 (p. 289), the **3a@(α-CyD)$_2$** and **(3a@β-CyD)$_2$** complexes are so structurally different that the generated carbene **39a** is directed toward different reaction pathways. The photolysis of the solid β-CyD complex predominantly gave adamantanone azine (**44a**), but none was formed within the α-CyD capsules (Scheme 10.11). This is due to the complete separation of the diazirine **3a** reactants from each other in the α-CyD complex. Instead, 2,4-didehydroadamantane (**46a**) and adamantane (**42a**) were formed as the only extractable C_{10} derivatives. Although **46a** forms from a 1,3-CH insertion within carbene **39a**, the pathway to **42a** is not conclusive. It most likely arises from (triplet) H-abstraction from either C–H or O–H bonds of the CyD. Indeed, oxidized CyDs were observed from UV-detected RP-HPLC chromatograms after the reaction.[17g] This redox behavior has been observed with other CyD ICs.[16b,c]

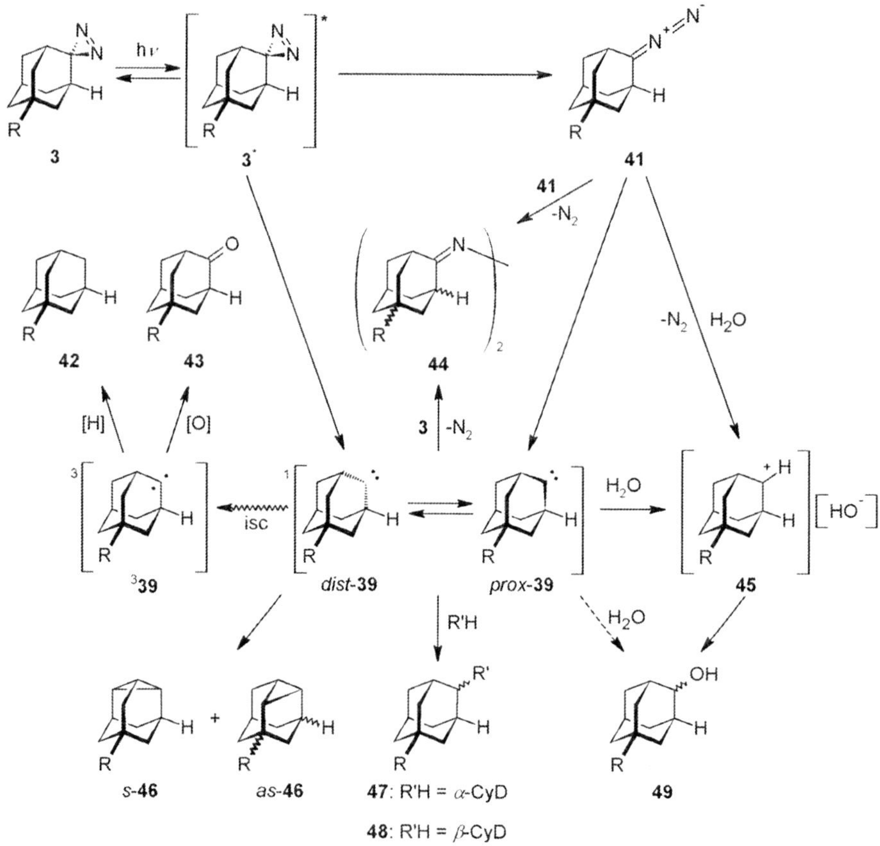

Scheme 10.10. [R = (**a**) H, (**b**) OH, (**c**) Br].

Homogeneous mixtures of diazirine **3a** with each CyD were also photolyzed in the solid state (Scheme 10.11). The results indicate that the exposed carbene precursors were outside the cavitands because much azine **44a** and adamantanone (**43a**) were formed. Bimolecular reaction of carbene **39a** with diazirine **3a** most likely forms an *N*-ylide,[59] which rearranges to azine **44a**. But a mechanism involving transient diazo compounds, which have been observed during the photolyses of supramolecular diazirine complexes,[17b,c,18] cannot be ruled out.[48b] The formation of ketone **43a** is suspected to derive from a radical reaction of carbene **39a** in its triplet state with molecular oxygen ($^3O_2$).

Dialkylcarbenes usually have triplet ground states. But the constraint placed upon the divalent carbon atom bridge in carbene **39a** by the skeletal framework reduces the pertinent bond angle from the typical value of ~130° to ca. 110°,[7,59b] which is characteristic of singlet carbenes. Thus, adamantylidene has a singlet ground state.[7,59b,104] The direct formation of ketone **43a** from $^3O_2$ is therefore spin forbidden. Although carbene **39a** is computed to have a low-lying triplet state (ΔE_{S-T} = 2.8–3.1 kcal/mol),[59b,104a] it was not detected during matrix-isolation

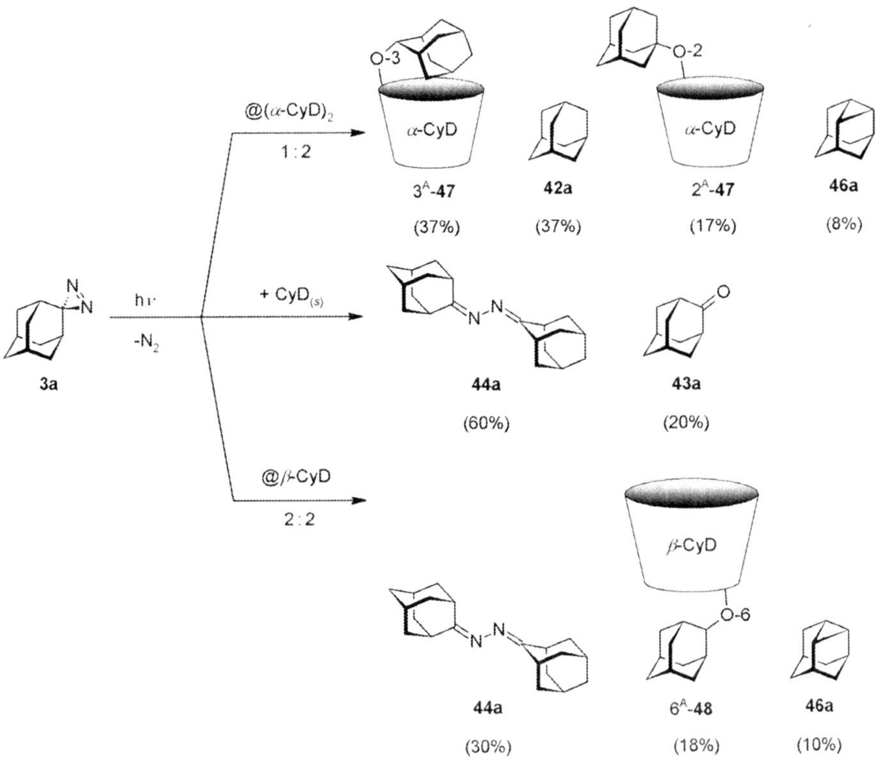

Scheme 10.11.[a–c] [a]The locants designated for the *x*-*O*-(2-adamantyl)-*m*-CyDs ($x = 2, 3, 6$; $m = \alpha, \beta$) refer to the carbon atom of the α-D-glucopyranose unit where O–H insertion occurred.[17g]
[b]FAB-MS and LC–MS indicated that x^A,y^n-di-*O*-(2-adamantyl)-β-CyDs ($x, y = 2, 3,$ or 6; $n =$ B, C, or D) were formed in 10% yield (not shown).[17a,g]
[c]$T =$ ca. 0 °C.[17g]

electron paramagnetic resonance (EPR) experiments.[7] Furthermore, the population of triplet carbene ^{3}**39** compared with singlet carbene ^{1}**39** is only about 2–3% at room temperature, given the aforementioned S_0–T_1 energy gap.[59b,104a] Therefore, if ^{3}**39** is indeed involved in the formation of ketone **43a**, then more triplet carbenes ^{3}**39** must be present to account for the large quantities of hydrocarbon **42a** that formed during the solid-state photolysis of **3a@(α-CyD)$_2$** (Scheme 10.11). This might be explained if the triplet state is stabilized by the α-CyD dimer.[15c,19,105] Therefore, it is not unreasonable to attribute the occurrence of H-abstraction to the triplet-state carbene.[13] However, molecular oxygen ($^3O_2$) efficiently traps triplet carbenes but only trace amounts of ketone **43a** were formed in aerated samples. This may also be attributable to the CyDs, which likely shield carbene **39a** from the external environment. Indeed, CyDs are purposely used to protect organic guests from intermolecular reaction with atmospheric oxygen.[106]

Substantial quantities of modified CyDs were isolated after reactions with both α- and β-CyD hosts (Scheme 10.11). They arose from innermolecular insertion reactions into the hydroxy groups of the oligosaccharides. Furthermore, the bond insertions were chemospecific[17d] because no C–H insertion products were found—even with the inward-pointing C-3–H and C-5–H bonds. In addition, O–H insertion was regioselective, reflecting the organization of the complexes (Fig. 10.12).[17g] Consequently, the inclusion of diazirines within CyDs may be exploited for the selective synthesis of monofunctionalized CyDs.[17d] Of course, unwanted innermolecular reactions will shorten carbene lifetimes thus hampering attempts to foster and study their intrinsic reactions.

Another study was performed using a hydrophobic resorcin[4]arene-based cavitand (Fig. 10.3).[20] Exchange of diazirine was slow on the NMR time scale, indicating that the generated carbene will likely react inside the cavity, which was large enough to accommodate even 4-azi-1-bromoadamantane (**3c**). The complex with **3c** has an association constant in DMSO $K_f = 6480\ M^{-1}$.[20] The crystal structure reveals some interactions between the benzylic C–H bonds inside the host and the halogen atom. The complex with bromo-substituted **3c** is 10 times more stable than that with unsubstituted **3a**. Nevertheless, this complex was photolyzed in the solid state only. Surprisingly, 5-bromoadamantan-2-ol (**49c**) was the main product with a yield of 74%, while only 25% 1- and 7-bromo-2,4-didehydroadamantane (*s*-**46c** >> *as*-**46c**) were formed.[20] The outcome of the reaction is unlike that of the irradiation of neat **3c**, which led only to azine **44c** formation (>95% yield). The resorcin[4]arene host is also a good microreactor for adamantylidenes because no innermolecular insertion of the carbene into the host was observed.

Even more exciting results were obtained with OA (Fig. 10.4),[21] which was also found to be large enough to enclose aziadamantane (**3a**) and the γ-substituted aziadamantanes 4-aziadamantan-1-ol (**3b**) and 4-azi-1-bromoadamantane (**3c**). The stoichiometry of each IC was derived from NMR titration, a comparison of diffusion coefficients, and by isothermal calorimetry (ITC). Each was found to be equimolar in host and guest, but the unsubstituted aziadamantane complex with OA was dimeric (i.e., (**3a@OA**)$_2$), as interpreted from their diffusion coefficients in water (Fig. 10.14): D(OA) = $0.19 \times 10^{-9}\ m^2/s$, D(**3b@OA**) = D(**3c@OA**) = $0.16 \times 10^{-9}\ m^2/s$, and D(**3a@OA**) = $0.12 \times 10^{-9}\ m^2/s$. The ITC measurements were made in a DMSO/1 mM $Na_2B_4O_{7(aq)}$ (2:3) buffer (pH = ~9) in order to fully dissolve the ICs. The results afforded the thermodynamic parameters and association equilibrium constants for each IC. Complexation was spontaneous, despite negative entropies attributable to a lessening of the guest's degrees of freedom, because this was easily compensated by exothermic enthalpy changes upon inclusion.[65] In fact, diazirine@OA binding was found to be rather strong, even in the presence of DMSO, due to the host's rigidity: K_f(**3b@OA**) = $5.5 \times 10^4\ M^{-1}$ and K_f(**3c@OA**) = $1.8 \times 10^6\ M^{-1}$.[21] One interesting feature was the reversed orientation of **3b** inside OA,[75] with respect to the other guests (Fig. 10.14). Evidence from 2-D NOESY spectra and molecular dynamics simulations (GROMACS, OPLS-AA force-field)

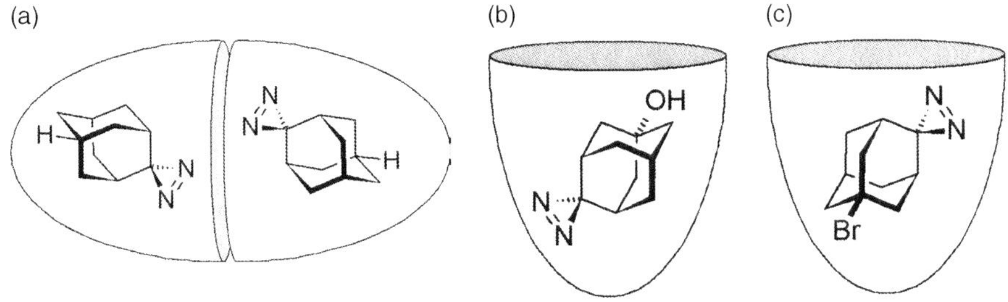

Figure 10.14. Orientation of γ-substituted aziadamantane guests within their OA hosts: (a) **(3a@OA)$_2$**, (b) **3b@OA**, and (c) **3c@OA**.

account for this variance. The reversed orientation, in which the hydroxy group of **3b** points toward the wider end of the OA host, is most probably due to hydrogen bonding interactions with the water molecules in the bulk phase. Moreover, this increased order explains why the magnitude of this IC's negative entropy is larger than found for the other γ-substituted aziadamantanes used in the study.

The large K_f values were assumed to be even larger in pure heavy water thereby ensuring that none of the carbene precursors would exist outside the OA cavitands during photolysis. Indeed, the signals of free and complexed molecules were distinct on the NMR time scale, which demonstrates that the complexes do not dissociate before the carbenes have exceeded their lifetimes. Thus, escape of the carbene from the host into the bulk solvent can be ruled out.

Photolysis of the parent and γ-substituted aziadamantanes within OAs led to very different reaction outcomes. Analysis of the structure of each complex in solution revealed the reason for the different product ratios. For example, unsubstituted **3a** yielded azine **44a** in 89% yield but the γ-substituted aziadamantanes did not form any azines.[21] This was ascribed to the dimeric nature of the **(3a@OA)$_2$** IC wherein two azi groups point toward each other in a head-to-head fashion (Fig. 10.14). Once adamantylidene is formed from one of the diazirines, it is immediately intercepted by another diazirine, presumably by the rapid formation of an ephemeral *N*-ylide that quickly rearranges to the azine. 4-Aziadamantan-1-ol (**3b**) produced 2,4-didehydroadamantan-7-ol (*s*-**46b**) plus 2,4-didehydroadamantan-1-ol (*as*-**46b**) in 92% yield.[21] This astonishing change is a direct consequence of the orientation of the guest. As noted, the hydroxy group of diazirine **3b** is positioned at the top of the cavitand facing the more polar part of the OA molecule and the aqueous solvent. Consequently, the azi functional group is deeply immersed within the bowl and the divalent carbon atom is well protected from the exterior surroundings upon generation of the carbene. With no other option, the carbene isomerizes to *s*- and *as*-**46b** by 1,3-CH insertions. Finally, diazirine **3c** produced **46c** in 37% yield and ketone **43c** in 63% yield.[21] This reaction of **3c@OA** IC also reflects its structure. The bromine atom is located at the bottom of the bowl and forces the azi group to be on top, thereby exposing it to the exterior environment. This situation leads to the formation of much ketone **43c**, which putatively results from the reaction of $^{3}O_2$ with the T_1 state of 5-bromo-2-adamantylidene (^{3}**39c**) (Scheme 10.10).

When these reactions were repeated after infusing the OA complexes onto a silica surface, more unexpected results were found.[21] Curiously, no azine was produced from the dimeric **(3a@OA)$_2$** IC. Instead, 2,4-didehydroadamantane was formed in 84% yield. Furthermore, 1,3-CH insertions predominated within the γ-substituted adamantylidenes as well with yields ranging from 80% to 95%. The higher fraction of substituted 2,4-didehydroadamantanes was accompanied by a reduction in the amount of the corresponding γ-substituted adamantanone. This was attributed to the reduced availability of molecular oxygen, which diffuses more slowly on SiO_2 than in D_2O.

Most recently, the investigation was extended to adamantanylidenes entrapped within cucurbiturils in aqueous solution.[107] Remarkably, no 2-adamantanol (**49a**) was obtained. This shows that adamantylidene (**39a**) is sufficiently protected from the bulk solvent. The 1:1 complexes with cucurbit[7]uril (CB[7]) have a particularly large association equilibrium constant ($K_f = 6.5 \times 10^5\,M^{-1}$ in DMSO/H_2O (3:2)) and the cucurbit[8]uril complexes are also kinetically stable on the NMR time scale. Their irradiation yields 2,4-didehydroadamantane (**46a**) almost exclusively (>90%).[107] These findings show that the CB hosts are relatively inert toward carbene **39a** despite their nitrogen atoms and carbonyl groups. The same study was conducted with γ-substituted adamantanylidenes **39b** and **39c**. Although the selectivity of the reactions was not particularly high, the aforementioned trend was still observed with yields of 1,3-insertion product **46** ranging from 49% to 66% in the CBs. The only by-products were the respective γ-substituted adamantanones **43**.

Aziadamantane (**3a**) was also included and photolyzed within FAU zeolites.[18] Although azine **44a** is observed for reactions in solution, its intermolecular formation was curtailed at relatively low occupancy levels within FAUs. However, FAU zeolites were found to be too reactive for the production of 2,4-didehydroadamantane (**46a**). The compound constitutes ~40% of the product mixture at 35% conversion, but its yield slowly decreased with prolonged irradiation. The reaction ultimately produced only 2-adamantanol (92%), adamantanone (6%), and adamantane (2%). It is likely that a protic mechanism transformed 1,3-CH insertion product **46a** into alcohol **49a** within the aluminosilicate, which is usually amenable to the formation of stable carbenium ions.[107]

10.6.6 3-Oxobicyclo[3.2.1]octan-8-ylidene

A remarkable rise in the chemoselectivity of 3-oxobicyclo[3.2.1]octan-8-ylidene (**51**) was achieved by encapsulating its diazirine precursor, 8-azibicyclo[3.2.1]octan-3-one (**50**), within an α-CyD dimer (Fig. 10.15).[17h] Carbene **51** can decay intramolecularly by three competing mechanisms, but only two pathways are practicable (Scheme 10.12). First, the tricyclic ketone tricyclo[3.3.0.0^{2,8}]octan-3-one (**52**) can be formed from a 1,3-CH insertion reaction

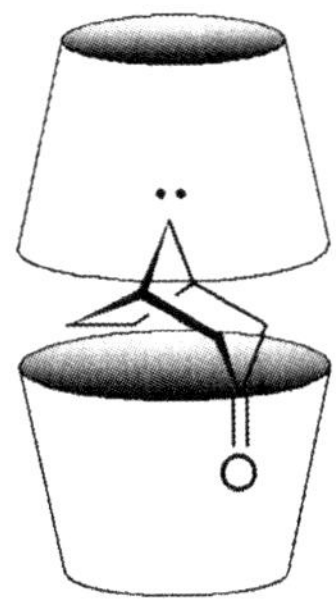

51@(α-CyD)$_2$

Figure 10.15. Encapsulation of 3-oxobicyclo[3.2.1]octan-8-ylidene (**51**) by a dimeric α-CyD assemblage markedly affects the selectivity of the carbene's reactions (Scheme 10.12).

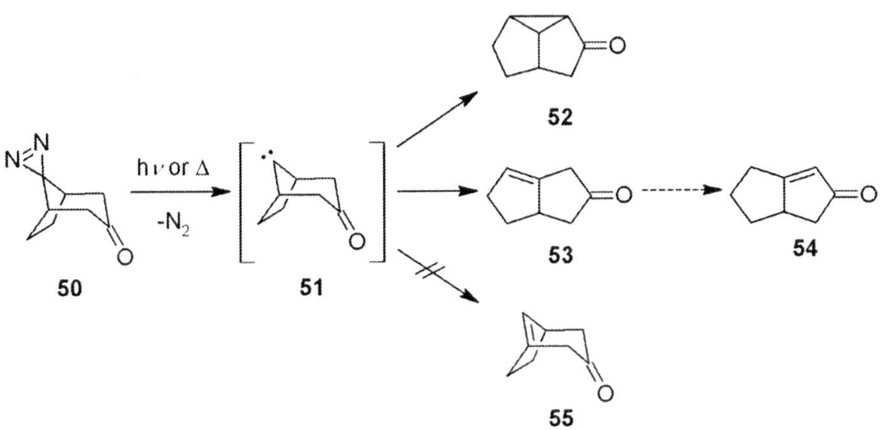

Scheme 10.12.

of the divalent carbon atom into either of the two *exo*-positioned C–H bonds vicinal to the carbonyl group, which likely induces the bridging divalent carbon to bend toward its side of the main ring.[108] Second, the alkenone bicyclo[3.3.0]-oct-5-en-3-one (**53**) will result from a 1,2-alkyl shift of either the C-1–C-2 or the C-4–C-5 bond toward the divalent carbon of **51**. Although characterized, this elusive product was found to readily rearrange to the more stable α,β-unsaturated ketone bicyclo[3.3.0]oct-1-en-3-one (**54**). Third, the strained anti-Bredt compound bicyclo[3.2.1]oct-1(8)-en-3-one (**55**) could arise from a 1,2-H shift in carbene **51**, but this simple rearrangement is unavailable due to orthogonal misalignment of either migrating C–H bond (HOMO) with the vacant p-orbital (LUMO) target. Remarkably, supramolecular inclusion of carbene **51** dramatically alters its chemistry. Whereas the ratio of constitutional isomers **52** and **53** from the pyrolysis of free diazirine **50** *in vacuo* is 0.9:1, it rises to 13.3:1

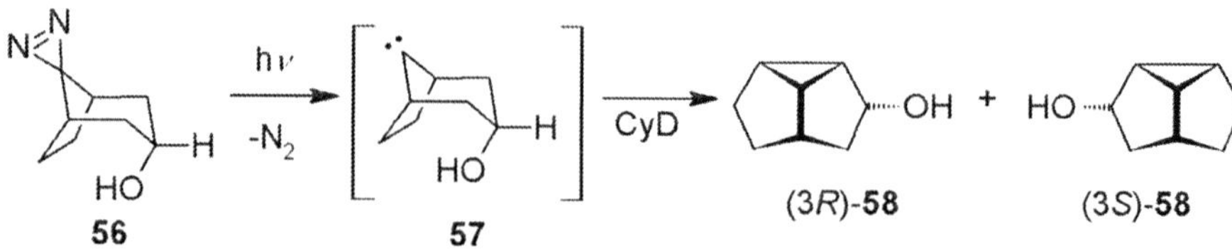

Scheme 10.13.

for the photolysis of **50@(α-CyD)$_2$** in the solid state. This represents a 15-fold increase in the chemoselectivity of carbene **51** for 1,3-CH insertion. The exact reason why carbene **51** substantially prefers 1,3-insertion over 1,2-alkyl migration within a confined space remains unclear, but calculating the energy barriers for the two competing reactions would be helpful. Indeed, extensive calculations need to be performed for carbenes entrapped within the microenvironments of host molecules.

10.6.7 *endo*-3-Hydroxybicyclo[3.2.1]octan-8-ylidene

One advantage of CyDs is that they belong to the chiral pool of natural products that are readily available. Their cavities can be used to perform reactions that generate chiral products from achiral reactants. Even fast-reacting carbene intermediates are susceptible to chiral induction. For example, photochemically generated *endo*-3-hydroxybicyclo[3.2.1]octan-8-ylidene (**57**) is prone to undergo a 1,3-CH insertion reaction giving *endo*-tricyclo[3.3.0.0^{2,8}]octan-3-ol (**58**). When this reaction was performed within β-CyD, a slight enantiomeric excess (ee = 4%) was obtained.[109] With the smaller and more rigid α-CyD, the effect was slightly higher (ee = 8%).[109] These results were determined from experiments at $T = 10\,°C$. Previous work on supramolecular photochirogenesis with ordinary neutral molecules showed that the enantiomeric excess can be increased significantly,[110] even up to values of 99%. A dual strategy was used that combined a chiral scaffold with a confining host.[111] Alternatively, temperature and pressure were varied (usually by lowering *T* and increasing *P*).[112] With the highly reactive carbene **57**, the enantiomeric excess of tricyclooctanol **58** was elevated to more than 40% at the temperature of liquid nitrogen (i.e., $T = -196\,°C$) (Mieusset, J.-L., Thiel, B., Abraham, M., Pačar, M., Brinker, U. H., University of Vienna, unpublished results).

endo-3-Hydroxybicyclo[3.2.1]octan-8-ylidene (**57**) was also generated within TRIMEB (cf. Section 10.5).[67,109] There were two reasons for examining permethylated β-CyD: (i) to avoid the undesired innermolecular products that are formed by ostensible O–H insertions into the upper and lower peripheries of natural CyD hosts and (ii) to enclose the guest more completely within a deeper cavity. Indeed, the first goal was achieved. TRIMEB proved to be inert toward innermolecular singlet carbene reactions; no insertion of the carbene into the protected host was observed.[109] However, the utility of the reaction

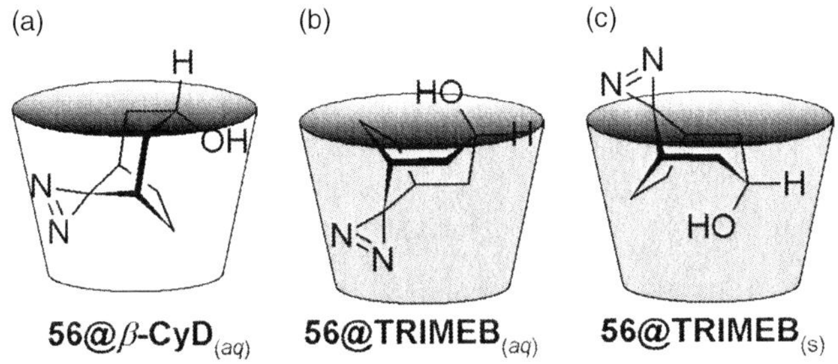

Figure 10.16. Different orientations of the 1:1 complexes of *endo*-8-azibicyclo[3.2.1]octan-3-ol (**56**) within (a) aqueous β-CyD, (b) aqueous TRIMEB, and (c) crystalline TRIMEB.[67]

in an aqueous solution was not improved due to the modified cavitand's flexibility. TRIMEB lacks the rigidifying intramolecular hydrogen bonds present in natural CyDs because it has no hydroxy groups. This lowers the association constant of the TRIMEB complex around 20-fold compared with the β-CyD IC.[67]

As depicted in Figure 10.16a versus b,[67] the orientation of *endo*-8-azibicyclo[3.2.1]octan-3-ol (**56**) within TRIMEB in solution is different from that in β-CyD. Both ICD fine structure and 2-D ROESY cross peaks suggest that the azi functional group of diazirine **56** is more deeply submerged within the permethylated CyD cavity than within that of the natural CyD. A single-crystal X-ray diffraction analysis of **56**@TRIMEB was obtained.[67] Interestingly, it showed a reversal in the orientation of diazirine **56** within the cavitand (Fig. 10.16c), which brings the polar hydroxy group of **56** deeper into the modified CyD and directs the apolar azi functionality toward the wider methoxy-lined perimeter of TRIMEB. This inverse hydrophobicity of TRIMEB compared with β-CyD has been investigated computationally.[113]

10.6.8 Ring Fragmentation and Expansion

10.6.8.1 Nortricyclylidene. Azinortricyclane (**59**) was used to form the CyD ICs **59@(α-CyD)$_2$** and **(59@β-CyD)$_2$**.[17c] Photolysis generated nortricyclylidene (**60**) within the ICs. Indeed, the results were in accord with these structures. No nortricyclanone azine (**62**) was formed upon photolysis of the sandwich-like **59@(α-CyD)$_2$** IC because the reactive guest was completely encapsulated. In contrast, bimolecular reaction of the guests within the **(59@β-CyD)$_2$** dimer produced azine **62** by virtue of the geminated assemblage. Of the two possible intramolecular products, quadricyclane (**61**) and 4-ethynylcyclopentene (**63**) (Scheme 10.14), only enyne **63** was observed after photolyses of the diazirine@CyD complexes. This is not surprising because the bilateral symmetry and endocyclic configuration of carbene **60** make it a prime candidate for cyclopropylcarbene fragmentation.[114–116] Control experiments with α- and β-CyD

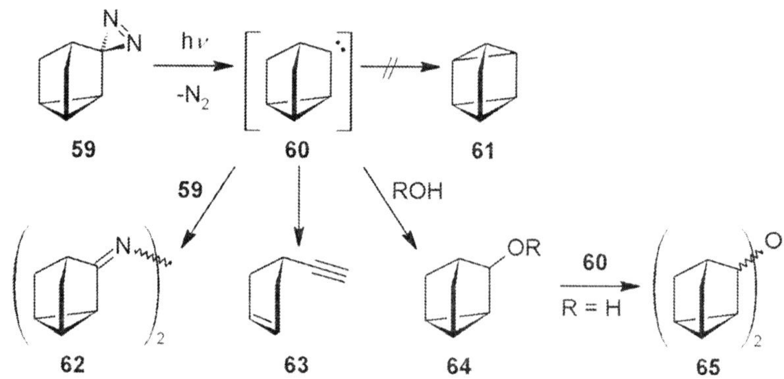

Scheme 10.14. (ROH = (a) α-CyD, (b) β-CyD, (c) H_2O)

complexes of **61** showed it to be stable under the reaction conditions. Thus, no **61** was formed from photolysis of the diazirine@CyD complexes. In addition to intramolecular fragmentation to **63**, innermolecular reaction of carbene **60** with the O–H bonds in both CyDs predominated, giving substituted CyDs **64a** and **64b**, which were detected by FAB-MS analysis.[17c]

Photolysis of **59@NaY** (‹s› = 0.25, cf. Section 10.2.6.1) yielded some **63** but no **61**, which was again found to be stable under the reaction conditions. The principal reaction is the hydrolysis of carbene **60** and the formation of 3-nortricyclanol (**64c**, 76%), which is also stable within NaY.[117] This demonstrates that a significant amount of water remains inside thermally activated NaY. The presence of bimolecularly derived azine **62** and ether **65** indicates some diffusion of transient carbene **60** within the lightly loaded FAU zeolite.

The absence of C–H insertion by carbene **60** is noteworthy because 2-norbornylidene (**66**), which merely lacks the secondary bridge of tricyclic **60**, undergoes C–H insertion to nortricyclane (**68**) almost exclusively (Scheme 10.15)[118,119] However, the 1,3-CH insertion **60** → **61** is also a 1,4-CH insertion. These reactions are too slow to compete with others due to improper orbital alignment.[86,119] Apparently, there is still enough space within the hosts' voids for the unraveling process; confinement of carbene **60** within CyDs and FAU zeolites neither hampers its fragmentation to enyne **63** nor enables 1,4-CH insertion to **61** to occur.

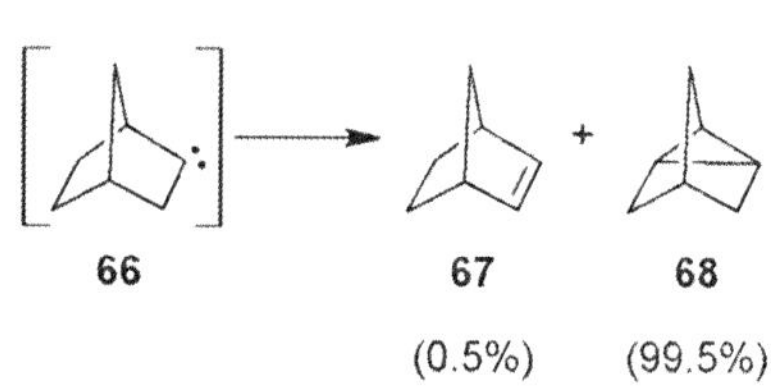

Scheme 10.15.

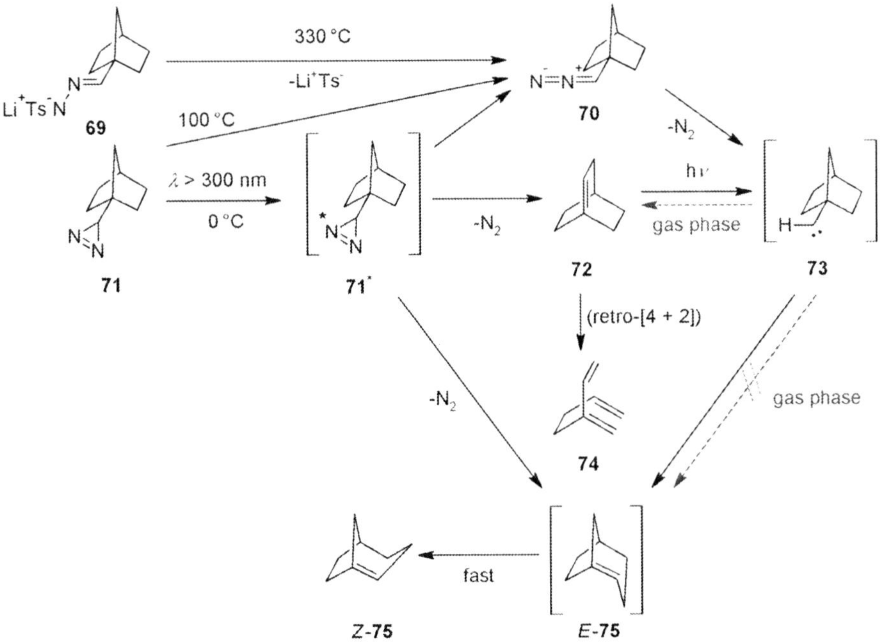

Scheme 10.16.

10.6.8.2 (Bicyclo[2.2.1]hept-1-yl)carbene. (Bicyclo[2.2.1]hept-1-yl)carbene (**73**)[101c,120] was generated during the photolysis of the corresponding incarcerated 3*H*-diazirine (**71**) (Scheme 10.16). A likely route involves diazoalkane **70**, which is the isomerization product of excited diazirine **71***. (Diazo compound **70** was independently generated by pyrolysis of the lithium tosylhydrazone salt **69**.) The ring-expansion products (*Z*)-bicyclo[3.2.1]oct-1-ene (*Z*-**75**) and bicyclo[2.2.2]oct-1-ene (**72**) were both observed.[121] These strained bridgehead bicycloalkenes were prevented from dimerization but not from reaction with dissolved oxygen. *Z*-**75** probably arose via rapid isomerization of its initially produced *E*-**75** isomer. For **72**, rearrangement through a retro-Diels–Alder reaction occurred yielding the acyclic molecule 3-methylenehepta-1,6-diene (**74**), which proceeded to escape from the HC host (Fig. 10.2; R = $O(CH_2)_4O$).

10.6.8.3 Phenylcarbene. The ring expansion of phenylcarbene (**77**), generated from diazirine **76**, to 1,2,4,6-cycloheptatetraene (**78**) has been extensively studied.[122] In addition to their characterization in a cryogenic matrix,[8] carbene **77** and strained allene **78** were generated within a HC, whereupon **77** immediately rearranged to **78**,[19a,b] as depicted in Scheme 10.17. Surprisingly, the usually transient **78** was found to be stable even at $T = 60\,°C$. Enantiomerization of asymmetric **78** putatively occurs through 2,4,6-cycloheptatrien-1-ylidene (**79**), which yields carbene insertion products (**80**; R = Cl, MeO, and EtO) when it is free in solution.[123] Indeed, the HC does prevent dimerization reactions but it also

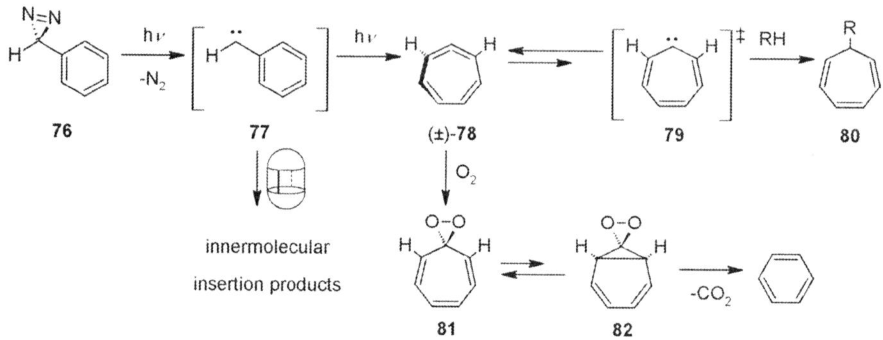

Scheme 10.17.

allows small molecules, such as HCl and O_2, to enter through its pores. For example, reaction of **79** with HCl produced 7-chlorocyclohepta-1,3,5-triene, and reaction with O_2 gave the corresponding dioxirane **81**, which decomposed, via valence tautomer **82** to benzene and CO_2 (Scheme 10.17).[19a,b] As mentioned earlier, cycloheptatetraene **78** also remains intact within the HC at $T = 60\,°C$. This is true even in the presence of methanol, which is small enough to infiltrate the host and react with incarcerated 7-chlorocyclohepta-1,3,5-triene. This seeming irregularity arises from the difficulty of protonating cycloheptatetraene **78** with a weak acid in the nonpolar inner phase of the HC.[19a,b]

10.6.9 Aryl(halo)carbenes

10.6.9.1 Fluoro(phenyl)carbene. Recently, an IC of fluoro(phenyl)carbene (**83**) with a HC was reported (Fig. 10.17).[124] The **83**@HC IC was metastable, however, because of the carbene's relatively high electrophilicity ($m = 0.89$).[42] Consequently, carbene **83** rapidly added to an aromatic moiety of the HC by cyclopropanation even at temperatures as low as –80 °C. This innermolecular

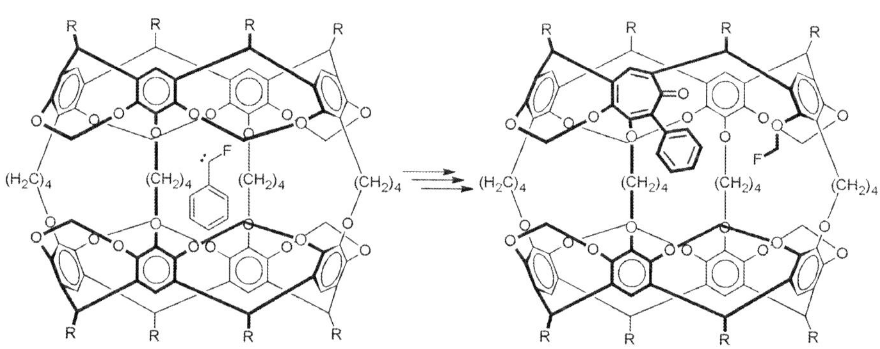

Figure 10.17. Innermolecular reaction of reactive fluoro(phenyl)carbene (**83**) included within a hemicarcerand led to ring scission of the host and tropone formation ($R = n\text{-}C_5H_{11}$).

reaction ultimately led to the tropone-embedded fluoro derivative of the HC (**84**) in 55% yield.[125] Kinetics experiments using LFP showed that the caged reactive intermediate forms a π-complex with the HC and decays by 10% in 4 μs (i.e., τ=38 μs).

10.6.9.2 Substituted Halo(phenyl)carbenes and O-Ylides. The reactivity of electrophilic carbenes, such as chloro(phenyl)carbene (**85a**) (Fig. 10.18) can be modulated in solution by the electron-donating ability of the solvent.[126–132] This is due to the formation of EDA complexes between the carbene and an ether or arene solvent. The degree of binding within the *O*-ylides or π-complexes thus formed can often be quite substantial, as with chloro(pentafluorophenyl) carbene (**85b**) dissolved in electron-rich trimethoxybenzene (K_f=3.21 × 10^5 M^{-1}).[132] Note that dialkylcarbenes behave differently, of course, because they belong to the group of *reactive-nucleophilic* carbenes (Fig. 10.8), which tend to donate electrons rather than accept them.[44]

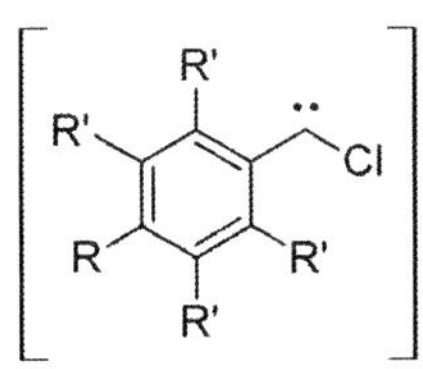

85a: R = R' = H

85b: R = R' = F

85c: R = NO_2; R' = H

85d: R = OCH_3; R' = H

Figure 10.18. The philicity of chloro(phenyl)carbene (**85a**) can be adjusted by functionalization of its phenyl ring.

Chloro(4-nitrophenyl)carbene (**85c**) is a typical electrophilic carbene that forms an *O*-ylide with tetrahydrofuran (THF). Thus, an *O*-ylide band was observed with the crown ether 18-crown-6 during LFP. When dibenzo-18-crown-6 was employed, both *O*-ylide and π-complexes were detected. The initially formed *O*-ylide decayed to the π-complex after 1500 ns (Fig. 10.19).[131] Not only do these findings help to explain innermolecular reactivity within HC hosts but they also may be applied to host-mediated reactions within zeolites (*vide infra*).

A study of the behavior of several *para*-substituted aryl(chloro)carbenes included within FAU zeolites revealed that the nature of interaction between guest and host depends on the philicity of the carbene.[133] The electron-rich chloro(4-methoxyphenyl)carbene (**85d**) tends to bind with Lewis acidic sites. In contrast, the electron-poor chloro(4-nitrophenyl)carbene (**85c**) is extremely short lived (τ< 100 ns) because it adds firmly to a nucleophilic [Si–O–Al]$^-$ bridge within the zeolite framework, creating *O*-ylidic adduct **86c**. This strong EDA complex then reacts with methanol to yield the typically

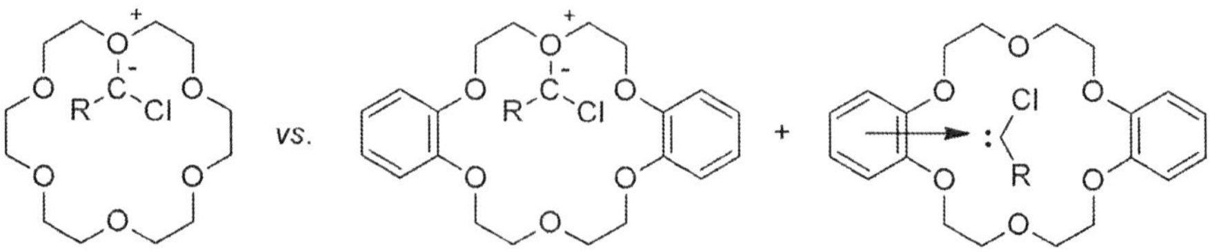

Figure 10.19. Chloro(4-nitrophenyl)carbene (**85c**, R = *p*-$C_6H_4NO_2$) not only forms *O*-ylides with the crown ethers 18-crown-6 and dibenzo-18-crown-6 but also a π-complex with the latter.

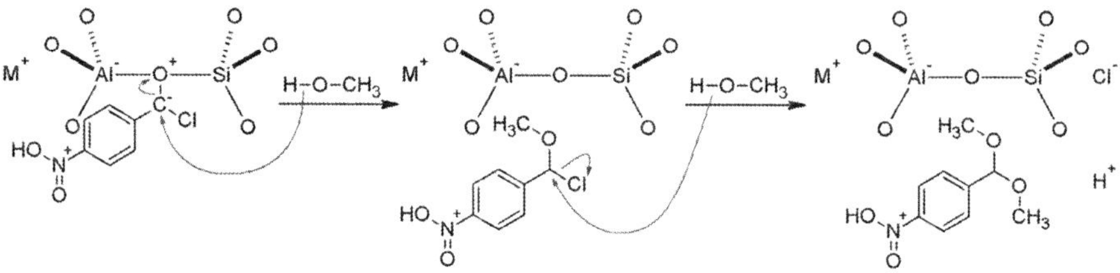

86c 87c 88c

Scheme 10.18.

reactive α-chloroether **87c**. Addition of methanol several hours later led to dimethyl acetal **88c** (Scheme 10.18).[133]

10.6.9.3 Chloro(phenyl)carbene. Photolysis of neat 3-chloro-3-phenyldiazirine (**89a**)[64,134] yields azine **90a** as the sole product (Scheme 10.19).[17f] In solution, chloro(phenyl)carbene (**85a**) also gives solvent insertion products.[135] In anhydrous methanol, transformation of the initially formed α-chloroether **87a** provides the corresponding dimethyl acetal **88a** exclusively. However, base is required to neutralize the HCl byproduct and quench subsequent hydrolysis, which leads to benzaldehyde. Finally, the ostensible carbene dimer **91a** is formed in minor amounts (up to 5%) only in nonpolar solvents.[17f]

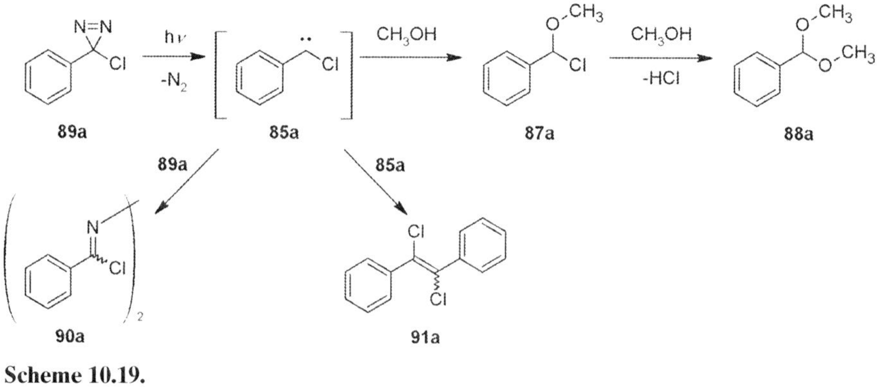

89a 85a 87a 88a 90a 91a

Scheme 10.19.

The structures of the diazirine@cyclodextrin complexes were found to be **89a@(α-CyD)$_2$** and **(89a@β-CyD)$_2$**.[17f] Photolysis of the solid **89a**@CyD ICs produced azine **90a**, benzaldehyde, and HCl. But RP-HPLC and FAB-MS revealed the presence of derivatized CyDs after the α- and β-CyD complexes of **89a** were photolyzed as slurries in alkaline buffer (pH=9.3).[17f] Their structures were not fully ascertained, but FAB-MS analysis ruled out $6^A,6^B$-*O*-(benzylidene)CyDs and related isomers.[136] Photolysis of **89a**@FAU ICs also gave azine **90a**, benzaldehyde, and HCl as the major products.[137] Formation of hydrolysis products from carbene **85a** demonstrated that a significant amount of chemisorbed water remained within thermally activated FAUs.

The first direct observation of a supramolecular carbene was achieved using a Y-type FAU zeolite.[138] The recording of transient diffuse reflectance UV/vis spectra during the LFP of 3-halo-3-phenyldiazirines at $\lambda=355$ nm showed intense absorption bands at $\lambda=320$ nm that were attributable to chloro(phenyl)- and bromo(phenyl)carbene, in accordance with established results obtained in solution.[139] In addition to this absorption band, further evidence was acquired from trapping experiments that generated a long-lived ylide. Although pyridine is routinely used for LFP experiments in solution, complications arise from its strong binding to the acidic sites of the zeolite. So, 4,4'-bipyridine was used because one of its nitrogen atoms remained free.

The lifetime of carbene **85a** was found to be sensitive to the alkali metal that served as the counterion in the cation-exchanged Y-type FAU zeolites[138] because of changes in the strength of the Lewis acidic and basic sites.[34] The rate of decay for the singlet carbene **85a** was slowest in CsY ($k=3.1\times10^5$ s^{-1}; $\tau=3.2$ μs) but increased inside RbY ($k=3.6\times10^5$ s^{-1}; $\tau=2.8$ μs) and KY ($k=5.2\times10^5$ s^{-1}; $\tau=1.9$ μs).[138] In NaY and LiY, the carbene could no longer be observed because it was too reactive. It has been suggested that the disappearance of the carbene may be caused by a reversible interaction with the zeolites.

From a synthetic viewpoint, slow thermolysis of **89** leads mainly to the formation of benzaldehyde, probably after reaction of the carbene with residual water. At high loading levels, however, 2,5-diphenyl-1,3,4-oxadiazole (**93a**)[140] becomes the main product (57% for $\langle s\rangle=2$ in NaY).[138] It results from the reaction between an intermediately formed azine and water (Scheme 10.20). Interestingly, oxadiazole **93a** does not form from azine **90a** in the absence of H_2O.[137,141] The heterocyclic compound was formed when **89a** was photolyzed in neat benzaldehyde.[137] Under these conditions, the corresponding diazo compound should be trapped by the aldehyde in a [3+2] dipolar addition. Subsequent elimination of HCl from the resulting Δ^3-1,3,4-oxadiazoline would give **93a**.

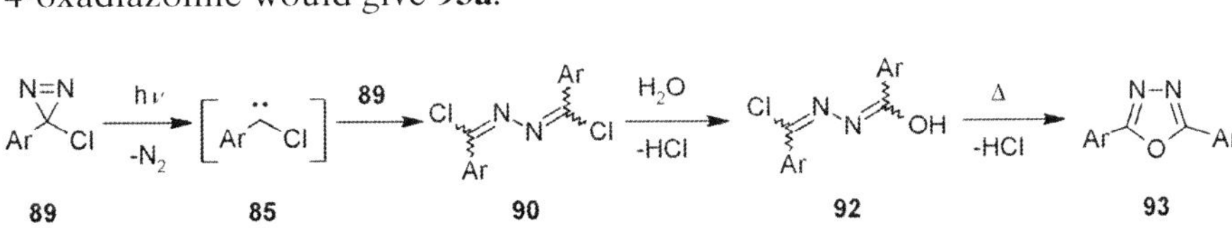

Scheme 10.20.

10.6.10 Fluoro(phenoxy)carbene

The extended lifetime of singlet fluoro(phenoxy)carbene (**95**) within a HC is most impressive. The carbene eschews innermolecular reaction with the host due to the carbene's intrinsic electronic stabilization, and carbene dimerization was restricted, of course. Thus, carbene **95**, formed by photolysis of diazirine **94**, was shown to persist in its less stable *cisoid* conformation for days at room temperature within a HC (Scheme 10.21).[19h] Eventually, acid-catalyzed hydrolysis of incarcerated **95** slowly led (via **96**) to formate **98**, while the reaction of incarcerated carbene **95** with the HF released from **96** gave difluoroether **97**.

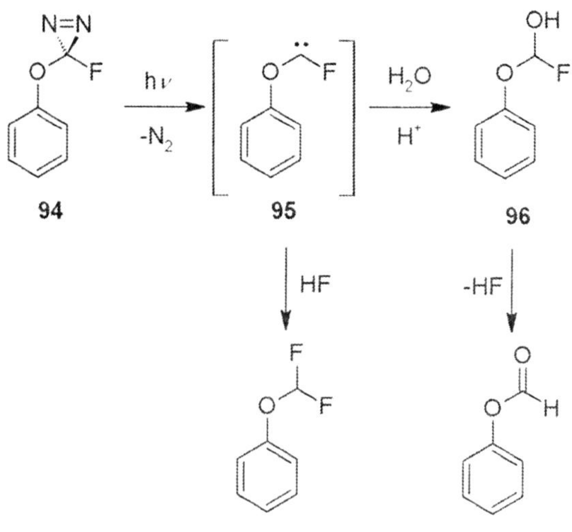

Scheme 10.21.

10.6.11 Persistent Triplet Carbenes

The triplet carbene bis(2,4,6-trichlorophenyl)carbene (**100**), formed by photolysis of diazo compound **99**, is intrinsically stabilized due to electron delocalization (Scheme 10.22; Fig. 10.20).[142] In addition, its large size sterically hinders it from intermolecular reaction. It is therefore a good candidate for observation within a zeolite. This endeavor was somewhat successful.[143] As noted, diazo compounds are sensitive to water, acids, and metals,[52] and FAU zeolites invariably have some proton-exchanged cationic sites.[144] Thus, it is not surprising that bis(2,4,6-trichlorophenyl)diazomethane (**99**) decomposed when included within the FAU NaY. In contrast, the aluminophosphate VPI-5 is a compatible host for diazo compounds because it is free of Brønsted–Lowry acidic sites (Fig. 10.7). Moreover, its empty spaces are big enough to

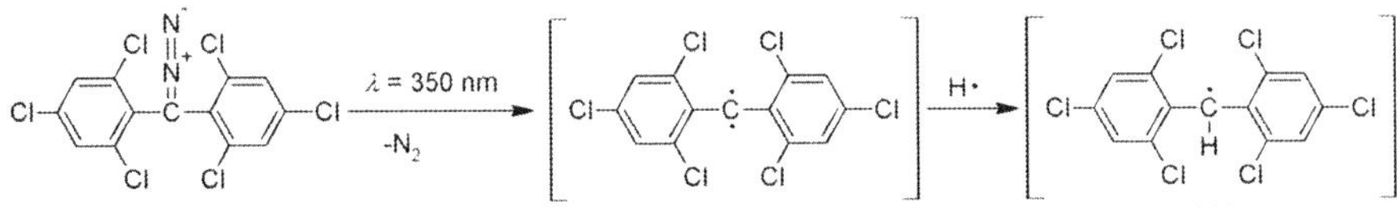

99 100 101

Scheme 10.22.

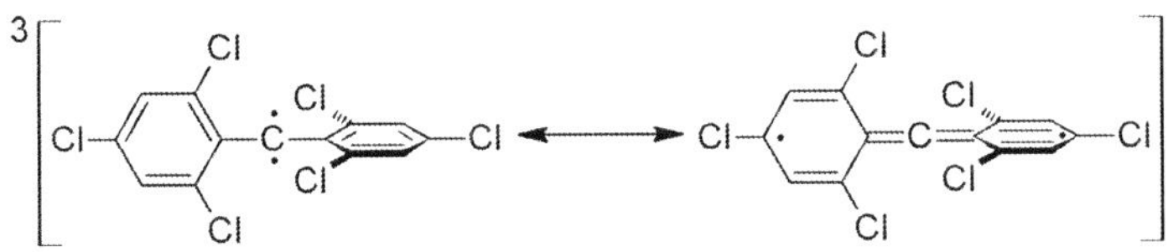

Figure 10.20. Bis(2,4,6-trichlorophenyl)carbene (**100**) was observed in the neutral zeolite VPI-5 under special conditions.

accommodate the relatively large diaryldiazo compound **99** (12 × 9 × 6 Å^3). Upon irradiation, however, the generated triplet diarylcarbene **100** decayed even faster than in solution. Triplet carbene **100** has a half-life ($t_{1/2}$) of 18 ms in room temperature benzene[145] but $t_{1/2}$ is less than 1 ns within the zeolite. This peculiar finding is due to multiple reflections of light within the VPI-5 powder. It causes the carbene to be promoted to its excited state and, since excited-state triplet carbenes are more efficient at abstracting hydrogen atoms than ground-state ones, the bis(2,4,6-trichlorophenyl)methyl radical (**101**) forms very rapidly (Scheme 10.22). In order to observe the triplet carbene, it was necessary to immerse the loaded VPI-5 in a solvent with the same refractive index. Hence, glycerol (n_D = 1.4730) was used. The zeolite-enclosed triplet carbene was detected by EPR spectroscopy under these conditions at T = −196 °C to −53 °C.[143]

10.7 CONCLUSION AND OUTLOOK

Many of the case studies described herein present compelling evidence that carbene reaction intermediates behave differently when they are generated in the supramolecular phase compared with other condensed phases. For example, the lifetime of fluoro(phenoxy)carbene (**95**) was dramatically extended when it was caged within a HC.[19h] To begin to achieve similar results with alkyl- and dialkylcarbenes, one must make prudent decisions. Rearrangements by facile 1,2-H shifts dramatically reduce the lifetimes of alkyl-substituted carbenes. Therefore, investigations of carbenes that feature a distinguishing, relatively slow intramolecular process (e.g., 1,3-CH insertion within adamantylidene (**39a**))[59b,104a,b] should be avidly pursued. However, a carbene might insert into the solvent or add to its nitrogenous precursor, resulting in unwanted azine formation. Nevertheless, the lifetime of carbene **39a** might be prolonged because τ(**39a**) already increases from 225 ns in aliphatic isooctane to 700 ns in aromatic benzene.[104a] This is likely attributable to the formation of a stable EDA complex with the latter solvent.

In any event, the altered product selectivities of carbene **39** indicate a strong dependence on which host compound is used for supramolecular complexation (cf. section 10.6.5). This remarkable phenomenon is accounted for by understanding the particular architecture of each guest@host complex. Host containers can offer a uniquely organized microenvironment wherein stoichiometry, free volume, guest orientation, and reaction partners can be controlled. Moreover, an optimized fit between the guest and the host cavity is needed to significantly reduce motion and prevent intermolecular reactions.

Diazirines have proven to be reliable sources for supramolecular carbenes. They fit well within many cavitand hosts and are easily reacted. However, unwanted innermolecular reactions (e.g., O–H insertion) sometimes deplete the supply of productive carbenes. This can lead to a faster rate of carbene decay due to incarceration, which is exactly opposite to the stated goal. Thus, the lifetimes of the photolytically generated carbenes may not be sufficiently prolonged despite complexation. In such cases, slower intramolecular reactions are difficult to foster, so more inert hosts, such as TRIMEB, HCs, and CB[7] should be considered. In the case of zeolites, the aluminosilicates are amphoteric hosts and can interfere with carbenes that exhibit a high degree of electro- or nucleophilicity.[133] Thus, aluminophosphate zeolites should be tested to observe new reactions because the $AlPO_4$ hosts have neutral frameworks.

As with any carbene reaction, supramolecular carbene reactions must be well planned. Each type of host vessel is unique and deserves careful inspection before being employed. Therefore, rudimentary concepts must be considered. For example, one must know how the carbene precursor is confined within the molecular container or how it interacts with the host. An exact knowledge of the structure of the guest@host complex is crucial for predicting the reaction's outcome. Unfortunately, crystals of these complexes suitable for X-ray diffraction analysis are hard to obtain. Therefore, other techniques, such as 2-D ROESY NMR and ICD, must be used to ascertain how the precursor is embedded within the host compound.

The supramolecular chemistry of carbenes continues to be a distinctive science. Although substantial progress has been made since its inception 25 years ago, challenging problems remain and need to be resolved. Computational exploration of supramolecular carbene chemistry will become more helpful in the coming years as the field continues to develop. All of this extra effort is well rewarded, however, because sometimes a dormant carbene reaction can be elicited by constraining the carbene intermediate.

ACKNOWLEDGMENTS

UHB is grateful to his coworkers at The State University of New York at Binghamton, where this research started, and to those at the University of Vienna for their efforts described in the works cited herein. Financial support from both The Petroleum Research Fund administered by the American Chemical Society and the *Fonds der wissenschaftlichen Forschung* in Austria is also appreciated. It was a great privilege to collaborate with Professor V. Ramamurthy, University of Miami. The authors are indebted to Dr. Daniel Krois (University of Vienna) for his important contributions to this chapter concerning diazirine@CyD complexes. UHB appreciates his introduction to the ICD method and its usefulness for studying cyclodextrin complexes.

SUGGESTED READING

Cram, D. J., Tanner, M. E., Thomas, R. "The Taming of Cyclobutadiene," *Angew. Chem. Int. Ed. Engl.* **1991**, *30*, 1024–1027.
Doyle, M. P. "The Selectivity of Carbenes Generated from Diazirines." In *Chemistry of Diazirines*. Liu, M. T. H., Ed. Vol. 2. CRC Press: Boca Raton, FL, **1987**, pp. 33–74.
Lehn, J.-M., Ed. *Comprehensive Supramolecular Chemistry*, 11 vols. Pergamon: New York, **1995**.
Ramamurthy, V., Ed. "Photoprocesses of Host–Guest Complexes in the Solid State." In *Photochemistry in Organized and Constrained Media*. VCH: New York, **1991**.

REFERENCES

1 (a) Hine, J. *Divalent Carbon*. Ronald Press: New York, **1964**. (b) Kirmse, W. *Carbene, Carbenoide, und Carbenanaloge*. Verlag Chemie: Weinheim, **1969**. (c) Kirmse, W. *Carbene Chemistry*, 2nd ed. Academic Press: New York, **1971**. (d) Jones, M., Jr., Moss, R. A., Eds. *Carbenes*, Vol. 1. Wiley: New York, **1973**. (e) Moss, R. A., Jones, M., Jr., Eds. *Carbenes*, Vol. 2. Wiley: New York, **1975**. (f) Jones, M., Jr. *Sci. Am.* **1976**, *234*(2), 101–113. (g) Jones, W. M., Brinker, U. H. "Some Pericyclic Reactions of Carbenes." In *Pericyclic Reactions*, Vol. 1. Marchand, A. P., Lehr, R. E., Eds. Academic Press: New York, **1977**. (h) Regitz, M., Ed. *Carbene (Carbenoide) Methoden der Organischen Chemie (Houben–Weyl)*, Vol. E19b. Thieme Verlag: Stuttgart, **1989**. (i) Brinker, U. H., Ed. *Advances in Carbene Chemistry*, Vol. 1. JAI Press: Greenwich, CT, **1994**. (j) Brinker, U. H., Ed. *Advances in Carbene Chemistry*, Vol. 2. JAI Press: Stamford, CT, **1998**. (k) Brinker, U. H., Ed. *Advances in Carbene Chemistry*, Vol. 3. Elsevier: Amsterdam, **2001**. (l) Helson, H. E., Jorgensen, W. L. *J. Org. Chem.* **1994**, *59*, 3841–3856. (m) Moss, R. A., Platz, M. S., Jones, M., Jr., Eds. *Reactive Intermediate Chemistry*. Wiley: New York, **2004**. (n) Bertrand, G., Ed. *Carbene Chemistry: From Fleeting Intermediates to Powerful Reagents*. Dekker: New York, **2002**.
2 Jones, M., Jr. *Organic Chemistry*. Norton: New York, **1997**, pp. 412, 1099.
3 Doering, W. v. E., Knox, L. H. *J. Am. Chem. Soc.* **1956**, *78*, 4947–4950.
4 Merrer, D. C., Moss, R. A. *Adv. Carbene Chem.* **2001**, *3*, 53–113.

5 Jackson, J. E., Platz, M. S. *Adv. Carbene Chem.* **1994**, *1*, 89–160.
6 (a) Bally, T. "Matrix Isolation." In *Reactive Intermediate Chemistry*. Moss, R. A., Platz, M. S., Jones, M., Jr., Eds. Wiley: New York, **2004**. (b) Sheridan, R. S. *Inter-Amer. Photochem. Soc. News*, **2000**, *23*(1), 39–48.(c) Sheridan, R. S. In *Organic Photochemistry and Photobiology*. Horspool, W. M., Ed. CRC Press: Boca Raton, FL, **1995**, pp. 999–1006.
7 Bally, T., Matzinger, S., Truttman, L., Platz, M. S., Morgan, S. *Angew. Chem. Int. Ed. Engl.* **1994**, *33*, 1964–1966.
8 Matzinger, S., Bally, T. *J. Phys. Chem. A* **2000**, *104*, 3544–3552.
9 Tomioka, H., Ozaki, Y., ., Izawa, Y. *Tetrahedron Lett.* **1982**, *23*, 1917–1920.
10 Tomioka, H., Ozaki, Y., Izawa, Y. *Chem. Lett.* **1982**, 843–846.
11 (a) Zuev, P. S. Sheridan, R. S. *J. Am. Chem. Soc.* **2001**, *123*, 12434–12435. (b) Sander, W. *Spectrochim. Acta, Part A*. **1987**, *43A*, 637–646. (c) Ganzer, G. A., Sheridan, R. S., Liu, M. T. H. *J. Am. Chem. Soc.* **1986**, *108*, 1517–1520.
12 Yao, G., Rempala, P., Bashore, C., Sheridan, R. S. *Tetrahedron Lett.* **1999**, *40*, 17–20.
13 Tomioka, H. "Triplet Carbenes." In *Reactive Intermediate Chemistry*. Moss, R. A., Platz, M. S., Jones, M., Jr., Eds. Wiley: New York, **2004**.
14 Tomioka, H. *Atom Tunneling Phenomena in Physics, Chemistry and Biology*, Springer Series on Atomic, Optical, and Plasma Physics **2004**, *36*, 147–172, *Chem. Abstr.* **2004**, 140, 303107.
15 (a) Brinker, U. H., Rosenberg, M. G. *Adv. Carbene Chem.* **1998**, *2*, 29–44. (b) Rosenberg, M. G., Brinker, U. H. *Adv. Phys. Org. Chem.* **2005**, *40*, 1–47. (c) Kirmse, W. *Angew. Chem. Int. Ed.* **2005**, *44*, 2476–2479. (d) Rosenberg, M. G., Brinker, U. H. *Eur. J. Org. Chem.* **2006**, 5423–5440. (e) Mieusset, J.-L., Brinker, U. H., Eds. "Encapsulation of Reactive Intermediates." In *Molecular Encapsulation: Organic Reactions in Constrained Systems*. Wiley: Chichester, **2010**.
16 (a) Abelt, C. J., Pleier, J. M. *J. Org. Chem.* **1988**, *53*, 2159–2162. (b) Abelt, C. J., Lokey, J. S., Smith, S. H. *Carbohydr. Res.* **1989**, *192*, 119–130. (c) Abelt, C. J. "Insertion Reactions of Cyclodextrin-Bound Carbenes," final report to the Petroleum Research Fund on Grant 22092–B4, College of William and Mary, Williamsburg, VA, **1992**. (d) Smith, S. H., Forrest, S. M., Williams, D. C., Jr., Cabell, M. F., Acquavella, M. F., Abelt, C. J. *Carbohydr. Res.* **1992**, *230*, 289–297. (e) Abelt, C. J. *Minutes 6th Int. Symp. Cyclodextrins*, **1992**, 649–654, *Chem. Abstr.* **1994**, *121*, 83794.
17 (a) Brinker, U. H., Buchkremer, R., Kolodziejczyk, M., Kupfer, R., Rosenberg, M. G., Poliks, M. D., Orlando, M., Gross, M. L. *Angew. Chem. Int. Ed. Engl.* **1993**, *32*, 1344–1345. (b) Kupfer, R., Brinker, U. H. *Liebigs Ann.* **1995**, 1721–1725. (c) Rosenberg, M. G., Kam, S. M., Brinker, U. H. *Tetrahedron Lett.* **1996**, *37*, 3235–3238. (d) Krois, D., Bobek, M. M., Werner, A., Kählig, H., Brinker, U. H. *Org. Lett.* **2000**, *2*, 315–318. (e) Rosenberg, M. G., Brinker, U. H. *J. Org. Chem.* **2001**, *66*, 1517–1522. (f) Rosenberg, M. G., Brinker, U. H. *J. Org. Chem.* **2003**, *68*, 4819–4832. (g) Krois, D., Brecker, L., Werner, A., Brinker, U. H. *Adv. Synth. Catal.* **2004**, *346*, 1367–1374. (h) Mieusset, J.-L., Thiel, B., Abraham, M., Pačar, M., Brinker, U. H. *Tetrahedron Lett.* **2013**, *54*, 681–683.
18 Kupfer, R., Poliks, M. D., Brinker, U. H. *J. Am. Chem. Soc.* **1994**, *116*, 7393–7398.
19 (a) Warmuth, R., Marvel, M. A. *Angew. Chem. Int. Ed.* **2000**, *39*, 1117–1119. (b) Warmuth, R., Marvel, M. A. *Chem.—Eur. J.* **2001**, *7*, 1209–1220. (c) Kemmis, C., Warmuth, R. *J. Supramol. Chem.* **2001**, *1*, 253–267. (d) Warmuth, R. *J. Am. Chem. Soc.* **2001**, *123*, 6955–6956. (e) Warmuth, R., Yoon, J. *Acc. Chem. Res.* **2001**, *34*, 95–105. (f) Warmuth, R. *Eur. J. Org. Chem.* **2001**, 423–437. (g) Kerdelhué, J.-L., Langenwalter, K. J.,

Warmuth, R. *J. Am. Chem. Soc.* **2003**, *125*, 973–986. (h) Liu, X., Chu, G., Moss, R. A., Sauers, R. R., Warmuth, R. *Angew. Chem. Int. Ed.* **2005**, *44*, 1994–1997.

20 Wagner, G., Knoll, W., Bobek, M. M., Brecker, L., van Herwijnen, H. W. G., Brinker, U. H. *Org. Lett.* **2010**, *12*, 332–335.

21 Gupta, S., Choudhury, R., Krois, D., Wagner, G., Brinker, U. H., Ramamurthy, V. *Org. Lett.* **2011**, *13*, 6074–6077.

22 Gupta, S., Choudhury, R., Krois, D., Brinker, U. H., Ramamurthy, V. *J. Org. Chem.* **2012**, *77*, 5155–5160.

23 (a) Villiers, A. C. R. *Hebd. Seances Acad. Sci.* **1891**, *112*, 536–538. (b) Schardinger, F. *Wien. Klin. Wochenschr.* **1904**, *17*, 207–209.

24 (a) Szejtli, J. *Chem. Rev.* **1998**, *98*, 1743–1753. (b) Saenger, W., Jacob, J., Gessler, K., Steiner, T., Hoffmann, D., Sanbe, H., Koizumi, K., Smith, S. M., Takaha, T. *Chem. Rev.* **1998**, *98*, 1787–1802. (c) Wenz, G. *Angew. Chem. Int. Ed. Engl.* **1994**, *33*, 803–822.(d) Saenger, W. *Angew. Chem. Int. Ed. Engl.* **1980**, *19*, 344–362.(e) Bender, M. L., Komiyama, M. *Cyclodextrin Chemistry*. Springer Verlag: Berlin, **1978**.

25 (a) Szejtli, J. "Helical and Cyclic Structures in Starch Chemistry." In *Biotechnology of Amylodextrin Oligosaccharides*. ACS Symposium Series 458. Friedman, R., Ed. American Chemical Society: Washington, DC, **1991**. (b) Kano, K., Tamiya, Y., Hashimoto, S. *J. Inclusion Phenom. Mol. Recognit. Chem.* **1992**, *13*, 287–293.

26 Behrend, R., Meyer, E., Rusche, F. *Justus Liebigs Ann. Chem.* **1905**, *339*, 1–37.

27 Freeman, W. A., Mock, W. L., Shih, N.-Y. *J. Am. Chem. Soc.* **1981**, *103*, 7367–7368.

28 (a) Masson, E., Ling, X., Joseph, R., Kyeremeh-Mensah, L., Lu, X. *RSC Adv.* **2012**, *2*, 1213–1247. (b) Lagona, J., Mukhopadhyay, P., Chakrabarti, S., Isaacs, L. *Angew. Chem. Int. Ed.* **2005**, *44*, 4844–4870. (c) Kim, K., Selvapalam, N., Ko, Y. H., Park, K. M., Kim, D., Kim, J. *Chem. Soc. Rev.* **2007**, *36*, 267–279.(d) Lee, J. W., Samal, S., Selvapalam, N., Kim, H.-J., Kim, K. *Acc. Chem. Res.* **2003**, *36*, 621–630.

29 Cronstedt, A. F. K. *Sven. Vetenskapsakad. Handl., Stockholm* **1756**, *17*, 120–123.

30 Manly, R., Holmes, J. *New Sci.* **1989**, *121*(1657), 39–43.

31 (a) Zhang, G., Liu, X., Thomas, J. K. *Radiat. Phys. Chem.* **1998**, *51*, 135–152. (b) Hashimoto, S. *J. Photochem. Photobiol. C: Rev.* **2003**, *4*, 19–49. (c) García, H., Roth, H. D. *Chem. Rev.* **2002**, *102*, 3947–4007. (d) Turro, N. J. *Acc. Chem. Res.* **2000**, *33*, 637–646.

32 Ramamurthy, V., Ed. "Photoprocesses of Organic Molecules Included in Zeolites." In *Photochemistry in Organized and Constrained Media*. VCH: New York, **1991**.

33 (a) Smoot, A. L., Lindquist, D. A. *J. Chem. Educ.* **1997**, *74*, 569–570. (b) Davis, P. J., van Bekkum, H., Coker, E. N. *J. Chem. Educ.* **1999**, *76*, 469–470.

34 Ramamurthy, V., Turro, N. J. *J. Inclusion Phenom. Mol. Recognit. Chem.* **1995**, *21*, 239–282.

35 Barthomeuf, D. *Catal. Rev.* **1996**, *52*, 521–612.

36 Jacobs, P. A., Beyer, H. K. *J. Phys. Chem.* **1979**, *83*, 1174–1177.

37 Loewenstein, W. *Am. Mineral.* **1954**, *39*, 92–96.

38 (a) Davis, M. E., Saldarriaga, C., Montes, C., Garces, J., Crowder, C. *Zeolites* **1988**, *8*, 362–366. (b) Annen, M. J., Young, D., Davis, M. E., Cavin, O. B., Hubbard, C. R. *J. Phys. Chem.* **1991**, *95*, 1380–1383. (c) McCusker, L. B., Baerlocher, Ch., Jahn, E., Bülow, M. *Zeolites* **1991**, *11*, 308–313. (d) Schmidt, W., Schüth, F., Reichert, H., Unger, K., Zibrowius, B. *Zeolites* **1992**, *12*, 2–8. (e) Rocha, J., Kolodziejski, W., He, H., Klinowski, J. *J. Am. Chem. Soc.* **1992**, *114*, 4884–4888. (f) He, H., Klinowski, J. *J. Phys. Chem.* **1994**, *98*, 1192–1197.

39 (a) Chateauneuf, J. E. *Res. Chem. Intermed.* **1994**, *20*, 249–263. (b) Peon, J., Polshakov, D., Kohler, B. *J. Am. Chem. Soc.* **2002**, *124*, 6428–6438.

40 (a) Kirmse, W. *Adv. Carbene Chem.* **2001**, *3*, 1–51. (b) Kirmse, W., Meinert, T. *J. Chem. Soc., Chem. Commun.* **1994**, 1065–1066. (c) Kirmse, W., Knöpfel, N., Loosen, K., Siegfried, R., Wroblowsky, H.-J. *Chem. Ber.* **1981**, *114*, 1187–1191.

41 (a) Kirmse, W. *Adv. Carbene Chem.* **1994**, *1*, 1–57. (b) Bethell, D., Newall, A. R., Stevens, G., Whittaker, D. *J. Chem. Soc., B* **1969**, 749–751. (c) Bethell, D., Newall, A. R., Whittaker, D. *J. Chem. Soc., B* **1971**, 23–31.

42 Moss, R. A. *Acc. Chem. Res.* **1980**, *13*, 58–64.

43 (a) Jones, W. M., LaBar, R. A., Brinker, U. H., Gebert, P. H. *J. Am. Chem. Soc.* **1977**, *99*, 6379–6391 (see footnote 27). (b) Schoeller, W. W., Brinker, U. H. *Z. Naturforsch.*, **1980**, *35B*, 475–476. (c) Moss, R. A. *Acc. Chem. Res.* **1989**, *22*, 15–21. (d) Chateauneuf, J. E., Johnson, R. P., Kirchhoff, M. M. *J. Am. Chem. Soc.* **1990**, *112*, 3217–3218. (e) Mendez, F., Garcia-Garibay, M. A. *J. Org. Chem.* **1999**, *64*, 7061–7066.

44 Mieusset, J.-L., Brinker, U. H. *J. Org. Chem.* **2008**, *73*, 1553–1558.

45 (a) Pascal, R. A., Jr., Mischke, S. *J. Org. Chem.* **1991**, *56*, 6954–6957. (b) Bach, R. D., Su, M.-D., Aldabbagh, E., Andrés, J. L., Schlegel, H. B. *J. Am. Chem. Soc.* **1993**, *115*, 10237–10246.

46 Gutsche, C. D., Bachman, G. L., Udell, W., Bäuerlein, S. *J. Am. Chem. Soc.* **1971**, *93*, 5172–5180.

47 Bauschlicher, C. W., Jr., Haber, K., Schaefer, H. F., III, Bender, C. F. *J. Am. Chem. Soc.* **1977**, *99*, 3610–3614.

48 (a) Korneev, S. M. *Eur. J. Org. Chem.* **2011**, 6153–6175. (b) Liu, M. T. H., Ed. *Chemistry of Diazirines*, Vols. 1 and 2. CRC Press: Boca Raton, FL, **1987**. (c) Schmitz, E. *Adv. Heterocycl. Chem.* **1963**, *2*, 83–130. (d) Schmitz, E. *Adv. Heterocycl. Chem.* **1979**, *24*, 63–107. (e) Schmitz, E. *Angew. Chem. Int. Ed. Engl.* **1964**, *3*, 333–341. (f) Schmitz, E. In *Methoden der Organischen Chemie (Houben–Weyl)*, Vol. E16c, Klamann, D., Ed. Thieme Verlag: Stuttgart, **1992**, pp. 678–728. (g) Church, R. F. R., Kende, A. S., Weiss, M. J. *J. Am. Chem. Soc.* **1965**, *87*, 2665–2671. (h) Jankowski, A., Paulsen, S. R. *Angew. Chem. Int. Ed. Engl.* **1964**, *3*, 229–230. (i) Frey, H. M., *Stevens*, I. D. R. *Proc. Chem. Soc., London* **1962**, 79.

49 Bamford, W. R., Stevens, T. S. *J. Chem. Soc.* **1952**, 4735–4740.

50 (a) Bobek, M. M., Krois, D., Brehmer, T. H., Giester, G., Wiberg, K. B., Brinker, U. H. *J. Org. Chem.* **2003**, *68*, 2129–2134. (b) Isaev, S. D., Kulik, N. I., Yurchenko, A. G. *Theor. Exp. Chem. (Transl. of Teor. Eksp. Khim.)* **1993**, *29*, 213–216. (c) Isaev, S. D., Zhalnina, G. G., Murzinova, Z. N., Lastovenko, S. I., Yurchenko, A. G. *J. Org. Chem. USSR (Transl. of Zh. Org. Khim.)* **1988**, *24*, 126–131.

51 Borrevang, P., Faarup, P., Hjort, J. U.S. Patent 3 538 084, **1970**.

52 (a) Regitz, M., Maas, G. *Diazo Compounds: Properties and Synthesis*. Academic Press: Orlando, FL, **1986**. (b) Patai, S., Ed. *The Chemistry of Diazonium and Diazo Groups*, Vols. 1 and 2. Wiley: New York, **1978**. (c) Baron, W. J., DeCamp, M. R., Hendrick, M. E., Jones, M., Jr., Levin, R. H., Sohn, M. B. "Carbenes from Diazo Compounds." In *Carbenes*, Vol. 1. Jones, M., Jr., Moss, R. A., Eds. Wiley: New York, **1973**.

53 Miller, J. B. *J. Org. Chem.* **1959**, *24*, 560–561.

54 Smith, L. I., Howard, K. L. *Org. Synth. Coll.* **1955**, *3*, 351–352.

55 Overberger, C. G., Anselme, J.-P. *Tetrahedron Lett.* **1963**, 1405–1408.

56 (a) Warmuth, R., Kerdelhué, J.-L., Sánchez Carrera, S., Langenwalter, K. J., Brown, N. *Angew. Chem. Int. Ed.* **2002**, *41*, 96–99. (b) Sánchez Carrera, S., Brown, N., Kerdelhué, J.-L., Langenwalter, K. J., Warmuth, R. *Eur. J. Org. Chem.* **2005**, 2239–2249.

57 Krois, D., Brinker, U. H. *J. Am. Chem. Soc.* **1998**, *120*, 11627–11632.
58 Krois, D., Brinker, U. H. "Circular Dichroism of Cyclodextrin Complexes." In *Cyclodextrins and Their Complexes*. Dodziuk, H., Ed. Wiley-VCH: Weinheim, **2006**.
59 (a) Shustov, G. V., Liu, M. T. H., Houk, K. N. *Can. J. Chem.* **1999**, *77*, 540–549. (b) Shustov, G. V., Liu, M. T. H. *Can. J. Chem.* **1998**, *76*, 851–861.
60 Bonneau, R., Liu, M. T. H. *J. Phys. Chem. A* **2000**, *104*, 4115–4120.
61 Çelebi, S., Leyva, S., Modarelli, D. A., Platz, M. S. *J. Am. Chem. Soc.* **1993**, *115*, 8613–8620.
62 (a) White, W. R., III, Platz, M. S. *J. Org. Chem.* **1992**, *57*, 2841–2846. (b) Modarelli, D. A., Morgan, S., Platz, M. S. *J. Am. Chem. Soc.* **1992**, *114*, 7034–7041. (c) Yamamoto, N., Bernardi, F., Bottoni, A., Olivucci, M., Robb, M. A., Wilsey, S. *J. Am. Chem. Soc.* **1994**, *116*, 2064–2074.
63 Wakahara, T., Niino, Y., Kato, T., Maeda, Y., Akasaka, T., Liu, M. T. H., Kobayashi, K., Nagase, S. *J. Am. Chem. Soc.* **2002**, *124*, 9465–9468.
64 Moss, R. A. *Acc. Chem. Res.* **2006**, *39*, 267–272.
65 Rekharsky, M. V., Inoue, Y. *Chem. Rev.* **1998**, *98*, 1875–1917.
66 Kano, K., Ishimura, T., Negi, S. *J. Inclusion Phenom. Mol. Recognit. Chem.* **1995**, *22*, 285–298.
67 Mieusset, J.-L., Krois, D., Pačar, M., Brecker, L., Giester, G., Brinker, U. H. *Org. Lett.* **2004**, *6*, 1967–1970.
68 (a) Nishijo, J., Nagai, M., Yasuda, M. *Carbohydr. Res.* **1993**, *245*, 43–56. (b) Gelb, R. I., Schwartz, L. M. J. *Inclusion Phenom. Mol. Recognit. Chem.* **1989**, *7*, 537–543. (c) Gelb, R. I., Raso, S., Alper, J. S. *Supramol. Chem.* **1995**, *4*, 279–285.
69 Cramer, F., Henglein F. M. *Chem. Ber.* **1957**, *90*, 2561–2571.
70 Tee, O. S., Gadosy, T. A., Giorgi, J. B. *Can. J. Chem.* **1996**, *74*, 736–744.
71 Isaev, S. D., Yurchenko, A. G., Stepanov, F. N., Kolyada, G. G., Novikov, S. S., Karpenko, N. F. *J. Org. Chem. USSR (Transl. of Zh. Org. Khim.)* **1973**, *9*, 745–748.
72 Sydnes, L. K., Brinker, U. H. In *Methoden der Organischen Chemie (Houben–Weyl)*, Vol. *E19b*, Regitz, M., Ed. Thieme Verlag: Stuttgart, **1989**, pp. 577–601.
73 Bobek, M. M., Giester, G., Kählig, H., Brinker, U. H. *Tetrahedron Lett.* **2000**, *41*, 5663–5667.
74 (a) Schneider, H.-J., Hacket, F., Rüdiger, V., Ikeda, H. *Chem. Rev.* **1998**, *98*, 1755–1785. (b) Rekharsky, M. V., Goldberg, R. N., Schwarz, F. P., Tewari, Y. B., Ross, P. D., Yamashoji, Y., Inoue, Y. *J. Am. Chem. Soc.* **1995**, *117*, 8830–8840.
75 (a) Alderfer, J. L., Eliseev, A. V. *J. Org. Chem.* **1997**, *62*, 8225–8226. (b) Komiyama, M., Hirai, H. *Chem. Lett.* **1980**, 1467–1470. (c) Mentzafos, D., Mavridis, I. M., Le Bas, G., Tsoucaris, G. *Acta Crystallogr., Sect. B* **1991**, *47*, 746–757. (d) Rüdiger, V., Eliseev, A., Simova, S., Schneider, H.-J., Blandamer, M. J., Cullis, P. M., Meyer, A. J. *J. Chem. Soc., Perkin Trans. 2* **1996**, 2119–2123. (e) Hamilton, J. A., Sabesan, M. N. *Acta Crystallogr., Sect. B* **1982**, *38*, 3063–3069.
76 Bobek, M. M., Brinker, U. H. *J. Am. Chem. Soc.* **2000**, *122*, 7430–7431.
77 Knoll, W., Bobek, M. M., Giester, G., Brinker, U. H. *Tetrahedron Lett.* **2001**, *42*, 9161–9165.
78 Sellner, B., Zifferer, G., Kornherr, A., Krois, D., Brinker, U. H. *J. Phys. Chem. B* **2008**, *112*, 710–714.
79 (a) Garces, F. O., Rao, V. P., Garcia-Garibay, M. A., Turro, N. J. *Supramol. Chem.* **1992**, *1*, 65–72. (b) Ripmeester, J. A. *J. Inclusion Phenom.* **1988**, *6*, 31–40. (c) Facey, G., Ripmeester, J. A. *J. Chem. Soc., Chem. Commun.* **1990**, 1585–1587.

80 (a) Isaev, S. D., Karpenko, N. F., Kolyada, G. G., Novikov, S. S., Yurchenko, A. G. *J. Org. Chem. USSR (Transl. of Zh. Org. Khim.)* **1978**, *14*, 708–712. (b) Isaev, S. D., Kolyada, G. G., Novikov, S. S., Yurchenko, A. G. *Bull. Acad. Sci. USSR, Div. Chem. Sci. (Transl. of Izv. Akad. Nauk SSSR, Ser. Khim.)* **1974**, *23*, 927.
81 Zifferer, G., Kornherr, A., Brinker, U. H. *Mol. Simul.* **2008**, *34*, 1177–1183.
82 Bobek, M. M., Krois, D., Brinker, U. H. *Org. Lett.* **2000**, *2*, 1999–2002.
83 Harata, K., Uedaira, H. *Bull. Chem. Soc. Jpn.* **1975**, *48*, 375–378.
84 Tinoco, I., Jr. *Adv. Chem. Phys.* **1962**, *4*, 113–160.
85 (a) Kodaka, M. *J. Phys. Chem. A* **1998**, *102*, 8101–8103. (b) Kodaka, M. *J. Am. Chem. Soc.* **1993**, *115*, 3702–3705.
86 Bradley, G. F., Evans, W. B. L., Stevens, I. D. R. *J. Chem. Soc., Perkin Trans. 2* **1977**, 1214–1220.
87 Sydnes, L. K., Brinker, U. H. In *Methoden der Organischen Chemie (Houben–Weyl)*, Vol. E19b, Regitz, M., Ed. Thieme Verlag: Stuttgart, **1989**, pp. 601–606.
88 (a) Wojnárovits, L., Szondy, T., Szekeres-Bursics, E., Földiák, G. *J. Photochem.* **1982**, *18*, 273–276. (b) Wojnárovits, L. *J. Chem. Soc., Perkin Trans. 2* **1984**, 1449–1451. (c) Cope A. C., Brown, M., Woo, G. L. *J. Am. Chem. Soc.* **1965**, *87*, 3107–3110.
89 The inherent migratory aptitude of a group depends mainly on three factors: (i) the intrinsic migratory aptitude, (ii) the bystander substituent at the migration origin, and (iii) the spectator substituent at the migration terminus. For reviews, see Reference 4, and Nickon, A. *Acc. Chem. Res.* **1993**, *26*, 84–89.
90 Wilt, J. W., Wagner, W. J. *J. Org. Chem.* **1964**, *29*, 2788–2789.
91 Sugawara, T., Tukada, H., Izuoka, A., Murata, S. *J. Am. Chem. Soc.* **1986**, *108*, 4272–4278.
92 (a) Shin, S. H., Keating, A. E., Garcia-Garibay, M. A. *J. Am. Chem. Soc.* **1996**, *118*, 7626–7627. (b) Shin, S. H., Cizmeciyan, D., Keating, A. E., Khan, S. I., Garcia-Garibay, M. A. *J. Am. Chem. Soc.* **1997**, *119*, 1859–1868. (c) Sanrame, C. N., Suhrada, C. P., Dang H., Garcia-Garibay, M. A. *J. Phys. Chem. A* **2003**, *107*, 3287–3294. (d) Garcia-Garibay, M. A. *Acc. Chem. Res.* **2003**, *36*, 491–498.
93 (a) Garcia-Garibay, M. A. *J. Am. Chem. Soc.* **1993**, *115*, 7011–7012. (b) Garcia-Garibay, M. A., Theroff, C., Shin, S. H., Jernelius, J. *Tetrahedron Lett.* **1993**, *52*, 8415–8418.
94 Garcia-Garibay, M. A., Shin, S., Sanrame, C. N. *Tetrahedron* **2000**, *56*, 6729–6737.
95 Garcia-Garibay, M. A., Dang, H. *Org. Biomol. Chem.* **2009**, *7*, 1106–1114.
96 Manet, I., Monti, S., Bortolus, P., Fagnoni, M., Albini, A. *Chem.—Eur. J.* **2005**, *11*, 4274–4282.
97 (a) Wynberg, H. *Chem. Rev.* **1960**, *60*, 169–184. (b) Wynberg, H., Meijer, E. W. *Org. React. (NY)* **1982**, *28*, 1–36. (c) Hassner, A., Stumer, C. *Organic Syntheses Based on Name Reactions and Unnamed Reactions*. Pergamon: Tarrytown, **1994**, p. 314. (d) Smith, M. B., March, J. *Advanced Organic Chemistry*, 5th ed. Wiley: New York, **2001**, pp. 716–717.
98 (a) Ohara, M., Fukuda, J. *Pharmazie* **1978**, *33*, 467. (b) Komiyama, M., Hirai, H. *Bull. Chem. Soc. Jpn.* **1981**, *54*, 2053–2056. (c) Komiyama, M., Hirai, H. *J. Am. Chem. Soc.* **1983**, *105*, 2018–2021.
99 Joshi, G. C., Singh, N., Pande, L. M. *Tetrahedron Lett.* **1972**, 1461–1464.
100 Tabushi, I., Yoshida, Z.-i., Takahashi, N. *J. Am. Chem. Soc.* **1971**, *93*, 1820.
101 (a) Conlin, R. T., Miller, R. D., Michl, J. *J. Am. Chem. Soc.* **1979**, *101*, 7637–7638. (b) Martella, D. J., Jones, M., Jr., Schleyer, P. v. R. *J. Am. Chem. Soc.* **1978**, *100*, 2896–2897. (c) Jones, M., Jr. *Adv. Carbene Chem.* **1998**, *2*, 77–96.

102 Isaev, S. D., Yurchenko, A. G., Stepanov, F. N., Kolyada, G. G., Novikov, S. S. *J. Org. Chem. USSR (Transl. of Zh. Org. Khim.)* **1973**, *9*, 436.

103 (a) Knoll, W., Kaneno, D., Bobek, M. M., Brecker, L., Rosenberg, M. G., Tomoda, S., Brinker, U. H. *J. Org. Chem.* **2012**, *77*, 1340–1360. (b) Knoll, W., Bobek, M. M., Kalchhauser, H., Rosenberg, M. G., Brinker, U. H. *Org. Lett.* **2003**, *5*, 2943–2946.

104 (a) Bonneau, R., Hellrung, B., Liu, M. T. H., Wirz, J. *J. Photochem. Photobiol., A* **1998**, *116*, 9–19. (b) Pezacki, J. P., Warkentin, J., Wood, P. D., Lusztyk, J., Yuzawa, T., Gudmundsdóttir, A. D., Morgan, S., Platz, M. S. *J. Photochem. Photobiol., A* **1998**, *116*, 1–7. (c) Morgan, S., Jackson, J. E., Platz, M. S. *J. Am. Chem. Soc.* **1991**, *113*, 2782–2783.

105 Wentrup, C. *Reactive Molecules*. Wiley: New York, **1984**, pp. 182–183.

106 (a) Pagington, J. S. *Chem. Br.* **1987**, *23*, 455–458. (b) Dodziuk, H. *Chem. Intell.* **1996**, *2*(2), 56. (c) Szejtli, J. *Chem. Intell.* **1999**, *5*(3), 38–45.

107 Walspurger, S., Sommer, J. "Reactions in Zeolites." In *Molecular Encapsulation: Organic Reactions in Constrained Systems*. Brinker, U. H., Mieusset, J.-L., Eds. Wiley: Chichester, **2010**.

108 For related examples, see Reference 103.

109 Mieusset, J.-L., Wagner, G., Su, K.-J., Steurer, M., Pačar, M., Abraham, M., Brinker, U. H. *Eur. J. Org. Chem.* **2009**, 5907–5912.

110 Hembury, G. A., Borovkov, V. V., Inoue, Y. *Chem. Rev.* **2008**, *108*, 1–73.

111 Yang, C., Ke, C., Liang, W., Fukuhara, G., Mori, T., Liu, Y., Inoue, Y. *J. Am. Chem. Soc.* **2011**, *133*, 13786–13789.

112 (a) Yang, C., Nakamura, A., Fukuhara, G., Origane, Y., Mori, T., Wada, T., Inoue, Y. *J. Org. Chem.* **2006**, *71*, 3126–3136. (b) Yang, C., Inoue, Y. "Supramolecular Photochirogenesis." In *Supramolecular Photochemistry: Controlling Photochemical Processes*. Ramamurthy, V., Inoue, Y., Eds. Wiley: Hoboken, **2011**.

113 Immel, S., Lichtenthaler, F. W. *Starch* **1996**, *48*, 225–232.

114 (a) Herges, R. *Angew. Chem. Int. Ed. Engl.* **1994**, *33*, 255–276. (b) Birney, D. M. *J. Am. Chem. Soc.* **2000**, *122*, 10917–10925.

115 (a) Shevlin, P. B., McKee, M. L. *J. Am. Chem. Soc.* **1989**, *111*, 519–524. (b) Moss, R. A., Liu, W., Krogh-Jespersen, K. *J. Phys. Chem.* **1993**, *97*, 13413–13418. (c) Moss, R. A. *Pure Appl. Chem.* **1995**, *67*, 741–747. (d) Huang, H., Platz, M. S. *Tetrahedron Lett.* **1996**, *37*, 8337–8340. (e) Huang, H., Platz, M. S. *J. Am. Chem. Soc.* **1998**, *120*, 5990–5999. (f) Ammann, J. R., Subramanian, R., Sheridan, R. S. *J. Am. Chem. Soc.* **1992**, *114*, 7592–7594. (g) Levashova, T. V., Semeikin, O. V., Balenkova, E. S. *J. Org. Chem. USSR (Transl. of Zh. Org. Khim.)* **1980**, *16*, 53–56. (h) Murray, R. K., Jr., Ford, T. M. *J. Org. Chem.* **1977**, *42*, 1806–1808. (i) Freeman, P. K., Pugh, J. K. *J. Am. Chem. Soc.* **1999**, *121*, 2269–2273. (j) Bergman, R. G., Rajadhyaksha, V. J. *J. Am. Chem. Soc.* **1970**, *92*, 2163–2164. (k) Wills, M. S. B., Danheiser, R. L. *J. Am. Chem. Soc.* **1998**, *120*, 9378–9379.

116 Diazirine RIES that mimics the result of cyclopropylcarbene fragmentation has been postulated. For examples, see the following: (a) References 115d,e. (b) Thamattoor, D. M., Jones, M., Jr., Pan, W., Shevlin, P. B. *Tetrahedron Lett.* **1996**, *37*, 8333–8336.

117 (a) Azbel', B. I., Gol'dshleger, N. F., Isakov, Ya. I., Épel'baum, E. T., Yampol'skii, Yu. Yu., Minachev, Kh. M. *Bull. Acad. Sci. USSR, Div. Chem. Sci. (Transl. of Izv. Akad. Nauk SSSR, Ser. Khim.)* **1986**, *35*, 1122–1125. (b) Azbel', B. I., Gol'dshleger, N. F., Isakov, Ya. I., Épel'baum, E. T., Yampol'skii, Yu. Yu., Khidekel', M. L., Minachev, Kh. M. *Dokl. Chem. (Transl. of Dokl. Akad. Nauk SSSR, Ser. Khim.)* **1984**, *276*, 197–199.

118 (a) Freeman, P. K., George, D. E., Rao, V. N. M. *J. Org. Chem.* **1964**, *29*, 1682–1684. (b) Kirmse, W., Meinert, T., Modarelli, D. A., Platz, M. S. *J. Am. Chem. Soc.* **1993**, *115*, 8918–8927.

119 Friedman, L., Shechter, H. *J. Am. Chem. Soc.* **1961**, *83*, 3159–3160.

120 Wolf, A. D., Jones, M., Jr. *J. Am. Chem. Soc.* **1973**, *95*, 8209–8210.

121 Roach, P., Warmuth, R. *Angew. Chem. Int. Ed.* **2003**, *42*, 3039–3042.

122 (a) Wong, M. W., Wentrup, C. *J. Org. Chem.* **1996**, *61*, 7022–7029. (b) Schreiner, P. R., Karney, W. L., Schleyer, P. v. R., Borden, W. T., Hamilton, T. P., Schaefer, H. F., III. *J. Org. Chem.* **1996**, *61*, 7030–7039. (c) McMahon, R. J., Abelt, C. J., Chapman, O. L., Johnson, J. W., Kreil, C. L., LeRoux, J.-P., Mooring, A. M., West, P. R. *J. Am. Chem. Soc.* **1987**, *109*, 2456–2469. (d) Jones, W. M., Joines, R. C., Myers, J. A., Mitsuhashi, T., Krajca, K. E., Waali, E. E., Davis, T. L., Turner, A. B. *J. Am. Chem. Soc.* **1973**, *95*, 826–835. (e) Geise, C. M., Hadad, C. M. *J. Org. Chem.* **2002**, *67*, 2532–2540.

123 (a) Harris, J. W., Jones, W. M. *J. Am. Chem. Soc.* **1982**, *104*, 7329–7330. (b) Kirmse, W., Loosen, K., Sluma, H.-D. *J. Am. Chem. Soc.* **1981**, *103*, 5935–5937. (c) Kirmse, W., Sluma, H.-D. *J. Org. Chem.* **1988**, *53*, 763–767.

124 Lu, Z., Moss, R. A., Warmuth, R., Krogh-Jespersen, K. *J. Phys. Chem. A* **2011**, *115*, 13799–13803.

125 Lu, Z., Moss, R. A., Sauers, R. R., Warmuth, R. *Org. Lett.* **2009**, *11*, 3866–3869.

126 Fukushima, M., Jones, M., Jr., Brinker, U. H. *Tetrahedron Lett.* **1982**, *23*, 3211–3214.

127 Ruck, R. T., Jones, M., Jr. *Tetrahedron Lett.* **1998**, *39*, 2277–2280.

128 Moss, R. A., Tian, J., Sauers, R. R., Krogh-Jespersen, K. *J. Am. Chem. Soc.* **2007**, *129*, 10019–10028.

129 (a) Moss, R. A. *J. Org. Chem.* **2010**, *75*, 5773–5783. (b) Moss, R. A., Tian, J., Sauers, R. R., Ess, D. H., Houk, K. N., Krogh-Jespersen, K. *J. Am. Chem. Soc.* **2007**, *129*, 5167–5174. (c) Moss, R. A., Wang, L., Odorisio, C. M., Zhang, M., Krogh-Jespersen, K. *J. Phys. Chem. A* **2010**, *114*, 209–217.

130 Moss, R. A., Wang, L., Odorisio, C. M., Krogh-Jespersen, K. *J. Am. Chem. Soc.* **2010**, *132*, 10677–10679.

131 Moss, R. A., Wang, L., Weintraub, E., Krogh-Jespersen, K. *J. Phys. Chem. A* **2008**, *112*, 4651–4659.

132 Wang, L., Moss, R. A., Krogh-Jespersen, K. *J. Phys. Chem. A* **2011**, *115*, 8113–8118.

133 Moya-Barrios, R., Cozens, F. L. *J. Am. Chem. Soc.* **2006**, *128*, 14836–14844.

134 (a) Graham, W. H. J. *Am. Chem. Soc.* **1965**, *87*, 4396–4397. (b) Padwa, A., Pulwer, M. J., Blacklock, T. J. *Org. Synth. Coll.* **1990**, *7*, 203–206. (c) Moss, R. A., Włostowska, J., Guo, W., Fedorynski, M., Springer, J. P., Hirshfield, J. M. *J. Org. Chem.* **1981**, *46*, 5048–5050.

135 Liu, M. T. H., Toriyama, K. *J. Phys. Chem.* **1972**, *76*, 797–801.

136 Yoshinaga, M. Jpn. Patent 05 32 704, 1993, *Chem. Abstr.* **1993**, *119*, 10738.

137 Rosenberg, M. G., Ph.D. Dissertation, The State University of New York at Binghamton, **2002**.

138 Moya-Barrios, R., Cozens, F. L. *Org. Lett.* **2004**, *6*, 881–884.

139 Moss, R. A., Turro, N. J. "Laser Flash Photolytic Studies of Arylhalocarbenes." In *Kinetics and Spectroscopy of Carbenes and Biradicals*. Platz, M. S., Ed. Plenum: New York, **1990**.

140 (a) Günther, E. *Ber. Dtsch. Chem. Ges.* **1888**, *21*, 516–518. (b) Günther, E. *Justus Liebigs Ann. Chem.* **1889**, *252*, 44–72. (c) Hayes, F. N., Rogers, B. S., Ott, D. G. *J. Am. Chem. Soc.* **1955**, *77*, 1850–1852. (d) Nesynov, E. P., Grekov, A. P. *Russ. Chem. Rev.*

(*Transl. of Usp. Khim.*) **1964**, *33*, 508–514.(e) Belen'kii, L. I., Brokhovetskii, D. B., Krayushkin, M. M. *Tetrahedron* **1991**, *47*, 447–456.
141 Stollé, R. *J. Prakt. Chem.* **1906**, *73*, 277–287.
142 Tomioka, H. *Adv. Carbene Chem.* **1998**, *2*, 175–214.
143 Hashimoto, S., Saitoh, M., Taira, N., Schmidt, W., Hirai, K., Tomioka, H. *J. Phys. Chem. B* **2005**, *109*, 20407–20414.
144 Breck, D. W. *Zeolite Molecular Sieves*. Krieger: Malabar, FL, **1984**.
145 (a) Tomioka, H., Hirai, K., Nakayama, T. *J. Am. Chem. Soc.* **1993**, *115*, 1285–1289. (b) Tomioka, H., Hirai, K., Fujii, C. *Acta Chem. Scand.* **1992**, *46*, 680–682.

PART 2

METAL CARBENES

11

MODERN LITHIUM CARBENOID CHEMISTRY

VITO CAPRIATI
Dipartimento di Farmacia, Università degli Studi di Bari "Aldo Moro", Consorzio Interuniversitario Nazionale Metodologie e Processi Innovativi di Sintesi, Bari, Italy

11.1 INTRODUCTION

What is meant by "carbenoid"? This is a tricky question, because of the historical evolution of the term over the years since its origin at the beginning of the 1960s. At that time, G. L. Closs and L. E. Closs first proposed the term "carbenoid molecule" in the course of studies aimed at finding new carbene precursors.[1] Thus, a "carbenoid species" was meant to be a synthetically useful version of a carbene. In 1964, Closs and Moss[2] suggested the appellation (as a noun) of "carbenoid" for those species capable of undergoing electrophilic reactions "qualitatively similar to those of carbenes without necessarily being free divalent carbene species."

In their textbook, *Modern Physical Organic Chemistry*, Anslyn and Dougherty define a carbenoid as a "carbene that is stabilized by complexation to a metal.[3] Such a definition immediately brings to mind the classic Simmons–Smith reagent ICH_2ZnI, generally known as a "zinc carbenoid," which stereospecifically delivers a methylene unit to both carbons of an alkene to form a cyclopropane. By tweaking the electronics of the ligand or metal center, however, a carbene-type ligand can also engage in alkylidene-like reactivity.[4] Lithium carbenoids (*vide infra*) are traditionally referred to as species showing a "chameleon-like" reactivity, that is being able not only to behave as

Contemporary Carbene Chemistry, First Edition. Edited by Robert A. Moss and Michael P. Doyle.

nucleophiles, but also as electrophiles in reactions typical for carbenes like additions to C–C double bonds, 1,2-alkyl(aryl) migrations, and C–H(D) insertions, to cite just a few. It is unfortunate that the term carbenoid is still used for distinct classes of molecules with a different reactivity, namely transition metal-bound carbenes and the aforementioned lithium carbenoids. Thus, to avoid confusion, it would be better in general, and in line with lithium derivatives, that the term "metal carbenoid" be restricted only to those ambiphilic organometallic compounds exhibiting "symptoms" of a split personality.[5]

On this basis, within the present chapter, the reactivity of α-lithiated organohalogens, ethers, and amines (today recognized as being lithium carbenoids), will be addressed with special emphasis on their structure, configurational stability, and applications to synthetic chemistry in light of the recent advances that have appeared in the literature over the past decade.[6] The role played by aggregation and solvation on the structure–reactivity relationship will be highlighted as well.

11.2 STRUCTURAL FEATURES OF LITHIUM CARBENOIDS

From a structural point of view, a lithium carbenoid is understood as a species bearing a lithium and a particular nucleofugal heterosubstituent X (X = Hal, OR, NR_2) (*vide infra*) on the same carbon atom able to stabilize a negative charge (Fig. 11.1).[7]

However, is the presence of a "generic" acceptor substituent at the anionic carbon sufficient to label the corresponding lithiated species as a carbenoid? Pioneering work in this area was performed by both the Seebach and the Boche groups. Seebach and coworkers, in particular, discovered that the ^{13}C chemical shift of the anionic carbon of α-lithiated halogens (Li/Hal) was always strongly deshielded up to 280 ppm, and the coupling constants with ^{1}H and ^{13}C nuclei attached to the carbenoid carbon atom were strongly decreased. At the same time, the ^{13}C–^{6}Li coupling constant, structurally independent, is almost constant and large (~17.0 Hz).[7a–c] Responsible to a large extent for the deshielding

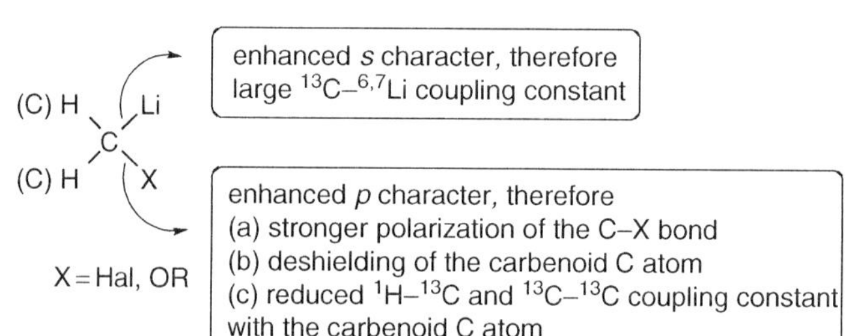

Figure 11.1. Summary of the NMR investigations of Li/Hal and Li/OR carbenoids performed by the Seebach and the Boche groups.

of such a carbon (which also causes a stronger polarization of the C–Hal bond) is most probably the energetically very favorable interaction of the high-lying σ_{C-Li} orbital with the low-lying σ^*_{C-X} orbital.[6b,c,7b,c] The greater the deshielding, the stronger the carbenoid character. Later on, Boche and coworkers also demonstrated the carbenoid nature of α-lithiated ethers (Li/OR).[6b,c,7f] Analogously to Li/Hal carbenoids, the ^{13}C NMR signal of the anionic carbon atom of Li/OR species proved to be downfield shifted compared to that of the corresponding nonlithiated species. This fact parallels the electrophilic reactivity observed for this class of compounds in reaction with nucleophiles.

A common structural feature also found in Li/Hal and Li/OR carbenoids is the *elongation* rather than the *shortening* of the C–X bond upon lithiation,[8] the latter situation being that usually found with "normal" acceptor substituents. This is the case, for example, with α-lithiated sulfones, sulfoximides, sulfoxides, thioethers, dithianes, nitriles, nitro compounds, and hydrazones.[9] The electronic structure of carbenoids has been mainly investigated by X-ray analysis and low-temperature multinuclear magnetic resonance studies, and corroborated by theoretical calculations.[6b,c,7f] Very recently, the carbenoid character of terminal and ring-fused lithiated aziridines, representatives of α-lithiated amines, has also come into play. This aspect will be discussed in Section 11.6.

11.3 LITHIUM HALIDE CARBENOIDS

11.3.1 Thermal and Kinetic Stability

From the data discussed in Section 11.2, it seems as though the electrophilic nature of a carbenoid species is strictly related to both the deshielding of the ^{13}C lithiated carbon atom in the corresponding NMR spectrum, and the C–X bond elongation compared to appropriate nonlithiated reference compounds. This is, however, not always true. As for Li/Hal carbenoids, they are usually known to be kinetically stable in the temperature range below −80 °C. At a higher temperature and in the absence of any trapping reagent, they generally decompose by α-elimination of LiX. In 2007, Le Floch and coworkers reported the case of a carbenoid which surprisingly proved to be stable in the solid state at room temperature and in solution up to 60 °C.[10] Li/Hal carbenoid **2** was synthesized by mild oxidation (hexachloroethane in Et_2O) of the geminal dianion **1** (Scheme 11.1). The most striking point for this species was the shielding of $\delta = 15.2$ ppm detected for the central lithiated ^{13}C atom on going from the hydrogenated species **3** ($\delta = 53.7$ ppm) (obtained upon quenching **2** with water) to **2** ($\delta = 38.5$ ppm). An X-ray analysis together with theoretical calculations revealed that (i) the C–Cl bond length in **2** (178.1(2) pm) is unusually slightly shortened compared to **3** (179.6(4) pm), (ii) the central carbon possesses a lone pair in a pure *p* orbital stabilized by negative hyperconjugation into the phosphorous antibonding orbitals,[11] and (iii) the lithium is efficiently stabilized by two thiophosphinoyl side groups. Most probably, the unusual thermal

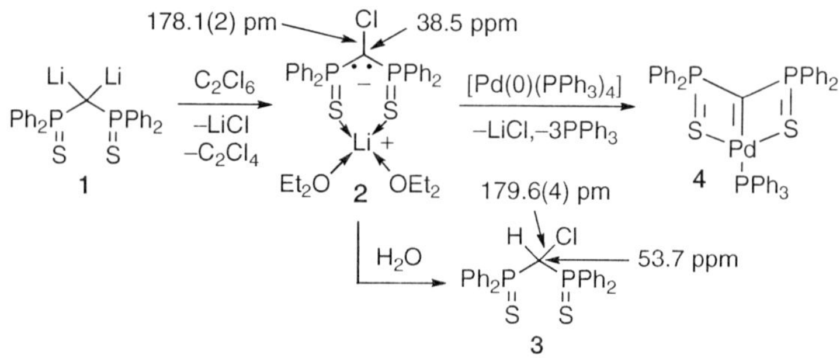

Scheme 11.1.

stability observed for **2** comes just from these last two features that prevent LiCl elimination. The electrophilic nature of **2** was, however, demonstrated showing that such a carbenoid is a convenient source of a carbene ligand. Indeed, upon treatment with electron-rich metal centers such as $[Pd(0)(PPh_3)_4]$, the formation of the carbene complex **4** smoothly took place, likely by an S_N2 attack promoted by the Pd(0) center which, in spite of the strong C–Cl bond, displaces the chloride anion of **2** (Scheme 11.1).

A remarkable thermal stability has also been reported for the ylene (phosphoranylidene) carbenoid **5** (Scheme 11.2) by Niecke and coworkers; in solution, it undergoes LiCl elimination only above −10 °C.[12] In such a case, the carbenoid center is a component of a delocalized four-electron–pπ-framework extending asymmetrically over three atoms because of the presence on the β-carbon of silyl groups more capable of stabilizing the incipient negative charge. This is accompanied by a significantly longer C–Cl bond (181.6(6) pm) with respect to the corresponding hydrogenated species **6** (171.2(8) pm) and a strong deshielding of the lithiated ^{13}C nucleus ($\Delta\delta = 67.8$ **5**/**6**). The carbenoid character of **5** is apparent in the LiCl elimination upon warming that leads to the λ^3-phospirene **7** (Scheme 11.2).

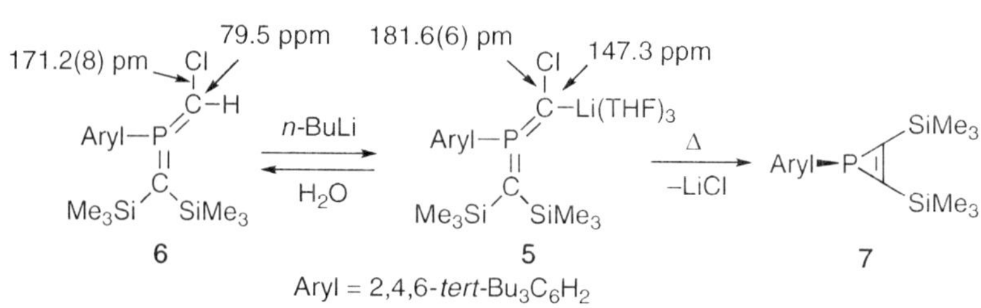

Scheme 11.2.

The lithiated vinyl iodide **9** (*Z*-isomer) is another example of a long-lived carbenoid. Generated by treatment of the corresponding 2-alkoxy-1,

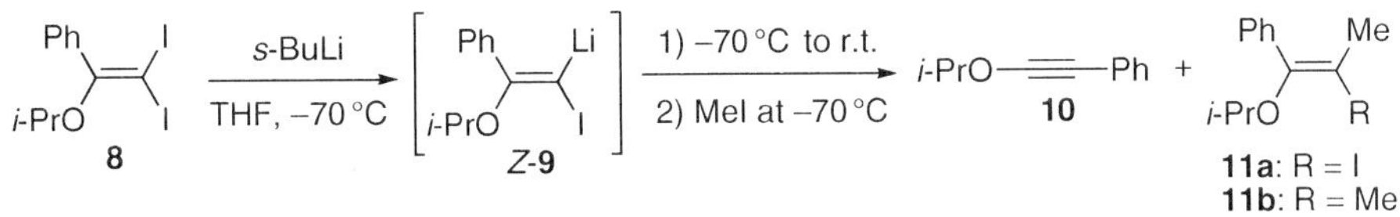

Scheme 11.3.

1-diiodoalkene **8** with *s*-BuLi in tetrahydrofuran (THF), it proved to be stable up to −20 °C (Scheme 11.3).[13a] Apparently, as the temperature is increased, the presence of the isopropyloxy group hampers both the phenyl migration through a Fritsch–Buttenberg–Wiechell (FBW) rearrangement and 1,5-CH insertion into the isopropyl CH_3 groups, typical reactions of alkylidenecarbenoids.[6c,14] Indeed, when **9** was warmed from −70 °C to room temperature with a large excess of *s*-BuLi in THF, quenching with MeI, upon cooling the mixture to −70 °C, afforded only 8% of the expected FBW rearrangement product **10** along with 39% of the two methylated derivatives **11a,b**.[13b] This indicates that more than one-third of carbenoid **9** survives under the warming conditions. Ab initio calculations confirmed that the trisolvated monomeric isomer *Z*-**9** is among the most stable ones, and that the isopropyloxy group does not take part in coordination of lithium.[15]

11.3.2 Configurational Stability and Stereochemistry of the Coupling Reactions with Electrophiles

Stereochemistry has always been a recurring theme when dealing with lithium carbenoids that have a stereogenic lithium-bearing carbon atom. Pioneering work by Köbrich[16] and Seyferth[17] with norcarane systems demonstrated that lithium chloro carbenoid **12a** undergoes epimerization only slowly at −80 °C to give **12b**, and at a rate comparable to its decomposition, whereas lithium bromo carbenoids **13a,b** proved to be completely configurationally stable at −95 °C in THF or THF/ether mixtures (Fig. 11.2). Likewise, Hoffmann and coworkers[18] showed that the lithiated bromocyclopropane **14a** is configurationally stable at −125 °C, although a rapid equilibration toward its epimer **14b** occurs at −78 °C before decomposition sets in at −60 °C (Fig. 11.2).

At the same time, some important observations were made. In the first place, the carbenoids **14a,b**, generated by a bromine–lithium exchange reaction from the corresponding dibromo derivative, equilibrate by a process mediated by the same precursor. Thus, only in the absence of the latter, are **14a,b** configurationally stable at the aforementioned temperature.[19] In the second place, the final diastereomeric ratio significantly depends on the method of generating the carbenoids and their trapping. These results show that carbenoids with a cyclopropane backbone are configurationally stable at low temperatures. One may argue that it is mainly the cyclopropane ring's strain that hampers equilibration. Open-chain Li/Hal carbenoids, however, are also surprisingly configurationally

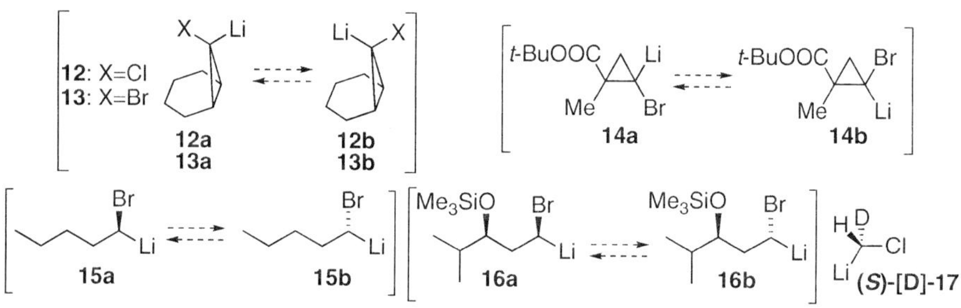

Figure 11.2. Selected examples of Li/Hal carbenoids configurationally stable at low temperatures.

stable. Chiral lithium bromocarbenoids **15a,b**, for example, showed a remarkable configurational stability at −120 °C on a timescale defined by their addition to electrophiles (macroscopic configurational stability). This was ascertained on the basis of the Hoffmann test, which is based on kinetic resolution (Fig. 11.2).[20] Similarly, diastereomeric lithium bromocarbenoids **16a,b** have been found to be configurationally stable at −110 °C but, again, only in the absence of their dibromo precursor (Fig. 11.2).[21] Very recently, Hammerschmidt and coworkers also ascertained that the enantiomerically enriched chloromethyllithium (*S*)-[D]-**17**, generated by tin–lithium exchange, is configurationally stable on both microscopic (*in situ* quenching) and macroscopic timescales up to its decomposition temperature, of −78 °C (Fig. 11.2).[22]

Reaction of *gem*-dihalogenocyclopropanes with alkyllithiums is a known way to obtain substituted allenes (Doering–Skattebøl–Moore reaction).[23] The mechanism was gradually uncovered, and proved to involve a carbene-like intermediate (either a Li/Hal carbenoid or a free cyclopropylidene), formed from an incipient 1-halo-1-lithiocyclopropane by lithium–halogen exchange. Substituent effects on the ring-opening of lithium bromocyclopropylidenoids to allenes have recently been computationally investigated.[24] The efficiency of the process was found to depend on the ease of departure of the bromide anion. Thus, whereas electron-withdrawing groups prevent such a ring-opening, stabilizing the corresponding cyclopropylidenes, electron-donating groups (+M) lower the barrier to allene formation. Interestingly, it was serendipitously found that treatment of *gem*-dihalogenospiropentanes **18** with MeLi at low temperature (−55 °C) favors the formation of unusual rearrangement products at the expense of the allenes **19** (Scheme 11.4).[25] Depending on their structure, spiropentanes **18** can react to generate either cyclobutene derivatives **20** or symmetric "dimeric" structures **21** (Scheme 11.4). Such unprecedented skeletal rearrangements are most likely the result of the carbenoid character of the spiropentylithium intermediates.

Using stereochemistry as a probe, it would be possible to distinguish between the carbenoid and carbene mechanisms because the carbenes are achiral, and their intervention inevitably leads to racemization. In contrast,

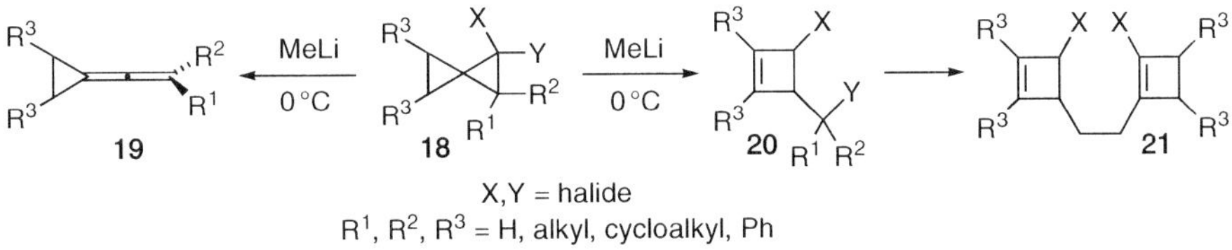

Scheme 11.4.

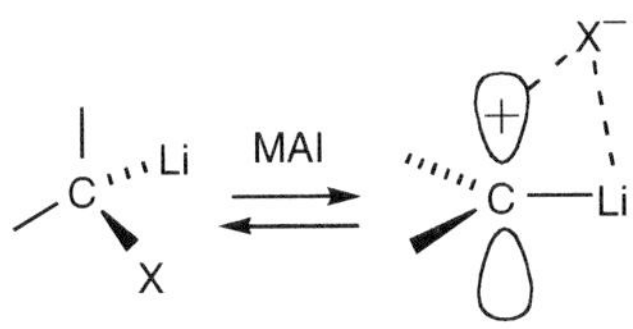

Figure 11.3. Metal-Assisted Ionization in a Li/Hal carbenoid.

carbenoids, according to their substitution, can be chiral. Walborsky and coworkers pioneered the study of the mechanism and the stereochemistry of reactions of chiral lithium carbenoids with nucleophiles.[26] In order to explain the stereochemical results of substitution reactions on chiral cyclopropyl and vinyl lithium carbenoids, they introduced the concept of "Metal-Assisted Ionization" (MAI) (Fig. 11.3). According to this model, the ionization of the carbon–halogen bond, assisted by the metal, allows the formation of a carbon cation–halogen anion tight ion pair which, due to the development of positive charge on the carbenoid carbon, is usually attacked from the backside by the incoming nucleophile with overall inversion of configuration. The rapid solvolysis that Li/Hal carbenoids undergo even at very low temperatures can be interpreted as due to MAI. Retention of configuration, however, was observed in the case of vinyl fluorocarbenoids.

The carbenoid-center stereochemistry is also an important parameter in carbenoid electrophilic reactivity. Early in the 1970s, Taylor and coworkers suggested that the reactivity differences between the two carbenoids α-chlorocyclopropyllithiums **22a,b**, which are epimeric at the carbenoid center only, result from their differing stereochemistry at C-8.[27] This, in turn, led to different electrophilic reaction pathways upon thermolysis: carbenoid **22a**, with exo Cl, cleanly gave a high yield of a product of intramolecular C–H insertion (**23**), whereas **22b**, with endo Cl, gave products of intermolecular reaction (**24** and **25**), provided that bromide ion was absent from the reaction mixture (Scheme 11.5). These experiments definitely ruled out the reversible formation of either a "free carbene" or a "carbene complex with an ionized C–Cl bond" as the reactive intermediate.

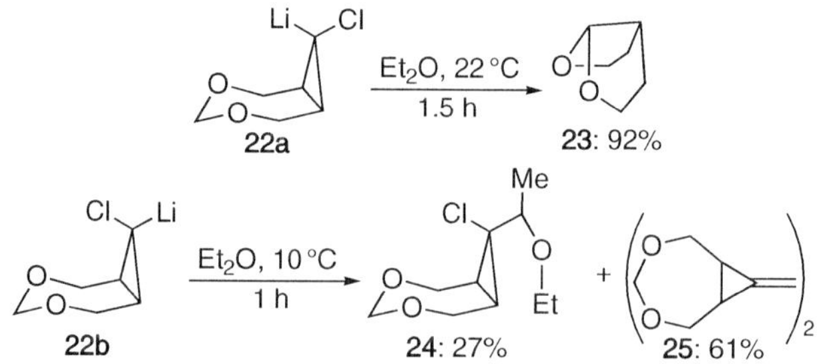

Scheme 11.5.

Using density functional theory (DFT), Zhao and coworkers recently studied the long-standing mechanistic dichotomy for the lithium-carbenoid-promoted cyclopropanation reaction, which has been shown to proceed solely through the methylene-transfer mechanism ("butterfly-type" transition state).[28a] Conversely, the carbometalation mechanism does not appear to make a significant contribution because of its higher activation free energy. This investigation also provided new insight into the intramolecular cyclopropanation mechanisms promoted by the two chiral carbenoids **26** and **27** (Scheme 11.6).[28b] Both Lewis base assistance and the configuration at the carbenoid carbon atom were found to be key factors in influencing the stereochemistry of cyclization. For carbenoid **26**, which is *R*-configured at the lithiated carbon, and Li-bridged, the silyloxy group directs the cyclization to form the bicyclohexane **28** as the sole stereoisomeric product. On the other hand, for epimeric carbenoid **27**, the most stable conformation is not Li-bridged. Due to the *S* configuration of the lithiated carbon, the Lewis base now cannot stereoselectively direct the cyclization reaction, and a mixture of the two bicyclohexanes **28** and **29** is obtained (Scheme 11.6).

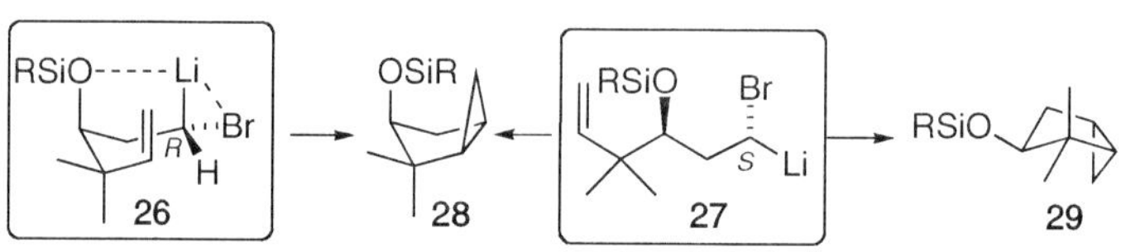

Scheme 11.6.

A seminal guiding contribution from Matteson's laboratory[29] inspired Blakemore and coworkers to design successful stereospecific insertions of enantioenriched chiral Li/Hal carbenoids into organometallic substrates.[30] This strategy allowed the first asymmetric homologation of boronic esters. Central to this synthesis was the preparation of the enantioenriched (*S*)-1-chloro-2-phenylethyllithium **31** from scalemic sulfoxide **30** (diastereomeric ratio (d.r.) > 99:1, enantiomeric excess (e.e.) >98%) by sulfoxide-ligand exchange (SLE).[31] Because of the chemical instability of Li/Cl carbenoid **31**, Barbier-type conditions were adopted for its generation. The addition of *n*-BuLi to a

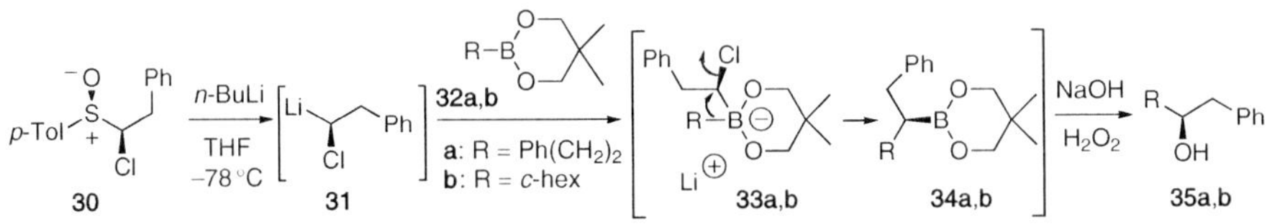

Scheme 11.7.

mixture of sulfoxide **30** and boronates **32a,b** in THF at −78 °C gave, after warming to 0 °C and oxidative workup, stereodefined *sec*-alcohols **35a,b** in very good yields (70–86%) and remarkable e.e. (87–96%), closely approaching that of the starting sulfoxide **30** (Scheme 11.7). The stereochemical outcome of these insertions and the excellent stereochemical fidelity are consistent with (i) a stereospecific SLE; (ii) configurational stability of the putative carbenoid **31**, under the conditions of its generation; and (iii) stereoretentive electrophilic substitution (S_E2) between boronates **32a,b** and the carbenoid **31**. A stereospecific 1,2-metalate rearrangement from the resulting ate-complexes **33a,b**, with concomitant inversion of configuration at the migratory terminus, finally furnishes the chiral boronates **34a,b**, which easily undergo oxidation to give chiral alcohols **35a,b** (Scheme 11.7).

The more robust pinacol boronate **36** could also be successfully chain extended, with excellent stereochemical fidelity, by two treatments with the enantioenriched chiral Li-carbenoid **31**, alternating with a homologation step with chloromethyllithium (generated via I/Li exchange from ICH_2Cl). Upon oxidation, this gave carbinol **37** (Scheme 11.8).[32a] A combination of *t*-BuLi with toluene as the solvent was optimal for effecting SLE. Iterative application of this so-called "stereospecific reagent-controlled homologation" (StReCH) process would, in principle, allow for the programmed assembly of stereodefined polysubstituted alkyl moieties with total control over stereochemistry and constitution.

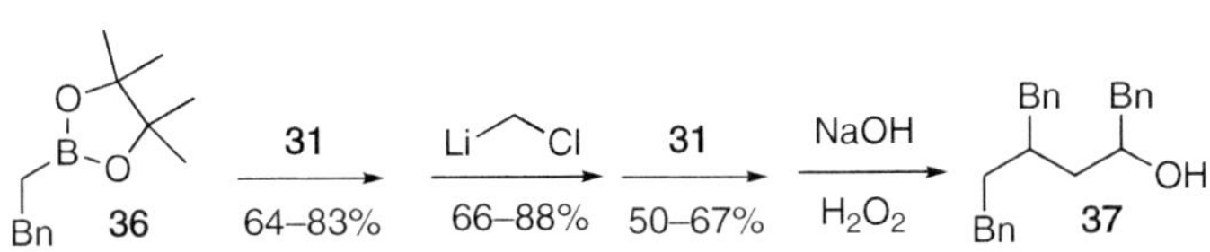

Scheme 11.8.

The StReCH strategy could also be iterated to access cyclic compounds, providing that the carbenoid building blocks bear functionalized substituents. For example, functionalized scalemic α-chloroalkyllithium **39**, generated by a PhLi-initiated SLE, has been employed by Blakemore and coworkers for sequential double StReCH of the commercially available pinacol boronate **38**. This process afforded an acyclic precursor (**40**) containing preset stereogenic centers amenable to subsequent stereospecific annulative transformations *en route* to the analgesic alkaloid (−)-epibatidine (**41**) (Scheme 11.9).[32b]

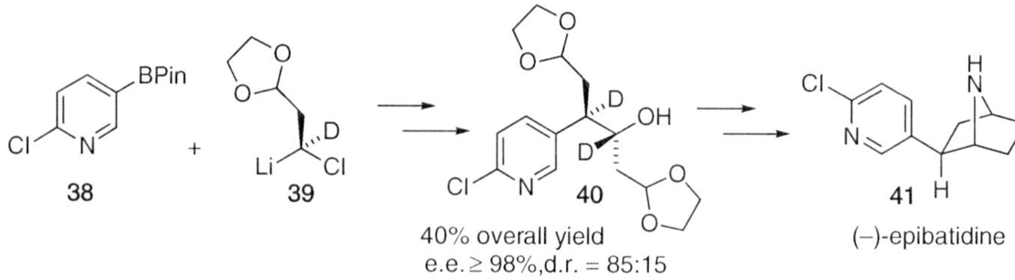

Scheme 11.9.

Polyfunctional organometallic reagents are valuable building blocks in organic synthesis for further elaborative transformations.[33] In particular, straightforward introduction of two metals into an organic molecule using an interelement compound of the kind M^1–M^2 is of great significance.[34] Shimizu, Hiyama, and coworkers have recently reported that the additions of bis(pinacolato) diboron (B–B) (**44**) and (dimethylphenylsilyl)(pinacolato)borane (B–Si) (**45**) to alkylidene-type carbenoids **43**, available from 1,1-dihaloalkenes or 1-haloalkenes **42**, stereoselectively affords, in one pot, the corresponding geminal difunctionalized derivatives **47**. These can then be easily converted into tri- or tetra-substituted alkenes **48** through various transition metal-catalyzed carbon–carbon bond transformations mediated by boron or silicon (Scheme 11.10).[35] The mechanism involves stereoselective formation of an intermediate borate complex **46**, followed by stereospecific 1,2-metalate rearrangement.

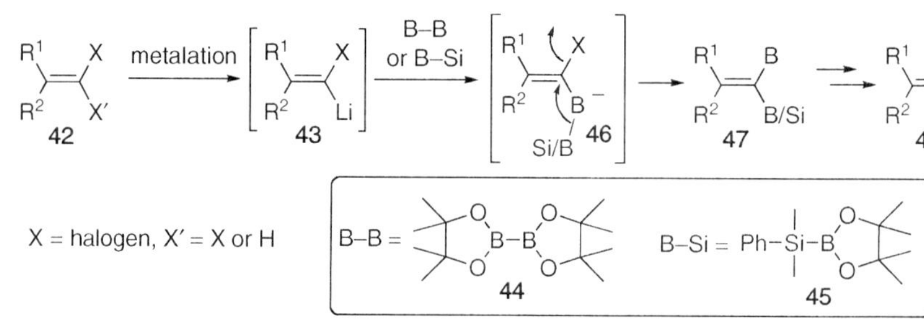

Scheme 11.10.

Interestingly, a palladium-catalyzed cross-coupling reaction of 2-aryl-1, 1-diboryl-1-alkenes **50** (easily prepared by *gem*-diborylation of 2-aryl-1,1-dibromo-1-alkenes **49** with bis(pinacolato)diboron **44**) with aryl iodides affords (*E*)-alkenylboronates **51** as single diastereomers. A further coupling reaction of **51** with another aryl iodide leads to stereodefined triarylated alkenes **52** as final products (Scheme 11.11).[36] Of note, both diastereomers of **52** can be prepared simply by changing the order of the aryl iodides employed.

Analogously, the synthesis of 1,1-diborylated cyclopropanes **54** could be accomplished by *gem*-diborylation of cyclopropylidene lithium carbenoids **53** with bis(pinacolato)diboron **44**. Cyclopropanes **54** proved to be useful precursors of

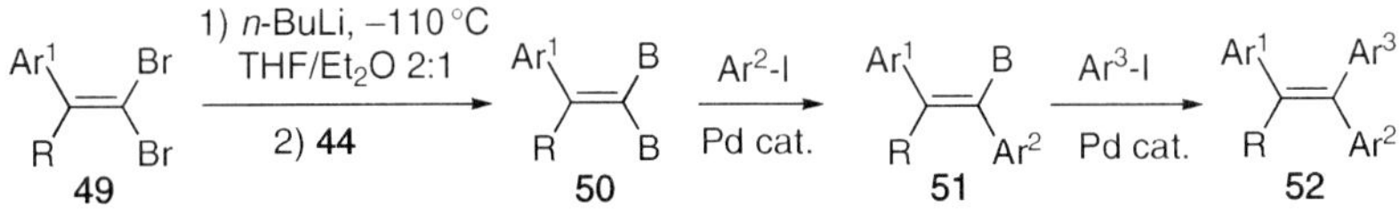

Scheme 11.11.

vic-diborylated methylenecyclopentenes **56**, obtained by a two-step approach that exploits (i) an allenylidene insertion into a carbon–boron bond of **54** and (ii) a ring-enlargement of the resulting allenylcyclopropanes **55**. Two different aryl groups could finally be installed stepwise on **56** via Pd-catalyzed cross-coupling reactions to give diarylated derivatives **57** stereoselectively (Scheme 11.12).[37]

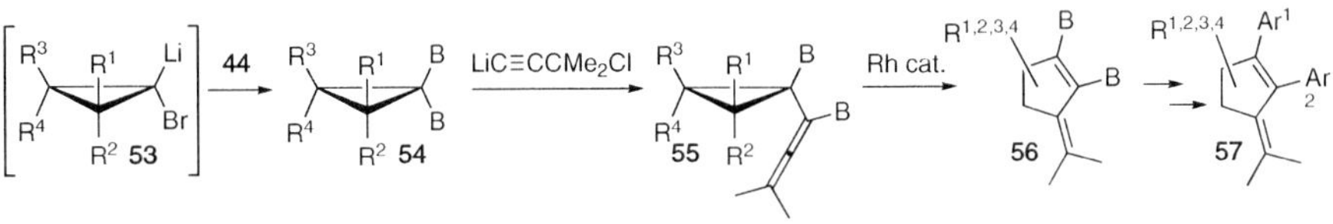

Scheme 11.12.

Treatment of 1,1-bis(pinacolatoboryl)ethene **58** with excess 1-bromo-1-lithioethene **59** furnished in high-yield 2,3-bis(pinacolatoboryl)-1,3-butadiene **61** via a 1,2-migration of the 1-borylethenyl group at the expense of the intermediate borate **60**. Palladium-catalyzed cross-coupling of the resulting diborylbutadiene **61** with electrophiles (e.g., aryl iodides, benzyl chloride, or alkenyl halides) gave 2,3-difunctionalized 1,3-butadienes **62**, which serve as useful intermediates for biologically active compounds and as functionalized monomers for polybutadienes (Scheme 11.13).[38] This strategy could be successfully iterated for the assembly of symmetrical and unsymmetrical dendralenes.

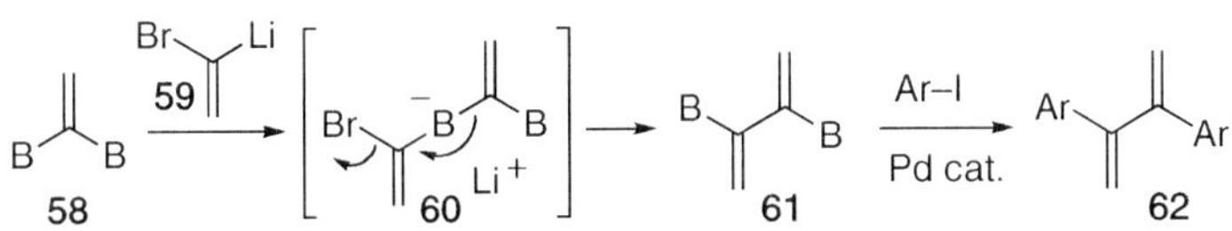

Scheme 11.13.

Li/Hal carbenoids can also act as viable partners in novel and selective carbon–carbon bond-forming reactions that proceed through a carbenoid-chain mechanism. Knorr and coworkers investigated the reactions of unactivated, terminal 2-(halomethylidene)indanes **63** with organolithium compounds (RLi) that lead to the corresponding vinylic substitution products **64** (Scheme 11.14).[39] It was clearly shown that despite the deceptively simple relationship between **63** and **64**, a direct nucleophilic attack of RLi at Cα of **63** (via the so-called addition–rotation–elimination mechanism)[40] does not occur; carbenoid chain processes are involved instead. The evidence supports the initial

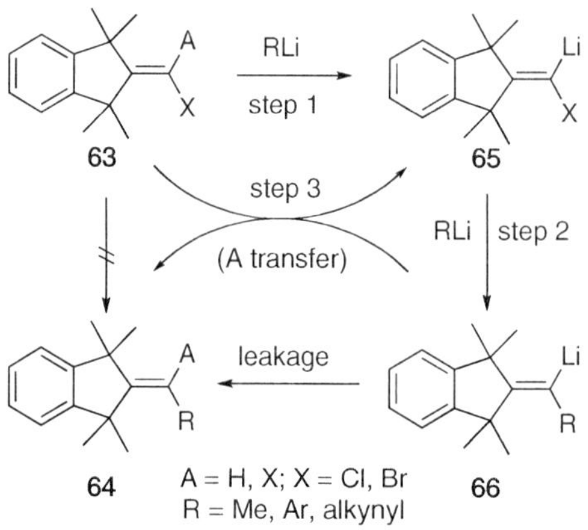

Scheme 11.14.

generation of Li/Hal carbenoid **65** (either by deprotonation (A = H) or by Hal/Li exchange (A = X) with RLi), followed by a vinylic substitution of **65** by RLi to give alkenyllithium **66**. The rate-limiting transfer of A from substrate **63** to **66** finally closes the chain cycle with both formation of product **64** and regeneration of Li/Hal carbenoid **65** (Scheme 11.14). For the carbenoid chain process to be effective, it is required that (i) step 2 occurs significantly faster than both the initiating step and carbenoid decomposition, and that (ii) step 3 is faster than both step 1 and the "leakage" reaction.

Different types of carbenoid chain processes have been reported by Harada and coworkers for (i) the cycloisomerization reaction involving ω-iodo and 1,ω-diiodo-1-alkynes in the presence of catalytic amount of lithium diisopropylamide (LDA) (or with alkynyllithium compounds), and for (ii) the alkylative carbocyclization reaction of iodoalkynes with organolithium compounds. In the former variant (Scheme 11.15, variant a), deprotonation or Li/Hal exchange at the expense of alkyne **67** gives lithium acetylide **68**, which then undergoes smooth *exo* cyclization at the β position of the acetylide moiety to generate the I,Li-alkylidene carbenoid **69**. The chain cycle is finally closed by protonation or iodination of carbenoid **69** by **67** to produce product **70** with simultaneous regeneration of acetylide **68**.[41] In the second variant (Scheme 11.15, variant b), an I/Li exchange reaction of iodoalkyne **71** with RLi furnishes acetylide **72** with a leaving group X. The subsequent reaction is proposed to involve *exo* cyclization of **72** followed by vinylic substitution of the resulting cycloalkylidene carbenoid **73** with RLi to give the alkenyllithium intermediate **74**. Finally, iodination of **74** gives product **75** with concomitant regeneration of the acetylide intermediate **72**.[42]

A novel alkylative elimination of lactone-derived *gem*-dichloroalkenes, which provides ready access to a variety of nonterminal alkynes, has recently been reported by Fürstner and coworkers *en route* to the first stereoselective synthesis of tulearin C.[43] An interesting solvent-tuned dicothomous behavior of dichloroolefins

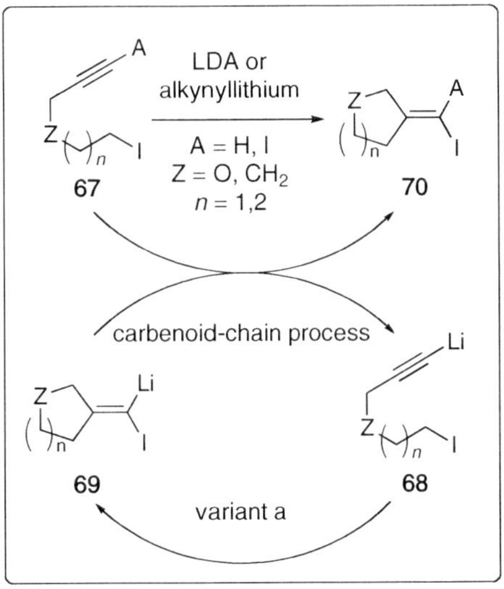

Scheme 11.15.

has been observed in the case of substrates derived from five-membered lactones. For example, alkyne **78** forms as the major product upon exposure of *gem*-dichloroalkene **76** to excess MeLi (5 equiv) in Et_2O at ambient temperature, whereas trisubstituted allene **80** prevails from the same reaction in THF (Scheme 11.16). It is assumed that a preliminary Li/Cl exchange, very likely followed by a vinylic nucleophilic substitution promoted by MeLi, and then by a ring-opening rearrangement of the putative alkenyllithium **77** to **78**, occurs faster in Et_2O than in THF. In THF, however, allylic deprotonation followed by a geminal disubstitution of chlorines in **79** seems to be favored, affording allene **80**.[43]

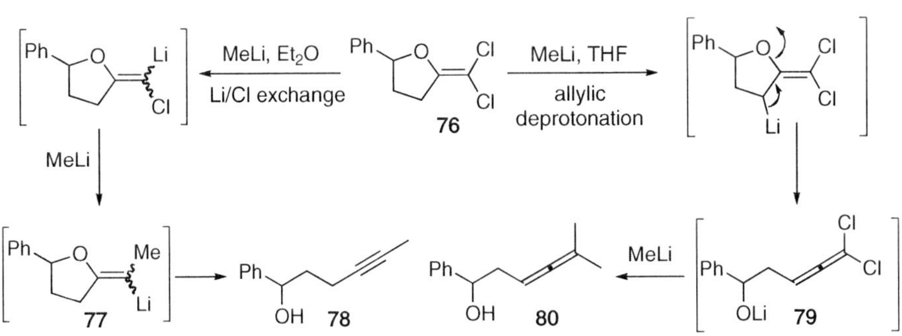

Scheme 11.16.

Whitby and coworkers have developed the use of carbenoid insertions in the elaboration of zirconacycles and investigated the relative stereochemistry.[44] Of note, the results obtained in the reactions of β-monosubstituted 1-chloro-1-lithioalkenes **81** (R^1 or R^2 = H) with organozirconocene chlorides **83** (X = Cl)

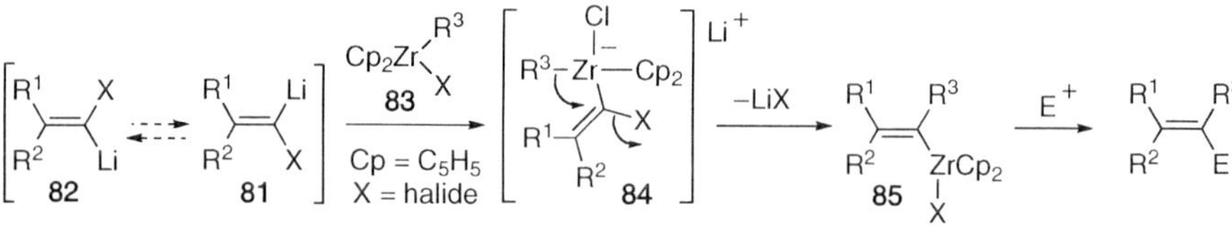

Scheme 11.17.

are consistent with a clean inversion of configuration at the carbenoid sp^2 center, strongly supporting a concerted stereospecific 1,2-metalate rearrangement of the putative intermediate ate complex **84**. The latter furnishes new, stereodefined organozirconium species **85**, which may be further elaborated (Scheme 11.17). On the other hand, poor stereocontrol results for the insertion of β,β-disubstituted 1-lithio-1-haloalkenes **81** (R^1, R^2 = alkyl). This loss of stereochemical integrity has been ascribed to an isomerization of the lithiated carbenoid **81** to **82** (most likely via an MAI mechanism) that occurs prior to 1,2-metalate rearrangement at the zirconium.[45]

An efficient multicomponent assembly of bicyclo[3.3.0]oct-2-enes and bicyclo[3.3.0]octanes, based on the tandem insertion of dihalocarbenoids and acetylides into zirconacycles, was serendipitously discovered by Whitby and coworkers.[46] Addition of lithium 2,2,6,6-tetramethylpiperidide (LTMP) (1.1 equiv) to a mixture of zirconacycle **86** and dichloromethane **87** (R^1 = H) (1.1 equiv) at −78 °C, followed by addition of lithium phenylacetylide **88** (R^2 = Ph) (2.2 equiv), afforded bicyclic compound **89** (R^1 = H, R^2 = Ph) in 72% yield upon protic workup (Scheme 11.18). This reaction could also be successfully extended to other substituted dihalocarbenoids and alkyl-substituted alkynes to give the expected carbocyclic adducts in reasonable yields (47–86%). Full details of this unusual reaction and related chemistry have only been recently described.[47] The proposed mechanism envisages, as a crucial step, a novel rearrangement that most likely involves addition of the organolithium species **90** to the β-position of the zirconium-alkyne complex **91** to give the alkylidene–zirconate species **92** that, after hydrolysis, furnishes **89** (Scheme 11.18).

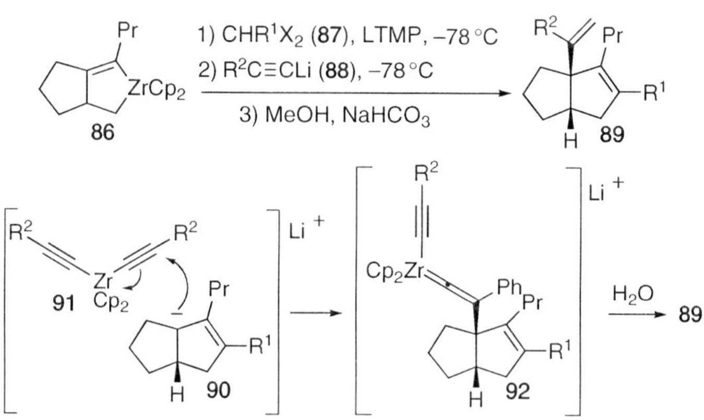

Scheme 11.18.

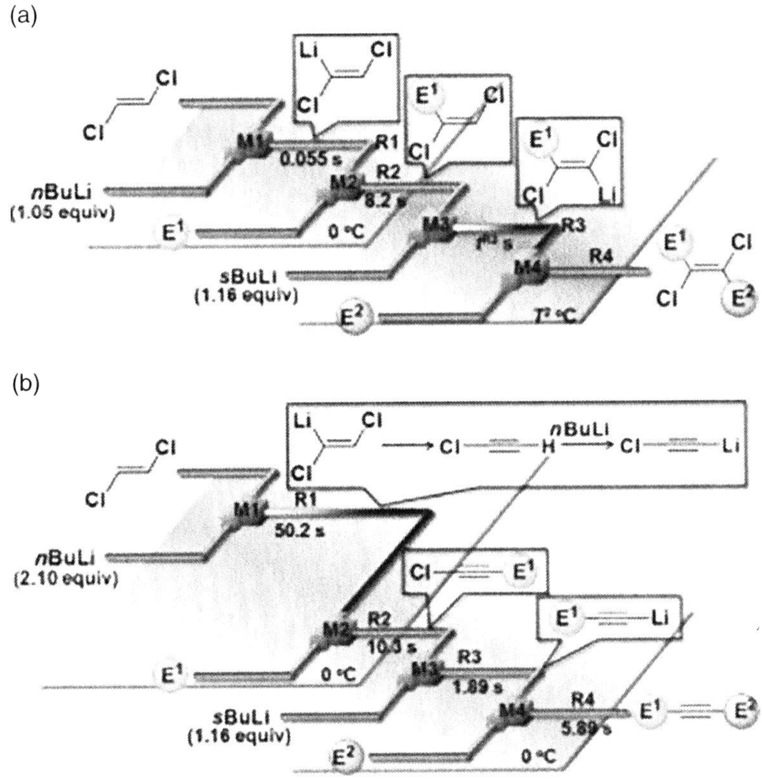

Figure 11.4. An integrated flow microreactor system for the selective synthesis either of disubstituted dichloroalkenes (a) or of nonsymmetrically disubstituted alkynes (b). (Reproduced with permission from Wiley-VCH, Weinheim, from Reference 49.)

As is evident from the examples described thus far, 2-halovinyllithium compounds are especially attractive intermediates in organic synthesis. One major drawback to their employment, however, is the competing β-elimination which usually occurs rapidly at temperatures higher than −78 °C. "Flash chemistry"[48] offers avenues to circumvent this problem. This is an exiciting new concept introduced by Yoshida: a continuous flow microreactors synthesis based on high-resolution control of the residence time, which is the length of time that the solution remains inside the reactor. This feature is quite effective, in particular, for conducting reactions involving short-lived highly reactive intermediates that can be generated and transferred to another location to be used in the next reaction before they decompose. With regard to halocarbenoids, Yoshida and coworkers have recently demonstrated that precise control of the residence time enables the reaction pathway of a 1,2-dichlorovinyllithium intermediate to be switched for the (stereo)selective synthesis either of disubstituted 1,2-dichloroethenes (Fig. 11.4.a) or of nonsymmetrically disubstituted alkynes employing temperatures up to 0 °C (Fig. 11.4.b).[49]

11.4 STRUCTURE–REACTIVITY RELATIONSHIPS

There is a prevalent misconception that organolithium compounds consist of a carbanion and a passive lithium cation. This is not an accurate description; in fact, the lithium cation has a central role in the chemistry of these complexes.[50] An important—but by no means unique—feature of organolithiums is the tendency they have to cluster in the solid state, in solution and even in the gas phase,[51] forming structures that can be diabolically complex, with the metal associated with more than one carbanion center. Lithium carbenoids are no exception. They too are variously aggregated in solution and understanding their reactivity is strongly connected to understanding their structure.[50] As to the latter point, Bauer, Winchester, and Schleyer, upon comparing the ^{1}J(C–Li) constants for a series of alkyllithiums of various oligomeric forms, realized that within relatively narrow limits the $^{13}C,^{6,7}Li$ coupling constants mainly depended on the state of aggregation, irrespective of the structure and hybridization of the carbanions.[52] Indirectly, this also suggests the C–Li bond to be largely ionic (80–90%), but to have a small nonneglectable covalent character. When the ^{6}Li isotope (nuclear spin quantum number $I = 1$, natural abundance $^{7}Li/^{6}Li \approx 92.5/7.5\%$) is chosen for the preparation of organolithiums, the empirically deduced rule is: $^{1}J_{13C-6Li} = 1/n\ (17 \pm 2)$ Hz, where n equals the number of lithium atoms *directly* bonded to carbon on the NMR timescale. This inverse dependence of ^{1}J(C–Li) on n was also corroborated by computational studies,[53] which simulated experimental data with high fidelity provided that the *aggregation* and/or the *solvation* of the organolithium species was taken into account. In particular, knowledge of the influence of solvation on both the reactant and transition structure stabilities is crucial as it helps in understanding solvent effects on reaction rates.[54] Later, an inverse dependence of ^{1}J(C–Li) from microsolvation of organolithium compounds was experimentally verified and quantified by Knorr and coworkers, who developed a novel expression for such a coupling constant: $^{1}J_{13C-6Li} = L[n(a+d)]^{-1}$, where $(a+d)$ is the total number of atoms in the coordination sphere of the lithium (anions and donor ligands), and L is the sensitivity factor.[55] The aforementioned empirical expression, at the same time, endows ^{1}J(C–Li) with a new dimension of analytical utility because it provides a new tool for assessing unknown microsolvation numbers of organolithiums provided that both their aggregation states and the sensitivity factors L are known.

Mixed (hetero) aggregates between two different lithium compounds are also quite common. They often exhibit a different structure and reactivity with respect to homo-aggregates of the parent compounds and can also have significant effects on the product distribution.[56] Particularly intriguing is the co-complexation provided by a lithium halide (e.g., lithium chloride) formed as a by-product, present as a contaminant, or deliberately added to the reaction mixture, in fine-tuning both the reactivity and selectivity of organolithiums.[57] DFT calculations have often been employed to examine structure, bonding, and solvation of lithium carbenoids and their hetero aggregates.[58] It has been suggested that mixed aggregates formation with halomethyllithium carbenoids may affect the activation barrier of carbenoid reactions, causing a change in the mechanism during the course of the same reaction.[58d] A systematic investigation of the performance of several DFT functionals using the reactions of aggregated halomethyllithium carbenoids in both the gas phase (or nonpolar solvents) and THF solvent has recently been carried out by Pratt and coworkers.[59]

For a certain lithium carbenoid, the presence of a heteroatom X (e.g., halogen, oxygen, nitrogen) in its organic framework generally, but not always,[60] allows it to form "bridged structures" with the heteroatom bridging the carbon–lithium bond. The point is that additional intramolecular coordination (particularly in the case of dimeric or higher aggregates), lowers the symmetry, and might induce geometrical distortion of the structure, so that the corresponding architecture differs from the most common type. In this context, NMR spectroscopic data suggestive of a certain "aggregation state" should always be treated with caution. The fact that the ^{13}C signal of a carbenoid is split by ^{6}Li, for example, into a triplet, does not necessarily mean that a monomeric species is present in solution; rather that there is only one ^{6}Li attached to each carbon atom. For example, with reference to $LiCH_2F$, calculations (MP2/6-31G(d)//3-21G) predict that the dimer **94** is 50 kcal/mol more stable than two bridged monomers **93**, the C–F bond being also much less elongated (156.6 pm) than that of the monomer (159.2 pm) (Fig. 11.5). In addition, two different and more elongated Li–C bonds were also found, displaying a length >250 pm, much too long for a C–Li bond.[7f] The features of this aggregate would also be in agreement with the triplet resonance of the carbenoid ^{13}C atom spectroscopically observed by Seebach and coworkers for $LiCH_2F$.[7d] As for the solvated structure **95**, parallel to the lengthening of the Li–F bonds, the corresponding C–F bonds are shortened when compared to the unsolvated species **94** (Fig. 11.5). Therefore, if C–X bond elongation is a significant criterion for "weighing" carbenoid character, then the more solvated a Li carbenoid, the more stable it becomes. Consider the following example. Li/Hal carbenoid **96** is a rare example of a carbenoid with a tetrahedral carbon atom characterized by X-ray crystallography by Boche and coworkers (Fig. 11.5).[61] C–Cl bond lengths are in nice agreement with the carbenoid nature of such a species in that both C–Cl bonds are ~10 pm longer than those in dichloromethane. In addition, the sum of the internal bond angles testify to a higher p-character in the C–H and C–Cl bonds resulting from the higher s-character in the C–Li bond.

To obtain information on the significance of solvation on the thermal stability and the properties of **96**, calculations were also performed on both the nonsolvated and tris-solvated species. For the former, bridged structure **97** was the most stable form. The Li-bridged C–Cl bond, in particular, proved to be significantly elongated with respect to the other C–Cl bond.

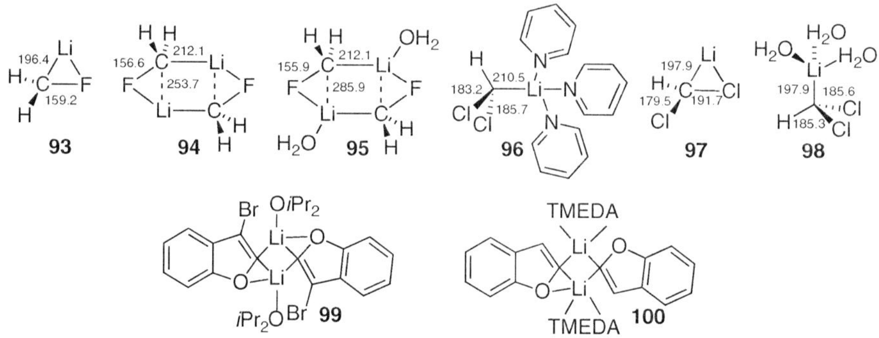

Figure 11.5. Bond lengths (pm), geometries, and essential features of representative Li/Hal and Li/OR carbenoids.

This leads to high reactivity, as experimentally observed in poorly coordinating solvents such as Et_2O. On the other hand, the tris-solvated species **98** is a better model for the solid state structure **96** because, upon solvation, the Li-bridged bond is broken, and both C–Cl bonds correspond closely to the values determined in **96** (Fig. 11.5).

Finally, with reference to the aforementioned "symmetry reduction" typical of a lithium carbenoid, the following examples related to Li/OR carbenoids are particularly enlightening. In the diisopropyl ether-complexed **99**, both carbenoid C–O bonds are bridged by lithium. Conversely, in the TMEDA-complexed **100**, only one C–O bond is Li-bridged (Fig. 11.5).[62] Such Li-bridging is also remarkable from a structural point of view as Li adopts an unusual pentacoordinate structure, whereas three and four are the usual coordination numbers of such a cation. In summary, the interplay of aggregation and solvation,[63] influencing the various bond lengths as well as the symmetry and geometry, have a strong effect in determining the so-called "carbenoid character" of a lithium carbenoid, and thus also its stability and reactivity (*vide infra*).

11.5 LITHIUM–OXYGEN CARBENOIDS

A perusal of the literature reveals that even in the first preparation of an α-lithiated ether described by Lüttringhaus in 1938,[64] the typical reaction exhibited by Li/Hal carbenoids, namely the substitution reaction with nucleophiles such as RLi, also occurs with α-lithiated ethers. Since then, many other examples indicate the generality of this displacement reaction: C–H insertion, cyclopropane formation with olefins, and the FBW rearrangement in the case of lithiated vinylic ethers.[6b,7f] This shows quite clearly that α-lithiated ethers are carbenoids. The comparatively facile breaking of the C–O bond parallels the remarkable elongation (7–10 pm) of this bond as found in the reported solid state structures.[62] ^{13}C NMR investigations corroborate this finding. Similarly to

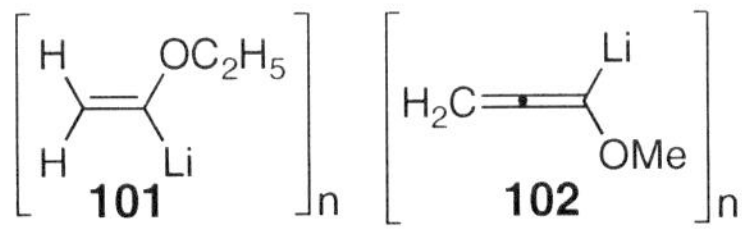

Figure 11.6. Examples of α-lithiated alkenyl ethers: α-ethoxyvinyllithium **101** and 1-methoxyallenyllithium **102**.

Li/Hal carbenoids, the ^{13}C signal of the carbenoid atom is markedly shifted to lower field compared to that of nonlithiated compounds, and quantum chemical calculations also nicely agree with the experimentally determined ^{13}C chemical shifts.[6b,7f]

Crystal data for Li/oxygen carbenoids often agree very well with data in solution, leading to the conclusion that bridged dimers are the preferred species. Exceptions, however, are known. α-Ethoxyvinyllithium **101** (Fig. 11.6), for example, a widely used synthetic reagent serving as an acyl anion equivalent, forms an unexpected self-assembled polymeric chain in the solid state with (distorted) tetrameric subunits linked by Li–C π interactions, and with two different lithium environments. This structure, however, is not retained in THF solution, where the seven-line ^{13}C multiplet observed due to ^{13}C–^{6}Li scalar coupling ($^{1}J(^{13}C–^{6}Li) = 5.0\,Hz$), is consistent with the α-lithiated carbon coupled with three equivalent lithium cations, as in the case of an isolated static tetramer.[65a] The carbenoid character of **101** is shown in an elongated Cα–O bond and lithium bridging.

A combined computational and ^{13}C NMR investigation was also recently used to determine the solution structures of a lithiated allenyl ether such as 1-methoxyallenyllithium **102** (Fig. 11.6) in both an apolar solvent (hexane) and THF. In hexane, the tetramer and the hexamer appear to be the major species. Conversely, the calculated and observed chemical shifts in THF were consistent with a dimer–tetramer equilibrium, with the former aggregate favored by lower temperatures and the latter favored by higher temperatures.[65b]

In general, the reactivity of a carbene complex (as an electrophile or a nucleophile) is known to be strongly influenced by both the nature of the metal and the electronic properties of the carbene substituents,[66] whereas the *reactivity of functionalized lithium carbenoids is determined by their structural features and can in principle be tuned by experimental conditions*. Lithiated epoxides, for example, are typical Li/oxygen carbenoids whose successful employment as synthons for asymmetric reactions has grown significantly over the last 10 years. They enjoy a very rich chemistry, substantially resulting from their peculiar "chameleon-like" character.[67] As nucleophiles, they can be captured by electrophiles to give more functionalized oxiranes and products derived from them. As electrophiles (typically in reactions with organolithiums and hindered lithium amides), they undergo 1,2-alkyl shifts to give enolates, eliminative dimerization, β-elimination, reductive alkylation, transannular C–H insertion (in particular in the case of medium-sized oxiranes), and intramolecular cyclopropanation onto a double bond. With reference to this last reaction, Hodgson and coworkers have recently reported an elegant and straightforward protocol for the stereocontrolled synthesis of (−)-cubelol (**105**) from (−)-menthone, exploiting an efficient intramolecular cyclopropanation of unsaturated terminal epoxide **103** which, upon lithiation with LTMP, stereospecifically gives tricyclic alcohol **104** (Scheme 11.19).[68] Despite stereochemical variation in the tethered alkene, the facial selectivity of cyclopropanation is controlled only by the epoxide stereochemistry. Similarly, the naturally occurring (−)-10-*epi*-cubelol (**106**) could also be prepared starting from (+)-menthone.

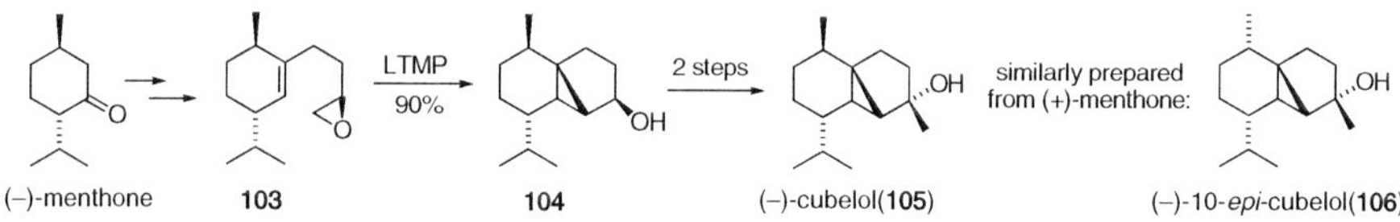

Scheme 11.19.

An unprecedented reaction was found to take place between hindered lithium amides and terminal epoxides, whereby the intermediate enamine **107**, retaining its ability to act as C-nucleophile, reacts with activated and simple alkyl halides providing α-alkylated aldehydes **108** as the final products (Scheme 11.20).[69] The scope of the reaction has been investigated and full mechanistic details were given.

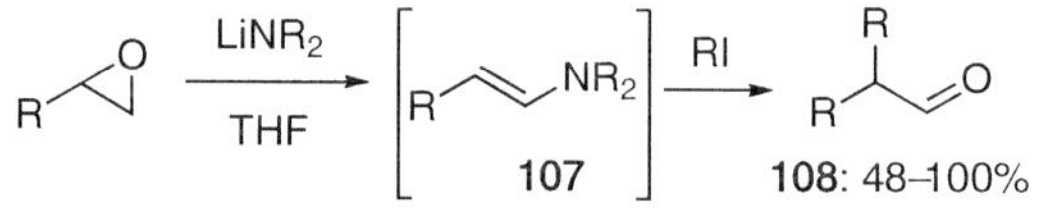

Scheme 11.20.

An in-depth multinuclear magnetic resonance investigation, supported by DFT calculations, was carried out by Capriati, Florio, and coworkers with the aim of investigating the "factors" responsible for the two facets of reactivity of a lithiated epoxide, that is the carbanionic and the carbene-like character. Employing a ^{13}C doubly enriched substrate, the dichotomous reactivity exhibited by α-lithiated styrene oxide was related to the different aggregation states favored in solution under certain experimental conditions.[70] Over a range of concentrations, 0.075–0.3 M, α-lithiated styrene oxide was found to be mainly present in THF at 173 K as a THF-solvated monomeric species (**109**) (with a tetrahedral arrangement around the carbanionic carbon), in equilibrium with a complex mixture of stereoisomeric dimers represented by **110** (Fig. 11.7). The reduced symmetry of some aggregates, due mainly to the partial breaking of some Cα–Li bonds, further complicates their NMR spectra. The natural bond orbital analysis suggests that an O-coordinated bridged dimer such as **110** has stronger carbenoid character than monomer **109**. This is also consistent with both the higher Cα–O bond elongation predicted by calculations, and the

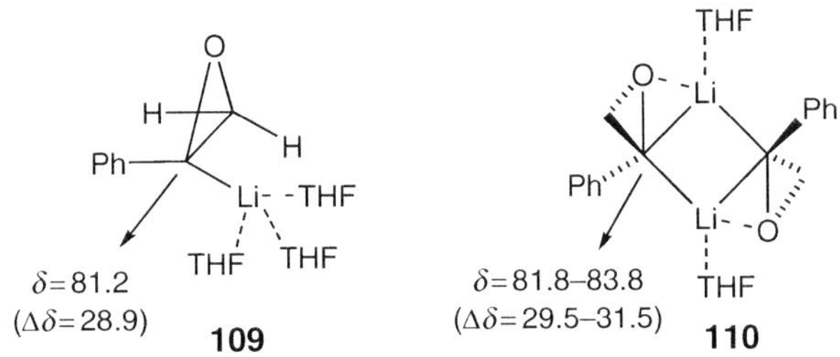

Figure 11.7. Selected aggregation states of lithiated styrene oxide (**109**, **110**); chemical shifts (ppm) of the carbenoid carbon atom determined in THF at 173 K (relative deshielding with respect to the nonlithiated precursor).

larger downfield ^{13}C shift observed for the carbenoid ^{13}C atom in the case of all O-coordinated disolvated dimers ($\Delta\delta = 29.5–31.5$) with respect to the monomer ($\Delta\delta = 28.9$) (Fig. 11.7).

A flow microreactor consisting of micromixers and microtubes provides an effective reactor for the preparative scale synthesis of substituted epoxides at a temperature higher than −78 °C. There is no significant decomposition or isomerization of the intermediate oxiranyllithiums because of the short residence time and efficient temperature control. The sequential and stereoselective introduction of various electrophiles onto 2,3-diphenyloxiranes could also be efficiently accomplished.[71]

An interesting aspect of the reactivity of α-lithiated oxiranes is the stereochemical outcome of their reactions with electrophiles. Many oxiranyllitiums are configurationally stable so that trapping with electrophiles usually occurs with complete retention of configuration at the lithiated carbon. Styrene oxide, for example, can be smoothly lithiated at −98 °C in THF and functionalized with a wide range of electrophiles, the absolute configuration at the benzylic carbon remaining completely unaffected.[67c] Racemization, however, is sometimes observed, depending upon the substitution at the oxiranyl skeleton and the employed experimental conditions.[72] In general, for certain chiral α-heterosubstituted organolithium compounds, configurational stability depends on many factors: the nature of the heteroatom and the solvent, the temperature, the substitution, and the geometry of the carbanionic carbon atom (planar or pyramidalized), as well as the aggregation and solvation, because of their influence on the structure–reactivity relationship. Several examples of retentive and inverting electrophilic substitutions (categorized as $S_E2$ret or $S_E2$inv in the Hughes–Ingold terminology) of configurationally stable organolithiums are known, as well as substitutions occurring via stepwise (single electron transfer) mechanisms. The latter lead to stereorandom couplings at the metal-bearing carbon. Common mechanistic pathways suggested for the enantiomerization of chiral organolithiums are: (i) a dissociative mechanism which proceeds via a solvent-separated ion pair derived from the contact ion pair (particularly effective in polar solvents with a strong donor capacity such as THF), (ii) a "conducted tour" mechanism, usually involving intramolecular Lewis base coordination, and (iii) epimerization through an aggregate.[73]

Oxazolinyloxiranyllithium **111** is a rare example of a Li/OR carbenoid (chemical shift of the carbenoid carbon atom, 93.2 ppm, $\Delta\delta = 36.3$) exhibiting an unusual thermal stability; in fact, it undergoes a successful deuterium incorporation (>98% D, 85% yield) even at 25 °C. In the concentration range of 0.08–0.3 M, it mainly exists as a monomeric η^3-aza-allyl-coordinated species rapidly equilibrating on the NMR timescale, with a complex mixture of diastereomeric oxazoline-bridged dimeric species. The bias that lithium has to be strongly coordinated by the iminic oxazoline moiety seems to be a key factor in either controlling the kinetic stability of such a lithium carbenoid, or in causing a fast racemization of **111** on the NMR timescale ($t_{1/2} = 6.05$ s, $\Delta G^{\ddagger}_{143\,K} = 8.8$ kcal/mol in 3:2 THF/Et_2O) (Fig. 11.8).[74a]

As for aryl-substituted α-lithiated styrene oxides, it has been ascertained that whereas electron-donating groups (e.g., methyl and methoxy) do not alter the configurational stability in THF and at low temperatures, electron-withdrawing groups (e.g., chloro and trifluoromethyl) trigger racemization in THF at a rate dependent on their position on the phenyl ring. In addition, for those oxiranyllithiums (**112–117**) proven to be configurationally unstable on the timescale of their reactions, a general trend became apparent in the presence of TMEDA: racemization rate was slowed in coordinating solvents (e.g., THF), but dramatically hindered in noncoordinating solvents (e.g., hexane). The calculated barriers to inversion and activation parameters (which have been collected in Fig. 11.8), suggest a subtle interplay of different mechanisms in the enantiomerization process. In particular, in the case of **112**, the presence of TMEDA has the interesting effect of considerably increasing the enthalpy of activation (C–Li bond breaking), which is consistent with a surprising and successful competition of this ligand with bulk THF.[74b-d]

α-Lithiated oxetanes also display a carbenoid character. The reactivity of 3,3-disubstituted oxetanes **118** toward alkyllithiums was first investigated by Klumpp and coworkers in the 1980s.[75] In an apolar solvent such as hexane, nucleophilic substitution at C-2 always competed with α-lithiation. Quenching of the reaction mixture with a deuterium source, however, provided no α-deuterated products. This is because α-lithiated derivatives **119**, being strongly electrophilic in the aforementioned reaction medium, were attacked by the same organolithium, as soon as they formed, to give only substitution products **120** (carbene-like character of lithiated oxetanes, Scheme 11.21). Capriati and coworkers have recently reported that the regioselective α-deprotonation of 2-phenyloxetane **121** (R = H) is feasible with *s*-BuLi as the base at −78 °C in THF. The carbanionic character of the corresponding α-lithiated species **122** can be fruitfully exploited for synthesizing phenyl-substituted derivatives **123** by reaction with different electrophiles, which are well-accommodated on the oxetane core (Scheme 11.21).[76] α-Lithiated intermediate **122**, however, proved to be configurationally unstable in both THF and hexane/TMEDA. The oxetane ring is also special for another reason. In view of its

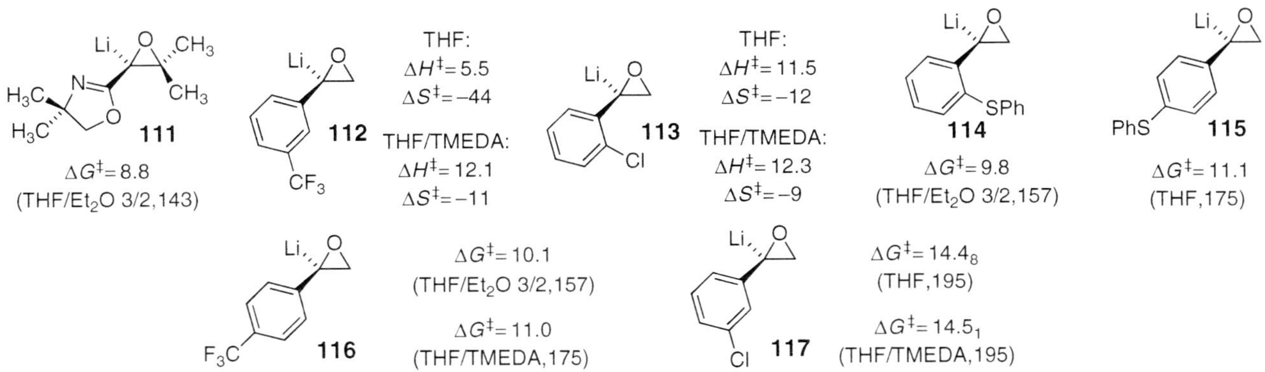

Figure 11.8. Enantiomerization barriers and activation parameters for representative oxiranyllithiums **111–117** (ΔG and ΔH (kcal/mol); ΔS (cal/mol·K)) (solvent, temperature in K).

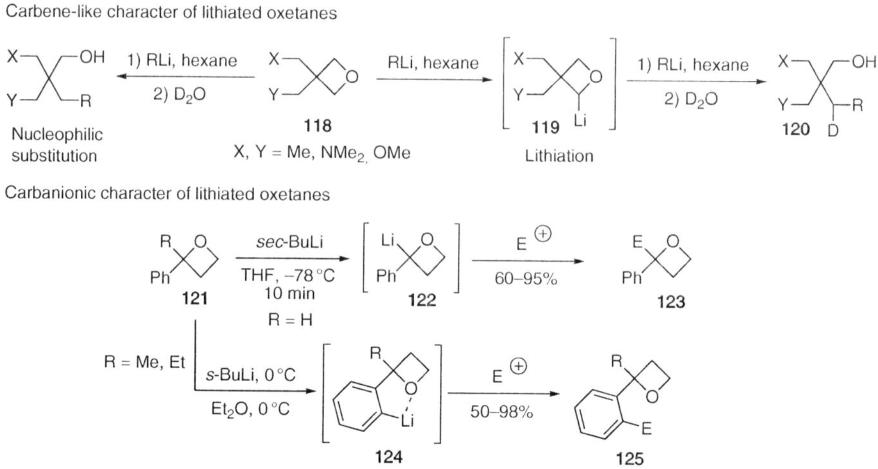

Scheme 11.21.

remarkable basicity (even higher than that of THF), 2-phenyloxetane surprisingly behaves as an effective *ortho*-directing group in the absence of a benzylic proton. The same reaction does not occur in the case of α-substituted styrene oxide. Under mild conditions (*s*-BuLi in Et_2O at 0 °C or even at room temperature), *ortho*-lithiation of **121** (R = Me, Et) proceeds smoothly, and the putative intermediate **124** can ultimately be trapped with a variety of structurally different electrophiles, thereby furnishing *ortho*-functionalized derivatives **125**. Of note, aryl and heteroaryl bromides could also be efficiently cross-coupled through a one-pot D*o*M-Suzuki–Miyaura reaction. The lithiation directing ability and the proton transfer mechanism have been investigated by competitive metalation and kinetic isotope effect studies.[77]

Carbene-like reactivity toward nucleophiles RLi has also been observed for O-alkyl-*N*,*N*-dialkylcarbamates. The general features of one of the rare crystal structures of a lithiated carbamate such as the carbamoyloxy-substituted **126**·(−)-sparteine, reported by Boche and coworkers, are in line with those of other Li/OR carbenoids. Once again, there is strong elongation of the carbenoid–oxygen bond (147.6 pm), with the lithium not bridging the adjacent oxygen atom, most probably because of a more favorable dipole-stabilized five-membered ring chelate. Such a structural feature is also retained in solution (Fig. 11.9).[6b,c]

Lithiation-borylation of optically active carbamates has recently spawned a major area of research in particular with reference to the construction of challenging quaternary stereogenic centers with high enantioselectivity. Aggarwal and coworkers developed a methodology for preparing tertiary boronic esters **129** with excellent enantiomeric ratio (e.r.) values from carbamates **127**, which can be easily obtained from the corresponding optically active secondary alcohols through homologation of a borate complex. The deprotonation of **127**

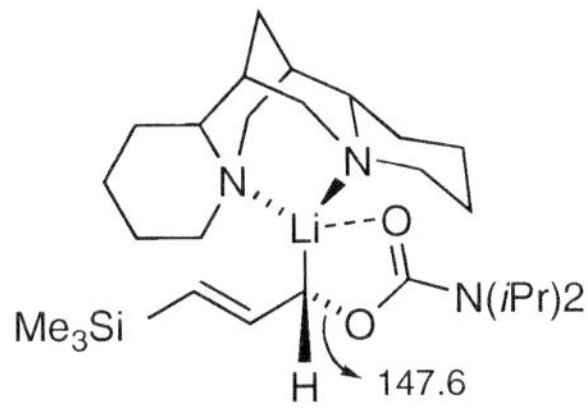

126 (–)-sparteine

Figure 11.9. Crystal structure of **126**·(–)-sparteine: essential features; bond length (pm).

with *s*-BuLi at low temperature, followed by treatment with a boronate such as RBpin, leads to the formation of the ate complex **128**. Upon warming of the reaction mixture to room temperature, **128** collapses, allowing the stereospecific 1,2-migration of the R substituent from the boron to the carbon, thereby giving **129** with concomitant elimination of the nucleofugal OCb group. The addition of $MgBr_2$/MeOH proved to be essential to prevent a considerable erosion in e.e. values that occurred with sterically demanding groups due to dissociation of the ate complex back to the starting lithiated carbamate, which is more prone to undergo racemization upon warming (Scheme 11.22).[78a,b] The application of this methodology, employing a mixed, unhindered borane, proved pivotal for the stereocontrolled synthesis of all the diastereomers of (+)-erogorgiaene, a marine diterpene that shows promising activity as an inhibitor of *Mycobacterium tuberculosis*.[78c]

Subjecting chiral tertiary boronic esters like **129** to a modified Matteson homologation, making use of $BrCH_2Li$ (obtained from CH_2Br_2 and *n*-BuLi) as an alternative reagent in place of the more common $ClCH_2Li$, permits their simple homologation to compounds with quaternary carbon centers bearing an α-primary alcohol moiety (**130**) (Scheme 11.22).[78d] Tertiary boronic esters **129** also react efficiently with 1-chloroallyllithium **131**, with both surprisingly high diastereoselectivity (24:1) and stereospecificity, to give chiral boronic esters **132** with two contiguous stereocenters. Such excellent diastereo- and enantioselection, starting from a racemic mixture, is consistent with a fast dynamic kinetic resolution of the intermediate Li/Cl carbenoid **131** during the reaction (Scheme 11.22).[78e] Homologation and vinylation of tertiary boronic esters is also feasible adopting a modified Zweifel vinylation protocol: treatment of optically active compounds **129** with 4 equiv of vinylmagnesium bromide, and then with I_2 and NaOMe, favored a complete and stereospecific conversion into vinylated adducts **133** with very high e.r. values. The synthetic utility of such a vinylation methodology was demonstrated by the highly enantioselective synthesis of (+)-(*S*)-sporochnol **134**, a monoterpene-substituted phenol isolated from the Caribbean marine alga *Sporochnus bolleanus*, which has been shown to exhibit significant feeding deterrence toward herbivorous fish (Scheme 11.22).[78e] Two general strategies for the effective preparation of both tertiary and the more challenging quaternary allylsilanes in high yield and

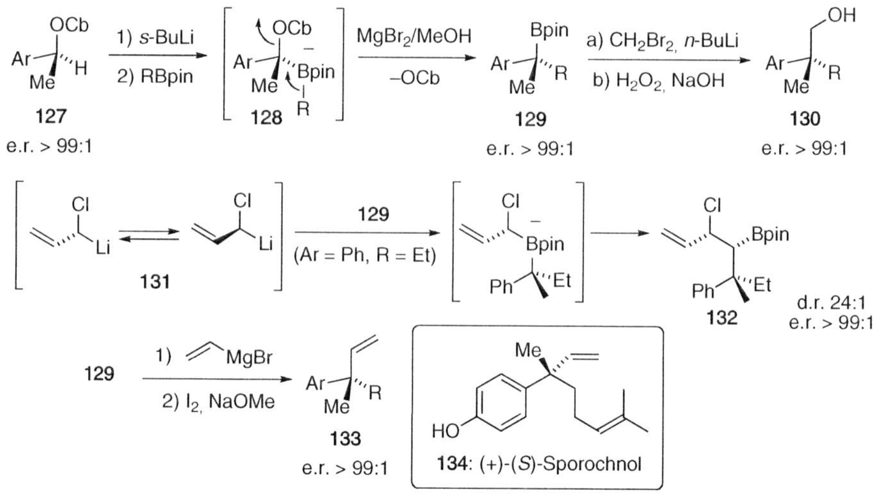

Scheme 11.22.

d.r. have also been reported.[78f] Further stereospecific transformations of tertiary boronic esters into tertiary alcohols, C-tertiary amines, and tertiary arylalkanes were discussed in a Concept article.[78g]

11.6 LITHIUM–NITROGEN CARBENOIDS

In Sections 11.3 and 11.5, the characteristic features of Li/Hal and Li/OR carbenoids were discussed; these include: (i) the electrophilic reactivity, as observed in reactions with nucleophiles; (ii) the deshielding of the ^{13}C atom in the NMR spectrum; and (iii) the C–X bond length in the corresponding solid state structures. What about α-lithiated amines? The few reported structures in the literature[6b,c] are consistent with a slight bond lengthening (2–5 pm) of the C–N bonds, which parallels the small deshielding of the anionic ^{13}C atom. However, until the beginning of 2000s, α-lithiated amines could not be strictly classified as "carbenoids" because reactions with nucleophiles were never reported and only "carbanion" behavior was known. The carbene-like reactivity of α-lithiated amines has only recently been recognized in the case of lithiated aziridines[79] bearing electron-withdrawing groups at the aziridine nitrogen, and under certain experimental conditions of base, solvent, and temperature. Between 2001 and 2005, the first insertions into both C–H bonds and organolithium reagents (alkylative ring opening) were independently described by O'Brien[80] and Müller[81] for *N*-sulfonyl aziridines. In particular, desymmetrization of *N*-sulfonyl aziridines **135** and **136** by *s*-BuLi in the presence of (−)-sparteine was found to enantiomerically generate enriched allylic (by insertion into adjacent β-CH bond) or bicyclic (by transannular CH insertion)

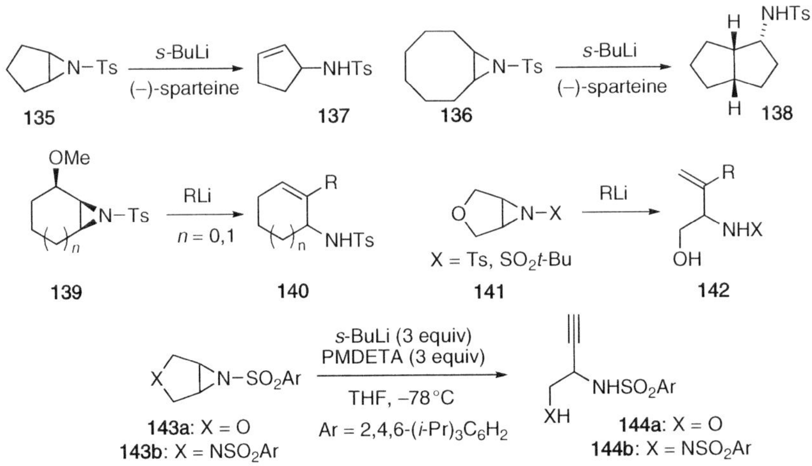

Scheme 11.23.

sulfonamides **137** and **138**, respectively. The latter is the typical reaction of medium-sized rings (Scheme 11.23).

Of note, in all cases investigated, the stereochemical outcome of the aziridine desymmetrization was opposite to that of the corresponding epoxides. In the presence of a β-alkoxy group, new routes to substituted cyclic allylic amines **140** and unsaturated amino alcohols **142** were established via organolithium-induced alkylative reduction of β-alkoxy *cis*-aziridines **139** and 2,5-dihydrofuran-derived aziridines **141**, respectively (Scheme 11.23).[82] Extension of this methodology to C_2-symmetric acyclic *cis*-aziridines broadened the synthetic applicability, also allowing the preparation of substituted allylic amino ethers in an enantioenriched form.[82b] A direct organolithium-mediated conversion of both dihydrofuran (X = O) and dihydropyrrole (X = NSO_2Ar) aziridines **143a,b** into alkynyl amino alcohols and diamines **144a,b**, respectively, was achieved by reacting *N*-2,4,6-triisopropylbenzenesulfonyl aziridines with 3 equiv of *s*-BuLi/PMDETA (Scheme 11.23).[83]

For eliminative dimerization reactions, *N*-Bus-protected terminal aziridines **145** were found to proceed smoothly by lithiation with 3 equiv of LTMP at −78 to 0 °C in THF/hexanes to give *N*-protected-2-ene-1,4-diamines **147**. Formation of the *E* isomer from enantiopure substrates is consistent with an initial *trans* lithiation at the terminal position of the aziridine ring to give intermediate **146**, followed by *syn* elimination with one unit of **146** reacting as an electrophile and another as a nucleophile. The uselfulness of this methodology was demonstrated in a concise and stereoselective synthesis of (*R*,*S*,*S*,*R*)-2,5-diamino-1,6-diphenylhexane-3,4-diol **148**, which is the core unit of many extremely potent HIV protease inhibitors (Scheme 11.22).[84a]

A variety of synthetically useful bicyclic amines of type **151** could be synthesized as single diastereomers in good to excellent yields by regio- and

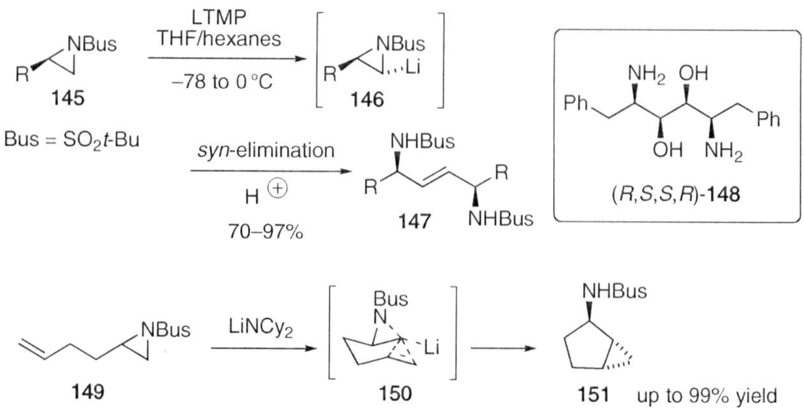

Scheme 11.24.

stereoselective deprotonation of bishomoallylic terminal *N*-Bus protected aziridines **149**. Lithium dicyclohexylamide ($LiNCy_2$) proved to be the best lithiating agent to promote a diastereoselective intramolecular cyclopropanation. However, slow addition of the aziridine to the base is crucial to avoiding a localized high concentration of the lithiated aziridine **150** which would allow dimerization to become competitive (Scheme 11.24). Cyclopropanation of enantiopure terminal aziridines furnishes chiral bicyclic amines with no loss of enantiopurity, thereby demonstrating the stereospecificity of the process.[84b] β-Amino alcohols (also bearing quaternary stereogenic centers) could be synthesized via a new stereocontrolled route described by Aggarwal and coworkers, employing lithiated *N*-Boc aziridines and boronic esters as key starting materials.[85]

Structural investigations on α-lithiated amines and aziridines are still in their infancy.[86] This knowledge may shed light on the role played by aggregation and solvation on structure–reactivity relationships for these reactive intermediates. This will stimulate the development of novel stereoselective applications in the field of organic synthesis and, eventually, unveil still undisclosed aspects of "carbenoid reactivity." For example, the α-lithiation-trapping sequence of *trans*-*N*-alkyl-2,3-diphenylaziridines of type **152** was found to take place with stereochemistry that dramatically depends on solvent coordinating ability, for example, inversion of configuration in THF and retention in toluene. Such opposite regioselectivity has recently been rationalized by Luisi, Florio, and coworkers by a detailed multinuclear spectroscopic investigation, suggesting involvement of two differently configured lithiated intermediates. A *trans*-configured homochiral dimeric aziridinyllithium, *trans*-**152-Li** tightly bonded to TMEDA, would be responsible for the formation of adducts **153** in toluene, whereas a *cis*-configured monomeric aziridinyllithium, *cis*-**152-Li**, would favor the formation of adducts **154** in THF (Scheme 11.25).[86b]

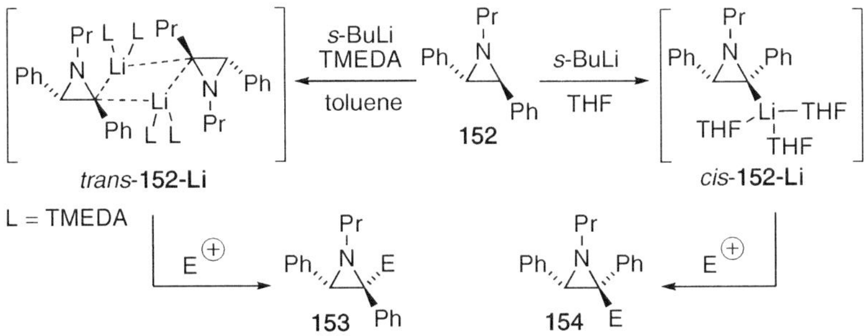

Scheme 11.25.

11.7 CONCLUSION AND OUTLOOK

Lithium carbenoids are today widely recognized as valuable synthons for the regio- and stereoselective functionalization of substituted carbon backbones. Much of the interest in their chemistry mainly derives from the intriguing complexity induced by their dichotomous carbanionic/carbene-like behavior, tunable by the experimental conditions that, in turn, can also influence their configurational stability/lability. Because of the predominantly ionic character of C–Li bonds, functionalized lithium carbenoids, like most common organolithium reagents, are variously aggregated in solution. However, the basic structural features of these aggregates may differ from those of the most common symmetric types owing to the presence of additional intra/intermolecular coordinations provided by the heteroatom (halogen, oxygen, nitrogen). Knowledge of these details is essential because it is known that the degree of aggregation and solvation has significant impact on the (ambiphilic) reactivity and selectivity of organolithium complexes. To fulfill this goal, there is a strong need to synergistically combine the potentials of computational chemistry, cryo-X-ray spectroscopic analysis, and modern NMR experiments. In particular, the recent development of quantum chemical methodology gives us an opportunity to accurately describe and interpret useful model systems, and to make reliable predictions, including solvation, with reasonable time and effort. This provides helpful support to the structural hypotheses stemming from the interpretation of spectral data furnished by nonroutine NMR techniques. Lithium is a nucleus that is very suitable for NMR spectroscopy, and "lithium spectra" (^{1}H, ^{6,7}Li, ^{13}C NMR) are very rich in information. They can provide insight not only into the solution structure of organolithium compounds, but also into their dynamic behavior in solution, and the mechanisms of their reactions.

Besides traditional methods based on the magnitude of ^{13}C–^{6,7}Li coupling constants, which depend on the state of aggregation,[52] structural studies of organolithiums can also be performed by means of diffusion-ordered spectroscopy (DOSY).[87] Williard and coworkers pioneered DOSY diffusion coefficient-formula weight analysis, and were thereby able to discriminate among different aggregation states of a certain organolithium compound by comparison with reference substances of defined molar masses.[88] The solution structures of 2-thienyllithium aggregates with various donor solvents were also solved by Stalke and coworkers, by applying $^{1}H,^{7}Li$ HOESY experiments and comparing the distance information provided by the $^{1}H,^{7}Li$ HOE buildup with the calculated $\Sigma r(H\cdots Li)^{-6}$ terms from the crystal structures.[89] A more straightforward method to identify the degree of organolithium aggregation, recently described by Stalke and John, is based on ^{7}Li residual quadrupolar couplings.[90]

Topological analysis of the charge densities from high-resolution X-ray diffraction data, bearing on the nature of chemical bonding and chemical reactivity, also provides access to the structure–reactivity relationship. Use of Bader's quantum mechanical "Atoms in Molecules" approach affords the possibility of unveiling "secondary interactions" (mainly noncovalent interactions) responsible for the geometries and the packing of organolithium aggregates.[91]

Major breakthroughs in the organolithium field can also be expected from a wider application of "flash chemistry" (*vide supra*, Section 11.3.2). A flow-microreactor approach, for example, was recently successfully applied to the protecting-group-free synthesis of substituted arylketones mediated by organolithiums, by greatly reducing the residence time (0.003 s or less),[92a] and also to the asymmetric carbolithiation of conjugated enynes to obtain enantioenriched chiral allenes.[92b]

In light of the explosive development of the chemistry of *N*-heterocyclic carbenes (NHCs), achieved in the last few years, with widespread application in catalysis and also beyond catalysis,[93] it is surprising that lithium carbene complexes (which cannot be defined as carbenoids) are still rare, given that "free" carbenes can be generated by deprotonation of the corresponding imidazolium salts with strong group I bases. To date, there are only a few structurally characterized Li-NHC adducts.[94] Selective functionalization of a NHC with lithium is an interesting and fascinating concept, worth investigating because, in principle, it may allow tuning of the electronic and steric properties of the carbene center. At the same time, Li-NHC adducts could also act as promising reagents for carbene transfer reactions.

More generally, an in-depth understanding of the reaction mechanisms in which lithium participates remains a great challenge. However, it is critical because it may open up new avenues toward the development of previously inconceivable reactions in synthetic chemistry. The ultimate goal is to learn more about mediation by lithium in transition-metal-free carbon–carbon and carbon–heteroatom bond-forming reactions. The recent results obtained in the case of lithiated carborane are very encouraging.[95] Now that the cat is out of the bag, stimulating new discoveries are expected. I am convinced that the resurgence of main-group chemistry has just begun!

SUGGESTED READING

Braun, M. "Lithium Carbenoids." In *The Chemistry of Organolithium Compounds*, Part 2. Rappoport, Z., Marek, I., Eds. John Wiley & Sons, Inc.: New York, **2004**, p. 829.

Boche, G., Lohrenz, J. C. W. "The Electrophilic Nature of Carbenoids, Nitrenoids, and Oxenoids," *Chem. Rev.* **2001**, *101*, 697.

Boche, G., Lohrenz, J. C. W., Opel, A. "From "Carbanions" to Carbenoids: The Structure of Lithiated Amines and Lithiated Ethers." In *Lithium Chemistry: A Theoretical and Experimental Overview*. Sapse, A.-M., Schleyer, P. v. R., Eds. John Wiley & Sons, Inc.: New York, **1995**, p. 195.

Claiden, J. *Organolithiums: Selectivity for Synthesis*. Pergamon: Amsterdam, **2002**.

Capriati, V., Florio, S. "Anatomy of Long-Lasting Love Affairs with Lithium Carbenoids: Past and Present Status and Future Prospects," *Chem. Eur. J.* **2010**, *16*, 4152.

Gawley, R. E., Siegel, J. S., Eds. "Topics in Stereochemistry." In *Stereochemical Aspects of Organolithium Compounds*, Vol. 26. Wiley-VCH: Weinheim, **2010**.

Knorr, R. "Alkylidenecarbene, Alkylidenecarbenoids, and Competing Species: Which is Responsible for Vinylic Nucleophilic Substitution, [1+2] Cycloadditions, 1,5-CH Insertions, and the Fritsch–Buttenberg–Wiechell Rearrangement?" *Chem. Rev.* **2004**, *104*, 3795.

Scott, H. K., Aggarwal, V. K. "Highly Enantioselective Synthesis of Tertiary Boronic Esters and their Stereospecific Conversion to Other Functional Groups and Quaternary Stereocentres," *Chem. Eur. J.* **2011**, *17*, 13124.

Stey, T., Stalke, D. "Lead Structures in Lithium Organic Chemistry." In *The Chemistry of Organolithium Compounds*, Part 1. Rappoport, Z., Marek, I., Eds. John Wiley & Sons, Inc.: New York, **2004**, p. 47.

Yoshida, J. *Flash Chemistry, Fast Organic Synthesis in Microsystems*. Wiley-Blackwell: Hoboken NJ, **2008**.

REFERENCES

1 Closs, G. L., Closs. L. E. *Angew. Chem.* **1962**, *74*, 431.

2 (a) Closs, G. L., Moss, R. A. *J. Am. Chem. Soc.* **1964**, *86*, 4042. (b) The term carbenoid as an adjective was, instead, first used L. Friedman and H. Shechter to describe a base-catalyzed decomposition of tosylhydrazones in aprotic solvents which, most probably, implies the release of a carbene fragment, see: Friedman, L., Shechter, H. *J. Am. Chem. Soc.* **1959**, *81*, 5512.

3 Anslyn, E. V., Dougherty, D. A. *Modern Physical Organic Chemistry*. University Science Books: Sausalito, CA, **2006**, p. 575.

4 Transition-metal carbenes and alkylidenes are sometimes considered, erroneously, similar species, for a survey, see: Mindiola, D. J., Scott, J. *Nat. Chem.* **2011**, *3*, 15.

5 Such a dicothomy also occurs in the case of singlet carbenes. In fact, whereas triplet carbenes exhibit radical-like reactivity, singlet carbenes are expected to show both electrophilic and nucleophilic behavior because of the lone pair and the empty *p* orbital, see: Bertrand, G. "Stable Singlet Carbenes." In *Reactive Intermediate Chemistry*. Moss, R. A., Platz, M. S., Jones, Jr., M., Eds. John Wiley & Sons: Hoboken, NJ, **2004**.

6 For early general reviews on lithium carbenoids, see: (a) Köbrich, G. *Angew. Chem. Int. Ed. Engl.* **1967**, *6*, 41. (b) Boche, G., Lohrenz, J. C. W., Opel, A. In *Lithium Chemistry: A Theoretical and Experimental Overview*. Sapse, A.-M., Schleyer, P. v. R., Eds. John Wiley & Sons, Inc.: New York, **1995**, p. 195. (c) Boche, G., Lohrenz, J. C. W. *Chem. Rev.* **2001**, *101*, 697. The reactivity of alkylidenecarbenoids has also been extensively reviewed: (d) Braun, M. *Angew. Chem. Int. Ed. Engl.* **1998**, *37*, 430. (e) Knorr, R. *Chem. Rev.* **2004**, *104*, 3795. (f) Braun, M. "Lithium Carbenoids." In *The Chemistry of Organolithium Compounds*, Vol. 2. Sapse, A.-M., Schleyer, P. v. R., Eds. John Wiley & Sons, Inc.: New York, **2004**, p. 829. For a recent review on α-lithiated α-chloroalkylheterocycles and their utility in synthetic organic chemistry, see: (g) Florio, S., Capriati, V., Luisi, R. *Curr. Org. Chem.* **2004**, *8*, 1529.

7 (a) Seebach, D., Siegel, H., Müllen, K., Hiltbrunner, K. *Angew. Chem. Int. Ed. Engl.* **1979**, *18*, 784. (b) Siegel, H., Hiltbrunner, K., Seebach, D. *Angew. Chem. Int. Ed. Engl.* **1979**, *18*, 785. (c) Seebach, D., Siegel, H., Gabriel, J., Hässig, R. *Helv. Chim. Acta* **1980**, *63*, 2046. (d) Seebach, D., Hässig, R., Gabriel, J. *Helv. Chim. Acta* **1983**, *66*, 308. (e) Seebach, D., Gabriel, J., Hässig, R. *Helv. Chim. Acta* **1984**, *67*, 1083. (f) Boche, G., Bosold, F., Lohrenz, J. C. W., Opel, A., Zulauf, P. *Chem. Ber.* **1993**, *126*, 1873.

8 In Li/Hal carbenoids, the elongation decreases (F>Cl>Br>I) with increasing atomic weight of the halogen, see: Ref. 6c.

9 Boche, G. *Angew. Chem. Int. Ed. Engl.* **1989**, *28*, 277.

10 Cantat, T., Jacques, X., Ricard, L., Le Goff, X. F., Mézailles, N., Le Floch, P. *Angew. Chem. Int. Ed.* **2007**, *46*, 5947.

11 For other examples dealing with the role played by hyperconjugation in stabilizing carbanions and in controlling chemical reactivity, see: Gaul, C., Arvidsson, P. I., Bauer, W., Gawley, R. E., Seebach, D. *Chem. Eur. J.* **2001**, *7*, 4117. (b) Nilsson Lill, S. O., Rauhut, G., Anders, E. *Chem. Eur. J.* **2003**, *9*, 3143.

12 Niecke, E., Becker, P., Nieger, M., Stalke, D., Schoeller, W. W. *Angew. Chem. Int. Ed. Engl.* **1995**, *34*, 1849.

13 (a) Barluenga, J., Rodríguez, M. A., Campos, P. J. *J. Am. Chem. Soc.* **1988**, *110*, 5567. (b) Maercher, A., Bös, B. *Main Group Met. Chem.* **1991**, *14*, 67.

14 Mechanistic aspects of alkyne formation in FBW rearrangements have been discussed in a recent paper, see: Bichler, P., Chalifoux, W. A., Eisler, S., Shi Shun, A. L. K., Chernick, E. T., Tykwinski, R. R. *Org. Lett.* **2009**, *11*, 519.

15 Barluenga, J., González, J. M., Llorente, I., Campos, P. J., Rodríguez, M. A., Thiel, W. *J. Organomet. Chem.* **1997**, *548*, 185.

16 Köbrich, G., Goyert, W. *Tetrahedron* **1968**, *24*, 4237.

17 (a) Seyferth, D., Lambert, Jr., R. L. *J. Organomet. Chem.* **1973**, *55*, C53. (b) Seyferth, D., Lambert, Jr., R. L., Massol, M. *J. Organomet. Chem.* **1975**, *88*, 255.

18 Schmidt, A., Köbrich, G., Hoffmann, R. W. *Chem. Ber.* **1991**, *124*, 1253.

19 Halogen-lithium exchange constitutes one of the most powerful methods for the preparation of organolithium compounds, although the lithium-metalloid exchange is still controversial, both radical and carbanionic mechanisms seem plausible according to the nature of the reactants and reaction conditions. For leading references, see: (a)Reich, H. J., Medina, M. A., Bowe, M. D. *J. Am. Chem. Soc.* **1992**, *114*, 11003. (b) Reich, H. J., Bevan M. J., Gudmundsson, B. Ö., Puckett, C. L. *Angew. Chem. Int. Ed.* **2002**, *41*, 3436. (c) Nudelman, N. S., García, G. V., Velurtas, S. *J. Phys. Org. Chem.* **2002**, *15*, 903.

20 Hoffmann, R. W., Julius, M., Chemla, F., Ruhland, T., Frenzen, G. *Tetrahedron* **1994**, *50*, 6049.

21 Hoffmann, R. W., Bewersdorf, M. *Chem. Ber.* **1991**, *124*, 1259.
22 Kapeller, D. C., Hammerschmidt, F. *J. Am. Chem. Soc.* **2008**, *130*, 2329.
23 Sydnes, I. K. *Chem. Rev.* **2003**, *103*, 1133.
24 Azizoglu, A., Balci, M., Mieusset, J.-L., Brinker, U. H. *J. Org. Chem.* **2008**, *73*, 8182.
25 Zefirov, N., Kuznetsova, T. S. *Synlett* **2011**, *16*, 2299.
26 (a) Duraisamy, M., Walborsky, H. M. *J. Am. Chem. Soc.* **1984**, *106*, 5035. (b) Topolski, M., Duraisamy, M., Rachón, J., Gawronski, J., Gawronska, K., Goedken, V., Walborsky, H. M. *J. Org. Chem.* **1993**, *58*, 546.
27 Taylor, K. G., Chaney, J., Deck, J. C. *J. Am. Chem. Soc.* **1976**, *98*, 4163.
28 (a) Ke, Z., Zhao, C., Phillips, D. L. *J. Org. Chem.* **2007**, *72*, 848. (b) Ke, Z., Zhou, Y., Gao, H., Zhao, C., Phillips, D. L. *Chem. Eur. J.* **2007**, *13*, 6274.
29 Matteson, D. S. *Tetrahedron* **1998**, *54*, 10555.
30 Blakemore, P. R., Marsden, S. P., Vater, H. D. *Org. Lett.* **2006**, *8*, 773.
31 (a) Satoh, T., Takano, K. *Tetrahedron* **1996**, *52*, 2349. (b) Hoffmann, R. W., Nell, P. G., Leo, R., Harms, K. *Chem. Eur. J.* **2000**, *6*, 3359.
32 (a) Blakemore, P. R., Burge, M. S. *J. Am. Chem. Soc.* **2007**, *129*, 3068. (b) Emerson, C. R., Zakharov, L. N., Blakemore, P. R. *Org. Lett.* **2011**, *13*, 1318.
33 Knochel, P., Ed. *Handbook of Functionalized Organometallics*, Vols. 1 and 2. Wiley-VCH, Weinheim, **2005**.
34 Special Volume entitled "The Chemistry of Interelement Linkage." *J. Organomet. Chem.* **2000**, *611*, 1–614.
35 (a) Hata, T., Kitagawa, H., Masai, H., Kurahashi, T., Shimizu, M., Hiyama, T. *Angew. Chem. Int. Ed.* **2001**, *40*, 790. For a survey, see: (b) Kurahashi, T., Hata, T., Masai, H., Kitagawa, H., Shimizu, M., Hiyama, T. *Tetrahedron* **2002**, *58*, 6381.
36 Shimizu, M., Nakamaki, C., Shimono, K., Schelper, M., Kurahashi, T., Hiyama, T. *J. Am. Chem. Soc.* **2005**, *127*, 12506.
37 (a) Shimizu, M., Schelper, M., Nagao, I., Shimono, K. *Chem. Lett.* **2006**, *35*, 1222. A highly stereoselective route to di- and trisubstituted cyclopropanes employing an 1,2-alkyl migration reaction of zincate carbenoids as intermediates has also been described: (b) Harada, T., Katsuhira, T., Hattori, K., Oku, A. *J. Org. Chem.* **1993**, *58*, 2958.
38 Shimizu, M., Kurahashi, T., Shimono, K., Tanaka, K., Nagao, I., Kiyomoto, S., Hiyama, T. *Chem. Asian J.* **2007**, *2*, 1400.
39 (a) Knorr, R., Pires, C., Behringer, C., Menke, T., Freudenreich, J., Rossmann, E. C., Böhrer, P. *J. Am. Chem. Soc.* **2006**, *128*, 14845. (b) Knorr, R., Pires, C., Freudenreich, J. *J. Org. Chem.* **2007**, *72*, 6084.
40 Rappoport, Z. *Acc. Chem. Res.* **1992**, *25*, 474.
41 Harada, T., Muramatsu, K., Mizunashi, K., Kitano, C., Imaoka, D., Fujiwara, T., Kataoka, H. *J. Org. Chem.* **2008**, *73*, 249.
42 Harada, T., Imaoka, D., Kitano, C., Kusukawa, T. *Chem. Eur. J.* **2010**, *16*, 9164.
43 Lehr, K., Mariz, R., Leseurre, L., Gabor, B., Fürstner, A. *Angew. Chem. Int. Ed.* **2011**, *50*, 11373.
44 (a) Fillery, S. F., Gordon, G. J., Luker, T. Whitby, R. J. *Pure Appl. Chem.* **1997**, *69*, 633. (b) Gordon, G. J., Whitby, R. J. *Chem. Comm.* **1997**, 1045. (c) Watkins, W. J., Whitby, R. J. *Tetrahedron Lett.* **1998**, *39*, 123.
45 Kasatkin, A., Whitby, R. *J. Am. Chem. Soc.* **1999**, *121*, 7039.
46 Thomas, E., Dixon, S., Whitby, R. J. *Angew. Chem. Int. Ed.* **2006**, *45*, 7070.
47 Stec, J., Thomas, E., Dixon, S., Whitby, R. J. *Chem. Eur. J.* **2011**, *17*, 4896.

48 (a) Yoshida, J. In *Flash Chemistry, Fast Organic Synthesis in Microsystems*. Wiley-Blackwell: Hoboken, NJ, **2008**. (b) Yoshida, J. *Chem. Rec.* **2010**, *10*, 332. (c) Yoshida, J., Kim, H., Nagaki, A. *ChemSusChem* **2011**, *4*, 331.
49 Nagaki, A., Matsuo, C., Kim, S., Saito, K., Miyazaki, A., Yoshida, Y. *Angew. Chem. Int. Ed.* **2012**, *51*, 3245.
50 For leading reviews, see: (a) Stey, T., Stalke, D. "Lead Structures in Lithium Organic Chemistry." In *The Chemistry of Organolithium Compounds*, Part 1. Rappoport, Z., Marek, I., Eds. John Wiley & Sons, Inc.: New York, **2004**, p. 47. (b) Gessner, V. H., Däschlein, C., Strohmann, C. *Chem. Eur. J.* **2009**, *15*, 3320.
51 (a) Williard, P. G. In *Comprehensive Organic Synthesis*, Vol. *1*. Pergamon: New York, **1991**, p. 1. (b) Setzer, W. N., Schleyer, P. v. R. *Adv. Organomet. Chem.* **1985**, *24*, 353. (c) Streitwieser, A., Bachrach, S. M., Dorigo, A., Schleyer, P. v. R. in *Lithium Chemistry: A Theoretical and Experimental Overview*. Sapse, A.-M., Schleyer, P. v. R., Eds. John Wiley & Sons, Inc.: New York, **1995**, p. 1. (d) Sapse, A.-M., Jain, D. C., Raghavachari, K. In *Lithium Chemistry: A Theoretical and Experimental Overview*. Sapse, A.-M., Schleyer, P. v. R., Eds. John Wiley & Sons, Inc.: New York, **1995**, p. 45.
52 Bauer, W., Winchester, W. R., Schleyer, *P. v. R. Organometallics* **1987**, *6*, 2371.
53 (a) Koizumi, T., Kikuchi, O. *Organometallics* **1995**, *14*, 987. (b) Parisel, O., Fressigné, C., Maddaluno, J., Glessner-Prettre, C. *J. Org. Chem.* **2003**, *68*, 1290.
54 Remenar, J. F., Collum, D. B. *J. Am. Chem. Soc.* **1997**, *119*, 5573.
55 Knorr, R., Menke, T., Ferchland, K., Mehlstäubl, J., Stephenson, D. S. *J. Am. Chem. Soc.* **2008**, *130*, 14179.
56 (a) Hall, P. L., Gilchrist, J. H., Collum, D. B. *J. Am. Chem. Soc.* **1991**, *113*, 9571. (b) Pratt, L. M. *Mini-Rev. Org. Chem.* **2004**, *1*, 209.(c) Gossage, R. A., Jastrzebski, J. T. B. H., van Koten, G. *Angew. Chem. Int. Ed. Engl.* **2005**, *44*, 1448.
57 (a) Stern, D., Finkelmeier, N., Stalke, D. *Chem. Comm.* **2011**, *47*, 2113. (b) Hevia, E., Mulvey, R. E. *Angew. Chem. Int. Ed.* **2011**, *50*, 6448.
58 (a) Pratt, L. M., Nguyên, N. V., Lê, L. T. *J. Org. Chem.* **2005**, *70*, 2294. (b) Pratt, L. M., Ramachandran, B. *J. Org. Chem.* **2005**, *70*, 7238. (c) Pratt, L. M., Thi, L., Truong, T. N. *J. Org. Chem.* **2005**, *70*, 8298. (d) Pratt, L. M., Merry, S., Nguyen, S. C., Quan, P., Thanh, B. T. *Tetrahedron* **2006**, *62*, 10821.
59 (a) Pratt, L. M., Phan, D. H. T., Tran, P. T. T., Nguyên, N. V. *Bull. Chem. Soc. Jpn.* **2007**, *80*, 1587. (b) Ramachandran, B., Kharidehal, P., Pratt, L. M., Voit, S., Okeke, F. N., Ewan, M. *J. Phys. Chem. A* **2010**, *114*, 8423.
60 Exceptions have been documented in the case of excellent solvation provided by solvents/co-solvents, or because of the formation of a favorable dipole-stabilized ring chelate, see: (a) Boche, G., Marsch, M., Müller, A., Harms, K. *Angew. Chem. Int. Ed. Engl.* **1993**, *32*, 1032. (b) Marsch, M., Harms, K., Zschage, O., Hoppe, D., Boche, G. *Angew. Chem. Int. Ed. Engl.* **1991**, *30*, 321.
61 Müller, A., Marsch, M., Harms, K., Lohrenz, J. C. W., Boche, G. *Angew. Chem. Int. Ed. Engl.* **1996**, *35*, 1518.
62 (a) Boche, G., Opel, A., Marsch, M., Harms, K., Haller, F., Lohrenz, J. C. W., Thümmler, C., Koch, W. *Chem. Ber.* **1992**, *125*, 2265. (b) Harder, S., Boersma, J., Brandsma, L., Kanters, J. A., Bauer, W., Pi, R., Schleyer, P. v. R., Schöllhorn, H., Thewalt, U. *Organometallics*, **1989**, *8*, 1688.
63 Saá J. M. *Helv. Chim. Acta* **2002**, *85*, 814, and references therein.
64 Lüttringhaus, A. V., Sääf, G. *Angew. Chem.* **1938**, *51*, 915.
65 (a) Sorger, K., Bauer, W., Schleyer, P. v. R., Stalke, D. *Angew. Chem. Int. Ed. Engl.* **1995**, *34*, 1594. (b) Dixon, D. D., Tius, M. A., Pratt, L. M. *J. Org. Chem.* **2009**, *74*, 5881.

66 Fischer-type carbenes, for example, are typical electrophilic complexes, whereas Schrock-type carbenes exhibit nucleophilic reactivity, for a survey, see: Dörwald, F. Z. in *Metal Carbenes in Organic Synthesis*. Wiley-VCH: Weinheim, **1999**.

67 For recent reviews on the reactivity of oxiranyllithiums, see: (a) Hodgson, D. M., Tomooka, K., Gras, E. *Top. Organomet. Chem.* **2003**, *5*, 217. (b) Chemla, F., Vranken, E. "Reactivity of Oxiranes with Organolithium Reagents." In *The Chemistry of Organolithium Compounds*, Vol. 2. Sapse, A.-M., Schleyer, P. v. R., Eds. John Wiley & Sons, Inc.: New York, **2004**, p. 1165. (c) Hodgson, D. M., Humphreys, P. G., Hughes, S. P. *Pure Appl. Chem.* **2007**, *79*, 269. (d) Capriati, V., Florio, S., Luisi, R. *Chem. Rev.* **2008**, *108*, 1918. (e) Capriati, V., Florio, S., Salomone A. "Oxiranyllithiums as Chiral Synthons for Asymmetric Synthesis." In *Topics in Stereochemistry*, Vol. 26: Stereochemical Aspects of Organolithium Compounds, Gawley, R. E., Siegel, J. S., Eds., Wiley-VCH, Weinheim, **2010**, p. 135.

68 Hodgson, D. M., Salik, S., Fox, D. J. *J. Org. Chem.* **2010**, *75*, 2157.

69 Hodgson, D. M., Bray, C. D., Kindon, N. D., Reynolds, N. J., Coote, S. J., Um, J. M., Houk, K. N. *J. Org. Chem.* **2009**, *74*, 1019.

70 Capriati, V., Florio, S., Perna, F. M., Salomone, A., Abbotto, A., Amedjkouh, M., Nilsson Lill, S. O. *Chem. Eur. J.* **2009**, *15*, 7958.

71 (a) Nagaki, A., Takizawa, E., Yoshida, J. *J. Am. Chem. Soc.* **2009**, *131*, 1654. (b) Nagaki, A., Takizawa, E., Yoshida, J. *Chem. Eur. J.* **2010**, *16*, 14149.

72 α-Lithiated styrene oxide, for example, loses its stereochemical integrity during the reaction with boronic esters in THF at −98 °C, but not in Et_2O at −115 °C, see: Vedrenne, E., Wallner, O. A., Vitale, M., Schmidt, F., Aggarwal, V. K. *Org. Lett.* **2009**, *11*, 165.

73 For leading references about the configurational stability and reactivity of chiral organolithiums, see: (a) Basu, A., Thayumanavan, S. *Angew. Chem. Int. Ed. Engl.* **2002**, *41*, 716. (b) Claiden, J. In *Organolithiums: Selectivity for Synthesis*, Pergamon: Amsterdam, **2002**. (c) Ott, H., Däschlein, D., Leusser, D., Schildbach, T., Seibel, D., Stalke, D., Strohmann C. *J. Am. Chem. Soc.* **2008**, *130*, 11901. (d) Gawley, R. E. "Overview of Carbanion Dynamics and Electrophilic Substitutions in Chiral Organolithium Compounds." In *Topics in Stereochemistry*, Vol. 26: Stereochemical Aspects of Organolithium Compounds. Gawley, R. E., Siegel, J. S., Eds. Wiley-VCH: Weinheim, **2010**, p. 93.

74 (a) Capriati, V., Florio, S., Luisi, R., Perna, F. M., Spina, A. *J. Org. Chem.* **2008**, *73*, 9552. (b) Capriati, V., Florio, S., Perna, F. M., Salomone, A. *Chem. Eur. J.* **2010**, *16*, 9778. (c) Perna, F. M., Salomone, A., Dammacco, M., Florio, S., Capriati, V. *Chem. Eur. J.* **2011**, *17*, 8216. (d) Interestingly, both (−)-sparteine and its (+)-surrogate were recently proven to be more effective than TMEDA in hampering the progress of racemization of electron-withdrawing substituted lithiated aryloxiranes, see, Mansueto, R., Perna, F. M., Salomone, A., Florio, S., Capriati, V. *Chem. Comm.* **2013**, *49*, 4911.

75 Schakel, M., Vrielink, J. J., Klumpp, G. W. *Tetrahedron Lett.* **1987**, *28*, 5747.

76 Coppi, D. I., Salomone, A., Perna, F. M., Capriati, V. *Chem. Comm.* **2011**, *47*, 9918.

77 Coppi, D. I., Salomone, A., Perna, F. M., Capriati, V. *Angew. Chem. Int. Ed.* **2012**, *51*, 7532.

78 (a) Stymiest, J. L., Bagutski, V., French, R. M., Aggarwal, V. K. *Nature* **2008**, *456*, 778. (b) Bagutski, V., French, R. M., Aggarwal, V. K. *Angew. Chem. Int. Ed.* **2010**, *49*, 5142. (c) Elford, T. G., Nave S., Sonawane, R. P., Aggarwal, V. K. *J. Am. Chem. Soc.* **2011**, *133*, 16798. (d) Dutheuil, G., Webster, M. P., Worthington, P. A., Aggarwal, V. K. *Angew. Chem. Int. Ed.* **2009**, *48*, 6317. (e) Sonawane, R. P., Jheengut, V., Rabalakos, C., Larouche-Gauthier, R., Scott, H. K., Aggarwal, V. K. *Angew. Chem. Int. Ed.* **2011**, *50*, 3760. (f) Aggarwal, V. K., Binanzer, M., de Ceglie, C., Gallanti, M., Glasspoole, B. W.,

Kendrick, S. J. F., Sonawane, R. P., Vázquez-Romero, A., Webster, M. P. *Org. Lett.* **2011**, *13*, 1490. (g) Scott, H. K., Aggarwal, V. K. *Chem. Eur. J.* **2011**, *17*, 13124.
79 For a recent review on the generation, reactivity and use in modern synthetic chemistry of aziridinyl anion, see: Florio, S., Luisi R. *Chem. Rev.* **2010**, *110*, 5128.
80 (a) O'Brien, P., Rosser, C. M., Caine, D. *Tetrahedron Lett.* **2003**, *44*, 6613. (b) O'Brien, P., Rosser, C. M., Caine, D. *Tetrahedron* **2003**, *59*, 9779. (c) Huang J., O'Brien P. *Chem. Comm.* **2005**, 5696.
81 (a) Müller, P., Nury, P. *Helv. Chem. Acta* **2001**, *84*, 662. (b) Müller, P., Riegert, D., Bernardinelli, G. *Helv. Chim. Acta* **2004**, *87*, 227.
82 (a) Rosser, C. M., Coote, S. C., Kirby, J. P., O'Brien, P., Caine, D. *Org. Lett.* **2004**, *6*, 4817. (b) Hodgson, D. M., Stefane, B., Miles, T. J., Witherington, J. *J. Org. Chem.* **2006**, *71*, 8510.
83 Huang, J., O'Brien, P. *Chem. Comm.* **2005**, 5696.
84 (a) Hodgson, D. M., Miles, S. M. *Angew. Chem. Int. Ed.* **2006**, *45*, 935. (b) Hodgson, D. M., Humphreys, P. G., Ward, J. G. *Org. Lett.* **2006**, *8*, 995.
85 Schmidt, F., Keller, F., Vedrenne, E., Aggarwal, V. K. *Angew. Chem. Int. Ed.* **2009**, *48*, 1149.
86 (a) Capriati, V., Florio, S., Luisi, R., Musio, B., Alkorta, I., Blanco, F., Elguero, J. *Struct. Chem.* **2008**, *19*, 785. (b) Capriati, V., Florio S., Luisi R., Mazzanti A., Musio, B. *J. Org. Chem.* **2008**, *73*, 3197.
87 (a) Morris, K. F., Johnson, C. S., Jr. *J. Am. Chem. Soc.* **1992**, *114*, 3141. (b) A combination of ^{7}Li and pulsed gradient spin-echo methods also proved to be useful for recognizing both ion pairing and aggregation of organolithiums in THF, see: Fernández, I., Martínez-Viviente, E., Breher, F., Pregosin, P. S. *Chem. Eur. J.* **2005**, *11*, 1495.
88 Li, D., Keresztes, I., Hopson, R., Williard, P. G. *Acc. Chem. Res.* **2009**, *42*, 270.
89 Granitzka, M., Pöppler, A.-C., Schwarze, E. K., Stern, D., Schulz, T., John, M., Herbst-Irmer, R., Pandey, S. K., Stalke, D. *J. Am. Chem. Soc.* **2012**, *134*, 1344.
90 Pöppler, A.-C., Keil, H., Stalke, D., John, M. *Angew. Chem. Int. Ed.* **2012**, *51*, 7843.
91 For a Concept article, see: Stalke, D. *Chem. Eur. J.* **2011**, *17*, 9264.
92 (a) Kim, H., Nagaki, A., Yoshida, J. *Nat. Commun.* **2011**, article number *264*. (b) Tomida, Y., Nagaki, A., Yoshida, J. *J. Am. Chem. Soc.* **2011**, *133*, 3744.
93 (a) Hermann, W. A. *Angew. Chem. Int. Ed.* **2002**, *41*, 1290. (b) Enders, D., Niemeier, O., Henseler, A. *Chem. Rev.* **2007**, *107*, 5606. (c) Kühl, O. In *Functionalized N-Heterocyclic Carbene Complexes*. Wiley: Chichester, **2010**. (d) Meres, L., Albrecht, M. *Chem. Soc. Rev.* **2010**, *39*, 1903.
94 (a) Shishkov, I. V., Rominger, F., Hofmann, P. *Organometallics* **2009**, *28*, 3532. (b) Wang, Y., Xie, Y., Abraham, M. Y., Wei, P., Schaefer III, H. F., Schleyer, P. v. R., Robinson, G. H. *J. Am. Chem. Soc.* **2010**, *132*, 14370. (c) Jana, A., Azhakar, R., Tavcar, G., Roesky, H. W., Objartel, I., Stalke, D. *Eur. J. Inorg. Chem.* **2011**, 3686.
95 (a) Popescu, A.-R., Musteti, A. D., Ferrer-Ugalde, A., Viñas, C., Núñez, R., Teixidor, F. *Chem. Eur. J.* **2012**, *18*, 3174. (b) Olid, D., Viñas, C., Teixidor, F. *Chem. Eur. J.* **2012**, *18*, 12936.

12

RHODIUM CARBENES

HUW M. L. DAVIES AND BRENDAN T. PARR
Department of Chemistry, Emory University, Atlanta, GA, USA

12.1 INTRODUCTION

The discovery of dirhodium teracarboxylates as highly effective catalysts for the decomposition of diazo compounds set forth a renaissance in the field of carbenoid chemistry. Since the pioneering work of Teyssie,[1,2] research in this area has experienced tremendous growth, especially with the development of chiral dirhodium catalysts capable of highly enantioselective transformations.[3–6] The state of the art of rhodium–carbenoid chemistry was comprehensively reviewed in the late 1990s in a book[7] and several key reviews.[8–12] Due to the extent of rhodium–carbenoid chemistry, many subsections of the field have been reviewed in recent years.[13–26] Rhodium–carbenoid intermediates are capable of a breadth of stereoselective transformations, including cyclopropanation,[20,23,24] cyclopropenantion,[23] C–H insertion,[15–19,23] and the generation of a wide variety of ylide intermediates[21,22,26–29] (Scheme 12.1). The scope of carbenoid reactions is very broad because many of the initial products/intermediates illustrated in Scheme 12.1 are often highly reactive, and thus, capable of undergoing a diverse range of elaborate cascade reactions.[21,22]

This chapter focuses on the most inspiring recent examples of rhodium–carbenoid chemistry, which we hope will illustrate the vitality of this exciting field. Due to space limitations, many useful transformations that follow well-established carbenoid protocols are not covered. Instead, the chapter concentrates on new directions in carbenoid chemistry because these are the areas most likely to inspire future growth. The chapter begins with an overview

Contemporary Carbene Chemistry, First Edition. Edited by Robert A. Moss and Michael P. Doyle.

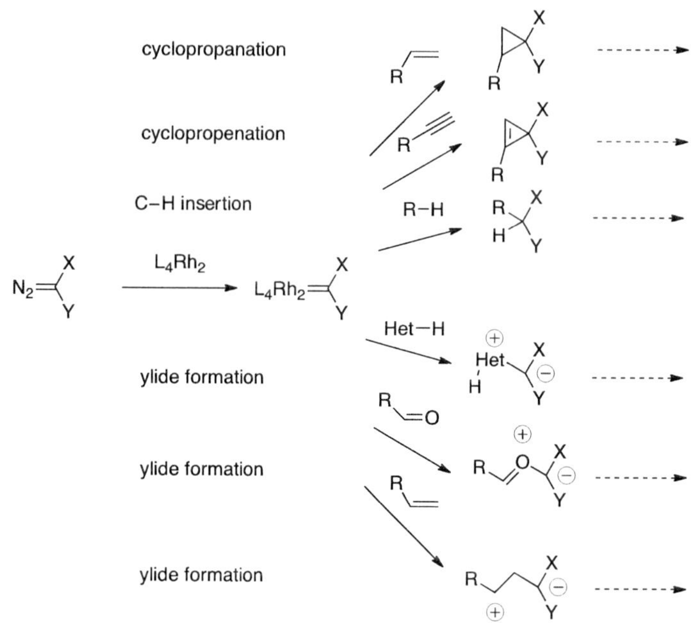

Scheme 12.1.

of the major classes of carbenoid intermediates and the most established chiral catalysts. This is followed by a discussion of enantioselective cyclopropanation and cyclopropenation, which illustrates the expanding range of carbenoids that are capable of highly enantioselective transformations. The next section describes the rapid growth of intermolecular C–H insertion into arguably one of the most effective strategies for stereoselective C–H functionalization at sp^3 C–H bonds. Another major topic is the recent advances in enantioselective cascade reactions of ylide intermediates. The final section describes new synthetic opportunities associated with vinylcarbenoid intermediates, in which reactions are initiated at the vinylogous position of the carbenoid rather than at the carbenoid center.

12.2 OVERVIEW OF RHODIUM–CARBENOID INTERMEDIATES AND CHIRAL CATALYSTS

The most common method for the synthesis of rhodium–carbenoid intermediates is by means of the metal-catalyzed extrusion of dinitrogen from diazo compounds.[7] In the last decade, it has become widely recognized that the nature of the functionality around the carbenoid has a major impact on the reaction outcome. Rhodium carbenoids are highly electrophilic and are commonly classified into three major groups: acceptor carbenoids (**1**), acceptor/acceptor carbenoids (**2**), and donor/acceptor carbenoids (**3**) (Scheme 12.2).[13,19,23,30]

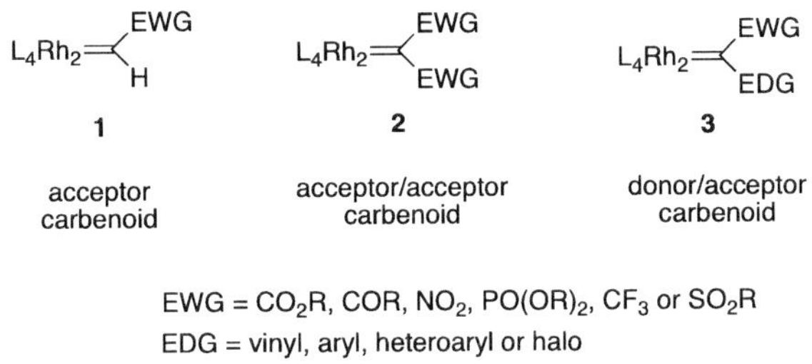

Scheme 12.2.

The acceptor carbenoids, such as the carbenoids derived from ethyl diazoacetate, have been the most extensively studied.[7] The acceptor/acceptor carbenoids are extremely electrophilic and the presence of two acceptor groups makes these carbenoids more prone to react *via* ylide intermediates compared to the other classes of carbenoids. Many of the most significant advances in recent years, especially for intermolecular reactions, have involved donor/acceptor carbenoids.[13,19,23,30] The reactivity of these carbenoids is attenuated by the presence of an electron-donating group, which has enabled the development of a number of highly selective transformations that are simply not possible with the other classes of rhodium–carbenoid intermediates.

The development of highly efficient chiral dirhodium catalysts has been a major factor in advancing the field of rhodium–carbenoid chemistry. The most notable chiral catalysts are dirhodium catalysts, such as Rh_2(5*S*-MEPY)$_4$ (**4**) and Rh_2(4*S*-MEOX)$_4$ (**5**)[6]; the phthalimido catalysts, such as Rh_2(*S*-PTTL)$_4$ (**6**),[31,32] Rh_2(*S*-TCPTTL)$_4$ (**7**),[33] Rh_2(*S*-TBPTTL)$_4$ (**8**),[34] Rh_2(*S*-PTAD)$_4$ (**9**),[35] and the related naphthylimido catalyst Rh_2(*S*-NTTL)$_4$ (**11**); and the prolinate catalysts,[4] such as Rh_2(*S*-DOSP)$_4$ (**12**)[36] and Rh_2(biTISP)$_2$ (**13**)[37] (Scheme 12.3). Recent developments in catalyst design that appear promising are the triarylcyclopropanecarboxylate catalyst Rh_2(*S*-BTPCP)$_4$[38] (**14**) and mixed ligand catalysts, such as the tetracarboxylate catalysts, Rh_2(*S*-PTTL)$_3$TPA (**15**)[39] and Rh_2(*S*-TCPTTL)$_3$OAc (**16**), and the monoacetate triscarboxamidate catalyst Rh_2(OAc)(DPTI)$_3$ (**17**). As a general guideline,[5] the dirhodium tetracarboxamidates tend to be the optimum rhodium catalysts for enantioselective reactions of acceptor carbenoids and the rhodium prolinates tend to be the best for donor/acceptor carbenoids, especially when the acceptor group is a methyl ester. On the other hand, the phthalimido catalysts have been shown to be capable of achieving high asymmetric induction with a variety of carbenoid types.

Even though it quickly became apparent that the dirhodium complexes were exceptional catalysts for the reactions of diazo compounds, early efforts at the development of effective chiral dirhodium catalysts were unsuccessful.[40] Low levels of enantioselectivity were observed for a series of dirhodium tetracarboxylate catalysts and it was considered that the presence of two potential carbene coordination sites plus the conformational mobility of the ligands and the bound carbenoid, would render the system unsuitable for achieving

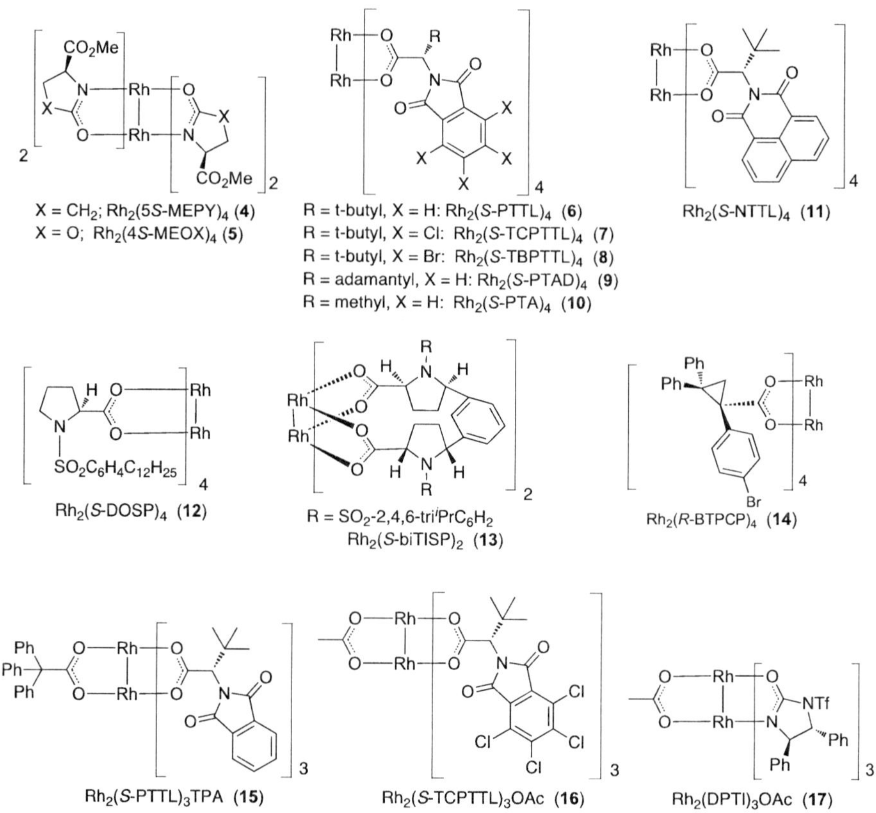

Scheme 12.3.

high asymmetric induction.[3] The dirhodium tetracarboxamidate complexes circumvent this problem because they adopt a C_2-symmetric arrangement with the amide ligands bound in an ααββ arrangement, leading to two identical faces of the catalysts (Fig. 12.1).[6] Early on in the development of chiral dirhodium tetracarboxylate catalysts, it was recognized that the arrangement of four identical ligands around the D_{4h}-symmetric dirhodium core could result in the complexes adopting a higher symmetry arrangement than that of the ligands themselves.[4,5,36] Even though there is considerable conformational mobility in many of these catalysts, the prolinate catalyst Rh_2(S-DOSP)$_4$ is considered to adopt a D_2-symmetric arrangement with the *N*-sulfonyl group adopting an αβαβ orientation,[36] and indeed, Rh_2(biTISP)$_2$ was specifically designed with bridging bis-carboxylate ligands, which force the complex to adopt a D_2-symmetric arrangement.[37] The phthalimido complexes were originally proposed to exist in a C_2-symmetric arrangement with an ααββ orientation of the ligands. However, more recent computational and X-ray crystallographic studies on the bulky phthalimido catalysts, Rh_2(*S*-PTTL)$_4$ and Rh_2(*S*-TCPTTL)$_4$,[33] reveal that these catalysts preferentially adopt an αααα-orientation. In the

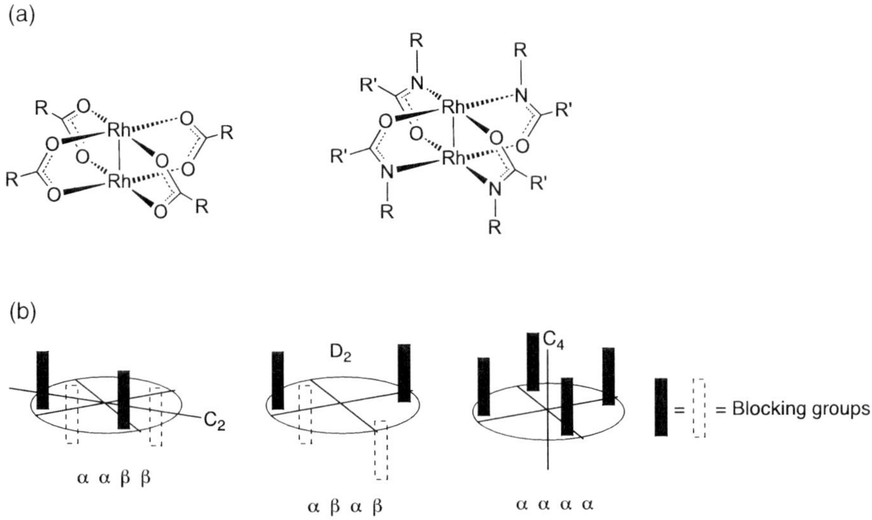

Figure 12.1. High-symmetry chiral dirhodium catalysts. (a) Dirhodium tetracarboxylate and tetracarboxamidate structures and (b) C_2, D_2 and C_4, symmetric arrangements of dirhodium complexes.

perfectly aligned αααα-orientation, the complex would be C_4-symmetric, but the structural studies revealed that the complex actually adopts a slightly distorted structure, making the coordinating face of the catalyst appear as a C_2-symmetric "chiral crown." Rh_2(*S*-PTTL)$_4$, however, is still considered to have considerable conformational mobility, whereas the tetrachloro analogue, Rh_2(*S*-TCPTTL)$_4$, is much more rigid.[33] Rh_2(*S*-BTPCP)$_4$ is a very bulky catalyst, which also preferentially adopts a D_2-symmetric arrangement, but when the carbenoid binds to the catalysts two of the aryl groups on the ligands need to slip to a more equatorial position to make room for the carbenoid.[38] It should be noted that the recent development of complexes containing mixed ligands[39,41,42] reveals that a highly symmetrical arrangement of the chiral ligands is not an absolute requirement for high levels of asymmetric induction.

12.3 ENANTIOSELECTIVE CYCLOPROPANATION

The standard reference reaction that has been used to evaluate new chiral catalysts has been intermolecular cyclopropanation of alkenes. By the late 1990s, the dirhodium tetracarboxamides had been demonstrated to be the optimum chiral rhodium catalysts for the cyclopropanation reactions of acceptor carbenoids, especially those derived from diazoacetates to form **18** (Scheme 12.4).[43] However, a range of other metal complexes, particularly copper, ruthenium, and cobalt complexes are broadly used for enantioselective cyclopropanations with acceptor carbenoids.[7,23] Enantioselective cyclopropanation of acceptor

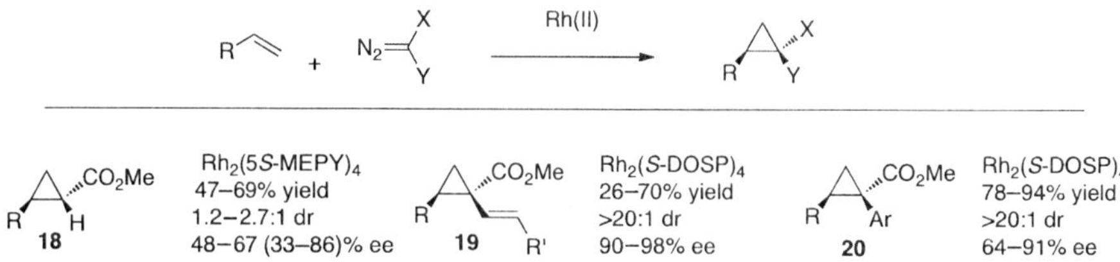

Scheme 12.4.

carbenoids are best achieved with nonrhodium catalysts. Rh_2(*S*-DOSP)$_4$ had been shown to be the most versatile catalyst for the cyclopropanation reactions of donor/acceptor carbenoids derived from methyl vinyldiazoacetates[36] and aryldiazoacetates,[44,45] to form **19** and **20**, respectively. A comprehensive review on intermolecular cyclopropanation by diazo compounds was published in 2001.[23] Scheme 12.4 highlights the optimum rhodium catalyst and the ranges of yields, diastereoselectivity and enantioselectivity for each class of carbenoid system.

In the last decade, more diverse carbenoid systems have been shown to be capable of highly enantioselective cyclopropanations. The adamantyl derived phthalimido catalyst Rh_2(*S*-PTAD)$_4$ gives high asymmetric induction with a range of donor/acceptor carbenoids, in which the typical methyl ester acceptor group has been replaced by dialkyl phosphonate,[35,46] cyano,[47] trifluoromethyl,[48] or keto[49] groups to form cyclopropanes **21–24** (Scheme 12.5). A further benefit of the cyclopropanation chemistry of donor/acceptor carbenoids is that the reactions are routinely highly diastereoselective. Rh_2(*S*-NTTL)$_4$ has been shown to be an effective chiral catalyst for the reactions of siloxyvinyldiazoacetates to form cyclopropanes **25**.[50] By using bulky catalysts, the typical 1,2-hydride shift plaguing reactions of alkyl-substituted diazoacetates can be averted,[51,52] and if Rh_2(*S*-PTTL)$_4$ is used as catalyst, cyclopropanes **26** are formed in a highly enantioselective manner.[53] Rh_2(*S*-BTPCP)$_4$ is another sterically encumbered catalyst that gives high asymmetric induction with *t*-butyl styryldiazoacetate to form **27**.[38] A further extension of the chemistry has been the use of bromodiazoacetates[54] and bromodiazophosphonates[55] to form **28** and **29**, respectively. Even though a bromo group would be normally considered inductively as an acceptor group, it is still able to stabilize the highly electrophilic carbenoid by donation through resonance.[56] Recently, enantioselective protocols were developed for acceptor/acceptor carbenoids. The Rh_2(*S*-NTTL)$_4$-catalyzed reactions of α-nitrilo diazoacetamides[57] and *p*-methoxyphenyl ketone derivative[33,58,59] gives rise to cyclopropanes **30** and **31**, respectively.

Even though diazo compounds have been used as carbenoid precursors in large-scale processes, there is always some concern associated with their safety.[60–63] Consequently, efforts have been made to develop safer, alternative carbenoid precursors. A promising carbenoid source has been *N*-sulfonyl-1,2,3-triazoles. Heating of the *N*-sulfonyltriazole **32** in the presence of styrene and Rh_2(*S*-NTTL)$_4$ (**11**) results in the formation of the cyclopropane **35** with high asymmetric induction (Scheme 12.6)[64] Presumably, the reaction proceeds *via*

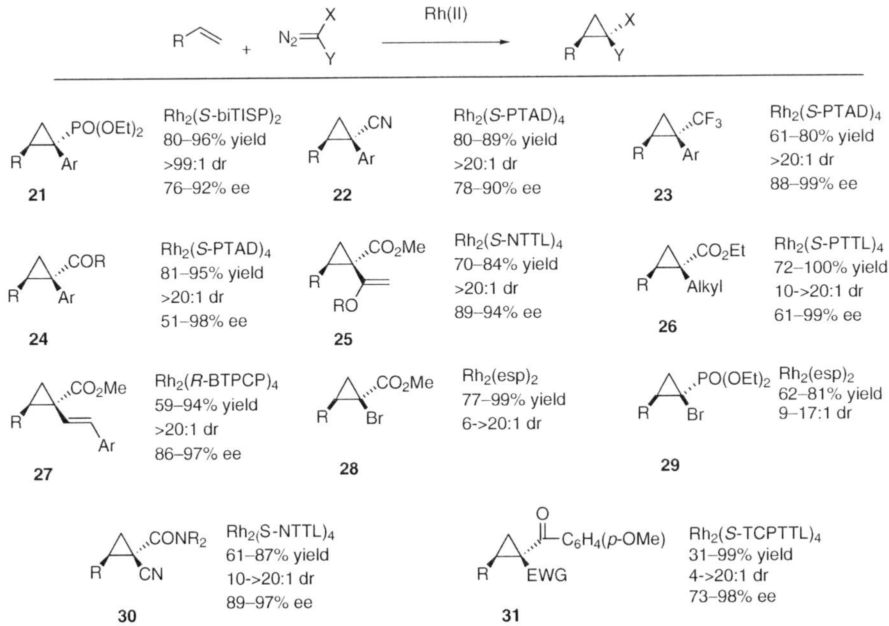

Scheme 12.5.

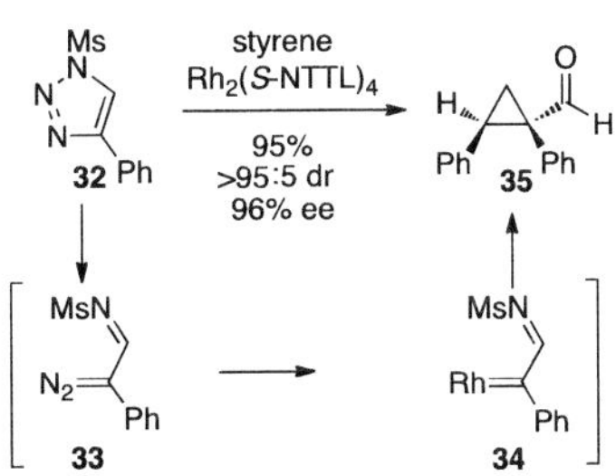

Scheme 12.6.

an electrocyclic ring opening to expose the imino diazo compound **33**, which then generates the iminocarbenoid intermediate **34**. Several recent studies show that *N*-sulfonyltriazoles undergo a variety of synthetically useful carbenoid transformations.[65–71]

The vast majority of reported rhodium-catalyzed cyclopropanations use a catalyst loading of 1 mol%.[23] Donor/acceptor carbenoids, however, are capable of effective transformations with much lower catalyst loadings because these carbenoids are more stabilized, and so are less likely to destroy the catalyst.[72] The $Rh_2(S\text{-PTAD})_4$-catalyzed cyclopropanation of cyclopentadiene (**37**) by *p*-methoxyphenyldiazoacetate **36** to form the bicycle **38** can be conducted with catalyst loadings as low as 0.00006 mol% when solvent-free conditions are used (Eq. 12.1).[73]

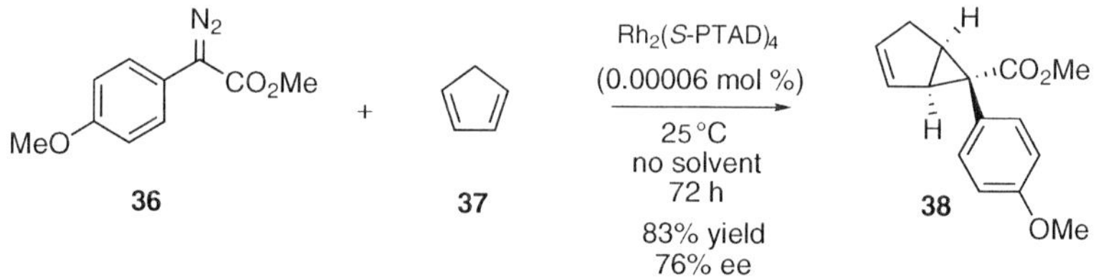

(12.1)

A number of technological advances have been made to increase the practicality of rhodium-catalyzed reactions.[74] Several strategies have been developed for immobilizing the catalysts so that they can be recovered and recycled. One of the simplest methods is a universal immobilization strategy, in which the catalysts are bound to a pyridine-derived, highly cross-linked, polystyrene resin.[75] A most impressive example is a silica-bound derivative of $Rh_2(S\text{-PTTL})_4$, which was recycled 100 times in an intramolecular C–H insertion without any depletion in the level of enantioselectivity.[76] Other technological advances have been the use of supercritical media[77] or ionic liquids,[78] which also allow for catalyst recovery and recycling.

12.4 CASCADE SEQUENCES INITIATED BY RHODIUM-CATALYZED CYCLOPROPANATION

Cyclopropanes are versatile intermediates in synthesis because they can undergo a variety of ring expansion reactions.[79,80] Rearrangement can be conducted thermally or by a second metal-catalyzed process. The cyclopropanation and subsequent rearrangement can be conducted sequentially or as a domino process. If the initial cyclopropanation is conducted asymmetrically, then the final product can often be obtained in an enantioenriched form.

Cyclopropanation by an acceptor/acceptor carbenoid followed by a thermal ring expansion was recently used in the synthesis of (+)-norrisolide (**43**) (Scheme 12.7).[81] Reaction of **39** with **40** generated the cyclopropane **41** which on thermolysis at 185°C provided the bicyclic system **42**, which was then converted in a series of steps to **43**.

An elegant application of a sequential cyclopropanation followed by an indium-catalyzed intramolecular Friedel–Crafts alkylation was recently described for the synthesis of (±)-deethyleburnaminone (**48**) (Scheme 12.8).[82] $Rh_2(esp)_2$-catalyzed reaction of the diazoacetoacetate **44** with the enamide **45** generated the cyclopropane **46**, which on treatment with indium(III) triflate, underwent cyclopropane ring opening, followed by cyclization to the tetracycle **47**, which could then be readily converted to **48**.

One of the synthetically useful features of rhodium(II)-catalyzed reactions of diazo compounds is the orthogonal reactivity compared to many other transition metal-catalyzed processes, such as palladium-catalyzed cross coupling processes.[83] Therefore, it is possible to design one-pot processes, in which

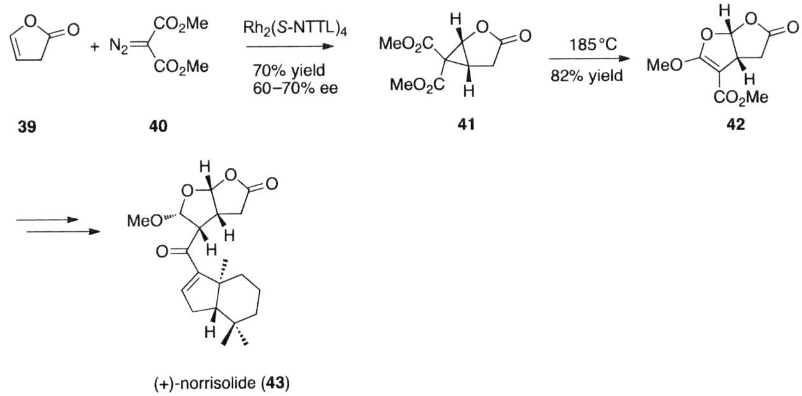

Scheme 12.7.

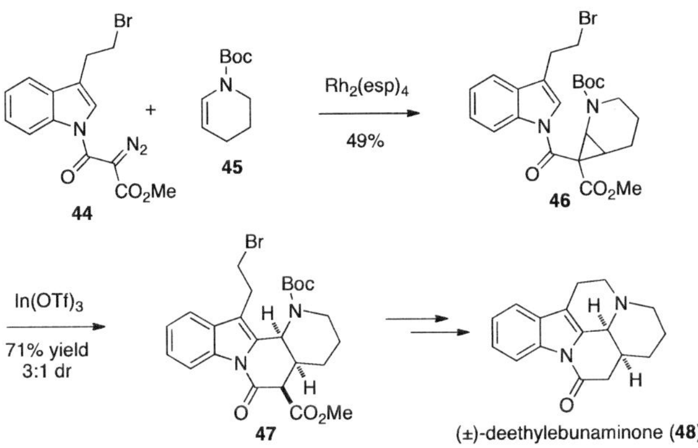

Scheme 12.8.

distinct catalysts are used for different steps in a sequential process. One example is combined rhodium- and gold-catalyzed processes for the synthesis of fused dihyrofuran **52** (Scheme 12.9).[84] Rhodium trifluoroacetate catalyzed the cyclopropanation between the diazoketone **49** and styrene to form **50**, which then underwent a gold-catalyzed ring expansion of the dihydrofuran **51**, followed by a gold-catalyzed electrocylization to form the fused dihydrofuran **52**.

When vinylcarbenoids are reacted with dienes, the resulting cyclopropane is a divinylcyclopropane, which is capable of undergoing a Cope rearrangement through a boat-like transition state, generating seven-membered rings containing up to three new stereogenic centers.[23] Extensive studies have been conducted to define the scope of this transformation, which is applicable to a wide variety of vinylcarbenoids and dienes. A recent example is the Rh_2(*S*-PTAD)$_4$-catalyzed reaction between the siloxyvinyldiazoacetate **54** and pyrrole **53** to form the tropane derivative **56**, presumably *via* the intermediacy of the divinylcyclopropane **55** (Scheme 12.10).[85]

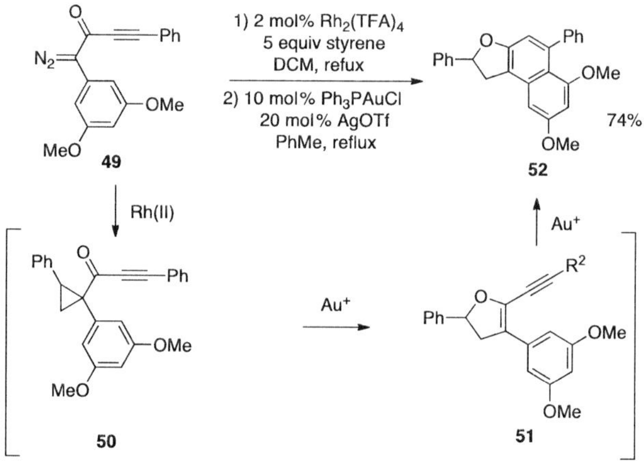

Scheme 12.9.

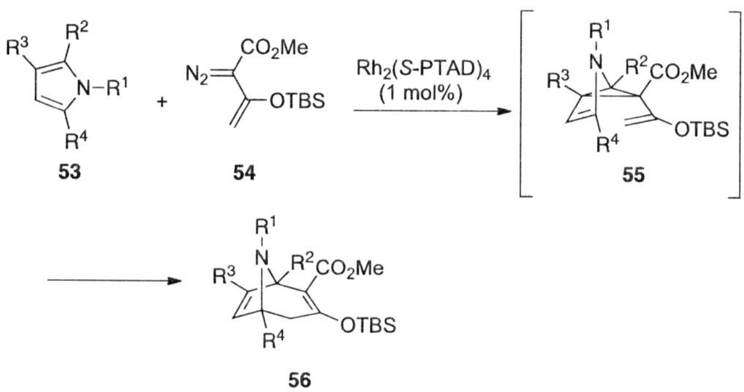

Scheme 12.10.

The tandem cyclopropanation/Cope rearrangement has been applied to the synthesis of a variety of natural products.[86] A recent example that illustrates the potential of this transformation is the total synthesis of 5-*epi*-vibsanin E (**60**) (Scheme 12.11).[87] $Rh_2(S\text{-PTAD})_4$-catalyzed reaction of the siloxyvinyldiazoacetate **54** with the triene **57** results in selective cyclopropanation of the monosubstituted alkene in **57**. A Cope rearrangement of the resulting divinylcyclopropane forms the cycloheptadiene **58**. The quaternary stereocenter generated in this step is used to control the stereochemistry during the formation of the tricyclic core **59** *en route* to 5-*epi*-vibsanin E (**60**).

Other notable applications of the tandem cyclopropanation/Cope rearrangement in total synthesis include the syntheses of (±)-tremulenolide (**61**),[88] (+)-barekol (**62**),[89] of (±)-isostemofoline (**63**),[90] (+)-frondosin B (**64**),[91] and (−)-englerin A (**65**)[92] (Scheme 12.12). In each structure, the bonds that are formed in the tandem cyclopropanation/Cope rearrangement are highlighted.

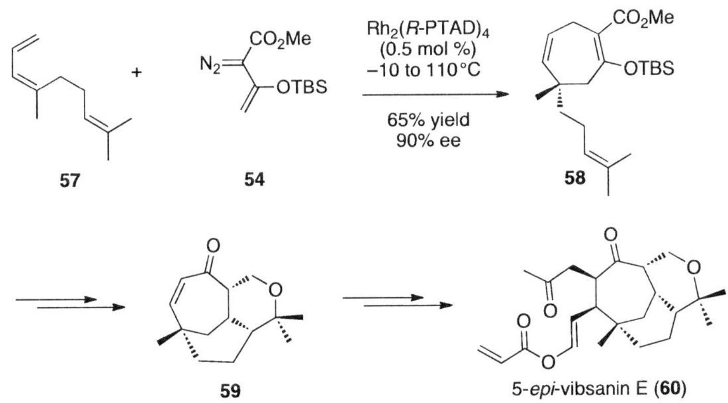

Scheme 12.11.

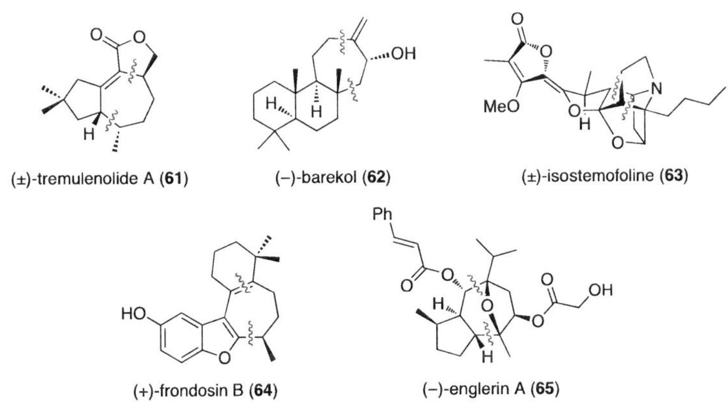

Scheme 12.12.

12.5 ENANTIOSELECTIVE CYCLOPROPENATION

The main advances in rhodium-catalyzed enantioselective cyclopropenation over the last decade have been directed toward broadening the range of carbenoids that undergo the reaction, as summarized in Scheme 12.13. Dirhodium tetracarboxamidate catalysts had previously been demonstrated by Doyle in the 1990s to be capable of highly enantioselective cyclopropenation by acceptor carbenoids to form **66**.[43] The mixed carboaxamidate/carboxylate $Rh_2(DPTI)_3(OAc)$ (**17**) was subsequently found to also be an efficacious chiral catalyst with acceptor carbenoids to form **67**.[41] $Rh_2(S\text{-DOSP})_4$ is an effective catalyst for cyclopropenation with methyl aryldiazoacetates[93] and styryldiazoacetates[94] to form **68** and **69**, respectively, although in the case of styryldiazoacetates, the reactions need to be conducted at low temperatures to avoid rearrangement of the vinylcyclopropenes to cyclopentadienes. $Rh_2(S\text{-PTAD})_4$ provides high

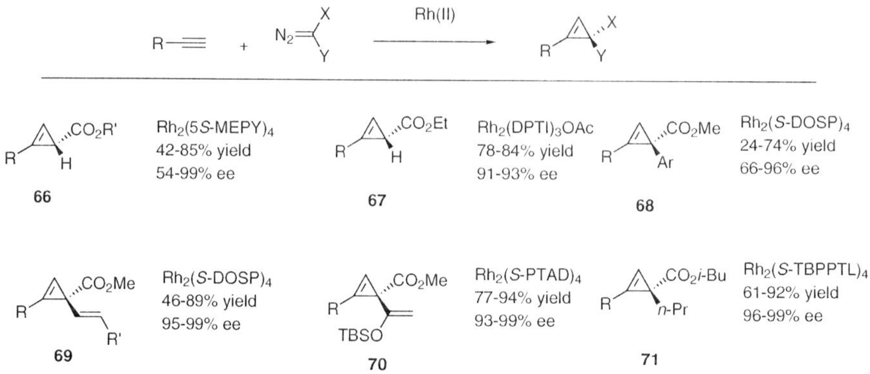

Scheme 12.13.

asymmetric induction in cyclopropenation with siloxyvinyldiazoacetate **54** to form **70**,[95] whereas, Rh_2(*S*-TBPTTL)$_4$ gave high asymmetric induction with *i*-butyl diazopentanoate to form **71**.[34]

The actual mechanism of the cyclopropenation has been the subject of controversy. The originally proposed mechanism by Doyle was cyclopropenation involving a concerted asynchronous process with alkyne approaching perpendicular to the face of the catalyst.[43] Corey proposed that the reaction is a stepwise process involving partial ligand dissociation with formation of a metalcyclobutene intermediate.[41] This latter hypothesis was refuted in a computational study by Singleton.[96] More recent computational studies by Davies[94] and Hashimoto[34] confirm that the cyclopropenation is a concerted asynchronous process. Moreover, the alkyne approaches at a slightly tilted angle of 18–33° from the perpendicular position, which allows for hydrogen bonding to occur between the alkyne C–H and a carboxylate ligand (Fig. 12.2). This model rationalizes the observed asymmetric induction and is consistent with the observation that rhodium-catalyzed cyclopropenation is not generally effective with internal alkynes.

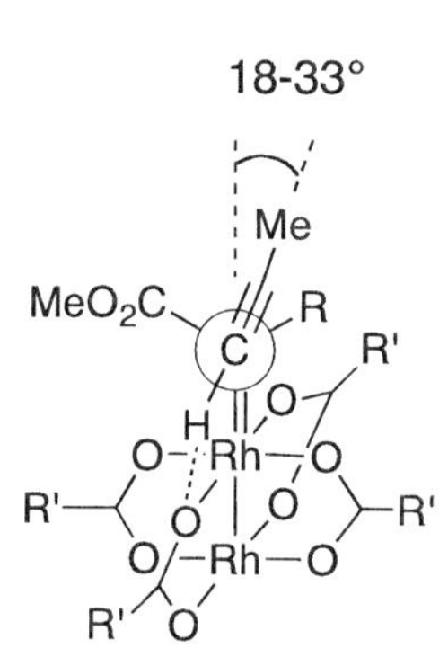

Figure 12.2. Transition state model for cyclopropenation.

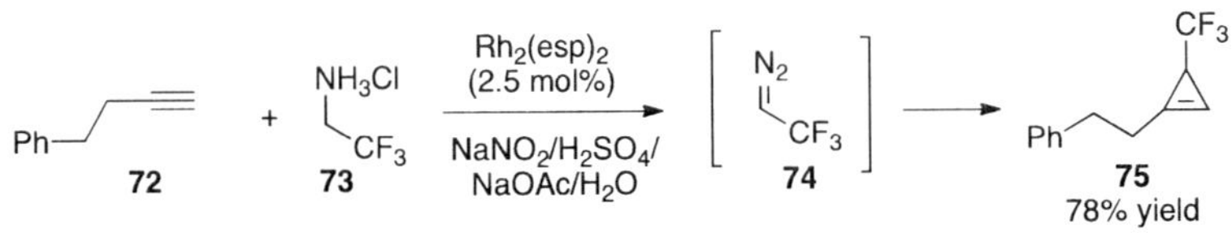

Scheme 12.14.

A particularly intriguing development is the *in situ* formation of the potentially explosive trifluoromethyldiazomethane (**74**) and its implementation in rhodium-catalyzed cyclopropenation in an aqueous environment (Scheme 12.14).[97] An illustrative example is the $Rh_2(esp)_4$-catalyzed reaction between trifluoroethylamine hydrochloride (**73**) and 4-phenyl-1-buytyne (**72**), which generated the trifluoromethylcyclopropene **75** in 78% yield.

Cyclopropenes often undergo facile ring expansion rearrangements and these rearrangements can be catalyzed by the dirhodium complexes.[98] In particular, the reactions of ketocarbenoids often preferentially form furan products either through ring expansion of initially formed cyclopropenes or *via* zwitterionic intermediates. A recent example is the reaction of ynamides with diazoacetoacetate, which generates aminofurans, which when appropriately functionalized will undergo a further intramolecular Diels–Alder cycloaddition.[99] Another interesting annulation example is the reaction of pyridotriazole **76** with alkyne **77** to form indolizine **79**, presumably proceeding through the carbenoid intermediate **78** (Scheme 12.15).[100,101]

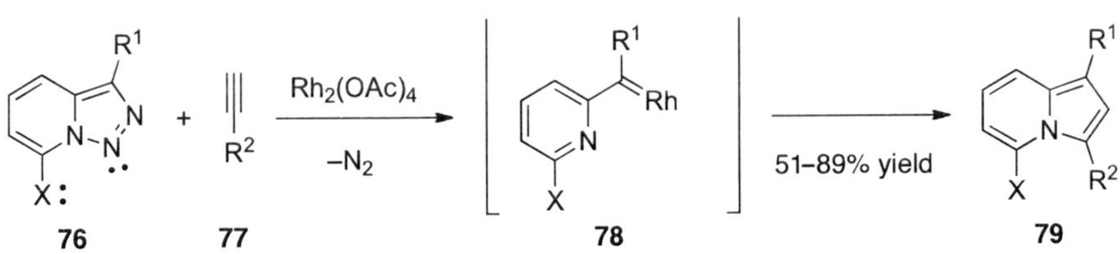

Scheme 12.15.

12.6 C–H FUNCTIONALIZATION BY CARBENOID-INDUCED C–H INSERTION

Carbenoid-induced C–H insertion is an attractive strategy for C–H functionalization.[16,18,19,30,102,103] Dirhodium catalysts were instrumental in the development of carbenoid C–H insertion.[104] By the mid-1990s the intramolecular version of this reaction was broadly developed.[9] Intramolecular C–H insertions preferentially formed five-membered rings as illustrated in the generic reaction of **80** to form **81** shown in Equation (12.2), but many examples for the selective formation of four- and six-membered rings were also known.[79] Site selectivity guidelines were becoming well-established; the order of reactivity decreased from methyne to methylene to methyl C–H bonds, whereas allylic and benzylic sites as well as sites adjacent to oxygen and nitrogen groups were particularly susceptible to C–H functionalization. The chiral dirhodium tetracarboxamidate catalysts enabled very high asymmetric induction for intramolecular C–H insertions of acceptor carbenoids,[105] whereas the *N*-phthalimido catalysts such as Rh_2(*S*-PTTL)$_4$ gave moderately high levels of enantioinduction for intramolecular reactions of acceptor/acceptor carbenoids.[106,107] Intramolecular C–H insertions by donor/acceptor carbenoids were first demonstrated in the early 2000s, and highly enantioselective examples have been demonstrated for both the prolinate and the phathlimido catalysts.[108,109]

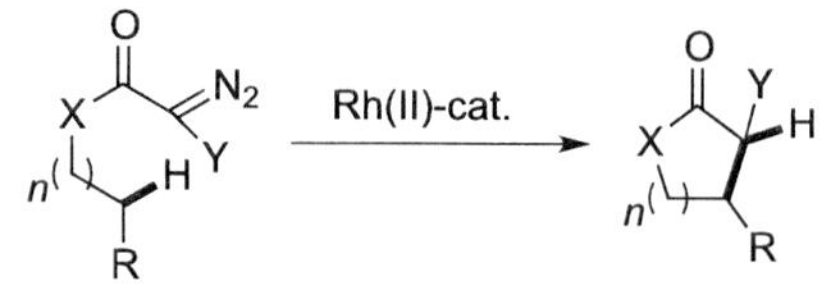

80 X = O or CH_2, *n* = 1-2 81
X = NR, *n* = 0
Y = H, COR, Ph

(12.2)

Intramolecular C–H insertion has been used extensively for the synthesis of natural products and pharmaceutical targets[110] and the applications of the established C–H insertions with acceptor carbenoids and acceptor/acceptor carbenoids have been extensively reviewed.[18,19,30,102,103] In recent years, a number of impressive examples of natural product synthesis using donor/acceptor carbenoids have appeared. One example is the total synthesis (+)-lithospermic acid (**84**) (Scheme 12.16).[111] In order to obtain a reasonably high level of asymmetric induction in the intramolecular C–H insertion, a combination of a chiral catalyst and auxiliary was required. Rh_2(*S*-DOSP)$_4$ catalyzed reaction of the aryldiazoacetate **82** generated the dihydrobenzofuran **83** in 85% yield and 8:1 dr, which was then converted to (+)-lithospermic acid (**84**).

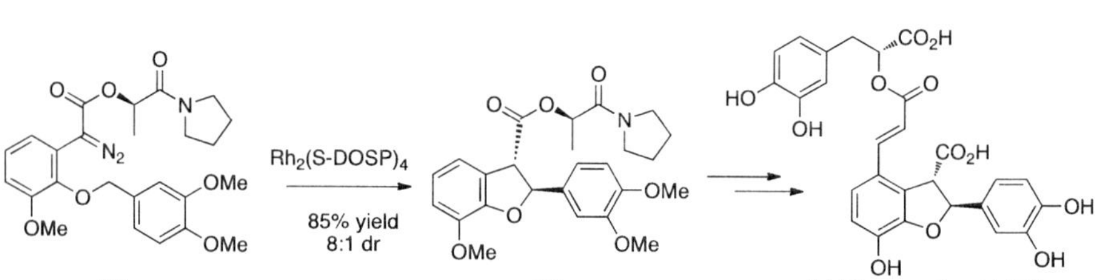

82 83 (+)-lithospermic acid (**84**)

Scheme 12.16.

Other notable recent applications of the intramolecular C–H insertion in total synthesis include the syntheses of (−)-hamigeran B (**85**),[112] (−)-ephedradine A (**86**),[113] (−)-serotobenine (**87**),[114] and (−)-tetrodotoxin (**88**)[115] (Scheme 12.17). In each structure, the bond that is formed during the intermolecular C–H insertion is highlighted.

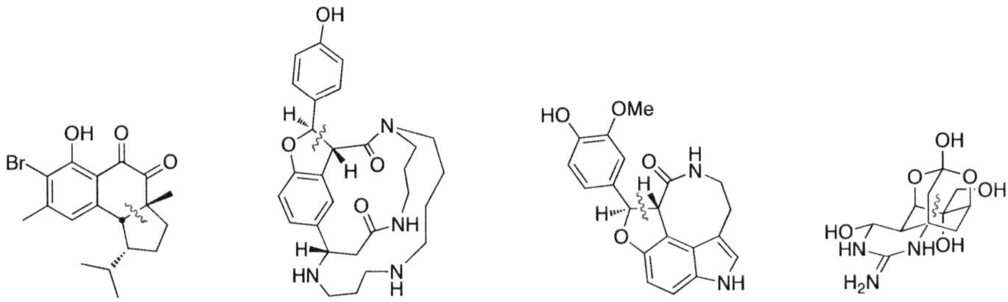

(–)-hamigeran B (**85**) (–)-ephedradine A (**86**) (–)-serotobenine (**87**) (–)-tetrodotoxin (**88**)

Scheme 12.17.

Even though the first examples of rhodium-catalyzed C–H insertions were intramolecular reactions, the development and synthetic application of intermolecular C–H insertions has lagged behind. Indeed, intermolecular C–H insertion in 1998 was still described as "mechanistically valuable, but not synthetically useful."[7] The challenge associated with intermolecular C–H insertions is control of site selectivity, because rhodium carbenoids are highly reactive and relatively unselective. Intermolecular C–H insertions with acceptor- and acceptor/acceptor carbenoids can be achieved with simple substrates such as cyclohexane and tetrahydrofuran, but when more complex substrates are implemented, mixtures tend to be formed.[1,2,104] The use of donor/acceptor carbenoids was the major breakthrough for intermolecular C–H insertions because they are still sufficiently electrophilic for reaction with C–H bonds, but due to their attenuated reactivity, the reactions can be highly regio- and stereoselective.[102] Representative examples of substrates that are capable of selective intermolecular C–H insertion with aryldiazoacetates are illustrated in Scheme 12.18.[102] Selective C–H functionalization can occur at a variety of C–H bonds from primary, secondary, or tertiary sites, from hydrocarbons to more activated bonds such as C–H bonds adjacent to oxygen or nitrogen or at allylic or benzylic sites.

Understanding of the mechanistic aspects of the C–H insertion process has greatly benefited from computational studies. The early proposals suggested that the reaction was a concerted, asynchronous process[105,116] or involved four-membered cyclic transition states.[116] Seminal computational studies by Nakamura on ethyl diazoacetate showed that the reaction was a concerted asynchronous process, initiated by what appeared to be a hydride transfer event from the C–H bond to the carbenoid.[117] Subsequent studies by Davies and Autschbach on donor/acceptor carbenoids also showed that the C–H insertion had considerable hydride transfer character and led to a model that rationalizes both the regiochemical and stereochemical outcome of the reaction (Fig. 12.3).[118] Due to the hydride transfer component of the mechanism, C–H bonds, in which positive charge build up at carbon is stabilized, are electronically favored for C–H insertion. This is counterbalanced by steric effects, because the rhodium carbenoid is sterically encumbered.

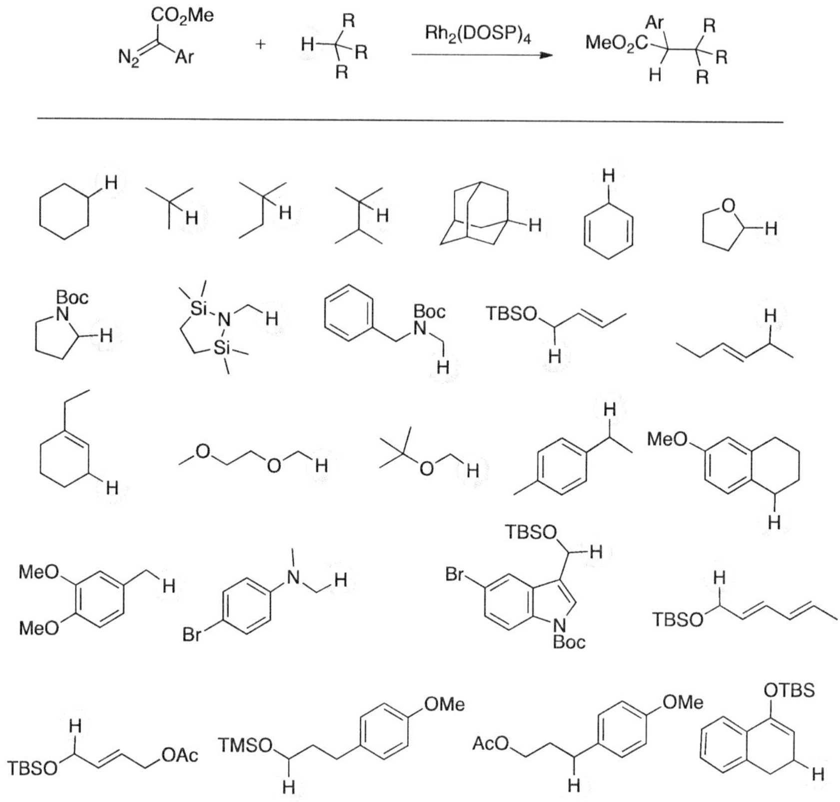

Scheme 12.18.

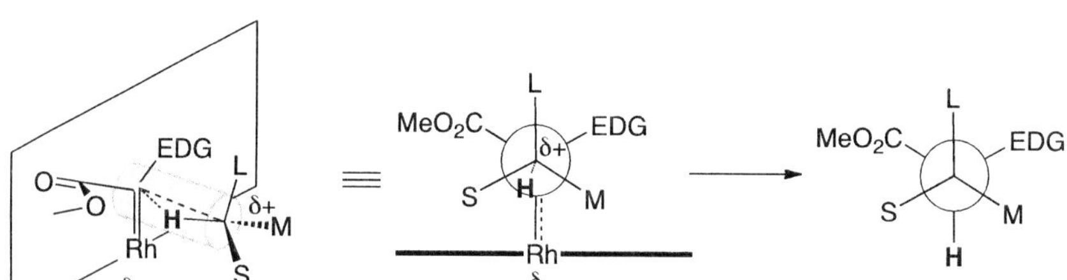

Figure 12.3. Transition state for C–H insertion. For color details, please see color plate section.

A review detailing the guiding principles for regioselective C–H functionalizations has recently been published,[102] and so this chapter will only give a limited overview of the scope of intermolecular C–H insertions. The predictive model shown in Figure 12.3 suggests that if the C–H functionalization occurs at a methylene site and there is appreciable size differentiation between the other two groups (i.e., L and M in Fig. 12.3), then the reaction can be highly diastereoselective. As Rh_2(*S*-DOSP)$_4$ routinely gives high asymmetric induction with

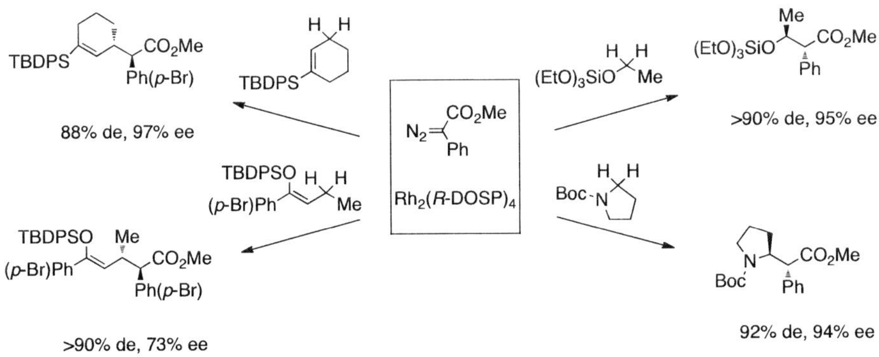

Scheme 12.19.

donor/acceptor carbenoids, then the reactions can also be highly enantioselective. Representative examples illustrating the levels of stereoselectivity that are attainable in intermolecular C–H insertions are shown in Scheme 12.19.

One of the most exciting aspects of C–H functionalization strategies is the opportunity for developing new retrosynthetic pathways for synthesis. As illustrated in Scheme 12.20, C–H insertion at sites adjacent to oxygen can be considered as strategic surrogates to the aldol reaction[119] and the Claisen condensation,[120] whereas C–H insertion at sites adjacent to nitrogen can be considered as strategic surrogates to the Mannich reaction.[121] Allylic C–H

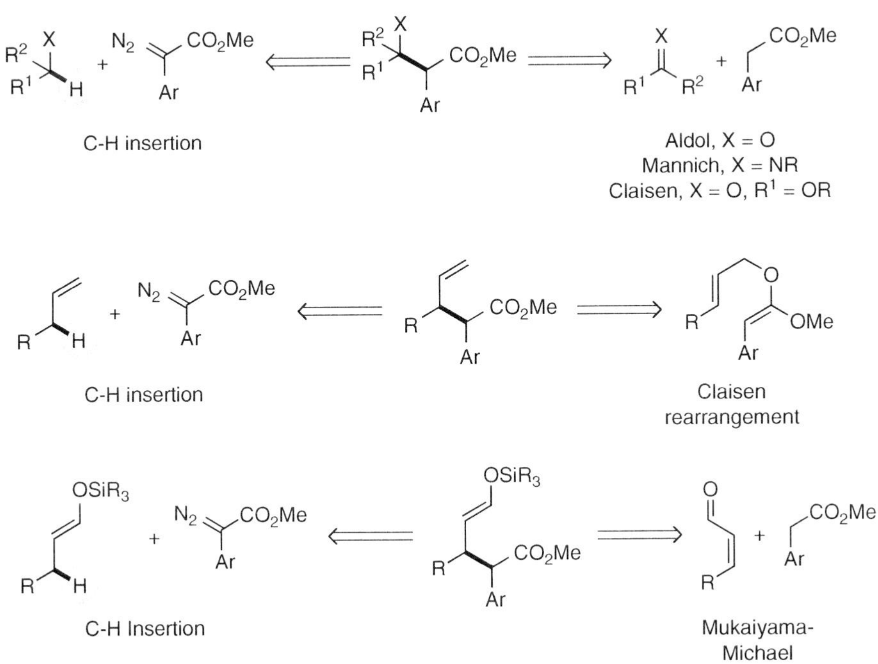

Scheme 12.20.

insertion generates the type of products that are typically formed in a Claisen rearrangement[122] or a Mukaiyama–Michael addition.[123]

Numerous examples have appeared in recent years of the application of intermolecular C–H insertion to the synthesis of pharmaceuticals and natural products.[102] Two recent applications are described in Scheme 12.21. Rh_2(*S*-DOSP)$_4$-catalyzed reaction of aryldiazoacetate **89** containing a chiral auxiliary with 1,4-cyclohexadiene (**90**) generated the C–H insertion product **91** in 88% yield and >39:1 dr, which was then converted to (+)-phenylkainic acid **92**.[124] Rh_2(*S*-DOSP)$_4$ catalyzed reaction of styryldiazoacetate **93** with allyl amine **94** generated the C–H insertion product **95** in 32% yield, 19:1 dr and 92% ee (cyclopropanation was a competing process).[125] This is an excellent example of the regioselectivity that is possible in C–H insertion with donor/acceptor carbenoids because the C–H insertion occurs at the *N*-ethyl site in preference to the *N*-allyl site, suggesting that steric effects are dominating over the electronic preference for this substrate. A subsequent metathesis transformation resulted in the formation of piperidine derivative **96**.

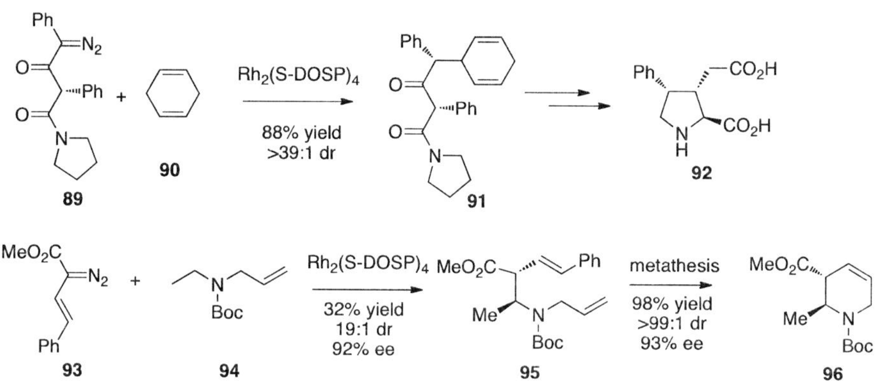

Scheme 12.21.

12.7 COMBINED C–H ACTIVATION/COPE REARRANGEMENT (CHCR)

The reaction between vinylcarbenoids (**97**) and allylic substrates (**98**) leads to unusual C–H functionalization products **99** by a transformation, which has been coined the CHCR (Scheme 12.22).[16] Computational studies reveal that the reaction begins by a hydride abstraction mechanism leading to an allyl cation and a rhodium-bound allyl anion.[126] Before any bond rotation is possible, C–C bond formation occurs to generate a product that is conceptually derived from a C–H insertion, which has been interrupted by a Cope rearrangement. The chair- and boat-like transition states are both energetically accessible and either can be induced depending on the functionality of the substrates. Cyclic substrates will tend to react through a chair-like transition

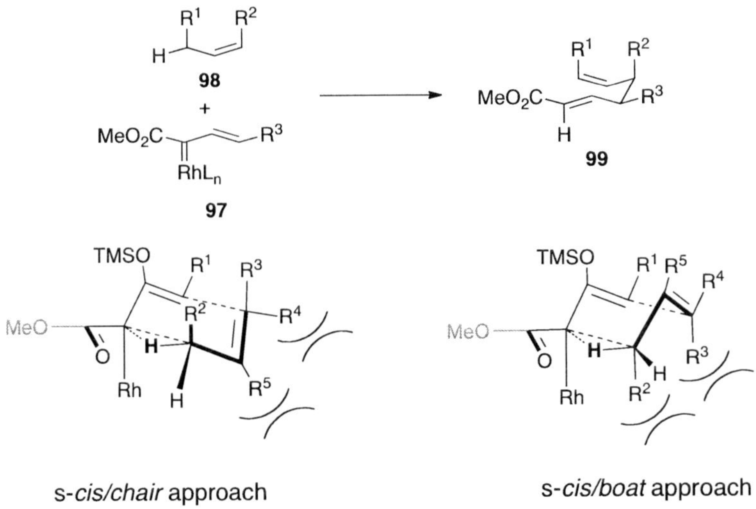

Scheme 12.22.

state, as otherwise the ring would point toward the wall of the catalyst. Acyclic substrates, especially when internally functionalized will tend to react through a boat-like transition state to avoid steric hindrance with the catalyst.

The CHCR reaction has been recently reviewed[16] and only a few highlights will be presented here. Cyclohexene,[127] cyclohexadiene,[128] and dihydronaphthalene[129] derivatives undergo the CHCR reaction through a chair-like transition state, resulting in very high diastereocontrol (>30:1 dr) and enantiocontrol (>97% ee). An early example is the Rh_2(*S*-DOSP)$_4$-catalyzed reaction of the styryldiazoacetate **100** with 1,3-cyclohexadiene (**101**), which generated the 1,4-cyclohexadiene **102** in 99% ee (Scheme 12.23). In a few steps, **102** can be converted to the pharmaceutical agent (+)-sertraline (**103**).[127]

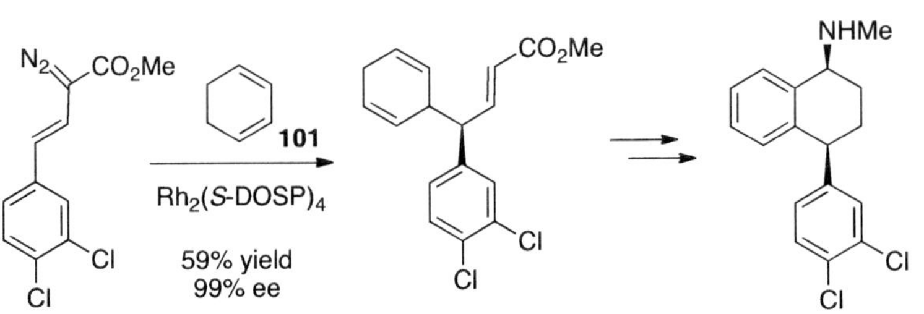

Scheme 12.23.

The CHCR reaction with 4-acetoxy-6,7-dihydroindoles is an interesting transformation because it generates chiral 4-substituted indoles.[130] The initially formed product of the CHCR reaction aromatizes upon facile elimination of acetic acid. An elaborate example of this transformation is the reaction of the indolylvinyldiazoacetate **104** with the dihydroindole **105**, which generates the tris indole **106** in 97% ee (Eq. 12.3).

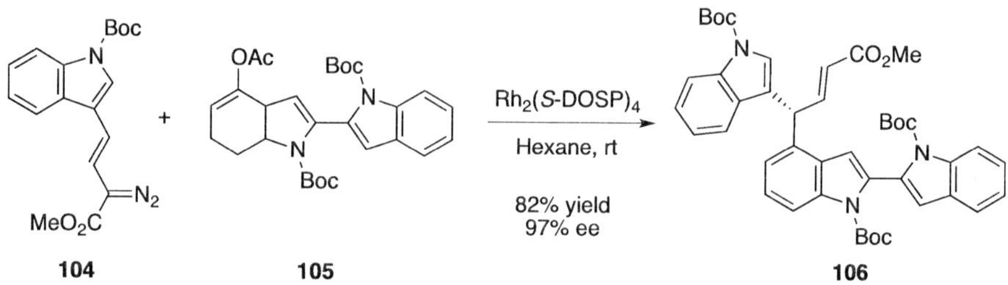

(12.3)

Appropriately functionalized 1,2-disubstituted cyclopentenes[131] and vinyl ethers[132] undergo the CHCR reaction through a boat-like transition state. An example of this is reaction of the vinyl ether **107** with styryldiazoacetate **93**, in which the CHCR product **108** is produced with very high diastereo- and enantiocontrol (Scheme 12.24).[132] From a strategic point of view, this transformation can be considered as a surrogate for the vinylogous Mukaiyama-aldol reaction between the aldehyde **109** and the diene **110**. The advantages of the carbenoid approach over the conventional disconnection are that it proceeds with very high stereocontrol and can be conducted with low catalyst loading (0.01 mol%).

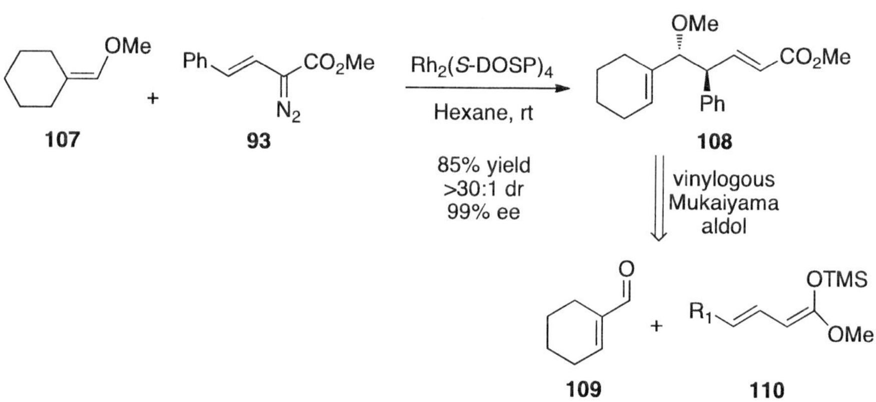

Scheme 12.24.

The CHCR reaction has been applied to the synthesis of a number of natural products.[133–135] The key step has been the enantiodivergent reaction with racemic dihydronaphthalenes.[133–135] An example is shown in the context of the total synthesis of (–)-colombiasin A (**114**) (Scheme 12.25).[134] Reaction of racemic dihydronaphthalene **111** with vinyldiazoacetate **112** generates a CHCR product and a cyclopropane, each derived from different enantiomers of **111**. After hydrogenation and ester reduction, the tetrahydronapthalene derivative **113** was isolated in 34% overall yield as a single diastereomer and in >95% ee.

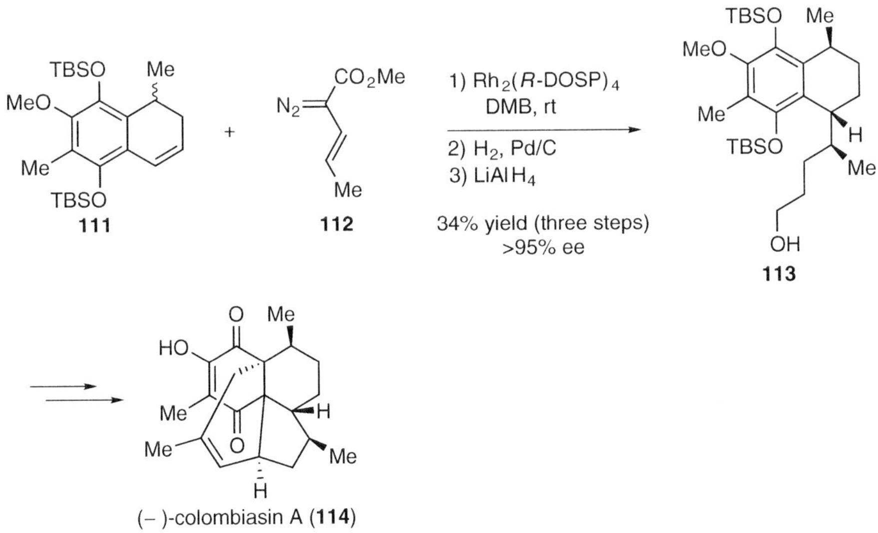

Scheme 12.25.

In this process, the three most challenging stereocenters in (–)-colombiasin A were controlled in one step.

12.8 FORMATION AND REACTIONS OF RHODIUM-BOUND YLIDES

Rhodium-bound ylides are generated when rhodium carbenoids react with oxygen, nitrogen, or sulfur Lewis bases.[7] Depending upon the substituents attached to the Lewis base, a number of subsequent inter- and intramolecular reaction pathways are also possible: proton transfer, [1,2]-alkyl shift (Stevens-type rearrangement), coupling of the ylide with electrophiles, [2,3]-sigmatropic rearrangement, [1,3]-dipolar cycloaddition, or simply isolation of the stable ylide. These reactions are of great interest to synthetic chemists, as the products of oxygen- and nitrogen-derived ylides are generally α-alkoxy and α-amino acids, respectively. Even though it may be reasonable to assume that many of the various chiral dirhodium catalysts would be capable of generating the rhodium-bound ylide **115** with high asymmetric induction, this alone is not sufficient to generate enantioenriched products (Scheme 12.26).[136] A proton transfer to generate an enol **116** or loss of the metal to generate the free ylide **117** would cause a loss of the chiral influence of the catalyst. If, however, the subsequent transformation were initiated from the rhodium-bound ylide, then it would be possible to generate products with retention of the asymmetric induction.

One of the classic reactions illustrating the importance of the subsequent steps in rhodium-bound ylide chemistry is O–H and N–H insertion.[137] Chiral rhodium catalysts routinely give racemic products because the reaction proceeds

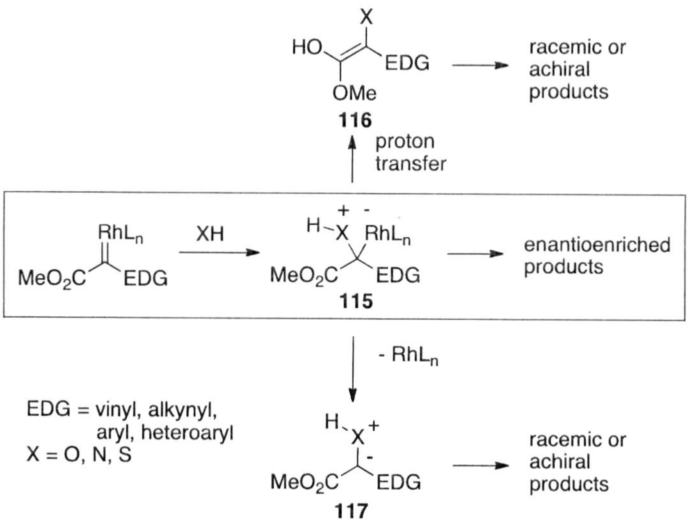

Scheme 12.26.

via the enol intermediate, which then tautomerizes to a keto product.[138] The effective chiral catalysts for enantioselective O–H and N–H insertions are all copper-based[139,140] because with copper catalysis the proton transfer from the metal bound ylide goes directly to the keto tautomer and does not involve enol intermediates. A clever solution for asymmetric induction in rhodium-catalyzed O–H insertion has been to use an achiral rhodium catalyst and a chiral protic acid to enantioselectively protonate the enol intermediate.[141] An example of this is the dirhodium triphenylacetate [$Rh_2(TPA_4)$]-catalyzed N–H insertion between aryldiazoacetate **118** and the carbamate **119** in the presence of the chiral spirocyclic phosphoric acid **120**, which generates the amino ester **121** in 92% ee (Scheme 12.27).[142] The reaction has been proposed to proceed *via* a

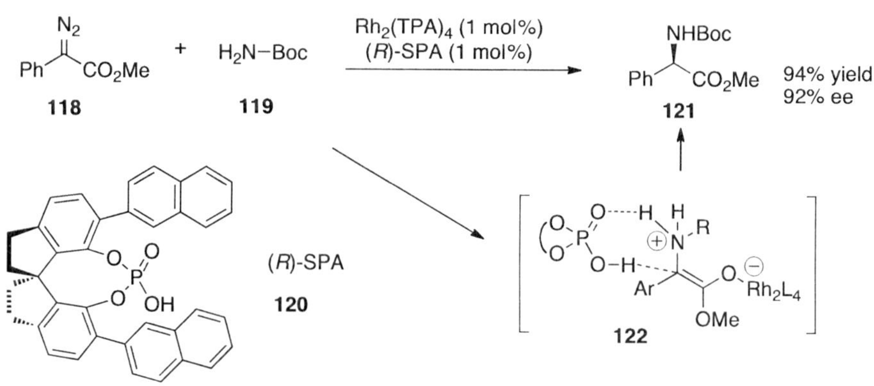

Scheme 12.27.

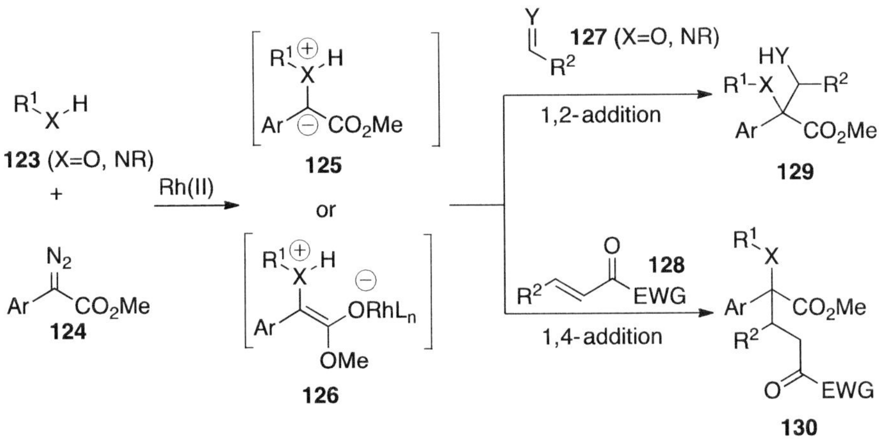

Scheme 12.28.

rhodium-bound enolate **122** rather than a free enol because the nature of the achiral catalyst impacts the level of asymmetric induction.

Numerous examples have been reported of three-component coupling where ylide intermediates are trapped by electrophilic reagents. The early studies describe the reaction of aryldiazoacetates **124** with a variety of ylide sources (**123**) and electrophilic partners (**127** and **128**) suitable for undergoing ylide formation followed by 1,2-[143–145] or 1,4-addition[146–148] processes (Scheme 12.28). Although aldehydes and *N*-aryl aldimines are most studied in the three-component coupling reaction, other highly electrophilic carbonyls such as α-ketoesters[149] and α-ketoamides[150] are compatible partners. Judicious screening of reaction conditions and substituents on the Lewis base and electrophile enabled the synthesis of hydroxy- and amino esters **129** and **130** in high yield and diastereocontrol. Mechanistically, there is still some uncertainty whether the reactive intermediate is the ylide, enol, or rhodium-bound enolate because all could be competent in the further reaction with electrophiles. The earlier studies suggested a mechanism involving metal-free ylide intermediate **125**, but recent publications suggest the intermediacy of rhodium-bound enolate **126**.[138]

A major advance in the three-component coupling has been to conduct the reaction in the presence of binaphthylphosphoric acids **131**. These act as chiral Brønsted acid catalysts in the coupling of the presumed rhodium-bound enolate with the electrophile (Scheme 12.29).[151–153] Highly enantioselective three-component coupling with *N*-aryl imines has been achieved with bulky alcohols using **131a** as catalyst to form **132**,[151,152] and with carbamates to form either diastereomeric series **133** and **134**.[153] When the chiral Brønsted acid is **131b**, **133** is strongly preferred; but when **131c** is used as catalyst, the diastereocontrol switches and **134** becomes the dominant product.

The tandem ylide formation/[1,4]-addition process was also rendered enantioselective by using a chiral Lewis acid. Combination of the aryldiazoacetates

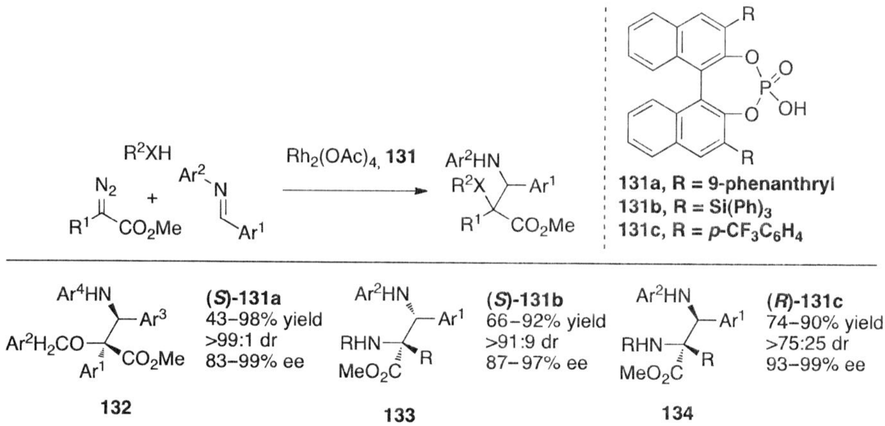

Scheme 12.29.

136 and water in the presence of a cinnamylimidazole Michael acceptor **137**, under the dual catalytic action of $Rh_2(OAc)_4$ and a chiral bisoxazoline Zn(II) complex **135** generated the three-component products **138** with high asymmetric induction (Scheme 12.30).[154] Presumably, the imidoacetate moiety is necessary for bidentate coordination to the chiral Lewis acid catalyst to achieve effective stereoinduction during the 1,4-addition.

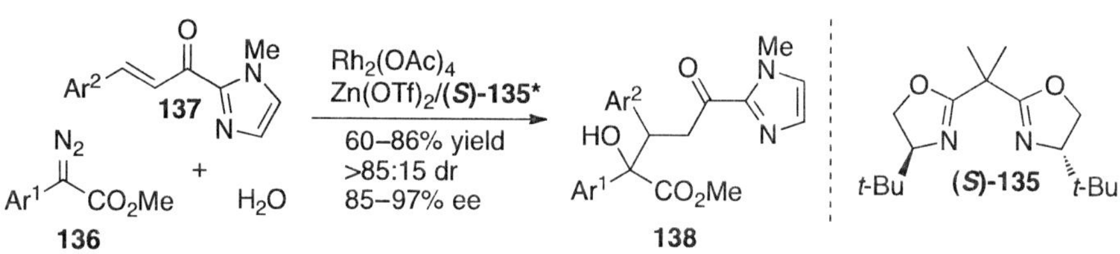

Scheme 12.30.

One of the most versatile reactions of rhodium carbenoids is the tandem cyclization/1,3-dipolar cycloaddition that has been popularized by Padwa.[22] This topic has been extensively reviewed and has been widely applied to the synthesis of complex natural products.[155] A recent example is the total synthesis of (±)-3H-epivincamine **142** (Scheme 12.31).[156] The key step is the reaction of **139**, which undergoes a cyclization to form an ylide intermediate that is engineered to participate in an intramolecular [1,3]-dipolar cycloaddition to form **141**. A relatively short number of synthetic steps are then required to convert **141–142**.

The [1,3]-diploar cycloaddition may proceed through a metal-bound or free ylide. Hence, the development of enantioselective versions of the tandem cyclization/[1,3]-dipolar cycloaddition has not been general because it depends on the active mechanism.[157–167] An impressive recent example of the cyclization/intermolecular cycloaddition is described in Scheme 12.32.[168] Rh_2(*S*-TCPTTL)$_4$ (**7**)-catalyzed reaction of the diazoacetoacetate **143** generated ylide **144**, which

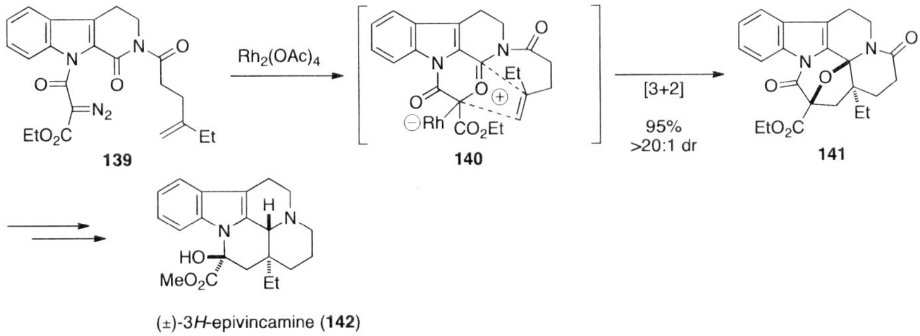

Scheme 12.31.

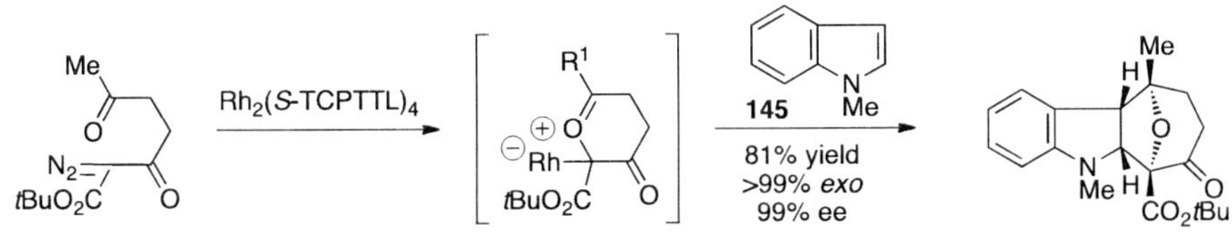

Scheme 12.32.

then reacted with the indole **145** to form the tetracycle **146** with exceptionally high stereocontrol. Even in this case, however, it is clear that the trapping agent and reaction conditions need to be carefully tuned to ensure that the rhodium-bound ylide undergoes the cycloaddition reaction. Although not fully determined, it is likely that the highly enantioselective cycloaddition is a stepwise process, involving electrophilic attack by the oxocarbenium ion of the rhodium-bound ylide **144** on the indole **145**, followed by cyclization to the final product **146**.

A well-established rearrangement of carbenoid-derived ylides is the [2,3]-sigmatropic rearrangement and numerous applications of this transformation to the synthesis of natural products are known.[27] An impressive recent example is the application of an intramolecular oxygen ylide formation/[2,3]-sigmatropic rearrangement to the synthesis of a series of natural products called the cladiellinins.[169] The application of this approach to one specific member of these natural products, (−)-cladiella-6,11-dien-3-ol (**150**) is illustrated in Scheme 12.33.[169] The key step is the metal-catalyzed reaction of the diazoketone **147** to form the *E* and *Z* isomers of the oxabicyclo[6.2.1]bicyclic system (**148** and **149**). When $Rh_2(TPA)_4$ is used as catalyst, the *E* isomer **148** is preferred by up to a 6.3:1 ratio, whereas copper catalysis favors the *Z* isomer **149**. Both **148** and **149** are useful precursors to various members of the cladiellinin class of natural products.

Several reports on the enantioselective version of the [2,3]-sigmatropic rearrangement have been reported since the seminal studies by Doyle in the 1990s.[170,171] Rhodium-catalyzed reaction of aryldiazoacetates with allyl sulfides have been reported to generate the [2,3]-sigmatropic rearrangement products

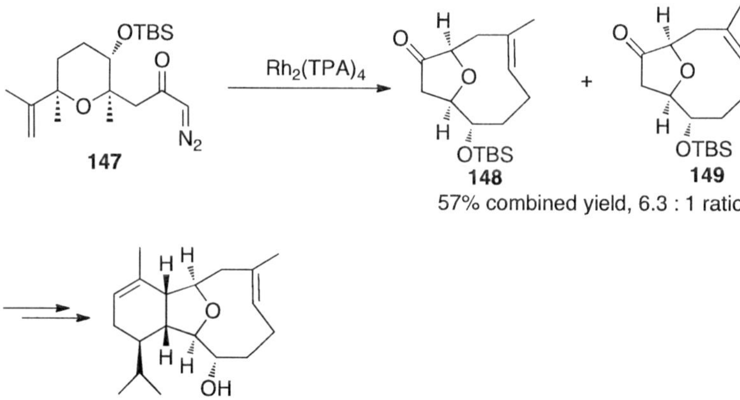

(–)-cladiella-6,11-dien-3-ol (**150**)

Scheme 12.33.

in moderate to high levels of enantioselectivity.[29] The Rh_2(*S*-DOSP)$_4$-catalyzed reaction between racemic allyl alcohols **151** and donor/acceptor carbenoids derived from either aryl- or styryldiazoacetates **152** generates [2,3]-sigmatropic rearrangement products **153** with very high levels of enantioselectivity (92–98% ee) (Scheme 12.34).[172] Normally, rhodium-catalyzed reactions of allyl alcohols with carbenoids would be expected to generate O–H insertion products.[173] This is indeed the case for the reaction with either an acceptor carbenoid or the reaction of donor/acceptor carbenoids with unsubstituted allyl alcohols.[172] When highly functionalized allyl alcohols and donor acceptor carbenoids react together, however, the intermediate rhodium-bound ylide **154** preferentially undergoes a [2,3]-sigmatropic rearrangement to form **153**. In these reactions, the enantioinduction during the formation of the rhodium-bound ylide is transferred to the product upon [2,3]-sigmatropic rearrangement. Control experiments showed that the enantioinduction is under catalyst control because the reaction of an enantioenriched isomer of the allyl alcohol had minimal effects on the asymmetric induction in the final product.

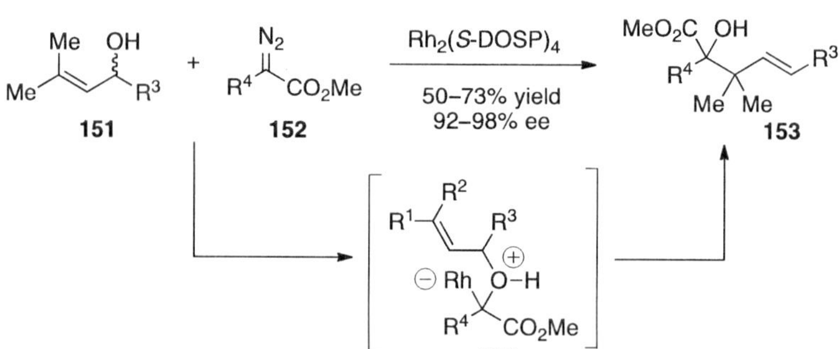

Scheme 12.34.

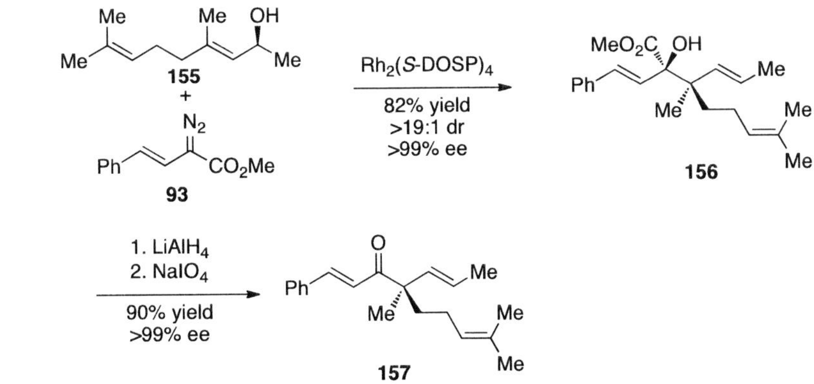

Scheme 12.35.

When the reaction is conducted on allyl alcohols with different vinyl terminus substituents, two stereogenic centers are generated during the [2,3]-sigmatropic rearrangement.[174] Since this new stereocenter is controlled by the chirality of the alcohol, any of the four possible stereoisomers can be obtained by proper choice of alcohol and catalyst. An example illustrating the effectiveness of this reaction is shown in Scheme 12.35.[174] Rh_2(*S*-DOSP)$_4$-catalyzed reaction between allyl alcohol **155** and styryldiazoaceate **93** generated the [2,3]-sigmatropic rearrangement product **156** in 82% yield, >19:1 dr and 99% ee. The utility of this process can be further demonstrated by converting **156** to the chiral ketone **157**.

Due to the high reactivity of rhodium carbenoids and the mild reaction conditions, often the products initially isolated are relatively unstable and prone to further rearrangement. This is graphically seen in the reaction of allyl alcohols **151** with vinyldiazoacetates **158**. When the reaction is conducted under forcing conditions (80–98 °C) in the presence of scandium triflate as a Lewis acid, an extensive series of rearrangements occur to form cyclopentanes **159** containing four new stereogenic centers (Scheme 12.36).[175] The 1,5-hexadien-3-ols **160**

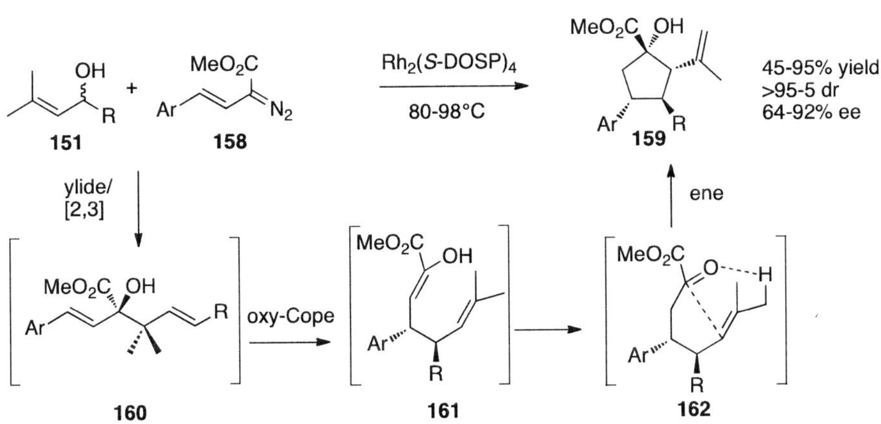

Scheme 12.36.

derived from the [2,3]-sigmatropic rearrangement undergo an oxy-Cope rearrangement to form **161** and, after a keto/enol tautomerization, **162** participates in an intramolecular ene reaction to form **159**.

Rhodium-bound ylides derived from tertiary propargyl alcohols **163** also undergo [2,3]-sigmatropic rearrangements to generate chiral allenes **164** (Scheme 12.37).[176] This type of behavior contrasts with the extensive studies by Wood and coworkers in the 1990s on the reaction of propargylic alcohols with diazoketones, which resulted in O–H insertion followed by either a [3,3]-sigmatropic rearrangement or a [2,3]-sigmatropic rearrangement of the transient propargyloxy enol ether.[177] Extensive computational studies reveal the critical reasons for these successful transformations.[176] The reaction of the carbenoid with the alcohol involves a two-point attachment (carbenoid–oxygen interaction and hydrogen bonding between the OH and the carboxylate ligand), which leads to a more ordered transition state and thus very high asymmetric induction. The resulting ylide **165** can undergo either a proton transfer to an enol or the [2,3]-sigmatropic rearrangement. The transition state of the [2,3]-sigmatropic rearrangement has considerable diradical character, which explains why ylides derived from donor/acceptor carbenoids and highly substituted propargylic alcohols would preferentially undergo this transformation.

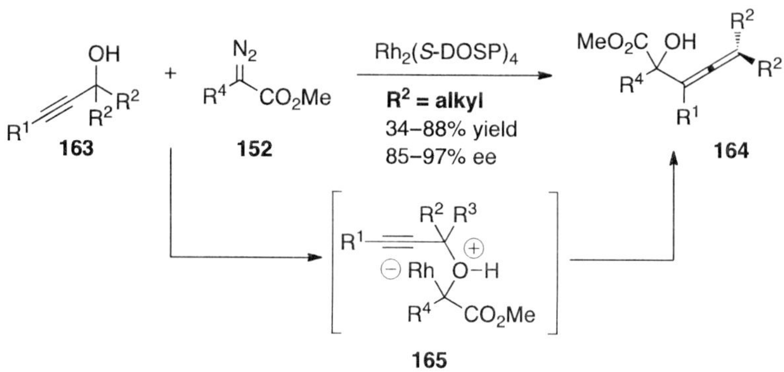

Scheme 12.37.

The reaction of donor/acceptor carbenoids with chiral tertiary propargylic alcohols has been shown to be prone to kinetic resolution due to considerable matched/mismatched character between the chiral catalyst and the chiral propargylic alcohol.[176] A matched reaction is the Rh_2(*S*-DOSP)$_4$-catalyzed reaction between styryldiazoacetate **93** and propargylic alcohol **166**, which generates the allene **167** in 70% yield, >19:1 dr and >99% ee (Scheme 12.38). Allenes such as **167** are useful intermediates because they readily cyclize to dihydrofurans, as illustrated in the silver-catalyzed rearrangement of **167** to **168**.

Ylide intermediates are also generated on reaction of rhodium carbenoids with electron-rich π-systems, where rather than undergoing cyclopropanation or cyclopropenation, a variety of other transformations can occur.[23,24] Recently,

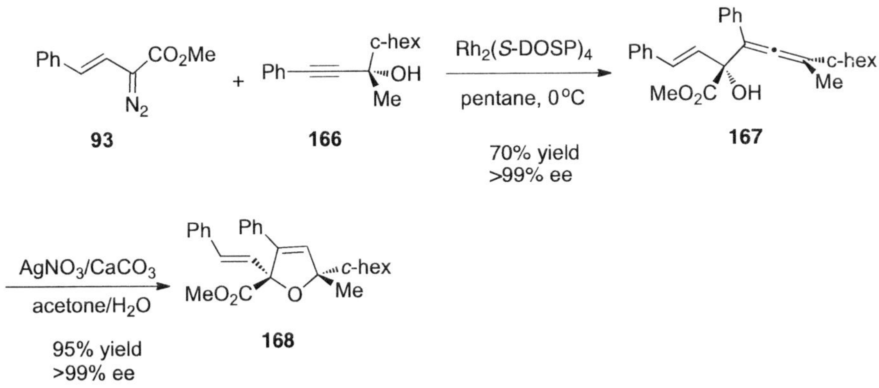

Scheme 12.38.

it has been shown that it is possible for these types of ylide transformations to be conducted with high asymmetric induction using chiral dirhodium catalysts. Rh_2(*S*-NTTL)$_4$-catalyzed reaction between 2-diazo-hexanoate **170** and indole **169** generates the alkylated indole **171** in 96% yield and 95% ee (Scheme 12.39).[178] A second example is the Rh_2(*S*-DOSP)$_4$-catalyzed reaction of styryldiazoacetate **93** with indole **169** to form the [3+2]-annulation product **172** in 68% yield and 97% ee.[179] Although the mechanistic interpretation of the asymmetric induction in these reactions is complex, they are both likely to involve the rhodium-bound

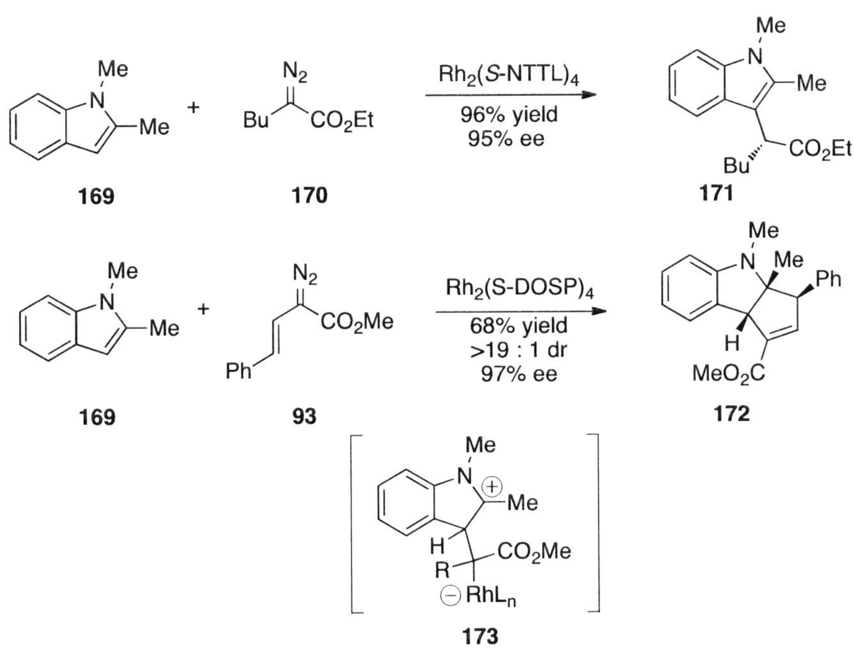

Scheme 12.39.

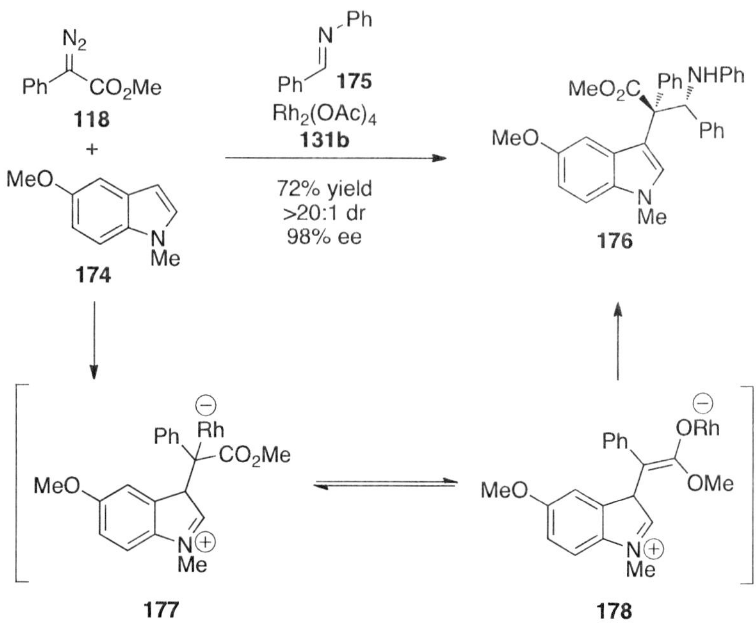

Scheme 12.40.

zwitterionic intermediate **173** or transition states that are structurally similar to **173**. These types of reactions offer exciting new opportunities for the design of novel enantioselective transformations of rhodium carbenoids.

Another interesting approach to the functionalization of the indole nucleus has been in its utilization as a carbon "Lewis base" in the three-component tandem ylide formation/electrophilic addition reaction (Scheme 12.40).[180] Reaction between the aryldiazoacetate **118**, indole **174**, and imine **175** generates the functionalized indole **176** in a highly stereoselective manner. Electrophilic attack of the rhodium carbenoid by indole C-3 generates the pseudo ylide intermediate **177**, which could also be envisioned as the rhodium enolate, zwitterionic structure **178**. In the presence of an aldimine electrophile, and chiral phosphoric acid **131b** as an activating agent, the rhodium-zwitterion undergoes a highly stereoselective Mannich reaction to furnish *iso*-tryptophan-type products **176**. It is unclear whether the chiral center generated during addition of the indole to the carbenoid is conserved, or whether the achiral rhodium enolate attacks the activated imine with phosphonate-induced enantiocontrol.

12.9 VINYLOGOUS REACTIONS OF RHODIUM–VINYLCARBENOIDS

Some sporadic reports in the 1990s indicated that vinylcarbenoids are capable of displaying electrophilic character at the vinylogous position in addition to the carbenoid site when the vinylogous terminus is unsubstituted.[181–184]

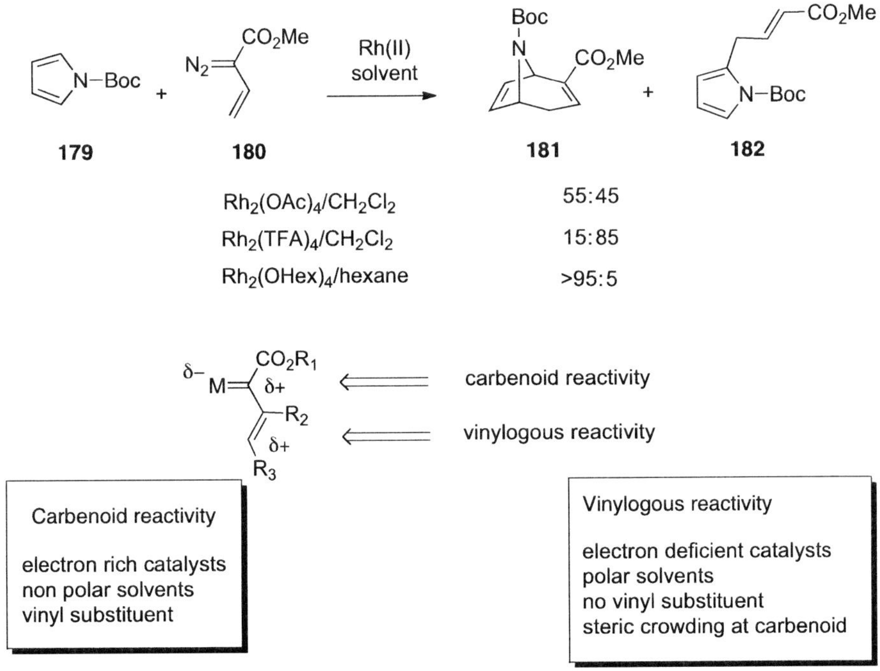

Scheme 12.41.

An example of this phenomenon is the reaction between pyrrole **179** and vinyldiazoacetate **180** (Scheme 12.41).[182] Instead of clean formation of the tropane **181**, a significant amount of the vinylogous alkylation product **182** could be formed. Vinylogous reaction to form **182** is preferred when the catalyst is electron deficient and relatively polar solvents are used. Subsequent studies revealed that steric influences around the carbenoid can drive the reaction toward the vinylogous addition.[183,184] A bulky ester group interferes with the reaction at the carbenoid site, as does an internal substituent, but this can also impede attack at the vinylogous position, causing the carbenoid to cyclize to a cyclopropene instead.

In recent years, there has been renewed interest in exploring the synthetic potential of vinylogous reactions of rhodium vinylcarbenoids. New factors have been discovered that enhance vinylogous reactions such that substituted vinylcarbenoids can be effective substrates. *Z*-Vinyldiazoacetates are more susceptible than *E*-vinyldiazoacetates at undergoing vinylogous reactions, and the outcome of the $Rh_2(esp)_2$-catalyzed reactions of *Z*-diazopentenoate **183** is very dependent on the steric environment of the substrate (Scheme 12.42).[185] When the reaction is conducted with indole **184** the normal carbenoid product **185** is formed but when the more crowded indole **169** is used, **186**, derived from the vinylogous reaction, is formed. The *Z* geometry of **186** is indicative that the vinylogous reaction occurred on the s-*trans* configuration **187** of the vinylcarbenoid.

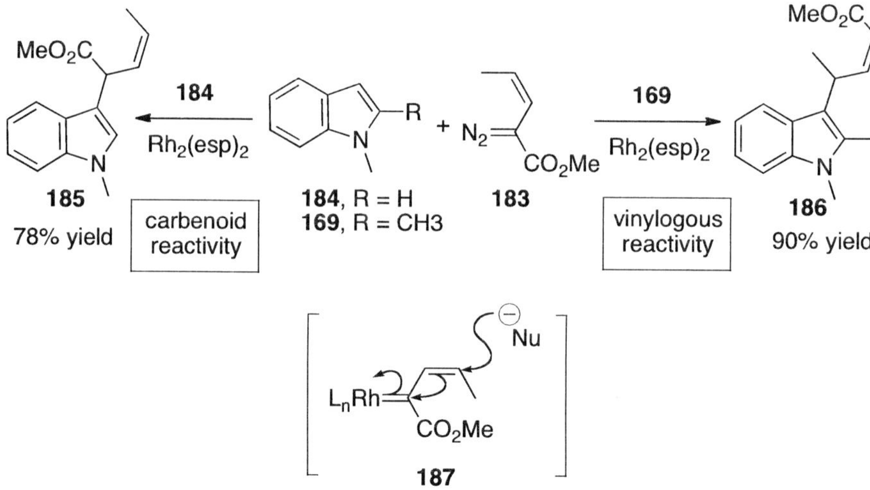

Scheme 12.42.

An alternative approach to enhance vinylogous reactions is to use bulky catalysts (Eq. 12.4).[186] Even though standard catalysts will preferentially result in normal carbenoid chemistry in the reactions of *E*-diazopentenoate **112**, when the bulky catalyst Rh_2(*S*-biTISP)$_2$ and substituted indoles **188** are used as substrates, the three-substituted indoles **189** are formed with high levels of enantioselectivity. The *Z* isomers of **189** are formed, which is indicative of the reaction occurring from the s-*trans* isomers of the vinylcarbenoid, and it is conceivable that one of the roles of the bulky catalysts is to favor the s-*trans* conformer of the carbenoid over the s-*cis* conformer.

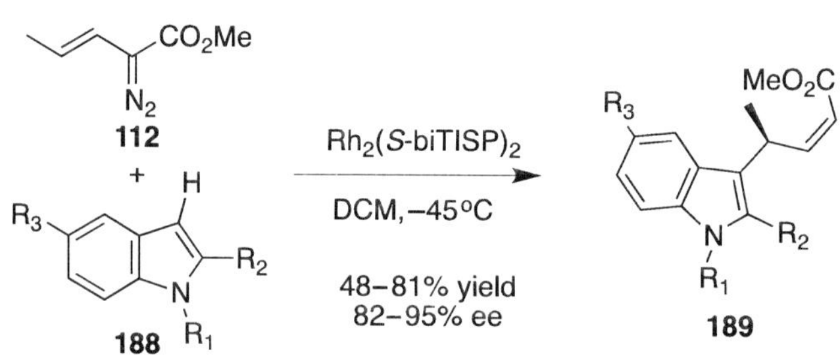

(12.4)

The carbenoid derived from siloxyvinyldiazoacetate **54** is also capable of reaction at the vinylogous position and can lead to some interesting cascade transformations (Scheme 12.43).[187] Reaction of **54** with nitrones **190** results in a formal [3+3]-cycloaddition to form **191**. The reaction is considered to proceed *via* the zwitterionic intermediate **193**, which is generated by vinylogous attack of the nitrone on the vinylcarbenoid **192**.

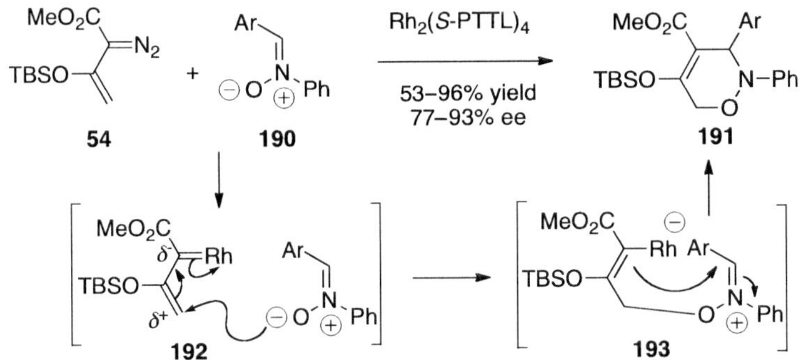

Scheme 12.43.

The reaction of the 2-diazobutenoate **178** with nitrones **194** results in the formation of an unusual tricyclic derivative **195** (Scheme 12.44).[188] It has been proposed that the reaction is initiated by a [1,3]-dipolar cycloaddition between the vinylcarbenoid and **194** to form **197**. An alternative possibility could be C–C bond formation to generate the zwitterion **196** by vinylogous reaction of **194** with the s-*trans* conformer of the vinylcarbenoid. As **196** has the improper geometry for a formal [3 + 3]-cycloaddition, it cyclizes to form **197**. A subsequent intramolecular cyclopropanation would generate the norcaradiene **198**, which after a 6π electrocyclic ring expansion followed by a [1,7]-sigmatropic rearrangement would generate the observed product **195**.

The vinylogous reaction of siloxyvinyldiazoacetates with silylenol ethers results in very different types of products in which a siloxy migration has occurred and an alkyne is formed. An example of this transformation is the

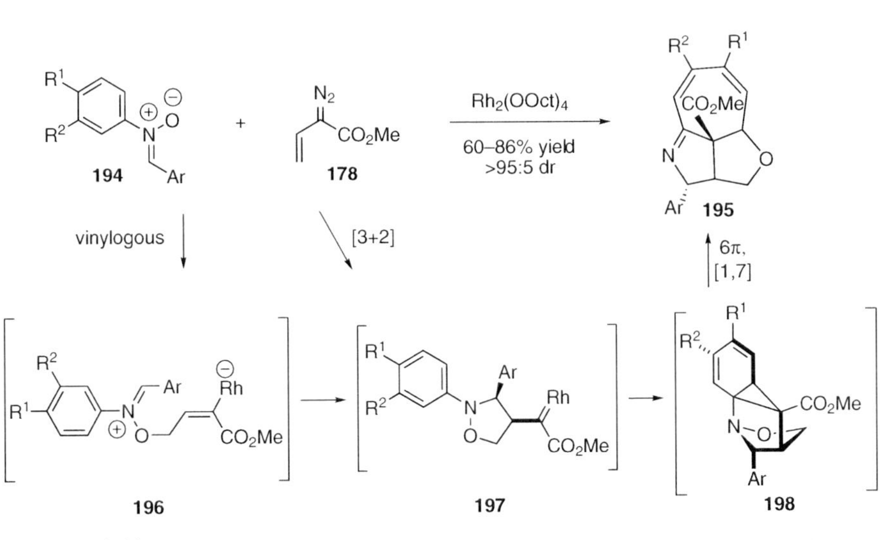

Scheme 12.44.

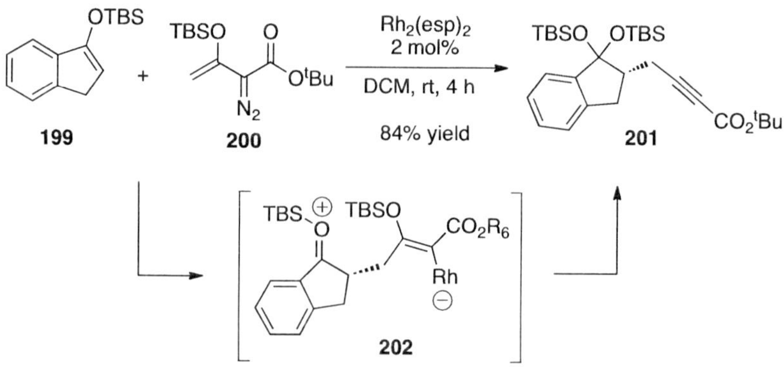

Scheme 12.45.

$Rh_2(esp)_2$-catalyzed reaction of the silyl enol ether **199** with the siloxyvinyldiazoacetate **200** to form the alkyne **201** (Scheme 12.45).[189] This reaction is general for a wide variety of cyclic enol ethers. The reaction is considered to involve zwitterionic intermediate **202**, which then undergoes a siloxy migration, either directly of *via* a cyclic intermediate to generate **201**.

The reaction with acyclic silyl enol ethers **203** is more complicated because two types of products can be formed with high levels of asymmetric induction (Scheme 12.46).[190] When the R group in **203** is small such as methyl, alkyne **204** is produced. However, when the R group in **204** is larger such as phenyl, the formal [3+2]-cycloaddition product **205** is formed. Both products are considered to be derived from zwitterionic intermediate **206** but the size of the R group influences the trajectory of approach of the enol ether to the carbenoid, and this in turn effects the orientation of the groups in **206**, leading to the different product outcomes.

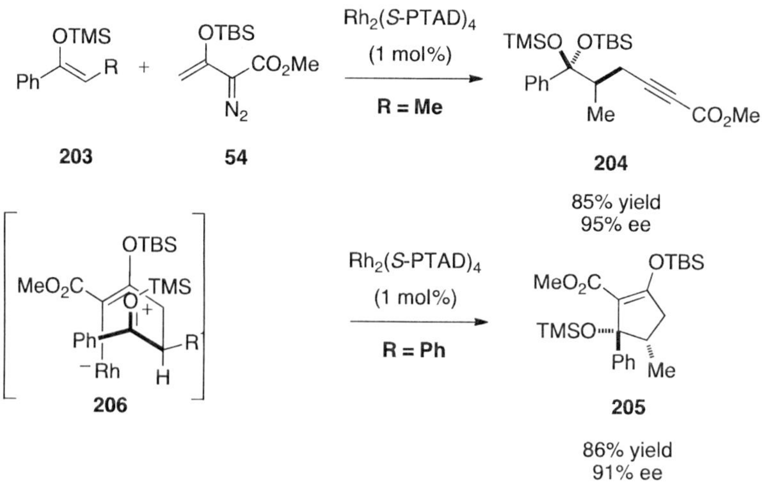

Scheme 12.46.

12.10 CONCLUSIONS AND FUTURE OUTLOOK

During the last decade, the range of enantioselective transformations of rhodium carbenoids has expanded tremendously. A series of highly efficient chiral catalysts are available and many of these are effective in a wide variety of transformations. Asymmetric cyclopropanation and cyclopropenation can now be conducted with several different classes of carbenoid intermediates. Intermolecular C–H insertion has developed into a very powerful strategy for C–H functionalization with the advent of donor/acceptor carbenoids, which make the C–H insertions much more selective. Ylide chemistry has continued to blossom, and now a much better understanding is available for the factors controlling high asymmetric induction in these reactions. A particularly rich area of research is the opportunity for cascade reactions in which the initial products can initiate a complex series of transformations. Several unusual types of organometallic intermediates are generated, such as new rhodium carbenoids, rhodium vinyl complexes, rhodium-bound zwitterionic intermediates and rhodium-bound ylides, and they have the potential of greatly increasing the versatility of the rhodium–carbenoid chemistry. Computational studies are generating great insights into the subtle controlling elements associated with these rhodium–carbenoid intermediates, which will enhance rational development of new types of transformations. The synthetic possibilities associated with rhodium carbenoids is clearly seen in the wide range of applications to total synthesis, and as the diverse chemistry continues to expand, it is likely that an even greater range of applications will be seen in the future.

SUGGESTED READING

Davies, H. M. L., Beckwith, R. E. J. "Catalytic Enantioselective C–H Activation by Means of Metal–Carbenoid-Induced C–H Insertion," *Chem. Rev.* **2003**, *103*, 2861.

Davies, H. M. L., Hedley, S. J. "Intermolecular Reactions of Electron-Rich Heterocycles with Copper and Rhodium Carbenoids," *Chem. Soc. Rev.* **2007**, *36*, 1109.

Davies, H. M. L., Lian, Y. "The Combined C–H Functionalization/Cope Rearrangement: Discovery and Applications in Organic Synthesis," *Acc. Chem. Res.* **2012**, *45*, 923.

Doyle, M. P. "Perspective on Dirhodium Carboxamidates as Catalysts," *J. Org. Chem.* **2006**, *71*, 9253–9260.

Doyle, M. P., McKervey, M. A., Ye, T. *Modern Catalytic Methods for Organic Synthesis with Diazo Compounds: From Cyclopropanes to Ylides*. Wiley-Interscience: New York, **1998**.

Doyle, M. P., Duffy, R., Ratnikov, M., Zhou, L. "Catalytic Carbene Insertion into C–H Bonds," *Chem. Rev.* **2010**, *110*, 704.

Hansen, J. H., Davies, H. M. L. "High Symmetry Dirhodium(II) Paddlewheel Complexes as Chiral Catalysts," *Coord. Chem. Rev.* **2008**, *252*, 545–555.

Lebel, H., Marcoux, J. F., Molinaro, C., Charette, A. B. "Stereoselective Cyclopropanation Reactions," *Chem. Rev.* **2003**, *103*, 977–1050.

Padwa, A. "Domino Reactions of Rhodium(II) Carbenoids for Alkaloid Synthesis," *Chem. Soc. Rev.* **2009**, *38*, 3072–3081.

Vanecko, J. A., Wan, H., West, F. G. "Recent Advances in the Stevens Rearrangement of Ammonium Ylides. Application to the Synthesis of Alkaloid Natural Products," *Tetrahedron* **2006**, *62*, 1043–1062.

REFERENCES

1 Hubert, A. J., Noels, A. F., Anciaux, A. J., Teyssie, P. *Synthesis* **1976**, *600*.
2 Anciaux, A. J., Hubert, A. J., Noels, A. F., Petiniot, N., Teyssie, P. *J. Org. Chem.* **1980**, *45*, 695.
3 Doyle, M. P. *Recl. Trav. Chim. Pay. B* **1991**, *110*, 305.
4 Davies, H. M. L. *Eur. J. Org. Chem.* **1999**, *2459*.
5 Hansen, J. H., Davies, H. M. L. *Coord. Chem. Rev.* **2008**, *252*, 545.
6 Doyle, M. P. *J. Org. Chem.* **2006**, *71*, 9253.
7 Doyle, M. P., McKervey, M. A., Ye, T. *Modern Catalytic Methods for Organic Synthesis with Diazo Compounds: From Cyclopropanes to Ylides*. Wiley-Interscience: New York, **1998**.
8 Ye, T., McKervey, M. A. *Chem. Rev.* **1994**, *94*, 1091.
9 Doyle, M. P., Forbes, D. C. *Chem. Rev.* **1998**, *98*, 911.
10 Padwa, A., Weingarten, M. D. *Chem. Rev.* **1996**, *96*, 223.
11 Adams, J., Spero, D. M. *Tetrahedron* **1991**, *47*, 1765.
12 Davies, H. M. L. *Tetrahedron* **1993**, *49*, 5203.
13 Davies, H. M. L., Walji, A. M. In *Modern Rhodium-Catalyzed Organic Reactions*, Evans, P., Ed. Blackwell Science Publ.: **2005**, pp. 301–340.
14 Doyle, M. P. In *Modern Rhodium-Catalyzed Organic Reactions*, Evans, P., Ed. Blackwell Science Publ.: **2005**, pp. 341–355.
15 Davies, H. M. L., Manning, J. R. *Nature* **2008**, *451*, 417.
16 Davies, H. M. L., Lian, Y. "The Combined C–H Functionalization/Cope Rearrangement: Discovery and Applications in Organic Synthesis," *Acc. Chem. Res.* **2012**, *45*, 923.
17 Davies, H. M. L., Loe, O. *Synthesis* **2004**, 2595.
18 Doyle, M. P., Duffy, R., Ratnikov, M., Zhou, L. "Catalytic Carbene Insertion into C–H Bonds," *Chem. Rev.* **2010**, *110*, 704.
19 Davies, H. M. L., Beckwith, R. E. J. "Catalytic Enantioselective C–H Activation by Means of Metal–Carbenoid-Induced C–H Insertion," *Chem. Rev.* **2003**, *103*, 2861.
20 Lebel, H., Marcoux, J. F., Molinaro, C., Charette, A. B. *Chem. Rev.* **2003**, *103*, 977.
21 Mehta, G., Muthusamy, S. *Tetrahedron* **2002**, *58*, 9477.
22 Padwa, A. *Chem. Soc. Rev.* **2009**, *38*, 3072.
23 Davies, H. M. L., Antoulinakis, E. G. *Org. Reacts.* **2001**, *57*, 1.
24 Davies, H. M. L., Hedley, S. J. "Intermolecular Reactions of Electron-Rich Heterocycles with Copper and Rhodium Carbenoids," *Chem. Soc. Rev.* **2007**, *36*, 1109.
25 Merlic, C. A., Zechman, A. L. *Synthesis* **2003**, 1137.
26 Hodgson, D. M., Pierard, F. Y. T. M., Stupple, P. A. *Chem. Soc. Rev.* **2001**, *30*, 50.
27 West, F. G. In *Modern Rhodium-Catalyzed Organic Reactions*, Evans, P., Ed. Blackwell Science Publ.: **2005**, pp. 417–431.
28 Vanecko, J. A., Wan, H., West, F. G. *Tetrahedron* **2006**, *62*, 1043.

29 Zhang, Y., Wang, J. *Coord. Chem. Rev.* **2010**, *254*, 941.
30 Davies, H. M. L., Pelphrey, P. *Org. React.* **2011**, *75*, 75.
31 Watanabe, N., Ogawa, T., Ohtake, Y., Ikegami, S., Hashimoto, S. *Synlett* **1996**, 85.
32 Hashimoto, S., Watanabe, N., Ikegami, S. *Tetrahedron Lett.* **1990**, *31*, 5173.
33 Lindsay, V. N. G., Lin, W., Charette, A. B. *J. Am. Chem. Soc.* **2009**, *131*, 16383.
34 Goto, T., Takeda, K., Shimada, N., Nambu, H., Anada, M., Shiro, M., Ando, K., Hashimoto, S. *Angew. Chem. Int. Ed.* **2011**, *50*, 6803.
35 Reddy, R. P., Lee, G. H., Davies, H. M. L. *Org. Lett.* **2006**, *8*, 3437.
36 Davies, H. M. L., Bruzinski, P. R., Lake, D. H., Kong, N., Fall, M. J. *J. Am. Chem. Soc.* **1996**, *118*, 6897.
37 Davies, H. M. L., Panaro, S. A. *Tetrahedron Lett.* **1999**, *40*, 5287.
38 Qin, C. M., Boyarskikh, V., Hansen, J. H., Hardcastle, K. I., Musaev, D. G., Davies, H. M. L. *J. Am. Chem. Soc.* **2011**, *133*, 19198.
39 Boruta, D. T., Dmitrenko, O., Yap, G. P. A., Fox, J. M. *Chem. Sci.* **2012**, *3*, 1589.
40 Brunner, H., Kluschanzoff, H., Wutz, K. *Bull. Soc. Chim. Belg.* **1989**, *98*, 63.
41 Lou, Y., Horikawa, M., Kloster, R. A., Hawryluk, N. A., Corey, E. J. *J. Am. Chem. Soc.* **2004**, *126*, 8916.
42 Lindsay, V. N. G., Charette, A. B. *ACS Catal.* **2012**, *2*, 1221.
43 Doyle, M. P., Protopopova, M., Muller, P., Ene, D., Shapiro, E. A. *J. Am. Chem. Soc.* **1994**, *116*, 8492.
44 Davies, H. M. L., Bruzinski, P. R., Fall, M. J. *Tetrahedron Lett.* **1996**, *37*, 4133.
45 Doyle, M. P., Zhou, Q. L., Charnsangavej, C., Longoria, M. A., McKervey, M. A., Garcia, C. F. *Tetrahedron Lett.* **1996**, *37*, 4129.
46 Davies, H. M. L., Lee, G. H. *Org. Lett.* **2004**, *6*, 2117.
47 Denton, J. R., Cheng, K., L. Davies, H. M. L. *Chem. Commun.* **2008**, 1238.
48 Denton, J. R., Sukumaran, D., Davies, H. M. L. *Org. Lett.* **2007**, *9*, 2625.
49 Denton, J. R., Davies, H. M. L. *Org. Lett.* **2009**, *11*, 787.
50 Muller, P., Bernardinelli, G., Allenbach, Y. F., Ferri, M., Flack, H. D. *Org. Lett.* **2004**, *6*, 1725.
51 Panne, P., DeAngelis, A., Fox, J. M. *Org. Lett.* **2008**, *10*, 2987.
52 Panne, P., Fox, J. M. J. *Am. Chem. Soc.* **2006**, *129*, 22.
53 DeAngelis, A., Dmitrenko, O., Yap, G. P. A., Fox, J. M. *J. Am. Chem. Soc.* **2009**, *131*, 7230.
54 Bonge, H. T., Pintea, B., Hansen, T. *Org. Biomol. Chem.* **2008**, *6*, 3670.
55 Schnaars, C., Hansen, T. *Org. Lett.* **2012**, *14*, 2794.
56 Bonge, H. T., Hansen, T. *J. Org. Chem.* **2010**, *75*, 2309.
57 Marcoux, D., Azzi, S., Charette, A. B. *J. Am. Chem. Soc.* **2009**, *131*, 6970.
58 Lindsay, V. N. G., Nicolas, C., Charette, A. B. *J. Am. Chem. Soc.* **2011**, *133*, 8972.
59 Marcoux, D., Lindsay, V. N. G., Charette, A. B. *Chem. Commun.* **2010**, *46*, 910.
60 Singh, A. K., Rao, M. N., Simpson, J. H., Li, W. S., Thornton, J. E., Kuehner, D. E., Kacsur, D. *J. Org. Proc. Res. Dev.* **2002**, *6*, 618.
61 Simpson, J. H., Godfrey, J., Fox, R., Kotnis, A., Kacsur, D., Hamm, J., Totelben, M., Rosso, V., Mueller, R., Delaney, E., Deshpande, R. P. *Tetrahedron-Asymmetry* **2003**, *14*, 3569.
62 Anthes, R., Bello, O., Benoit, S., Chen, C. K., Corbett, E., Corbett, R. M., DelMonte, A. J., Gingras, S., Livingston, R., Sausker, J., Soumeillant, M. *Org. Proc. Res. Dev.* **2008**, *12*, 168.
63 Anthes, R., Benoit, S., Chen, C. K., Corbett, E. A., Corbett, R. M., DelMonte, A. J., Gingras, S., Livingston, R. C., Pendri, Y., Sausker, J., Soumeillant, M. *Org. Proc. Res. Dev.* **2008**, *12*, 178.

64 Chuprakov, S., Kwok, S. W., Zhang, L., Lercher, L., Fokin, V. V. *J. Am. Chem. Soc.* **2009**, *131*, 18034.
65 Horneff, T., Chuprakov, S., Chernyak, N., Gevorgyan, V., Fokin, V. V. *J. Am. Chem. Soc.* **2008**, *130*, 14972.
66 Grimster, N., Zhang, L., Fokin, V. V. *J. Am. Chem. Soc.* **2010**, *132*, 2510.
67 Chuprakov, S., Malik, J. A., Zibinsky, M., Fokin, V. V. *J. Am. Chem. Soc.* **2011**, *133*, 10352.
68 Culhane, J. C., Fokin, V. V. *Org. Lett.* **2011**, *13*, 4578.
69 Zibinsky, M., Fokin, V. V. *Org. Lett.* **2011**, *13*, 4870.
70 Selander, N., Fokin, V. V. *J. Am. Chem. Soc.* **2012**, *134*, 2477.
71 Selander, N., Worrell, B. T., Chuprakov, S., Velaparthi, S., Fokin, V. V. *J. Am. Chem. Soc.* **2012**, *134*, 14670.
72 Davies, H. M. L., Venkataramani, C. *Org. Lett.* **2003**, *5*, 1403.
73 Pelphrey, P., Hansen, J., Davies, H. M. L. *Chem. Sci.* **2010**, *1*, 254.
74 Candeias, N. R., Afonso, C. A. M., Gois, P. M. P. *Org. Biomol. Chem.* **2012**, *10*, 3357.
75 Davies, H. M. L., Walji, A. M. *Org. Lett.* **2005**, *7*, 2941.
76 Takeda, K., Oohara, T., Anada, M., Nambu, H., Hashimoto, S. *Angew. Chem. Int. Ed.* **2010**, *49*, 6979.
77 Jessop, P. G., Ikariya, T., Noyori, R. *Chem. Rev.* **1999**, *99*, 475.
78 Fraile, J. M., Garcia, J. I., Herrerias, C. I., Mayoral, J. A., Carrie, D., Vaultier, M. *Tetrahedron-Asymmetry* **2001**, *12*, 1891.
79 Tang, P., Qin, Y. *Synthesis* **2012**, *44*, 2969.
80 Reissig, H. U., Zimmer, R. *Chem. Rev.* **2003**, *103*, 1151.
81 Granger, K., Snapper, M. L. *Eur. J. Org. Chem.* **2012**, 2308.
82 Patil, D. V., Cavitt, M. A., France, S. *Heterocycles* **2012**, *84*, 1363.
83 Ni, A. W., France, J. E., Davies, H. M. L . *J.Org. Chem.* **2006**, *71*, 5594.
84 Wang, H., Denton, J. R., Davies, H. M. L. *Org. Lett.* **2011**, *13*, 4316.
85 Reddy, R. P., Davies, H. M. L. *J. Am. Chem. Soc.* **2007**, *129*, 10312.
86 Davies, H. M. L., Denton, J. R. *Chem. Soc. Rev.* **2009**, *38*, 3061.
87 Schwartz, B. D., Denton, J. R., Lian, Y., Davies, H. M. L., Williams, C. M. *J. Am. Chem. Soc.* **2009**, *131*, 8329.
88 Davies, H. M. L., Doan, B. D. *J. Org. Chem.* **1998**, *63*, 657.
89 Lian, Y. J., Miller, L. C., Born, S., Sarpong, R., Davies, H. M. L. *J. Am. Chem. Soc.* **2010**, *132*, 12422.
90 Kende, A. S., Smalley, T. L., Huang, H. *J. Am. Chem. Soc.* **1999**, *121*, 7431.
91 Olson, J. P., Davies, H. M. L. *Org. Lett.* *12*, **1144**.
92 Xu, J., Caro-Diaz, E. J. E., Theodorakis, E. A. *Org. Lett.* **2011**, *12*, 3708.
93 Davies, H. M. L., Lee, G. H. *Org. Lett.* **2004**, *6*, 1233.
94 Briones, J. F., Hansen, J., Hardcastle, K. I., Autschbach, J., Davies, H. M. L. *J. Am. Chem. Soc.* **2010**, *132*, 17211.
95 Briones, J. F., Davies, H. M. L. *Tetrahedron* **2011**, *67*, 4313.
96 Nowlan, D. T., Singleton, D. A. *J. Am. Chem. Soc.* **2005**, *127*, 6190.
97 Morandi, B., Carreira, E. M. *Angew. Chem. Int. Ed.* **2010**, *49*, 4294.
98 Muller, P., Pautex, N., Doyle, M. P., Bagheri, V. *Helv. Chim. Acta* **1990**, *73*, 1233.
99 Li, H. Y., Hsung, R. P. *Org. Lett.* **2009**, *11*, 4462.
100 Chuprakov, S., Hwang, F. W., Gevorgyan, V. *Angew. Chem. Int. Ed.* **2007**, *46*, 4757.
101 Chuprakov, S., Gevorgyan, V. *Org. Lett.* **2007**, *9*, 4463.
102 Davies, H. M. L., Morton, D. *Chem. Soc. Rev.* **2011**, *40*, 1857.
103 Doyle, M. P., Liu, Y., Ratnikov, M. *Org. Reacts.* **2013**, *in press*.

104 Demonceau, A., Noels, A. F., Hubert, A. J., Teyssie, P. *Chem. Commun.* **1981**, 688.
105 Doyle, M. P., Westrum, L. J., Wolthuis, W. N. E., See, M. M., Boone, W. P., Bagheri, V., Pearson, M. M. *J. Am. Chem. Soc.* **1993**, *115*, 958.
106 Anada, M., Hashimoto, S. *Tetrahedron Lett.* **1998**, *39*, 79.
107 Anada, M., Mita, O., Watanabe, H., Kitagaki, S., Hashimoto, S. *Synlett* **1999**, 1775.
108 Davies, H. M. L., Grazini, M. V. A., Aouad, E. *Org. Lett.* **2001**, *3*, 1475.
109 Saito, H., Oishi, H., Kitagaki, S., Nakamura, S., Anada, M., Hashimoto, S. *Org. Lett.* **2002**, *4*, 3887.
110 Doyle, M. P., Ratnikov, M., Liu, Y. *Org. Biomol. Chem.* **2011**, *9*, 4007.
111 Wang, D. H., Yu, J. Q. *J. Am. Chem. Soc.* **2011**, *133*, 5767.
112 Taber, D. F., Tian, W. W. *J. Org. Chem.* **2008**, *73*, 7560.
113 Kurosawa, W., Kobayashi, H., Kan, T., Fukuyama, T. *Tetrahedron* **2004**, *60*, 9615.
114 Koizumi, Y., Kobayashi, H., Wakimoto, T., Furuta, T., Fukuyama, T., Kan, T. *J. Am. Chem. Soc.* **2008**, *130*, 16854.
115 Hinman, A., Du Bois, J. *J. Am. Chem. Soc.* **2003**, *125*, 11510.
116 Taber, D. F., You, K. K., Rheingold, A. L. *J. Am. Chem. Soc.* **1996**, *118*, 547.
117 Nakamura, E., Yoshikai, N., Yamanaka, M. *J. Am. Chem. Soc.* **2002**, *124*, 7181.
118 Hansen, J. H., Autschbach, J., Davies, H. M. L. *J. Org. Chem.* **2009**, *74*, 6555.
119 Davies, H. M. L., Beckwith, R. E. J., Antoulinakis, E. G., Jin, Q. H. *J. Org. Chem.* **2003**, *68*, 6126.
120 Davies, H. M. L., Yang, J., Nikolai, J. *J. Organomet. Chem.* **2005**, *690*, 6111.
121 Davies, H. M. L., Venkataramani, C., Hansen, T., Hopper, D. W. *J. Am. Chem. Soc.* **2003**, *125*, 6462.
122 Davies, H. M. L., Ren, P. D., Jin, Q. H. *Org. Lett.* **2001**, *3*, 3587.
123 Davies, H. M. L., Ren, P. D. *J. Am. Chem. Soc.* **2001**, *123*, 2070.
124 Higashi, T., Isobe, Y., Ouchi, H., Suzuki, H., Okazaki, Y., Asakawa, T., Furuta, T., Wakimoto, T., Kan, T. *Org. Lett.* **2011**, *13*, 1089.
125 McMills, M. C., Humes, R. J., Pavlyuk, O. M. *Tetrahedron Lett.* **2012**, *53*, 849.
126 Hansen, J. H., Gregg, T. M., Ovalles, S. R., Lian, Y., Autschbach, J., Davies, H. M. L. *J. Am. Chem. Soc.* **2011**, *133*, 5076.
127 Davies, H. M. L., Stafford, D. G., Hansen, T. *Org. Lett.* **1999**, *1*, 233.
128 Davies, H. M. L., Jin, Q. *Proc. Natl. Acad. Sci. USA* **2004**, *101*, 5472.
129 Davies, H. M. L., Jin, Q. *J. Am. Chem. Soc.* **2004**, *126*, 10862.
130 Davies, H. M. L., Manning, J. R. *J. Am. Chem. Soc.* **2006**, *128*, 1060.
131 Lian, Y., Hardcastle, K. I., Davies, H. M. L. *Angew. Chem. Int. Ed.* **2011**, *50*, 9370.
132 Lian, Y., Davies, H. M. L. *J. Am. Chem. Soc.* **2011**, *133*, 11940.
133 Davies, H. M. L., Walji, A. M. *Angew. Chem. Int. Ed.* **2005**, *44*, 1733.
134 Davies, H. M. L., Dai, X., Long, M. S. *J. Am. Chem. Soc.* **2006**, *128*, 2485.
135 Dai, X., Wan, Z., Kerr, R. G., Davies, H. M. L. *J. Org. Chem.* **2007**, *72*, 1895.
136 Li, Z., Davies, H. M. L. *J. Am. Chem. Soc.* **2009**, *132*, 396.
137 Miller, D. J., Moody, C. J. *Tetrahedron* **1995**, *51*, 10811.
138 Liang, Y., Zhou, H., Yu, Z.-X. *J. Am. Chem. Soc.* **2009**, *131*, 17783.
139 Diaz-Requejo, M. M., Belderrain, T. R., Nicasio, M. C., Perez, P. J. *Dalton Trans.* **2006**, 5559.
140 Zhu, S. F., Zhou, Q. L. *Acc. Chem. Res.* **2012**, *45*, 1365.
141 Saito, H., Uchiyama, T., Miyake, M., Anada, M., Hashimoto, S., Takabatake, T., Miyairi, S. *Heterocycles* **2010**, *81*, 1149.
142 Xu, B., Zhu, S.-F., Xie, X.-L., Shen, J.-J., Zhou, Q.-L. *Angew. Chem. Int. Ed.* **2011**, *50*, 11483.

143 Wang, Y., Zhu, Y., Chen, Z., Mi, A., Hu, W., Doyle, M. P. *Org. Lett.* **2003**, *5*, 3923.
144 Lu, C.-D., Liu, H., Chen, Z.-Y., Hu, W.-H., Mi, A.-Q. *Org. Lett.* **2004**, *7*, 83.
145 Lu, C.-D., Liu, H., Chen, Z.-Y., Hu, W.-H., Mi, A.-Q. *Chem. Commun.* **2005**, 2624.
146 Guo, X., Huang, H., Yang, L., Hu, W. *Org. Lett.* **2007**, *9*, 4721.
147 Guo, X., Yue, Y., Hu, G., Zhou, J., Zhao, Y., Yang, L., Hu, W. *Synlett* **2009**, *2009*, 2109.
148 Xu, X., Han, X., Yang, L., Hu, W. *Chem. Eur. J.* **2009**, *15*, 12604.
149 Zhu, Y., Zhai, C., Yue, Y., Yang, L., Hu, W. *Chem. Commun.* **2009**, 1362.
150 Huang, H., Guo, X., Hu, W. *Angew. Chem. Int. Ed.* **2007**, *46*, 1337.
151 Hu, W., Xu, X., Zhou, J., Liu, W.-J., Huang, H., Hu, J., Yang, L., Gong, L.-Z. *J. Am. Chem. Soc.* **2008**, *130*, 7782.
152 Xu, X., Zhou, J., Yang, L., Hu, W. *Chem. Commun.* **2008**, 6564.
153 Jiang, J., Xu, H.-D., Xi, J.-B., Ren, B.-Y., Lv, F.-P., Guo, X., Jiang, L.-Q., Zhang, Z.-Y., Hu, W.-H. *J. Am. Chem. Soc.* **2011**, *133*, 8428.
154 Guan, X.-Y., Yang, L.-P., Hu, W. *Angew. Chem. Int. Ed.* **2010**, *49*, 2190.
155 Padwa, A. *Tetrahedron* **2011**, *67*, 8057.
156 England, D. B., Padwa, A. *J. Org. Chem.* **2008**, *73*, 2792.
157 Kitagaki, S., Anada, M., Kataoka, O., Matsuno, K., Umeda, C., Watanabe, N., Hashimoto, *S.-i. J. Am. Chem. Soc.* **1999**, *121*, 1417.
158 Kitagaki, S., Yasugahira, M., Anada, M., Nakajima, M., Hashimoto, S. *Tetrahedron Lett.* **2000**, *41*, 5931.
159 Tsutsui, H., Shimada, N., Abe, T., Anada, M., Nakajima, M., Nakamura, S., Nambu, H., Hashimoto, S. *Adv. Synth. Catal.* **2007**, *349*, 521.
160 Hodgson, D. M., Stupple, P. A., Johnstone, C. *Tetrahedron Lett.* **1997**, *38*, 6471.
161 Hodgson, D.M., Stupple, P.A., Johnstone, C. *Chem. Commun.* **1999**, 2185.
162 Hodgson, D. M., Stupple, P. A., Pierard, F. Y. T. M., Labande, A. H., Johnstone, C. *Chem. Eur. J.* **2001**, *7*, 4465.
163 Hodgson, D. M., Glen, R., Grant, G. H., Redgrave, A. J. *J. Org. Chem.* **2002**, *68*, 581.
164 Hodgson, D. M., Labande, A. H., Pierard, F. Y. T. M. *Synlett* **2003**, *2003*, 0059.
165 Hodgson, D. M., Labande, A. H., Pierard, F. Y. T. M., Castro, M. A. E. *J. Org. Chem.* **2003**, *68*, 6153.
166 Hodgson, D. M., Selden, D. A., Dossetter, A. G. *Tetrahedron-Asymmetry* **2003**, *14*, 3841.
167 Hodgson, D. M., Brockl, T., Glen, R., Labande, A. H., Selden, D. A., Dossetter, A. G., Redgrave, A. *J. Proc. Natl. Acad. Sci. USA* **2004**, *101*, 5450.
168 Shimada, N., Oohara, T., Krishnamurthi, J., Nambu, H., Hashimoto, S. *Org. Lett.* **2011**, *13*, 6284.
169 Clark, J. S., Berger, R., Hayes, S. T., Thomas, L. H., Morrison, A. J., Gobbi, L. *Angew. Chem. Int. Ed.* **2010**, *49*, 9867.
170 Jaber, D. M., Burgin, R. N., Helper, M., Zavalij, P. Y., Doyle, M. P. *Org. Lett.* **2012**, *14*, 1676.
171 Doyle, M. P., Ene, D. G., Forbes, D. C., Tedrow, J. S. *Tetrahedron Lett.* **1997**, *38*, 4367.
172 Li, Z. J., Davies, H. M. L. *J. Am. Chem. Soc.* **2010**, *132*, 396.
173 Wood, J. L., Moniz, G. A., Pflum, D. A., Stoltz, B. M., Holubec, A. A., Dietrich, H. J. *J. Am. Chem. Soc.* **1999**, *121*, 1748.
174 Li, Z. J., Parr, B. T., Davies, H. M. L. *J. Am. Chem. Soc.* **2012**, *134*, 10942.
175 Parr, B. T., Li, Z., Davies, H. M. L. *Chem. Sci.* **2011**, *2*, 2378.
176 Li, Z., Boyarskikh, V., Hansen, J. H., Autschbach, J., Musaev, D. G., Davies, H. M. L. *J. Am. Chem. Soc.* **2012**, *134*, 15497.

177 Moniz, G. A., Wood, J. L. *J. Am. Chem. Soc.* **2001**, *123*, 5095.
178 DeAngelis, A., Shurtleff, V. W., Dmitrenko, O., Fox, J. M. *J. Am. Chem. Soc.* **2011**, *133*, 1650.
179 Lian, Y., Davies, H. M. L. *J. Am. Chem. Soc.* **2010**, *132*, 440.
180 Qiu, H., Li, M., Jiang, L.-Q., Lv, F.-P., Zan, L., Zhai, C.-W., Doyle, M. P., Hu, W.-H. *Nat. Chem.* **2012**, *4*, 733.
181 Davies, H. M. L., Saikali, E., Clark, T. J., Chee, E. H. *Tetrahedron Lett.* **1990**, *31*, 6299.
182 Davies, H. M. L., Saikali, E., Young, W. B. *J. Org. Chem.* **1991**, *56*, 5696.
183 Davies, H. M. L., Smith, H. D., Hu, B. H., Klenzak, S. M., Hegner, F. J. *J. Org. Chem.* **1992**, *57*, 6900.
184 Davies, H. M. L., Hu, B. H., Saikali, E., Bruzinski, P. R. *J. Org. Chem.* **1994**, *59*, 4535.
185 Lian, Y., Davies, H. M. L. *Org. Lett.* **2010**, *12*, 924.
186 Lian, Y., Davies, H. M. L. *Org. Lett.* **2012**, *14*, 1934.
187 Wang, X., Xu, X., Zavalij, P. Y., Doyle, M. P. *J. Am. Chem. Soc.* **2011**, *133*, 16402.
188 Wang, X., Abrahams, Q. M., Zavalij, P. Y., Doyle, M. P. *Angew. Chem. Int. Ed.* **2012**, *51*, 5907.
189 Valette, D., Lian, Y., Haydek, J. P., Hardcastle, K. I., Davies, H. M. L. *Angew. Chem. Int. Ed.* **2012**, *51*, 8636.
190 Smith, A. G., Davies, H. M. L. *J. Am. Chem. Soc.* **2012**, *134*, 18241.

13

RUTHENIUM CARBENES

Steven T. Diver and Jonathan M. French
Department of Chemistry, University at Buffalo, The State University of New York, Amherst, NY, USA

13.1 INTRODUCTION

Alkene metathesis with ruthenium carbenes has become one of the most widely used synthetic methods employing a transition metal catalyst. Alkene metathesis is promoted by the ruthenium carbene complexes developed by Grubbs and by the molybdenum carbene complexes of Schrock. In this chapter, we focus on the chemistry of ruthenium carbenes. Ruthenium carbenes, collectively called the Grubbs complexes, are tolerant of organic functional groups and display remarkable selectivity for unsaturated reactants such as alkenes. With the Grubbs catalysts, chemoselective targeting of alkenes is possible within a complex framework of organic functional groups. That the ruthenium carbenes are able to cleave and reassemble stable carbon–carbon double bonds but largely ignore other Lewis bases, shows remarkable poise that is synthetically valuable. The importance of the ruthenium carbene catalysts and the unique chemoselectivity that they offer in organic and materials synthesis was recognized in the award of the Nobel Prize in Chemistry (2005) to Professors Grubbs, Schrock, and Chauvin. In the last decade, there have been major advances in catalyst development, and mechanistic understanding and applications continue to swell. We will give some mechanistic background and detail of two recent advances in mechanistic understanding, the identification of ruthenacyclobutane intermediates and conformational changes of the ruthenium carbene, derived from early Density Functional Theory (DFT)

Contemporary Carbene Chemistry, First Edition. Edited by Robert A. Moss and Michael P. Doyle.

calculations and confirmed through experiment. The next two major areas are the development of phosphine-free ruthenium carbene complexes and greater refinement of our understanding of ruthenium carbene decomposition pathways. We will move onto issues of selectivity in alkene metathesis: enantioselectivity, site selectivity achievable through relay metathesis, and *Z*-selectivity in cross alkene metathesis. Last, the applications are overviewed in three subsections: natural product synthesis, applications in aqueous media, and the use of metathesis as a means for creating diverse nonnatural structures (diversity-oriented synthesis). Finally, we will provide a brief conclusion and outlook on this amazing field of chemistry.

13.2 IMPROVED MECHANISTIC UNDERSTANDING

The alkene metathesis mechanism is well understood for the ruthenium carbene complexes **Ru1** and **Ru2** (Scheme 13.1). The other available catalysts undergo initiation with different rates but operate by the same mechanism, proceeding through common intermediates. The landmark kinetic experimental work by Grubbs' group yielded the mechanistic sequence of events with identification of the rate-limiting step. The basic scheme of catalysis applies to the more active second generation Grubbs complex, **Ru2**, though a different and somewhat surprising kinetic profile was found. Since these early kinetic studies, DFT calculations have played an increasingly important role in mechanistic study, addressing some of the implications of the Grubbs work. Calculations provide a detailed view of the intermediates and the energy landscape of catalysis. The improved quality of the calculations and the number of studies directed towards understanding reactivity trends using calculations has made them more attractive and widely accepted. The recent experimental support for ruthenacyclobutane intermediates has further refined our understanding of catalysis, providing insight into the remarkable qualities of ruthenium carbenes to engage uniquely in alkene metathesis rather than, for example, cyclopropanation. In this section, the greater interplay of theory and experiment will be seen. The basic mechanism of alkene metathesis, key intermediates, and the improved reactivity profile of **Ru2** will be discussed.

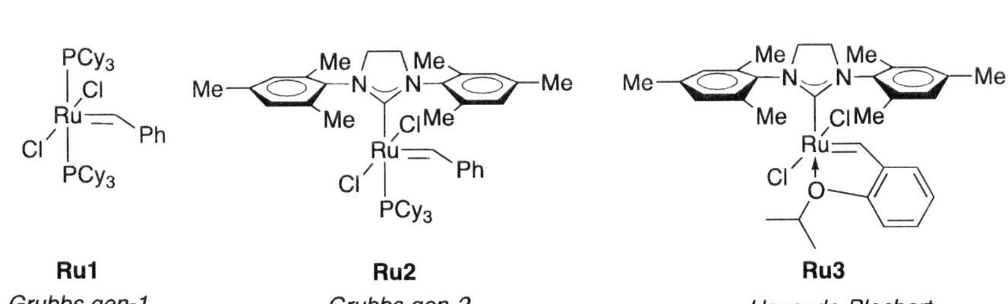

Scheme 13.1.

Two of the top 10 highlights are intermingled: (i) improved mechanistic understanding through identification of ruthenacyclobutane intermediates and (ii) the use of DFT calculations to help improve understanding of processes such as conformational changes of the ruthenium carbene.

13.2.1 Background

The catalytic cycle of alkene metathesis proceeds first through an initiation step to make an active ruthenium carbene and is then followed by cross metathesis. The Grubbs catalysts **Ru1** and **Ru2** are 16-valence electron complexes. To become active in alkene metathesis, these "precatalysts" must initiate to form a new metal carbene capable of entering the catalytic cycle. The initiation is a two-step process where the phosphine must be replaced with an alkene, and a new carbene forms (Eq. 13.1). The initiation process results in loss of styrene. After phosphine loss, the remaining steps in Equation (13.1) result in creation of an active carbene. The high-energy alkene π-complex passes to an intermediate ruthenacyclobutane via a cycloaddition-like process. Fragmentation of the ruthenacyclobutane produces a new π-complex which will dissociate styrene and bind another alkene reactant. The conversion of one ruthenium carbene to another via the metallacycle mechanism is called transalkylidenation. In this chapter, we refer to it as a carbene ligand exchange and in this case CHPh is exchanged for CHR, as depicted in Equation (13.1).

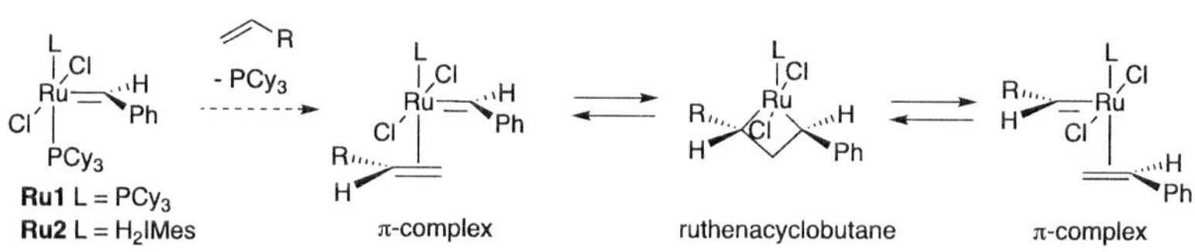

(13.1)

Rate studies indicate that the catalyst becomes activated by a dissociative mechanism. For **Ru1**, mechanistic studies showed a rate depression in the presence of excess phosphine and found that the catalyst rate dependence was half order.[1] These observations suggest a dissociative process. Detailed kinetic analysis led Grubbs and coworkers[1] to estimate that 90%–95% of alkene metathesis proceeded by a dissociative pathway based on an Eyring plot for degenerate tricyclohexylphosphine exchange. Both **Ru1** and **Ru2** were found to have positive entropies of activation in the range expected for a dissociative ligand exchange mechanism.

Once initiated, the alkene metathesis follows the catalytic cycle depicted in Scheme 13.2. The metal carbene is shown in its active form at the top of the cycle. When active, it will turn over many times before being diverted to an inactive, phosphine-bound resting state **A**. (Decomposition is another alternative, considered next.) In catalysis, a second alkene binds, forms the metallacycle, which breaks down to form a new alkene complex. Dissociation of the alkene

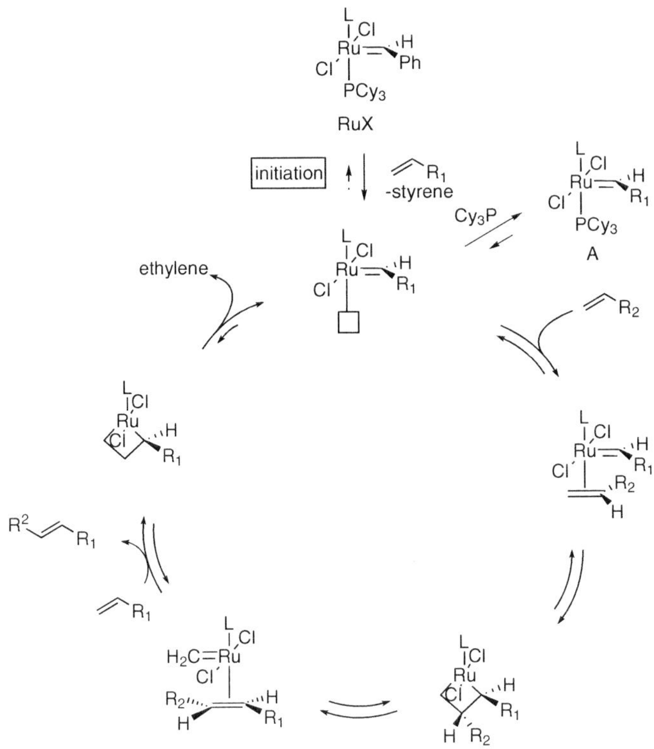

Scheme 13.2.

and replacement by excess, reactive 1-alkene provides a monosubstituted ruthenacyclobutane which fragments to give ethylene and the reactive metal carbene. All elementary steps are reversible. In practice, the serial equilibria of the catalytic cycle are stressed by mass action when the ethylene is driven out of solution by the refluxing action of solvent. Ultimately, this loss of ethylene is used to drive the entire catalytic cycle forward to give the desired cross metathesis product, a 1,2-disubstituted alkene. Though alkene complexation is shown through "bottom binding," it is also possible that alkene binding can occur from the side as well. Side binding is important in the *Z*-selective chelated ruthenium carbenes discussed in Section 13.4.3.

The second-generation Grubbs carbene **Ru2** is more active than **Ru1** due to the presence of a stronger σ-donating ligand. The **Ru2** catalyst bears a dihydroimidazole carbene ligand, an *N*-heterocyclic carbene (NHC). To enter the catalytic cycle of Scheme 13.2, the "lower" Cy_3P must be lost. This dissociated phosphine is oriented *trans* to the NHC ligand. Since phosphine dissociation controls entry of an active catalyst into the cycle, it was tempting to explain the greater reactivity due to a *trans*-labilizing effect in the parent 16-electron complex. This hypothesis predicted that **Ru2** would dissociate the Cy_3P more easily (and faster) than the diphosphine complex **Ru1**. However, Grubbs and coworkers[2,3] demonstrated surprisingly that **Ru2** has a higher alkene affinity which accounts for its much higher activity in alkene metathesis. The phosphine exchange rate is an order of magnitude faster for the less active catalyst **Ru1** (Scheme 13.3). Based on a plot of $1/k_{obs}$ versus $[Cy_3P]/[alkene]$, k_{-1} and the quotient k_{-1}/k_2 could be determined.

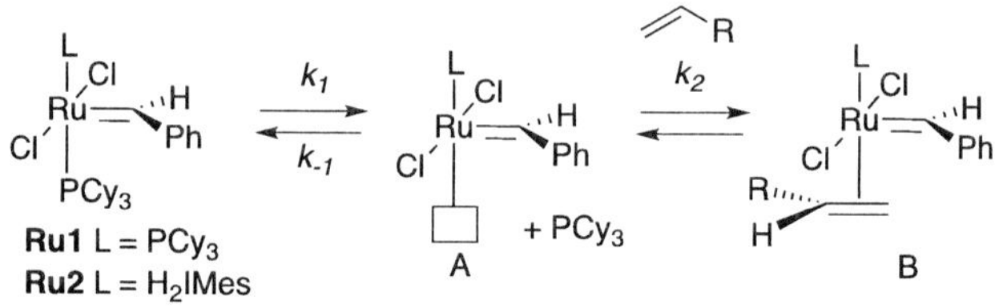

(13.2)

	k_{-1} (80 °C)	ΔG≠	ΔH≠	ΔS≠	k_{-1}/k_2 (50 °C)
Ru1	9.6 ± 0.2 s^{-1}	19.88	23.6	12	1.3×10^4
Ru2	0.13 ± 0.01 s^{-1}	23.0	27	13	1.25

ΔG,ΔH in kcal/mole; ΔS in e.u.

Scheme 13.3.

In the kinetic model of Equation (13.2), the ratio k_{-1}/k_2 represents the partitioning of the 14-electron intermediate between entering catalysis or reverting to an inactive precatalyst state. This assumes that the intermediate in Equation (13.2) irreversibly enters catalysis. The assumption that all subsequent intermediates are energetically downhill from the 14-electron intermediate was supported based on a DFT calculation by Cavallo.[4] The second generation carbene **Ru2** has a much greater affinity for alkene binding versus **Ru1**. For **Ru2** by itself, it has roughly equal probability to revert or to enter the catalytic cycle. In contrast, **Ru1** prefers to revert to the 16-electron inactive precatalyst state 10,000:1 versus alkene metathesis. The second generation Grubbs catalyst bonds more strongly to alkenes. Stronger backbonding to the alkene lowest unoccupied molecular orbital (LUMO) occurs in **Ru2** due to greater electron density on the ruthenium atom *vis-à-vis* the NHC ligand. This more than compensates for the slower phosphine dissociation rate, improving catalyst performance by a factor of 1000.

The greater reactivity of the second generation Grubbs catalyst has allowed more difficult alkene metatheses to be performed successfully. The increased reactivity has allowed the use of less reactive alkenes and more forcing reaction conditions. Though alkene metathesis emerged as a synthetic method due to its prowess performing ring-closing reactions, in recent years, selective cross metathesis continues to pose a challenge due to both reactivity and selectivity considerations. For example, in cross alkene metathesis involving an electron-deficient alkene partner, **Ru2** is superior.

One example is shown in Equation (13.3).[5] In addition, cross reaction with geminal alkenes works well (Eq. 13.4).[6] The greater reactivity of **Ru2** permits the use of 1,3-dienes as reactants (Eq. 13.5), whereas **Ru1** does not react with conjugated dienes.[7]

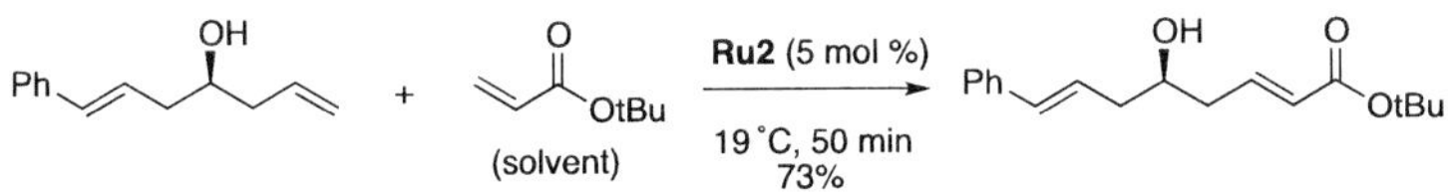

(13.3)

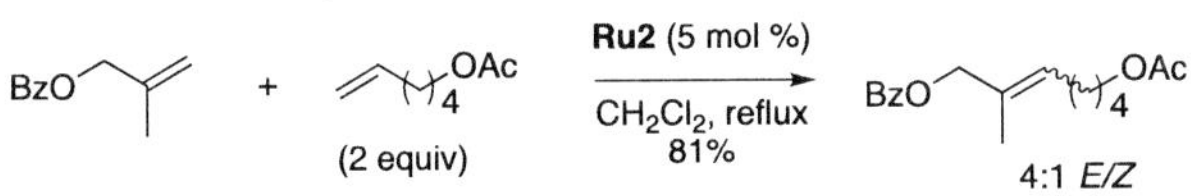

(13.4)

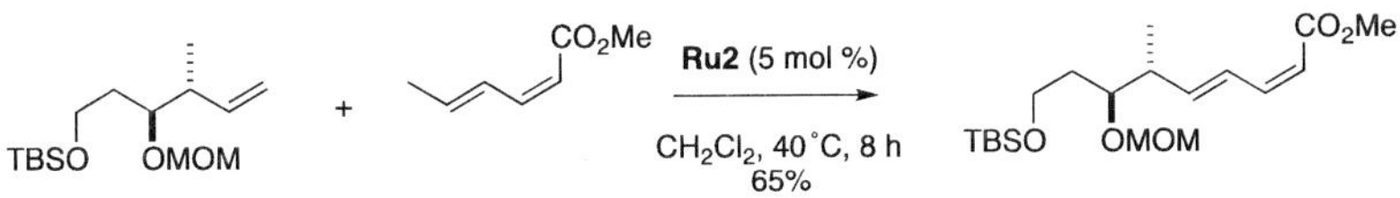

(13.5)

Selective cross metathesis occurs between different alkenes if they have a different reactivity towards the Grubbs catalyst. This can be predicted on the basis of the Grubbs empirical model for cross alkene metathesis (Scheme 13.4).[6] For the most selective combination, one alkene should react with the catalyst and the other should not. Type I alkenes, such as 1-hexene, are the most reactive. They cross react well with alkenes in less reactive categories, such as type III or type IV alkenes. With only one reactive ruthenium carbene produced, both the desired cross metathesis (Eq. 13.6) and the self-metathesis (Eq. 13.7) takes place. However, the homodimer is reactive to further metathesis, often referred to as "secondary metathesis." Secondary metathesis is a consecutive alkene metathesis occurring on the primary products of cross metathesis, which achieves equilibrium control. The homodimer reacts again with the catalyst **Ru2** and repopulates the active metal carbene which partitions again between homodimer and cross product.

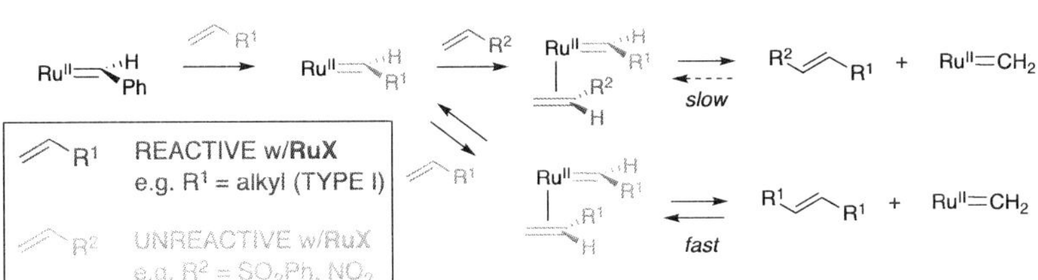

(13.6)

(13.7)

Scheme 13.4.

An excellent example of reversibility (e.g., secondary metathesis) can be found in the synthetic studies towards the cylindrocyclophanes by Smith and

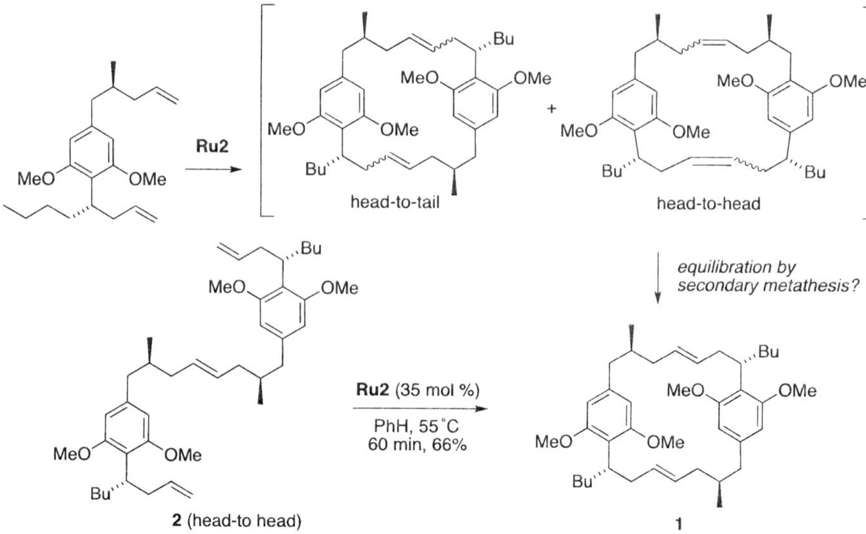

Scheme 13.5.

coworkers.[8,9] The assembly of the head-to-tail macrocycle **1** was achieved through stepwise alkene metathesis. However, a direct homodimerization-type of cross metathesis offered a more direct solution to the target molecule from a single diene precursor (Scheme 13.5). Head-to-head cross metathesis would assemble an undesired product, which was predicted to be less stable than the desired product on the basis of molecular modeling. The alkenes are reactive, type I alkenes and capable of secondary metathesis (equilibrium control). Thus, it was predicted that head-to-head intermediates would form but then undergo reversible secondary metathesis, thereby equilibrating to the more stable and desired head-to-tail product **1**.[9] To test this hypothesis, a model head-to-head dimer **2** was found to equilibrate to the desired product. This study in natural product total synthesis nicely demonstrates how reversible "homodimerization" can be used to simplify organic synthesis on the basis of the cross metathesis selectivity model and the principles of kinetic and thermodynamic control.

13.2.2 Identification of Ruthenacyclobutane Intermediates in Alkene Metathesis

A significant advance in mechanistic studies came with the identification of metallacycle intermediates involved in alkene metathesis. The ruthenacycle intermediate in alkene metathesis is a high energy species. The ruthenacyclobutane is high energy since it is an electron-deficient, 14-electron complex with the ruthenium atom formally in the Ru(IV) oxidation state. In considering the mechanism of carbene exchange, the geometry of the ruthenacyclobutanes was puzzling as it related to the forming of metal carbenes on either side of the equilibrium process (Eq. 13.8). If the intermediate ruthenacyclobutane ruptured, it would produce a carbene trans to the ligand L (Cy_3P or H_2IMes). Furthermore, microscopic reversibility was hard to reconcile because the reverse reaction from the alternate ruthenium carbene produced a different ruthenacyclobutane. A more refined picture of these intermediates was needed. Greater insight was provided in the early twenty-first century largely from DFT calculations by Chen[10,11] and Cavallo,[4] which resolved the issue of microscopic reversibility (*vide infra*).

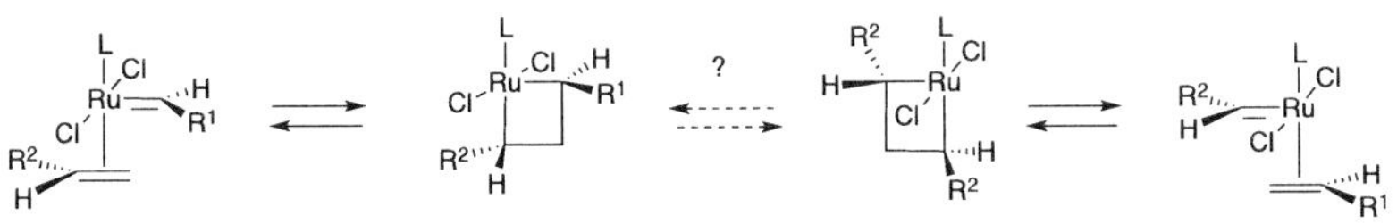

(13.8)

The geometry and existence of the ruthenacyclobutane were established by the direct observation of ruthenacyclobutane intermediates. Piers and Romero[12] focused on low temperature NMR studies which detected the intermediate using catalyst **Ru4** (Eq. 13.9).[13] The four H_α protons appeared at +6.6 ppm. The two H_b protons appeared at −2.6 ppm. The ^1H NMR also revealed that the mesityl groups of the NHC ligand were equivalent. Taken together with the equivalencies in the H_a and H_b proton environments, a symmetrical structure with overall C_{2v} symmetry was suggested. Similarly, Wenzel and Grubbs characterized a substituted ruthenacyclobutane spectroscopically at −95 °C using an assortment of NMR techniques.[14]

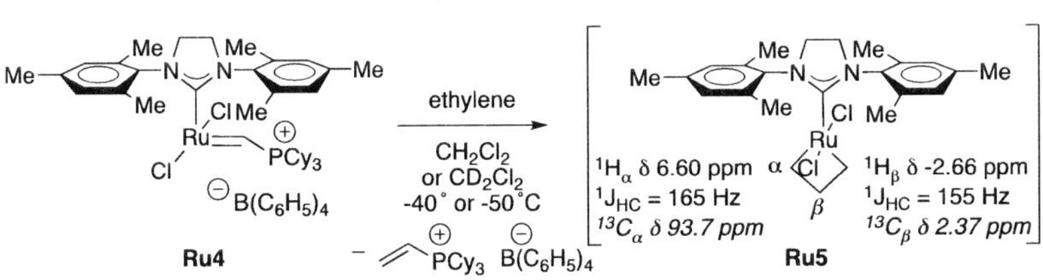

(13.9)

The chemical shifts and coupling constants suggested a vertical compression of the four-membered ring due to a strong transannular agostic interaction. There was a significant difference between the $^{13}C_\alpha$ and $^{13}C_\beta$ chemical shifts, significantly greater than the differences between the same two carbons on a variety of other metallacycles previously characterized.[15] Additionally, the $^{13}C_\alpha$–$^{13}C_\beta$ one bond coupling constant was found to be small ($^1J_{CC}$ = 15 Hz) as compared to normal values of $^1J_{CC}$ (30–39 Hz). Small $^1J_{CC}$ values are a primary indicator for (C–C)···M agostic interactions.[16] This interaction provides deeper

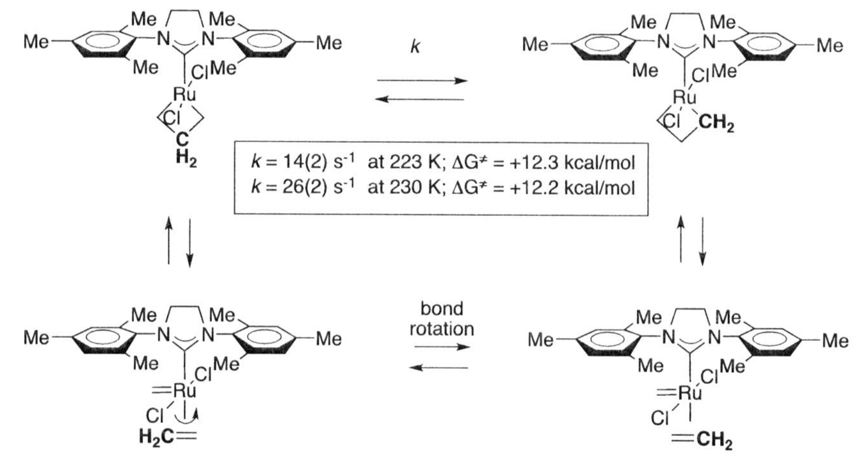

Scheme 13.6.

insight into how the Grubbs catalyst is able to split relatively stable C–C bonds—structural evidence pointing to a β-scission reaction *en route* to the η^2-bound metal carbene–alkene complex.

Both Grubbs' and Piers' groups determined that the α and β-carbons readily exchange positions. The exchange of these carbons was attributed to an equilibrium with the η^2-bound methylidenes (Scheme 13.6). Based on exchange spectroscopy, the transition state barrier was determined to be $\Delta G^{\neq} = 12.2$ kcal/mol. The exchange process provides experimental evidence linking metallacyclobutane intermediates with their *higher energy* π-alkene complexes.

13.2.3 Conformational Dynamics "Windshield Wipering"

In considering that the conformation of the Ru=C bond changes during catalysis, Chen designed a remarkable catalyst that displays alternating selectivity in its reactions with alkenes.[17] Conformational insight came from theoretical and experimental perspectives. Density Functional Theory calculations by Chen[10,11] and independent calculations by Cavallo[4] gave a theoretical framework for the conformational changes taking place as the metathesis reaction progressed through metallacyclic intermediates. Importantly, the geometry of the computed intermediate was validated by the observation of the ruthenacyclobutane intermediate by Piers and Romero.[13]

The orientation of the ruthenium carbene bond alternates each time it passes through a ruthenacyclobutane. Chen colorfully described this conformational change as "windshield wipering" because the flipping of the carbene back and forth in Equation (13.10) resembles the motion of a windshield wiper.[17]

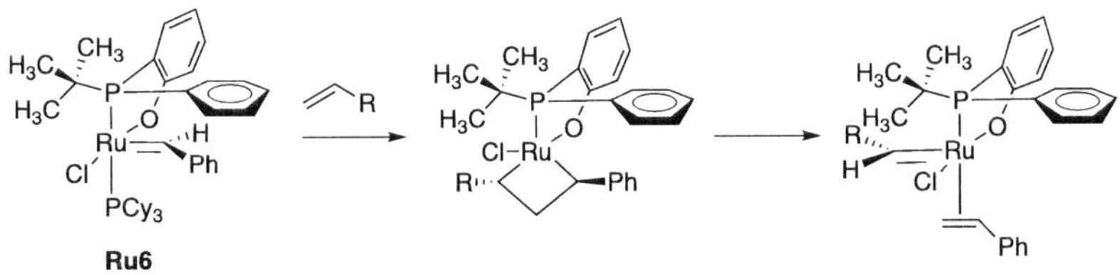

(13.10)

Due to its unsymmetric, bidentate phosphine ligand, Chen's catalyst **Ru6** has two distinct reactivity modes associated with its alternating carbene conformations. This design enabled the selective synthesis of an alternating copolymer composed of norbornene and cyclooctene subunits. In the copolymerization, different reactivity was used to produce selectivity in the alkene binding steps. In its starting conformation, the benzylidene proved to be unreactive towards polymerization of cyclooctene, but it was found to react with norbornene, which has pronounced reactivity due to its ring strain (steps I, II in Scheme 13.7). When the carbene was flipped into its alternate conformation, steric strain between the alkylidene substituent and the *tert*-butyl group would cause this carbene to be highly reactive (step IV). The higher reactivity led Chen and Bornand to infer that reaction with a "low strain" alkene such as cyclooctene would be possible.[17] In Scheme 13.7, the Ph moiety on the carbene initiator is replaced with a growing polymer on each turn of the catalytic cycle, which is summarized as P in the Scheme.

13.3 CATALYST DEVELOPMENT

In the past decade, catalyst development has resulted in better catalysts for applications in organic synthesis. In particular, with less reactive substrates, there are greater options for catalyst selection. This has improved applications to include more demanding cases. In general, the area that has seen the most growth is that of phosphine-free catalysts. The chelating ether design that originated from Hoveyda's lab has proven highly versatile. A catalyst is more desirable when it has a rapid initiation profile, when it can be used at higher temperatures, and when it is easily handled (and shelf stable). In this section, a few different phosphine-free catalysts are discussed with emphasis on the Hoveyda catalyst superfamily.

13.3.1 Development of Phosphine-Free Catalysts

As mechanistic understanding improved, it became apparent that though phosphine ligands were useful for the synthesis and shelf stability of the Grubbs catalyst, they were not necessarily desirable in the metathesis reaction. First, mechanistic studies showed that excess phosphine ligand depressed the rate of alkene metathesis. Second, though mechanistic studies explained why

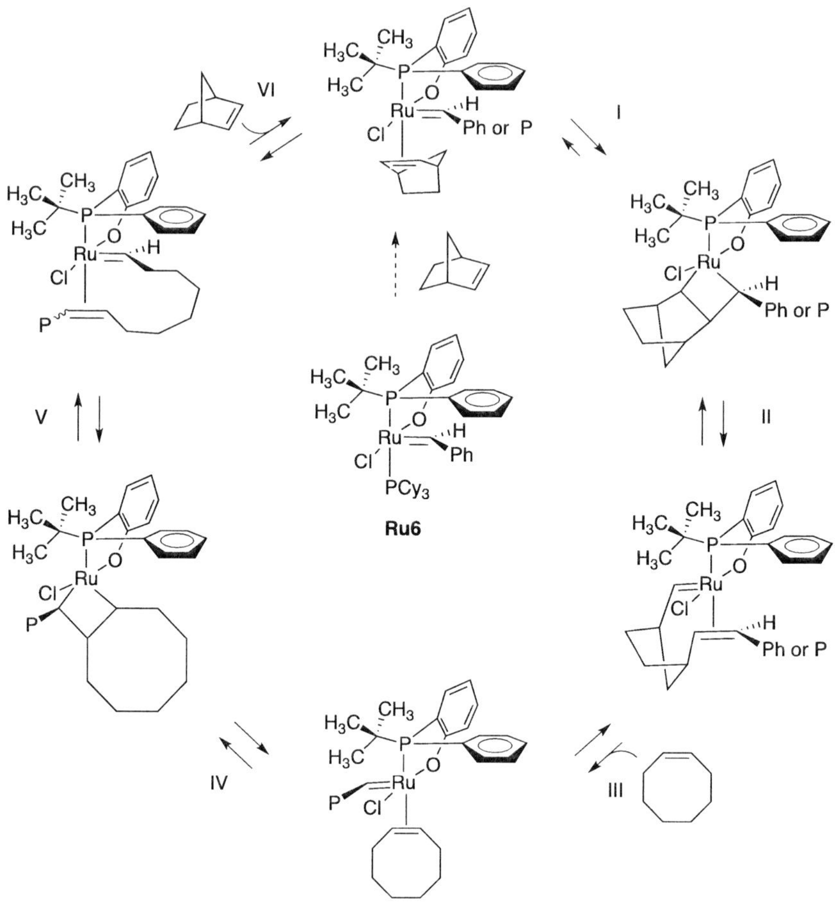

Scheme 13.7.

there was greater reactivity for **Ru2**, the dissociated tricyclohexylphosphine still competed with alkene binding. Third, the key intermediate in alkene metathesis is vulnerable to poisoning by the phosphine ligand.[2] This intermediate $(H_2IMes)Cl_2Ru{=}CH_2$ is required for cross alkene metathesis of 1-alkenes and it is kinetically deactivated by Cy_3P. Last, there has been continuous catalyst development since 1999,[18] largely due to the tunability (and modular synthesis) of the NHC ligand. As a result, catalyst development efforts sought better, faster-initiating catalysts that abandoned the phosphine ligand while still seeking shelf stability. Focus on improved initiation rate led many investigators to consider phosphine-free ruthenium carbene catalysts. This section will focus on this group of catalysts. Other catalyst development efforts directed toward *Z*-selective catalysts (Section 13.4.3) and water-soluble catalysts (Section 13.5.2) are described next.

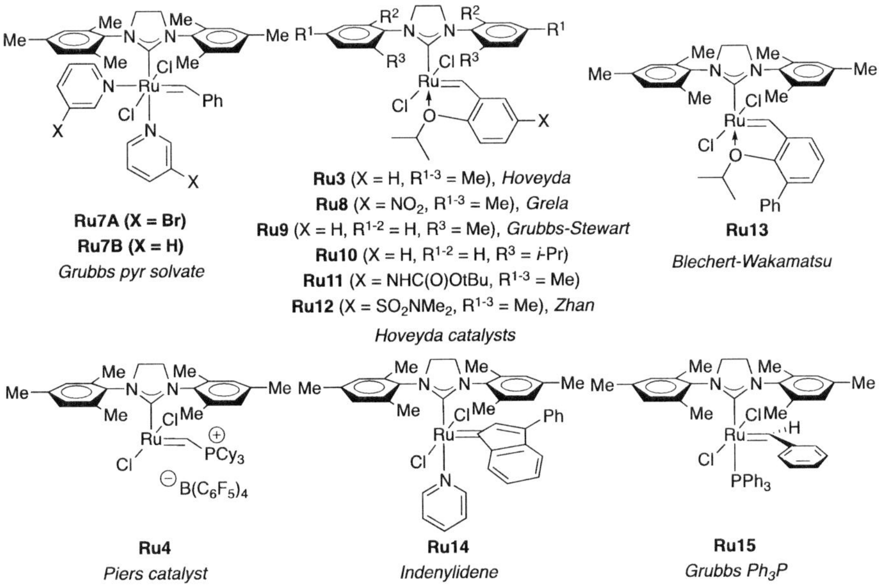

Scheme 13.8.

Several phosphine-free catalysts have emerged from this appraisal. They are illustrated in Scheme 13.8. Replacement of the phosphine by the more labile pyridine ligand gave the solvate **Ru7A**.[19] This is commonly known as the Grubbs pyridine solvate and has been used with both 5-bromopyridine and pyridine. Placing a coordinating group on the arylidene ligand gives a chelating ether which replaced the phosphine.[20] The aromatic ring can then be tuned electronically and sterically by additional substitution on the aromatic ring. In some cases, substitution had dramatic results. The Grela catalyst **Ru8** has a *p*-nitro group.[21] **Ru9** is known as the Grubbs–Stewart catalyst,[22] designed for sterically demanding ring-closing metathesis (RCM). **Ru10**,[22] **Ru11**,[23] and **Ru12**[24] are similar. **Ru13** uses steric compression to weaken the dative bond, known as the Blechert–Wakamatsu catalyst.[25] This family of catalysts is based on the catalyst design originated by Hoveyda and derived from the Hoveyda–Blechert catalyst (also called the Hoveyda–Grubbs catalyst). Other phosphine-free catalysts include **Ru4A** (Piers' catalyst)[13] and **Ru14** which has an indenylidene carbene.[26,27]

13.3.1.1 Pyridine Solvates. The pyridine complexes **Ru7A,B** are very fast initiators that promote difficult alkene metathesis. The rates of initiation for several complexes are provided in Table 13.1.[19] Alkene cross metathesis of electron-deficient alkenes such as acrylonitrile are challenging. The product yields given in Table 13.3 for the cross alkene metathesis between acrylonitrile and allylbenzene provide an estimate of catalytic effectiveness in a difficult cross alkene metathesis.

TABLE 13.1. Comparison of Initiation Rates and Cross Metathesis (CM) Yields[a]

Entry	Catalyst	k_{init} ($\times 10^{-3}$ s^{-1} at 5 °C)[b]	k_{rel}	Yield[c]
1	**Ru2**	0.0032 ± 0.0006	1	21
2	**Ru15**	1.8 ± 0.1	563	35
3	**Ru3**	2.6 ± 0.1	812	68
4	**Ru7B**	>200[d]	$>6.25 \times 10^4$	26
5	**Ru7A**	>4000[d]	$>1.25 \times 10^6$	67

[a]Reference 19.
[b]Measured by reaction with ethyl vinyl ether.
[c]Yield of CM product between allylbenzene and acrylonitrile.
[d]Lower limit estimate.

The pyridine complexes **Ru7A,B** proved to be versatile synthetic intermediates to make other ruthenium carbene complexes. Complexes **Ru7A,B** are easily prepared from **Ru2** using an excess of the appropriate pyridine. The pyridine ligands are rapidly displaced by different phosphines and readily undergo halide ion exchange (e.g., to make the diiodide monopyridine solvate).[28] Sponsler and coworkers[29] used the pyridine solvates to synthesize ruthenium alkylidenes from **Ru7A** (Eq. 13.11). These alkylidene complexes were found to initiate 10–20 times faster than **Ru2**.

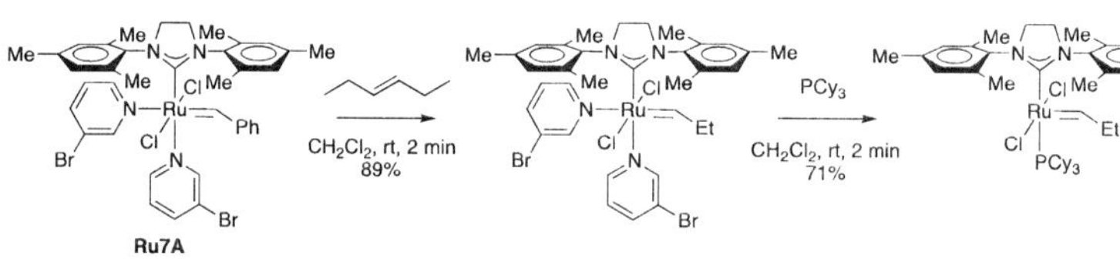

(13.11)

13.3.1.2 Hoveyda Catalyst Superfamily. The chelating ether motif is characteristic of the Hoveyda catalyst superfamily. This ligand design has become widely used because it offers a stable precatalyst and has proven to be a versatile platform for substitution. The chelating ether design originated from Hoveyda and coworkers.[20] In this work, the goal was to have the dissociated ligand rebound and reattach to the ruthenium atom after the alkene reactants were consumed (Barrett's boomerang idea).[30] The placement of a chelating ether on the arylidene moiety meant that the ligand used to protect what would become the open valence of the metal departs when the catalyst initiated, leaving the ruthenium catalyst turned on, in a 14-electron configuration. This bold idea meant that the catalyst would be "switched on" in the presence of alkene and could be "switched off" when the alkene was consumed by rebinding the bidentate styrene ligand, *o*-(isopropoxy)styrene. Contemporaneously, **Ru2** was reported by Grubbs' group.[18] The new chelating ether design was quickly modified to incorporate the NHC ligand, resulting in the Hoveyda–Blechert complex.[31,32] The synthesis of **Ru3** was reported by Blechert almost simultaneously.[33]

The Hoveyda catalyst is well-suited to ring-closing metathesis involving electron-deficient alkenes (Eq. 13.12).[34] Hoveyda recognized that electron-deficient carbenes derived from electron-poor alkenes were more susceptible to phosphine binding, which would deactivate ruthenium carbene intermediates.[35] The phosphine-free system **Ru3** avoids this problem.

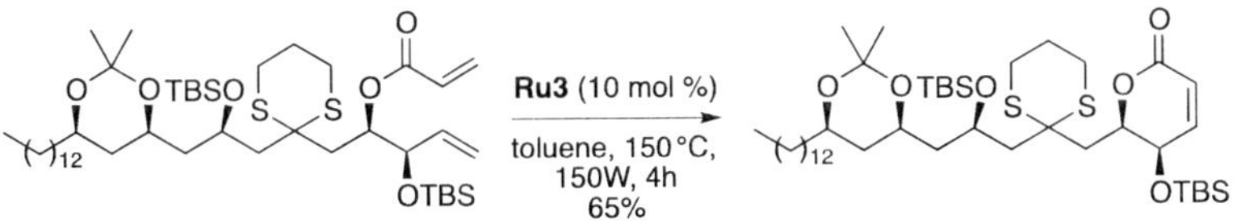

(13.12)

Tetrasubstituted alkenes can be closed with a Hoveyda-type complex. By deleting an ortho methyl group, the catalyst **Ru9** becomes more capable for handling highly substituted carbenes. This is known as the Grubbs–Stewart catalyst. The reactivity is sufficient to break into unreactive geminal alkenes.[22,36] Currently, however a tetrasubstituted alkene cannot be made by cross alkene metathesis. A nice application of catalyst **Ru9** can be found in Stolz and coworkers total synthesis of elatol (Eq. 13.13).[37]

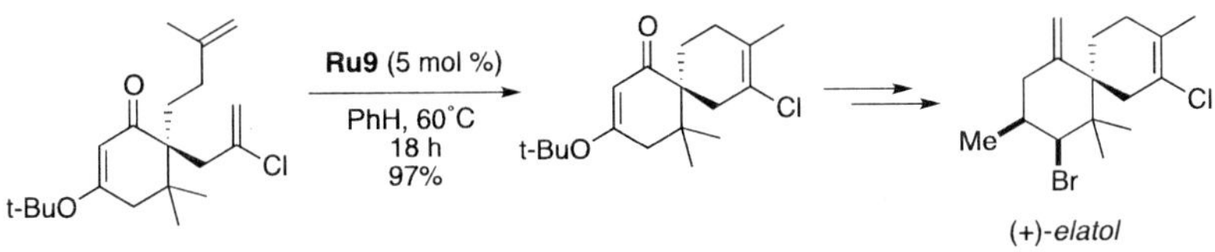

(13.13)

13.3.1.3 Piers' Catalyst. Another unique phosphine-free catalyst can be found in the work of Piers.[13] A ruthenium carbide[38] was protonated with strong inorganic acid. This promoted a 1,2-shift of the Cy_3P group to the carbyne carbon, producing a **Ru4** (Eq. 13.14). This clever approach removes the phosphine from the coordination sphere of the metal and upon initiation with a suitable alkene is lost as a vinylphosphonium salt.

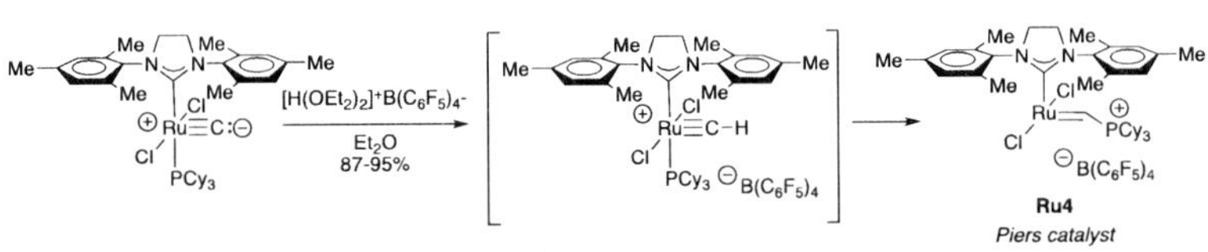

(13.14)

In addition to being a good catalyst for promoting alkene metathesis, the Piers catalyst has proven significant in mechanistic studies.[12,14,15] As detailed earlier, the Piers catalyst was used to identify a ruthenacyclobutane intermediate.

13.3.2 Decomposition of the Grubbs' Complex

One of the most important facets of ruthenium carbene-catalyzed alkene metathesis is the fate of the Grubbs complex. The fate of the catalyst is significant as it reveals fundamental reactivity. Decomposition presents a limitation to efficiency and may lead to unwanted side reactions. Early initiators suffered significant decomposition during catalysis. The more robust second generation Grubbs carbenes experience less decomposition during simple alkene metathesis, which allows them to be used at lower catalyst loading.

Decomposition can be classified into three major categories. First, there is an intrinsic decomposition of ruthenium carbenes. This intrinsic decomposition is due to the metal carbene complex reacting by itself or with itself. One example is thermal decomposition—the carbene complexes decompose on heating. Second, there are decomposition pathways that arise due to interactions with functional groups within specific substrates. These positioned functional groups are found in complex organic molecules. Third, decomposition might be caused by an external promoter, an agent that is not one of the reactants. In this section, decomposition of ruthenium carbenes relevant to alkene metathesis will be summarized.

13.3.2.1 Decomposition of the Methylidene. Of all the ruthenium carbenes, decomposition of the ruthenium methylidene is most significant. The ruthenium methylidene is significant because it is the propagating species in alkene metathesis involving 1-alkenes (α-olefins). The Ru=CH_2 species is formed in metatheses with either the first- or second-generation Grubbs catalysts.

The ruthenium methylidene decomposes thermally to produce a dinuclear ruthenium hydride (Eq. 13.15). Heating the complex **Ru16** in benzene for an extended time period produced the dinuclear complex **Ru17** in 46% yield (Scheme 13.9).[39] The temperature of this reaction is comparable to that used in routine cross alkene metatheses. A likely mechanism for decomposition is illustrated in Scheme 13.9.

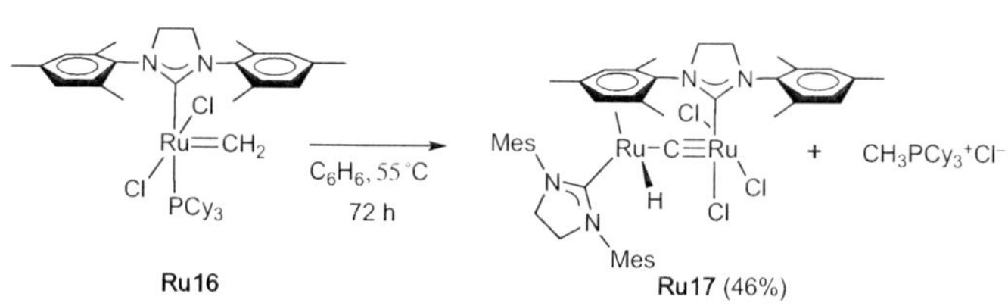

(13.15)

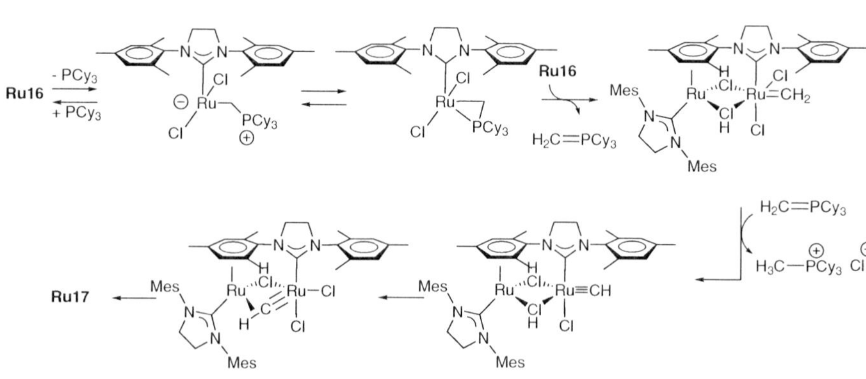

Scheme 13.9.

The dinuclear hydride **Ru17** causes alkene isomerization. Grubbs identified alkene isomerization under nominal ring-closing metathesis conditions (Eq. 13.16).[40] Grubbs demonstrated that the hydride **Ru17** caused alkene isomerization and this unwanted process was effectively suppressed by using benzoquinone as an additive. Benzoquinone is believed to oxidize ruthenium hydride intermediates.[40] Use of benzoquinone has been widely used as a means of preventing undesired alkene isomerization during a metathesis.[41–46] For example, a cross ene-yne metathesis with alkenols possessing free hydroxyl groups results in a side reaction caused by ruthenium hydride.[47] This process could be eliminated using benzoquinone as an additive, giving the ene-yne metathesis product in high yield (Eq. 13.17).[48]

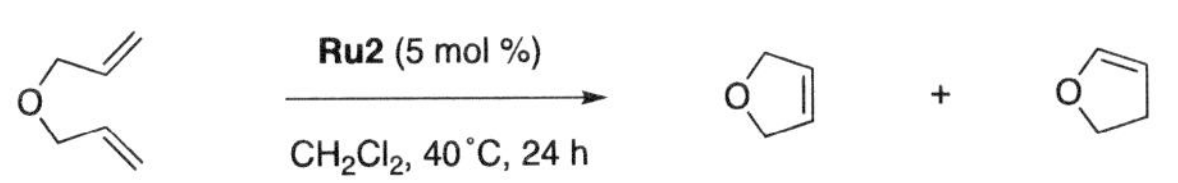

(13.16)

no additive	< 5%	> 95%
HOAc (10 mol %)	> 95%	none
benzoquinone (10 mol %)	> 95%	none

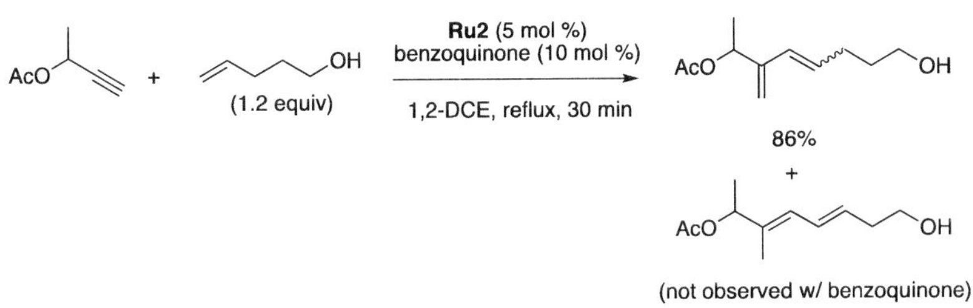

(13.17)

Under an atmosphere of ethylene, a different decomposition pathway occurs via CH bond insertion (Scheme 13.10).[49] Complex **Ru15** reacts with ethylene at room temperature to give a dinuclear intermediate methylene **Ru18**. On cooling to −40 °C, the expected ruthenacyclobutane **Ru19** is obtained by a reversible ethylene-induced bridge-splitting reaction. After extended reaction time at ambient temperature, decomposition gives a benzylic C–H insertion product **Ru20** is obtained along with $MePPh_3Cl$.

13.3.2.2 In the Presence of Air. Benzylic C–H insertion has also been identified during large-scale preparation of **Ru2**. In the large-scale synthesis, trace impurities of a Ru alkyl complex were observed (Eq. 13.18). The generation of this by-product arises from air and water, present as trace impurities in the solvent and could be eliminated by sparging solvents used for the crystallization step. Relatedly, a net C–H insertion was observed into the ortho position of a phenyl group in non-symmetrical Hoveyda-type catalysts. Unlike most second-generation Grubbs complexes, this carbene lacked ortho

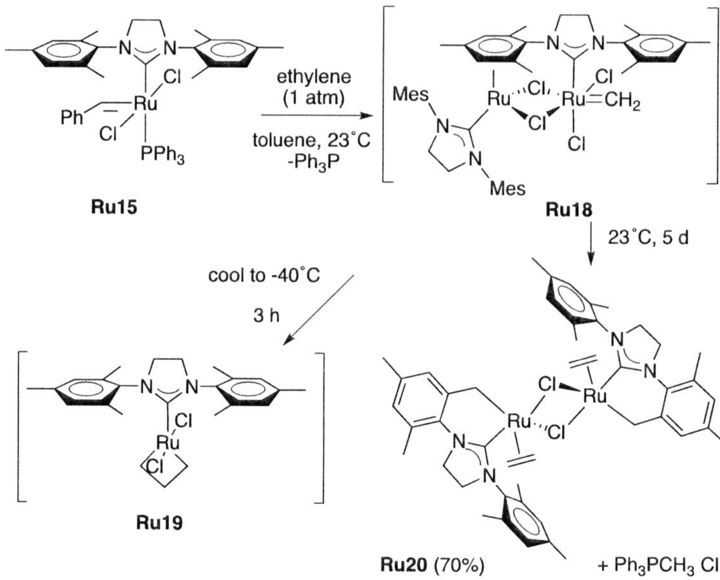

Scheme 13.10.

methyl groups. In the presence of air, Blechert found that a C–H insertion took place, giving a bis-chelated ruthenium carbene (Eq. 13.19).[50] Interestingly, the carbene complex was not metathesis active.

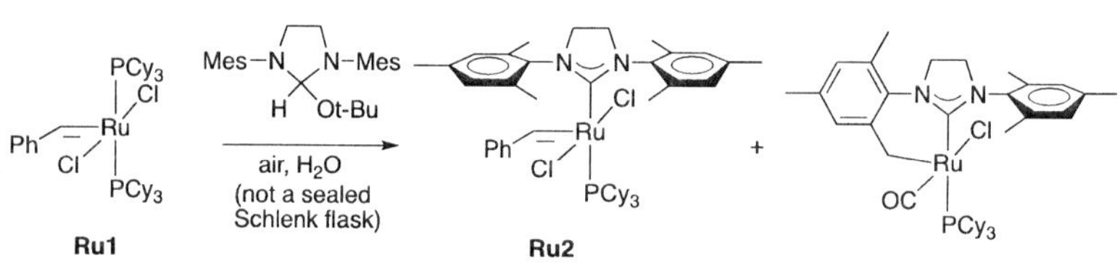

(13.18)

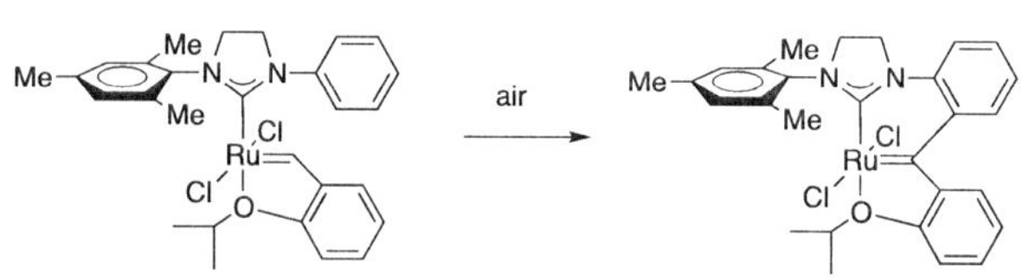

(13.19)

13.3.2.3 Thermal Decomposition. A unique β-hydride elimination pathway was proposed for ruthenacyclobutane intermediates on the basis of calculations and verified experimentally. The direct observation of ruthenacyclobutanes suggested that these high-energy intermediates might be vulnerable to non-metathesis reactions, leading to carbene decomposition. By DFT calculations, van Rensburg and coworkers found a β-hydride elimination transition state leading to propene.[51] Reductive elimination occurs to form the η^2-propene complex (Eq. 13.20). The activation barrier for the decomposition was comparable to that of phosphine dissociation for **Ru2**.

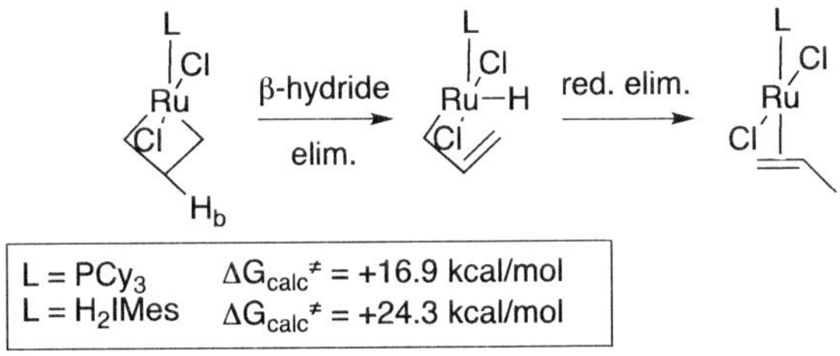

(13.20)

The observed formation of propene from **Ru1** or **Ru2** and ethylene supported the β-hydride elimination hypothesis. When **Ru1** and **Ru2** complexes reacted with ethylene, propene was observed along with variable quantities of higher alkenes. Propene had also been observed in reactions of η^3-vinyl carbene complexes with ethylene.[52,53]

Vinyl carbenes appear to decompose thermally in a dimerization process related to the decomposition of the methylidene.[54] Attempted purification of a bis-triphenylphosphine vinyl carbene complex led to the identification of a dinuclear monocarbene complex (Eq. 13.21).[55] The dinuclear complex features different ligand substitution on the ruthenium(II) centers, which are linked by three bridging chloride ligands. The presence of a *cis/trans* mixture of a triene provided mass balance, demonstrating that carbene dimerization has taken place.

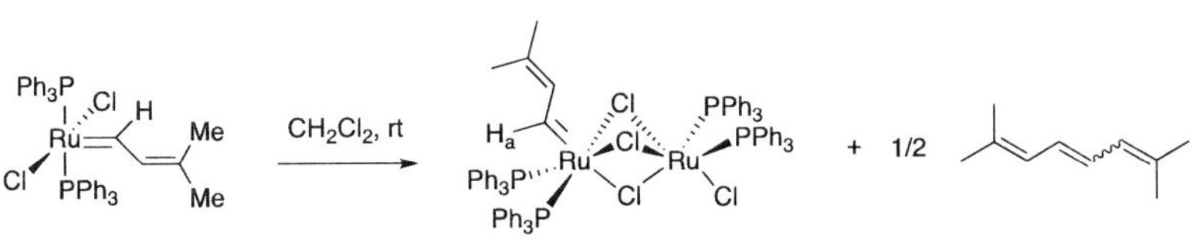

(13.21)

13.3.2.4 Via Unique Carbene Substitution. When **Ru2** was reacted with a methylenecyclopropane, a unique decomposition pathway leading to a terminal carbide was found. In both the first- and second generation ligand environments, a cheleotropic extrusion of dimethylfumarate occurs in the intermediate cyclopropyl alkylidene (Eq. 13.22).[38] The carbide **Ru21B** serves as the synthetic precursor to the Piers catalyst **Ru4**.[13]

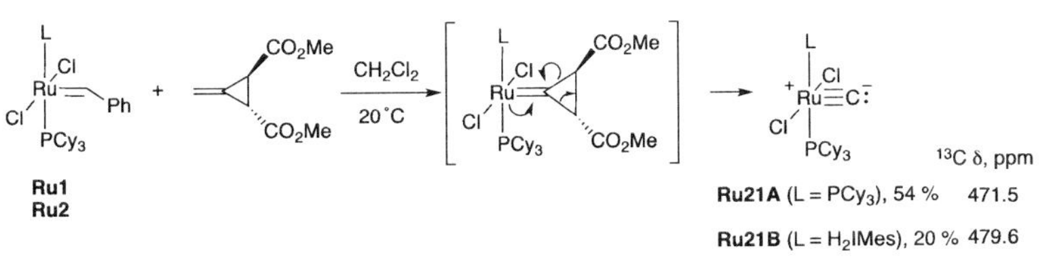

(13.22)

13.3.2.5 In the Presence of Alcohols: Formation of Ruthenium Hydrides. In the presence of alcohols, ruthenium carbenes decompose to hydridocarbonyls. This reaction represents a potential concern during synthetic applications of organic substrates containing free hydroxyl groups. Ruthenium hydrides bring about double bond isomerization, a side reaction accompanying even facile metathesis reactions such as RCM. The first generation Grubbs complex **Ru1** was shown to form hydride **Ru22** in the presence of a primary alcohol (methanol or 1-propanol) and base (Eq. 13.23).[56] The second generation complex **Ru2** similarly forms a hydride by reaction with methanol.[57] Conversion of carbene complexes to their Fischer carbenes followed by heating results in conversion to the corresponding hydride.[58] Excellent yields can be obtained via trimethylsilylenol ethers using the procedure of Arisawa and coworkers (Eq. 13.24).[59]

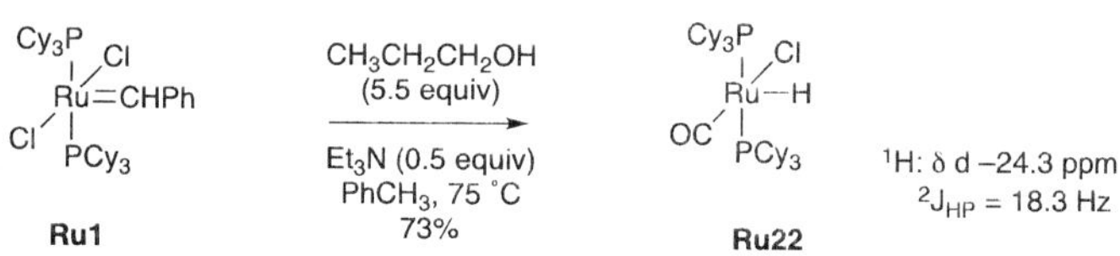

(13.23)

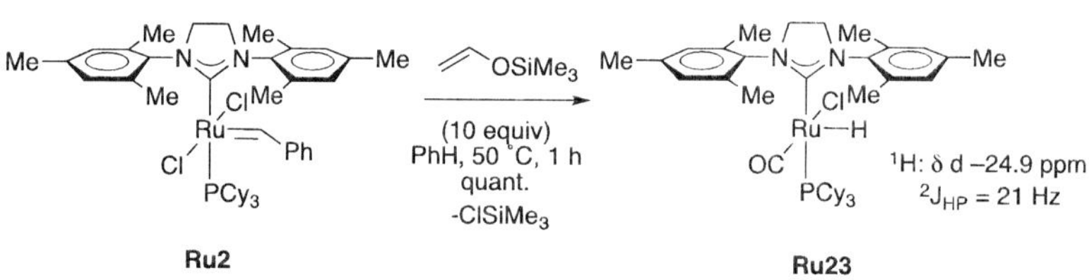

(13.24)

The *in situ* conversion of a ruthenium carbene to a hydride has been used effectively in tandem RCM/isomerization. This process was developed independently by work of Schmidt[60] and Snapper.[61] For example, upon completion of the RCM step, Snapper and coworkers[61] converted the carbene to a hydride by addition of dilute hydrogen gas,[62] which triggered ruthenium hydride formation. The uncharacterized hydride promoted an alkene isomerization *in situ* to form a more stable endocyclic enol ether (Eq. 13.25).

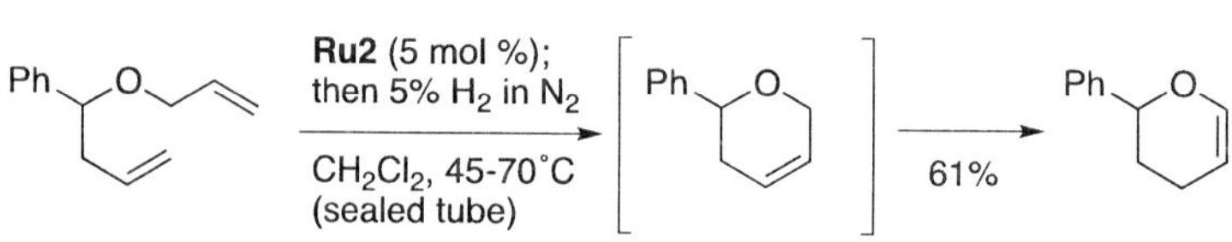

(13.25)

13.3.2.6 In the Presence of Donor/Acceptor Ligands. Carbon monoxide and isocyanides cause an unusual cyclopropanation/ring expansion for the **Ru2** carbene complex (Scheme 13.11).[63] The coordination of an isocyanide causes a

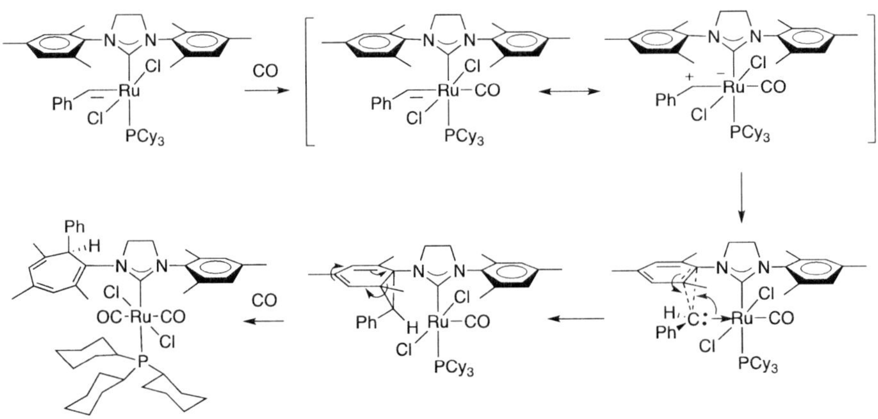

Scheme 13.11.

dramatic change in the electronic properties of the carbene, making it more electron deficient. This results in a cyclopropanation reaction of the adjacent aromatic ring (Buchner reaction). The intermediate cyclopropane then undergoes 6π electrocyclization to provide the observed cycloheptatriene.

The commercially available, polar isocyanide KO_2CCH_2NC has been used as a quenching and "clean-up" agent added at the end of an alkene metathesis reaction.[64] The addition of KO_2CCH_2NC quenches catalyst activity and transforms the ruthenium complex into a polar complex, easily removed by filtration through a short plug of silica gel. Catalyst activity is quenched because the carbene reacts internally in a very rapid reaction analogous to that depicted in Scheme 13.11. How the electronic nature of the carbene affects the through-space interaction with the H_2IMes group has been calculated by Cavallo giving additional mechanistic insight into this unusual process.[65] Many different carbene complexes showed effective quenching including **Ru3**, **Ru7B**, and **Ru8**.[64] Though **Ru1** does not have aromatic rings, the isocyanide causes quenching of activity of **Ru1**.[66]

In the past decade, decomposition pathways available to the ruthenium carbenes have become better understood. Significant focus on the ruthenium methylidene has been warranted due to its prominence as a propagating carbene in alkene metathesis. The intermediacy of electron-poor reactive intermediates leads to C–H bond insertion into the ortho benzylic positions of the H_2IMes group. Thermal decomposition may proceed through carbene dimerization or via β-hydride elimination of ruthenacyclobutane intermediates. The presence of alcohols or enol ethers can be used to trigger the formation of ruthenium hydrides. In applications where this may be problematic, benzoquinone can be used to intercept the hydride species. External promoters such as isocyanides can be used to stop metathesis reactions, to quench carbene intermediates, and to aid in the purification of products of alkene metathesis. Future catalyst design strategies will use this mechanistic understanding to design more robust catalysts that will extend the utility of the ruthenium carbenes even further.

13.4 ACHIEVING SELECTIVITY IN ALKENE METATHESIS

13.4.1 Enantioselective Metathesis with Ruthenium Carbenes

Catalytic asymmetric synthesis is valuable because it allows for the efficient synthesis of highly enantiomerically pure materials. Enantioselective metathesis requires a chiral catalyst, and there have been two different approaches used to achieve this goal. Chiral NHC backbone substituents have been used to induce a gearing effect of the *N*-aryl rings, causing an *anti*-disposition of the ortho substituents. Biaryl alkoxide ligands have been attached to one nitrogen atom of the NHC, creating two points of binding to the chiral element.

The first successful application of enantioselective synthesis with a chiral ruthenium carbene was the desymmetrization by RCM on to a symmetrical triene (Scheme 13.12). The backbone substituents enforce an *anti*-projection of the ortho isopropyl substituents.[67] The C_2-symmetric catalyst ensures that each side of the catalyst has identical environments. Enantioselectivity of the desymmetrization step was significantly influenced by alkene configuration and the halide. In the best case, the alkene undergoing RCM bore *cis*-methyl groups and the catalyst bears iodide substituents. A suggested transition state explaining the observed sense of enantioselection is shown where the alkene displays "side-on" binding and the halides are pinched into a *cis*-configuration. Though beneficial to enantioselectivity, the presence of iodides slowed reaction rates significantly.

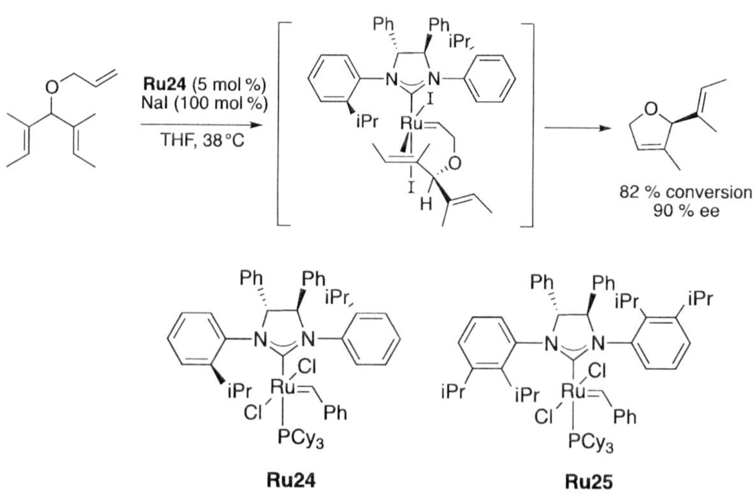

(13.26)

Scheme 13.12.

With additional substitution at the meta position of the ring as in **Ru25**, the enantioselectivity matched that of the earlier catalyst, without the need to replace the chlorides with iodides.[68] This was a significant improvement because the bulky iodide ligands resulted in diminished activity of the catalyst. A similar design strategy using an unsymmetrically substituted NHC has also been employed with modest enantioselectivity.[69,70]

A bidentate ligand environment offers another catalyst design for asymmetric ring-opening/cross metathesis (AROCM).[71] The two points of attachment created by the binaphthyl ring serve to lock catalyst **Ru26** into a rigid conformation (Scheme 13.13). This catalyst and a series of related catalysts were able to perform AROCM with high enantioselectivities (Eq. 13.27).[72] Due to the presence of the aryloxide, the reactivity of **Ru26** did not match monodentate catalysts developed by Grubbs. Similar to the work of Grubbs,[67] the observed enantioselectivity could be improved by introducing an iodide in place of the chloride ligand.[73] Catalyst **Ru27** showed improved reactivity and greater longevity.[74]

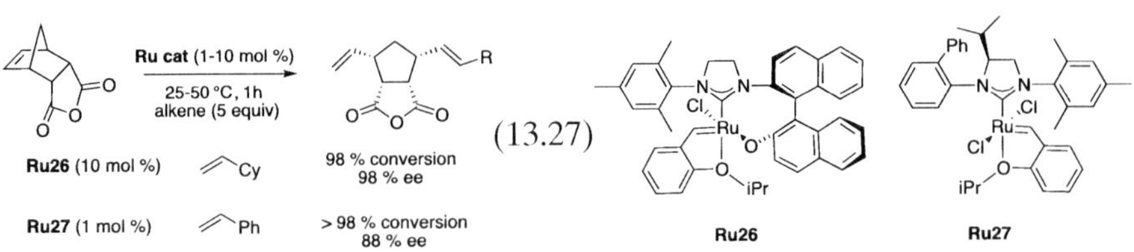

Scheme 13.13.

The "windshield wipering" conformational behavior has had a dramatic impact on development of enantioselective ruthenium carbene catalysts. There are two carbene states that possess reactivity and for unsymmetric chiral catalysts (see **Ru28**), these have different reactivity and will display different stereoselectivity. In Equation (13.28), a highly selective AROCM was described. In this dramatic example, Hoveyda and coworkers found that a chiral Fischer carbene catalyst could sample the two different carbene conformations in a preequilibrium (Eq. 13.29), which preceded a higher activation barrier for ring-opening cross metathesis (Scheme 13.14).[75] This study also utilized DFT calculations to render the energy surface of the catalytic reaction, finding higher energy ROM, and leading to a Curtin–Hammett situation. A related ring opening with styrenes provides the opposite enantiomer and with *E*-selectivity. This example demonstrates how chiral catalyst development led to a highly selective ring-opening cross metathesis and how a culmination of mechanistic knowledge coupled with DFT calculations led to a rationale for a sophisticated metathesis reaction.

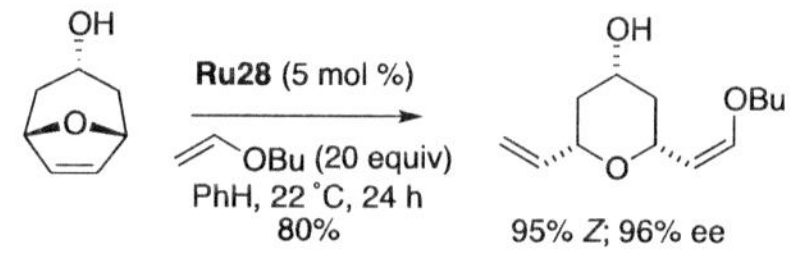

(13.28)

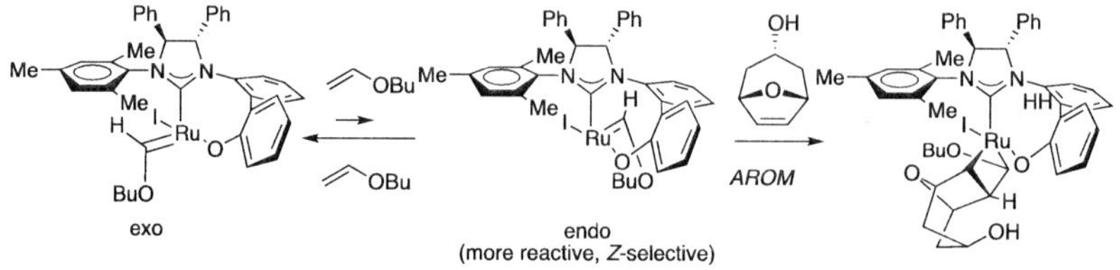

(13.29)

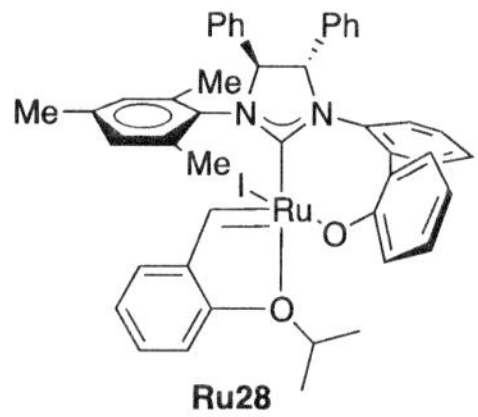

Scheme 13.14

There are only a handful of examples of enantioselective ruthenium carbene metathesis reactions. They are highly enantioselective for a few limited desymmetrization applications. They also demonstrate that a chiral NHC or a chiral bidentate ligand can be used to create a chiral environment for asymmetric catalysis. More commonly documented examples of asymmetric synthesis rely on the stereogenic-at-molybdenum–based systems developed in the labs of Schrock and Hoveyda.[20,76–80]

13.4.2 Site Selectivity Using Relay Ring-Closing Metathesis

Relay metathesis was developed to influence site selectivity in a substrate containing multiple sites of unsaturation possessing similar reactivity. The relay ring-closing metathesis (RRCM) strategy involves the installation of an alkene tether (relay element) that is used to direct catalyst initiation. This alkene tether directs the reaction to one alkene over another, allowing for selective ruthenium carbene formation. Site-selective placement of carbene reactivity directs the next metathesis reaction. This allows for "choreographed movement" of ruthenium carbenes. The relay metathesis is useful in controlling the directionality of a metathesis cascade as well as inducing the activation of an otherwise unreactive alkene. The relay concept was introduced by Hoye and coworkers.[81] This concept was embraced quickly after it was discovered in the arena of natural product synthesis[82,83] where localization of reactivity is *sine qua non*.

Relay ring-closing metathesis was developed to address site-selective initiation of ruthenium carbene reactivity. This process was particularly important to control tandem ring-closing metathesis where there are two possible products depending on which alkene reacts first (Eq. 13.30).[84] With no directing relay element, a 1:2 mixture results. Similar examples lead to random initiation and result in a statistical mixture of ring-closed products.[85,86] Installing the tether and directing the metathesis sequence from left-to-right afforded a 4.7:1 mixture and right-to-left afforded a 1:7 mixture. RRCM was used to selectively direct catalyst initiation at a specific alkene, thereby directing the metathesis sequence to preferentially form one product.

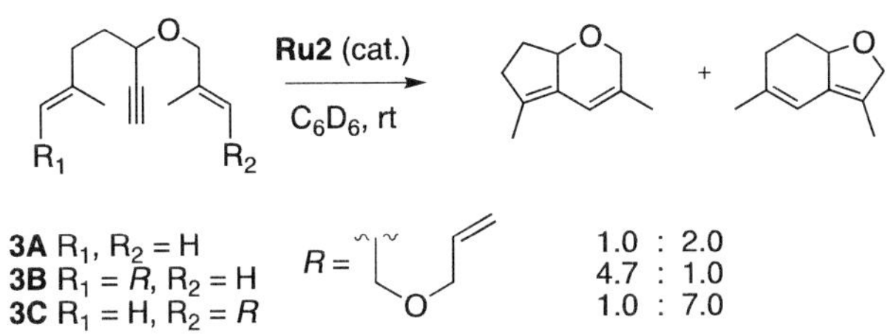

(13.30)

A cascading relay ring closing metathesis sequence was also used to construct the bicyclic core of (–)-englerin A (Eq. 13.31).[87] The mixture of carbonate diastereomers was subjected to the highly active Grubbs–Stewart catalyst, both diastereomers reacting to afford the desired bicyclic ring system. The major *cis*-diastereomer was carried forward in the synthesis.

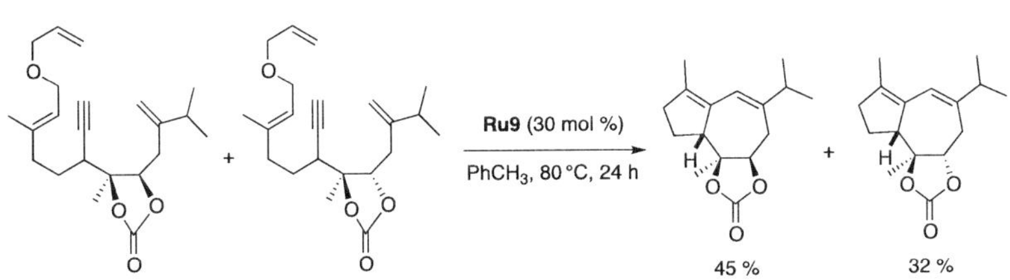

(13.31)

Relay metathesis has been used to control ring-closing alkene metathesis in natural product synthesis. For example, the total synthesis of (–)-archazolid was accomplished using a relay ring closing cross metathesis to afford the macrolide ring in 27% yield with **Ru2** (Eq. 13.32).[88]

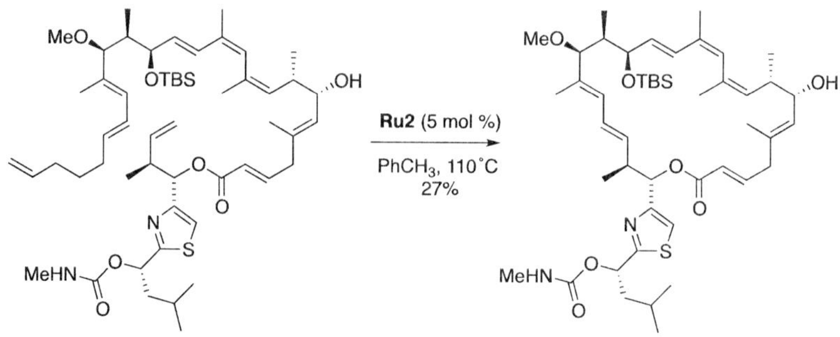

(13.32)

In a demanding environment, a relay ring-closure strategy failed to precede macroring closure (Eq. 13.33). The Trost group used relay ring closing metathesis in the last step of their synthesis of the anticancer antibiotic bryostatin.[89] In this case, the relay linker became incorporated in the large ring. The desired ruthenium carbene is difficult to form due to the adjacent quaternary center, which forestalls the relay ring closure step.

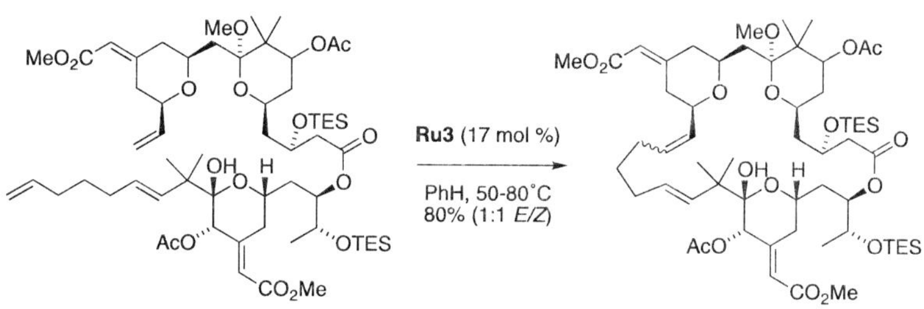

(13.33)

When multiple possibilities of ring-closing metathesis exist, the relay element may be used to influence the sequence of ring closures (Eq. 13.34). Hoye has evaluated several different alkene tethers for this process, finding different preferences depending on the relay element.[90] In this case, the polyene bearing the dimethylmalonate tether favored the RCM process shown primarily due to a Thorpe–Ingold effect.[90]

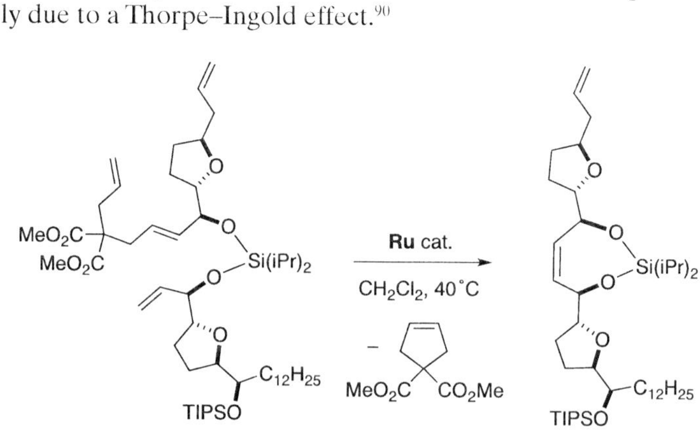

(13.34)

Lee has combined ring closing enyne metathesis reactions with a 1,3-metallotropic shift to synthesize (−)-tricholomenyn A.[91] In Equation (13.35), a relay element was used to start the cascade. After the initial ring-closing metathesis, a ring-closing enyne metathesis forms the six-membered ring and is followed by the 1,3-metalotropic shift. This strategy was also employed for the total synthesis of (+)-asperpentyn and (−)-harveyone.

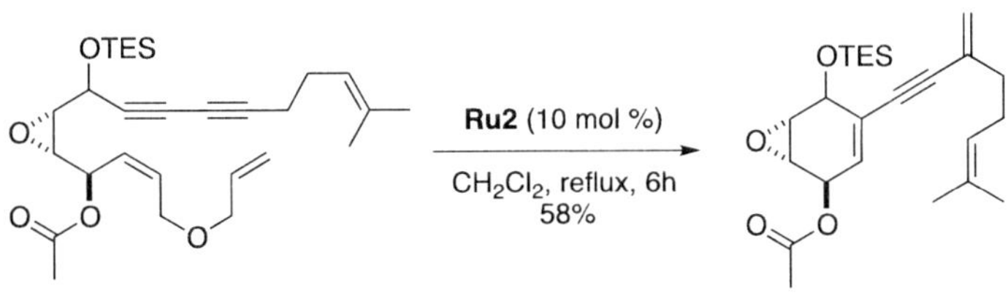

(13.35)

The relay metathesis strategy has been used in cross metathesis to obtain reactivity from geminally substituted alkenes.[92] A similar reaction had been used earlier.[93] In the examples earlier, relay metathesis was used to direct reactivity to a single site in a reactant. In this case, the relay was used to direct reactivity to an unreactive geminally-substituted alkene, setting the stage for an intermolecular ene-yne metathesis (Eq. 13.36). This extends the concept of relay metathesis beyond site selectivity.

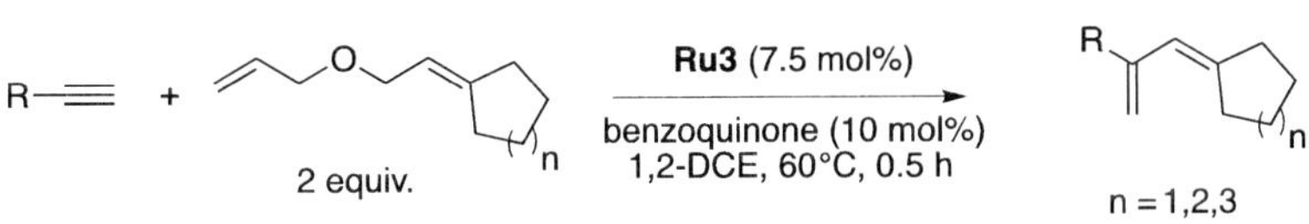

(13.36)

Relay RCM has had a significant impact on both alkene and ene-yne metathesis. It offers a unique strategy for the site-selective activation of alkenes or to direct catalyst reactivity to sterically hindered or otherwise unreactive alkenes. It has also allowed for the development of directed or "choreographed" movement across a substrate with more than one center of reactivity; this allows for selective product formation. Due to the utility of this method and the strong desire to localize ruthenium carbene reactivity in the synthesis of complex molecules, relay metathesis will continue to find broad application in organic syntheses.

13.4.3 *Z*-Selective Alkene Metathesis

One of the last remaining synthetic goals employing ruthenium-based metathesis is that of *Z*-selective alkene metathesis. With the traditional Grubbs catalysts such as **Ru2**, the *E*-isomer tends to predominate. To obtain a *Z*-selective metathesis, the catalyst would need to favor the formation of the *Z*-isomer and not cause secondary metathesis so that kinetic control can be maintained. Though some substrate-based solutions exist, the best *general* solution for this catalytic reaction would be a *Z*-selective catalyst. Recent efforts of the Grubbs' group have produced *Z*-selective ruthenium carbene catalysts, achieving catalytic stereoselective *Z*-alkene synthesis. Though continued development will be needed to achieve more robust catalysts, it is anticipated that these catalysts will be widely used for alkene synthesis.

Young and coworkers devised a method by which a silyl group could be used to direct the stereochemistry of a macrocyclative RCM (Eq. 13.37). The resulting RCM would be *E*-selective and stereospecific protodesilylation would afford the *Z*-alkene.[94] The silyl protecting group also enables regiospecific functionalization. The method is applicable to a variety of ring sizes. The silyl group limits by-product formation and prevents the desired product from succumbing to secondary metathesis. As a result, macrocycle formation is nonreversible.

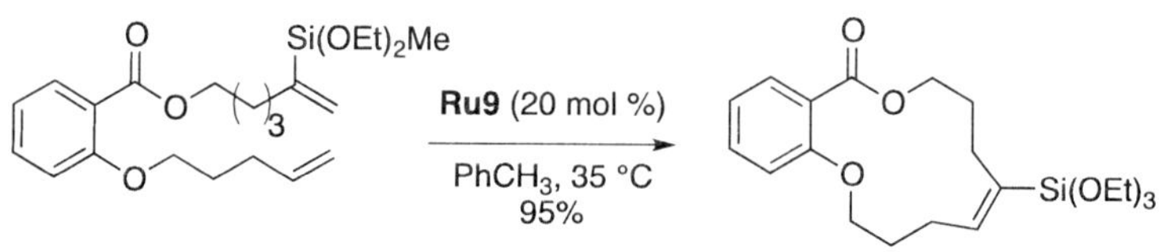

(13.37)

13.4.3.1 Z-Selective Ruthenium Carbene Catalysts. Replacement of the halide ligand with a sulfonate resulted in modest improvements in *Z*-selectivity. Though the NHC ligand is easily modified, its substitution is remote from the reactive center, making it difficult to influence *Z*-selectivity. For *Z*-selectivity, not only must the ligand environment control selectivity, the catalyst must not equilibrate the kinetically formed product to the more stable *E*-isomer. Chen introduced sulfonates in his selective copolymerization catalyst **Ru29** which resulted in increased *Z*-selectivity in ring-opening metathesis polymerization (ROMP).[93] Grubbs also found that the *E/Z* ratio could be decreased when using bulkier sulfonates to replace a chloride ligand (Eq. 13.38).[96] In the cross metathesis of allylbenzene with *cis*-1,4-diacetoxy-2-butene (a standard cross metathesis test for selectivity),[97] the *E/Z* ratio was found to be 2.7:1 as compared to 10:1 with **Ru3**. Importantly, modification of the halides led to the discovery of a novel catalyst.

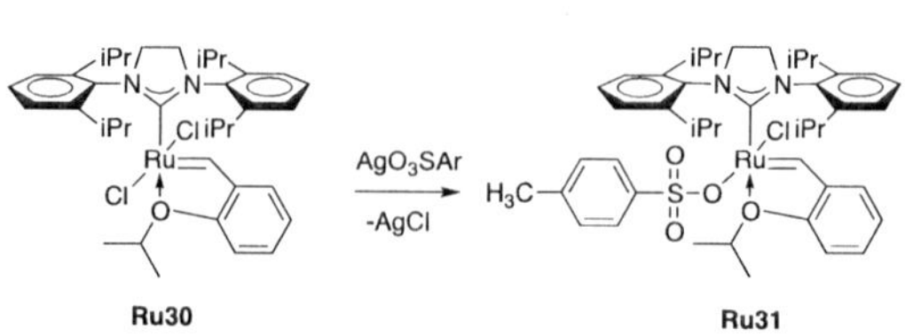

(13.38)

When pivalate was introduced as an anionic ligand, it triggered a C–H insertion into the ortho methyl of the mesityl ring (Eq. 13.39).[98] C–H insertion was also attempted with the **Ru3** variant containing an adamantyl group (Eq. 13.40). Both catalysts were found to catalyze RCM of diethyldiallylmalonate. In the standard cross metathesis between allylbenzene and *cis*-1,4-diacetoxy-2-butene, both catalysts gave good *Z*-selectivity. For example, the adamantly-chelated catalyst **Ru34** provided an *E/Z* ratio of 0.12, which corresponds to almost 90% *Z*-isomer (Eq. 13.41). Similarly, a wide variety of terminal alkenes resulted in highly *Z*-selective homodimerizations.[99] These new catalysts proved less active than the parent **Ru3** catalyst; however, they demonstrated remarkable and distinctive *Z*-selectivity.

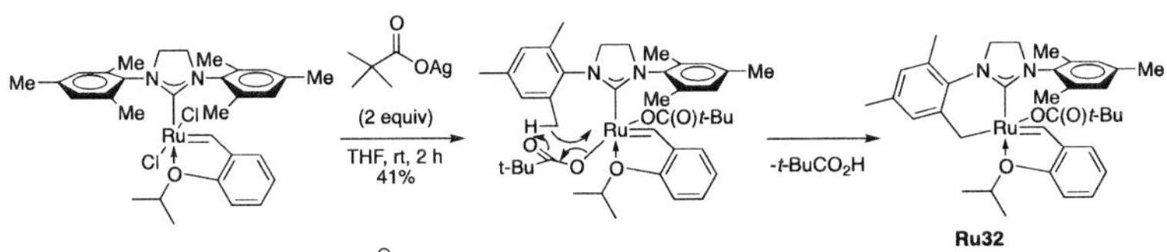

(13.39)

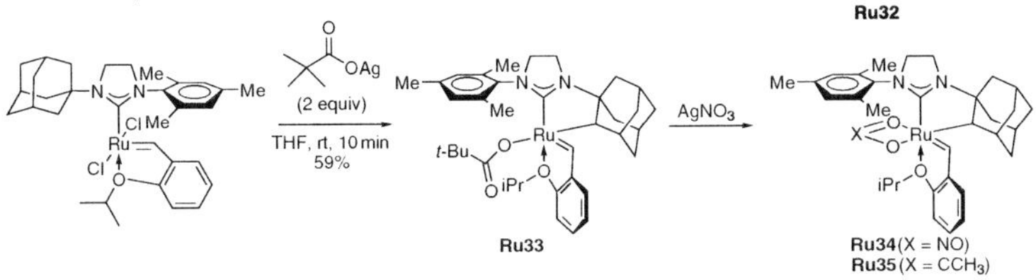

(13.40)

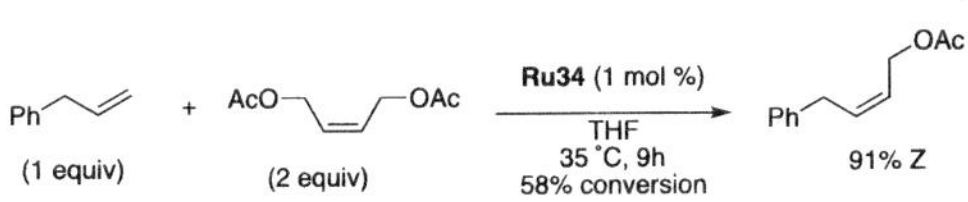

(13.41)

Density functional theory calculations of the homodimerization of ethylene and propene were used to probe the source of the *Z*-selectivity exhibited by this family of chelated catalysts.[100] The adamantyl-chelated catalyst **Ru34/35** favors a side bound approach of the alkene.[100] This conformation allows the NHC ligand to influence the orientation of substituents on the forming ruthenacyclobutane (Fig. 13.1a). The side bound approach also benefits from electronic interactions with the ruthenium center. In this orientation, the π^* orbital of the alkylidene is perpendicular to that of the NHC; this allows two different *d* orbitals from ruthenium to participate in back donation, one with the alkylidene and the other with NHC (Fig. 13.1a). In the bottom bound conformation, the π^* orbital of the NHC and alkylidene are in the same plane allowing only one *d* orbital of ruthenium to participate in the back donation (Fig. 13.1b). The computational study of the homodimerization of propene revealed a lower energy transition state leading to the *Z*-isomer. The side bound Z-transition states direct the metallacycle substituents (the methyl groups in Fig. 13.1c) away from the bulky ligand due to steric strain.

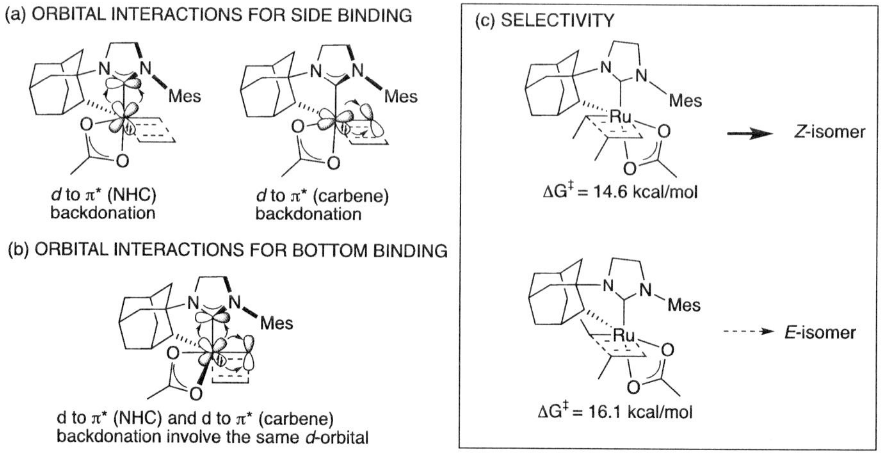

Figure 13.1. Rationalized *Z*-selectivity of **Ru35** based on DFT calculations.[98] (a) Orbital interactions for side binding, (b) orbital interactions for bottom binding, and (c) selectivity.

The biggest breakthrough in *Z*-selective alkene metathesis was the discovery of the chelated ruthenium carbenes catalysts. Continued development of these catalysts will offer powerful tools for synthetic organic chemistry.

13.5 APPLICATIONS

13.5.1 Selected Examples of Alkene Metathesis in the Total Synthesis of Natural Products

The chemoselective Grubbs ruthenium carbenes have proven to be highly effective catalysts that promote alkene metathesis in natural product total synthesis. Typically, amines and free hydroxyl groups are avoided. Most applications have been ring-closing methods to access macrocyclic rings. These have benefitted in some cases from the use of relay metathesis to control site selectivity. Enyne metathesis and alkene-1,3-diene metathesis are also commonly used for ring closure. Enyne metathesis has proven particularly useful for multiple ring synthesis by cascade metathesis. Industrial applications are also described in the last section.

Demonstrating the chemoselectivity of these catalysts, **Ru2** could be used for the RCM of the polyene with the desired chemoselection (Eq. 13.42).[101] This chemoselectivity is impressive in that no ring truncation products were observed which would have resulted from nonselective catalyst initiation. This macrocycle-forming RCM was a key step in the synthesis of lejimalide A-D.

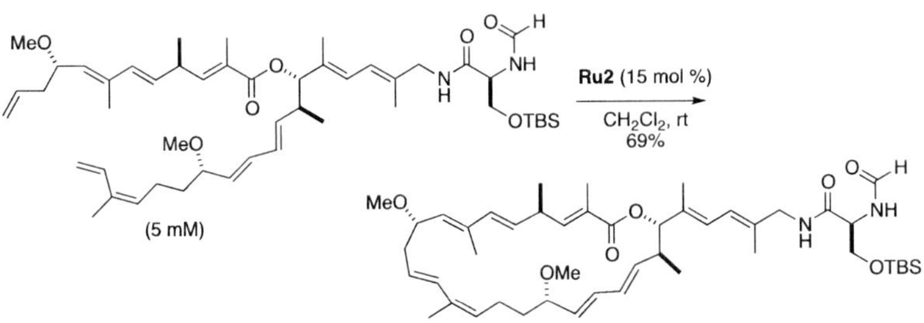

(13.42)

The 27-membered macrocycle of woodrosin I, could be formed through the RCM of the tether connecting the glycoside backbone of this natural product (Eq. 13.43).[102] Though protected, the four glucose subunits present a highly oxygen-rich environment that does not diminish catalysis by the ruthenium carbene **Ru36**.

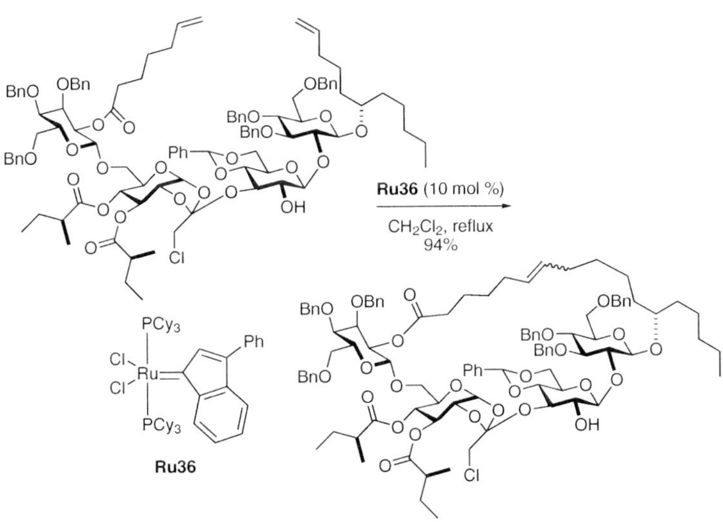

(13.43)

The "orthogonal" reactivity of ruthenium carbenes allows selective reactions in settings replete with coordinating heteroatoms. Hirama and coworkers nicely demonstrated this chemoselectivity of **Ru1** in a late stage RCM of ciguatoxin CTX3C (Eq. 13.44).[103]

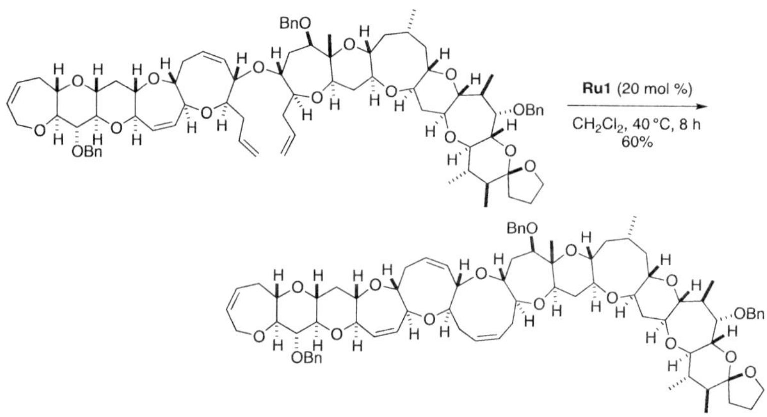

(13.44)

Alkaloids pose a challenge for synthesis with transition metal catalysts due to the potential for coordination to the metal. Chemoselective ring-closing alkene metathesis in the presence of an alkyne is comparatively rare. A recent example of both forms of chemoselectivity can be found in RCM to construct the eight-membered ring of manzamine A (Eq. 13.45).[104] The metathesis required high loading. In complex molecule synthesis, high loadings are not uncommon due to decomposition caused by the dense array of organic functional groups. In this challenging example, the addition of *p*-methoxyphenol was used to limit enyne metathesis with the internal alkyne (a side reaction) that partly accounted for the diminished yield.

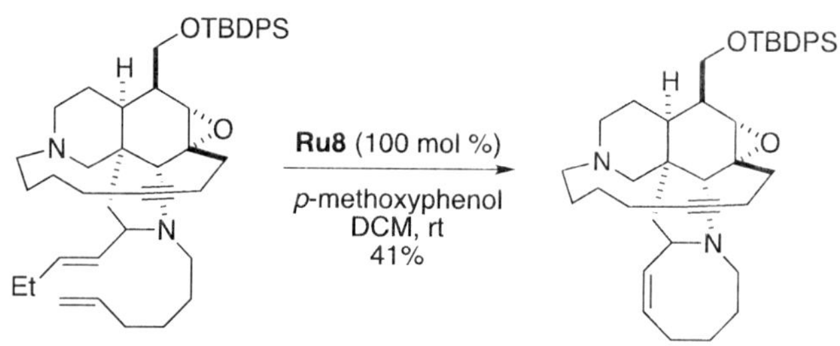

(13.45)

Cross alkene metathesis has become increasingly used in fragment coupling. Fragment coupling offers a means to unite two complex molecules in convergent total synthesis. Alkene metathesis can be used at late stages, as elegantly shown in the synthesis of norhalichondrin B by Phillips and coworkers (Eq. 13.46).[105] The Grubbs–Stewart catalyst[36] is used to join the chiral allylic alcohol with the α,β-unsaturated ketone in good yield. Alkene metathesis is a catalytic protocol that achieves the same objective as Wittig olefination but without prior activation of the fragments undergoing coupling. This is a good example that shows how far the alkene metathesis has progressed, capitalizing on reactivity of the catalyst and the predicted cross reactivity that comes from the Grubbs model.

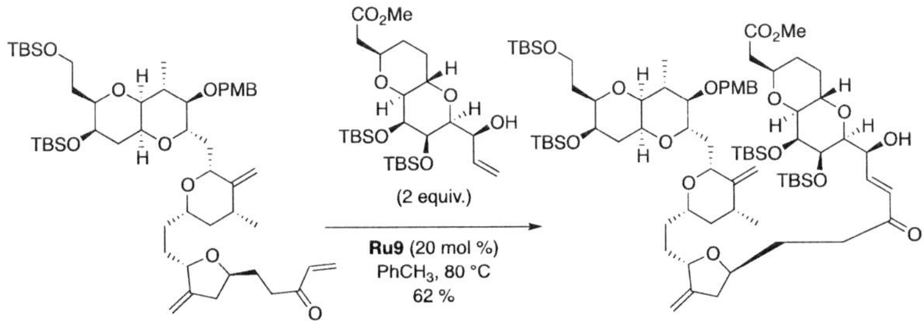

(13.46)

Highly functionalized synthetic intermediates undergo ring-closing metathesis with good results. In ring-closing metathesis of 1,3-dienes, there is a risk of truncation giving rise to smaller ring by-products. Danishefsky used alkene–diene metathesis to assemble the macrocyclic core of radicicol (Eq. 13.47).[106] The RCM occurred selectively to preserve the 1,3-diene moiety.

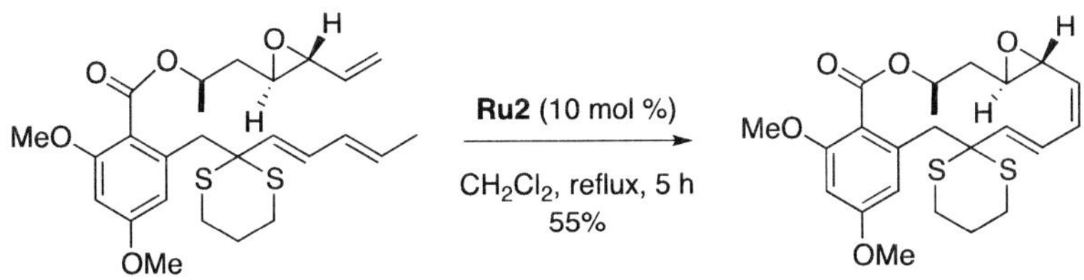

(13.47)

13.5.1.1 Tandem Reactions. Metathesis cascades have the potential to rapidly create polycyclic ring systems. A sound strategy uses the ring strain of a bicyclic molecule to drive the ring-opening metathesis step. Phillips' group used a two-directional tandem ring-opening cross metathesis reaction to create the five- and seven-membered ring of the core of (+)-cyanthiwigin U (Eq. 13.48).[107] A similar tandem reaction was applied to the total synthesis of (+)-cylindramide A (Eq. 13.49).[108]

(13.48)

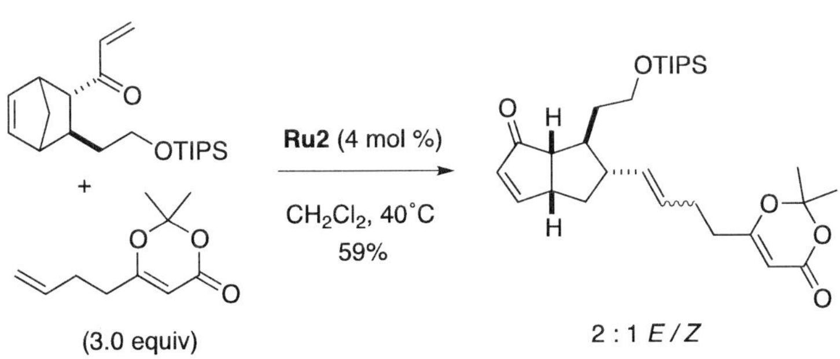

(13.49)

Blechert developed a carbocycle-to-heterocycle ring rearrangement used to produce several heterocyclic natural products. *En route* to the natural product (+)-astrophylline, a two-directional ring-opening cross metathesis was used to construct the piperidine rings in 82% yield (Eq. 13.50).[109] This is similar to the work of Wallace.[110,111] Blechert has used ring rearrangement metathesis for the synthesis of alkaloids such as (−)-anaferine,[112] (−)-swainsonine,[113] (−)-halosaline,[114] (+)-dihydrocuscohygrine,[115] and *ent*-lepadin F and G.[116]

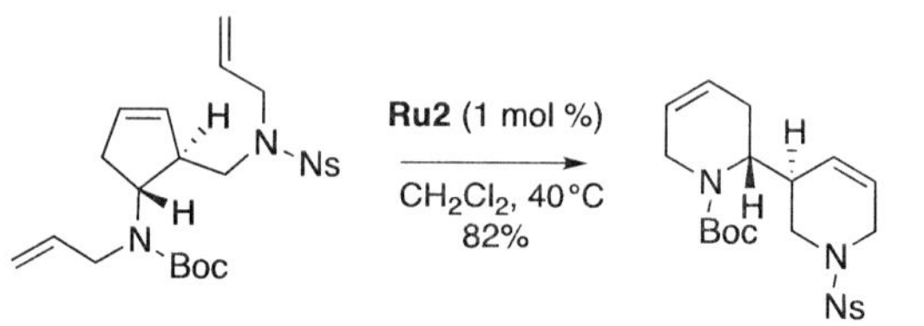

(13.50)

The 5-7-6 linear tricyclic core of guanacastepene was accessed through tandem ring-closing enyne metathesis (Eq. 13.51).[117] This cascade is initiated at the 1,3-diene, which is more reactive than the trisubstituted alkene. The core structure is a late stage intermediate used earlier by Danishefsky to complete the total synthesis of guanacastepene.[118,119]

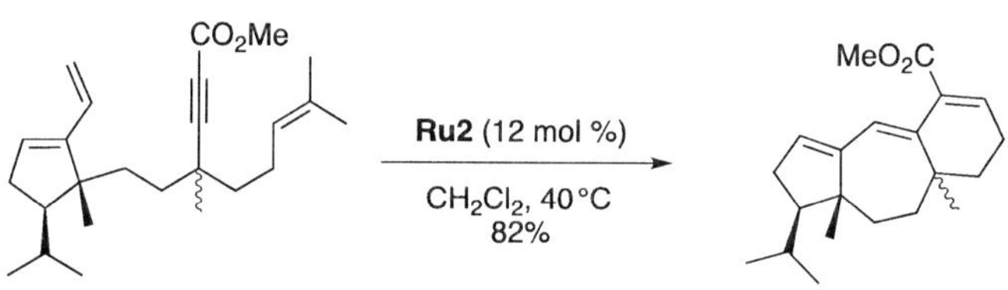

(13.51)

Temporary silyl ethers offer an effective synthetic strategy to preorganize unsaturated elements for a ring-closing metathesis cascade. Lee used the temporary siloxane tether to initiate a selective ene-yne metathesis sequence to construct the *E*,*Z*-diene in the intermediate toward the total synthesis of (−)-cochleamycin A.[120] The reaction sequence initiates on the least sterically hindered alkene (Eq. 13.52). A similar use of a siloxane linker can be found in Movassaghi's total synthesis of (−)-acylfulvene and (−)-irofulven.[121] The temporary siloxy tether was used to direct the cascading reaction sequence which afforded the desired triol after silicon group removal (Eq. 13.53). Completion of the synthesis required another ring-closing alkene metathesis to give acylfulvene (Eq. 13.54).

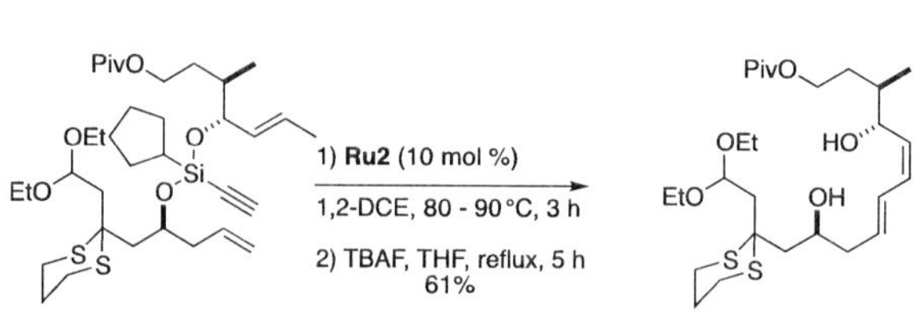

(13.52)

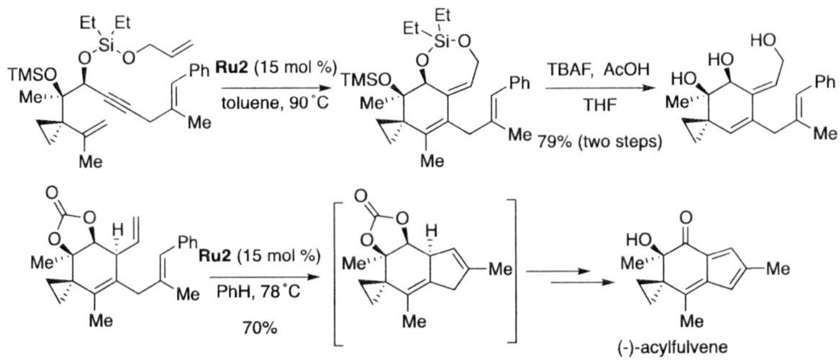

(13.53)

(13.54)

13.5.1.2 Industrial Applications. The industrial synthesis of the Hepatitis C Virus (HCV) NS3 protease inhibitor BILN 2061 utilized a macro ring-closing alkene metathesis. In early work, the Boehringer–Ingelheim group found an unusual epimerization was occurring at the vinyl cyclopropane (see asterisk in Eq. 13.55); it could be limited using Hoveyda-type catalysts.[122] RCM with a Grubbs catalyst resulted in epimerization of the allylic cyclopropane facilitated by the labile tricyclohexyl phosphine ligand.[123] The high dilution conditions were a limiting factor in scale-up efforts needed for animal testing. The process group was able to successfully increase the concentration to 0.2 M, significantly improving throughput in the critical metathesis step.[124] Similarly, a Merck group used a bold double ring-closing metathesis to form a conformationally constrained HCV protease inhibitor that was conceived through rational drug design (Eq. 13.56).[125]

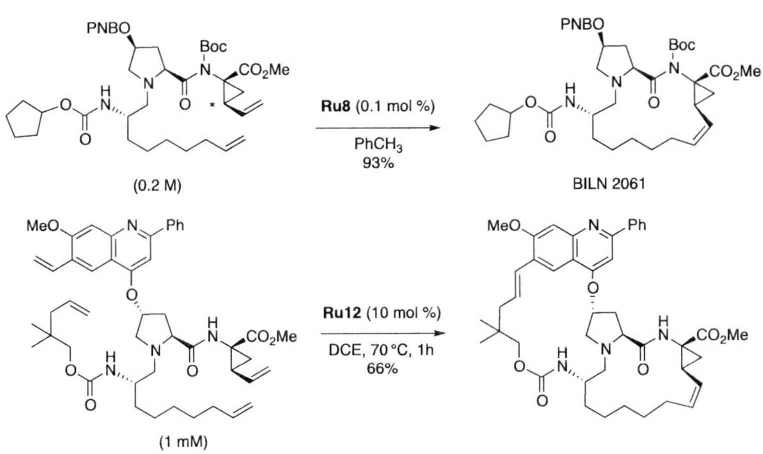

(13.55)

(13.56)

Ring-closing metathesis to form small and medium rings is reliable and has been used industrially for the synthesis of a dehydroazepane. The process group from GlaxoSmithKline (GSK) optimized the synthesis of SB-462795, a known inhibitor of cathepsin K, via a ring-closing metathesis reaction. A low loading of **Ru3** was found to be the optimal catalyst (Eq. 13.57).[126] However, yield varied from batch to batch. Through a meticulous screen of potential impurities in the starting material, the group identified 1,8-diazabicyclo[5.4.0] undec-7-ene (DBU) and a urea coproduct from previous steps that had a deleterious effect on the yield of the metathesis reaction. This example demonstrates that low catalyst loadings can be achieved during scale-up efforts with careful attention to purity of reactants.

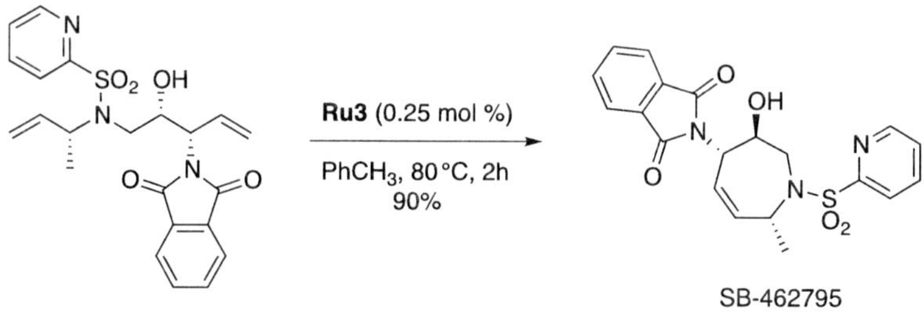

(13.57)

A macro ring-closing RCM and subsequent hydrogenation were key steps in the Merck process group synthesis of the HCV protease inhibitor vaniprevior (MK-7009).[127] Either the carboxylic acid or the methyl ester could be used with equal results. It was determined that slow addition of the catalyst and diene substrate combined with use of 2,6-dichlorobenzoquinone helped prevent olefin isomerization. Olefin isomerization led to an undesired 19-membered lactam by-product arising from ring truncation (Eq. 13.58).

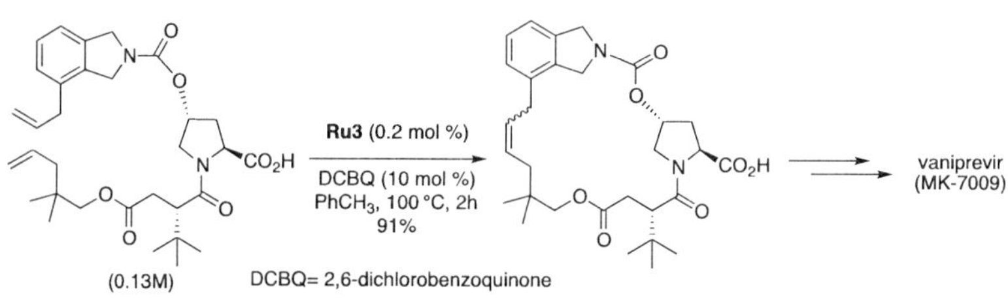

(13.58)

13.5.2 Applications of Metathesis in Aqueous Systems

Though there are several transition metal-based catalysts for alkene metathesis, a unique advantage of the ruthenium-based catalysts is their tolerance to polar functional groups. This tolerance of diverse functional groups has allowed for the deployment of this family of catalysts in polar and protic solvents. Much of the effort to conduct metathesis in water has been directed to modifications of the ligand sphere of the catalyst or modifying the reaction conditions with the addition of surfactants. This has allowed for the use of these catalysts in the chemoselective modification of biologically relevant substrates such as proteins and glycosides. In this section, we will not recapitulate catalyst development but simply focus on aqueous applications.

The first example of heterogeneous metathesis in a polar–protic solvent was the living ROMP of several substituted norbornenes as well as 7-oxanorbornenes. Grubbs and coworkers demonstrated that **Ru1** could catalyze the ROMP of these monomers in water and the yield could be improved with the addition of dodecyltrimethylammonium bromide (DTAB), a cationic emulsifier.[128] This method was used to construct block copolymers. This new method was utilized in early work by Kiessling and coworkers who used living ROMP to create neoglycopolymers (Eq. 13.59) to study multivalent carbohydrate–protein interactions.[129]

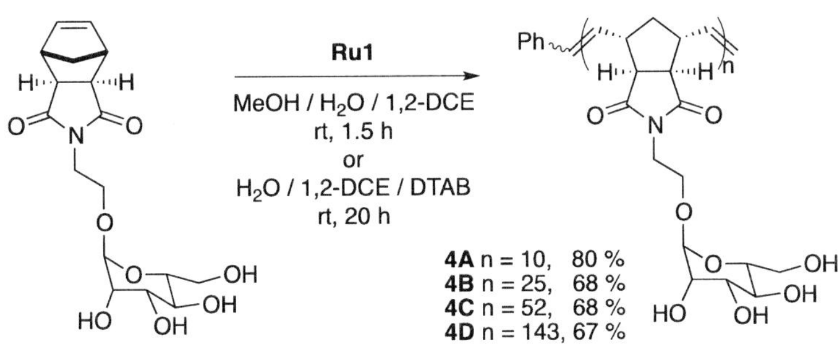

(13.59)

Water-soluble catalysts have been used to promote cross- and ring-closing metathesis. Blechert showed that allylic and homoallylic alcohols could be homodimerized in methanol–water cosolvent (Eq. 13.60).[130,131] Addition of a surfactant has also been used successfully to aid solubility as developed by Lipshutz.[132] Ring-closing metatheses (intramolecular reactions) usually work better than cross metatheses in water cosolvents. For instance, using the water-soluble catalyst **Ru38**, dienyl amine hydrochlorides prove to be good substrates for RCM (Eq. 13.61).[133]

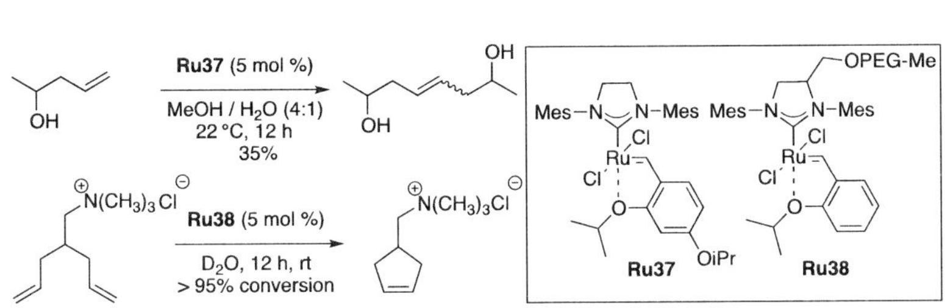

(13.60)

(13.61)

Cross metathesis was performed in a heterogeneous system for the synthesis of a library of vancomycin dimers.[134] The library of these dimers were made through **Ru1**-catalyzed cross metathesis (CM) in water with different length tethers appended to the aminoglycoside portion of vancomycin (Eq. 13.62). The divalent products were then tested for their biological activity relative to monomer of vancomycin.

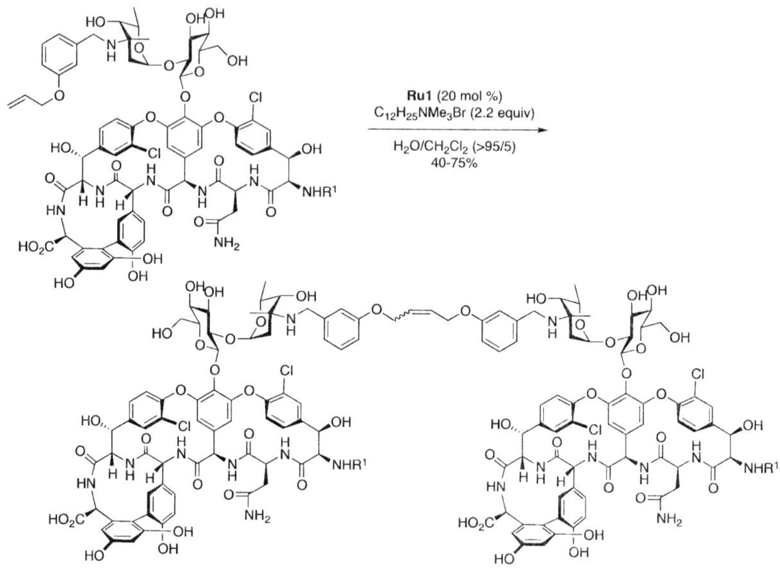

(13.62)

Cross metathesis on allylic sulfides has been conducted in aqueous cosolvents. In an effort to develop methods to modify peptides, Davis and coworkers studied the CM of several alkene-containing amino acids and discovered that *S*-allylcysteine underwent successful cross metathesis with allyl alcohol (Eq. 13.63).[135] It was determined that the allylic thioether was a requirement for successful CM.

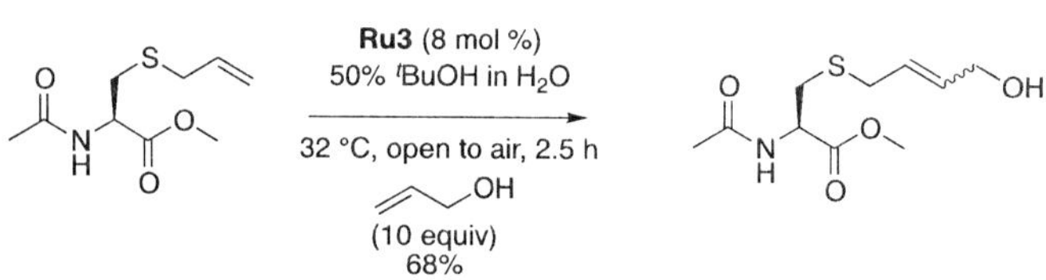

(13.63)

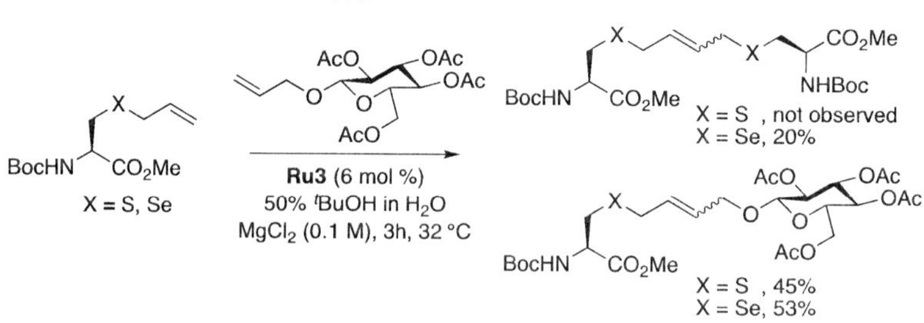

(13.64)

Cross alkene metathesis was used to modify a cysteine mutant of the serine protease subtilisin *Bacillus lentus* (SBL). The model study is shown in Equation (13.64). Once the cysteine residue was allylated,[136] the modified protein could then successfully undergo CM with allyl alcohol. These reactions required the presence of 100 mM of $MgCl_2$ to prevent chelation of the catalyst with the substrate. The procedure was recently extended to allyl selenides for coupling to several different glycosides and aminoglycosides (Eq. 13.64).[137] These conditions have also been used to conjugate a fluorescent probe containing an allyl thioether to an alkene-functionalized amino acid.[138]

Schultz used similar conditions to effect an RCM on the surface of a protein.[139] The Schultz group developed plasmids that could be used to genetically encode alkene containing non-natural amino acids into an amino acid chain within *Escherichia coli* containing an orthogonal tRNA/amino-acyl–tRNA synthetase pair. Using this technology, alkene-containing amino acids could be site-selectively incorporated into peptides. For instance, two *O*-crotylserine amino acids were incorporated along the peptide 10 amino acids apart from placing them in close proximity on the exterior of the protein (Eq. 13.65).

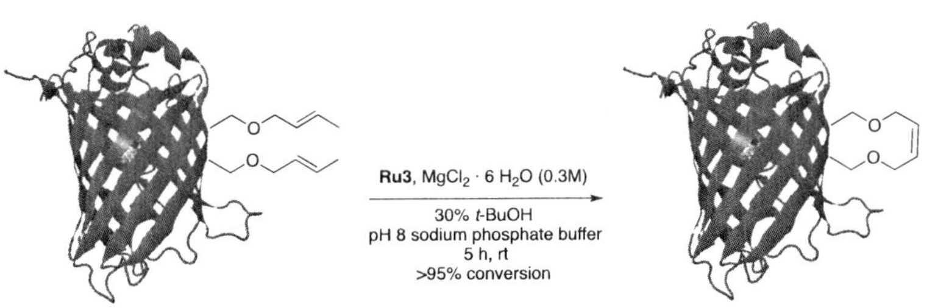

(13.65)

13.5.3 Applications in Diversity-Oriented Synthesis

Diversity-oriented synthesis (DOS) aims to generate a large number of small molecules with a wide range of skeletal and stereochemical diversity. Robust chemistry is employed to quickly create libraries of compounds that represent a wide and diverse range of chemical space. Metathesis chemistry has been used prominently in DOS because it offers orthogonal reactivity to that of organic functional groups, it is robust, and because unsaturated products lend themselves to further transformation. Enyne metathesis can also be rationally cascaded, thus offering a powerful chemical means to build molecular scaffolds in a single catalytic transformation.

Metathesis methods can be combined with distinct bond-forming processes to rapidly form complex molecular scaffolds. This is a key requirement for diversity-oriented synthesis. Banti and North nicely demonstrated this concept with a doubly armed norbornene scaffold (Eq. 13.66).[140] Under an atmosphere of ethylene, a two-directional synthesis occurred by a ring-opening/double ring-closing metathesis sequence. The resulting bis(1,3-diene) was trapped with maleic anhydride in a Diels–Alder reaction. In this simple sequence, seven rings were formed from a single bicyclic molecule.

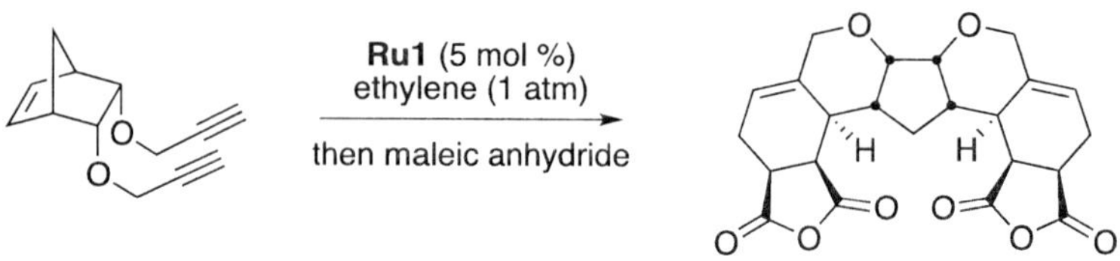

(13.66)

Schreiber was the first to realize the potential of Grubbs carbenes in multicomponent reactions that could be used to quickly generate molecular diversity. The first example involved the combination of a Ugi four component coupling reaction, followed by an intramolecular Diels–Alder reaction, and finally a ring-opening-closing metathesis to finish the sequence.[141] This concept was later expanded upon with a family of substrates capable of CM or ring-opening metathesis (ROM)–RCM reactions (Scheme 13.15).[142] Interestingly, the course of the metathesis depended on the stereochemical relationship of the unsaturated components in the reactant.

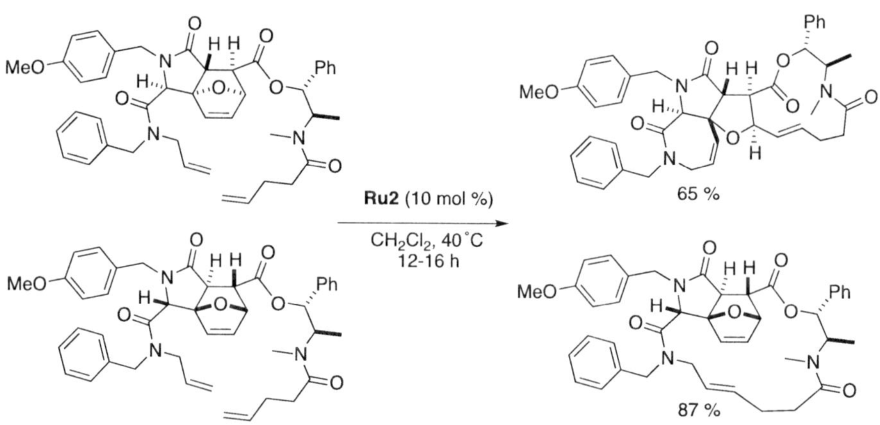

Scheme 13.15.

Ring-closing mode selectivity can be used to introduce an element of diversity in library synthesis. The scaffold of chiral amino propargylic alcohols lends itself to a ring-closing enyne metathesis. Depending on the propargylic protecting group, either the six- or seven-membered nitrogen heterocycle was formed (Eqs. 13.67 and 13.68).[143] It is interesting to note that the six-membered *exo* product is the expected product of the reaction, yet it was only obtained when the propargylic alcohol was acetate protected. The free propargylic alcohol resulted in the formation of the unexpected *endo* regioisomer. A similar strategy was applied by Hanson to generate a library of structurally diverse sultams starting from a chiral vinyl sulfonamide.[144]

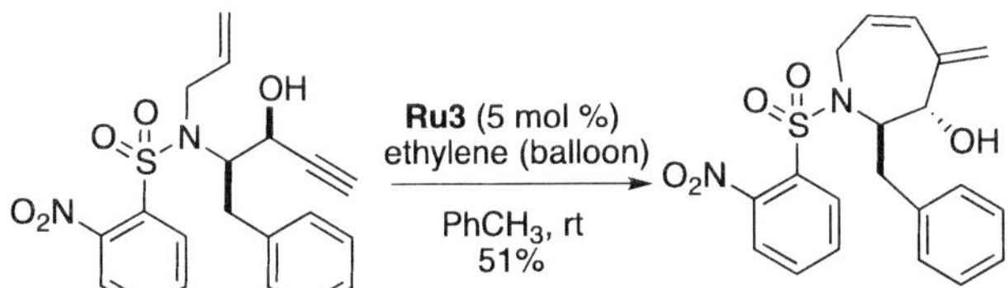

(13.67)

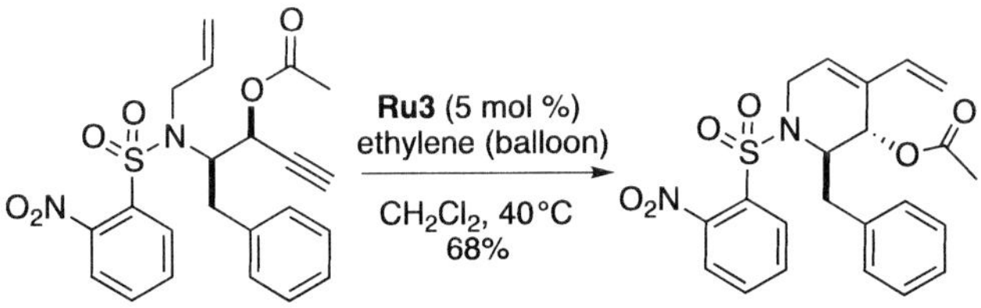

(13.68)

Double ring-closing enyne metathesis can be used to form diverse heterocycles in polycyclic frameworks. In one key step, the Grubbs catalyst **Ru1** performs chemoselective ring-closure (Scheme 13.16). Ethylene was used to assist the ring closure, known as Mori's conditions.[145] Using this approach, coupled with the additional versatility of a Diels–Alder cycloaddition, Schreiber and coworkers constructed a library of structurally diverse heterocycles.[146]

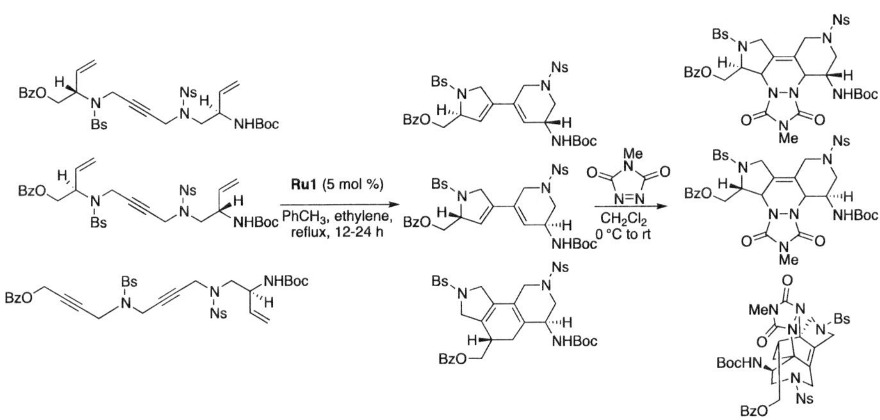

Scheme 13.16.

With an unsymmetrically substituted norbornene core, Spring and coworkers executed a two-directional synthesis of diverse polycyclic structures (Eq. 13.69).[147] In this instance, greater selectivity in the initial ring closure was obtained with **Ru1**. On the other hand, **Ru2** resulted in four different products as well as product loss due to competing ROMP of the norbornene. Subsequent to ring-opening alkene metathesis sequence, **Ru2** was introduced to catalyze the ring-closing enyne metathesis which furnished the 1,3-diene. In the final step, the 1,3-diene was trapped with *N*-ethylmaleimide resulting in formation of a complex polycyclic structure.

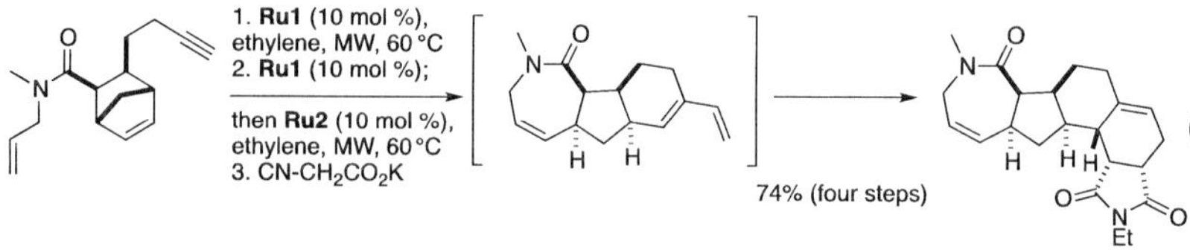

(13.69)

Ring rearrangement metathesis has been used to furnish a dienophile–diene duo leading to *in situ* intramolecular Diels–Alder reaction (Eq. 13.70).[148] The metathesis sequence generated an intermediate butenolide which serves as dienophile in cycloaddition with the 1,3-diene previously formed by ring-closing enyne metathesis.

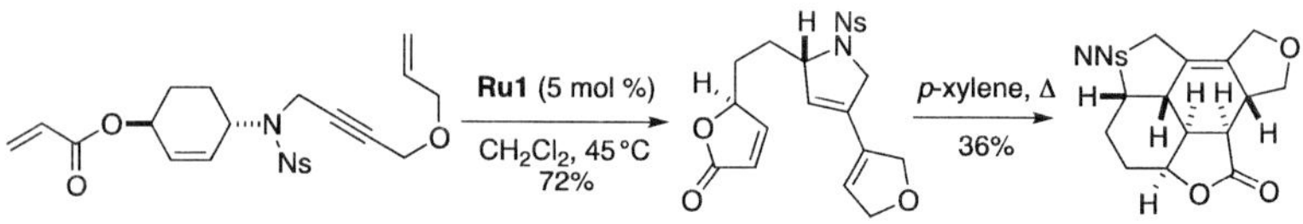

(13.70)

The chemoselectivity of the Grubbs' carbenes has allowed for their use in diversity-oriented synthesis because it allows for the facile construction of molecular scaffolds by metathetical union of alkenes and alkynes. Stereochemistry may play a determining role in dictating the mode of ring closure. The juxtaposition of the unsaturated elements can be used to access polycyclic molecules and heterocyclic ring systems. Cascade sequences have proven to be powerful methods for a rapid increase in structural complexity. For cascades involving enyne metathesis, the 1,3-diene makes a willing partner for cycloaddition reactions.

13.6 CONCLUSION AND OUTLOOK

Alkene metathesis with the Grubbs ruthenium carbenes has become an indispensible tool for the synthesis of complex molecules of all sizes. Because these carbene complexes are so chemoselective for alkenes, they offer a high degree of functional group tolerance. Significant progress has been made in understanding the origins of alkene metathesis reactivity for ruthenium carbene complexes. Built from solid kinetic mechanistic studies, a mechanistic scenario has been validated through characterization of reactive intermediates and through extensive computational studies. The ruthenacyclobutane intermediate validates the Chauvin mechanism and explains the conformational change that the ruthenium carbene undergoes, the so-called "windshield wipering." Catalyst development led to some of the key mechanistic insights and continues largely in the area of phosphine-free ruthenium carbene complexes known collectively as the Hoveyda–Grubbs superfamily of catalysts. Major advances have been made in selective reactions using ruthenium carbene complexes. Though limited in range of application as compared to the chiral molybdenum systems, enantioselectivity is possible with chiral ruthenium carbenes. Bidentate ligand systems have proven very successful in asymmetric applications and as a means to control Z-selectivity in alkene cross metathesis. As cascading metatheses have become more popular because of their complexity-building potential, the problem of site-selectively arose. To address this problem, relay metathesis was developed as a means of starting, or choreographing, the cascade. Cascades have become key steps in many natural product total syntheses and in diversity-oriented synthesis, especially when using enyne metathesis. Applications in complex molecule total synthesis have continued, and there is increasing use of relay metathesis, cascades, and selective cross alkene metathesis. Ring-closing applications are still widely used in synthesis particularly for macrocyclic ring closure. They are generally successful. Ring-closing metathesis has played a prominent role in the synthesis of protease inhibitors, illustrating that metathesis is efficient and amenable to scale-up, useful in process chemistry. Applications of metathesis chemistry to proteins and biopolymers have defined the need for aqueous-soluble catalysts. In protein chemistry, it is possible to use alkene metathesis as a bioorthogonal strategy for attaching molecular probes to alkene-containing catalysts. Though still quite limited, these applications are expected to grow as catalyst development further improves solubility and reactivity in aqueous media.

Future developments will utilize selective ruthenium carbene catalysts in aqueous chemistry and for Z-selective applications in natural product synthesis. Continued application of metathesis chemistry to library synthesis is also expected because of the potential of cascades to rapidly and reliably generate skeletal diversity. Catalyst development will and should continue. The widespread use of the Grubbs ruthenium carbenes has been largely due to their ease of handling and shelf stability. New catalysts with increased selectivity usually suffer from poor reactivity; this will need to be balanced with increased metathesis reactivity. At the same time, it is highly desirable for new catalysts to be bench stable. These are not easy goals as they often run in conflict. Because of the need of high turnover catalysts in industrial applications such as renewable energy, we expect further improvements will be sought in both catalyst structure and the use of additives. Some of the most selective catalysts feature bidentate binding to one of the positions occupied by a chloride ligand. New catalyst environments may lead to unexpected deactivation pathways. Mechanistic studies will continue to contribute to our understanding of ruthenium carbene reactivity.

SUGGESTED READING

Bieniek, M., Michrowska, A., Usanov, D. L., Grela, K. "In an Attempt to Provide a User's Guide to the Galaxy of Benzylidene, Alkoxybenzylidene, and Indenylidene Ruthenium Olefin Metathesis Catalysts," *Chem. Eur. J.* **2008**, *14*, 806–818.

Boeda, F., Clavier, H., Nolan, S. P. "Ruthenium-Indenylidene Complexes: Powerful Tools for Metathesis Transformations," *Chem. Commun.* **2008**, 2726–2740.

Buchmeiser, M. R. "Polymer-Supported Well-Defined Metathesis Catalysts," *Chem. Rev.* **2008**, *109*, 303–321.

Burtscher, D., Grela, K. "Aqueous Olefin Metathesis," *Angew. Chem. Int. Ed.* **2009**, *48*, 442–454.

Diver, S. T., Giessert, A. J. "Enyne Metathesis," *Chem. Rev.* **2004**, *104*, 1317–1382.

Hoye, T. R., Jeon, J. "Metathesis Involving a Relay and Applications in Natural Product Synthesis." In *Metathesis in Natural Product Synthesis. Strategies, Substrates, and Catalysts*. Cossy, J., Arseniyadis, S., Meyer, C., Eds. Wiley: Weinheim, **2010**, pp. 261–285.

Li, J., Lee, D. "Enyne-Metathesis-Based Tandem Processes," *Eur. J. Org. Chem.* **2011**, 4269–4287.

Nicolaou, K. C., Bulger, P. G., Sarlah, D. "Metathesis Reactions in Total Synthesis," *Angew. Chem. Int. Ed.* **2005**, *44*, 4490–4527.

Porta, M., Blechert, S. "Cascade Metathesis in Natural Product Synthesis." In *Metathesis in Natural Product Synthesis*, Cossy, J., Arseniyadis, S., Meyer, C., Eds. Wiley-VCH Verlag GmbH & Co. KGaA: Weinheim, **2010**, pp. 313–341.

Trnka, T. M., Grubbs, R. H. "The Development of L_2X_2RuCHR Olefin Metathesis Catalysts: An Organometallic Success Story," *Acc. Chem. Res.* **2000**, *34*, 18–29.

Vougioukalakis, G. C., Grubbs, R. H. "Ruthenium-Based Heterocyclic Carbene-Coordinated Olefin Metathesis Catalysts," *Chem. Rev.* **2009**, *110*, 1746–1787.

REFERENCES

1 Dias, E. L., Nguyen, S. T., Grubbs, R. H. *J. Am. Chem. Soc.* **1997**, *119*, 3887–3897.
2 Sanford, M. S., Love, J. A., Grubbs, R. H. *J. Am. Chem. Soc.* **2001**, *123*, 6543–6554.
3 Sanford, M. S., Ulman, M., Grubbs, R. H. *J. Am. Chem. Soc.* **2001**, *123*, 749–750.
4 Cavallo, L. J. *Am. Chem. Soc.* **2002**, *124*, 8965–8973.
5 Lautens, M., Maddess, M. L. *Org. Lett.* **2004**, *6*, 1883–1886.
6 Chatterjee, A. K., Choi, T.-L., Sanders, D. P., Grubbs, R. H. *J. Am. Chem. Soc.* **2003**, *125*, 11360–11370.
7 Moura-Letts, G., Curran, D. P. *Org. Lett.* **2006**, *9*, 5–8.
8 Smith, A. B., Kozmin, S. A., Paone, D. V. J. *Am. Chem. Soc.* **1999**, *121*, 7423–7424.
9 Smith, A. B., Adams, C. M., Kozmin, S. A., Paone, D. V. *J. Am. Chem. Soc.* **2001**, *123*, 5925–5937.
10 Adlhart, C., Chen, P. *Angew. Chem. Int. Ed.* **2002**, *41*, 4484–4487.
11 Adlhart, C., Chen, P. *J. Am. Chem. Soc.* **2004**, *126*, 3496–3510.
12 Romero, P. E., Piers, W. E. J. *Am. Chem. Soc.* **2005**, *127*, 5032–5033.
13 Romero, P. E., Piers, W. E., McDonald, R. *Angew. Chem. Int. Ed.* **2004**, *43*, 6161–6165.
14 Wenzel, A. G., Grubbs, R. H. *J. Am. Chem. Soc.* **2006**, *128*, 16048–16049.
15 Romero, P. E., Piers, W. E. *J. Am. Chem. Soc.* **2007**, *129*, 1698–1704.
16 Harvey, B. G., Mayne, C. L., Arif, A. M., Ernst, R. D. J. *Am. Chem. Soc.* **2005**, *127*, 16426–16435.
17 Bornand, M., Chen, P. *Angew. Chem. Int. Ed.* **2005**, *44*, 7909–7911.
18 Scholl, M., Ding, S., Lee, C. W., Grubbs, R. H. *Org. Lett.* **1999**, *1*, 953–956.

19 Love, J. A., Morgan, J. P., Trnka, T. M., Grubbs, R. H. *Angew. Chem. Int. Ed.* **2002**, *41*, 4035–4037.
20 Kingsbury, J. S., Harrity, J. P. A., Bonitatebus, P. J., Hoveyda, A. H. *J. Am. Chem. Soc.* **1999**, *121*, 791–799.
21 Michrowska, A., Bujok, R., Harutyunyan, S., Sashuk, V., Dolgnos, G., Grela, K. *J. Am. Chem. Soc.* **2004**, *126*, 9318–9325.
22 Stewart, I. C., Ung, T., Pletnev, A. A., Berlin, J. M., Grubbs, R. H., Schrodi, Y. *Org. Lett.* **2007**, *9*, 1589–1592.
23 Rix, D., Caijo, F., Laurent, I., Boeda, F., Clavier, H., Nolan, S. P., Mauduit, M. *J. Org. Chem.* **2008**, *73*, 4225–4228.
24 Zhan, Z.-Y. (Peop. Rep. China) Application: WO2006-CN1551, WO2007003135, **2007**, p. 61 pp.
25 Wakamatsu, H., Blechert, S. *Angew. Chem. Int. Ed.* **2002**, *41*, 794–796.
26 Fuerstner, A., Guth, O., Duffels, A., Seidel, G., Liebl, M., Gabor, B., Mynott, R. *Chem. – Eur. J.* **2001**, *7*, 4811–4820.
27 Monsaert, S., Drozdzak, R., Dragutan, V., Dragutan, I., Verpoort, F. *Eur. J. Inorg. Chem.* **2008**, 432–440.
28 Sanford, M. S., Love, J. A., Grubbs, R. H. *Organometallics* **2001**, *20*, 5314–5318.
29 Williams, J. E., Harner, M. J., Sponsler, M. B. *Organometallics* **2005**, *24*, 2013–2015.
30 Ahmed, M., Barrett, A. G. M., Braddock, D. C., Cramp, S. M., Procopiou, P. A. *Tetrahedron Lett.* **1999**, *40*, 8657–8662.
31 Garber, S. B., Kingsbury, J. S., Gray, B. L., Hoveyda, A. H. *J. Am. Chem. Soc.* **2000**, *122*, 8168–8179.
32 Stragies, R., Voigtmann, U., Blechert, S. *Tetrahedron Lett.* **2000**, *41*, 5465–5468.
33 Gessler, S., Randl, S., Blechert, S. *Tetrahedron Lett.* **2000**, *41*, 9973–9976.
34 Dong, L., Gordon, V. A., Grange, R. L., Johns, J., Parsons, P. G., Porzelle, A., Reddell, P., Schill, H., Williams, C. M. *J. Am. Chem. Soc.* **2008**, *130*, 15262–15263.
35 Hoveyda, A. H., Gillingham, D. G., Van Veldhuizen, J. J., Kataoka, O., Garber, S. B., Kingsbury, J. S., Harrity, J. P. A. *Org. Biomol. Chem.* **2004**, *2*, 8–23.
36 Stewart, I. C., Douglas, C. J., Grubbs, R. H. *Org. Lett.* **2008**, *10*, 441–444.
37 White, D. E., Stewart, I. C., Grubbs, R. H., Stoltz, B. M. *J. Am. Chem. Soc.* **2007**, *130*, 810–811.
38 Carlson, R. G., Gile, M. A., Heppert, J. A., Mason, M. H., Powell, D. R., Velde, D. V., Vilain, J. M. *J. Am. Chem. Soc.* **2002**, *124*, 1580–1581.
39 Hong, S. H., Day, M. W., Grubbs, R. H. J. *Am. Chem. Soc.* **2004**, *126*, 7414–7415.
40 Hong, S. H., Sanders, D. P., Lee, C. W., Grubbs, R. H. *J. Am. Chem. Soc.* **2005**, *127*, 17160–17161.
41 Gallenkamp, D., Fürstner, A. *J. Am. Chem. Soc.* **2011**, *133*, 9232–9235.
42 Marcaurelle, L. A., Comer, E., Dandapani, S., Duvall, J. R., Gerard, B., Kesavan, S., Lee, M. D., Liu, H., Lowe, J. T., Marie, J.-C., Mulrooney, C. A., Pandya, B. A., Rowley, A., Ryba, T. D., Suh, B.-C., Wei, J., Young, D. W., Akella, L. B., Ross, N. T., Zhang, Y.-L., Fass, D. M., Reis, S. A., Zhao, W.-N., Haggarty, S. J., Palmer, M., Foley, M. A. *J. Am. Chem. Soc.* **2010**, *132*, 16962–16976.
43 Nicolaou, K. C., Tria, G. S., Edmonds, D. J., Kar, M. *J. Am. Chem. Soc.* **2009**, *131*, 15909–15917.
44 Raju, R., Allen, L. J., Le, T., Taylor, C. D., Howell, A. R. *Org. Lett.* **2007**, *9*, 1699–1701.
45 Toya, H., Okano, K., Takasu, K., Ihara, M., Takahashi, A., Tanaka, H., Tokuyama, H. *Org. Lett.* **2010**, *12*, 5196–5199.
46 Winbush, S. M., Roush, W. R. *Org. Lett.* **2010**, *12*, 4344–4347.

47 Clark, D. A., Clark, J. R., Diver, S. T. *Org. Lett.* **2008**, *10*, 2055–2058.
48 Clark, J. R., Diver, S. T. *Org. Lett.* **2011**, *13*, 2896–2899.
49 Hong, S. H., Wenzel, A. G., Salguero, T. T., Day, M. W., Grubbs, R. H. *J. Am. Chem. Soc.* **2007**, *129*, 7961–7968.
50 Vehlow, K., Gessler, S., Blechert, S. *Angew. Chem. Int. Ed.* **2007**, *46*, 8082–8085.
51 van Rensburg, W. J., Steynberg, P. J., Meyer, W. H., Kirk, M. M., Forman, G. S. *J. Am. Chem. Soc.* **2004**, *126*, 14332–14333.
52 Trnka, T. M. *Catalysts for Olefin Metathesis: Ruthenium Alkylidene Complexes with Phosphine and N-Heterocyclic Carbene Ligands*, Ph.D Dissertation. California Institute of Technology, **2003**.
53 Trnka, T. M., Day, M. W., Grubbs, R. H. *Organometallics* **2001**, *20*, 3845–3847.
54 Ulman, M., Grubbs, R. H. *J. Org. Chem.* **1999**, *64*, 7202–7207.
55 Amoroso, D., Yap, G. P. A., Fogg, D. E. *Organometallics* **2002**, *21*, 3335–3343.
56 Dinger, M. B., Mol, J. C. *Eur. J. Inorg. Chem*. **2003**, 2827–2833.
57 Trnka, T. M., Morgan, J. P., Sanford, M. S., Wilhelm, T. E., Scholl, M., Choi, T.-L., Ding, S., Day, M. W., Grubbs, R. H. *J. Am. Chem. Soc.* **2003**, *125*, 2546–2558.
58 Louie, J., Grubbs, R. H. *Organometallics* **2002**, *21*, 2153–2164.
59 Arisawa, M., Terada, Y., Takahashi, K., Nakagawa, M., Nishida, A. *J. Org. Chem.* **2006**, *71*, 4255–4261.
60 Schmidt, B. *J. Org. Chem.* **2004**, *69*, 7672–7687.
61 Sutton, A. E., Seigal, B. A., Finnegan, D. F., Snapper, M. L. *J. Am. Chem. Soc.* **2002**, *124*, 13390–13391.
62 Louie, J., Bielawski, C. W., Grubbs, R. H. *J. Am. Chem. Soc.* **2001**, *123*, 11312–11313.
63 Galan, B. R., Gembicky, M., Dominiak, P. M., Keister, J. B., Diver, S. T. *J. Am. Chem. Soc.* **2005**, *127*, 15702–15703.
64 Galan, B. R., Kalbarczyk, K. P., Szczepankiewicz, S., Keister, J. B., Diver, S. T. *Org. Lett.* **2007**, *9*, 1203–1206.
65 Poater, A., Ragone, F., Correa, A., Cavallo, L. *J. Am. Chem. Soc.* **2009**, *131*, 9000–9006.
66 Galan, B. R., Pitak, M., Keister, J. B., Diver, S. T. *Organometallics* **2008**, *27*, 3630–3632.
67 Seiders, J. T., Ward, W. D., Grubbs, R. H. *Org. Lett.* **2001**, *3*, 3225–3228.
68 Funk, T. W., Berlin, J. M., Grubbs, R. H. *J. Am. Chem. Soc.* **2006**, *128*, 1840–1846.
69 Fournier, P.-A., Collins, S. K. *Organometallics* **2007**, *26*, 2945–2949.
70 Stenne, B., Timperio, J., Savoie, J., Dudding, T., Collins, S. K. *Org. Lett.* **2010**, *12*, 2032–2035.
71 Van Velduizen, J. J., Garber, S. B., Kingsbury, J. S., Hoveyda, A. H. *J. Am. Chem. Soc.* **2002**, *124*, 4954–4955.
72 Van Veldhuizen, J. J., Gillingham, D. G., Garber, S. B., Katoka, O., Hoveyda, A. H. *J. Am. Chem. Soc.* **2003**, *125*, 12502–12508.
73 Gillingham, D. G., Kataoka, O., Garber, S. B., Hoveyda, A. H. *J. Am. Chem. Soc.* **2004**, *126*, 12288–12290.
74 Tiede, S., Berger, A., Schlesiger, D., Rost, D., Luhl, A., Blechert, S. *Angew. Chem. Int. Ed.* **2010**, *49*, 3972–3975.
75 Khan, R. K. M., O'Brien, R. V., Torker, S., Li, B., Hoveyda, A. H. *J. Am. Chem. Soc.* **2012**, *134*, 12774–12779.
76 Malcolmson, S. J., Meek, S. J., Sattely, E. S., Schrock, R. R., Hoveyda, A. H. *Nature* **2008**, *456*, 933–937.

77 Ibrahem, I., Yu, M., Schrock, R. R., Hoveyda, A. H. *J. Am. Chem. Soc.* **2009**, *131*, 3844–3845.
78 Sattely, E. S., Meek, S. J., Malcolmson, S. J., Schrock, R. R., Hoveyda, A. H. *J. Am. Chem. Soc.* **2009**, *131*, 943–953.
79 Harvey, J. S., Malcolmson, S. J., Dunne, K. S., Meek, S. J., Thompson, A. L., Schrock, R. R., Hoveyda, A. H., Gouverneur, V. *Angew. Chem. Int. Ed.* **2009**, *48*, 762–766.
80 Hoveyda, A. H., Malcolmson, S. J., Meek, S. J., Zhugralin, A. R. *Angew. Chem. Int. Ed.* **2010**, *49*, 34–44.
81 Hoye, T. R., Jeffrey, C. S., Tennakoon, M. A., Wang, J., Zhao, H. *J. Am. Chem. Soc.* **2004**, *126*, 10210–10211.
82 Wallace, D. J. *Angew. Chem. Int. Ed.* **2005**, *44*, 1912–1915.
83 Hoye, T. R., Jeon, J. "Metathesis Involving a Relay and Applications in Natural Product Synthesis." In *Metathesis in Natural Product Synthesis. Strategies, Substrates, and Catalysts*. Cossy, J., Arseniyadis, S., Meyer, C., Eds. Wiley: Weinheim, **2010**, pp. 261–285.
84 Hoye, T. R., Jeffrey, C. S., Tennakoon, M. A., Wang, J., Zhao, H. *J. Am. Chem. Soc.* **2004**, *126*, 10210–10211.
85 Wu, C.-J., Madhushaw, R. J., Liu, R.-S. J. *Org. Chem.* **2003**, *68*, 7889–7892.
86 Huang, J., Xiong, H., Hsung, R. P., Rameshkumar, C., Mulder, J. A., Grebe, T. P. *Org. Lett.* **2002**, *4*, 2417–2420.
87 Lee, J., Parker, K. A. *Org. Lett.* **2012**, *14*, 2682–2685.
88 Roethie, P. A., Chen, I. T., Trauner, D. *J. Am. Chem. Soc.* **2007**, *129*, 8960–8961.
89 Trost, B. M., Yang, H., Thiel, O. R., Frontier, A. J., Brindle, C. S. J. *Am. Chem. Soc.* **2007**, *129*, 2206–2207.
90 Hoye, T. R., Jeon, J., Tennakoon, M. A. *Angew. Chem. Int. Ed.* **2011**, *50*, 2141–2143.
91 Li, J., Park, S., Miller, R. L., Lee, D. *Org. Lett.* **2009**, *11*, 571–574.
92 Clark, J. R., French, J. M., Jecs, E., Diver, S. T. *Org. Lett.* **2012**, *14*, 4178–4181.
93 Clark, J. R., French, J. M., Diver, S. T. *J. Org. Chem.* **2012**, *77*, 1599–1604.
94 Wang, Y., Jimenez, M., Hansen, A. S., Raiber, E.-A., Schreiber, S. L., Young, D. W. J. *Am. Chem. Soc.* **2011**, *133*, 9196–9199.
95 Torker, S., Muller, A., Chen, P. *Angew. Chem. Int. Ed.* **2010**, *49*, 3762–3766.
96 Teo, R., Grubbs, R. H. *Organometallics* **2010**, *29*, 6045–6050.
97 Ritter, T., Hejl, A., Wenzel, A. G., Funk, T. W., Grubbs, R. H. *Organometallics* **2006**, *25*, 5740–5745.
98 Endo, K., Grubbs, R. H. *J. Am. Chem. Soc.* **2011**, *133*, 8525–8527.
99 Keitz, B. K., Endo, K., Herbert, M. B., Grubbs, R. H. *J. Am. Chem. Soc.* **2011**, *133*, 9686–9688.
100 Liu, P., Xu, X., Dong, X., Keitz, B. K., Herbert, M. B., Grubbs, R. H., Houk, K. N. *J. Am. Chem. Soc.* **2012**, *134*, 1464–1467.
101 Furstner, A., Nevado, C., Waser, M., Tremblay, M., Chevrier, C., Teply, F., Aissa, C., Moulin, E., Muller, O. *J. Am. Chem. Soc.* **2007**, *129*, 9150–9161.
102 Fürstner, A., Jeanjean, F., Razon, P. *Angew. Chem. Int. Ed.* **2002**, *41*, 2097–2101.
103 Hirama, M., Oishi, T., Uehara, H., Inoue, M., Maruyama, M., Oguri, H., Satake, M. *Science* **2001**, *294*, 1904–1907.
104 Toma, T., Kita, Y., Fukuyama, T. *J. Am. Chem. Soc.* **2010**, *132*, 10233–10235.
105 Jackson, K. L., Henderson, J. A., Motoyoshi, H., Phillips, A. J. *Angew. Chem. Int. Ed.* **2009**, *48*, 2346–2350.
106 Garbaccio, R. M., Stachel, S. J., Baeschlin, D. K., Danishefsky, S. J. *J. Am. Chem. Soc.* **2001**, *123*, 10903–10908.

107 Pfeiffer, M. W. B., Phillips, A. J. *J. Am. Chem. Soc.* **2005**, *127*, 5334–5335.
108 Hart, A. C., Phillips, A. J. *J. Am. Chem. Soc.* **2006**, *128*, 1094–1095.
109 Schaudt, M., Blechert, S. *J. Org. Chem.* **2003**, *68*, 2913–2920.
110 Wallace, D. J. *Tetrahedron Lett.* **2003**, *44*, 2145–2148.
111 Wallace, D. J. *J. Mol. Cat. A: Chem.* **2006**, *254*, 78–84.
112 Blechert, S., Stapper, C. *Eur. J. Org. Chem.* **2002**, *2002*, 2855–2858.
113 Buschmann, N., Ruckert, A., Blechert, S. *J. Org. Chem.* **2002**, *67*, 4325–4329.
114 Stragies, R., Blechert, S. *Tetrahedron* **1999**, *55*, 8179–8188.
115 Strapper, C., Blechert, S. *J. Org. Chem.* **2002**, *67*, 6456–6460.
116 Niethe, A., Fischer, D., Blechert, S. *J. Org. Chem.* **2008**, *73*, 3088–3093.
117 Boyer, F.-D., Hanna, I., Ricard, L. *Org. Lett.* **2004**, *6*, 1817–1820.
118 Tan, D. S., Dudley, G. B., Danishefsky, S. J. *Angew. Chem. Int. Ed.* **2002**, *41*, 2185–2188.
119 Lin, S., Dudley, G. B., Tan, D. S., Danishefsky, S. J. *Angew. Chem. Int. Ed.* **2002**, *41*, 2188–2191.
120 Mukherjee, S., Lee, D. *Org. Lett.* **2009**, *11*, 2916–2919.
121 Movassaghi, M., Piizzi, G., Siegel, D. S., Piersanti, G. *Angew. Chem. Int. Ed.* **2006**, *45*, 5859–5863.
122 Yee, N. K., Farina, V., Houpis, I. N., Haddad, N., Frutos, R. P., Gallou, F., Wang, X.-J., Wei, X., Simpson, R. D., Feng, X., Fuchs, V., Xu, Y., Tan, J., Zhang, L., Xu, J., Smith-Keenan, L. L., Vitous, J., Ridges, M. D., Spinelli, E. M., Johnson, M., Donsbach, K., Nicola, T., Brenner, M., Winter, E., Kreye, P., Samstag, W. *J. Org. Chem.* **2006**, *71*, 7133–7145.
123 Zeng, X., Wei, X., Farina, V., Napolitano, E., Xu, Y., Zhang, L., Haddad, N., Yee, N. K., Grinberg, N., Shen, S., Senanayake, C. H. J. *Org. Chem.* **2006**, *71*, 8864–8875.
124 Shu, C., Zeng, X., Hao, M.-H., Wei, X., Yee, N. K., Busacca, C. A., Han, Z., Farina, V., Senanayake, C. H. *Org. Lett.* **2008**, *10*, 1303–1306.
125 McCauley, J. A., Rudd, M. T., Nguyen, K. T., McIntyre, C. J., Romano, J. J., Bush, K. J., Varga, S. L., Ross, C. W., Carroll, S. S., DiMuzio, J., Stahlhut, M. W., Olsen, D. B., Lyle, T. A., Vacca, J. P., Liverton, N. J. *Angew. Chem. Int. Ed.* **2008**, *47*, 9104–9107.
126 Wang, H., Goodman, S. N., Dai, Q., Stockdale, G. W., Clark, W. M. *Org. Proc. Res. Dev.* **2008**, *12*, 226–234.
127 Kong, J., Chen, C.-Y., Balsells-Padros, J., Cao, Y., Dunn, R. F., Dolman, S. J., Janey, J., Li, H., Zacuto, M. J. *J. Org. Chem.* **2012**, *77*, 3820–3828.
128 Lynn, D. M., Kanaoka, S., Grubbs, R. H. *J. Am. Chem. Soc.* **1996**, *118*, 784–790.
129 Kanai, M., Mortell, K. H., Kiessling, L. L. *J. Am. Chem. Soc.* **1997**, *119*, 9931.
130 Binder, J. B., Blank, J. J., Raines, R. T. *Org. Lett.* **2007**, *9*, 4885–4888.
131 Jordan, J. P., Grubbs, R. H. *Angew. Chem. Int. Ed.* **2007**, *46*, 5152–5155.
132 Lipshutz, B. H., Aguinaldo, G. T., Ghorai, S., Voigtritter, K. *Org. Lett.* **2008**, *10*, 1325–1328.
133 Hong, S. H., Grubbs, R. H. *J. Am. Chem. Soc.* **2006**, *128*, 3508–3509.
134 Nicolaou, K. C., Hughes, R., Cho, S. Y., Winssinger, N., Labischinski, H., Endermann, R. *Chem. Eur. J.* **2001**, *7*, 3824–3843.
135 Lin, Y. A., Chalker, J. M., Floyd, N., Bernardes, G. J. L., Davis, B. G. *J. Am. Chem. Soc.* **2008**, *130*, 9642–9643.
136 Bernardes, G. J. L., Chalker, J. M., Errey, J. C., Davis, B. G. *J. Am. Chem. Soc.* **2008**, *130*, 5052–5053.
137 Lin, Y. A., Chalker, J. M., Davis, B. G. *J. Am. Chem. Soc.* **2010**, *132*, 16805–16811.
138 Alam, J., Keller, T. H., Loh, T.-P. *J. Am. Chem. Soc.* **2010**, *132*, 9546–9548.

139 Ai, H.-W., Shen, W., Brustad, E., Schultz, P. G. *Angew. Chem. Int. Ed.* **2010**, *49*, 935–937.
140 Banti, D., North, M. *Tetrahedron Lett.* **2002**, *43*, 1561–1564.
141 Lee, D., Sello, J. K., Schreiber, S. L. *Org. Lett.* **2000**, *2*, 709–712.
142 Sello, J. K., Andreana, P. R., Lee, D., Schreiber, S. L. *Org. Lett.* **2003**, *5*, 4125–4127.
143 Pizzirani, D., Kaya, T., Clemons, P. A., Schreiber, S. L. *Org. Lett.* **2010**, *12*, 2822–2825.
144 Zhou, A., Rayabarapu, D., Hanson, P. R. *Org. Lett.* **2009**, *11*, 531–534.
145 Mori, M., Sakakibara, N., Kinoshita, A. J. *Org. Chem.* **1998**, *63*, 6082–6083.
146 Spiegel, D. A., Schroeder, F. C., Duvall, J. R., Schreiber, S. L. *J. Am. Chem. Soc.* **2006**, *128*, 14766–14767.
147 Spandl, R. J., Rudyk, H., Spring, D. R. *Chem. Commun.* **2008**, 3001–3003.
148 O'Leary-Steele, C., Pendersen, P. J., James, T., Lanyon-Hogg, T., Leach, S., Hayes, J., Nelson, A. *Chem. Eur. J.* **2010**, *16*, 9563–9571.

14

NUCLEOPHILIC CARBENES OF THE CHROMIUM TRIAD

ZACHARY J. TONZETICH
Department of Chemistry, University of Texas at San Antonio, San Antonio, TX, USA

14.1 INTRODUCTION

The discovery of the first isolable metal carbenes by E. O. Fischer in 1964 involved metals of the chromium triad,[1] and to this day Group 6 carbenes continue to be the subject of extensive investigation due, in large part, to their use as catalysts for olefin metathesis. The nature of the carbene ligand in Group 6 complexes can generally be classified as nucleophilic or electrophilic. Nucleophilic carbenes or *alkylidenes* (also *Schrock carbenes*) are distinguished from their electrophilic counterparts or *Fischer carbenes* foremost by their reactivity but also by the formal oxidation states of the metals to which they bind. Fischer carbenes are generally found with low oxidation state metals in tandem with π-acidic ligands such as CO. Alkylidenes, by contrast, are typically found with metals in their highest oxidation states (formally d^0). The bonding in both types of ligands is usually discussed in terms of the spin states of the hypothetical free carbenes. Whereas Fischer carbenes are considered singlets with low-lying π* orbitals and a carbon-based LUMO, alkylidenes are generally considered as triplet species containing unpaired electrons in orbitals available for both σ- and π-bonding with the metal (Scheme 14.1). Accordingly, the bonding in alkylidene compounds is more akin to a traditional double bond. The HOMO of these compounds is the metal–carbon π-bond, which is

Contemporary Carbene Chemistry, First Edition. Edited by Robert A. Moss and Michael P. Doyle.

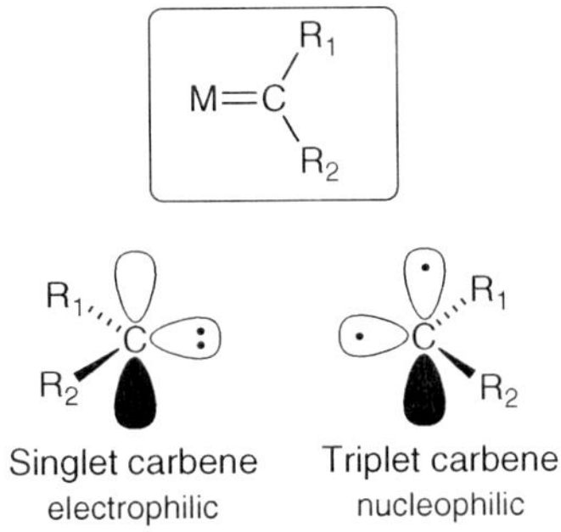

Scheme 14.1.

polarized towards carbon giving rise to the nucleophilic character. Although intermediate cases can exist, theoretical work has demonstrated the substantial differences in bonding between electrophilic and nucleophilic carbenes of the Group 6 metals,[2,3] especially in comparison to their carbyne counterparts.[4]

14.1.1 Discovery

Ten years after Fischer's initial discovery, the first nucleophilic carbene complex was prepared by Schrock through reaction of Mowat and Wilkinson's $[Ta(CH_2{}^tBu)_3Cl_2]$ compound[5] with 2 equiv of neopentyl lithium (Eq. 14.1).[6] The initial product of this reaction, $[Ta(CH_2{}^tBu)_5]$, is unstable and immediately undergoes α-hydrogen abstraction to generate neopentane and alkylidene **1**.[7,8] Schrock's original report also described the first nucleophilic carbyne complex by demonstrating that chemically reversible deprotonation can take place at the carbene carbon atom. α-Hydrogen abstraction as a means of generating nucleophilic carbenes has now been demonstrated with a variety of transition metals and remains a very common procedure for the preparation of alkylidene complexes.

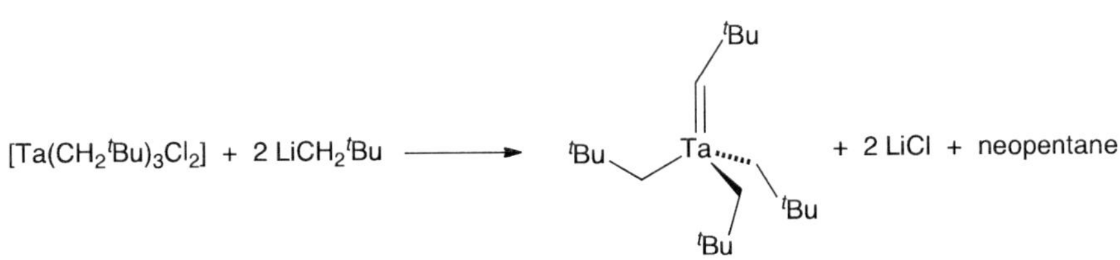

1

(14.1)

Early impetus for the exploration of nucleophilic carbene complexes came from a desire to create well-defined olefin metathesis catalysts. This desire prompted an exploration of other metals, and several years after the first report of nucleophilic carbenes with tantalum, Schrock also disclosed the first alkylidene compounds of a Group 6 metal, tungsten.[9] These tungsten alkylidenes were formed via α-hydrogen abstraction upon addition of phosphines to the trialkylalkylidyne compound, $[W(\equiv C^tBu)(CH_2{}^tBu)_3]$. As these species were inactive for the metathesis of olefins, alkylidene compounds containing other supporting ligands were investigated. Noteworthy is the subsequent preparation of tungsten alkylidene **2**, which contains the strongly π-donating oxo ligand (Eq. 14.2).[10] Interestingly, oxo-alkylidene **2** was not synthesized via α-hydrogen abstraction but rather through an alkylidene transfer reaction from tantalum complex **3**.[11] Activation of **2** with $AlCl_3$ afforded catalyst species active toward olefin metathesis.

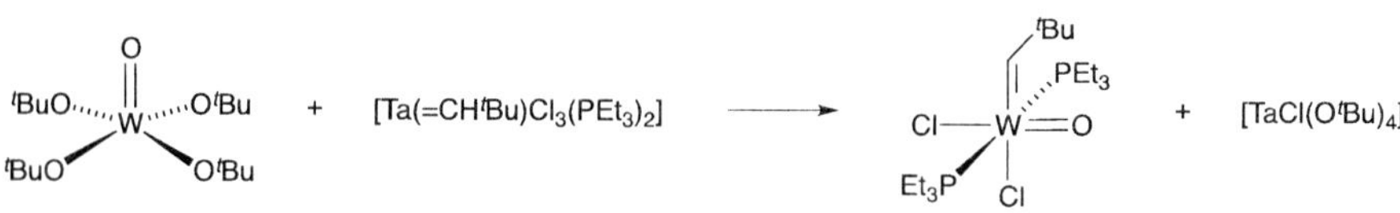

4 **3** **2**

(14.2)

It was recognized early on that the presence of the oxo ligand was beneficial for activity in olefin metathesis catalysis,[12,13] but its small size encouraged coordinative saturation about the metal inhibiting reactivity. In an effort to decrease the coordination number of tungsten alkylidene species and thereby obviate the need for activators such as $AlCl_3$, the imido ligand (NR^{2-}) was examined by Schrock as a sterically encumbering alternative to the oxo ligand.[14] Eventual replacement of the oxo with the imido ligand afforded a family of tungsten complexes active toward the olefin metathesis reaction in the absence of cocatalysts.[15] Sluggish reactivity and functional group intolerance with these early tungsten catalysts prompted an examination of their molybdenum congeners, and the first examples of catalytically active Mo(VI) alkylidenes were reported in 1987.[16] The initial syntheses of these Mo alkylidenes involved proton transfer from nitrogen to carbon within an alkylidyne compound mediated by catalytic amounts of Et_3N (Eq. 14.3). The discovery of compound **6** was in many respects a watershed moment in the development of nucleophilic carbene chemistry, setting the stage for two and a half decades of vigorous research into improved olefin metathesis catalysts.

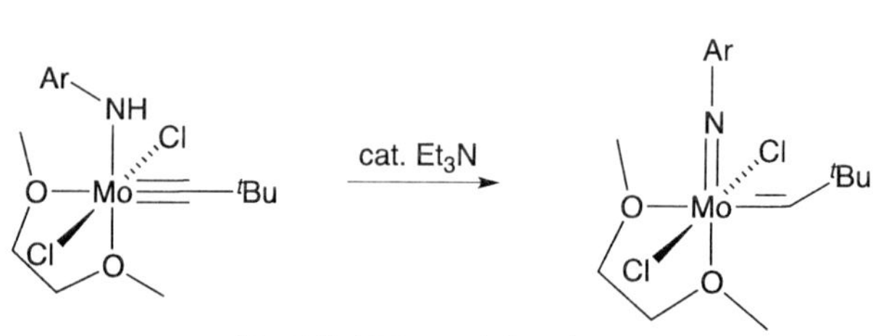

Ar = 2,6-diisopropylphenyl

5 **6** (14.3)

14.1.2 Spectroscopic Properties

The diamagnetism of nearly all carbene compounds has made NMR spectroscopy the preferred tool for their characterization in solution. NMR features of the alkylidene ligand in Group 6 complexes are especially diagnostic. The carbene carbon atom (herein denoted C_α) generally appears at ^{13}C chemical shift values between 250 and 330 ppm. Primary alkylidenes (containing at least one H atom substituent on C_α) also display diagnostic NMR features, with H_α chemical shift values between 10 and 15 ppm. In compounds containing additional strongly π-bonding ligands (oxo, imido, or alkylidyne), the possibility for alkylidene rotational isomers can exist. These isomers are termed *syn* and *anti*, referring to the disposition of the highest priority group on C_α with respect to the multiply bound ligand (Fig. 14.1). Interconversion of these isomers requires disruption of the M–C π-bond, which can be a high-energy process if the orthogonal $d\pi$ orbital used to stabilize the transition state is unavailable due to bonding with ancillary ligands. $^1J_{CH}$ coupling constants of primary alkylidenes can be very indicative of the rotational isomer. For oxo- and imido-alkylidene complexes of the Group 6 metals, this coupling constant is attenuated for the *syn* isomer due to an agnostic interaction with the metal (Fig. 14.1).[17]

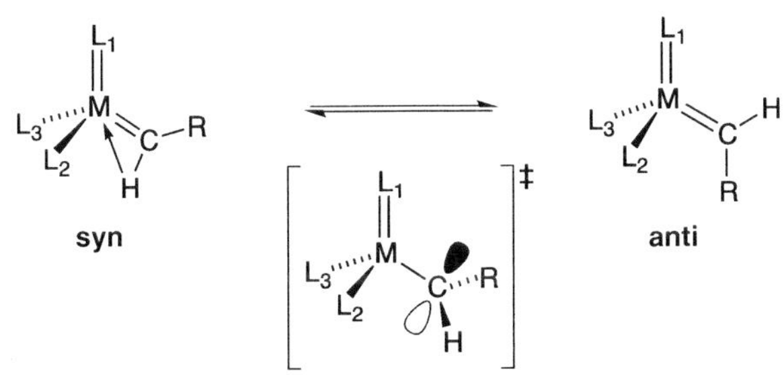

Figure 14.1. Alkylidene rotational isomerization in compounds containing a multiply bound ligand (L_1).

14.1.3 Metathesis Reactivity

Many nucleophilic carbene complexes of the Group 6 metals react rapidly with organic functional groups that feature π-bonds (Scheme 14.2). These so-called metathesis reactions proceed through the Chauvin mechanism, whereby a metallacyclic intermediate is formed through a formal [2+2] cycloaddition of the unsaturated reactant to the metal alkylidene.[18] Cycloreversion of this intermediate can reform the starting materials or lead to new metathesis products depending upon the thermodynamics of the reaction. This type of reactivity is most useful for the metathesis of olefins and has been successfully realized for a series of reactions that make use of convenient driving forces such as release of ring strain or production of volatile by-products. In cases of polarized π-bonds, the more electrophilic atom is found to bind to the carbene carbon atom consistent with its nucleophilic character. This reactivity is exemplified by the reaction of nucleophilic carbenes with aldehydes or ketones, which proceeds in a "Wittig-like" fashion to produce a metal-oxo compound and an alkene (Scheme 14.2, X = O, Y = CR'_2).

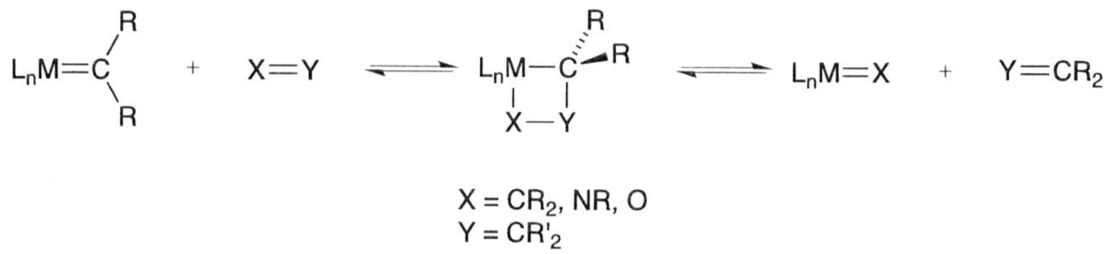

Scheme 14.2.

A prominent decomposition mode of nucleophilic carbenes is bimolecular coupling of alkylidenes to afford an olefin and a bimetallic species (Scheme 14.3). Less sterically encumbered alkylidenes are especially prone to these decomposition pathways and therefore primary alkylidenes usually feature bulky substituents such as *tert*-butyl, CMe_2Ph (termed *neophyl*), or trimethylsilyl. Intermediate alkylidenes formed during olefin metathesis reactions, especially methylidenes, typically do not posses such steric bulk, and the propensity for these intermediates to undergo coupling can be a major limit to catalytic activity. In addition to alkylidene coupling, intermediate metallacyclobutane species may also decompose via β-hydrogen elimination to afford reduced metal species after subsequent reductive elimination. These reduced metal centers can coordinate olefins leading to formation of alkene complexes. Although such complexes have been demonstrated to initiate metathesis reactions,[19] their activity is significantly less than that of the respective carbene compounds. Therefore, bulky supporting ligands are commonly used in an effort to avoid decomposition pathways that destroy the alkylidene ligand.

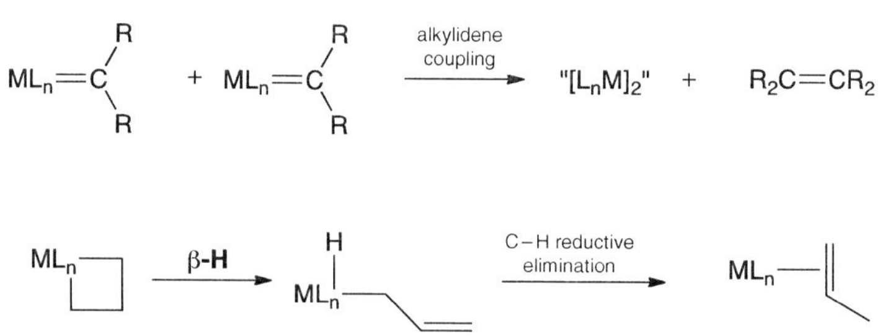

Scheme 14.3.

The well-defined nature of Group 6 nucleophilic carbene compounds has allowed for detailed mechanistic studies of the olefin metathesis reaction, which has resulted in ever-more finely tuned catalyst structures. In this chapter, recent developments in the chemistry of Group 6 alkylidenes will be highlighted, particularly those that relate to the olefin metathesis reaction. Extensive reviews on metal alkylidene chemistry and the development of the olefin metathesis reaction have been published elsewhere and the reader is directed to the Suggested Readings at the end of the chapter for a complete account of past work.

14.2 CHROMIUM CARBENES

Among the Group 6 metals, chromium displays the fewest number of examples of nucleophilic carbene complexes. This fact is due in part to the susceptibility of high-oxidation state chromium complexes to undergo reduction.[20] Despite these difficulties, chromium carbenes have been investigated by several research groups. Much of the interest in nucleophilic alkylidenes of chromium stems from their possible involvement in olefin polymerization reactions, in particular the Phillips catalyst system,[21] and several promising compounds have been prepared in recent years that demonstrate notable differences with their better studied Mo and W counterparts.

14.2.1 Cr(VI) Carbenes

The first and only example of molecular hexavalent chromium alkylidene compounds were reported by Gibson in 1996.[22] The synthetic route to these compounds involves alkylation of a diimido Cr(VI) precursor with neopentyl Grignard to give diimido-dineopentyl species **7**. Previous work by Schaverien and Orpen had identified a number of similar diimido-dialkyl species, but their attempts to observe a Cr(VI) alkylidene by a variety of methods were unsuccessful.[23] The Cr(VI) alkylidene **8** is formed by α-H abstraction from the diimido-dineopentyl species **7** and can be trapped by addition of tetrahydrofuran (THF) (Eq. 14.4). Subsequent displacement of the THF ligand by PMe_3 affords the phosphine-ligated derivative **9**. Although these species were identified in solution, they were never subjected to X-ray crystallographic analysis and so no metric parameters for a Cr(VI) alkylidene exist to date.

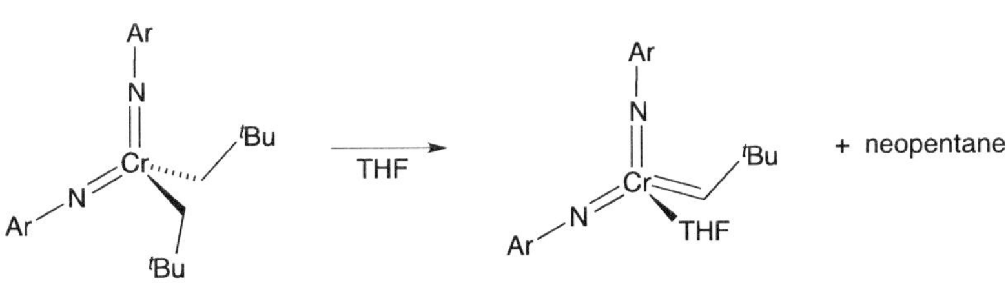

7 Ar = 2,6-diisopropylphenyl **8** (14.4)

Compound **7** is capable of undergoing α-H abstraction in the absence of added base to give a highly reactive three-coordinate alkylidene compound **10** as evidenced by its reactivity with benzene-d_6 to afford neopentane and the new

Cr(VI) neopentyl-phenyl complex **11** that displays deuterium incorporation into the neopentyl methylene protons (Eq. 14.5). Such C–H activation chemistry is not unique among Group 6 nucleophilic carbenes and has been observed in a series of Mo and W alkylidenes containing cyclopentadienyl ligands (see Section 14.3.3). The ^{13}C and ^{1}H NMR parameters of **8** and **9** are in line with those observed in Mo and W chemistry. Curiously, compounds **8**, **9**, and **10** demonstrated no reactivity towards olefins. This observation may be linked to the presence of the second imido ligand. Such a motif has not been reported in Mo and W alkylidene chemistry and is likely incompatible with metallacycle formation.

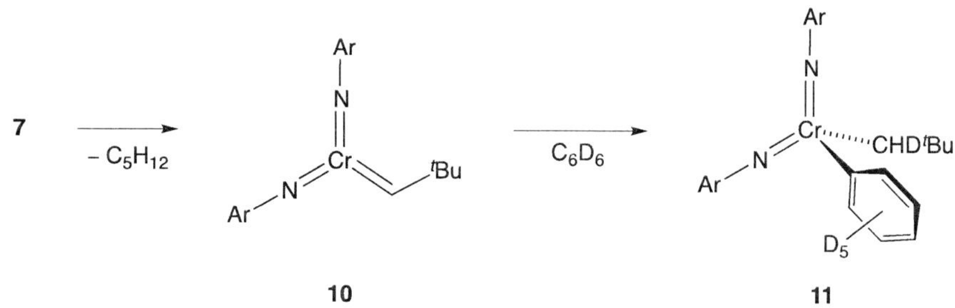

(14.5)

Sundermeyer has also reported an example of a high-oxidation state carbene complex of chromium containing a phosphonio-alkylidene.[24] Like **7**, this species was prepared from a diimido-dichloride Cr(VI) precursor. Addition of methylene triphenylphosphorane to this precursor afforded the phosphonio-alkylidene compound **12** (Eq. 14.6). Compound **12** may be viewed as a zwitterionic Cr(VI) alkylidene, although the presence of the PPh_3 group argues against its formulation as a true nucleophilic carbene. Nevertheless, **12** possesses NMR features consistent with other Group 6 alkylidene compounds and displays a high barrier to alkylidene rotation. Furthermore, **12** was demonstrated to react with several unsaturated molecules such as diphenylketene to give products consistent with an initial metathetical [2+2] addition. Unfortunately, as with compounds **8** and **9**, compound **12** was not subjected to crystallographic study. Analogous Mo and W phosphonio alkylidenes were also prepared in similar fashion and displayed comparable reactivity to **12**.

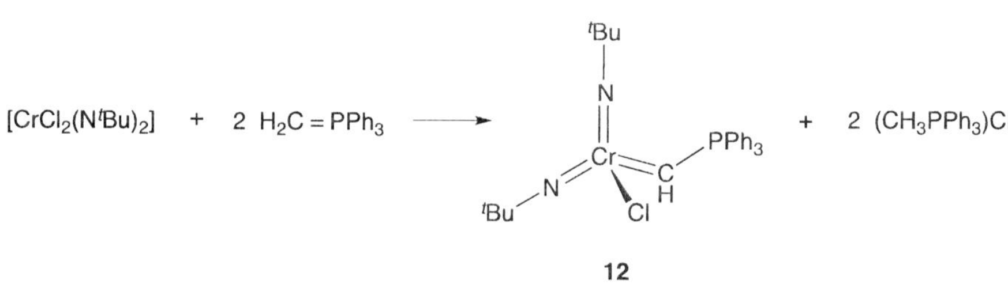

(14.6)

14.2.2 Nucleophilic Cr(III) Carbenes

Theopold has demonstrated that α-hydrogen abstraction can occur for a variety of low-coordinate trimethylsilylmethyl complexes of Cr(III).[25] For example, compound **13** was found to decompose upon thermolysis to the bridging alkylidene complex **14** with concomitant loss of tetramethylsilane (Scheme 14.4). Formation of compound **14** is consistent with a mechanism involving transient formation of the Cr(III) terminal alkylidene **15**. This reaction pathway, however, competes with γ-hydrogen elimination to afford the metallacycle **16**. Attempts to trap the putative alkylidene **15** with a variety of chelating ligands led to the formation of non-alkylidene-containing compounds including an activated bipyridine complex.[26,27]

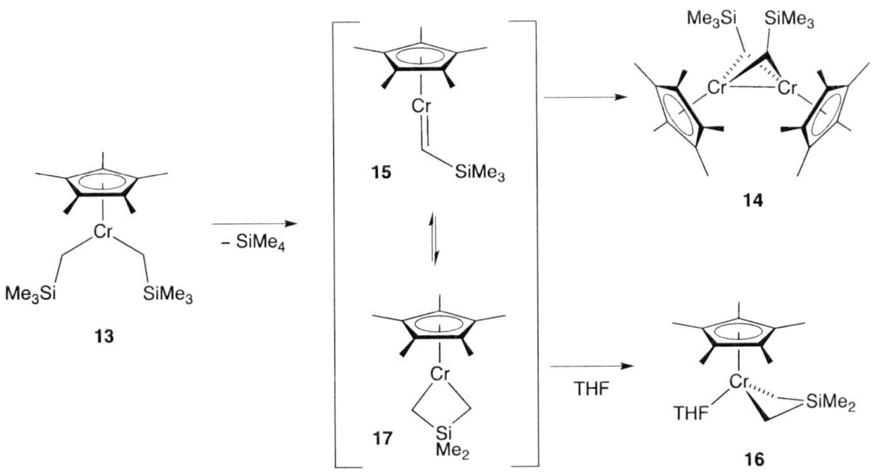

Scheme 14.4.

14.2.3 Supported Cr(IV) Carbenes

In addition to homogeneous versions of mid-oxidation state chromium alkylidenes, recent work has also examined the preparation of heterogeneous analogs by grafting of Cr(IV) precursors to silica. Work by Scott has demonstrated the formation of silica-bound Cr(IV) alkylidenes **18** and **19** by thermolysis of dialkyl precursors **20** and **21**.[28] Kinetics and deuterium-labeling experiments demonstrated that alkylidene formation from **20** and **21** proceeds by α-H elimination to form a Cr(VI) alkyl-alkylidene-hydride compound followed by reductive elimination of neopentane or $SiMe_4$ to generate the corresponding Cr(IV) alkylidene (Eq. 14.7).[29]

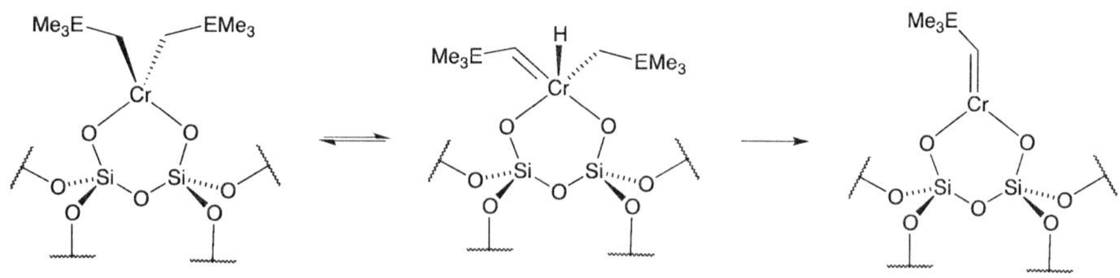

E = C, **20**
Si, **21**

E = C, **18**
Si, **19**

(14.7)

Alkylidene **18** reacts with acetone and styrene in a metathetical fashion to generate the expected olefins ($Me_2C{=}CH^tBu$ and $H_2C{=}CH^tBu$, respectively), consistent with its formulation as a nucleophilic Cr carbene.[30] Somewhat surprisingly, alkylidene **18** does not react with simple olefins such as ethylene to give the products of olefin metathesis, instead leading to polymerization/oligomerization. Mechanistic work on the polymerization reaction has ruled out the involvement of alkylidenes (via a Green–Rooney mechanism) in favor of a mechanism involving insertion of incoming olefin into a Cr(IV)-alkyl bond.[31–34] The extent to which these well-defined alkylidenes resemble possible species in the Phillips process is not known, but similarities in the polymerization mechanisms argue for their involvement. It will be of interest to see whether future work can identify a role for supported Cr carbenes as catalysts for olefin metathesis.

14.3 MOLYBDENUM CARBENES

In contrast to chromium, nucleophilic carbene complexes of molybdenum are numerous, thanks in large part to now well-established synthetic protocols. The prototypical Mo(VI) alkylidene compound, **23**, contains an imido and two alkoxide ligands. These four-coordinate species are prepared from a general triflate precursor, **22**, using a synthetic sequence outlined in Scheme 14.5.[35] Substitution of

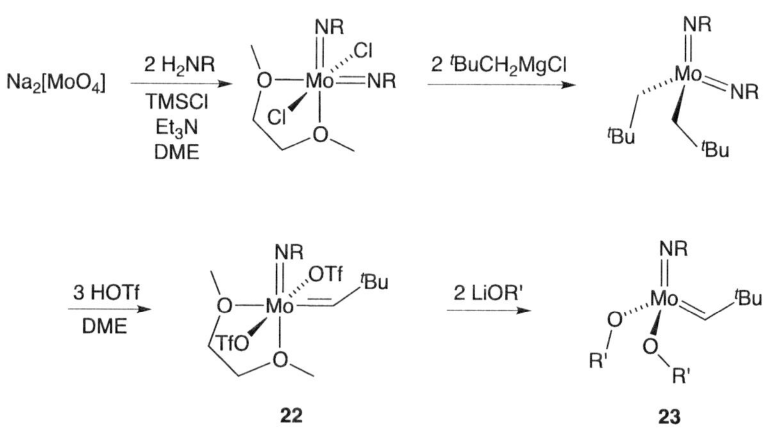

22

23

Scheme 14.5.

the triflate ligands with various alkoxides results in compounds active towards a variety of metathesis reactions including ring-opening metathesis polymerization (ROMP), ring-closing metathesis (RCM), and certain cross-metatheses (CM). In addition to alkoxide substitution, the general synthesis of **22** allows for the modification of the imido group starting from different amine precursors, although certain very bulky anilines are incompatible with this general synthesis.[36] The majority of contemporary work in the area of Mo carbene chemistry has elaborated upon the traditional imido-alkylidene-bis-alkoxide archetype and demonstrated that modifications to the classic catalyst framework can yield compounds with very different reactivities and selectivities in olefin metathesis reactions.

14.3.1 Mo(VI) Imido-Alkylidene-bis-Alkoxide Compounds

Examples of Schrock-type imido-alkylidene-bis-alkoxide (e.g., **23**) catalysts first appeared in the late 1980s. These four-coordinate species can feature a variety of different bulky alkoxide and imido ligands and are active catalysts for a variety of metathesis reactions such as ROMP and RCM.[37,38] The alkoxide ligands that feature most prominently in classical versions of Schrock catalysts are the electron-withdrawing hexafluoro-*tert*-butoxide (OR_{F6}) and the electron-donating *tert*-butoxide. Alkylidene rotational isomerization and olefin binding with these species has been the subject of much investigation. In a notable study, Mo complexes featuring alkoxide ligands that are more electron donating such as *tert*-butoxide were demonstrated to undergo alkylidene rotation at much higher rates than those containing electron-withdrawing ligands such as OR_{F6}.[17] Examples of five-coordinate imido-alkylidene-bis-alkoxide species can be obtained by addition of simple Lewis bases (Scheme 14.6).

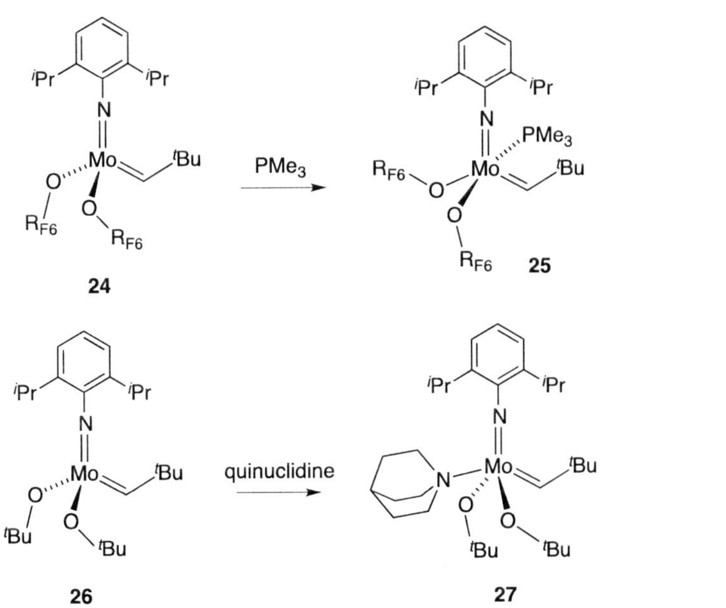

Scheme 14.6.

Coordination of bases to the CNO face of the molecule is generally considered to be a good model of olefin approach to the metal carbene although examples of base coordination to the NOO face are also known. These five-coordinate species are generally unreactive in olefin metathesis reactions due to their more electronically and coordinatively saturated nature.

The product of olefin addition to the alkylidene is a metallacyclobutane complex (Scheme 14.2). These species are generally unstable in molybdenum chemistry, although recent work has produced several isolable examples (see Section 14.3.2).[39] By contrast, metallacyclobutane complexes are the resting states during olefin metathesis with tungsten catalysts. With imido-alkylidenes of Mo and W, the metallacycles have been identified to possess two possible geometries, square pyramidal (SP) and trigonal bipyramidal (TBP). In either geometry, the imido ligand occupies an axial position. Recent experimental and computational work both favor a mechanism for olefin metathesis involving metallacyclobutanes possessing TBP geometries.[40,41]

A major advance in the olefin metathesis reaction was the introduction of asymmetry into the traditional catalyst architecture by use of chiral alkoxide or aryloxide ligands. Early work by Heppert demonstrated that racemic mixtures of C_2-symmetric binaphtholate ligands could be used to generate asymmetric tungsten alkylidenes.[42] These alkylidene-tetrakis-alkoxide species, structurally related to older catalysts first reported by Osborn,[43,44] did not prove to be very active for olefin metathesis in the absence of Lewis acids and displayed only modest stereoselectivity in the ROMP of norbornene. More successful asymmetric catalysts containing racemic and enantiomerically pure diolate ligands were reported later by Schrock and found to polymerize several substituted norbornadienes in a stereospecific fashion.[45] Subsequent work by Grubbs, and Schrock and Hoveyda described the application of such chiral catalysts (Scheme 14.7) to asymmetric RCM reactions.[46,47] These complexes have since proven successful in a number of different asymmetric olefin metathesis reactions,[48] and their fundamental reactivity with a variety of simple olefins has also been examined in detail.[49,50]

An interesting metathesis polymerization reaction effected by imido-alkylidene-bis-alkoxides is the cyclopolymerization of α,ω-heptadiynes. The

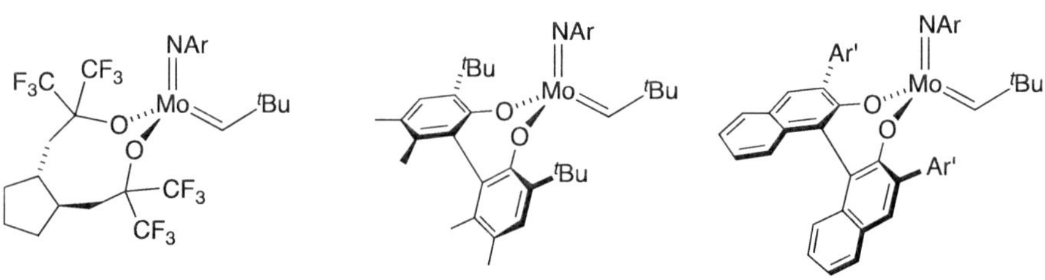

28 **29** **30**

Scheme 14.7.

living polymerization of diethyldipropargylmalonate by the prototypical Schrock catalyst **24** was first demonstrated in 1992, although examples using classical "black box" catalysts had been reported previously.[51] The photophysical properties of these polymers are intriguing given that they are soluble versions of substituted polyacetylenes. The cyclopolymerization reaction is capable of producing either five- or six-membered rings in the polymer backbone. The formation of these rings arises from α- versus β-addition of the initial alkyne group to the Mo alkylidene (Scheme 14.8).[52,53]

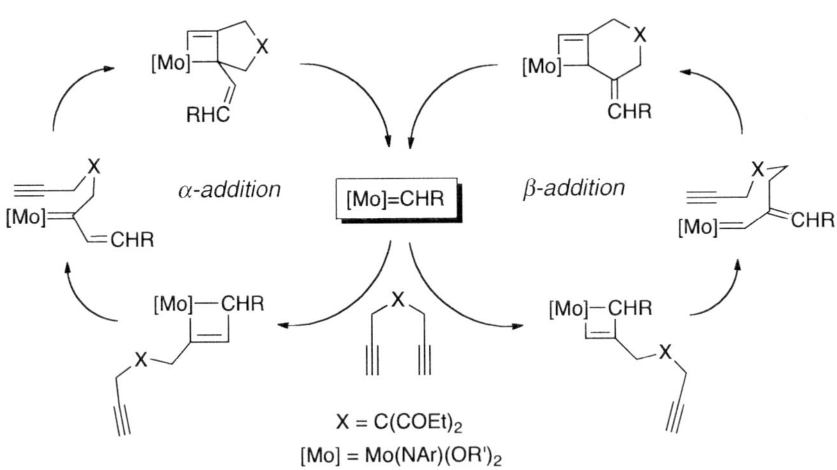

Scheme 14.8.

Use of catalyst **24** produces polymers with a mixture of five- and six-membered rings. In contrast, use of catalyst **26** produces polymers containing 99% five-membered rings.[54] Replacing the alkoxide ligand in the prototypical catalyst by carboxylate ligands gives rise to polymers containing 99% six-membered rings.[55,56] Recently, modification of the alkylidene ligand has lead to new compounds featuring carbene ligands that mimic the growing polymer chain (Fig. 14.2).[57] Stoichiometric reactions of these alkylidenes with various aldehydes generates polyenes that correspond to oligomeric fragments of the heptadiyne polymer.[58] These rational syntheses have allowed for an in-depth examination of the photophysical properties of polyene fragments.[59,60]

Other metathetical reactions of prototypical Mo imido-alkylidene-bis-alkoxide compounds with non-olefins include their reaction with phospha-alkynes to afford substituted phospha-metallacyclobutenes[61,62] and their reaction with imines to generate bis-imido species.[63,64] Addition of methylene triphenylphosphorane to the prototypical alkylidene **24** does not result in metathetical reactivity but rather in deprotonation to afford anionic alkylidyne **31** (Eq. 14.8).[65] This reaction is unique in that the alkylidene phosphorane is basic enough to deprotonate the alkylidene ligand but too poor of a

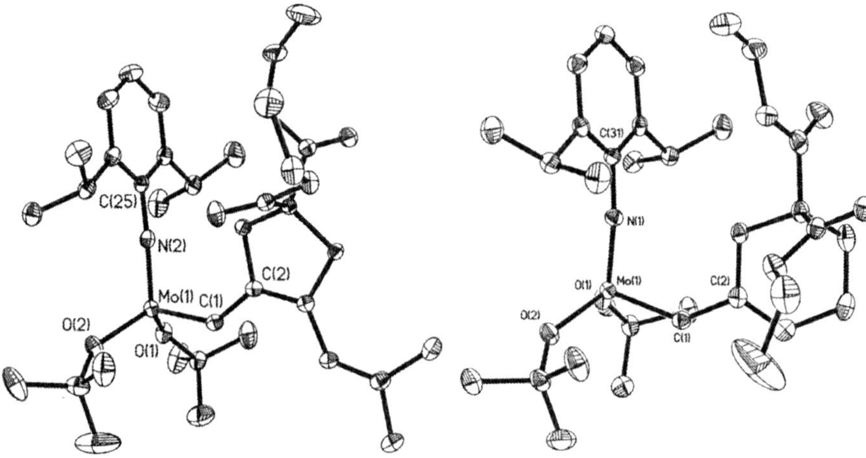

Figure 14.2. Structures of [Mo(=CH[N])(NAr)(O^{t}Bu)$_2$] (ring size (N)=5 and 6) featuring polymer-like alkylidenes. Reproduced with permission from Reference 57. Copyright 2006 American Chemical Society.

nucleophile to displace the alkoxide or form a simple base adduct. Anionic alkylidyne **31** features a substantially bent imido group consistent with less π-bonding between Mo and N due to the presence of the alkylidyne ligand. This species was not active for alkyne metathesis reactions but was found to display nucleophilic character at the imido nitrogen and could be reversibly protonated by addition of acid.

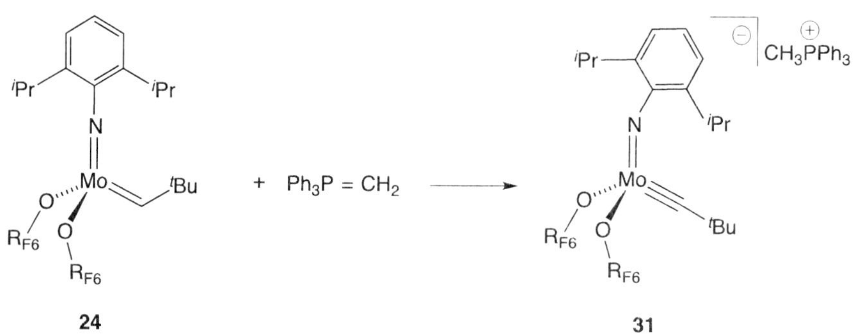

(14.8)

In related chemistry to that shown in Equation (14.8), Veige has demonstrated that deprotonation of a chelating bis-phenoxide ligand with methylene triphenylphosphorane can give rise to anionic alkylidene **32** (Eq. 14.9).[66] In this case, deprotonation occurs preferentially at the ancillary phenyl group as opposed to the alkylidene ligand. The metathesis reactivity of **32** was not reported, although it is unlikely that such a species would prove effective in catalysis due to its anionic five-coordinate nature.

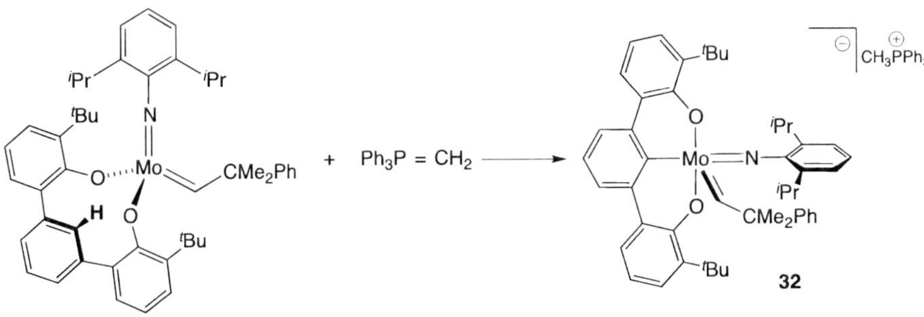

32

(14.9)

The sensitivity of most imido-alkylidene-bis-alkoxide compounds of Mo(VI) toward ambient conditions has prompted several investigations into stabilized alternatives. Odom has described a series of Mo alkylidenes containing tethers to either the imido ligand or an internal Lewis basic functionality (Scheme 14.9). The imido-tethered alkylidene **33** was structurally characterized, but no metathesis reactivity was reported.[67,68] The imino-alkylidene **34**, prepared by ring-opening of cyclooctyne, was competent for polymerization of norbornene when activated with $AlCl_3$.[69] Similar compounds featuring tungsten were also reported and shown to be active for carbonyl olefination.[70]

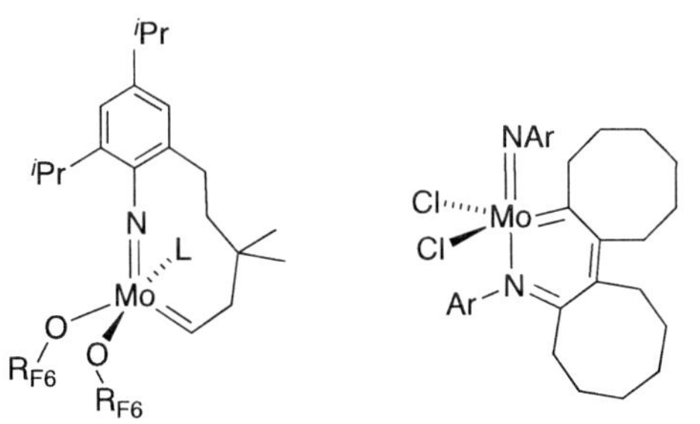

L = quinuclidine

33

Ar = 2,6-diisopropylphenyl

34

Scheme 14.9.

Fürstner has demonstrated that air-stable Schrock-type catalysts can be prepared by employing 2,2′-bipyridine or 1,10-phenanthroline as ligands with traditional four-coordinate alkylidenes (Eq. 14.10).[71] Heating of these compounds in the presence of $ZnCl_2$ regenerates the active catalysts, which are then capable of effecting olefin metathesis reactions. The presence of the bpy or phen ligands gives rise to six-coordinate 18-electron complexes such as **35**, accounting for the marked increase in stability.

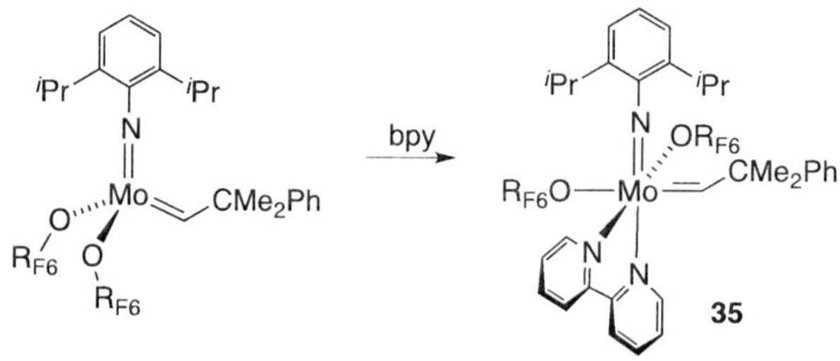

35

(14.10)

Although alkoxide ligands have played a dominant role in the development of Mo imido-alkylidene catalysts, examples not containing alkoxide ligands have also been investigated. Boncella prepared several early examples employing a chelating diamide ligand.[72] These compounds could be prepared from dialkyl precursors by thermolysis at 80 °C in the presence of PMe_3 without requiring precursor **22** (Eq. 14.11). Five-coordinate **36** is unreactive towards normal olefins, but dialkyl precursors of the type shown in Equation (14.11) were found to polymerize norbornene when thermolyzed prior to the addition of monomer. The lack of reactivity of these compounds is consistent with observations with other alkylidene-bis-amides. The π-basic nature of the amide ligands renders the alkylidene complex too electron rich to react with olefins in most cases.

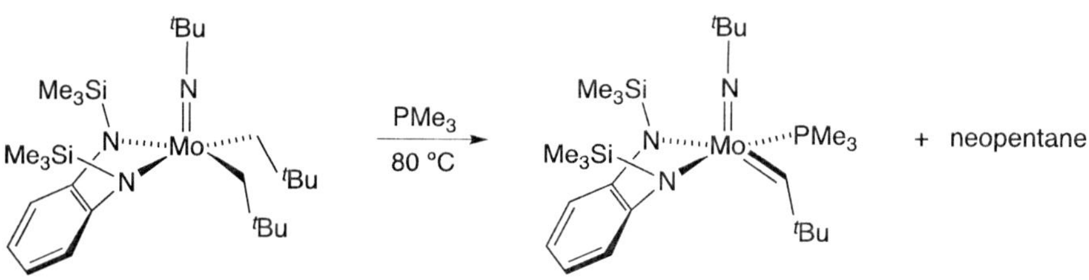

36

(14.11)

Cationic alkylidenes are an intriguing catalyst target given the precedent for such compounds to serve as very active catalysts.[73] Additional work by Boncella described a family of neutral and cationic imido-alkylidene compounds of Mo and W containing the tris-pryazolylborate (Tp) ligand.[74] The Tp ligand was introduced by a salt metathesis reaction of KTp with precursor **22**. In all cases, Tp-containing species were six-coordinate and demonstrated improved stability toward ambient conditions when compared to traditional four-coordinate Mo(VI) alkylidenes. Unfortunately, the compounds were unreactive towards olefins unless activated with $AlCl_3$.

More recently, cationic versions of imido alkylidene complexes possessing coordination numbers less than six have been prepared. These species contain β-diketonate and β-diketiminate ligands and can be isolated as five-coordinate solvent adducts (Scheme 14.10) upon salt metathesis with $NaBAr^f_4$ (Ar^f = 3,5-bis-(trifluoromethyl)phenyl).[75]

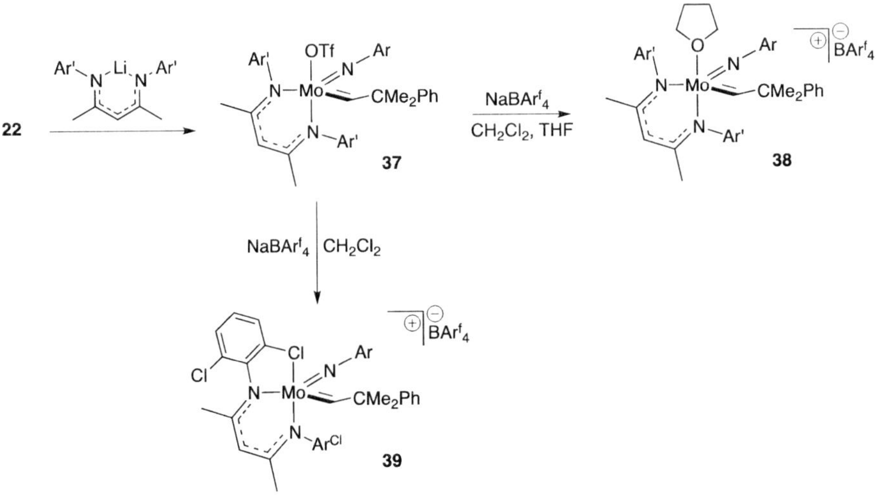

Scheme 14.10.

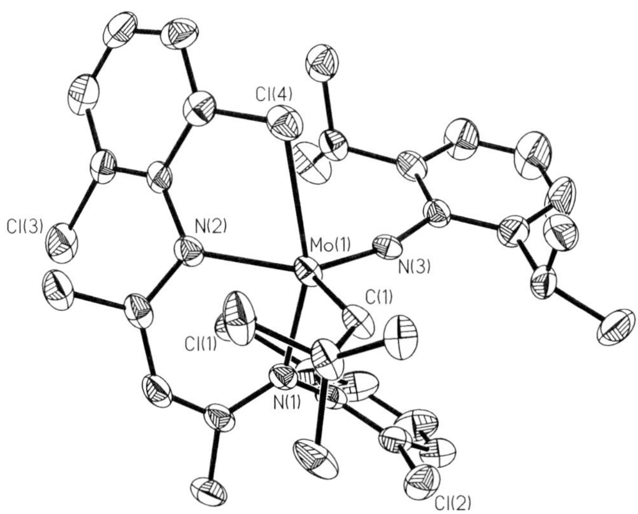

Figure 14.3. Structure of cationic alkylidene **39**. Adapted from Reference 75.

In the case of alkylidene **37**, which contains 2,6-dichlorophenyl substituents on the nitrogen atoms of the β-diketiminate ligand, a solvent-free cationic species could be isolated with an intramolecular M–Cl interaction (Fig. 14.3). Interestingly, this species exists as predominantly the *trans* isomer in contrast to related cationic alkylidenes such as **38**. Cationic species bearing β-diketonate or β-diketiminate ligands were not effective metathesis catalysts and displayed reactions with ethylene consistent with intermediate metallacyclobutanes being unstable with respect to β-H elimination.[76] Cationic complexes featuring pyrrolide ligands have also been reported but appear to suffer from the same instability as their diketonate and diketiminate counterparts.[77]

14.3.2 Mo(VI) Imido-Alkylidene Pyrrolide Compounds

In recent years, the primary focus in Mo alkylidene chemistry has moved away from catalysts containing two alkoxide ligands and centered on compounds that feature a pyrrolide ligand. Pyrrolide compounds of Mo imido-alkylidenes were first prepared with the intent of serving as precursors to bis-alkoxide species (e.g., **23**) through protonolysis of the Mo–N_{pyr} bond.[78] Earlier work had examined imido-alkylidenes featuring both dialkyl[79,80] and diamide ligands;[81] however, each of these systems proved undesirable for protonolysis reactions with various alcohols yielding in most instances mixed alkyl/alkoxide or amide/alkoxide species that were not very active towards olefin metathesis reactions. In contrast, use of pyrrolide complexes was found to result in smoother alcoholysis reactions demonstrating the utility of these species as universal precatalysts. Such species have also now been examined in combination with bipyridine ligands to afford air-stable versions of precatalysts analogous to compound **35**.[82]

Pyrrolide complexes are prepared by salt metathesis reactions of various pyrrolide salts with precursor **22**. In the case of unsubstituted pyrrole, the resulting imido-alkylidene complex forms a dimeric structure in the solid state with a single pyrrolic η^5-interaction per dimer (Scheme 14.11).[78] Use of substituted pyrrolides affords monomeric species that also feature a single η^5-interaction with the pyrrolide.[83] Work with these substituted pyrrolides, in particular 2,5-dimethylpyrrolide, revealed that mixed alkoxide/pyrrolide species could be isolated with select alcohols.[84] These so-called mono-alkoxide-pyrrolide (MAP) alkylidenes were demonstrated to be catalytically active themselves. In many

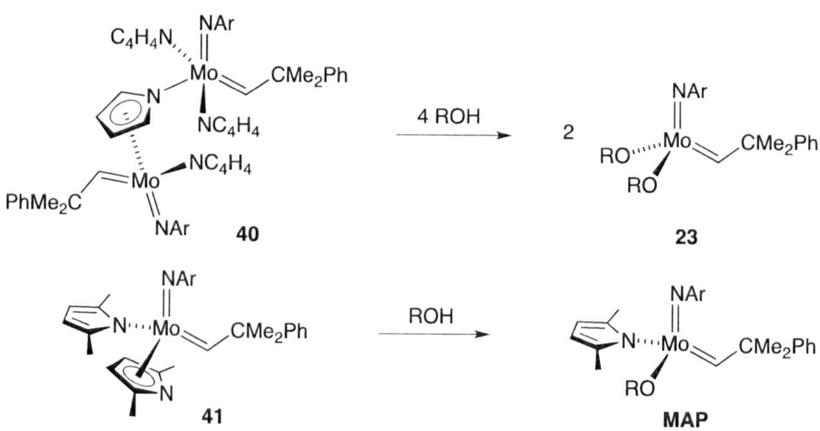

Scheme 14.11.

cases, this reactivity surpasses that of traditional bis-alkoxide-ligated olefin metathesis catalysts both in terms of activity and selectivity. In addition, these species are chiral at the metal as opposed to simply containing a chiral auxiliary group. Thus, the two faces of the alkylidene are differentiated to a greater extent than in traditional imido-alkylidenes featuring asymmetric diolates (e.g., **29**).

Theoretical work has highlighted the advantages of these so-called "stereogenic-at-metal" catalysts in promoting productive metathesis. Calculations by Eisenstein demonstrate that asymmetry at the metal favors the distortion from tetrahedral geometry required for binding of olefin and formation of the intermediate metallacyclobutane.[40] As a result of the differing electronics of pyrrolide versus alkoxide, MAP catalysts are rendered even more efficient than their bis-alkoxide progenitors. Furthermore, the pyrrolide ligand imparts added stability to intermediate metallacyclobutanes disfavoring decomposition via β-H elimination (Scheme 14.3).[85]

Mono-alkoxide-pyrrolide catalysts of molybdenum such as those shown in Scheme 14.12 have allowed for new types of olefin metathesis reactions hitherto unrealized in Group 6 chemistry. Enyne metathesis was reported using a series of MAP catalysts including **42**, which features a bulky 2,6-diisopropylphenoxide ligand (Eq. 14.12).[84] The enyne reaction is not possible with traditional imido-alkylidene-bis-alkoxide catalysts due to uncontrolled polymerization of the alkyne moiety. Variation of the alkoxide ligand leads to enyne products consistent with differing selectivities for initial α- or β-addition of the alkyne to the alkylidene species (Scheme 14.8).

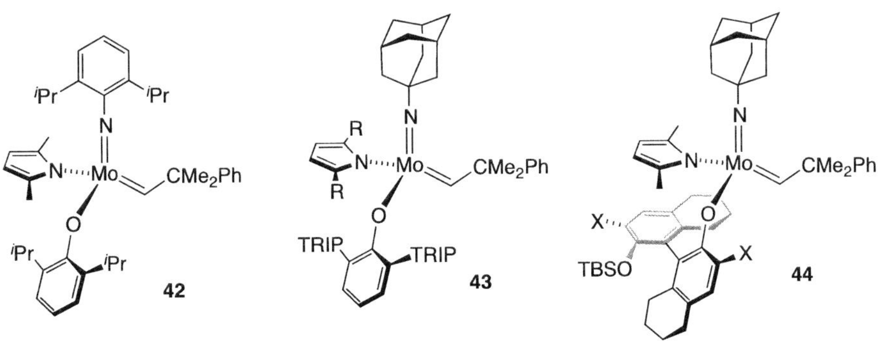

Scheme 14.12.

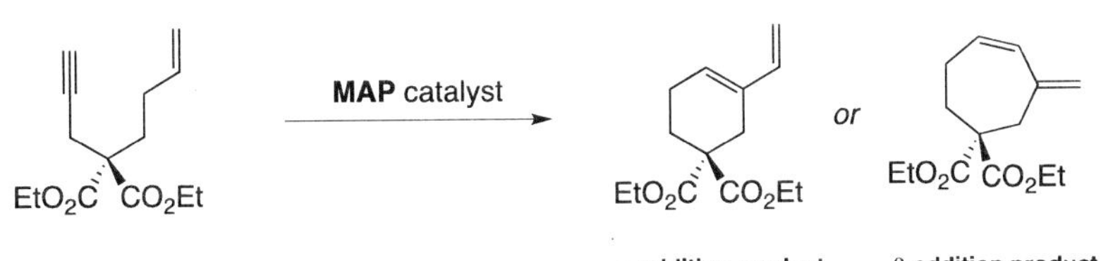

(14.12)

A long-standing problem in the field of olefin metathesis is the regioselective metathesis of olefins to give pure *E* or *Z* products.[86] Here as well, MAP catalysts have provided a breakthrough, allowing for the first *Z*-selective metathesis reactions to be carried out.[87] This selectivity has also allowed for the observation of new polymer microstructures. Catalyst **43**, which contains a small 1-adamantyl imido ligand coupled with the very bulky 2,6-bis(2,4,6-triisopropylphenyl)phenoxide (HIPTO), was found to polymerize norbornene derivatives in a *Z*-selective fashion to give all *cis*-syndiotactic polymer.[88] This polymer microstructure was unknown with all previous bis-alkoxide catalysts. An analog to catalyst **43** featuring a bulkier imido ligand, 2,6-diisopropylphenylimido, was also reported to effect the homocoupling of (*E*)-1,3-dienes to *E*,*Z*,*E*-trienes in high yield with *Z*-selectivities >95%.[89]

The use of enantiomerically pure phenoxide ligands in MAP chemistry has resulted in new examples of asymmetric RCM and CM reactions. Mono-protected biphenolate ligands have figured prominently in these systems. In addition to being stereogenic-at-metal, MAP catalysts containing enantiomerically pure alkoxide ligands possess a second stereocenter (e.g., **44**). As a result, diastereomeric metal complexes are generated upon protonation of bis-pyrrolide precursors with the enantiomerically pure alcohol. As expected, these diastereomeric catalysts are not formed in equal ratios. However, in certain instances the olefin metathesis reactivity of the two diastereomers is very different, and asymmetric metathesis reactions can be effected with very high enantiomeric excesses and regioselectivities by simply using a mixture of catalyst diastereomers. In fact, rapid epimerization at the metal center through degenerate metathesis with ethylene (the by-product of metathesis reactions) has been invoked as a necessary means of producing the reactive diastereomer in enantioselective reactions.[41]

Mono-alkoxide-pyrrolide catalysts featuring mono-protected biphenolate ligands have been exploited to great extent in natural product synthesis,[90] allowing for the preparation of a variety of key alkene linkages.[91–93] Functionalized olefins have also proven successful as substrates for enantioselective metathesis with MAP catalysts.[94] Metathesis reactions in the presence of ethylene using catalyst **44** and derivatives thereof can produce terminal olefins through a reaction constituting a formal "ethenolysis" of the internal alkene.[95] Remarkably, when mixtures of *E* and *Z* olefins were subjected to this ethenolysis reaction, the *Z* isomers were transformed, leaving the *E* olefins unchanged.[96] As with the ROMP reactions discussed earlier, the *Z*-selectivity with these catalyst systems is proposed to arise from a combination of a small imido group and a large alkoxide. Such a combination is believed to force the substituents in the metallacyclobutane intermediate to position themselves on one face of the MoC_3 ring resulting in exclusively *Z*-olefins upon cycloreversion (Scheme 14.13).[87]

The use of pyrrolide ligands in Mo(VI) imido-alkylidene chemistry has also permitted the isolation of traditionally unstable species such as methylene compounds and metallacyclobutanes.[39,95] The aforementioned bulky alkoxide ligands derived from terphenyl groups have been particularly instrumental in this chemistry. As a result, examples of Mo methylidenes and unsubstituted metallacyclobutanes have

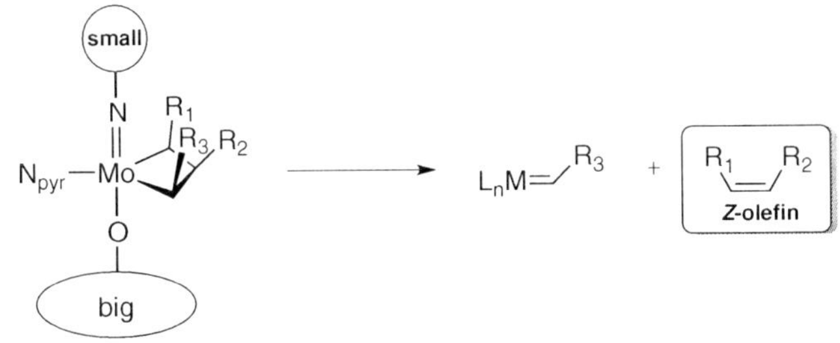

Scheme 14.13.

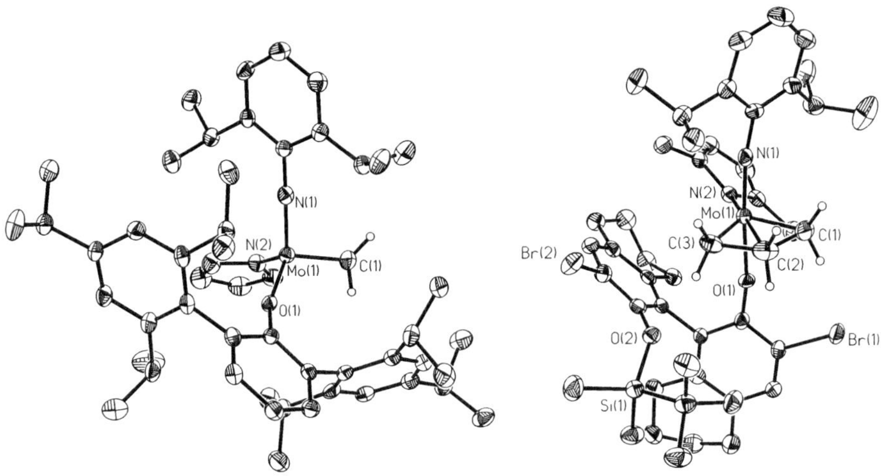

Figure 14.4. Structures of Mo methylene (left) and metallacyclobutane (right) complexes featuring pyrrolide ligands. Adapted from References 39 and 95.

been isolated and subjected to X-ray diffraction analysis (Fig. 14.4). These species have also allowed for unprecedented depth in spectroscopic studies concerning olefin binding and exchange with alkylidene/metallacycle species. Findings from these studies demonstrate that the stability imparted by pyrrolide ligands to intermediate methylidene species is key to their catalytic efficiency.

Pyrrolide complexes of Mo imido-alkylidenes are now the preferred catalysts for many metathesis reactions. The high degree of modularity afforded by traditional Mo catalysts coupled with the favorable attributes of the pyrrolide ligand promises to reveal an even greater number of new and selective metathesis reactions in coming years.

14.3.3 Other Nucleophilic Mo Carbenes

Although less common, nucleophilic carbenes of molybdenum not containing imido ligands have been prepared. Legzdins has described the chemistry of a family of Mo alkylidene compounds containing cyclopentadienyl and nitrosyl

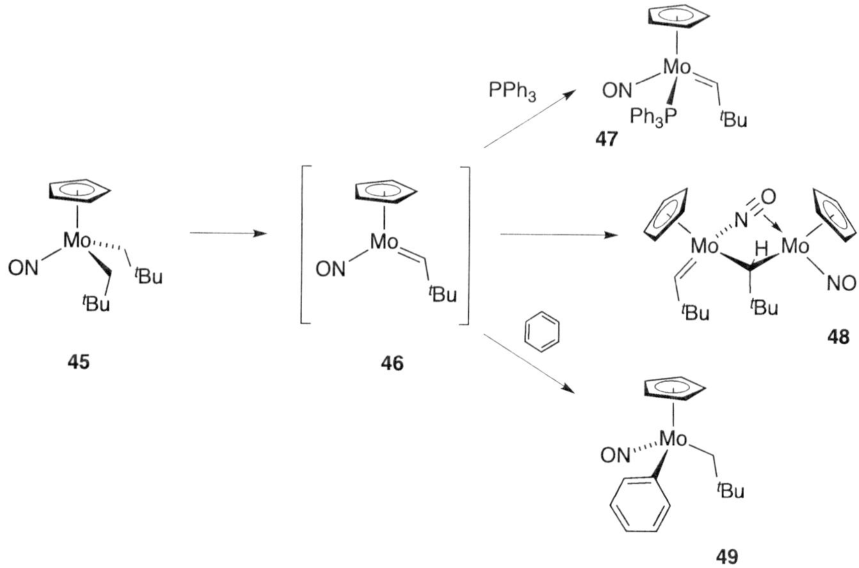

Scheme 14.14.

ligands (Scheme 14.14).[97] These species do not contain Mo in its highest formal oxidation state and therefore do not conform to the traditional mold of nucleophilic carbene complexes. However, such compounds are formed via α-hydrogen abstraction and can participate in C–H activation reactions reminiscent of high-oxidation state Cr alkylidenes (see Section 14.2.1). Both these observations argue for their consideration as truly nucleophilic carbenes. As shown in Scheme 14.14, dineopentyl molybdenum complex **45** undergoes thermally induced α-H abstraction near room temperature leading to the putative three-coordinate alkylidene **46** that can undergo a variety of different reactions including trapping by phosphine (**47**), dimerization (**48**), and C–H activation (**49**).[98,99] The reactivity of an analog of compound **46** containing the bulkier Cp* ligand has also been investigated with olefins. This work has established that [2+2] cycloaddition to form a molybdacyclobutane is the first step in the reaction of such alkylidenes with olefins. Productive metathesis from the metallacycle does not occur, however, and C–H activation reactions (β-hydrogen abstraction) take place leading to formation of π-allyl and diene complexes.[100,101] The tungsten congener of **46** has also been examined in detail and shown to display similar C–H activation chemistry.[102] Of note with tungsten, both *syn* and *anti* rotational isomers of the alkylidene ligand (with respect to Cp) can be identified, suggesting a further commonality with the high-oxidation state alkylidenes discussed earlier.[103]

Schrock has reported alkyne-alkylidene compound **51** supported by a diamido-amine ligand.[104] This species is prepared by α-hydrogen abstraction from Mo(IV) dialkyl **50** upon treatment with 2-butyne. Viewing the alkyne ligand as a metallacyclopropene results in a formal oxidation state for Mo of

6+. In this view, alkylidene **51** is related to base adduct **25** discussed earlier. Reactivity of dialkyl **50** with other oxidants such as cyclohexene sulfide and hexachloroethane afforded non-alkylidene species **52** and **53**, which were proposed to result from initial α-hydrogen abstraction to from a Mo(VI) alkylidene, followed by alkylidene coupling (Scheme 14.15).

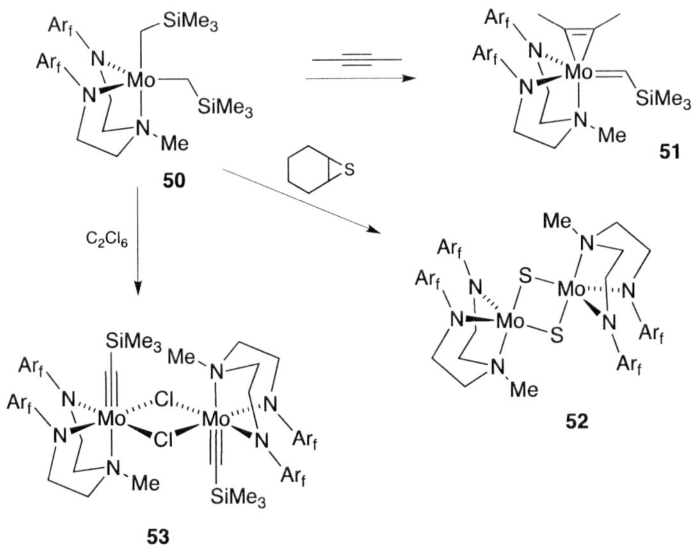

Scheme 14.15.

14.3.4 Supported Mo Alkylidenes

Heterogeneous versions of Mo(VI) alkylidene catalysts have been prepared by several research groups.[105,106] Early work by Schrock and Hoveyda described the free-radical polymerization of a styrene-appended biphenolate ligand which, after attachment to Mo(VI) with precursor **22**, afforded a polystyrene-supported metathesis catalyst.[107] These catalysts demonstrated comparable activity and stereoselectivity to the parent homogeneous catalyst **29** but had limited recyclability. In further work, Buchmeiser demonstrated that polynorbornene-appended biphenolate ligands could be synthesized by a Ru-catalyzed ROMP reaction prior to metallation with precursor **22**.[108] Mo catalysts bearing these polynorbornene-derived ligands displayed similar activity to those based on the polystyrene ligands and could be further integrated into polymer monoliths. These monoliths could then be conveniently sectioned to generate individual "catalyst pieces" for use in high-throughput screening.

In a related study, Schrock and Hoveyda demonstrated that ROMP of a diolate-appended norbornene group could result by formation of an active catalyst *in situ*.[109] Formation of these polymer-encapsulated catalysts (**55**) occurred upon substitution of the triflate ligand in **22** with the diolate precursor **54**

(Eq. 14.13). As with other polymer-appended catalysts, these polynorbornene-encapsulated catalysts were active for a variety of metathesis reactions, demonstrating comparable activity and selectivity to the homogeneous systems.

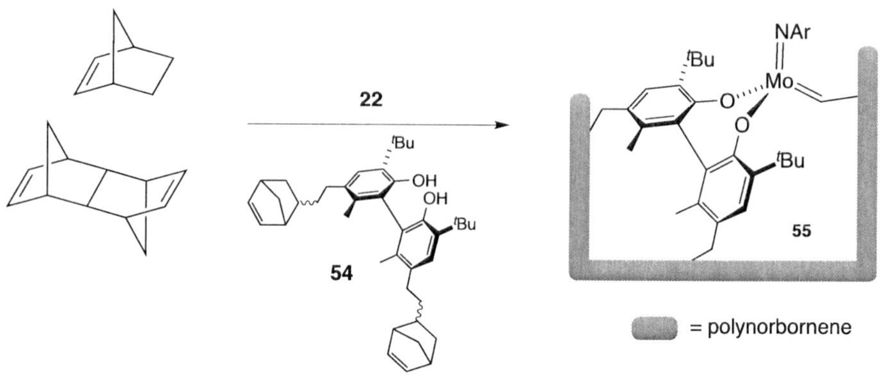

(14.13)

Buchmeiser has also reported a series of polymer-supported Mo alkylidenes by engineering of the imido ligand.[110] This strategy allows greater modification of the catalyst architecture because it does not limit the choice of alkoxide ligand. Alkylidene **56** was prepared by the general synthesis shown in Scheme 14.5 starting from an aniline containing a halo-alkyl chain in the *para* position of the benzene ring. Tethering of this alkyl chain to a polymer through a sulfonate ester linkage produced supported catalysts **57** and **58** (Scheme 14.16). As with catalysts of type **55**, these polymer-encapsulated catalysts demonstrated modest stability to metal leaching and proved efficient in asymmetric RCM reactions in the case of **58**.

In addition to polymer supports, more recent work has examined the use of silica as a platform for Mo and W alkylidene catalysts. Basset and Copéret have described in detail a series of alkylidene compounds covalently linked to partially dehydrated silica. These species are prepared by protonolysis reactions employing imido-alkylidene precursors featuring basic ligands (Eq. 14.14).[111]

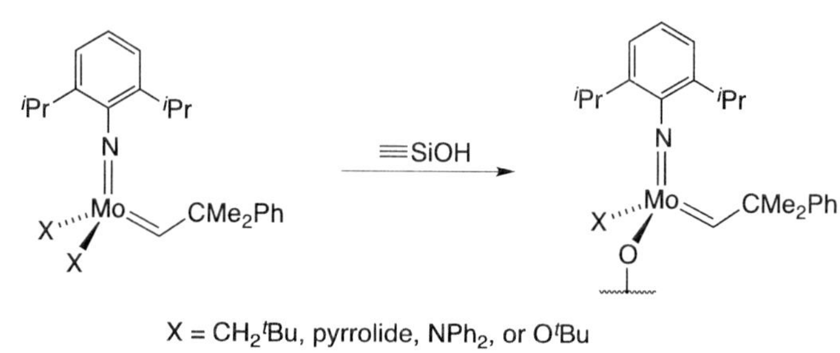

(14.14)

A variety of different alkylidene species have been reported by this method and examined for metathesis activity with terminal and internal olefins. With

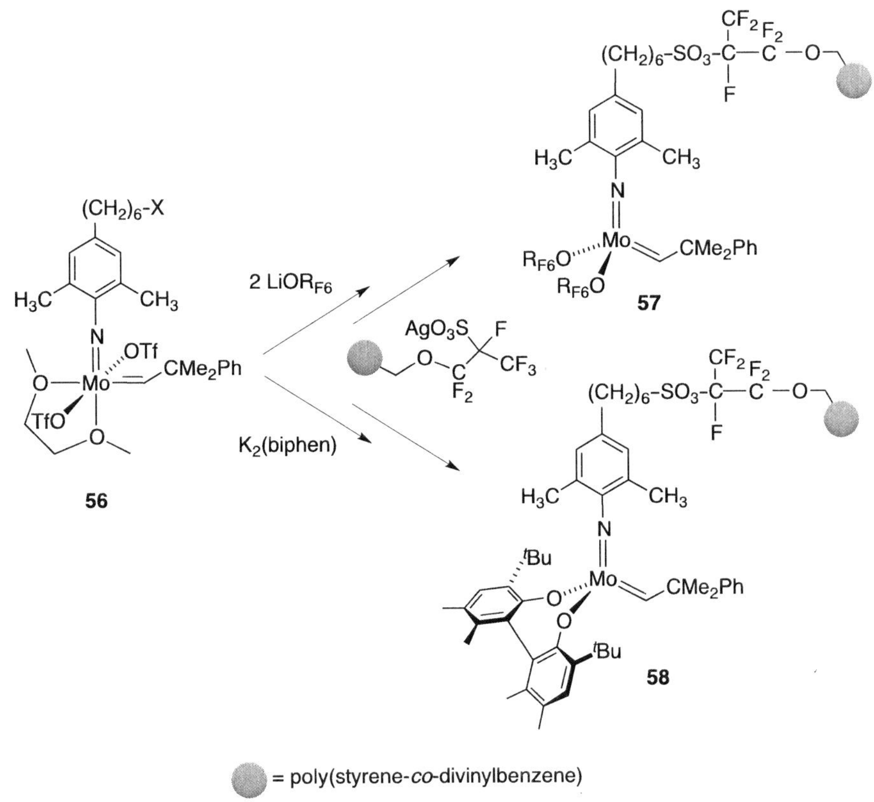

Scheme 14.16.

all precursors except X = 2,5-dimethylpyrrolide, protonolysis only removes one of the X-type ligands due to the limited availability of ≡SiOH groups within the local environment of alkylidene complex. With the 2,5-dimethylpyrrolide derivative, protonation of both the pyrrolide and alkylidene ligands is observed leading to a mixture of active and inactive species, respectively. Despite this fact, silica-supported catalysts containing pyrrolide ligands give rise to very efficient catalysts.[85] Amide and alkoxide ligands have also proven effective and show superior performance to those catalysts containing alkyl ligands (e.g., X = $CH_2{}^tBu$ in Eq. 14.14).[112,113] Solid-state NMR studies of the analogous supported tungsten catalysts have identified important intermediates including methylene species and unsubstituted metallacyclobutanes.[114,115] All studies to date support a mechanism for the supported catalysts that is analogous to that believed to be operative in solution with homogeneous alkylidene compounds.[116] Work by Trunschke has also demonstrated that heterogeneous precursors such as molybdenum oxide dispersed on silica can lead to catalysts related to those shown in Equation (14.14) in the presence of propylene.[117]

14.4 TUNGSTEN CARBENES

Nucleophilic carbene complexes of Group 6 were first realized in tungsten chemistry and this heaviest congener can boast the largest variety of alkylidene complexes reported to date. Due to their higher metal–ligand bond strengths, tungsten complexes are typically preferable to molybdenum for fundamental studies of alkylidenes and related species such as metallacyclobutanes and olefin complexes. Despite this fact, tungsten olefin metathesis catalysts have generally been neglected in favor of their molybdenum counterparts. For many years, this preference for Mo was ascribed to the greater stability of intermediate W metallacyclobutane species, which are slow to release olefin thereby limiting catalytic activity. In addition, W(VI) imido-alkylidene-bis-alkoxides are less tolerant than the corresponding Mo compounds toward certain functional groups such as esters. Together, these factors provided one of the principle motivations for development of Mo olefin metathesis catalysts.[118] Despite these perceived shortcomings, tungsten alkylidene chemistry has witnessed a resurgence in recent years due to new catalyst designs, in particular, those incorporating a pyrrolide ligand. Such progress has demonstrated that tungsten catalysts can be both efficient and tolerant of certain functional groups. In this section, focus will be placed on recent developments in tungsten alkylidene chemistry. Classical tungsten alkylidenes have been the subject of several extensive reviews.[119]

14.4.1 W(VI) Imido-Alkylidene-bis-Alkoxide Compounds

Imido-alkylidenes of tungsten have traditionally been examined in tandem with their molybdenum counterparts as catalysts for olefin metathesis. As a result, many of the compounds discussed in the previous sections have analogs containing W in place of Mo. The synthesis of tungsten imido-alkylidenes can be accomplished by a preparative route analogous to that in Scheme 14.5 employing WO_2Cl_2 as a starting material instead of molybdate. This route, however, is not amenable to the synthesis of alkylimidos of tungsten. Recent work by Schrock has identified an alternative route to alkylimido-alkylidenes of tungsten employing the reaction sequence in Scheme 14.17.[120] In this scenario, addition of a silylated amine directly to WCl_6 affords the chloride-bridged dimeric bis-imido complex **59**, first reported by Nielson.[121] Alkylation of this species with neopentyl or neophyl Grignard leads to bis-imido-dialkyl species **60**. As observed with select bulky imido ligands in Mo chemistry, these dialkyl species are not suitable for reactions with triflic acid as shown in Scheme 14.4. Instead, dialkyl **60** must be reacted with pyridinium chloride to afford the desired alkylidene **61**. Compound **61** serves as a precursor for various tungsten alkylimido-alkylidenes featuring alkoxide and pyrrolide ligands, a catalyst type heretofore unavailable through the synthetic pathway in Scheme 14.5.

In another interesting modification to the traditional catalyst synthesis, pentafluorophenylimido groups were introduced to tungsten by a metathesis

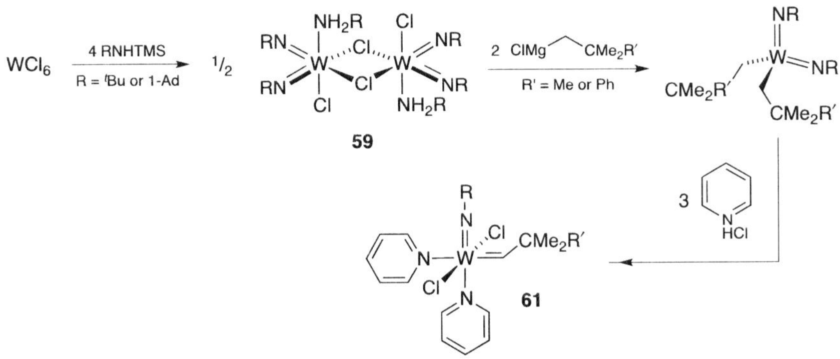

59

Scheme 14.17.

reaction between WO_2Cl_2(DME) and a sulfinylamine (Eq. 14.15).[122] This reaction, reported by Sundermeyer, generates SO_2 as a by-product, which must be removed under a nitrogen stream during reaction.[24] The resulting bis-imido complex **62** can then be used in the traditional reaction sequence to afford tungsten catalysts bearing the pentafluorophenylimido ligand.

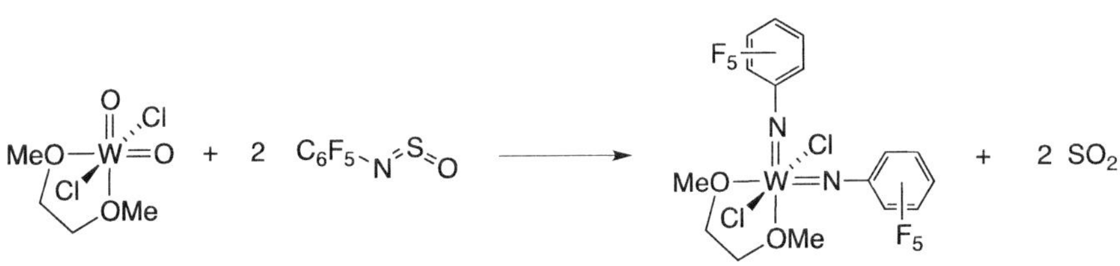

62

(14.15)

14.4.2 W(VI) Imido-Alkylidene Pyrrolide Compounds

As with molybdenum, pyrrolide ligands have featured prominently in the contemporary chemistry of tungsten imido-alkylidenes. Like their Mo counterparts, W bis-pyrrolide complexes can be used as precursors to bis-alkoxides or MAP species by treatment with various alcohols. Additionally, bis-pyrrolide supported tungsten alkylidenes also display unique reactivity (Scheme 14.18). Compound **63** was found to react directly with ethylene to afford methylidene **64**, which could be isolated and crystallized.[123] This species displays reversible degenerate metathesis in the presence of excess ethylene demonstrating that certain bis-pyrrolide complexes can react with olefins. Compound **63** can also be protonated with ($HNMe_2Ph$)(BAr^f_4) to afford a cationic alkylidene complex. Interestingly, protonation of **63** occurs at the pyrrolide carbon to afford the imine-bound **66**. In contrast to complex **63**, complex **67** containing the unsubstituted pyrrolide ligand was found to undergo bimolecular coupling in the presence of ethylene to afford compound **68** (Eq. 14.16). In this system, the

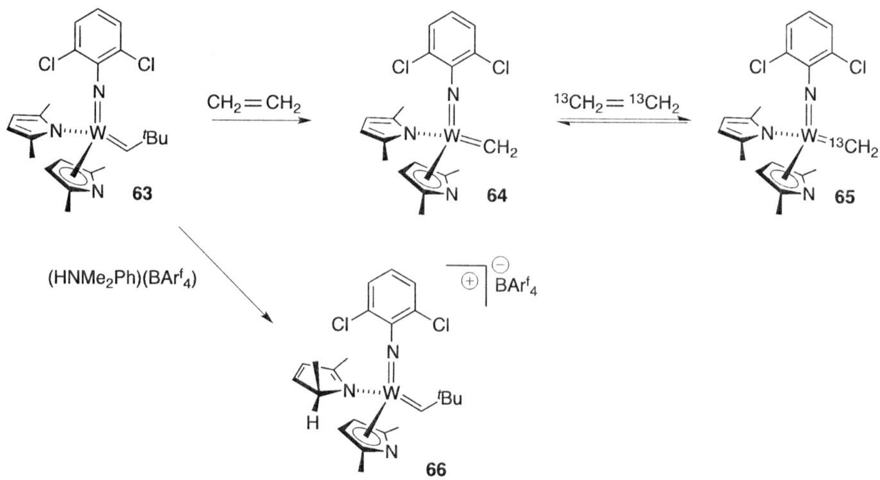

Scheme 14.18.

decreased steric bulk of the parent pyrrolide ligand is not sufficient to stabilize the resulting methylidene species.

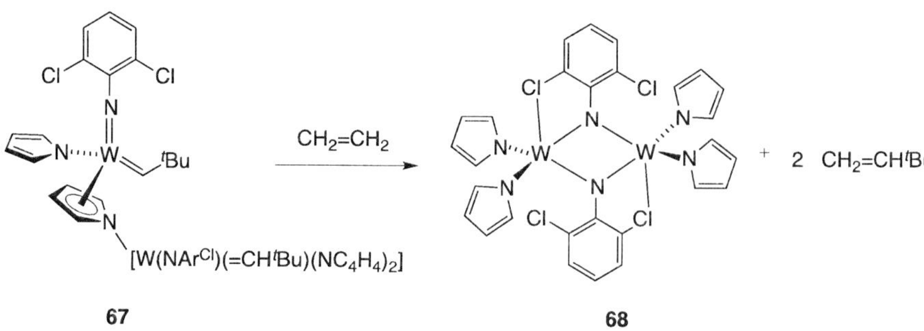

(14.16)

Mono-alkoxide-pyrrolide complexes of tungsten have also been the subject of recent investigations due to their catalytic activity in olefin metathesis reactions. Several studies have addressed the role of olefin binding and exchange to afford metallacyclobutanes.[39,124,125] As observed with traditional imido-alkylidene-bis-alkoxides, tungsten MAP species display much slower rates of metallacyclobutane break up than their molybdenum counterparts. Interestingly, however, methylidenes of tungsten MAP complexes were found to display much higher rates of alkylidene rotation than their Mo analogs.[39]

The homocoupling of terminal olefins has been especially successful with tungsten MAP catalysts (Eq. 14.17).[126] Reaction of several tungsten MAP catalysts with a variety of terminal olefins was found to give the homocoupled products in good yield with high *Z*-selectivity.[127] As with Mo MAP compounds, the unique combination of a large alkoxide ligand and a small imido group is believed to be responsible for the observed selectivity.

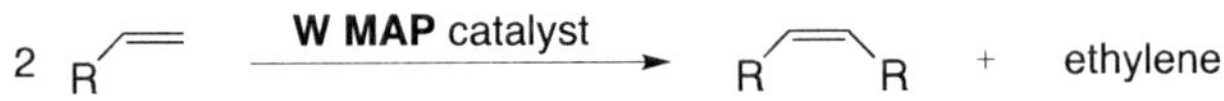

(14.17)

Tungsten MAP compounds have also proven effective as catalysts for the enyne reaction.[128] In this regard, tungsten catalysts actually outperform their molybdenum congeners in terms of regio- and enantioselectivity. For example, tungsten catalyst **69** (Scheme 14.19) was found to give twofold better selectivity for the β-addition product (Scheme 14.8) than the analogous molybdenum catalyst. Chiral versions of tungsten MAP catalysts, such as **70**, were also found to give dramatically better enantiomeric excesses in enyne reactions of prochiral substrates.

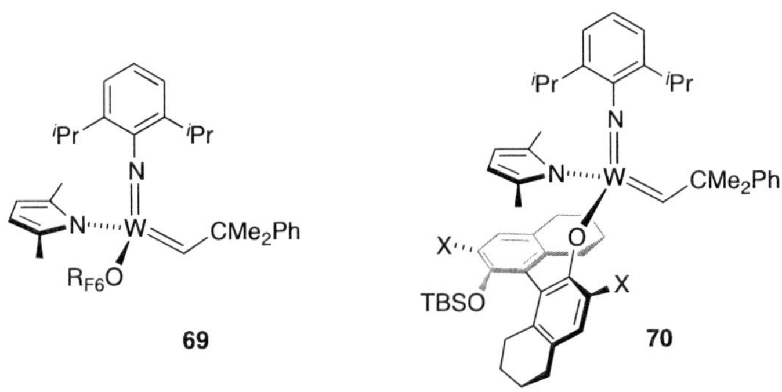

Scheme 14.19.

Beyond imido-alkylidenes, the pyrrolide ligand has provided new directions in the chemistry of tungsten oxo-alkylidenes. These species were prominent in the early development of tungsten alkylidene chemistry, but as mentioned, the small size of the oxo ligand was ill-suited to the stabilization of low-coordinate alkylidenes. Recently, however, four-coordinate oxo-alkylidenes of tungsten have been prepared by using a combination of very bulky phenoxide ligands and pyrrolides.[129] Compound **73** was prepared by successive salt metathesis reactions from precursor **71** to afford the sole example of a four-coordinate tungsten oxo-alkylidene (Eq. 14.18). Compound **74**, a derivative of complex **72** containing the slightly smaller 2,6-bis(2,4,6-trimethylphenyl)phenoxide (OHMT) ligand, was active for the *Z*-selective homocoupling of several terminal olefins and for the formation of *cis*, *syndiotactic*-polynorbornenes. Addition of $B(C_6F_5)_3$ to **74** resulted in a more active catalyst for 1-octene homocoupling, but the resulting coupled olefin (7-tetradecene) displayed a 4:1 mixture of *E* to *Z* isomers. Stoichiometric reactions of **74** with 2 equiv of borane afforded the borane-appended oxo-alkylidene **75** that was crystallized and shown to display a very long B–O bond distance (Fig. 14.5).

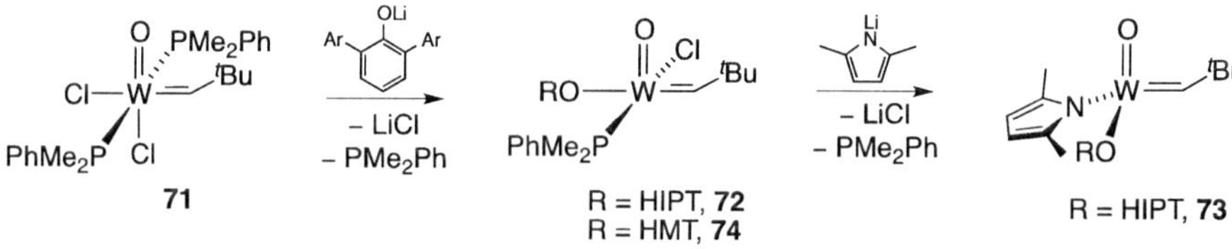

(14.18)

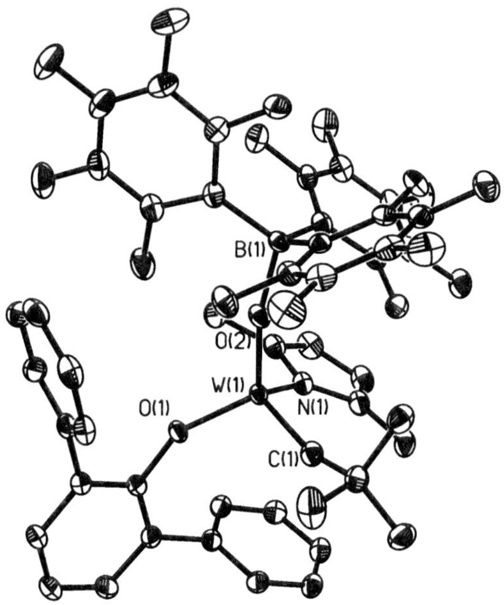

Figure 14.5. Solid-state structure of compound **75**. Adapted from Reference 129.

14.4.3 Other Nucleophilic Tungsten Carbenes

Unlike nucleophilic molybdenum carbenes, many examples of such species in tungsten chemistry are found without an imido (or oxo) ligand. Prior to the development of the synthetic sequence displayed in Scheme 14.5, imido-alkylidene-bis-alkoxides of tungsten (**77**) were prepared by a synthetic sequence involving alkylidyne precursor **76** (Scheme 14.20).[15] Addition of dimethylphosphinoethane (dmpe) to **76** results in formation of the well-known alkyl-alkylidene-alkylidyne complex **78**.[130]

In the absence of added bases, alkylidyne **76** is known to undergo relatively rapid scrambling of the alkyl and alkylidyne ligands.[131] This scrambling was originally proposed to occur through a bis-alkylidene-dialkyl intermediate.[132] Work over the last decade by Xue has described this chemistry in detail for a variety of hydrocarbyl- and silyl-derived ligands.[133–135] These studies have unequivocally demonstrated the intermediacy of bis-alkylidene species as intermediates in such scrambling reactions and revealed the kinetic and thermodynamic factors

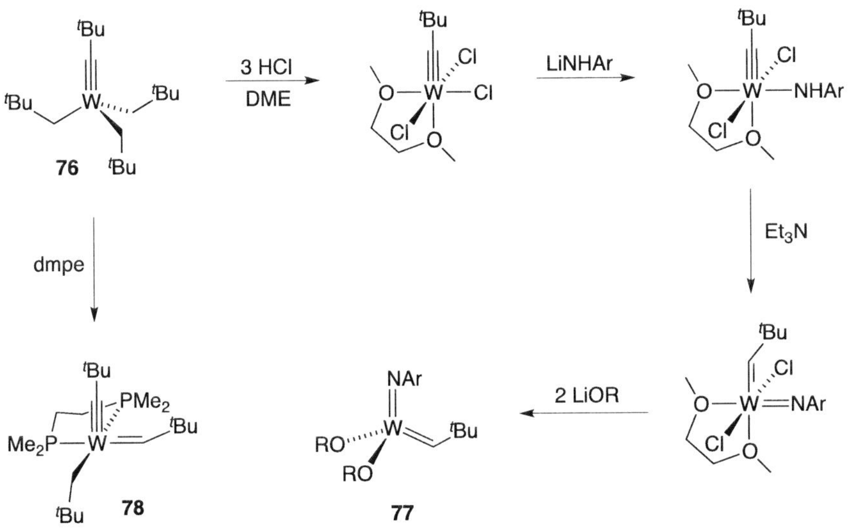

Scheme 14.20.

governing the ligand exchange. As a prototypical example, alkylidyne **79** was found to transform into an equilibrium mixture of alkylidyne **80** and bis-alkylidene **81** upon addition of PMe_3 (Eq. 14.19).[135] Compound **80** displays two alkylidene resonances by ^1H NMR spectroscopy consistent with a *syn*,*anti* conformation of the two alkylidene ligands with respect to each other. The solid-state structure of **80** shows two shortened W-C contacts also consistent with its formulation as a bis-alkylidene, although the structure suffers from disorder precluding a rigorous examination of the metric parameters.[133] Addition of a second equivalent of PMe_3 to **80/81** affords the alkyl-alkylidene-alkylidyne compound similar to **78** but with a *trans* disposition of the two phosphine ligands.[136] Formation of the alkylidene-alkylidyne species was demonstrated to occur by rate-limiting α-H abstraction prior to coordination of phosphine.

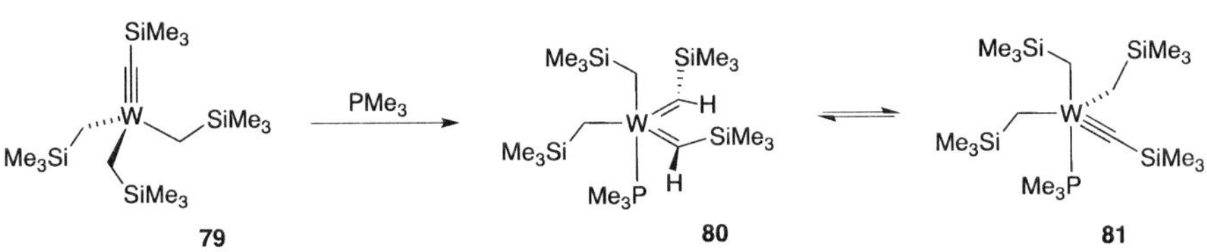

(14.19)

Replacement of one of the neopentyl ligands in compound **76** with a bulky *tert*-butyldiphenylsilyl ligand affords a species that exists in equilibrium as a four-coordinate bis-alkylidene/alkylidyne (**82/83** in Scheme 14.21). In the case of compound **82**, however, NMR spectroscopy is consistent with the *anti*, *anti*

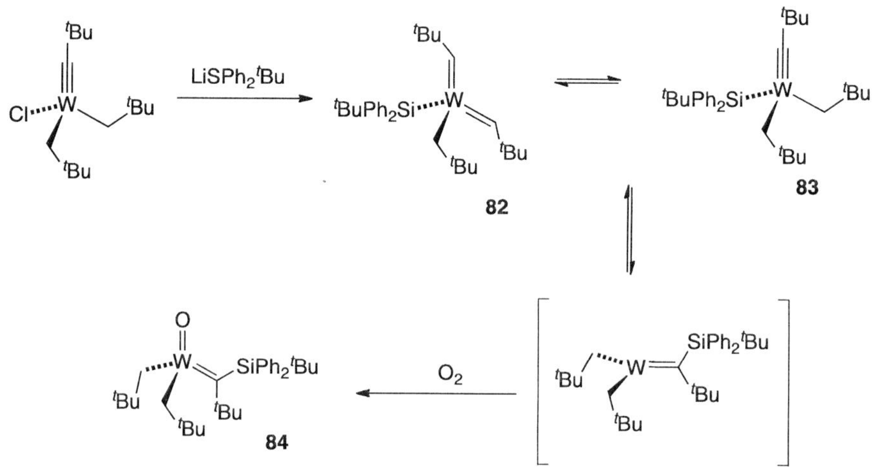

Scheme 14.21.

conformation of the bis-alkylidene.[134] Interestingly, reaction of **82/83** with molecular oxygen affords W(VI) oxo-alkylidene **84**. Computational studies of this reaction favor a pathway involving silyl migration to form a transient triplet W(IV) alkylidene prior to trapping by O_2.

In an effort to prepare highly reactive alkylidyne species, Veige has described a family of tungsten alkylidene compounds containing a trianionic OCO pincer ligand.[137] These species are prepared from alkylidyne precursors by addition of diphenol **85** (Scheme 14.22). Reaction of **85** with alkylidyne **86**[138] occurs

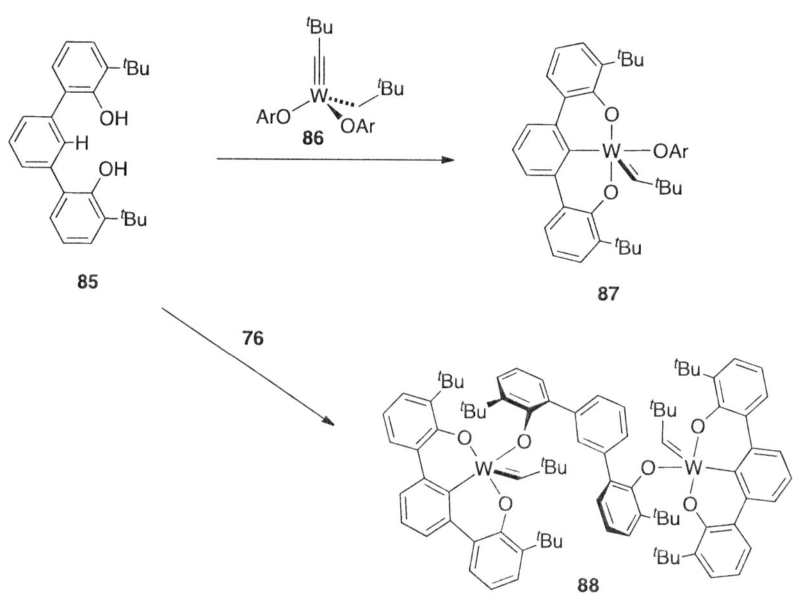

Scheme 14.22.

smoothly to yield alkylidene **87**, which possesses a five-coordinate distorted square pyramidal geometry with the alkylidene ligand occupying the axial position. Use of alkylidyne **76** in place of **86** results in complete consumption of all three alkyl ligands and formation of bimetallic species **88**, in which the pincer ligand now bridges two tungsten centers.

Replacement of the OCO pincer ligand in Scheme 14.22 with the more basic trianion derived from diol **89** affords the tungsten alkylidene species **91** after reaction with alkylidyne **90** (Scheme 14.23).[139] Alkylidene **91** can be deprotonated by methylene triphenylphosphorane to afford the nucleophilic alkylidyne **92**. Compound **92** reacts with electrophiles at carbon, which stands in contrast to previous anionic alkylidynes such as **31** that demonstrate nucleophilic reactivity at the ancillary ligands. Thus, reaction of **92** with methyl triflate occurs at the alkylidyne α-carbon to produce a rare example of a secondary alkylidene **93**. With the bulkier electrophile, TMSOTf, initial electrophilic attack is proposed to occur at the oxygen atom of the *tert*-butoxide ligand triggering isobutylene expulsion and formation of alkylidene **94**. Despite such interesting nucleophilic reactivity, no olefin metathesis activity was observed for any of the pincer-ligated alkylidene species in Schemes 14.22 and 14.23. A lack of such

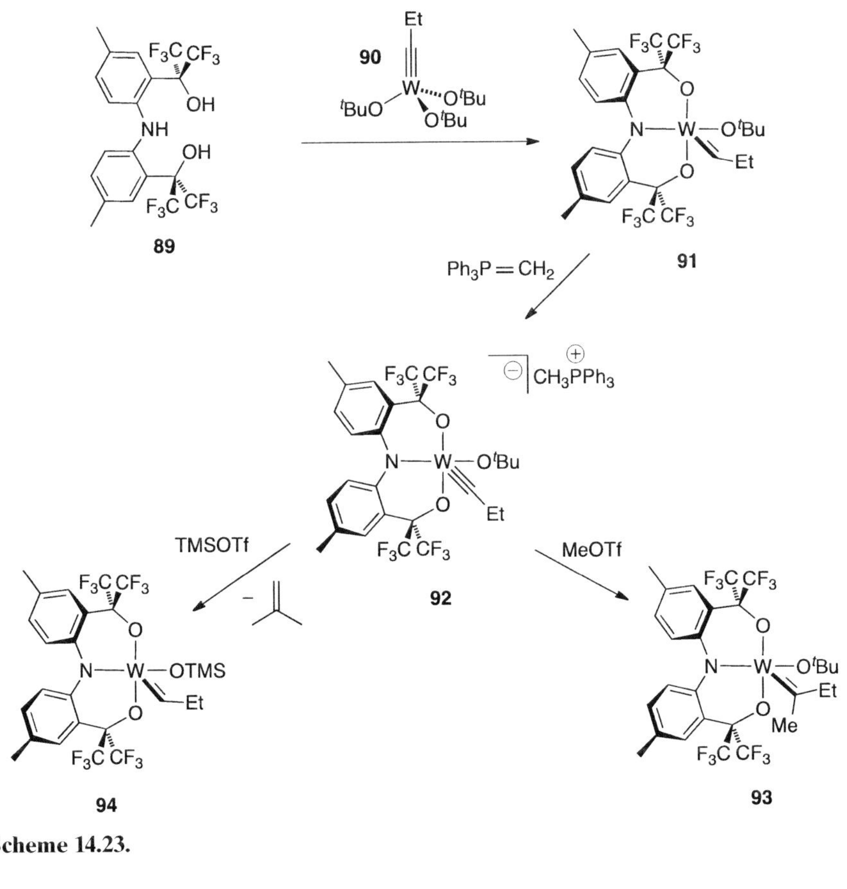

Scheme 14.23.

reactivity is not surprising given the five-coordinate nature of these complexes and the absence of an available coordination site adjacent to the alkylidene ligand for olefin binding.

14.5 CONCLUSIONS AND OUTLOOK

The large amount of recent work with Mo and W alkylidene compounds attests to the continuing interest in nucleophilic carbenes of the Group 6 metals despite a history spanning three decades. The driving force continues to be the olefin metathesis reaction in all its forms. New catalyst designs, especially those involving pyrrolide ligands have featured most prominently in recent years providing new examples of catalyst architectures that are uniquely selective for a variety of metathesis reactions in both small molecule synthesis and polymerization. In addition, these compounds have allowed for an unprecedented look into the intimate mechanism of the olefin metathesis reaction, establishing the elementary processes responsible for reactivity and selectivity within these systems. The ability to fine tune the steric and electronic requirements of the supporting ligands within Mo and W imido-alkylidene complexes is a major advantage and will no doubt continue to be a source of future exploration.

Looking forward, one exciting application of nucleophilic Group 6 carbenes still to be fully developed is the catalytic tandem alkane dehydrogenation/olefin metathesis reaction.[140] This reaction involves pairing of an iridium pincer catalyst[141] with a Group 6 alkylidene[142] to promote the net metathesis of unfunctionalized alkanes (Scheme 14.24). The compatibility of these two catalyst systems is key to the overall process and exemplifies the beneficial aspects of orthogonal reactivity.[143]

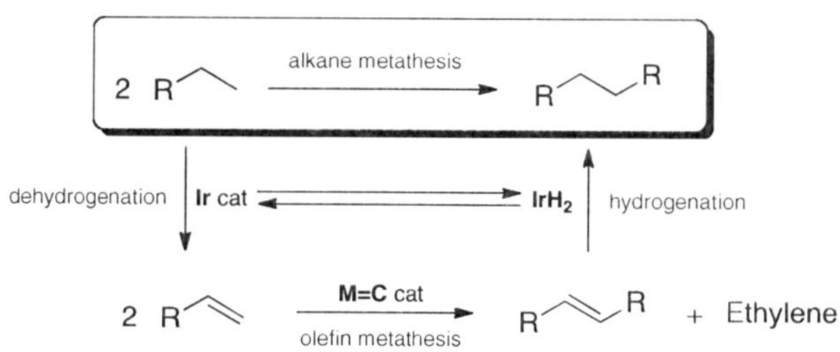

Scheme 14.24.

The fundamental reactivity of Group 6 carbenes has also provided fruitful ground for research in recent years. Continuing work in this area is warranted to discover new reactivity pathways and to further establish the utility of nucleophilic carbenes in reactions other than olefin metathesis. In this respect, it is hoped that future investigators will pick back up the gauntlet of chromium chemistry to help reveal how nucleophilic carbenes of this lightest Group 6 metal compare with the well-established reactivity of their heavier congeners.

SUGGESTED READING

Feldman, J., Schrock, R. R. "Recent Advances in the Chemistry of d^0 Alkylidene and Metallacyclobutane Complexes," *Prog. Inorg. Chem.* **1991**, *39*, 1–74.
Nugent, W. A., Mayer, J. M. *Metal-Ligand Multiple Bonds*. John Wiley & Sons, Inc.: New York, **1988**.
Schrock, R. R. "High Oxidation State Multiple Metal-Carbon Bonds," *Chem. Rev.* **2002**, *102*, 145–180.
Schrock, R. R."The Discovery and Development of High Oxidation State Mo and W Imido Alkylidene Complexes for Alkene Metathesis." In *Handbook of Metathesis*. Robert H. Grubbs, Ed. John Wiley & Sons, Inc.: New York, **2003**.
Schrock, R. R., Hoveyda, A. H. "Discovery and Development of Molybdenum and Tungsten Imido Alkylidene Complexes as Efficient Olefin Metathesis Catalysts for use in Organic Synthesis," *Angew. Chem. Int. Ed.* **2003**, *42*, 4592–4633.
Schrock, R. R. "Multiple Metal-Carbon Bonds for Catalytic Metathesis Reactions," *Angew. Chem. Int. Ed.* **2006**, *45*, 3748–3759 (Nobel Prize lecture).
Schrock, R. R. "Recent Advances in High Oxidation State Mo and W Imido Alkylidene Chemistry," *Chem. Rev.* **2009**, *109*, 3211–3226.

REFERENCES

1 Fischer, E. O., Maasböl, A. *Angew. Chem. Int. Ed.* **1964**, *3*, 580–581.
2 Taylor, T. E., Hall, M. B. *J. Am. Chem. Soc.* **1984**, *106*, 1576–1584.
3 Vyboishchikov, S. F., Frenking, G. *Chem.—Eur. J.* **1998**, *4*, 1428–1438.
4 Vyboishchikov, S. F., Frenking, G. *Chem.—Eur. J.* **1998**, *4*, 1439–1448.
5 Mowat, W., Wilkinson, G. *J. Chem. Soc., Dalton Trans.* **1973**, 1120–1124.
6 Schrock, R. R. *J. Am. Chem. Soc.* **1974**, *96*, 6796–6797.
7 Li, L., Hung, M., Xue, Z. *J. Am. Chem. Soc.* **1995**, *117*, 12746–12750.
8 Abbott, J. K. C., Li, L., Xue, Z.-L. *J. Am. Chem. Soc.* **2009**, *131*, 8246–8251.
9 Clark, D. N., Schrock, R. R. *J. Am. Chem. Soc.* **1978**, *100*, 6774–6776.
10 Schrock, R., Rocklage, S., Wengrovius, J., Rupprecht, G., Fellmann, J. *J. Mol. Catal.* **1980**, *8*, 73–83.
11 Wengrovius, J. H., Schrock, R. R. *Organometallics* **1982**, *1*, 148–155.
12 Mocella, M. T., Rovner, R., Muetterties, E. L. *J. Am. Chem. Soc.* **1976**, *98*, 4689–4690.
13 Kress, J., Wesolek, M., Le Ny, J.-P., Osborn, J. A. *J. Chem. Soc., Chem. Commun.* **1981**, 1039–1040.
14 Pedersen, S. F., Schrock, R. R. *J. Am. Chem. Soc.* **1982**, *104*, 7483–7491.
15 Schaverien, C. J., Dewan, J. C., Schrock, R. R. *J. Am. Chem. Soc.* **1986**, *108*, 2771–2773.
16 Murdzek, J. S., Schrock, R. R. *Organometallics* **1987**, *6*, 1373–1374.
17 Oskam, J. H., Schrock, R. R. *J. Am. Chem. Soc.* **1993**, *115*, 11831–11845.
18 Chauvin, Y., Commereuc, D., Cruypelinck, D. *Makromol. Chem.* **1976**, *177*, 2637–2646.
19 Schrock, R. R., Lopez, L. P. H., Hafer, J., Singh, R., Sinha, A., Müller, P. *Organometallics* **2005**, *24*, 5211–5213.
20 Danopoulos, A. A., Leung, W.-H., Wilkinson, G., Hussain-Bates, B., Hursthouse, M. B. *Polyhedron* **1990**, *9*, 2625–2634.

21 Groppo, E., Lamberti, C., Bordiga, S., Spoto, G., Zecchina, A. *Chem. Rev.* **2005**, *105*, 115–184.

22 Coles, M. P., Gibson, V. C., Clegg, W., Elsegood, M. R. J., Porrelli, P. A. *Chem. Commun.* **1996**, 1963–1964.

23 Meijboom, N., Schaverien, C. J., Orpen, A. G. *Organometallics* **1990**, *9*, 774–782.

24 Sundermeyer, J., Weber, K., Pritzkow, H. *Angew. Chem. Int. Ed.* **1993**, *32*, 731–733.

25 Heintz, R. A., Leelasubcharoen, S., Liable-Sands, L. M., Rheingold, A. L., Theopold, K. H. *Organometallics* **1998**, *17*, 5477–5485.

26 Leelasubcharoen, S., Lehmann, M. W., Theopold, K. H., Evans, D. H. *J. Electrochem. Soc.* **2001**, *148*, E118–E120.

27 Leelasubcharoen, S., Lam, K.-C., Concolino, T. E., Rheingold, A. L., Theopold, K. H. *Organometallics* **2001**, *20*, 182–187.

28 Amor Nait Ajjou, J., Scott, S. L., Paquet, V. *J. Am. Chem. Soc.* **1998**, *120*, 415–416.

29 Amor Nait Ajjou, J., Rice, G. L., Scott, S. L. *J. Am. Chem. Soc.* **1998**, *120*, 13436–13443.

30 Beaudoin, M. C., Womiloju, O., Fu, A., Ajjou, J. A. N., Rice, G. L., Scott, S. L. *J. Mol. Catal. A: Chem.* **2002**, *190*, 159–169.

31 Amor Nait Ajjou, J., Scott, S. L. *J. Am. Chem. Soc.* **2000**, *122*, 8968–8976.

32 Scott, S. L., Amor Nait Ajjou, *J. Chem. Eng. Sci.* **2001**, *56*, 4155–4168.

33 Liu, B., Nakatani, H., Terano, M. *J. Mol. Catal. A.* **2003**, *201*, 189–197.

34 McGuinness, D. S., Davies, N. W., Horne, J., Ivanov, I. *Organometallics* **2010**, *29*, 6111–6116.

35 Oskam, J. H., Fox, H. H., Yap, K. B., McConville, D. H., O'Dell, R., Lichtenstein, B. J., Schrock, R. R. *J. Organomet. Chem.* **1993**, *459*, 185–198.

36 Gerber, L. C. H., Schrock, R. R., Müller, P., Takase, M. K. *J. Am. Chem. Soc.* **2011**, *133*, 18142–18144.

37 Fu, G. C., Grubbs, R. H. *J. Am. Chem. Soc.* **1992**, *114*, 7324–7325.

38 Fu, G. C., Grubbs, R. H. *J. Am. Chem. Soc.* **1992**, *114*, 5426–5427.

39 Schrock, R. R., Jiang, A. J., Marinescu, S. C., Simpson, J. H., Müller, P. *Organometallics* **2010**, *29*, 5241–5251.

40 Poater, A., Solans-Monfort, X., Clot, E., Copéret, C., Eisenstein, O. *J. Am. Chem. Soc.* **2007**, *129*, 8207–8216.

41 Meek, S. J., Malcolmson, S. J., Li, B., Schrock, R. R., Hoveyda, A. H. *J. Am. Chem. Soc.* **2009**, *131*, 16407–16409.

42 Heppert, J. A., Dietz, S. D., Eilerts, N. W., Henning, R. W., Morton, M. D., Takusagawa, F., Kaul, F. A. *Organometallics* **1993**, *12*, 2565–2572.

43 Kress, J., Osborn, J. A. *J. Am. Chem. Soc.* **1987**, *109*, 3953–3960.

44 Youinou, M. T., Kress, J., Fischer, J., Aguero, A., Osborn, J. A. *J. Am. Chem. Soc.* **1988**, *110*, 1488–1493.

45 McConville, D. H., Wolf, J. R., Schrock, R. R. *J. Am. Chem. Soc.* **1993**, *115*, 4413–4414.

46 Fujimura, O., Grubbs, R. H. *J. Am. Chem. Soc.* **1996**, *118*, 2499–2500.

47 Alexander, J. B., La, D. S., Cefalo, D. R., Hoveyda, A. H., Schrock, R. R. *J. Am. Chem. Soc.* **1998**, *120*, 4041–4042.

48 Hoveyda, A. H., Schrock, R. R. *Chem. – Eur. J.* **2001**, *7*, 945–950.

49 Tsang, W. C. P., Jamieson, J. Y., Aeilts, S. L., Hultzsch, K. C., Schrock, R. R., Hoveyda, A. H. *Organometallics* **2004**, *23*, 1997–2007.

50 Tsang, W. C. P., Hultzsch, K. C., Alexander, J. B., Bonitatebus, P. J., Schrock, R. R., Hoveyda, A. H. *J. Am. Chem. Soc.* **2003**, *125*, 2652–2666.

51 Fox, H. H., Schrock, R. R. *Organometallics* **1992**, *11*, 2763–2765.
52 Sheng, Y.-H., Wu, Y.-D. *J. Am. Chem. Soc.* **2001**, *123*, 6662–6668.
53 Sheng, Y., Wu, Y.-D., Leszczynski, J. *Organometallics* **2004**, *23*, 3189–3196.
54 Anders, U., Nuyken, O., Buchmeiser, M. R., Wurst, K. *Macromolecules* **2002**, *35*, 9029–9038.
55 Schattenmann, F. J., Schrock, R. R., Davis, W. M. *J. Am. Chem. Soc.* **1996**, *118*, 3295–3296.
56 Schrock, R. R., Tonzetich, Z. J., Lichtscheidl, A. G., Müller, P., Schattenmann, F. J. *Organometallics* **2008**, *27*, 3986–3995.
57 Adamchuk, J., Schrock, R. R., Tonzetich, Z. J., Müller, P. *Organometallics* **2006**, *25*, 2364–2373.
58 Czekelius, C., Hafer, J., Tonzetich, Z. J., Schrock, R. R., Christensen, R. L., Müller, P. *J. Am. Chem. Soc.* **2006**, *128*, 16664–16675.
59 Scriban, C., Schrock, R. R., Müller, P. *Organometallics* **2008**, *27*, 6202–6214.
60 Scriban, C., Amagai, B. S., Stemmler, E. A., Christensen, R. L., Schrock, R. R. *J. Am. Chem. Soc.* **2009**, *131*, 13441–13452.
61 Jamison, G. M., Saunders, R. S., Wheeler, D. R., McClain, M. D., Loy, D. A., Ziller, J. W. *Organometallics* **1996**, *15*, 16–18.
62 Jamison, G. M., Saunders, R. S., Wheeler, D. R., Alam, T. M., McClain, M. D., Loy, D. A., Ziller, J. W. *Organometallics* **1996**, *15*, 3244–3246.
63 Cantrell, G. K., Geib, S. J., Meyer, T. Y. *Organometallics* **1999**, *18*, 4250–4252.
64 Cantrell, G. K., Geib, S. J., Meyer, T. Y. *Organometallics* **2000**, *19*, 3562–3568.
65 Tonzetich, Z. J., Schrock, R, R., Müller, P. *Organometallics* **2006**, *25*, 4301–4306.
66 Jan, M. T., Sarkar, S., Kuppuswamy, S., Ghiviriga, I., Abboud, K. A., Veige, A. S. *J. Organomet. Chem.* **2011**, *696*, 4079–4089.
67 Ciszewski, J. T., Xie, B., Cao, C., Odom, A. L. *Dalton Trans.* **2003**, 4226–4227.
68 Lokare, K. S., Staples, R. J., Odom, A. L. *Organometallics* **2008**, *27*, 5130–5138.
69 Lokare, K. S., Ciszewski, J. T., Odom, A. L. *Organometallics* **2004**, *23*, 5386–5388.
70 Lokare, K. S., Odom, A. L. *Inorg. Chem.* **2008**, *47*, 11191–11196.
71 Heppekausen, J., Fürstner, A. *Angew. Chem. Int. Ed.* **2011**, *50*, 7829–7832.
72 Ortiz, C. G., Abboud, K. A., Boncella, J. M. *Organometallics* **1999**, *18*, 4253–4260.
73 Kress, J., Osborn, J. A. *J. Am. Chem. Soc.* **1983**, *105*, 6346–6347.
74 Vaughan, W. M., Abboud, K. A., Boncella, J. M. *Organometallics* **1995**, *14*, 1567–1577.
75 Tonzetich, Z. J., Jiang, A. J., Schrock, R. R., Müller, P. *Organometallics* **2006**, *25*, 4725–4727.
76 Tonzetich, Z. J., Jiang, A. J., Schrock, R. R., Müller, P. *Organometallics* **2007**, *26*, 3771–3783.
77 Jiang, A. J., Schrock, R. R., Müller, P. *Organometallics* **2008**, *27*, 4428–4438.
78 Hock, A. S., Schrock, R. R., Hoveyda, A. H. *J. Am. Chem. Soc.* **2006**, *128*, 16373–16375.
79 Sinha, A., Schrock, R. R. *Organometallics* **2004**, *23*, 1643–1645.
80 Sinha, A., Lopez, L. P. H., Schrock, R. R., Hock, A. S., Müller, P. *Organometallics* **2006**, *25*, 1412–1423.
81 Sinha, A., Schrock, R. R., Müller, P., Hoveyda, A. H. *Organometallics* **2006**, *25*, 4621–4626.
82 Lichtscheidl, A. G., Ng, V. W. L., Müller, P., Takase, M. K., Schrock, R. R., Malcolmson, S. J., Meek, S. J., Li, B., Kiesewetter, E. T., Hoveyda, A. H. *Organometallics* **2012**, *31*, 4558–4564.

83 Marinescu, S. C., Singh, R., Hock, A. S., Wampler, K. M., Schrock, R. R., Müller, P. *Organometallics* **2008**, *27*, 6570–6578.
84 Singh, R., Schrock, R. R., Müller, P., Hoveyda, A. H. *J. Am. Chem. Soc.* **2007**, *129*, 12654–12655.
85 Solans-Monfort, X., Copéret, C., Eisenstein, O. *J. Am. Chem. Soc.* **2010**, *132*, 7750–7757.
86 Córdova, A., Rios, R. *Angew. Chem. Int. Ed.* **2009**, *48*, 8827–8831.
87 Ibrahem, I., Yu, M., Schrock, R. R., Hoveyda, A. H. *J. Am. Chem. Soc.* **2009**, *131*, 3844–3845.
88 Flook, M. M., Jiang, A. J., Schrock, R. R., Müller, P., Hoveyda, A. H. *J. Am. Chem. Soc.* **2009**, *131*, 7962–7963.
89 Townsend, E. M., Schrock, R. R., Hoveyda, A. H. *J. Am. Chem. Soc.* **2012**, *134*, 11334–11337.
90 Hoveyda, A. H., Malcolmson, S. J., Meek, S. J., Zhugralin, A. R. *Angew. Chem. Int. Ed.* **2010**, *49*, 34–44.
91 Malcolmson, S. J., Meek, S. J., Sattely, E. S., Schrock, R. R., Hoveyda, A. H. *Nature* **2008**, *456*, 933–937.
92 Yu, M., Wang, C., Kyle, A. F., Jakubec, P., Dixon, D. J., Schrock, R. R., Hoveyda, A. H. *Nature* **2011**, *479*, 88–93.
93 Meek, S. J., O'Brien, R. V., Llaveria, J., Schrock, R. R., Hoveyda, A. H. *Nature* **2011**, *471*, 461–466.
94 Yu, M., Ibrahem, I., Hasegawa, M., Schrock, R. R., Hoveyda, A. H. *J. Am. Chem. Soc.* **2012**, *134*, 2788–2799.
95 Marinescu, S. C., Schrock, R. R., Müller, P., Hoveyda, A. H. *J. Am. Chem. Soc.* **2009**, *131*, 10840–10841.
96 Marinescu, S. C., Levine, D. S., Zhao, Y., Schrock, R. R., Hoveyda, A. H. *J. Am. Chem. Soc.* **2011**, *133*, 11512–11514.
97 Pamplin, C. B., Legzdins, P. *Acc. Chem. Res.* **2003**, *36*, 223–233.
98 Legzdins, P., Rettig, S. J., Veltheer, J. E., Batchelor, R. J., Einstein, F. W. B. *Organometallics* **1993**, *12*, 3575–3585.
99 Wada, K., Pamplin, C. B., Legzdins, P., Patrick, B. O., Tsyba, I., Bau, R. *J. Am. Chem. Soc.* **2003**, *125*, 7035–7048.
100 Graham, P. M., Buschhaus, M. S. A., Legzdins, P. *J. Am. Chem. Soc.* **2006**, *128*, 9038–9039.
101 Graham, P. M., Buschhaus, M. S. A., Pamplin, C. B., Legzdins, P. *Organometallics* **2008**, *27*, 2840–2851.
102 Tran, E., Legzdins, P. *J. Am. Chem. Soc.* **1997**, *119*, 5071–5072.
103 Adams, C. S., Legzdins, P., Tran, E. *J. Am. Chem. Soc.* **2001**, *123*, 612–624.
104 Hock, A. S., Schrock, R. R. *Chem. Asian J.* **2007**, *2*, 867–874.
105 Buchmeiser, M. R. *Chem. Rev.* **2009**, *109*, 303–321.
106 Rendón, N., Blanc, F., Copéret, C. *Coord. Chem. Rev.* **2009**, *253*, 2015–2020.
107 Hultzsch, K. C., Jernelius, J. A., Hoveyda, A. H., Schrock, R. R. *Angew. Chem. Int. Ed.* **2002**, *41*, 589–593.
108 Kroll, R. M., Schuler, N., Lubbad, S., Buchmeiser, M. R. *Chem. Commun.* **2003**, 2742–2743.
109 Dolman, S. J., Hultzsch, K. C., Pezet, F., Teng, X., Hoveyda, A. H., Schrock, R. R. *J. Am. Chem. Soc.* **2004**, *126*, 10945–10953.
110 Wang, D., Kröll, R., Mayr, M., Wurst, K., Buchmeiser, M. R. *Adv. Synth. Catal.* **2006**, *348*, 1567–1579.

111 Blanc, F., Copéret, C., Thivolle-Cazat, J., Basset, J.-M., Lesage, A., Emsley, L., Sinha, A., Schrock, R. R. *Angew. Chem. Int. Ed.* **2006**, *45*, 1216–1220.
112 Blanc, F., Berthoud, R., Salameh, A., Basset, J.-M., Copéret, C., Singh, R., Schrock, R. R. *J. Am. Chem. Soc.* **2007**, *129*, 8434–8435.
113 Blanc, F., Rendon, N., Berthoud, R., Basset, J.-M., Copéret, C., Tonzetich, Z. J., Schrock, R. R. *Dalton Trans.* **2008**, 3156–3158.
114 Blanc, F., Copéret, C., Thivolle-Cazat, J., Basset, J.-M., Lesage, A., Emsley, L., Sinha, A., Schrock, R. R. *Inorg. Chem.* **2006**, *45*, 9587–9592.
115 Blanc, F., Basset, J.-M., Copéret, C., Sinha, A., Tonzetich, Z. J., Schrock, R. R., Solans-Monfort, X., Clot, E., Eisenstein, O., Lesage, A., Emsley, L. *J. Am. Chem. Soc.* **2008**, *130*, 5886–5900.
116 Rascón, F., Copéret, C. *J. Organomet. Chem.* **2011**, *696*, 4121–4131.
117 Amakawa, K., Wrabetz, S., Kröhnert, J., Tzolova-Müller, G., Schlögl, R., Trunschke, A. *J. Am. Chem. Soc.* **2012**, *134*, 11462–11473.
118 Schrock, R. R. *J. Organomet. Chem.* **1986**, *300*, 249–262.
119 Schrock, R. R. *Chem. Rev.* **2002**, *102*, 145–180.
120 Jeong, H., Axtell, J. C., Török, B., Schrock, R. R., Müller, P. *Organometallics* **2012**, *31*, 6522–6525.
121 Nielson, A. J. *Polyhedron* **1987**, *6*, 1657–1667.
122 Yuan, J., Schrock, R. R., Müller, P., Axtell, J. C., Dobereiner, G. E. *Organometallics* **2012**, *31*, 4650–4653.
123 Kreickmann, T., Arndt, S., Schrock, R. R., Müller, P. *Organometallics* **2007**, *26*, 5702–5711.
124 Flook, M. M., Börner, J., Kilyanek, S. M., Gerber, L. C. H., Schrock, R. R. *Organometallics* **2012**, *31*, 6231–6243.
125 Yuan, J., Townsend, E. M., Schrock, R. R., Goldman, A. S., Müller, P., Takase, M. K. *Adv. Synth. Catal.* **2011**, *353*, 1985–1992.
126 Jiang, A. J., Zhao, Y., Schrock, R. R., Hoveyda, A. H. *J. Am. Chem. Soc.* **2009**, *131*, 16630–16631.
127 Marinescu, S. C., Schrock, R. R., Müller, P., Takase, M. K., Hoveyda, A. H. *Organometallics* **2011**, *30*, 1780–1782.
128 Zhao, Y., Hoveyda, A. H., Schrock, R. R. *Org. Lett.* **2011**, *13*, 784–787.
129 Peryshkov, D. V., Schrock, R. R., Takase, M. K., Müller, P., Hoveyda, A. H. *J. Am. Chem. Soc.* **2011**, *133*, 20754–20757.
130 Schrock, R. R. *Acc. Chem. Res.* **1986**, *19*, 342–348.
131 Caulton, K. G., Chisholm, M. H., Streib, W. E., Xue, Z. *J. Am. Chem. Soc.* **1991**, *113*, 6082–6090.
132 Choi, S.-H., Lin, Z., Xue, Z. *Organometallics* **1999**, *18*, 5488–5495.
133 Morton, L. A., Wang, R., Yu, X., Campana, C. F., Guzei, I. A., Yap, G. P. A., Xue, Z.-L. *Organometallics* **2006**, *25*, 427–434.
134 Chen, T., Zhang, X.-H., Wang, C., Chen, S., Wu, Z., Li, L., Sorasaenee, K. R., Diminnie, J. B., Pan, H., Guzei, I. A., Rheingold, A. L., Wu, Y.-D., Xue, Z.-L. *Organometallics* **2005**, *24*, 1214–1224.
135 Morton, L. A., Zhang, X.-H., Wang, R., Lin, Z., Wu, Y.-D., Xue, Z.-L. *J. Am. Chem. Soc.* **2004**, *126*, 10208–10209.
136 Morton, L. A., Chen, S., Qiu, H., Xue, Z.-L. *J. Am. Chem. Soc.* **2007**, *129*, 7277–7283.
137 Kuppuswamy, S., Peloquin, A. J., Ghiviriga, I., Abboud, K. A., Veige, A. S. *Organometallics* **2010**, *29*, 4227–4233.

138 Tonzetich, Z. J., Lam, Y. C., Müller, P., Schrock, R. R. *Organometallics* **2007**, *26*, 475–477.
139 O'Reilly, M. E., Ghiviriga, I., Abboud, K. A., Veige, A. S. *J. Am. Chem. Soc.* **2012**, *134*, 11185–11195.
140 Goldman, A. S., Roy, A. H., Huang, Z., Ahuja, R., Schinski, W., Brookhart, M. *Science* **2006**, *312*, 257–261.
141 Choi, J., MacArthur, A. H. R., Brookhart, M., Goldman, A. S. *Chem. Rev.* **2011**, *111*, 1761–1779.
142 Bailey, B. C., Schrock, R. R., Kundu, S., Goldman, A. S., Huang, Z., Brookhart, M. *Organometallics* **2008**, *28*, 355–360.
143 Haibach, M. C., Kundu, S., Brookhart, M., Goldman, A. S. *Acc. Chem. Res.* **2012**, *45*, 947–958.

15

COBALT-MEDIATED CARBENE TRANSFER REACTIONS

XIN CUI AND X. PETER ZHANG
Department of Chemistry, University of South Florida, Tampa, FL, USA

15.1 INTRODUCTION

Carbenes in their free form are highly reactive species and can undergo many types of organic transformations. Due to the lack of selectivity, however, these transformations have found limited synthetic utility. Coordination to transition metal ions restrains and regulates the high reactivity of free carbenes. The resulting transition metal carbene complexes have been shown to have more controllable reactivity and are enabling regio- and stereoselective carbene transformations.[1] Extensive studies have been carried out to develop effective catalytic systems for selective carbene transfer reactions with the use of different transition metals.[2] Although rhodium represents the most common transition metal for these studies, its first-row cousin cobalt has also been demonstrated to involve formation of carbene complexes for effective carbene transfers, mainly through the decomposition of diazo reagents. During the past decades, a number of Co-catalyzed carbene transfer reactions have been reported.[3] Most efforts have been devoted to the development of asymmetric cyclopropanation of olefins with diazo reagents as carbene sources.[4] Cobalt-based catalytic systems have also been shown to catalyze other selective carbene transformations, such as cyclopropenation,[5] carbonylation,[6] and olefination reactions.[7]

Contemporary Carbene Chemistry, First Edition. Edited by Robert A. Moss and Michael P. Doyle.

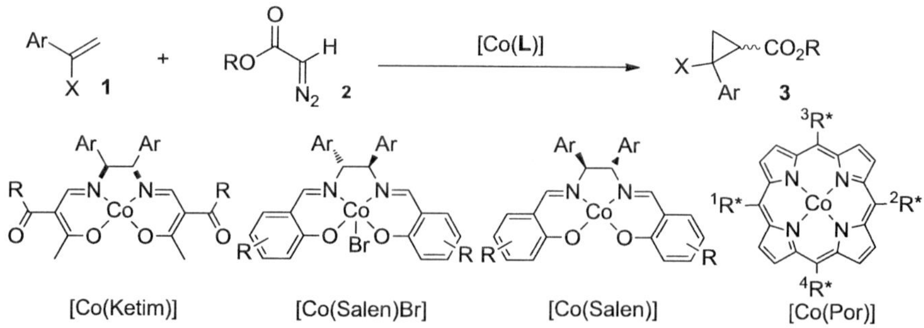

Figure 15.1. Asymmetric cyclopropanation catalyzed by Co-based chiral catalysts.

Early reports were exclusively focused on the decomposition of diazoacetates (**2**) with cobalt complexes, in either Co(II) or Co(III) oxidation states, for cyclopropanation of styrene derivatives (**1**) (Fig. 15.1).[3d,8] Effective catalysts have included the cobalt complexes of ketoiminato compounds ([Co(II)(Ketim)]), *N*,*N*'-bis(salicylidene)ethylenediamines ([Co(III)(Salen)Br] and [Co(II)(Salen)]), and porphyrins ([Co(II)(Por)]). These Co-based chiral catalysts were shown to be as active for asymmetric cyclopropanation using diazoacetates as were other metal-based catalytic systems, achieving high levels of reactivity and stereocontrol. Similar to other catalytic systems, the olefin substrate scope of asymmetric cyclopropanation by these cobalt catalysts was largely limited to styrene derivatives and related terminal olefins with comparable electron richness. However, the uniqueness of cobalt in catalytic carbene transfers was not significantly demonstrated until a report by Zhang and coworkers on [Co(Por)]-catalyzed asymmetric cyclopropanation of electron-deficient olefins (**4**) (Fig. 15.2).[9] Electron-deficient olefins such as

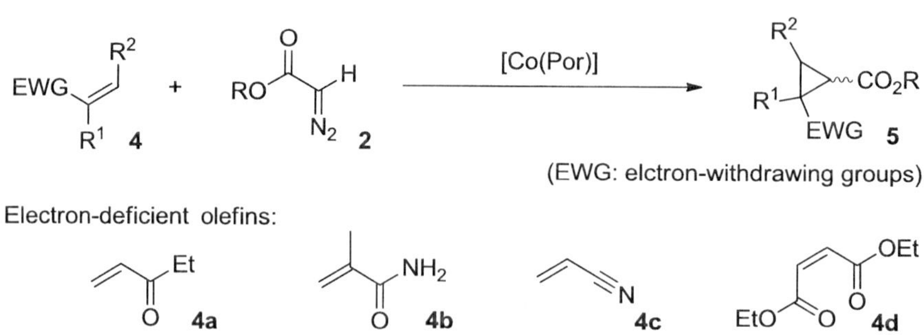

Figure 15.2. Asymmetric cyclopropanation of electron-deficient olefins by [Co(Por)].

α,β-unsaturated carbonyl compounds and nitriles (**4a–d**) represent a class of substrates that are challenging for cyclopropanation reactions for most metal-based catalytic systems, presumably due to the electrophilic nature of the carbene intermediates (like Fischer-type) in their catalytic cycles. The exceptional reactivity toward electron-deficient olefins by [Co(Por)] suggests that the corresponding Co-carbene intermediate may have different reactivity characteristics and most likely possess a certain degree of nucleophilicity.

In addition to the broadened scope of olefin substrates, cobalt-based catalysts have since been reported to catalyze asymmetric cycloaddition reactions by effectively decomposing different types of diazo reagents, including those that are known to be challenging carbene sources for catalytic carbene transfers. For example, Carreira and coworkers reported that chiral [Co(Salen)] complexes could catalyze the decomposition of trifluoromethyl diazomethane (**6**) for asymmetric olefin cyclopropanation, resulting in the formation of enantiomerically enriched trifluoromethylcyclopropanes (**8**) (Scheme 15.1).[10] Chiral [Co(Por)] complexes were shown by Zhang and coworkers for the first time to be effective catalysts to decompose sulfonyldiazomethanes (**7**) for asymmetric synthesis of cyclopropyl sulfones (**8**) from olefins (**1**) with both excellent diastereoselectivity and enantioselectivity (Scheme 15.1).[11]

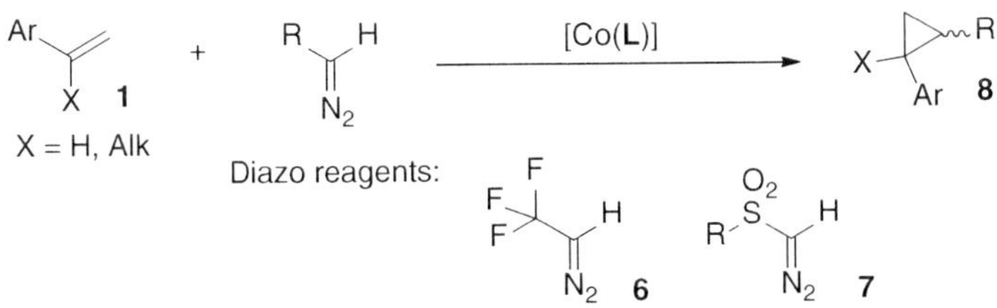

Scheme 15.1. Asymmetric cyclopropanation with non-carbonyl diazo reagents.

Diazo compounds bearing two electron-withdrawing groups, namely acceptor/acceptor-substituted diazo reagents, typically display both low reactivity and poor stereocontrol in transition metal-catalyzed carbene transfer reactions.[2a,b] As another distinct catalytic property of the Co-based system, these acceptor/acceptor-substituted diazo reagents could be effectively activated by Zhang's cobalt(II) complexes of D_2-symmetric chiral porphyrins [Co(D_2-Por*)] as carbene sources for selective carbene transfer reactions. The [Co(D_2-Por*)] catalysts were shown to undergo highly selective asymmetric cyclopropanation reactions of varied olefins with both α-nitrodiazoacetates (**10**) and α-cyanodiazoacetates (**13**) (Fig. 15.3).[12] Later, [Co(D_2-Por*)]-based catalytic system was also successfully applied by Zhang and coworkers to alkyne substrates (**12**) for enantioselective cyclopropenation reactions with acceptor/acceptor-substituted diazo **13** (Fig. 15.3).[5] These results have unlocked the door toward the general use of acceptor/acceptor-substituted diazo reagents for catalytic carbene transfers.

This chapter presents an account of the development of Co-catalyzed carbene transfer processes with presumed involvement of cobalt-carbene intermediates from decomposition of diazo reagents. Its focus is on the reactivity and stereoselectivity of Co-based catalytic systems for carbene transfer reactions in relation to their potential applications in organic synthesis. Several unique aspects of Co-based catalytic processes are highlighted in comparison

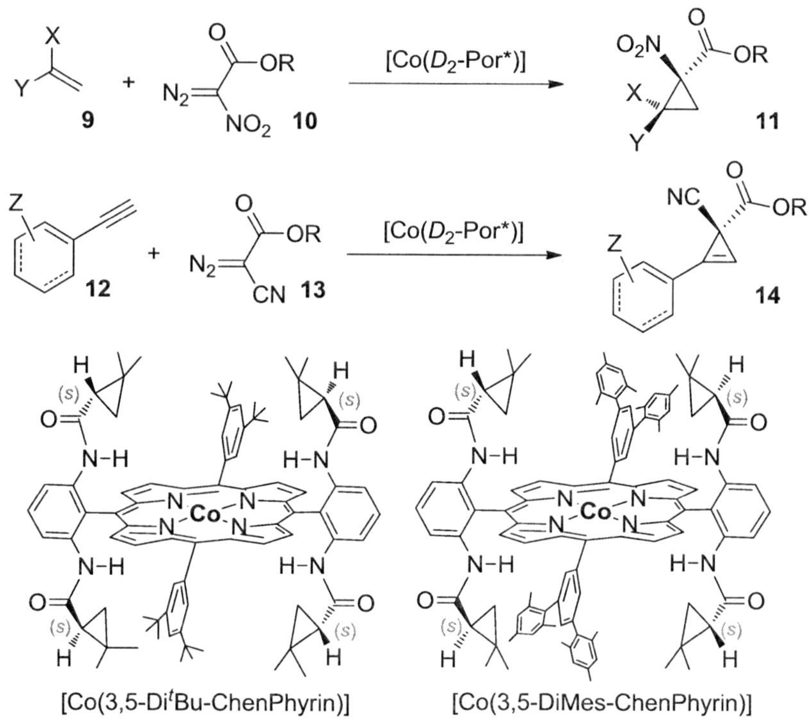

Figure 15.3. Asymmetric cyclopropanation and cyclopropenation with acceptor/acceptor-substituted diazo reagents.

with other metal-based systems. The chapter is structured to describe the several types of highly selective carbene transfer reactions that have been demonstrated with Co-based catalysts, including cyclopropanation, cyclopropenation, olefination, and carbonylation reactions. Mechanistic topics that are closely related to these transformations will also be briefly discussed. Within each type of carbene transfer reaction, discussion is organized based on the structure of the ligands, which plays a decisive role in the control of reactivity and selectivity. The chapter ends with discussion about the remaining challenges associated with Co-catalyzed carbene transfers and future research directions.

15.2 COBALT-CATALYZED CYCLOPROPANATION REACTIONS

15.2.1 Cobalt Dioximato and Cobalt Ketoiminato Catalysts

The earliest examples of cobalt-catalyzed enantioselective carbene transfers were reported in 1970s by Otsuka, Nakamura, and coworkers.[13] They showed that the cobalt(II) complex of camphor-based dioximato [Co(**Cqd**)] could catalyze olefin cyclopropanation with diazoacetates (**2**), affording the desired

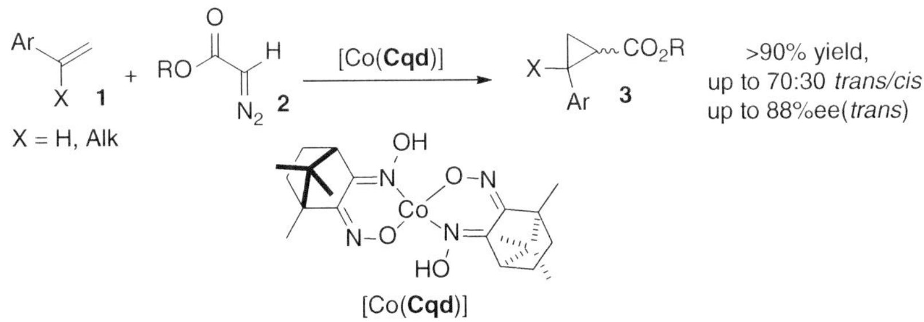

Scheme 15.2. Asymmetric cyclopropanation by dioximato Co(II) complex.

cyclopropanes (**3**) in high yields and good enantioselectivities (up to 88% optical purity) (Scheme 15.2).[13] Although moderate diastereoselectivity was a major issue for the catalytic system, these pioneering reports demonstrated for the first time that cobalt(II) complexes supported by chiral ligands were capable of catalyzing asymmetric cyclopropanation of olefins with high levels of enantiocontrol and their catalytic performance could be as equally effective as their copper, ruthenium, and rhodium counterparts.

In 1999, Yamada and coworkers reported the development of a structurally well-defined cobalt(II) complex of a tetradentate ketoiminato ligand **K1** [Co(**K1**)] as a catalyst for enantioselective cyclopropanation of styrene derivatives (**1**) with diazoacetate (**2**) (Scheme 15.3).[14] Given that the catalytic reaction could occur in the coordinating solvent THF, different Lewis bases were added as potential axial donor ligands for further tuning of the reactivity and selectivity of [Co(**K1**)]-catalyzed cyclopropanation. Among a series of *N*-heterocycles tested, addition of *N*-methylimidazole (NMI) resulted in significant improvement of asymmetric olefin cyclopropanation with *tert*-butyl diazoacetate (**2a**) by [Co(**K1**)], affording the corresponding cyclopropanes **15** in high yields with up to 84% ee but with moderate diastereoselectivity.

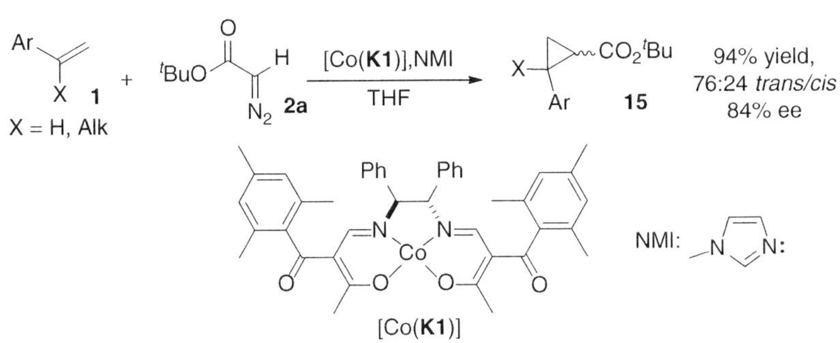

Scheme 15.3. Asymmetric cyclopropanation by ketoiminato Co(II) complex.

To further improve diastereoselectivity and enantioselectivity of the Co(II)-catalyzed cyclopropanation, Yamada and coworkers introduced new chiral ketoiminato ligands **K2** and **K3** which increased the steric bulk on the ethylenedi-

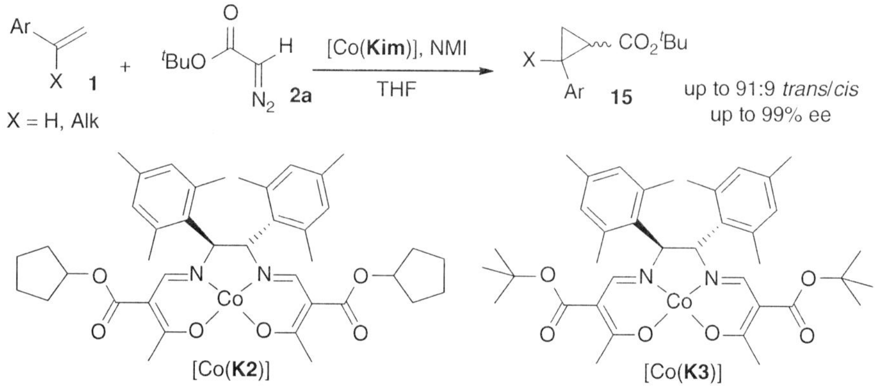

Scheme 15.4. Asymmetric cyclopropanation by new ketoiminato cobalt(II) complexes.

amine bridge while replacing the ketone substituents of **K1** with bulky ester groups (Scheme 15.4).[8,15] The complexes [Co(**K2**)] and [Co(**K3**)] proved to be highly selective catalysts for the reaction, generating *trans*-cyclopropane **15** with stereocontrol of up to 91:9 diastereomeric ratio and 99% ee (Scheme 15.4).

Further studies by Yamada and coworkers disclosed a remarkable acceleration effect of aqueous or alcoholic solvents on Co-catalyzed cyclopropanation reactions.[16] Using a protic solvent system, high stereoselectivities for the formation of products **3** could be achieved without the need of any additives. In addition to styrene derivatives, 1,1-disubstituted olefins, such as 1,1-diphenylethylene (**3b**), could also be smoothly cyclopropanated by [Co(**K2**)] with diazoacetates (**2**) with high enantioselectivity (Fig. 15.4).

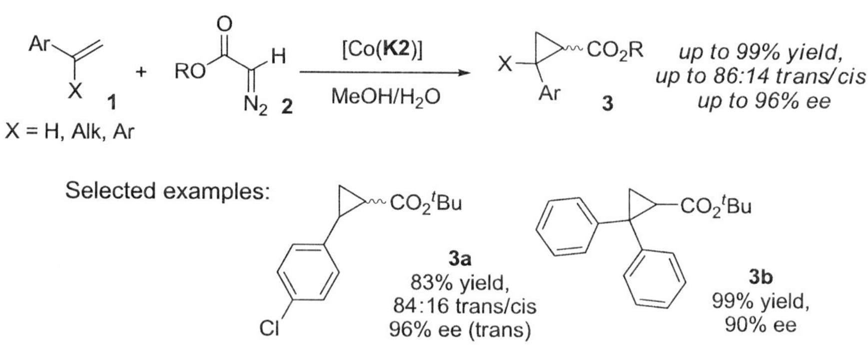

Figure 15.4. Acceleration effect of protic solvents on asymmetric cyclopropanation.

15.2.2 Cobalt Salen Catalysts

Since 1997, Katsuki and coworkers have reported a series of [Co(Salen)]-based catalytic systems for asymmetric cyclopropanation with diazoacetates. In addition to the steric and electronic nature of the chiral salen ligands, they demonstrated that the stereoselectivity of the [Co(Salen)]-catalyzed cyclopropanation reaction could be influenced by the oxidation state of the cobalt ion as well as the axial donor ligand. By using suitable [Co(Salen)] complexes as catalysts, asymmetric cyclopropanation could result in the formation of either *trans*- or *cis*-cyclopropane products. For example, when the Co(III) complex [Co(**S1**)Br] was employed as the catalyst, asymmetric cyclopropanation of styrene (**1a**) with *tert*-butyl diazoacetate **2a** was found to be *trans*-selective (*trans*:*cis* = 96:4), generating *trans*-**15a** in 80% yield and 93% ee (Scheme 15.5).[17]

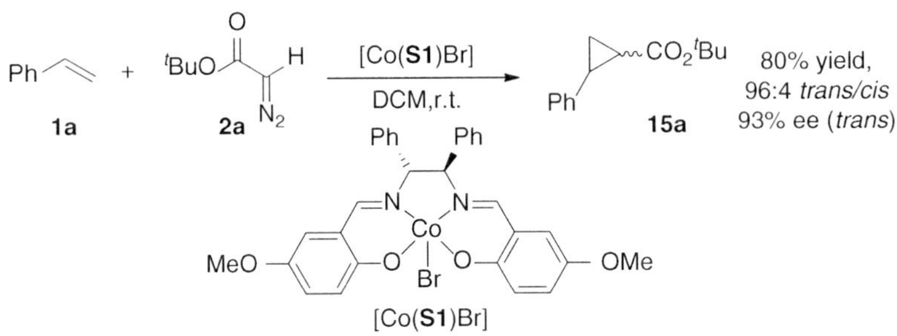

Scheme 15.5. *trans*-Selective cyclopropanation by [Co(III)(Salen)].

Remarkably, the same reaction could be made *cis*-selective when catalyzed by the Co(II) complex of Katsuki's salen ligand **S2** ([Co(**S2**)]) in THF in the presence of NMI as the additive. As shown in Scheme 15.6, the [Co(**S2**)]-catalyzed reaction of *tert*-butyl diazoacetates with styrene produced cyclopropane **15a** in the *cis*-dominant form in high yield while retaining excellent asymmetric induction.[17a,18] The different sense of diastereoselectivity for this catalytic cyclopropanation reaction between [Co(III)(**S1**)Br)] and [Co(II)(**S2**)] raises an interesting question: what is the major factor that is responsible for the switch of diastereoselectivity? Among the several differences between the two catalysts are the oxidation state of the metal ion, the chiral environment of the salen ligand, and the nature of

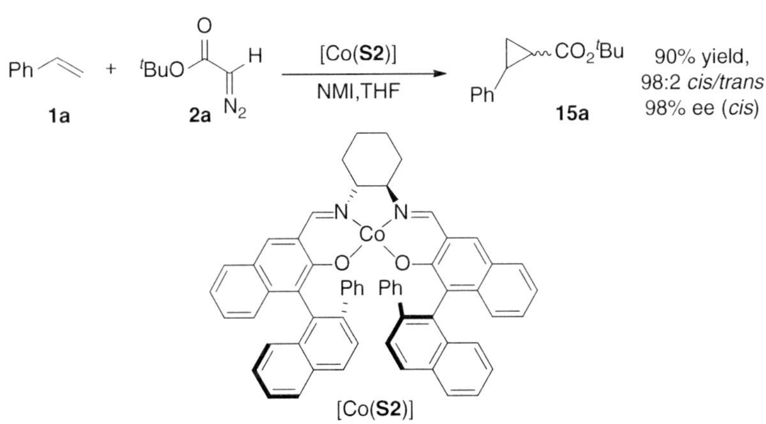

Scheme 15.6. *cis*-Selective cyclopropanation by [Co(II)(Salen)].

the fifth coordinating ligand. While the use of Co(II) metal ion and related axial ligand ensured sufficient catalytic activity with this sterically more hindered **S2** ligand, the bulky β-phenyl-containing naphthalene unit in **S2** was believed to be the main contributing factor for the observed diastereoselectivity switch as it would force the formation of an intermediate with a totally different orientation toward the formation of *cis*-cyclopropane products.

The notion that the steric environment of salen ligand other than the oxidation state of cobalt ion determines the sense of diastereoselectivity was nicely illustrated with the reported asymmetric cyclopropanation of *p*-chlorostyrene (**1b**) with [Co(**S3**)] (Scheme 15.7)[18a] to produce *trans*-cyclopropanecarboxylate **3a** in high yield and excellent enantioselectivity. Comparing the two catalytic reactions by [Co(**S2**)] and [Co(**S3**)] (Schemes 15.6 and 15.7), it is clear that only the change of the β-phenyl groups in ligand **S2** to the β-methyl groups in ligand **S3** resulted in the switch of *cis*-diastereoselectivity to *trans*-diastereoselectivity. These results highlight the great flexibility of [Co(II)(Salen)]-based catalytic systems for controlling the diastereoselectivity of olefin cyclopropanation, permitting the preparation of either *cis*- or *trans*-cyclopropanes through simply tuning the ligand steric environment.

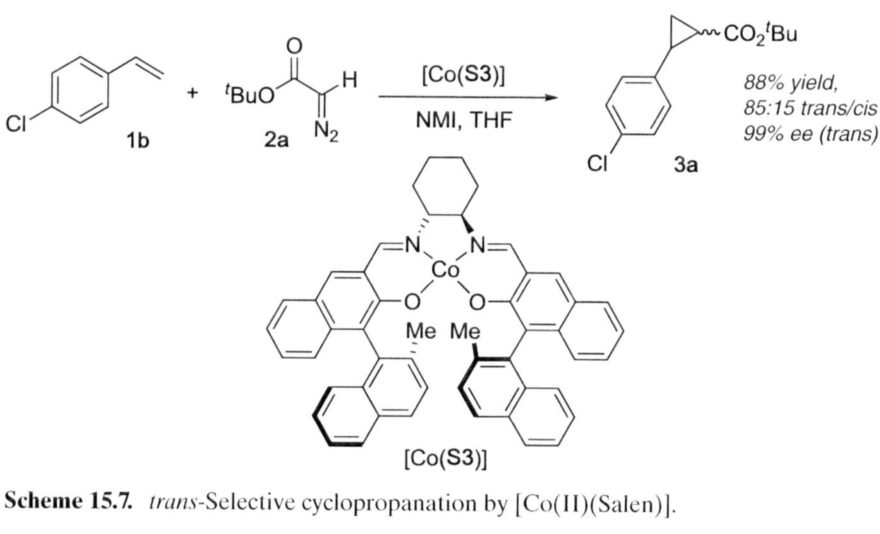

Scheme 15.7. *trans*-Selective cyclopropanation by [Co(II)(Salen)].

As [Co(II)(Salen)] catalytic systems usually requires axial ligands as an extra component to ensure high reactivity and stereoselectivity, Katsuki and coworkers reported the design and synthesis of new chiral salen ligands bearing a pendent imidazole functionality as the fifth coordinating group as exemplified with the **S2**-derived pentadentate chiral ligand **S4** (Fig. 15.5).[19] This type of ligand is expected to intramolecularly coordinate to the metal center and should exert a large trans effect. As expected, the Co(II) complex of the pentadentate ligand **S4** [Co(**S4**)] was shown to be an effective catalyst for highly

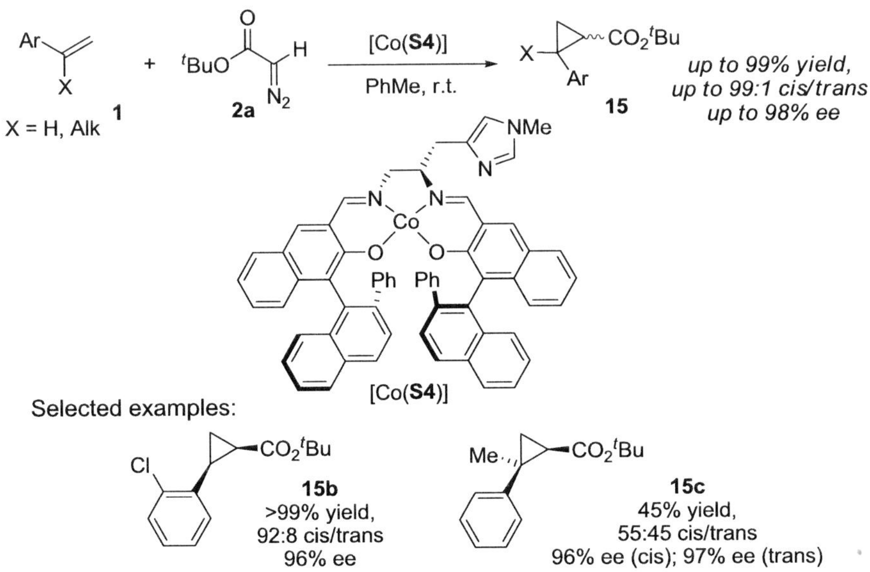

Figure 15.5. Asymmetric cyclopropanation by [Co(II)(pentadentate-Salen)].

asymmetric cyclopropanation of terminal alkenes (**1**) with diazoacetates without addition of extra ligands, suggesting intramolecular coordination of the pendent imidazole to the Co(II) center during the catalytic process (Fig. 15.5).[19] Like the [Co(**S2**)]/NMI-based-catalyzed system, asymmetric cyclopropanation by [Co(**S4**)] gave the unusual *cis*-diastereoselectivity.

In addition to intermolecular cyclopropanation, [Co(II)(Salen)] complexes were shown by Katsuki and coworkers to be also capable of catalyzing intramolecular cyclopropanation of alkenyl diazoacetates in a highly stereoselective manner (Fig. 15.6). For example, intramolecular cyclopropanation of

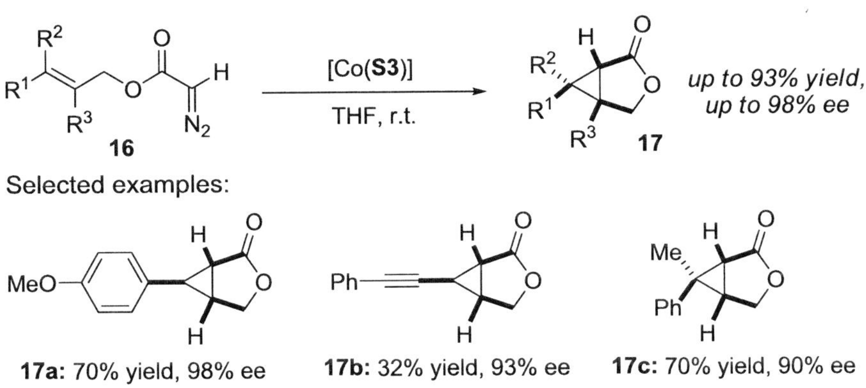

Figure 15.6. [Co(II)(Salen)]-catalyzed asymmetric cyclopropanation.

alkenyl diazoacetates (**16**) could be catalyzed by [Co(**S3**)] to afford the desired 3-oxabicycl[3.1.0]hexan-2-one products (**17**) with complete diastereocontrol in up to 93% yield and 98% ee.[17a,20]

Through the design and synthesis of a series of chiral [Co(Salen)] catalysts, Katsuki and co-workers have developed asymmetric cyclopropanation systems in both inter- and intramolecular manners, resulting in the production of chiral cyclopropane derivatives with high level of efficiency and stereoselectivity. So far, their development of *cis*-selective catalysts remains one of the few systems from which cyclopropane products in the atypical *cis*-configuration can be obtained with high diastereomeric excess.

[Co(II)(Salen)]-catalyzed cyclopropanation has been extended to employ other types of diazo reagents in addition to diazoacetates. Carreira and coworkers recently reported a [Co(II)(**S5**)]-catalyzed system for the highly enantioselective cyclopropanation of styrene derivatives (**1**) with trifluoromethyl diazomethane (**6**) (Scheme 15.8).[10] The Co(II)-catalyzed cyclopropanation occurred in acidic aqueous media by a practical procedure for *in situ* generation of the non-carbonyl diazo reagent **6** in smooth reaction of $CF_3CH_2NH_2$ with $NaNO_2$, which indicated the high effectiveness of Co(II)-based catalysts under eco-friendly conditions.[21] In the presence of Ph_3As, the one-pot catalytic process by [Co(II)(**S5**)] provided a high-yielding synthesis of trifluoromethyl cyclopropane derivatives **18** in nearly complete stereoselectivity (Scheme 15.8).

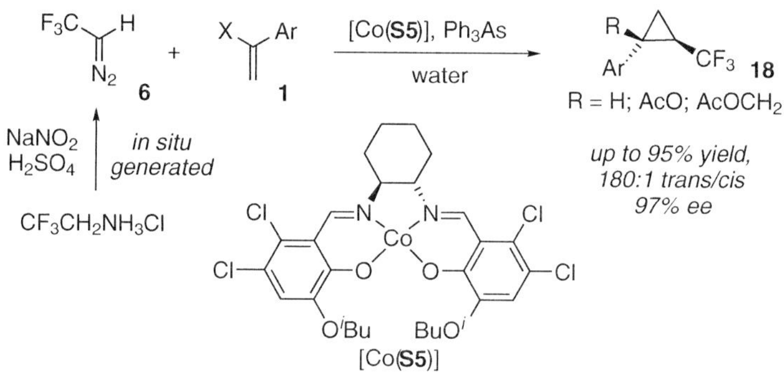

Scheme 15.8. [Co(II)(Salen)]-catalyzed asymmetric cyclopropanation with trifluoromethyldiazomethane.

With chiral cyclic salen derivatives **S6** and **S7** as supporting ligands, Gao and coworkers reported that the resulting dinuclear Co(II) and Co(III) complexes ([Co_2(**S6**)] and [Co_2(**S7**)]) are effective catalysts for asymmetric cyclopropanation of styrenes with diazoacetates (Fig. 15.7).[22] It is believed that both the Co centers on the same macrocyclic ligand platform might be involved with the generation of Co-carbene intermediates, which could subsequently reorient as a result of their interaction. This postulated model of the intermediate might transfer the chirality through highly ordered ligand–carbene–carbene interactions rather than the common ligand–carbene interactions in mononuclear

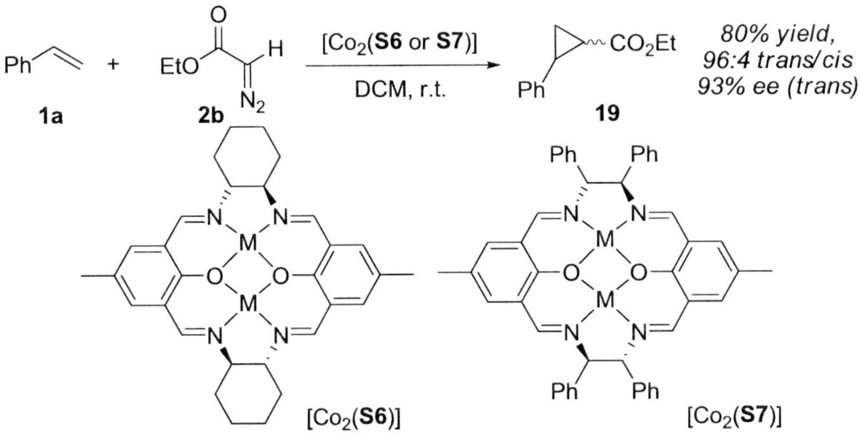

[Co_2(II)(**S6**)]	90% yield,	72:28 trans/cis	86% ee (trans)	92% ee (cis)
[Co_2(III)(**S6**)]	92% yield,	74:26 trans/cis	88% ee (trans)	94% ee (cis)
[Co_2(II)(**S7**)]	93% yield,	70:30 trans/cis	89% ee (trans)	92% ee (cis)
[Co_2(III)(**S7**)]	94% yield,	76:24 trans/cis	94% ee (trans)	91% ee (cis)

Figure 15.7. Chiral dinuclear Co catalysts for asymmetric cyclopropanation.

catalyst systems. The high yields and promising stereocontrol from uses of these structurally unique catalysts may encourage the development of more diverse chiral ligands for Co(II)-catalyzed transformations.

15.2.3 Cobalt Porphyrin Catalysts

Porphyrins, whose structures are found in many biologically relevant molecules such as enzymes, chlorophylls, and other heme proteins, serve as supporting ligands for many metal-catalyzed transformations. This class of tetradentate macrocyclic nitrogen ligands is well known to form metal complexes with excellent thermal and metal coordination stability.[23] Especially for stereoselective reactions, the highly rigid and stable chiral environments provided by the aromatic backbone of optically active porphyrins have advantages in reaching high levels of stereocontrol. Since the first demonstration of stereocontrol in cyclopropanation reactions with rhodium porphyrins in 1980,[24] porphyrin complexes of several metals, including iron, ruthenium, and osmium, have been found to catalyze cyclopropanation reactions.[25] Despite the apparent periodic relationship to the other catalytic metals and low cost of cobalt precursors, it was not until two decades later that [Co(Por)] was independently disclosed by the Cenini[26] and Zhang[27] groups as efficient cyclopropanation catalysts. In 2003, Cenini and coworkers reported the catalytic capability of cobalt(II) tetraphenylporphyrin ([Co(TPP)]) for cyclopropanation with diazoacetates and carried out kinetic and mechanistic studies of the catalytic process.[26] In the same year, Zhang and coworkers demonstrated that [Co(Por)] complexes are superior catalysts for stereoselective cyclopropanation in comparison with other known metalloporphyrin catalysts.[27] They showed that [Co(Por)]-based catalytic system can operate with alkenes as limiting reagents without the need for slow addition of diazo regents. It was further shown that both the diastereoselectivities and enantioselectivities of [Co(Por)]-catalyzed cyclopropanation reactions can be tuned by the use of different porphyrin ligands. With chiral camphanic acid-based porphyrin **P1** as the ligand, asymmetric cyclopropanation of styrene (**1a**) with ethyl diazoacetate (**2b**) could be catalyzed by [Co(**P1**)] to form cyclopropane **19** as *cis*-dominant product in good yield with enantioselectivities of 77% ee and 62% ee for *cis*- and *trans*-cyclopropane, respectively (Scheme 15.9).[27] Although stereoselectivity was limited in this initial catalytic system, the results provided the first indication that asymmetric induction in catalytic cyclopropanation by cobalt(II) porphyrins was possible.

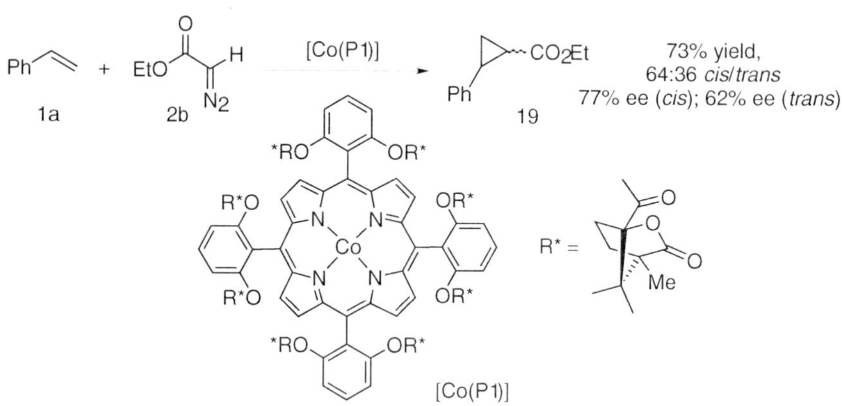

Scheme 15.9. Co(II) complex of chiral camphanic acid-based porphyrin for asymmetric cyclopropanation.

Subsequent investigations were focused on the design of new chiral porphyrin ligands to improve the Co(II)-catalyzed cyclopropanation. Early efforts were made to synthesize chiral porphyrins with chiral substituents at the *meso*-positions (Fig. 15.8).[28] A series of *meso*-chiral porphyrins of this type were synthesized, including the novel *meso*-(*R*)-BINOL-substituted chiral porphyrin **P2** (Fig. 15.8).[28b] Cobalt complexes of these *meso*-chiral porphyrins were shown to be effective catalysts for cyclopropanation of styrene with ethyl diazoacetate, producing the desired cyclopropane in excellent yields but with limited asymmetric induction.[28] In contrast to the *meso*-chiral porphyrins with a flexible chiral environment oriented away from the reaction center, *meso*-chiral porphyrins with chiral moieties attached at the *ortho*-positions of *meso*-aryl groups were believed to enable more effective asymmetric induction (Fig. 15.9).[27,29] In 2004, Zhang and coworkers reported the synthesis and evaluation of a new class of D_2-symmetric chiral Co(II) porphyrins ([Co(D_2-Por*)])

Chirality at *meso*-positions:

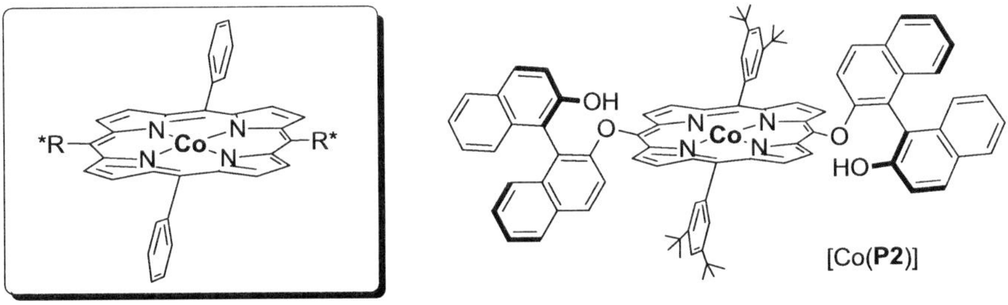

Figure 15.8. [Co(Por)] catalysts with chiral functionalities at *meso*-positions.

Chirality at *ortho*-aryl positions:

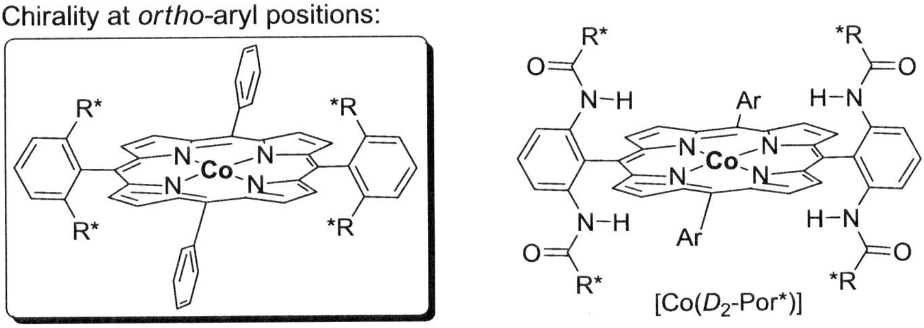

Figure 15.9. [Co(Por)] catalysts with chiral functionalities at *ortho*-aryl positions.

as catalysts for asymmetric cyclopropanation (Fig. 15.9).[29a] Through modular construction of chiral porphyrins via palladium-catalyzed multiple amidation reactions of bromoporphyrin synthons with chiral amides,[29a] a "toolbox" of [Co(D_2-Por*)] catalysts ([Co(**P3-8**)]) with varied chiral (R*) and non-chiral (Ar) substituents were constructed and applied for various asymmetric cyclopropanation and related group transfer reactions (Fig. 15.10).

Using styrene derivatives as model substrates, Zhang and coworkers discovered that the Co(II) complex of the D_2-symmetric chiral cyclopropyl-containing porphyrin 3,5-Di^{t}Bu-ChenPhyrin ([Co(**P3**)]) was an effective catalyst for asymmetric cyclopropanation with diazoacetates. In the presence of 4-dimethylaminopyridine as an axial ligand, for example, [Co(**P3**)] effectively cyclopropanated styrene (**1a**) with diazoacetate **2a** resulting in the formation of the desired cyclopropane product **15a** in high yield with complete *trans*-diastereoselectivity and excellent enantioselectivity (Scheme 15.10).[29a,30]

Dimerization of diazo compounds is known as a common side reaction in transition metal-mediated carbene transfers. To minimize the dimerization side reaction, the use of excess olefin and slow addition of the diazo reagents are usually required.[2f,4] Notably, the [Co(Por)]-based asymmetric cyclopropanation reaction could operate effectively in a one-pot fashion with alkenes as

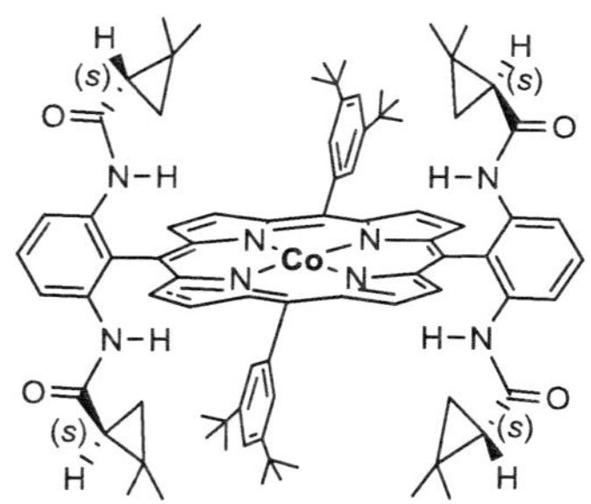

[Co(**P3**)]
(**P3** = 3,5-Di^{t}Bu-ChenPhyrin)

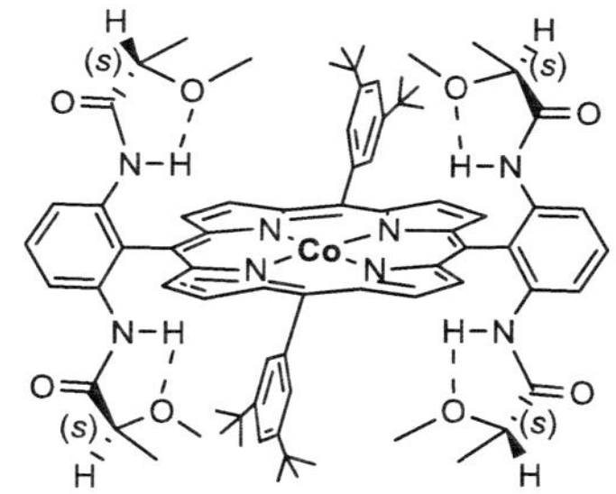

[Co(**P5**)]
(**P5** = 3,5-Di^{t}Bu-RuppelPhyrin)

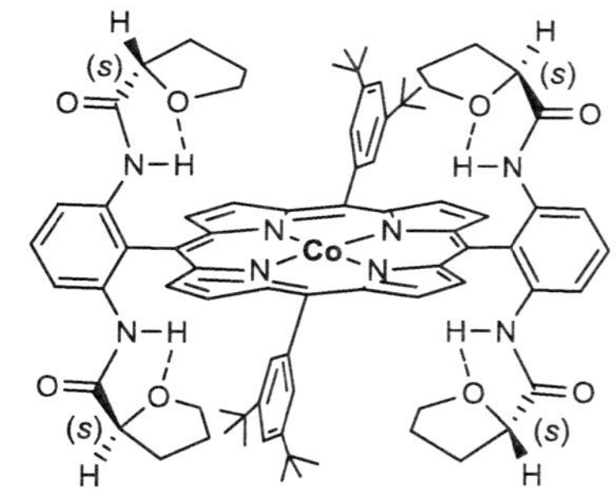

[Co(**P7**)]
(**P7** = 3,5-Di^{t}Bu-ZhuPhyrin)

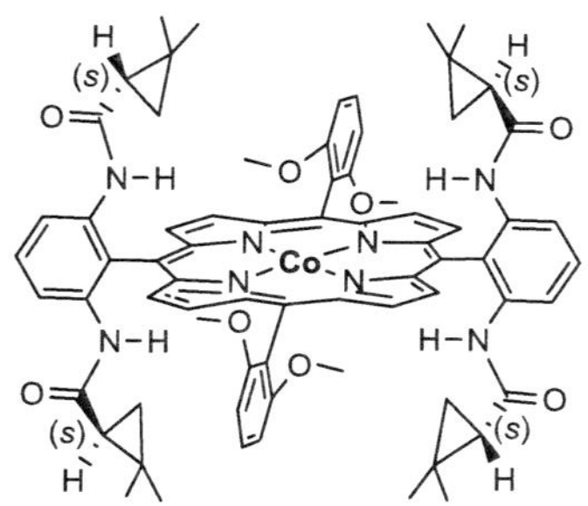

[Co(**P4**)]
(**P4** = 2,6-DiMeO-ChenPhyrin)

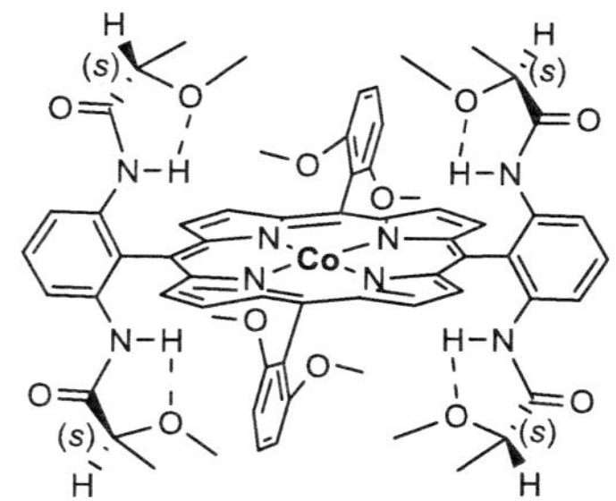

[Co(**P6**)]
(**P6** = 2,6-DiMeO-RuppelPhyrin)

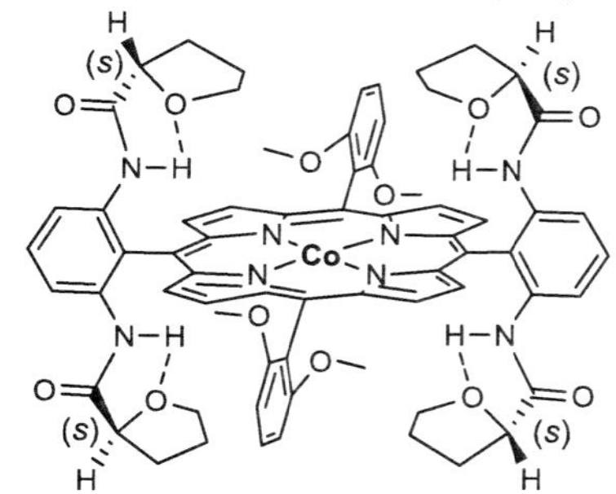

[Co(**P8**)]
(**P8** = 2,6-DiMeO-ZhuPhyrin)

Figure 15.10. Structures of D_2-symmetric chiral cobalt(II) porphyrins.

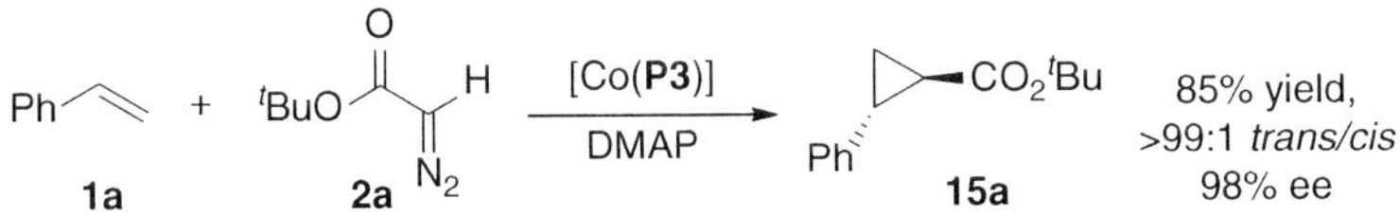

Scheme 15.10. [Co(D_2-Por*)]-catalyzed asymmetric cyclopropanation with diazoacetates.

limiting regents and did not require the slow addition of diazo reagents without significant formation of dimerization side products. The unique characteristics and superior performance of [Co(Por)] catalysts in asymmetric cyclopropanation were further demonstrated through comparative studies with the same chiral porphyrin **P3** complexes of other Group VIIIB metals, such as the Fe(III) (Fig. 15.11).[31] For example, while [Co(**P3**)] could affect highly asymmetric cyclopropanation with stoichiometric amounts of styrenes, 10 equiv of styrenes were required for [Fe(**P3**)Cl]-catalyzed reactions in order to minimize the dimerization side reaction, and even with this excess of alkene cyclopropanes **3** were formed in low yields with poor enantioselectivity. These results indicate [Co(Por)] catalysts have high inherent selectivity toward cyclopropanation, suggesting that Co-carbene species distinctively favor addition to olefins rather than dimerization with another diazo molecule.

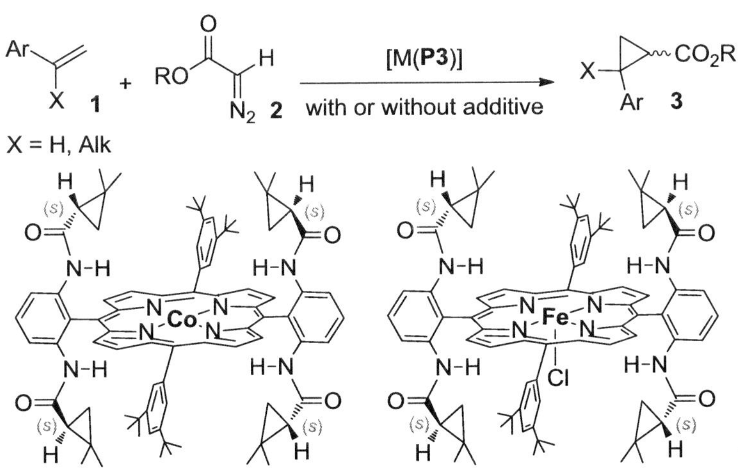

Figure 15.11. Comparison of [M(D_2-Por*)]-catalyzed cyclopropanation.

As a synthetic application of [Co(**P3**)]-catalyzed asymmetric cyclopropanation, Zhang and coworkers developed an enantioselective synthesis of cyclopropyl carboxamides by reacting preformed cyclopropyl chiral synthons with various amines (Fig. 15.12).[32] For example, cyclopropane succinimidyl ester **21** could be

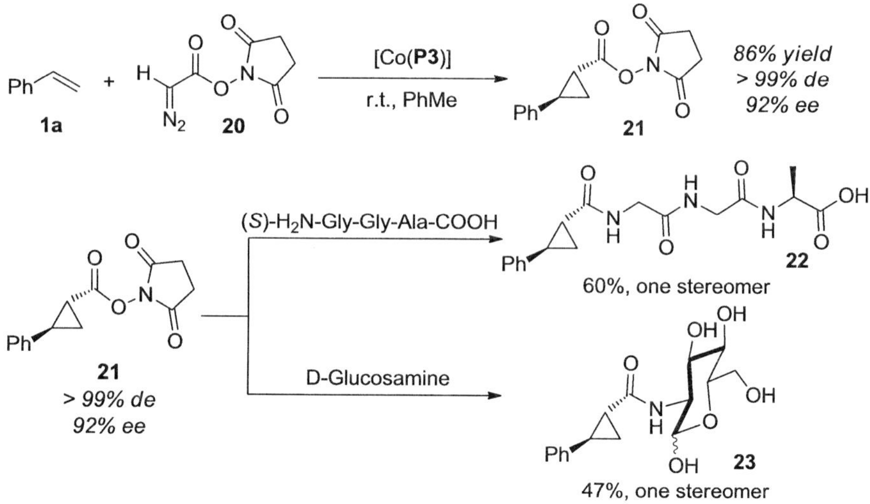

Figure 15.12. Synthetic application of [Co(D_2-Por*)]-catalyzed cyclopropanation.

prepared in high yield with excellent diastereo- and enantioselectivity through Co(II)-catalyzed asymmetric cyclopropanation of styrene with succinimidyl diazoacetate (N_2CHCO_2Su, **20**). The hydroxylsuccinimide ester functionality in the enantioenriched chiral synthon **21** reacted with a broad range of different amines such as the unprotected tripeptide (*S*)–H_2N–Gly–Ala–COOH and D-glucosamine under mild conditions to produce the corresponding cyclopropyl tripeptide **22** and cyclopropyl carboxamido sugar **23**. In addition to its versatility, this postderivatization strategy features a high degree of functional group tolerance.

While asymmetric cyclopropanation reactions of styrene derivatives and electron-rich olefins are common with the vast majority of catalytic systems, cyclopropanation of electron-deficient olefins containing electron-withdrawing groups has proven to be rare and challenging. This is attributed to the electrophilic nature of the metal-carbene intermediates involved in metal-catalyzed cyclopropanation.[2b,d,33] In 2007, by using [Co(**P3**)] as the catalyst, Zhang and coworkers reported the first transition metal-catalyzed asymmetric cyclopropanation of electron-deficient olefins (**4**), including unsaturated esters, amides, ketones, and nitriles.[9] With the absence of dimerization of diazo compounds, the catalytic reactions can be performed under mild conditions in a one-pot protocol using olefins as the limiting reagents, forming the desired electrophvilic cyclopropane derivatives (**5**) in high yields and stereoselectivities. In most cases, both excellent diastereo- and enantioselectivities were achieved (Fig. 15.13).

While the majority of catalytic cyclopropanation systems have employed diazocarbonyl compounds, mostly diazoacetates, as carbene sources, metal-catalyzed asymmetric cyclopropanation with other types of diazo reagents has been less developed.[2b,d,33a] To explore the effectiveness of [Co(Por)]-based catalytic systems for asymmetric cyclopropanation with diazosulfones, Zhang and coworkers

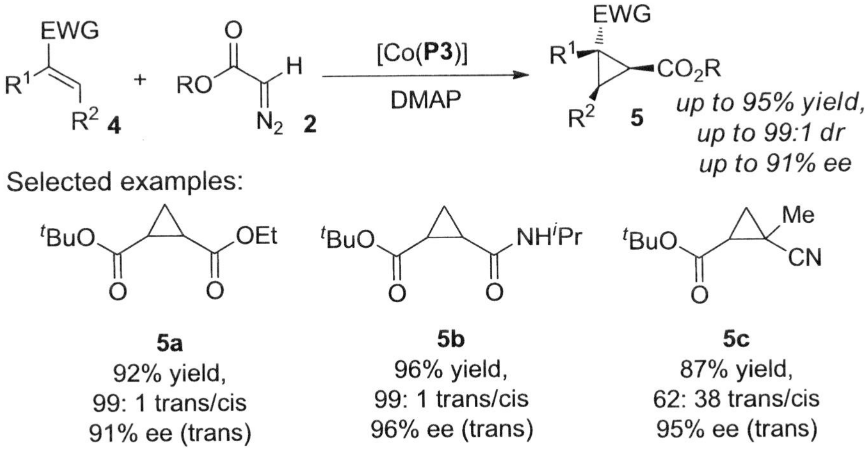

Figure 15.13. [Co(D_2-Por*)]-catalyzed cyclopropanation of electron-deficient olefins.

designed and synthesized a new D_2-symmetric chiral porphyrin 2,6-DiMeO-ZhuPhyrin (**P8**) with high rigidity and polarity of the chiral environment, which results from both intramolecular hydrogen bonding interactions and the use of the cyclic structure of (*S*)-2-tetrahydofurnancarboxamide (Scheme 15.11).[11] The Co(II) complex of this supporting ligand [Co(**P8**)] was shown to effectively catalyze asymmetric olefin cyclopropanation with diazosulfone **7a**. The catalytic system can be generally applied to various aromatic olefins as well as electron-deficient olefins, leading to the formation of the corresponding cyclopropyl sulfones (**25**) in high yields with high stereoselectivities (Scheme 15.11).

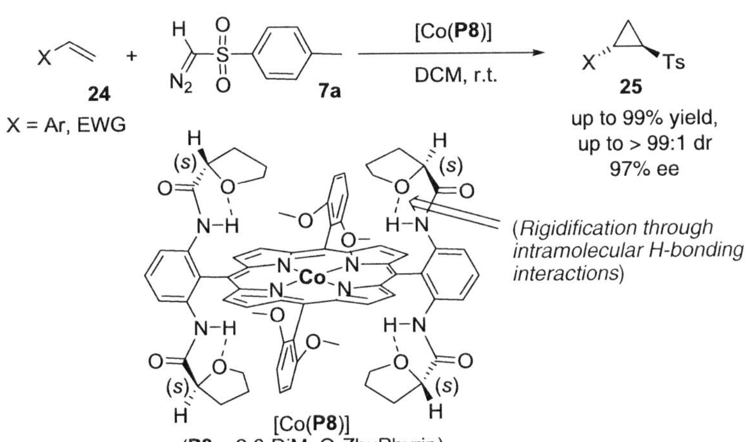

Scheme 15.11. [Co(D_2-Por*)]-catalyzed cyclopropanation with sulfonyldiazomethane.

In addition to tosyldiazomethane (**7a**), diazosulfones with different electronic properties, such as 4-methoxybenzenesulfonyldiazomethane (**7b**)

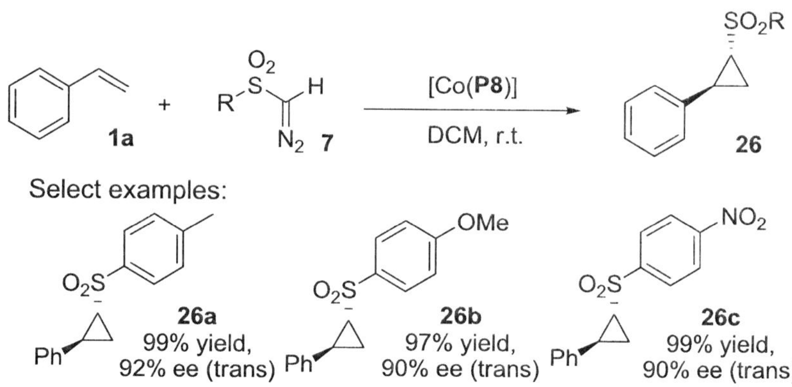

Figure 15.14. [Co(D_2-Por*)]-catalyzed cyclopropanation with varied sulfonyldiazomethanes.

and 4-nitrobenzenesulfonyldiazomethane (**7c**) were also suitable carbene sources for [Co(**P8**)]-catalyzed asymmetric cyclopropanation, affording enantiomerically enriched cyclopropyl sulfones **26b** and **26c** in high yields and excellent *trans*-diastereoselectivity (Fig. 15.14).[34]

As another highlight, the [Co(D_2-Por*)] catalytic system was shown to achieve effective asymmetric cyclopropanation with acceptor/acceptor-substituted diazo reagents, a class of carbene sources that are known for their inherent low reactivity and perceived poor enantioselectivity.[2b–d,f] [Co(**P3**)] was demonstrated to be an effective catalyst for asymmetric cyclopropanation with the acceptor/acceptor-substituted diazo reagent α-nitrodiazoacetates (**10**).[12a,35] The unique reactivity of [Co(**P3**)]-based catalytic system toward this challenging diazo reagent is general and suitable for electron-sufficient, -neutral, and -deficient alkenes, forming the corresponding cyclopropyl nitroesters (**11**) in high yields. In addition to high diastereo- and enantioselectivity, the catalytic process exhibits the atypical *Z*-selectivity (Fig. 15.15).

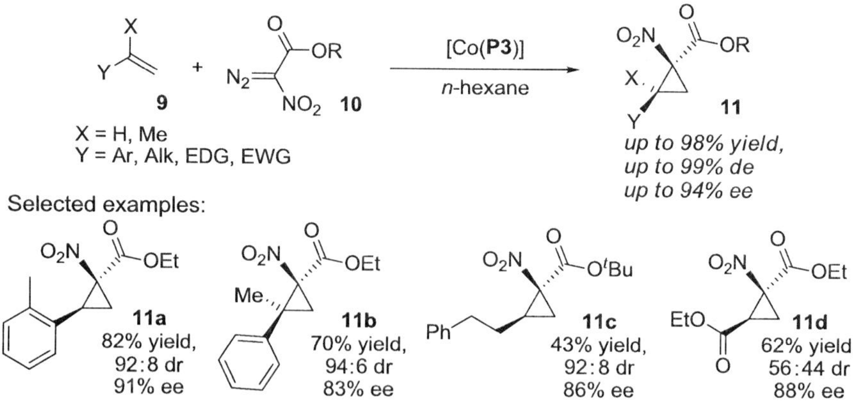

Figure 15.15. [Co(D_2-Por*)]-catalyzed cyclopropanation with α-nitrodiazoacetates.

The effectiveness of [Co(**P3**)] toward asymmetric cyclopropanation with diazo compound **10** was rationalized as a consequence of two potential N–H···O hydrogen-bonding interactions between two of the chiral amide N–H moieties on the **P3** ligand with both the N=O (–NO_2 group) and the C=O (–CO_2R group) units of the carbene moiety, respectively, in a proposed metallocarbene intermediate **27** (Fig. 15.16). These double H-bonding interactions would promote carbene formation from the less reactive diazo reagent and rigidify the intermediate towards its subsequent reaction with the olefin substrate, which would lead to a more effective and selective catalytic process.

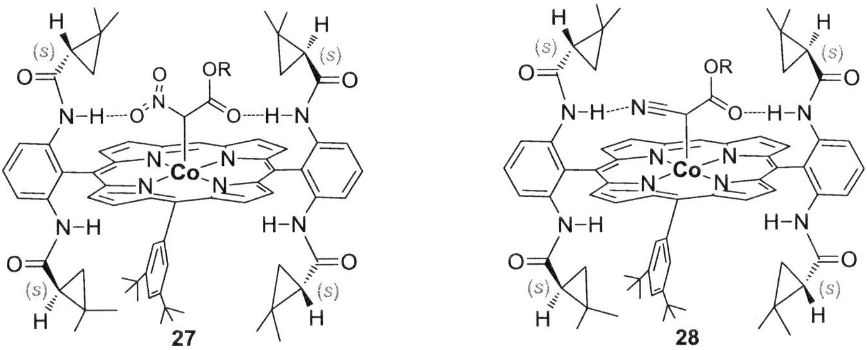

Figure 15.16. Potential double H-bonding in postulated carbene complex of [Co(**P3**)].

By applying the double H-bonding hypothesis of Figure 15.16 to another type of acceptor/acceptor-substituted diazo reagent α-cyanodiazoacetates (**13**), whose cyano group is normally considered as a stronger hydrogen bond acceptor than a nitro group, Zhang and coworkers showed that [Co(**P3**)] is also an effective catalyst for olefin cyclopropanation with diazo compound **13** in even higher stereoselectivity.[12b] This Co(II)-based catalytic system is also suitable for both aromatic and aliphatic olefins with varied electronic properties, affording densely functionalized cyclopropane products **29** with high optical purity (Fig. 15.17).

In comparison with significant advances in asymmetric intermolecular cyclopropanation, asymmetric intramolecular cyclopropanation, especially its scope with acceptor/acceptor-substituted diazo reagents, has been relatively underdeveloped.[2b,d,36] After demonstrations of the unusual effectiveness of [Co(D_2-Por*)] catalysts for a number of challenging asymmetric intermolecular cyclopropanation reactions, Zhang and coworkers reported a catalytic system for asymmetric intramolecular cyclopropanation that is based on the Co(II) complex of D_2-symmetric chiral porphyrin 3,5-Di^{t}Bu-QingPhyrin ([Co(**P9**)])

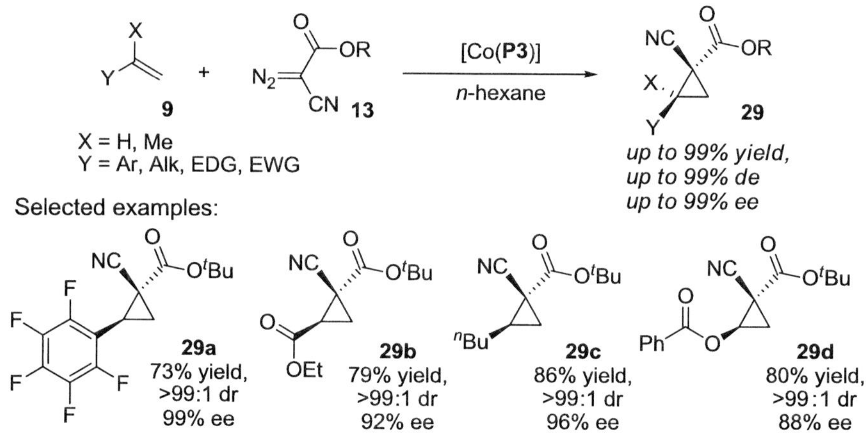

Figure 15.17. [Co(D_2-Por*)]-catalyzed cyclopropanation with α-cyanodiazoacetates.

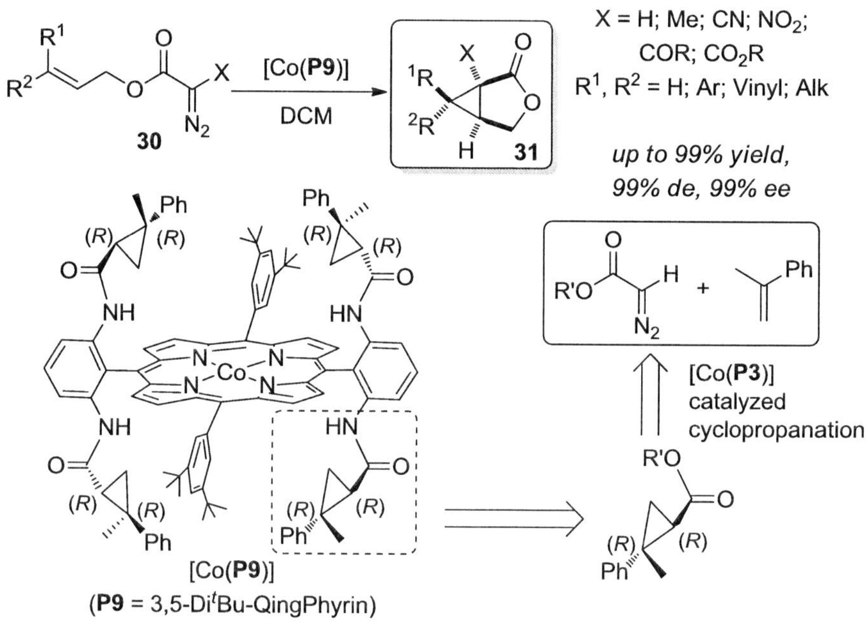

Figure 15.18. [Co(D_2-Por*)]-catalyzed intramolecular asymmetric cyclopropanation.

(Fig. 15.18).[37] In contrast to several prior asymmetric intramolecular catalytic systems that mainly focused on unsaturated acceptor-substituted diazo reagents,[2b,d,38] [Co(**P9**)] proved to be a general and effective catalyst for asymmetric intramolecular cyclopropanation of allylic diazoacetates (**30**) with various α-substituents, especially including the challenging electron-withdrawing groups, affording the corresponding 3-oxabicyclo[3.1.0]hexan-2-one derivatives (**31**) in high yields with excellent diastereoselectivity and enantioselectivity.

Interestingly, the preparation of the cyclopropanecarboxamide unit with two contiguous chiral centers in the D_2-symmetric chiral porphyrin ligand **P9** was based on [Co(**P3**)]-catalyzed asymmetric intermolecular cyclopropanation (Fig. 15.18), although [Co(**P3**)] itself was shown to be a less selective catalyst for asymmetric intramolecular cyclopropanation.[31] This iterative approach has led to the development of a new generation of [Co(D_2-Por*)] catalysts with more diverse chiral environments that may find new applications in Co(II)-catalyzed carbene transfer reactions.

The superior reactivity and stereoselectivity shown by [Co(D_2-Por*)]-based catalytic system for asymmetric cyclopropanation is likely the combined result of the +2 oxidation state of the cobalt metal ion and the D_2-symmetry of the chiral amidoporphyrin ligand. As aforementioned, the Fe(III) complex of D_2-symmetric chiral porphyrin **P3**, [Fe(**P3**)Cl], is a much inferior catalyst for asymmetric cyclopropanation than the Co(II) complex of the same ligand, [Co(**P3**)] (Fig. 15.11). Cobalt(III) complexes of D_2-symmetric chiral porphyrins and related ligands also exhibit a different catalytic carbene transfer profile from their corresponding Co(II) complexes (Kim, C.-S. J., H.-L., Zhang, X. P. Unpublished Results). This combined effect is further illustrated through comparison of asymmetric cyclopropanation reactions with Co(II) complexes of other chiral amidoporphyrins having different symmetry. Gallo, Rose, and coworkers described their studies on asymmetric cyclopropanation by Co(II) complexes of a series of C_2-symmetric chiral binaphthyl tetraamidoporphyrins ([Co(**P10**)]) and A_3B diamidoporphyrins ([Co(**P11**)]) (Fig. 15.19).[29b,c] Although interesting trends on the relationship

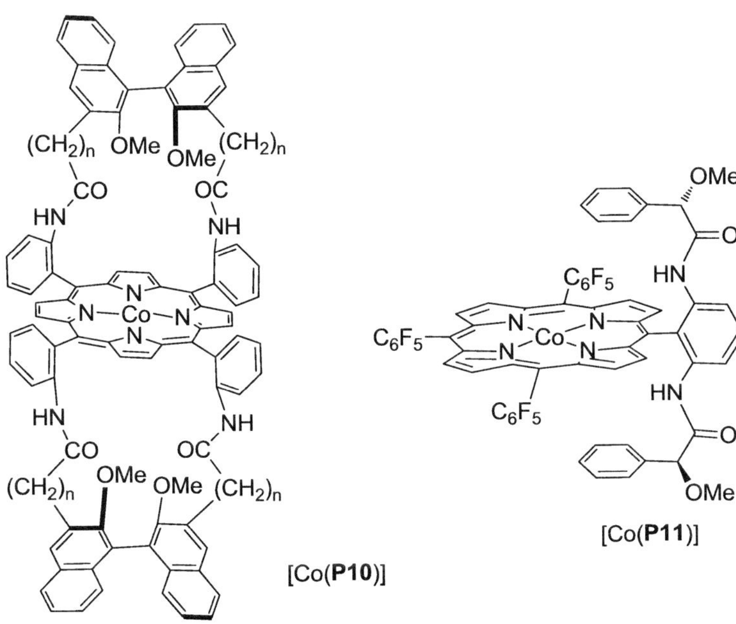

[Co(**P10**)]

[Co(**P11**)]

Figure 15.19. Co(II) complexes of C_2-symmetric chiral amidoporphyrins.

between catalytic reactivity and ligand structure were observed, the catalytic performances of these C_2-symmetric chiral amidoporphyrins for Co(II)-catalyzed asymmetric cyclopropanation are generally not comparable to those of Zhang's D_2-symmetric chiral amidoporphyrins.

15.2.4 Cobalt Catalysts with Other Supporting Ligands

The ligand families shown earlier have demonstrated great structural tunability for pursuing high reactivity and selectivity in Co(II)-based cyclopropanation reactions. In addition to skeletons, such as those of salen and porphyrin, other classes of chiral ligands have also been reported to support cobalt-catalyzed cyclopropanation reactions. Jommi and coworkers designed and synthesized a chiral N_2S_2 tetradentate ligand and studied its Co(II) complex [Co(**N2S2**)] as a catalyst for asymmetric cyclopropanation. Remarkably, 1-octene (**32**) was successfully cyclopropanated with ethyl diazoacetate (**2b**) by [Co(**N2S2**)] to produce *trans*-ethyl 2-hexylcyclopropanecarboxylate (**33**) in 97% ee, although with moderate diastereoselectivity (Scheme 15.12).[39]

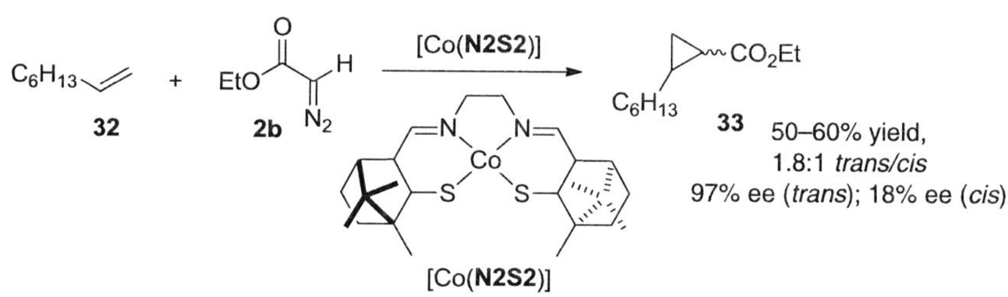

Scheme 15.12. Asymmetric cyclopropanation by Co(II) complex of chiral N_2S_2 tetradentate ligand.

Gade and coworkers developed a series of chiral pincer-type ligands based on bis(2-pyridylimino)isoindole (bpi) for the cobalt(II)-catalyzed asymmetric inter- and intramolecular cyclopropanation of alkenes.[40] One representative example is the **Carbpi** ligand, which is derived from (+)-2-carene with a tetraphenylated bpi backbone (Fig. 15.20). The structure of the **Carbpi** ligand consists of the chiral bicyclic alkane unit as the stereodirecting element and the bulky tetraphenylbenzene unit as a protective hedge, which is expected to block the metal center from backside attack and force the substrate to approach along the chiral bicyclic alkane unit. The judiciously designed catalyst [Co(**Carbpi**)] resulted in excellent stereoselectivities in both intermolecular and intramolecular cobalt-catalyzed cyclopropanation reactions.

Besides cobalt complexes with synthetic ligands, the naturally occurring vitamin B_{12} derivatives, which contain the Co(III) complex with a chiral corrin macrocycle, could be directly used to catalyze asymmetric cyclopropanation reactions of alkenes with ethyl diazoacetate (**2b**) (Scheme 15.13).[41] Among several common derivatives, aquocobalamin (B_{12a}) was shown to be the most

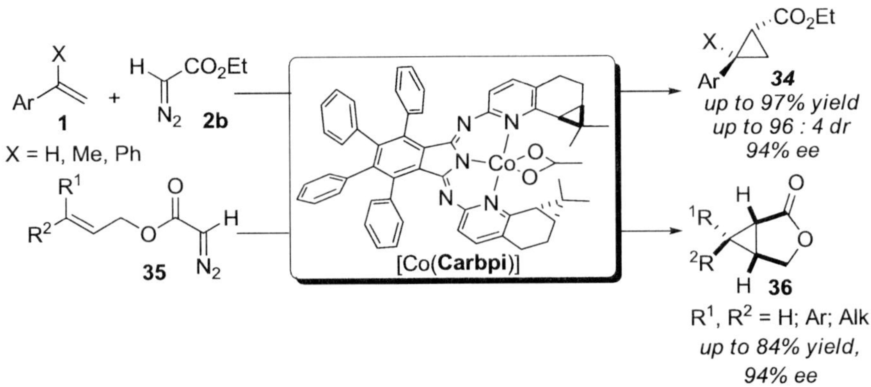

Figure 15.20. Co(II) complex of pincer-type ligand catalyzed cyclopropanation.

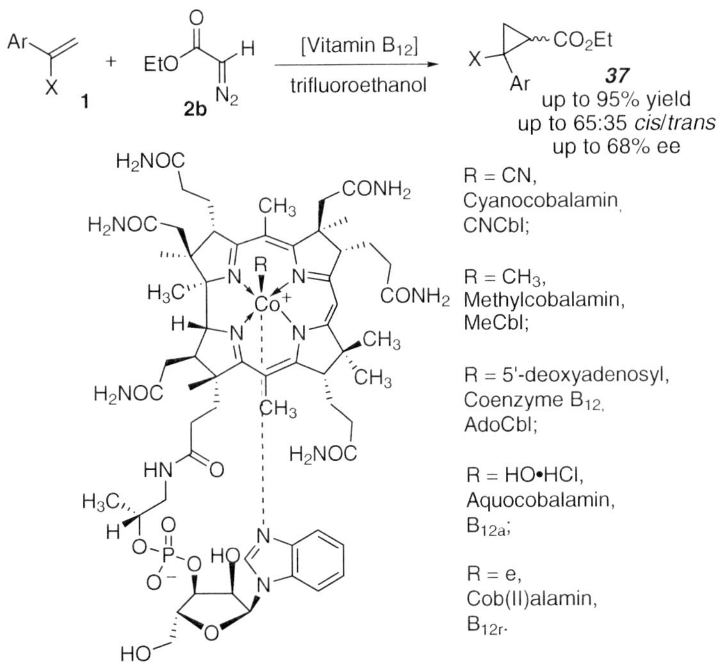

Scheme 15.13. Asymmetric cyclopropanation catalyzed by Vitamin B_{12} derivatives.

effective catalyst for cyclopropanation of a variety of alkenes, providing *cis*-dominant cyclopropanes **37** in excellent yields and moderate enantioselectivity. Experimental data suggested that the catalytically active species might be Co(II) species that was generated possibly via *in situ* reduction of B_{12a} by ethyl diazoacetate **2b**.

Through developments by a number of research groups during the past decades, cobalt-based catalytic carbene transfer systems have grown to become one of the most effective and general catalytic systems for asymmetric cyclopropanation of alkenes. A number of effective chiral ligands have been developed to support Co-catalyzed asymmetric cyclopropanation and to successfully address several long-standing challenges in the field. In addition to those documented earlier, an increasing number of new cobalt catalysts, including cobalt carboporphyrin,[42] cobalt terpyridine,[43] and heterogeneous cobalt catalysts,[44] have also been reported to have catalytic capability for olefin cyclopropanation. These new developments have certainly enriched the chemistry of cobalt carbenes.

15.2.5 Mechanistic Studies

It is evident from earlier discussion that the cyclopropanation reactivity profile of Co-based catalytic systems is distinctly different from those of the widely studies Rh_2- and other metal-based systems. Between two different oxidation states, Co(II) complexes are generally much more versatile and effective catalysts for olefin cyclopropanation than are Co(III) complexes. There have been several reports on the underlying mechanism of Co-catalyzed cyclopropanation, including the nature of cobalt–carbene intermediates involved in the catalytic cycle.[27,45]

While certain Co(III) complexes are believed to form the typical Fischer-type carbene from reaction with diazo reagents and then undergo an electrophilic addition to alkenes,[45a,e,h,i] early theoretical and FT-IR studies on [Co(II)(Salen)] and [Co(II)(Ketim)] systems by Yamada and coworkers suggested that these Co(II) complexes likely formed corresponding carbene species with a single Co–C bond and have substantial unpaired spin density at the carbene carbon.[45b,j] A cyclopropanation mechanism based on a similar Co(III)-carbene intermediate with Co–C single bond and carbon-based radical character was proposed by Zhang and coworkers in their original report on the discovery of [Co(Por)]-catalyzed olefin cyclopropanation.[27]

In 2010, de Bruin, Zhang, and coworkers reported a comprehensive study on the mechanism of [Co(Por)]-catalyzed cyclopropanation of olefins with diazoacetates (**2**).[45g] Density functional theory (DFT) calculations in combination with electron paramagnetic resonance (EPR) and high-resolution mass spectrometry (HRMS) experiments provided a detailed picture of an unprecedented stepwise radical reaction mechanism of the Co(II)-catalyzed cyclopropanation (Fig. 15.21). Due to the strong ligand field and planar coordination environment of porphyrins, [Co(II)(Por)] complexes are known to adopt a low-spin ground state with the $(dxy)^2(dxz,yz)^4(dz^2)^1$ electron configuration as well-defined metalloradicals. DFT calculations suggested the Co(II)-based metalloradical is capable of activating diazo reagents to generate an unusual Co(III)-carbene radical (**B**) as the key intermediate of the catalytic cycle. The net process may be described as a

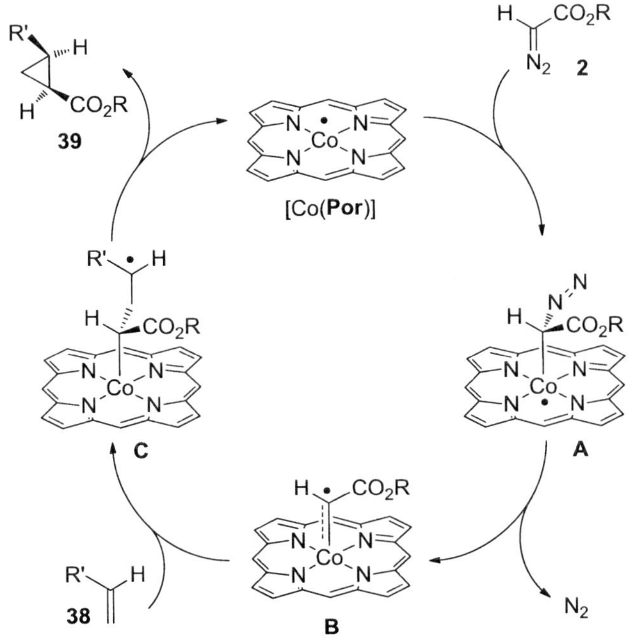

Figure 15.21. Proposed stepwise metalloradical cyclopropanation (MRC).

radical transfer where the radical character is transferred from the Co-center to the α-carbon atom of the original diazo reagent. Different from the common electrophilic Fischer-type carbene intermediates, the new radical carbene intermediate **B**, which was detected experimentally using EPR and HRMS techniques, is effectively a [Co(Por)]-supported carbon-based radical.[45g,d,f] Like carbon free radicals, the Co(III)-carbene radical **B** is capable of undergoing common radical reactions such as radical addition and substitution. Accordingly, radical addition of the α-radical **B** to the C=C bond of olefin substrate (**38**) results in further radical transfer to form the γ-radical intermediate **C**. Because of relatively weak bond dissociation energy of Co–C bonds, the γ-radical **C** readily undergoes ring-closure reaction via intramolecular radical substitution with almost no barrier, resulting in the formation of the desired cyclopropane product (**39**) while regenerating the metalloradical catalyst (Fig. 15.21). Although both the α-radical **B** and γ-radical **C** are essentially carbon-based radicals, they are not "free" as their reactivity and selectivity are well regulated by the electronic, steric, and chiral environments of the covalently attached [Co(Por)] moiety. This proposed one-electron radical mechanism is consistent with the experimental results and explains the unique reactivity and selectivity profiles of the Co(II)-catalyzed cyclopropanation.

In 2011, Zhang, de Bruin, and coworkers reported more definite experimental evidences for the existence of the cobalt(III)-carbene radicals.[45c] In

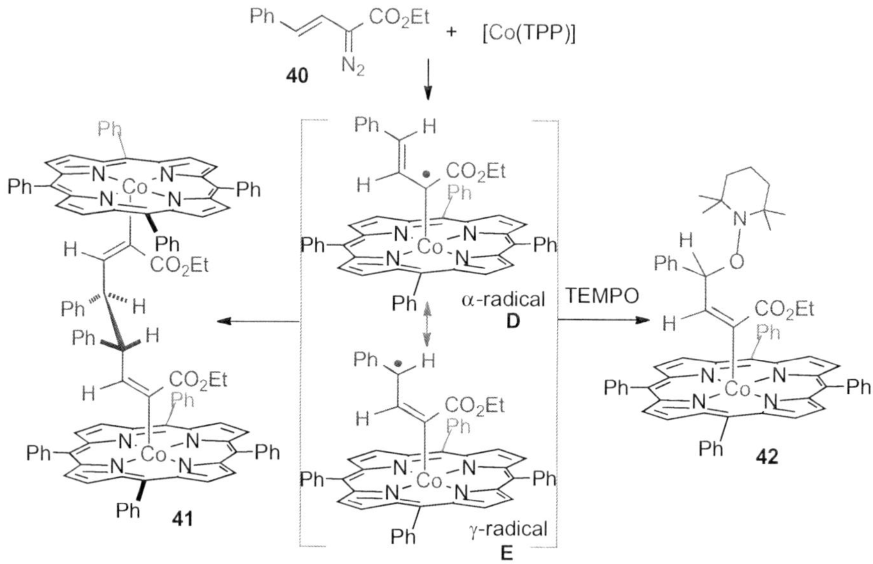

Figure 15.22. Experimental evidences for the existence of the cobalt(III)-carbene radicals.

the absence of olefin substrates, the reaction of [Co(TPP)] with ethyl styryldiazoacetate (**40**) was found to result in the formation of a new metalloporphyrin complex in high yield that was identified as a dinuclear cobalt(III) porphyrin complex (**41**), whose structure was established by single-crystal X-ray diffraction (Fig. 15.22). The formation of dimeric complex **41** was believed to arise from radical dimerization of the initially generated cobalt(III)-carbene radical that exists as a hybrid of α-radical (**D**) and γ-radical (**E**) allylic resonance forms. In addition to homodimerization, the cobalt(III)-vinylcarbene radical intermediate could be effectively trapped by the stable radical TEMPO, leading to the heterodimerization product **42** (Fig. 15.22). These results were taken as convincing further evidence for the existence of the cobalt(III)-carbene radicals that were proposed as key intermediates in the radical-type mechanism of [Co(II)(Por)]-catalyzed cyclopropanation.

Unlike the behavior of the electrophilic Fischer-type carbenes, further mechanistic study by DFT computations showed that cobalt(III)-carbene radical **43** (Fig. 15.23) could possess a certain degree of nucleophilicity.[45d] This finding explains well the unusual catalytic capability of [Co(Por)] catalysts for cyclopropanation of electron-deficient olefins. More broadly, the theoretical study demonstrated that the charge of the carbene radical carbon in cobalt(III)-carbene radicals can be fine-tuned through the introduction of different α-carbon R substituents, from slightly positive to highly negative (Fig. 15.23).[45d] Accordingly, the Co(II)-based radical process may have a potential to be

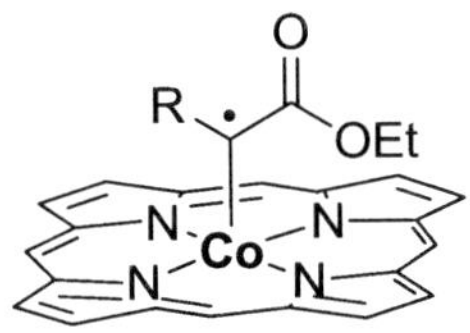

43

Figure 15.23. Electronically tunable cobalt(III)-carbene radicals.

developed into a general cyclopropanation reaction that is suitable to a wide range of olefins with varied electronic properties.

15.3 OTHER COBALT-CATALYZED CARBENE TRANSFER REACTIONS

Compared to the substantial advancements in Co-catalyzed cyclopropanation, the developments of other carbene transfer reactions by Co-based catalysts are still in their early stages. Although less developed, the potential of these Co-catalyzed carbene transformations in organic synthesis, such as asymmetric cyclopropenation,[5] carbonylation,[6] and olefination reactions,[7] has already been shown by several research groups in their recent reports.

15.3.1 Cyclopropenation

Encouraged by the unique catalytic activity of [Co(D_2-Por*)] in asymmetric cyclopropanation of alkenes with acceptor/acceptor-substituted diazo reagents, Zhang and coworkers subsequently developed a [Co(D_2-Por*)]-catalyzed system for enantioselective cyclopropenation with acceptor/acceptor-substituted diazo reagents.[5] Although the Co(II) complex of 3,5-Di^{t}Bu-ChenPhyrin ([Co(**P3**)]) is highly effective for asymmetric cyclopropanation, its catalytic performance for asymmetric cyclopropenation was found to be inferior. Exchange of the aliphatic *tert*-butyl group in **P3** by the aromatic mesityl group generated a new D_2-symmetric chiral porphyrin **P12**, the Co(II) complex of which ([Co(**P12**)]) proved to be a highly effective catalyst for asymmetric cyclopropenation of aryl/vinyl alkynes with both α-cyanodiazoacetamides and α-cyanodiazoacetates (Fig. 15.24). This [Co(**P12**)]-catalyzed cyclopropenation reaction is suitable for a wide range of terminal aromatic and related conjugated alkynes (**44**) with varied steric and electronic properties, providing the corresponding trisubstituted cyclopropenes **46** in high yields with excellent enantiocontrol of the all-carbon quaternary stereogenic centers. Moreover, substrates bearing aldehyde, phenol, and amino groups could be successfully converted to cyclopropene products (**46a–d**) without protection. The high degree of functional group tolerance of this catalysis is believed to have close relevance to the radical pathway of Co(II)-based metalloradical catalysis.

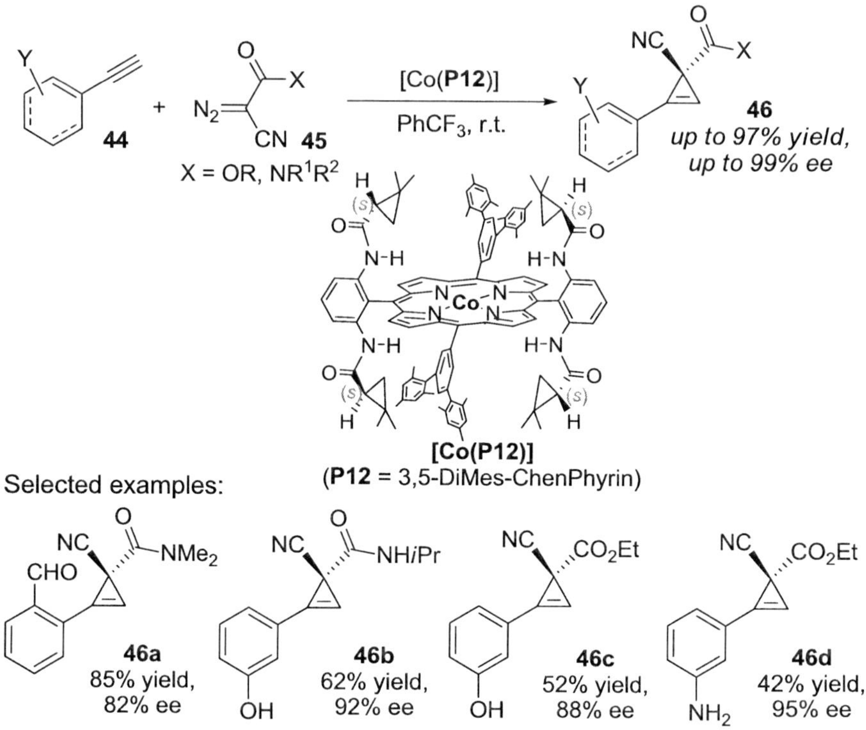

Figure 15.24. [Co(D_2-Por*)]-catalyzed enantioselective cyclopropenation of alkynes with acceptor/acceptor-substituted diazo reagents.

15.3.2 Carbonylation

Ungvary and coworkers carried out systematic studies on carbonylation reaction of diazoacetates with octacarbonyl dicobalt complex [$Co_2(CO)_8$].[6a,b] In the presence of carbon monoxide, [$Co_2(CO)_8$] could effectively catalyze the selective transformation of diazoacetates to the corresponding ketenes. For example, effective dediazotation of ethyl diazoacetate (**2b**) under a CO atmosphere by the dinuclear cobalt(0) complex [$Co_2(CO)_8$] resulted in the formation of ethoxycarbonylketene (**47**) (Scheme 15.14) in high yield.[6c–e] The *mono*- and *di*-ethoxycarbonylcarbene-bridged dicobalt carbonyl complexes **48a** and **48b** were identified as active intermediates for the catalytic carbonylation reaction.

Due to the high reactivity of ketenes, *in situ* formed **47** was directly converted into malonic acid derivatives **49–51** in the presence of alcohol, phenol, amine, or diazoacetate derivatives (Scheme 15.15).[6f,g] When *N*-*tert*-butylbenzylimine was added, direct reaction with ketene **47** resulted in cyclization to produce β-lactam **52** in 95% yield.[6h]

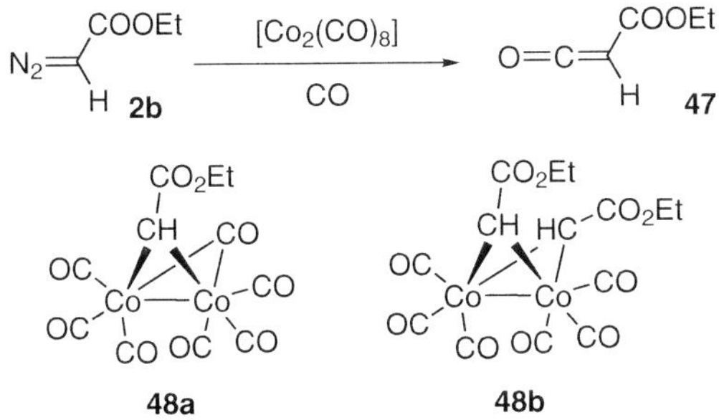

Scheme 15.14. [$Co_2(CO)_8$]-catalyzed carbonylation of diazoacetates.

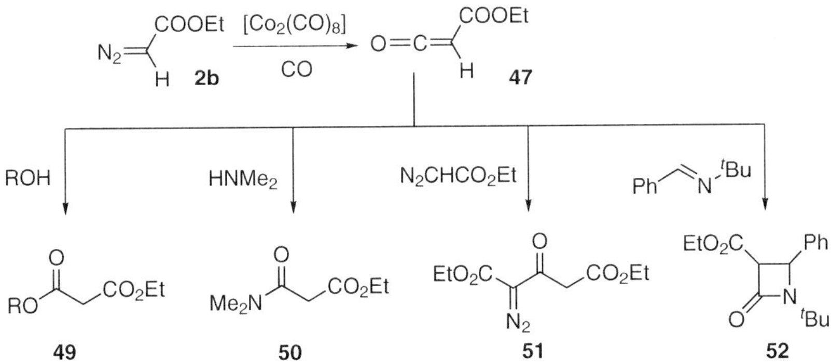

Scheme 15.15. Synthetic applications of [$Co_2(CO)_8$]-catalyzed carbonylation.

15.3.3 Olefination

Several transition metal complexes, including those of Mo, Re, Ru, Rh, and Fe, have been known to catalyze selective olefination of aldehydes with suitable diazo reagents in the presence of tertiary phosphines. Commercially available [Co(TPP)] was also shown to be an effective catalyst for selective olefination of carbonyl compounds (**53**) with ethyl diazoacetate (**2b**) in the presence of triphenylphosphine.[7] Due to the minimized dimerization of diazo reagents in Co-based systems, the catalytic olefination reactions of aldehydes could be carried out in a one-time protocol without the slow addition of the diazo **2b**, affording olefins **54a** and **54b** in high yields and excellent (*E*)-selectivity (Scheme 15.16). The catalytic reaction has a wide scope and is applicable to a variety of aldehydes including electron-neutral, -poor, -rich, sterically demanding, and functionalized aromatic aldehydes as well as α,β-unsaturated, cyclic, and aliphatic aldehydes. Trifluoromethyl ketones are also suitable substrates for the olefination reaction, leading to the stereoselective synthesis of β-trifluoromethyl-α,β-unsaturated esters. Under similar conditions, however, unactivated simple ketones were ineffective substrates for olefination. But in

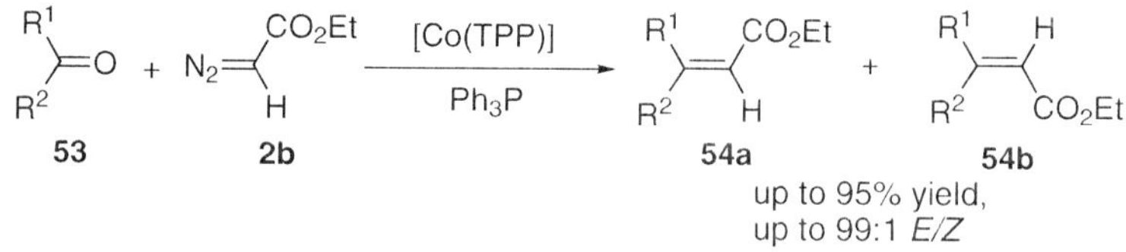

Scheme 15.16. [Co(TPP)]-catalyzed olefination reactions.

the presence of substoichiometric amounts of benzoic acids, these reactions could be significantly facilitated to reach better yields.[7]

15.4 CONCLUSION AND OUTLOOK

Supported by chiral salen and porphyrin ligands, Co(II)-based metalloradical catalysts have been developed to become one of the major classes of catalytic systems for asymmetric cyclopropanation of alkenes. Co(II)-catalyzed cyclopropanation has been shown to possess a unique catalytic reactivity profile which is attributed to its underlying radical mechanism with the unusual Co(III)-carbene radical as the key intermediate. In combination with suitable chiral ligands, Co(II)-based metalloradical catalysts have been shown to be highly effective for asymmetric cyclopropanation of a wide range of alkene substrates with different classes of diazo reagents as carbene sources, affording corresponding cyclopropane derivatives in high yields and with excellent diastereoselectivity and enantioselectivity. In particular, Co(II)-based catalytic systems have unprecedented catalytic activity toward electron-deficient olefins. In addition to acceptor-substituted diazo reagents with both carbonyl and non-carbonyl α-substituents, Co(II)-based metalloradical catalysts have displayed unusual capability to activate acceptor/acceptor-substituted diazo reagents as carbene sources for asymmetric cyclopropanation. By changing the ligand environment, a diastereoselectivity switch from the common *trans*-selective to the rare *cis*-selective stereocontrol has been realized in Co(II)-catalyzed asymmetric cyclopropanation. Intramolecular versions of asymmetric cyclopropanation by Co(II)-based catalysts have also been successfully developed.

These demonstrated achievements will undoubtedly stimulate further development of Co(II)-based catalytic systems for addressing other challenging issues in asymmetric cyclopropanation and related carbene transfer processes. Cobalt(II)-catalyzed asymmetric cyclopropanation will continue to see further expansion of substrate scope to include other difficult olefin substrates, including aliphatic and internal olefins. In parallel, the Co(II)-based catalytic system will be expanded to employ other acceptor/acceptor-substituted diazo reagents as carbene sources for stereoselective cyclopropanation. Among them, dicarbonyldiazo compounds such as α-ketodiazoacetates, α-formyldiazoacetates, and α-diazomalonates will be attractive but challenging candidates. It is noted that

Co(II)-based metalloradical cyclopropanation has yet to be demonstrated with donor/acceptor-substituted diazo reagents. Like cyclopropanation, Co(II)-catalyzed asymmetric cyclopropenation will be further developed for alkyne substrates beyond terminal aryl and vinyl alkynes and with diazo reagents other than α-cyanodiazoacetates and α-cyanodiazoacetamides.

The Co(II)-based metalloradical system will be explored for other important catalytic carbene transfer processes. Among them, an exciting possibility is Co(II)-catalyzed stereoselective radical C–H alkylation via formal carbene insertion. Asymmetric C–H insertion reactions have been successfully demonstrated with Rh_2- and other metal-based catalytic systems using acceptor- and donor/acceptor-substituted diazo reagents as carbene sources.[2a,c,46] Due to the electrophilic nature of the metal carbene intermediates, these existing catalytic insertion systems are limited to electron-rich C–H bonds. As demonstrated for asymmetric cyclopropanation, it is conceivable that a Co(II)-based metalloradical system for C–H alkylation would have a much broader scope of both C–H substrates and carbene sources.

Evidence has accumulated to suggest that both trivalent and divalent cobalt complexes supported by N_2O_2- and N_4-tetradentate ligands have the capability to activate diazo reagents to generate cobalt–carbene intermediates that can be catalytically transferred to substrates. Between cobalt complexes in their +2 and +3 oxidation states, the resulting cobalt–carbene species appear to differ fundamentally. Similar to the carbene intermediates involved in catalytic carbene transfer processes by many other metals, the corresponding cobalt carbenes from d^6 Co(III) complexes are most likely electrophilic Fischer-type carbenes (Fig. 15.25). On the other hand, convincing evidence has been attained to show that Co(II) complexes, as metalloradicals with d^7 open-shell electronic structure, tend to form a new class of metal carbenes with unusual M–C single bond and carbon-based radical character (Fig. 15.25).[45] Like carbon free radicals, these so-called "metal carbene radicals" can undergo common radical reactions but with potential control of reactivity and stereoselectivity by the attached metal ligand complex. Given that cobalt complexes in +1 oxidation state are also well known, it is tempting to speculate the possible existence of corresponding cobalt carbenes from d^8 Co(I) complexes. If possible, would they behave more like Fisher-type or Schrock-type carbenes (Fig. 15.25)? This

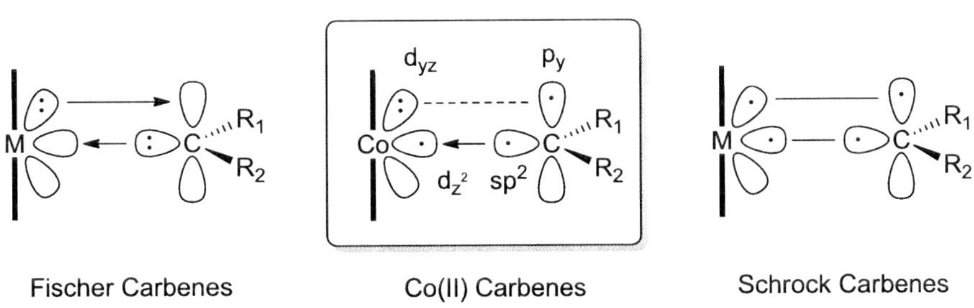

Figure 15.25. Bonding interactions of metallocarbene species.

and related fundamental questions may open totally new research avenues for further development of cobalt carbene chemistry.

ACKNOWLEDGMENTS

We are grateful for financial support by the National Science Foundation (CHE-1152767) and the National Institutes of Health (R01-GM098777).

SUGGESTED READING

Caballero, A., Prieto, A., Diaz-Requejo, M. M., Perez, P. J. "Metal-Catalyzed Olefin Cyclopropanation with Ethyl Diazoacetate: Control of the Diastereoselectivity," *Eur. J. Inorg. Chem.* **2009**, *1137*–1144.

Doyle, M. P. "Exceptional Selectivity in Cyclopropanation Reactions Catalyzed by Chiral Cobalt(II)-Porphyrin Catalysts," *Angew. Chem. Int. Ed.* **2009**, *48*, 850–852.

Doyle, M. P., McKervey, M. A., Ye, T. *Modern Catalytic Methods for Organic Synthesis with Diazo Compounds: From Cyclopropanes to Ylides*. Wiley: New York, **1998**.

Dzik, W. I., Xu, X., Zhang, X. P., Reek, J. N. H., de Bruin, B. "'Carbene Radicals' in Co^{II}(por)-Catalyzed Olefin Cyclopropanation," *J. Am. Chem. Soc.* **2010**, *132*, 10891–10902.

Intrieri, D., Caselli, A., Gallo, E. "Cyclopropanation Reactions Mediated by Group 9 Metal Porphyrin Complexes," *Eur. J. Inorg. Chem.* **2011**, 5071–5081.

Katsuki, T. "Chiral Metallosalen Complexes: Structures and Catalyst Tuning for Asymmetric Epoxidation and Cyclopropanation," *Adv. Synth. Catal.* **2002**, *344*, 131–147.

Lebel, H., Marcoux, J. F., Molinaro, C., Charette, A. B. "Stereoselective Cyclopropanation Reactions," *Chem. Rev.* **2003**, *103*, 977–1050.

Moss, R. A., Platz, M., Jones, M. *Reactive Intermediate Chemistry*. Wiley-Interscience: Hoboken, NJ, **2004**.

Spessard, G. O., Miessler, G. L. *Organometallic Chemistry*. 2nd ed. Oxford University Press: New York, **2010**.

Yamada, T., Ikeno, T., Ohtsuka, Y., Kezuka, S., Sato, M., Iwakura, I. "Manganese and Cobalt 3-Oxobutylideneaminato Complexes: Design and Application for Enantioselective Reactions," *Sci. Technol. Adv. Mat.* **2006**, *7*, 184–196.

REFERENCES

1 (a) Spessard, G. O., Miessler, G. L. *Organometallic Chemistry*, 2nd ed. Oxford University Press: New York, **2010**. (b) Moss, R. A., Platz, M., Jones, M. *Reactive Intermediate Chemistry*. Wiley-Interscience: Hoboken, NJ, **2004**. (c) Schubert, U. *Advances in Metal Carbene Chemistry*. Kluwer Academic Publishers: Dordrecht, the Netherlands and Boston, **1989**.

2 (a) Doyle, M. P., Duffy, R., Ratnikov, M., Zhou, L. *Chem. Rev.* **2010**, *110*, 704–724. (b) Lebel, H., Marcoux, J. F., Molinaro, C., Charette, A. B. *Chem. Rev.* **2003**, *103*, 977–1050.

(c) Davies, H. M. L., Beckwith, R. E. *J. Chem. Rev*. **2003**, *103*, 2861–2903 (d) Doyle, M. P., Forbes, D. C. *Chem. Rev*. **1998**, *98*, 911–935. (e) Doyle, M. P., McKervey, M. A., Ye, T. *Modern Catalytic Methods for Organic Synthesis with Diazo Compounds: From Cyclopropanes to Ylides*. Wiley: New York, **1998**. (f) Doyle, M. P. *Chem. Rev.* **1986**, *86*, 919–939.

3 (a) Intrieri, D., Caselli, A., Gallo, E. Eur. J. Inorg. Chem. **2011**, 5071–5081. (b) Yamada, T. *J. Synth. Org. Chem. Jpn*. **2010**, *68*, 966–972. (c) Hess, W., Treutwein, J., Hilt, G. Synthesis-Stuttgart **2008**, 3537–3562.(d) Katsuki, T. *Adv. Synth. Catal.* **2002**, *344*, 131–147.

4 Caballero, A., Prieto, A., Diaz-Requejo, M. M., Perez, P. J. *Eur. J. Inorg. Chem.* **2009**, 1137–1144.

5 Cui, X., Xu, X., Lu, H. J., Zhu, S. F., Wojtas, L., Zhang, X. P. *J. Am. Chem. Soc.* **2011**, *133*, 3304–3307.

6 (a) Ungvari, N., Fordos, E., Kegl, T., Ungvary, F. *Inorg. Chim. Acta* **2009**, *362*, 1333–1342. (b) Tuba, R., Fordos, E., Ungvary, F. *J. Mol. Catal. A: Chem*. **2005**, *236*, 113–118. (c) Kegl, T., Ungvary, F. *Lett. Org. Chem*. **2010**, *7*, 634–644.(d) Kegl, T., Ungvary, F. *J. Organomet. Chem.* **2007**, *692*, 1825–1833. (e) Fordos, E., Ungvari, N., Kegl, T., Ungvary, F. *Eur. J. Inorg. Chem*. **2006**, 1875–1880. (f) Ungvari, N., Fordos, E., Kegl, T., Ungvary, F. *Inorg. Chim. Acta* **2010**, *363*, 2016–2028. (g) Tuba, R., Ungvary, F. *J. Mol. Catal. A: Chem*. **2003**, *203*, 59–67. (h) Fordos, E., Tuba, R., Parkanyi, L., Kegl, T., Ungvary, F. *Eur. J. Org. Chem*. *2009*, 1994–2002.

7 Lee, M. Y., Chen, Y., Zhang, X. P. *Organometallics* **2003**, *22*, 4905–4909.

8 Yamada, T., Ikeno, T., Ohtsuka, Y., Kezuka, S., Sato, M., Iwakura, I. *Sci. Technol. Adv. Mat.* **2006**, *7*, 184–196.

9 Chen, Y., Ruppel, J. V., Zhang, X. P. *J. Am. Chem. Soc.* **2007**, *129*, 12074–12075.

10 Morandi, B., Mariampillai, B., Carreira, E. M. *Angew. Chem. Int. Ed.* **2011**, *50*, 1101–1104.

11 Zhu, S. F., Ruppel, J. V., Lu, H. J., Wojtas, L., Zhang, X. P. *J. Am. Chem. Soc.* **2008**, *130*, 5042–5043.

12 (a) Zhu, S. F., Perman, J. A., Zhang, X. P. *Angew. Chem. Int. Ed*. **2008**, *47*, 8460–8463. (b) Zhu, S. F., Xu, X., Perman, J. A., Zhang, X. P. *J. Am. Chem. Soc.* **2010**, *132*, 12796–12799.

13 (a) Tatsuno, Y., Konishi, A., Nakamura, A., Otsuka, S. *J. Chem. Soc.—Chem. Commun*. **1974**, 588–589. (b) Nakamura, A., Konishi, A., Tatsuno, Y., Otsuka, S. *J. Am. Chem. Soc.* **1978**, *100*, 3443–3448.(c) Nakamura, A., Konishi, A., Tsujitani, R., Kudo, M., Otsuka, S. *J. Am. Chem. Soc.* **1978**, *100*, 3449–3461.(d) Nakamura, A. *Pure Appl. Chem.* **1978**, *50*, 37–42.

14 Yamada, T., Ikeno, T., Sekino, H., Sato, M. *Chem. Lett*. **1999**, 719–720.

15 (a) Ikeno, T., Sato, M., Yamada, T. *Chem. Lett*. **1999**, 1345–1346. (b) Ikeno, T., Sato, M., Iwakura, I., Kokura, A., Nagata, T., Yamada, T. *J. Synth. Org. Chem. Jpn.* **2008**, *66*, 110–123.

16 (a) Ikeno, T., Nishizuka, A., Sato, M., Yamada, T. *Synlett* **2001**, 406–408. (b) Ikeno, T., Sato, M., Sekino, H., Nishizuka, A., Yamada, T. *Bull. Chem. Soc. Jpn*. **2001**, *74*, 2139–2150. (c) Ikeno, T., Iwakura, I., Yamada, T. *Bull. Chem. Soc. Jpn.* **2001**, *74*, 2151–2160.

17 (a) Uchida, T., Katsuki, T. *Synthesis-Stuttgart* **2006**, 1715–1723. (b) Fukuda, T., Katsuki, T. *Tetrahedron* **1997**, *53*, 7201–7208.(c) Itoh, K., Fukuda, T., Kitajima, H., Katsuki, T. *J. Synth. Org. Chem. Jpn*. **1997**, *55*, 764–773;(d) Fukuda, T., Katsuki, T. *Synlett* **1995**, 825–826.

18 (a) Niimi, T., Uchida, T., Irie, R., Katsuki, T. *Adv. Synth. Catal.* **2001**, *343*, 79–88. (b) Niimi, T., Uchida, T., Irie, R., Katsuki, T. *Tetrahedron Lett.* **2000**, *41*, 3647–3651.

19 Shitama, H., Katsuki, T. *Chem. – Eur. J.* **2007**, *13*, 4849–4858.

20 (a) Saha, B., Uchida, T., Katsuki, T. *Tetrahedron: Asymmetry* **2003**, *14*, 823–836. (b) Uchida, T., Saha, B., Katsuki, T. *Tetrahedron Lett.* **2001**, *42*, 2521–2524.

21 Wurz, R. P., Charette, A. B. *Org. Lett.* **2002**, *4*, 4531–4533.

22 Gao, J., Woolley, F. R., Zingaro, R. A. *Org. Biomol. Chem.* **2005**, *3*, 2126–2128.

23 Kadish, K. M., Smith, K. M., Guilard, R. *The Porphyrin Handbook*. Academic Press: San Diego,CA, **2000**.

24 Callot, H. J., Piechocki, C. *Tetrahedron Lett.* **1980**, *21*, 3489–3492.

25 Ruppel, J. V., Fields, K. B., Snyder, N. L., Zhang, X. P. "Metalloporphyrin-Catalyzed Asymmetric Atom/Group Transfer Reactions." In *Handbook of Porphyrin Science*. Kadish, K. M., Smith, K. M., Guilard, R., Eds. World Scientific: Hackensack, NJ; London, **2010**, Vol. *10*, pp. 1–84.

26 Penoni, A., Wanke, R., Tollari, S., Gallo, E., Musella, D., Ragaini, F., Demartin, F., Cenini, S. *Eur. J. Inorg. Chem.* **2003**, 1452–1460.

27 Huang, L. Y., Chen, Y., Gao, G. Y., Zhang, X. P. *J. Org. Chem.* **2003**, *68*, 8179–8184.

28 (a) Caselli, A., Gallo, E., Ragaini, F., Ricatto, F., Abbiati, G., Cenini, S. *Inorg. Chim. Acta* **2006**, *359*, 2924–2932. (b) Chen, Y., Gao, G. Y., Zhang, X. P. *Tetrahedron Lett.* **2005**, *46*, 4965–4969.

29 (a) Chen, Y., Fields, K. B., Zhang, X. P. *J. Am. Chem. Soc.* **2004**, *126*, 14718–14719. (b) Fantauzzi, S., Gallo, E., Rose, E., Raoul, N., Caselli, A., Issa, S., Ragaini, F., Cenini, S. *Organometallics* **2008**, *27*, 6143–6151. (c) Raoul, N., Gallo, E., Rose, E. *J. Porphyrins Phthalocyanines* **2011**, *15*, 602–611.

30 Chen, Y., Zhang, X. P. *Synthesis-Stuttgart* **2006**, 1697–1700.

31 Chen, Y., Zhang, X. P. *J. Org. Chem.* **2007**, *72*, 5931–5934.

32 Ruppel, J. V., Gauthier, T. J., Snyder, N. L., Perman, J. A., Zhang, X. P. *Org. Lett.* **2009**, *11*, 2273–2276.

33 (a) Padwa, A., Krumpe, K. E. *Tetrahedron* **1992**, *48*, 5385–5453. (b) Pietruszka, *J. Chem. Rev.* **2003**, *103*, 1051–1070.

34 Zhu, S. F., Cui, X., Zhang, X. P. *Eur. J. Inorg. Chem.* **2012**, 430–434.

35 Doyle, M. P. *Angew. Chem. Int. Ed.* **2009**, *48*, 850–852.

36 Pellissier, H. *Tetrahedron* **2008**, *64*, 7041–7095.

37 Xu, X., Lu, H. J., Ruppel, J. V., Cui, X., de Mesa, S. L., Wojtas, L., Zhang, X. P. *J. Am. Chem. Soc.* **2011**, *133*, 15292–15295.

38 (a) Xu, Z.-J., Fang, R., Zhao, C., Huang, J.-S., Li, G.-Y., Zhu, N., Che, C.-M. *J. Am. Chem. Soc.* **2009**, *131*, 4405–4417. (b) Abu-Elfotoh, A. M., Phomkeona, K., Shibatomi, K., Iwasa, S. *Angew. Chem. Int. Ed.* **2010**, *49*, 8439–8443. (c) Ito, J., Ujiie, S., Nishiyama, H. *Chem. – Eur. J.* **2010**, *16*, 4986–4990.

39 Jommi, G., Pagliarin, R., Rizzi, G., Sisti, M. *Synlett* **1993**, 833–834.

40 Langlotz, B. K., Wadepohl, H., Gade, L. H. *Angew. Chem. Int. Ed.* **2008**, *47*, 4670–4674.

41 Chen, Y., Zhang, X. P. *J. Org. Chem.* **2004**, *69*, 2431–2435.

42 Fields, K. B., Engle, J. T., Sripothongnak, S., Kim, C., Zhang, X. P., Ziegler, C. *J. Chem. Commun.* **2011**, *47*, 749–751.

43 Yeung, C. T., Sham, K. C., Lee, W. S., Wong, W. T., Wong, W. Y., Kwong, H. L. *Inorg. Chim. Acta* **2009**, *362*, 3267–3273.

44 (a) Boldini, I., Guillemot, G., Caselli, A., Proust, A., Gallo, E. *Adv. Synth. Catal.* **2010**, *352*, 2365–2370. (b) Caselli, A., Buonomenna, M. G., de Baldironi, F., Laera, L.,

Fantauzzi, S., Ragaini, F., Gallo, E., Golemme, G., Cenini, S., Drioli, E. *J. Mol. Catal. A: Chem.* **2010**, *317*, 72–80.

45 (a) Albrecht, M., Maji, P., Hausl, C., Monney, A., Muller-Bunz, H. *Inorg. Chim. Acta* **2012**, *380*, 90–95. (b) Aldajaei, J. T., Gronert, S. *Int. J. Mass Spectrom.* **2012**, *316*, 68–75. (c) Lu, H. J., Dzik, W. I., Xu, X., Wojtas, L., de Bruin, B., Zhang, X. P. *J. Am. Chem. Soc.* **2011**, *133*, 8518–8521. (d) Belof, J. L., Cioce, C. R., Xu, X., Zhang, X. P., Space, B., Woodcock, H. L. *Organometallics* **2011**, *30*, 2739–2746. (e) Chattopadhyay, P., Matsuo, T., Tsuji, T., Ohbayashi, J., Hayashi, T. *Organometallics* **2011**, *30*, 1869–1873. (f) Dzik, W. I., Zhang, X. P., de Bruin, B. *Inorg. Chem.* **2011**, *50*, 9896–9903. (g) Dzik, W. I., Xu, X., Zhang, X. P., Reek, J. N. H., de Bruin, B. *J. Am. Chem. Soc.* **2010**, *132*, 10891–10902. (h) Koehn, S. K., Gronert, S., Aldajael, J. T. *Org. Lett.* **2010**, *12*, 676–679. (i) Iwakura, I., Tanaka, H., Ikeno, T., Yamada, T. *Chem. Lett.* **2004**, *33*, 140–141. (j) Ikeno, T., Iwakura, I., Yamada, T. *J. Am. Chem. Soc.* **2002**, *124*, 15152–15153.

46 (a) Davies, H. M. L., Lian, Y. *Acc. Chem. Res.* **2012**, *45*, 923–935. (b) Davies, H. M. L., Morton, D. *Chem. Soc. Rev.* **2011**, *40*, 1857–1869. (c) Davies, H. M. L., Dick, A. R. In *C–H Activation*; Yu, J. Q., Shi, Z., Eds. Springer: Heidelberg, NY, **2010**, Vol. *292*, pp. 303–345. (d) Davies, H. M. L., Manning, J. R. *Nature* **2008**, *451*, 417–424. (e) Davies, H. M. L., Antoulinakis, E. G. *J. Organomet. Chem.* **2001**, *617*, 47–55. (f) Slattery, C. N., Ford, A., Maguire, A. R. *Tetrahedron* **2010**, *66*, 6681–6705.

16

GOLD CARBENES

LIMING ZHANG
Department of Chemistry and Biochemistry, University of California, Santa Barbara, CA, USA

16.1 INTRODUCTION

Homogeneous gold catalysis[1] has experienced explosive development from little known at the turn of the new millennium to a broadly recognized area in organic synthesis due to its rich chemistry and fascinating reactivities. With synergistic efforts by many research groups around the globe, its rapid rise to "stardom" has little to do with surprises against the long-held perception that gold is an inert metal and of little use in homogeneous catalysis, but squarely rests on the unique reactivities that gold complexes bestow on various substrates with C–C π bonds, especially alkynes and allenes.

An outstanding, though not necessarily unique, feature of gold(I) complexes and, to a less extent, of Au(III) complexes is their softness as Lewis acids and thereby their potent capabilities in activating π systems toward efficient attack by various nucleophiles. This reactivity mode serves most often as the entry point into gold catalysis, generating an alkenyl gold intermediate **A** in the case of an alkyne substrate (as shown in Scheme 16.1). While the Au–C(sp^2) bond in this intermediate can be transformed by an approaching electrophile (i.e., the α approach), an alternative electrophilic attack at the alkene end distal to gold (i.e., the β approach) would generate a carbocation with a gold complex directly attached (i.e., **B**). By invoking back donation by the gold center, a gold carbene (i.e., **C**) becomes a mesomeric isomer of likely significant contribution to the overall structure. This intermediate and intermediates of this nature are rich in

Contemporary Carbene Chemistry, First Edition. Edited by Robert A. Moss and Michael P. Doyle.

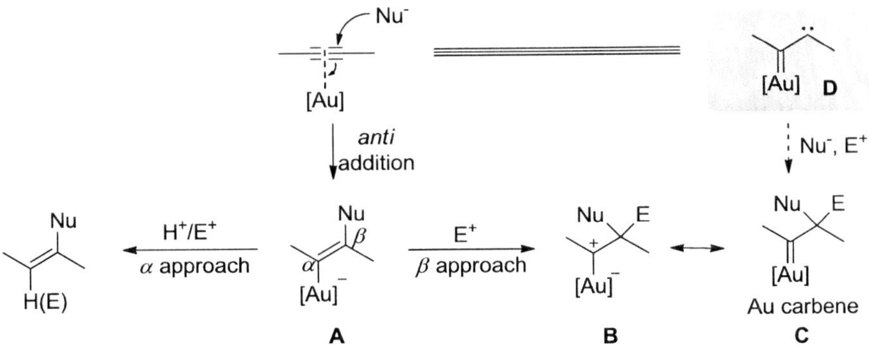

Scheme 16.1.

reactivities. Due to their ready access via a variety of intramolecular or intermolecular approaches following the route outlined in Scheme 16.1, a large array of synthetically highly useful reactions have been developed based on these versatile intermediates.

Notably, the overall process of generating the gold carbene **C** from an alkyne can be viewed formally as a hypothetical and perhaps provocative α-carbene gold carbene **D** that reacts consecutively with a nucleophile and an electrophile at its free carbene center. Hence, the gold-coordinated alkyne, if en route to a gold carbene intermediate, is formally equivalent to **D** (Scheme 16.1)! This formalism, while perhaps of little theoretical usage, is practically useful as a mnemonic model that can unify most of the strategies of gold carbene formation including many to be discussed in this chapter. In addition, it could serve as a simple model to inspire further reaction discovery.

With allenes as substrates, a generic reaction pathway similar to that in Scheme 16.1 could also lead to the formation of a gold carbene intermediates (i.e., **E**) (Scheme 16.2); however, the difference is that the nucleophile and the electrophile are in a 1,3-relationship instead of geminal. In this case, the overall process can be viewed as a 1,3-dipolar gold carbene (i.e., **F**) that undergoes consecutive reactions at the charged carbon centers. As a result, the formal equivalency between a gold-coordinated allene and a 1,3-dipolar gold carbene can be conceived. This formalism, while useful for envisioning reaction outcomes and helpful with designing new reactions, may not just

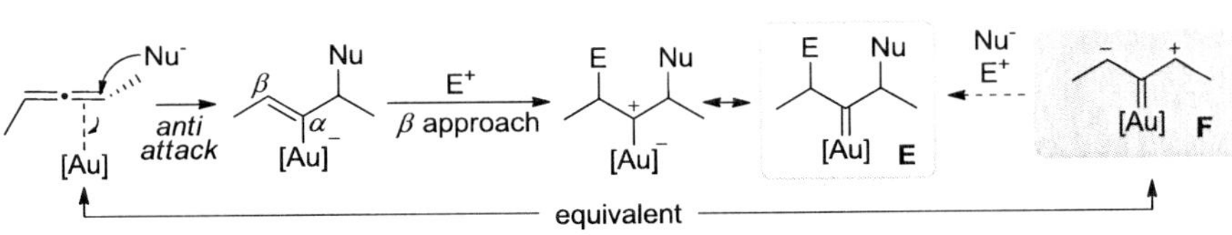

Scheme 16.2.

be used for mnemonic purposes; instead, as shown in Section 16.3.5, gold carbenes of type **F** could be generated as real reactive species en route to concerted pericyclic reactions.

With the general strategy for the generation of gold carbenes outlined, this chapter will first examine the bonding nature of the gold carbene moiety and then discuss generations and reactivities of different types of gold carbenes. Due to the large number of published works involving gold carbene intermediates, it is neither possible nor the intention of the author to cover the topic in a comprehensive manner. Instead, representative examples are chosen to highlight the uniqueness of different approaches employed, and comments, wherever appropriate, are provided to offer perspectives. Throughout the chapter, the term "gold carbene" is preferred over "gold carbenoid" and "gold-stabilized carbocation" for simplicity. In addition, gold *N*-heterocyclic carbene complexes are not covered as they are, in general, not reactive intermediates but mostly stable isolable compounds.[2]

16.2 NATURE OF THE Au–CARBON DOUBLE BOND

16.2.1 Experimental Studies

The resonance structures **B** and **C** represents two extremes of a gold carbene, with the former indicating no back donation from the gold metal and the latter suggesting a complete back bonding. Although **B** and **C** most likely define between them a continuum that covers gold carbenes with a variety of substitution patterns, it is still tempting to probe which extreme a gold carbene leans more toward. Moreover, this understanding can predict and/or rationalize whether a gold carbene reacts as an electrophilic center, as in the case of **B**, or undergoes cyclopropanations and C–H insertions, as in the case of **C**. In this context, Fürstner estimated the rotation barrier of C2–C3 bond in the alkenyl gold complex **1** by ^{13}C NMR[3] to be around or <7.2 kcal, which is in the same range as that of the C2–C3 bond in 1,3-butadiene (Fig. 16.1). This data suggests little double bond character between C2 and C3 in **1** and hence minimal contribution by the gold carbene resonance form **1′**. However, the necessity of the two γ-positioned oxygens in **1** for its smooth generation makes a liberal interpretation of a weak back donation by gold open to further examination.

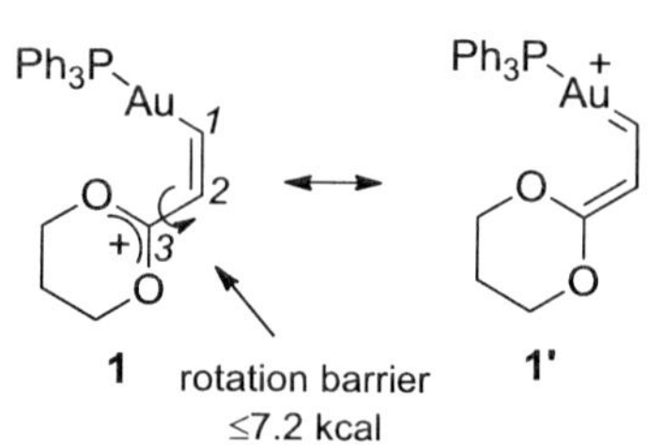

Figure 16.1. The rotation barrier of the alkenyl gold complex **1** estimated by ^{13}C NMR.

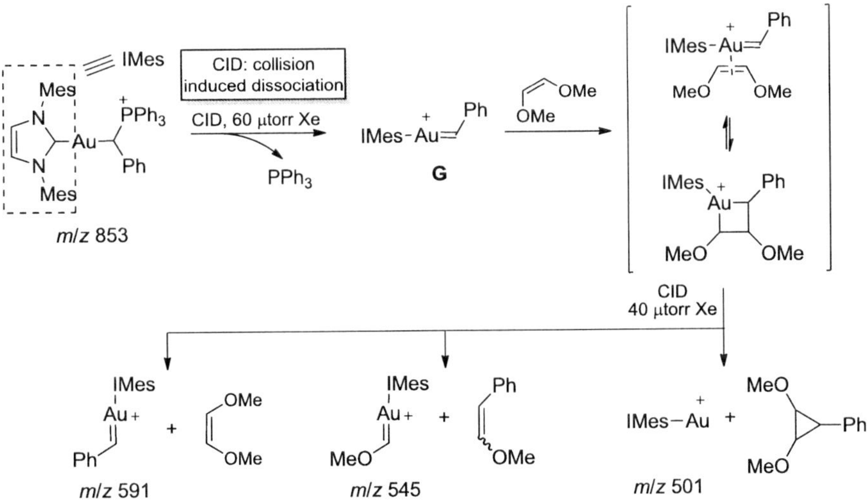

Scheme 16.3.

Chen and coworkers generated gold benzylidene complexes (e.g., **G**, Scheme 16.3) via gas-phase synthesis and examined their reactivities with *cis*-1,2-dimethoxyethylene and *cis*-3-hexene.[4] The typical reactions associated with Ru and Rh carbenoids such as cyclopropanations and metathesis are observed, indicating the carbene nature of **G** and hence meaningful back donation by the gold center.

16.2.2 DFT Calculations

Density functional theory (DFT) calculations using M06 functional have been performed on gold carbenes to understand the nature of bonding between the metal and the carbene center. In this theoretical study by Goddard and Toste,[5] the alkenyl gold carbene/gold-substituted allylic cation system, previously examined experimentally by Fürstner,[3] is computed with substituents of different electronic natures at the positions γ to the gold moiety. As shown in Figure 16.2, comparing the structures in the top row, the structures in the bottom row, due to their access to gold carbene mesomeric isomers, have larger A values in each case, which are defined as the ratio of the length of the C2–C3 bond over that of the C1–C2 bond; moreover, the ΔA increases as the X group becomes less electron donating and more electron withdrawing. These trends are consistent with the gold center being able to back donate to stabilize α carbocations in the form of gold carbon double bonds, and the extent of this stabilization increases as the cation becomes less stable. Additional calculations also predict that more σ-donating ligands on the metal enhance the contribution of gold carbene to the overall structure.

By examining the rotation barrier (Fig. 16.3a), the Me_3PAu moiety appears to be comparable to a MeO group in stabilizing an allylic cation. Additional computations of the gold carbene **H** (Fig. 16.3b) reveal that the natural bond order between Au and C is 1.14, 88% of which is surprisingly of π character.

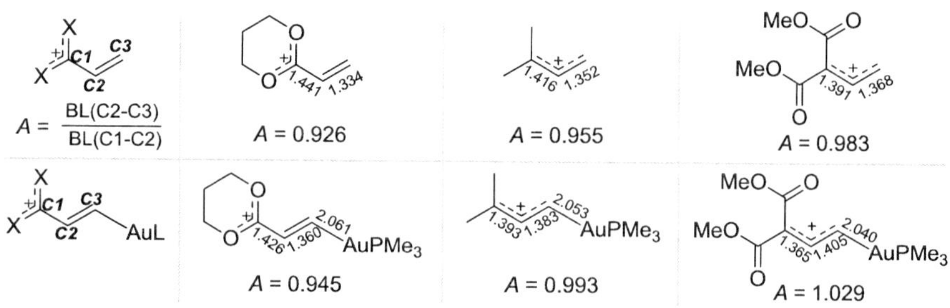

Figure 16.2. Comparison of bond lengths of different substituted allyl cations via DFT (M06) calculations.

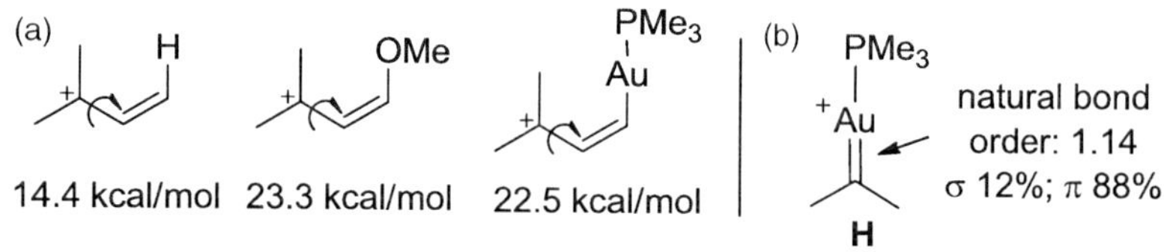

Figure 16.3. (a) Calculated activation energies of bond rotation and (b) the bond type of gold carbene.

This result indicates that the carbene behaves as a weak σ donor and a strong π acceptor and gold can be an effective back donator. As the carbene center becomes more electron deficient, the gold carbene bond would likely have increasing π nature in the expense of the σ counterpart.

16.3 GENERATION AND REACTIONS OF GOLD CARBENES

16.3.1 Gold Carbenes with Dative or Alkyl Substituents

16.3.1.1 Via Enyne Cycloisomerization. Gold-catalyzed enyne cycloisomerization[1f,6] is perhaps the most studied transformation in homogenous gold catalysis, especially in the early days. In his pioneer work in this area published in 2004,[7] Echavarren reported that cationic gold(I) complexes are highly alkynophilic catalysts for endo or exo cyclization of 1,6-enynes (Scheme 16.4). The regioselectivity of the cyclization is mostly controlled by the Z group. With Z as carbon-based groups, the preferred initial step is an exo cyclization, leading to the exocyclic gold carbene I upon a concerted cyclopropane ring construction. Density functional theory calculations of this system established that the structure of this carbene resembles the canonical form I′ as the fusing C–C bond is long (1.728 Å, Y = H, L = PH_3). The mnemonic model of equating a gold-activated alkyne to an α-carbene gold carbene (see Scheme 16.1) is applicable as the formation of I can be rationalized as an intramolecular cyclopropanation by a hypothetical biscarbene intermediate (i.e., J). When Z is *p*-toluenesulfonylimino (NTs), the initially formed intermediate is the endocyclic gold carbene K. Again, its formation can be formally

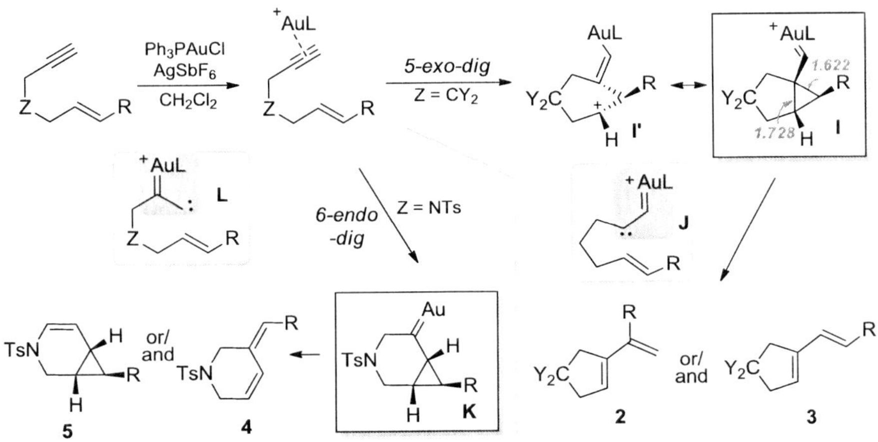

Scheme 16.4.

attributed to an intramolecular cyclopropanation by a regioisomeric α-carbene gold carbene (i.e., L). These two pathways can lead to diene products 2–4 via Wagner–Meerwein type skeleton rearrangements or to bicyclic alkene 5 via a formal carbene 1,2-C–H insertion by **K**. While the latter reveals carbene-type reactivities, a more confirming example is the intramolecular cyclopropanation shown in Equation (16.1). In this reaction, the gold carbene intermediate of type I exhibits the typical reactivities of carbenes. Later, the same group demonstrated that the cyclopropanation can be achieved in intermolecular fashions.[8]

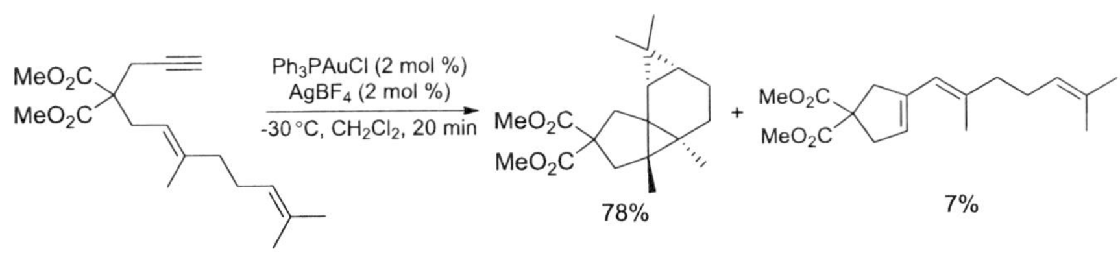

(16.1)

Similar to 1,6-enynes, 1,5-enynes also proceed in a stereospecific manner through cyclopropanated gold carbene intermediates[9] (e.g., **N**) in the presence of a gold catalyst (Scheme 16.5); however, the exo cyclization mode is not possible due to the severe strain associated with the bicyclo[2.1.0]pentane skeleton. The formation of **N** can again be rationalized via an intramolecular cyclopropanation by a hypothetical α-carbene gold carbene (i.e., **M**).

The gold carbene **N** can undergo facile 1,2-C(sp^3)–H insertions,[9] a chemoselective 1,2-acyloxy migration[9a,10] if there is an α-acyloxy group, or even insertion into a β-C(sp^3)–H bond if no α-hydrogens are available.[11] Skeleton rearrangements involving **N** are reported with alkynes terminated with aryl[12] or siloxy[13] groups.

This generation of gold carbenes from alkynes by engaging tethered alkenes has been expanded by Echavarren and coworkers to intermolecular cases.[14]

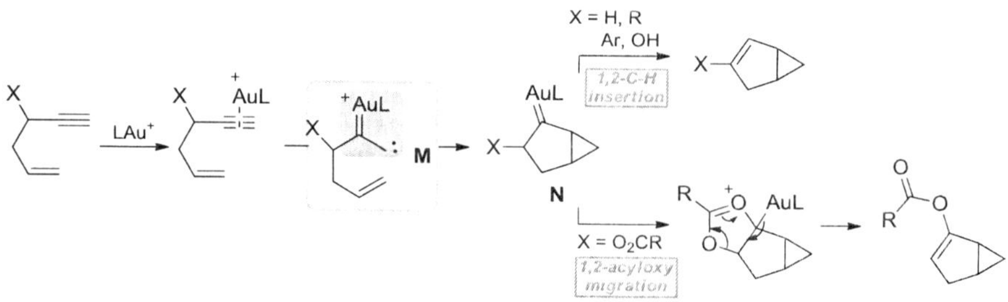

Scheme 16.5.

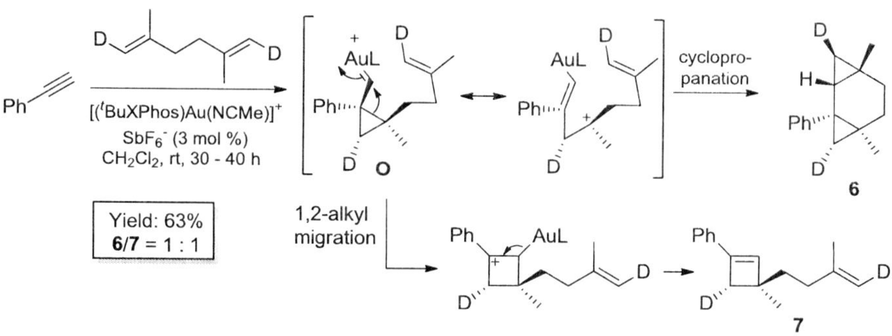

Scheme 16.6.

As shown in Scheme 16.6, in the presence of a sterically hindered gold(I) catalyst, phenylacetylene reacts with 2,5-dimethyl-1,5-hexadiene to yield the tricyclic biscyclopropane **6** and the cyclobutene **7** in a combined 63% yield. The reaction outcome can be explained by invoking the cyclopropyl gold carbene **O**. Considering the rich reactivities in enyne cycloisomerization, this intermolecular variant should open new opportunities for methodology development and much work awaits to be done.

16.3.1.2 Via 1,2-Acyloxy Migration of Propargyl Carboxylates. Another efficient approach to generating gold carbenes is a gold-promoted 1,2-acyloxy migration of propargylic carboxylates. Scheme 16.7 outlines this process along with the competing 3,3-rearrangement.[1j] The initial step for the carbene formation is a gold-promoted 5-exo-dig cyclization, which is followed by ring opening to form the vinyl gold carbene **P**. This transformation can also be viewed as the hypothetical gold carbene **Q** undergoes a carbene-promoted acyloxy migration. A 6-endo-dig cyclization diverts the reaction towards 3,3-rearrangement. Although calculations[15] and experiments[16] have established the reversible nature of both processes, the reaction pathways and their eventual reaction outcomes are mostly dependent on the substrate structures. Carbene formation via a 1,2-acyloxy migration is preferred with a terminal alkyne that has the propargylic center fully substituted or substituted by an aryl group.

This approach to vinyl gold carbenes of type **P** was initially reported by Uemura[17] using $AuCl_3$ as the catalyst and later extensively studied by Toste using

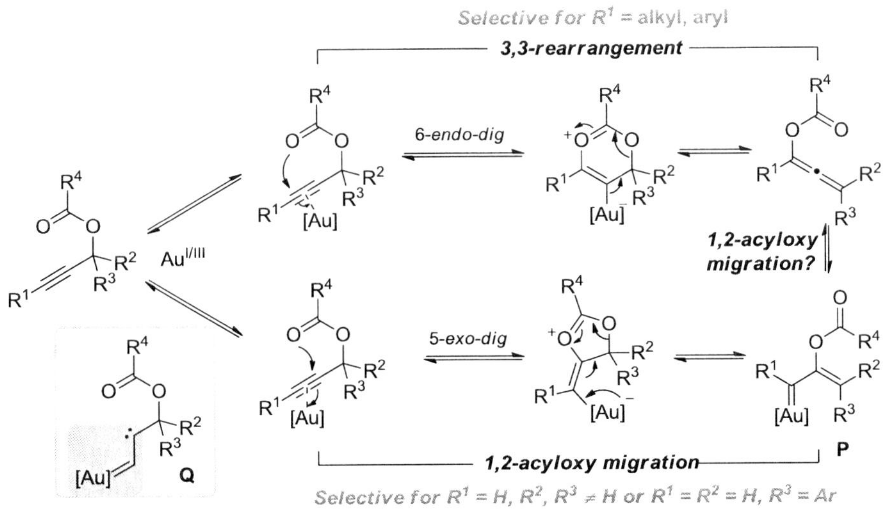

Scheme 16.7.

cationic gold(I) complexes as much more effective catalysts. Two representing examples are shown in Equations (16.2) and (16.3). In the former example, the vinyl gold carbene cyclopropanates diastereomeric alkenes stereospecifically, affording cyclopropyl enol pivalates in good yields.[18] In the latter case, the carbene behaves initially as an electrophile and then cyclizes as a nucleophile at its γ-position, highlighting a reactivity mode equivalent to a 1,3-dipole and different from carbenes.[19] A similar phenomenon is reported by Liu.[20]

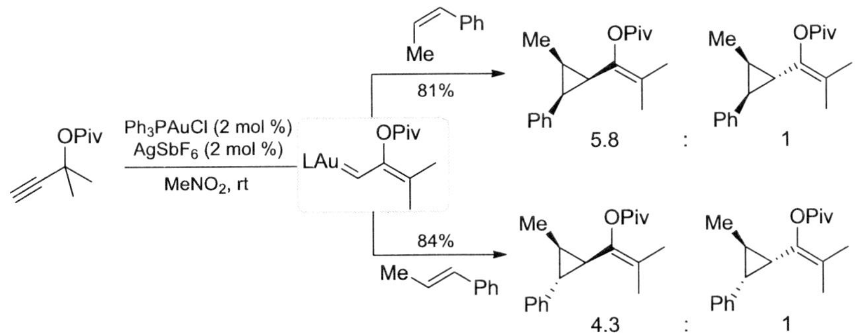

(16.2)

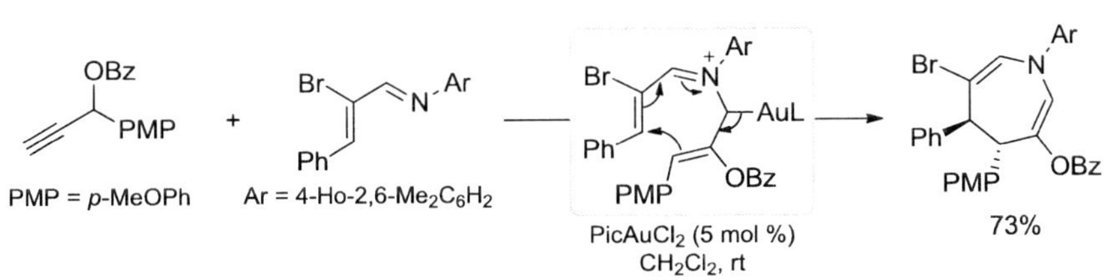

(16.3)

16.3.1.3 From Retrocyclopropanation. Echavarren in 2011[21] reported a novel generation of gold carbenes without using alkyne substrates. It relies on a gold-promoted retrocyclopropanation and is driven by the restoration of aromaticity. As shown in Equation (16.4), this strategy manifests as a retro-Buchner reaction from a cycloheptatriene and the gold carbene generated can cyclopropanate external alkenes or tether alkenes efficiently.

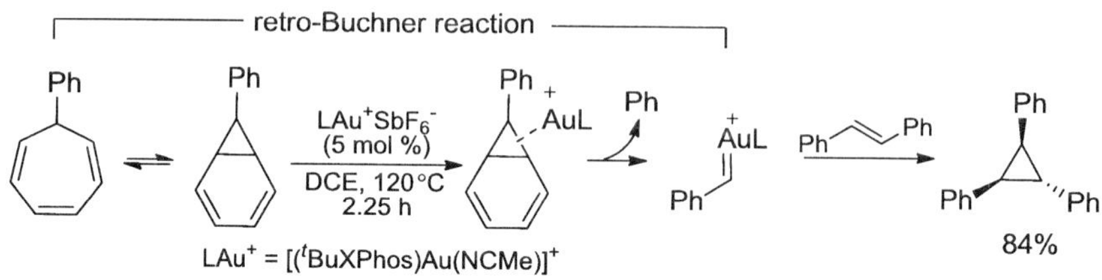

(16.4)

16.3.2 Generation and Reactions of α-Oxo Gold Carbenes

16.3.2.1 Via Decomposition of α-Diazo Esters. Dediazotization of diazo compounds is a well-applied strategy in the generation of metal carbenoids.[22] Díaz-Requejo, Nolan, Pérez, and coworkers[23] examined the decomposition of ethyl diazoacetate using IPrAuCl/NaBAr$^F_4$ [Ar^F=3,5-(CF_3)$_2$Ph] in the presence of various substrates as solvent. As shown in Equation (16.5), the presumed gold carbene intermediate reacts with styrene via competing cyclopropanation and the Friedel–Crafts reactions, with toluene via the Buchner reaction and the Friedel–Crafts reaction, with ethanol via OH insertion, and notably with alkanes by C–H bond functionalization.[24] Although no dimerization of the gold carbene **R** is detected in the first three cases, diethyl maleate and fumarate are formed in the last case, presumably due to the kinetically slow nature of the C–H insertions.

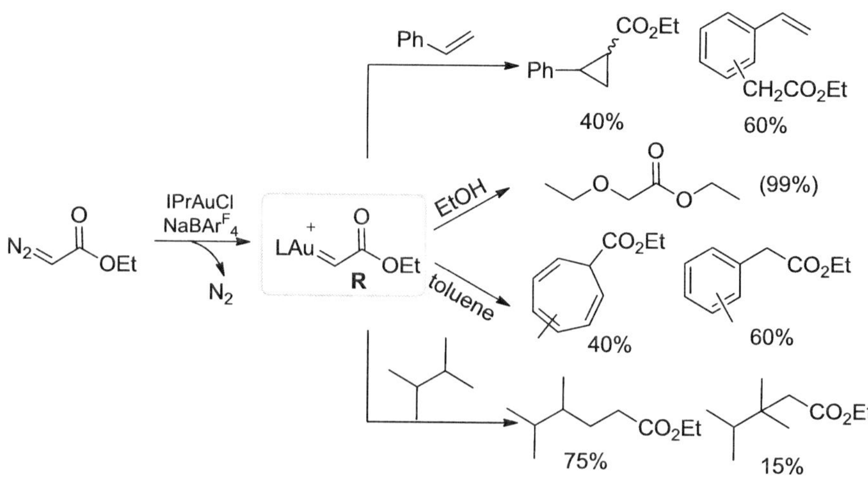

(16.5)

Notably, a recent study by Houk and Zhang showed that decomposition of α-diazo methyl ketones might not lead to the formation of gold carbene intermediates as the reaction undergoes the Wolff rearrangement chemoselectively.[25]

16.3.2.2 Via Intramolecular Redox Processes. If both the nucleophilic group and the electrophilic group in Scheme 16.1 were oxygen, the gold carbene **C** would become an α-oxo gold carbene. In this scenario, an oxidant would be required and, moreover, it needs to be nucleophilic as the first step is a nucleophilic attack. By using the mnemonic model **D**, this strategy can be viewed as its carbene center being oxidized (Scheme 16.8).

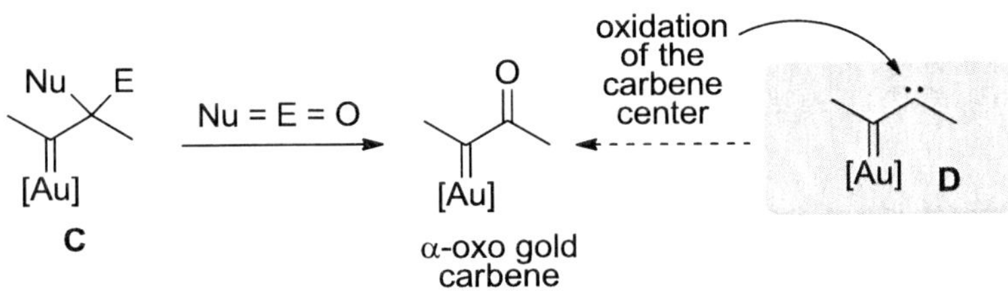

Scheme 16.8.

To realize this strategy, Toste[26] and Zhang[27] independently reported in early 2007 that an appropriately tethered sulfinyl group could serve as the requisite oxidant, leading to the formation of tetrahydrobenzothiepinones (Scheme 16.9). The proposed reaction mechanism entails a gold-promoted cyclization by the sulfinyl oxygen followed by a heterolytic fragmentation of the S–O bond to form α-oxo gold carbene **T**. This intermediate would then undergo Friedel–Crafts-type cyclization by the electron-deficient carbene moiety to afford the product. However, the facile formation of the thiepinone ring raises suspicion. Alternatively, a 3,3-rearrangement by the intermediate **S**, similar to the intermolecular case,[28]

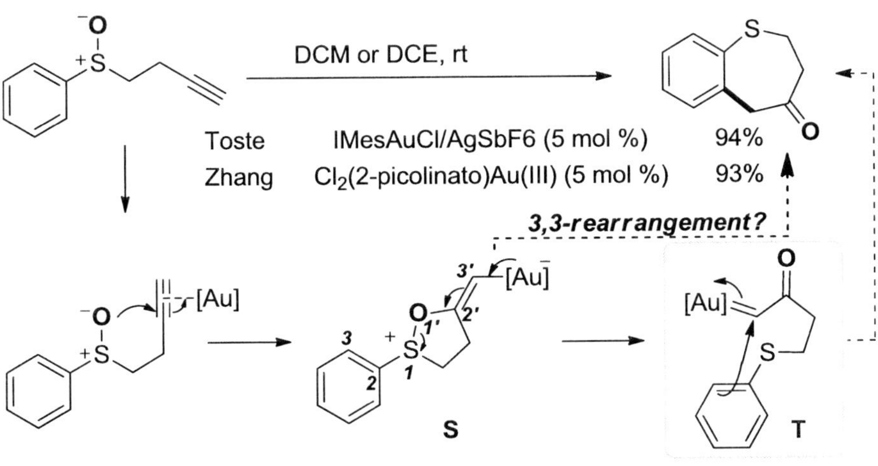

Scheme 16.9.

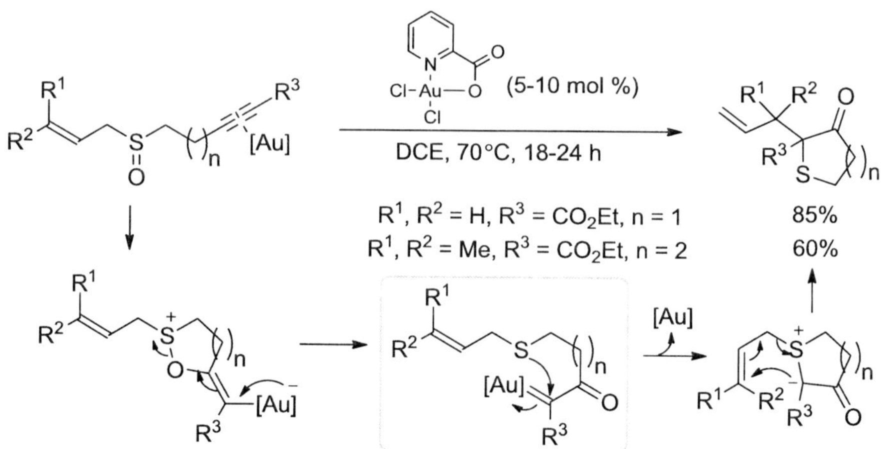

Scheme 16.10

could explain the facile formation of the seven-membered ring. However, some other reactions in both works do support the generation of α-oxo gold carbenes.

Davies reported that this gold carbene can be trapped by the tethered nascent sulfide, therefore yielding a sulfur ylide. With an allyl group substituted on the sulfur atom, a 2,3-sigmatropic rearrangement could ensue under the thermo conditions to afford substituted dihydrothiophenones and dihydrothiopyranones (Scheme 16.10).[29] In some cases, $PtCl_2$ is a better catalyst.

Instead of sulfoxides, nitrones could serve as the internal nucleophilic oxidants. As shown in Scheme 16.10, the α-oxo gold carbene **U** is generated upon intramolecular delivery of oxygen from the nitrone moiety to the C–C triple bond. This intermediate, however, does not undergo 1,2-C–H insertion or cyclopropanation; instead, due to its likely highly electrophilic nature, the carbene moiety reacts with the nascent imine moiety to form an azomethine ylide (e.g., **V**), which then undergoes [3+2] cycloaddition to form the polycyclic product (Scheme 16.11).[30] Notably, gold(III) works the best in this reaction.

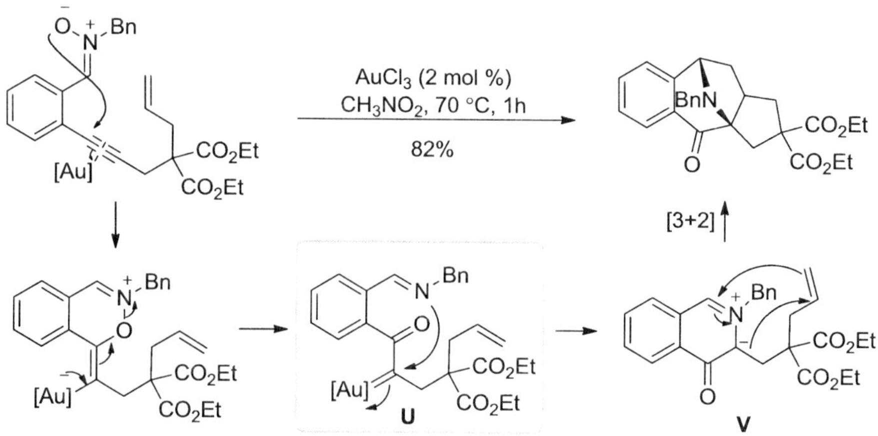

Scheme 16.11.

By extension of his original sulfoxide chemistry (Scheme 16.7),[27] Zhang reported that the use of a tertiary amine *N*-oxide as the nucleophilic internal oxidant and developed efficient syntheses of piperidine-4-ones[31] and azepan-4-ones[32] in two-step [4+2] and [5+2] annulations, respectively (Scheme 16.12). However, DFT calculations do not support the intermediacy of gold carbenes but rather a retro-aza-ene mechanism.[25]

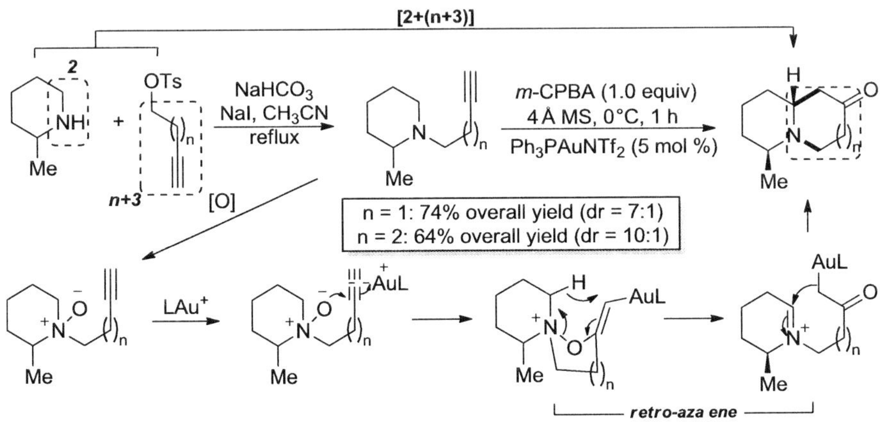

Scheme 16.12.

Surprisingly, Hashmi[33] and Liu[34] independently reported that a disubstituted epoxide moiety can act as an internal oxidant to generate α-oxo gold carbene intermediates. Interestingly, Liu showed that the gold carbene (e.g., **W**) undergoes stepwise [3+2] cycloaddition with electron-rich alkenes, forming dihydrofurans products. These compounds annulate with the remaining alkene to form an additional ring under acidic conditions (Scheme 16.13).[35]

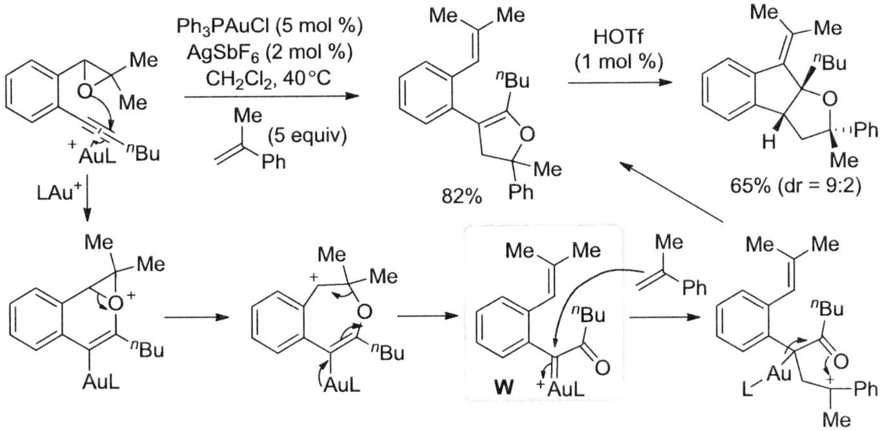

Scheme 16.13.

The latest variation on types of internal oxidants was reported by Liu, where a nitro group is used. Different from the nitrone case,[30] a nitroso ylide is formed as the intermediate, which undergoes [3 + 2] cycloadditions to furnish a bridged product (Eq. 16.6).[36]

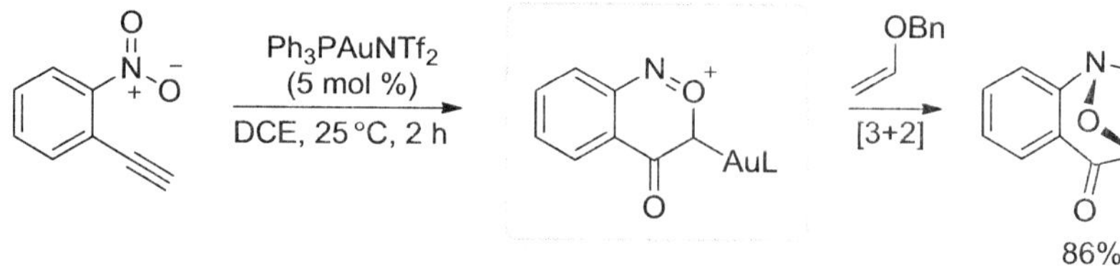

(16.6)

16.3.2.3 Via Intermolecular Alkyne Oxidation. A challenging yet natural extension of the intramolecular alkyne oxidation strategy is intermolecular oxidation (Scheme 16.14). Compared to the dediazotization approach,[23,37] this strategy would make benign and readily available alkynes effective surrogates of hazardous and potential explosive diazo ketones, thereby offering significant benefit on operational safety and reaction scale. In addition, this strategy would improve step economy and synthetic efficiency as diazo ketones are often prepared from precursors in more than one linear step.

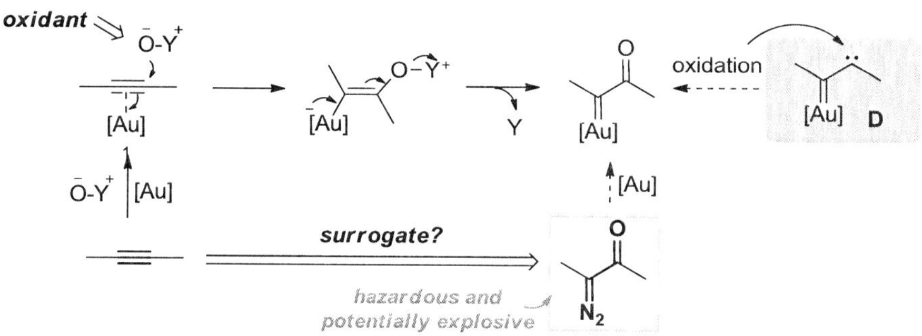

Scheme 16.14.

However, initial attempts to achieve this intermolecular oxidation using aryl sulfoxides was not successful but revealed an interesting 3,3-rearrangement process, leading to selective *ortho*-functionalization of the arene ring (Scheme 16.15).[28] This was confirmed by a later study by Liu,[38] where alkynylcyclopropanes underwent oxidative ring expansion to furnish synthetically useful cyclobuten-1-yl ketones or imides.

The first implementation of Scheme 16.13 was realized by Zhang using pyridine *N*-oxides as the oxidant.[39] As shown in Equation (16.7), the terminal C–C triple bond in the ring-fused homopropargyl alcohol **1** is oxidized into the α-oxo gold carbene moiety in **X**, which undergoes a facile intramolecular O–H insertion to afford the bicyclic dihydrofuranone **8**. The fact that the strained

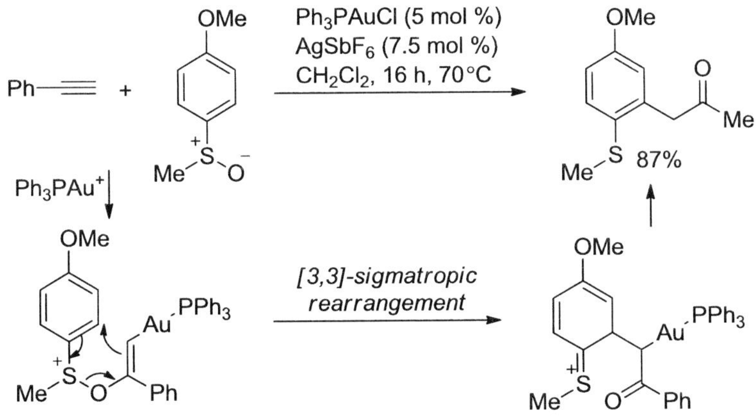

Scheme 16.15.

5,6-*trans*-fused bicycle is formed with ease suggests a highly reactive precursor and a really early transition state, consistent with the formation of the carbene **X**. Of note, 1.2 equiv. of MsOH is needed to prevent the pyridine byproduct from coordinating to and hence deactivating the cationic gold complex.

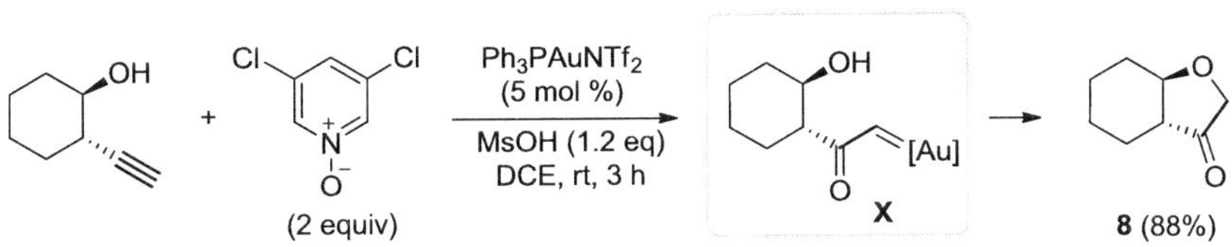

(16.7)

A further demonstration of this intermolecular oxidation is a highly efficient synthesis of strained oxetan-3-ones from readily available secondary propargyl alcohols developed by the same group (Eq. 16.8).[40] With a tertiary propargylic alcohol as the substrate, the alkyne terminus needs to be functionalized with an electron-withdrawing group in order to prevent ionization at the propargylic position under the acidic reaction conditions.

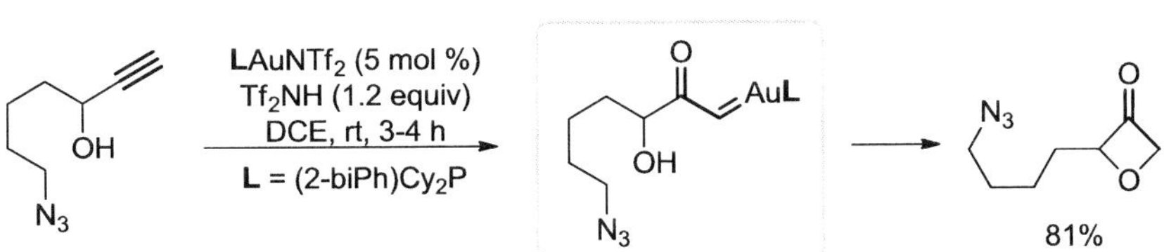

(16.8)

The same strategy is applied to an efficient synthesis of strained azetidine-3-ones from propargylic amides (Eq. 16.9). Importantly, no acid additive is required in this reaction; as a result, a range of sensitive functional groups can be tolerated. Steric hindrance appears to be beneficial as bulky BrettPhos is the optimal ligand on gold and sterically congested 2,6-dibromopyridine *N*-oxide works the best as the oxidant. Since chiral *N*-propargylsulfinamides can be prepared in excellent diastereoselectivities using chiral sulfinyl imine chemistry,[41] their oxidations by *m*-CPBA provide chiral sulfonamides for the gold catalysis without additional manipulation of the nitrogen protection group. With no racemization during the gold catalysis, azetidine-3-ones can be prepared readily with high enantiomeric excess (ee).

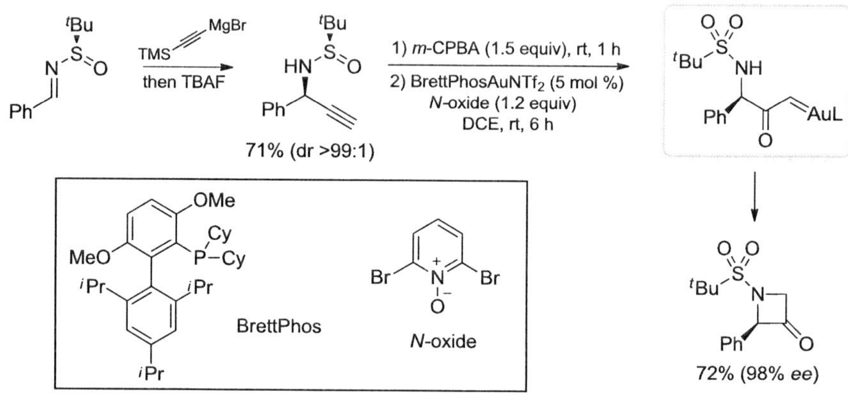

(16.9)

Besides intramolecular trapping the gold carbene moiety with heteronucleophiles, intramolecular cyclopropanations, one of the typical reactions of metal carbenes/carbenoids, was realized by Liu[42] and J. Zhang group[43] separately. An example is shown in Equation (16.10). An alternative mechanism involving sequential enyne cycloisomerization and oxidation of the gold carbene intermediate was ruled out.

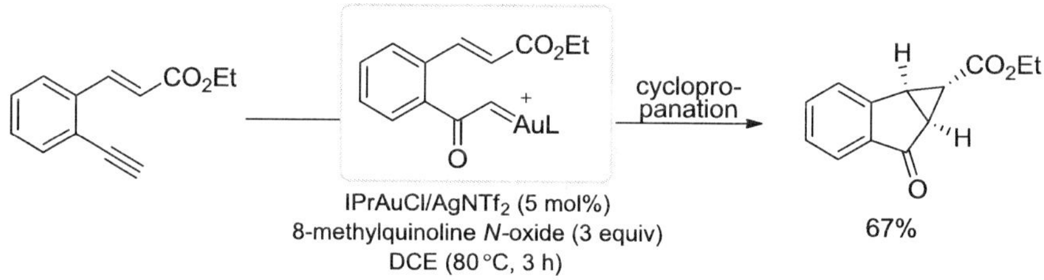

(16.10)

C–H insertion is another characteristic reaction of metal carbenes, and 1,2-C–H insertions can be readily realized in the case of α-oxo gold carbenes. For example, internal α-oxo gold carbenes, generated from internal alkynes, undergo facile 1,2-C–H insertions to furnish enone products with excellent *E*-selectivities (Eq. 16.11).[44] The stereoselectivity is, however, opposite to the *Z*-selectivity observed with Rh carbenoids,[45] highlighting the importance of

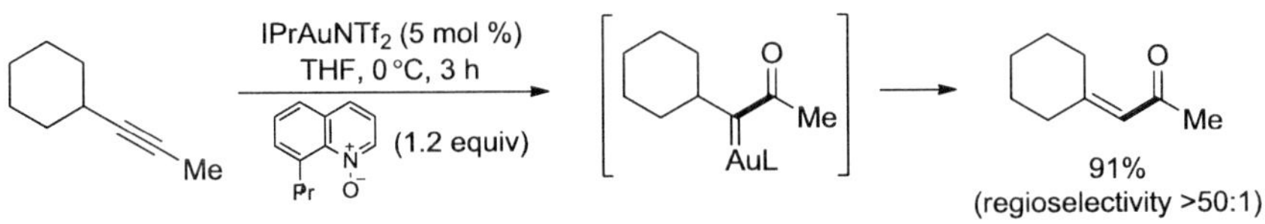

(16.11)

the metal center to the carbene reactivities and suggesting different reaction mechanisms. Of synthetic significance is that this chemistry permits the revelation of a versatile enone moiety from a propargyl moiety with predictably good to excellent regioselectivities by exploiting steric or electronic difference between the two alkyne ends. Notably, this study by Zhang showed for the first time that 8-alkylquinoline *N*-oxides are effective external oxidant and in most cases perform better than derivatized pyridine *N*-oxides and, moreover, permit the avoidance of acid additives. As a result, acid-sensitive functional groups such as OTBS and OTHP are tolerated in this reaction.

Long-range C–H functionalizations by α-oxo gold carbenes are, however, not attributed to concerted insertions but to stepwise processes such as the Friedel–Crafts-type reactions[46] initiated by the highly electrophilic carbene centers. Notably, Liu and coworkers reported a formal C(sp^3)–H insertion reaction that affords synthetically useful cyclopentenones and indanones (Scheme 16.16).[47] The authors prefer a 1,5-hydride migration of the initial addition product **Y**; however, such a process involving the gold carbene appears also possible.

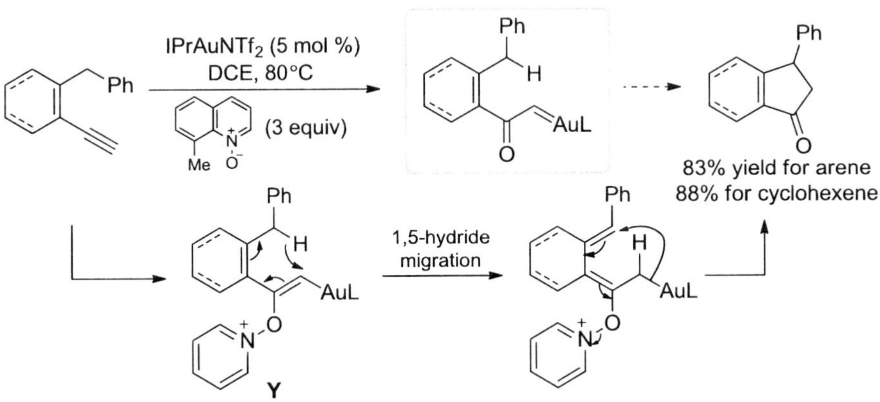

Scheme 16.16.

The reactivities of α-oxo gold carbenes generated via alkyne oxidation can be mostly attributed to their high electrophilicities, which is different from those of their related rhodium counterparts. For example, the highly electron-deficient nature of the carbene center becomes a major issue in reaction development as it tends to react with minimal chemoselectivity. The participation of even typically inert solvents becomes a possibility. In fact, chloride abstraction from dichloroethane was recently developed into a synthetically useful synthesis of chloromethyl ketones by Xiang and Zhang (Eq. 16.12).[48]

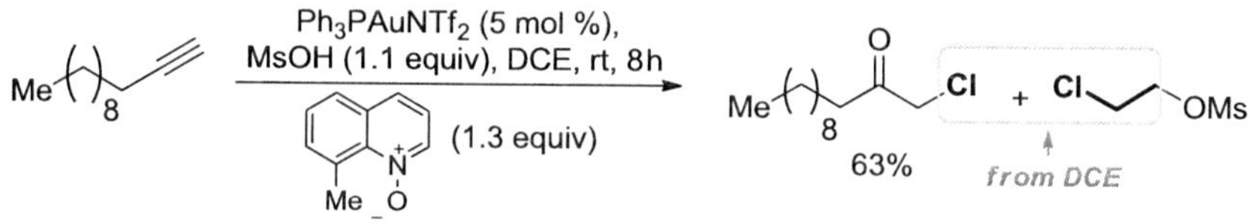

(16.12)

16.3.3 Generation and Reactions of α-Imino Gold Carbenes

If the carbene center of the hypothetical intermediate **D** is nitrenated, an α-imino gold carbene (i.e., **Z**) could be generated (Scheme 16.17). In theory, a nucleophilic nitrene transfer reagent would realize this transformation, therefore making alkynes surrogates of α-diazo imines. Notably, this type of diazo compounds is difficult to access due to their readily attained cyclizations into 1,2,3-triazoles,[49] let alone the typical hazards associated with diazo compounds. As a result, the reactivities of α-imino metal carbenes[50] have rarely been explored, except when the R group is a strong electron-withdrawing group (Scheme 16.17). In this exception, 1,2,3-triazole could reversibly ring open to provide a diazo species for the formation of an α-imino Rh carbene in the presence of a Rh catalyst.[51]

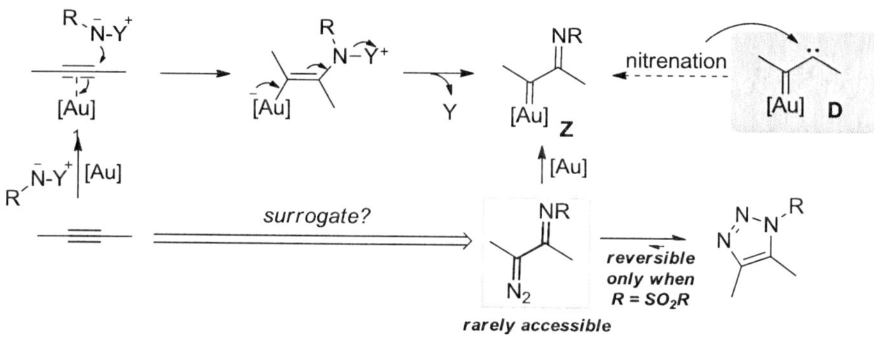

Scheme 16.17.

16.3.3.1 Via Intermolecular Nitrene Transfer. Zhang and coworkers demonstrated that, indeed, α-imino gold carbenes can be generated via gold-promoted nitrenation of alkynes.[52] With the nitrogen counterpart of pyridine *N*-oxides, iminopyridinium ylides as the nitrene-transfer reagent, reactive ynamides can be transformed into amidines in good yields albeit with low diastereoselectivities (Eq. 16.13). Davies later reported a related reaction, offering a synthesis of trisubstituted oxazoles.[53] Unfortunately, typical alkynes are not good substrates. Notwithstanding, the feasibility of this strategy should open opportunities to study the reactivities of rarely unexplored α-imino gold carbenes in an efficient manner.

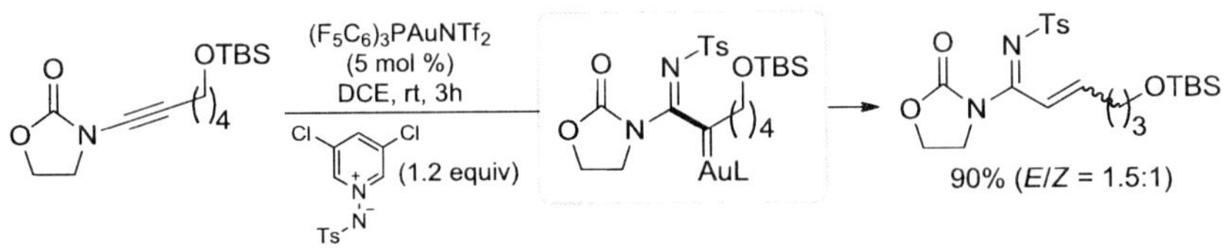

(16.13)

16.3.3.2 Via Intramolecular Nitrene Transfer. The related intramolecular version has been successfully implemented using readily accessible azido groups as the tethered nitrene precursors. Though not originally conceived as an approach to the generation of α-imino gold carbenes, the gold-catalyzed intramolecular acetylenic Schmidt reaction of homopropargylic azide, reported by Toste in 2005, proceeds through such an intermediate (i.e., **AA**) via an initial gold-promoted 5-endo-dig cyclization by the azido group, eventually leading to the formation of a pyrrole product (Eq. 16.14).[54] A similar chemistry was later realized by Hiroya using $PtCl_4$ as the catalyst.[55]

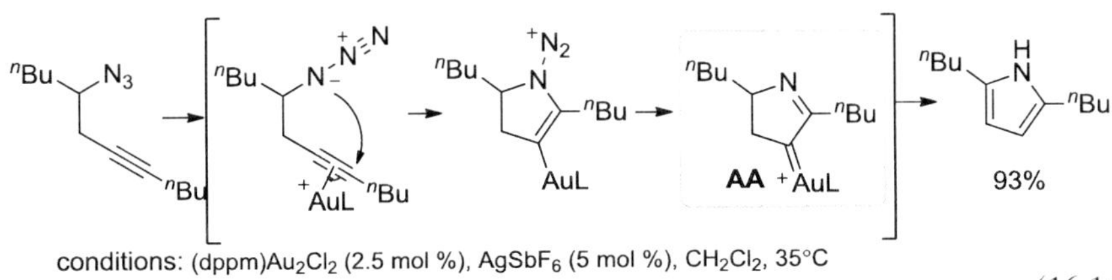

(16.14)

Gagosz[56] and Zhang[57] independently reported that the homopropargyl azides in Scheme 16.18 can be fused to a benzene ring. The α-imino gold carbene **AB** generated upon gold catalysis behaves as an indole precursor that is highly electrophilic at its 3-position. Considering that, typically, the indole 3-position is nucleophilic, this intermediate offers reactivity umpolung and hence a unique approach to the preparation of indole derivatives. In Gagosz's chemistry, the initial trapping of **AB** by an allyl alcohol is coupled with a likely gold-promoted Claisen rearrangement under mild heating to afford a pseudoindoxyl product efficiently. In Zhang's work, bisindole products can be readily prepared.

While all the earlier examples invoke initial 5-endo-dig cyclization by the azido group, a recent study by Zhang and coworkers[58] utilizes electronic bias to realize a highly regioselective 5-*exo*-dig cyclization. As shown in Equation (16.15), the α-imino gold carbene **AC** generated in this way has an electron-deficient alkene directly attached and can undergo a 4π electrocyclization to eventually form the central pyrrole ring.

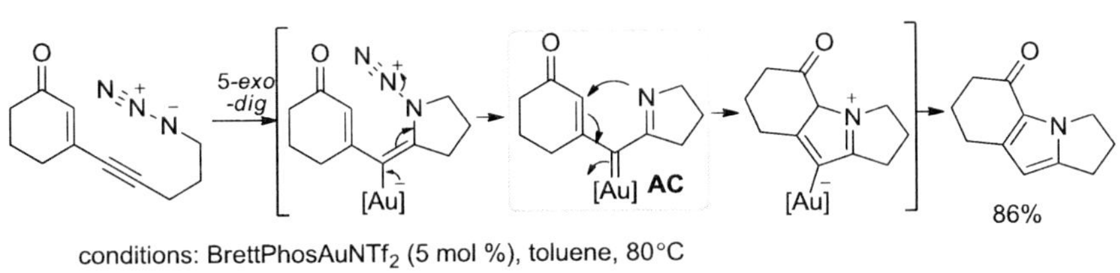

(16.15)

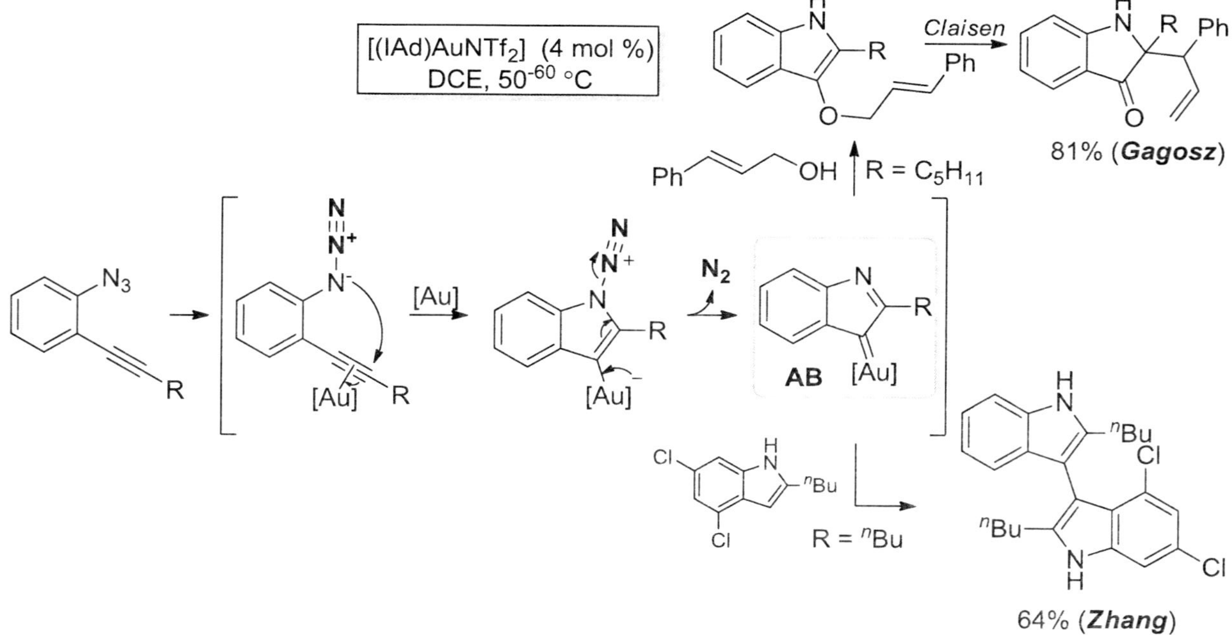

[(IAd)AuNTf2] (4 mol %)
DCE, 50-60 °C
N3
R
[Au]
N2
AB
R = C5H11
Ph
OH
Claisen
81% (Gagosz)
R = nBu
Cl
nBu
64% (Zhang)

Scheme 16.18.

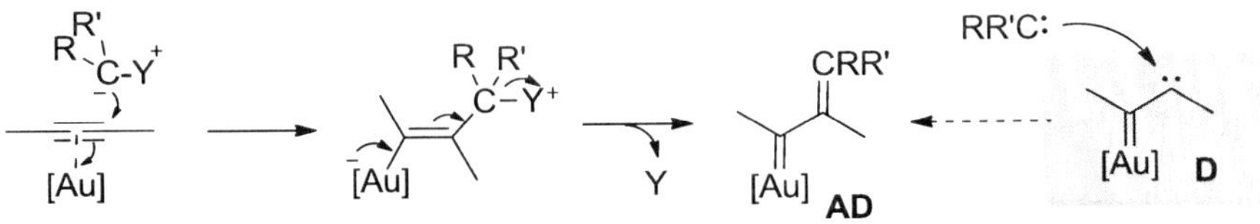

Scheme 16.19.

16.3.4 Generation and Reactivities of Gold Carbenes via Carbene Transfer

Besides oxidation and nitrene transfer, carbene transfer to alkynes under gold catalysis could generate α-alkenyl gold carbenes (i.e., **AD**, Scheme 16.19). This strategy was first realized by Skrydstrup by using sulfur ylides as the carbene-transfer reagents (Eq. 16.16).[59] In the reaction, the ylide behaves first as a nucleophile and then as an electrophile en route to the formation of the gold carbene **AE**, which can also be viewed formally as the combination of an incoming carbene with the carbene center of the hypothetical intermediate **D**. Cyclization of **AE** leads to the formation of 2,4-disubstituted furans. The overall process is a modular [3+2] cycloaddition.

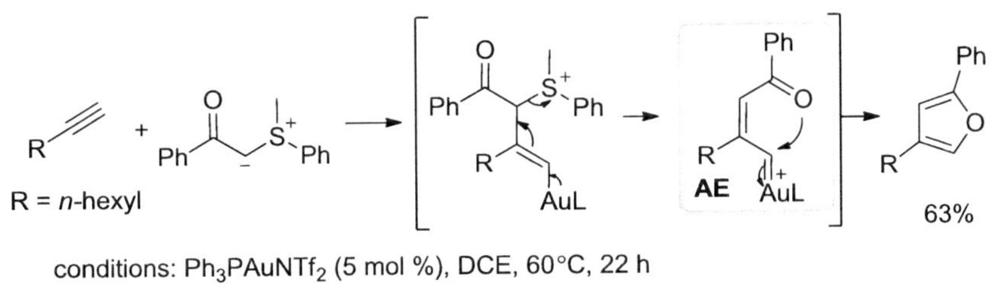

(16.16)

16.3.5 Generation and Reactivities of 1,3-Dipolar Gold Carbenes from Allenes

Besides alkynes, allenes are also preferred substrates for gold catalysis. The formalism outlined in Scheme 16.2 describes the 1,2-diene unit as a precursor to 1,3-dipolar gold carbenes. Zhang in his original work,[60] however, showed that 1,3-dipolar gold carbenes can be generated from allenes as actual reactive species in gold catalysis. As shown in Scheme 16.20, the initial activation of the allene moiety by gold generates a relatively stable oxonium species **AF**, which is in resonance with the dipolar gold carbene **AF'**. This 1,3-dipole then undergoes an intramolecular concerted [3+2] cycloaddition, which is substantiated by its observed stereospecificity. The subsequent cyclopropane ring opening and protodeauration afford the spiro product with the phenyl group *trans* to the carbonyl group, retaining the spacial relationship found in the enone substrate.

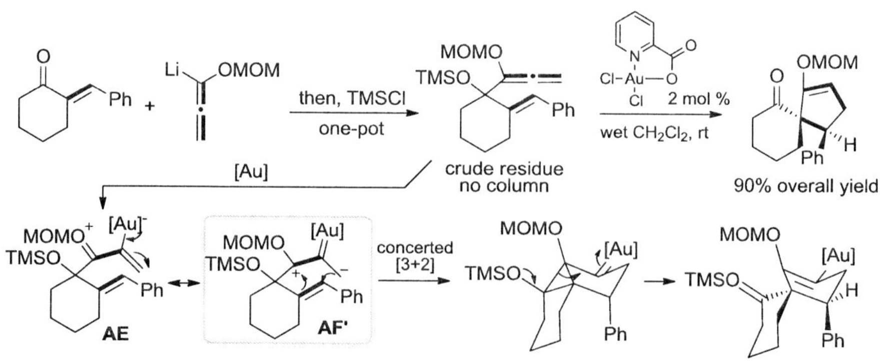

Scheme 16.20.

Considering the synthetic versatility associated with this type of carbenes, it is reasonable to anticipate many synthetic opportunities by utilizing this strategy. However, there have been so far only a few successes in this area. One study reported by Mascareñas, initially accomplished by $PtCl_2$[61] and later by Au(I),[63] is an intramolecular [4+3] cycloaddition between an allene and a 1,3-diene (Eq. 16.17). The concerted nature of this seven-membered ring formation process is established by experiments and the DFT calculations support the intermediacy of a metal allyl cation **AG**[63] and hence its resonance form, the 1,3-dipolar gold carbenes **AGn**. Toste later published a similar [4+3] cycloaddition and, in addition, one example of likely concerted [3+2] cycloaddition between an allene and a tethered alkene (Eq. 16.18).[63]

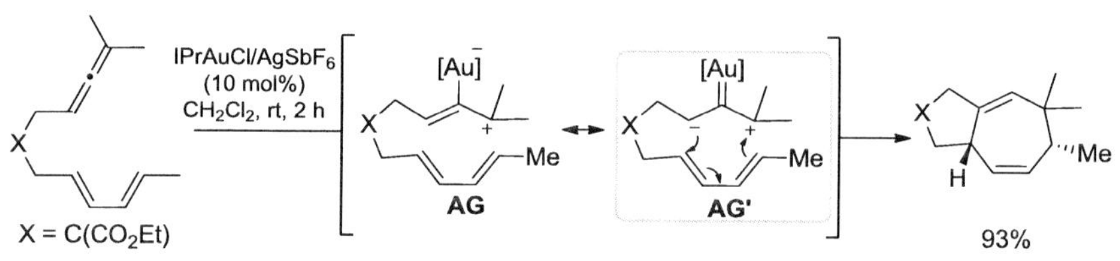

(16.17)

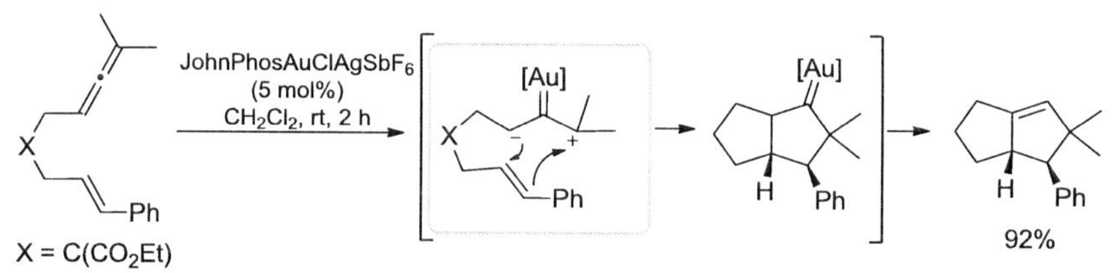

(16.18)

A relevant study by Gagosz, which predates the 1,3-dipolar gold carbene concept, can be formally viewed as an intramolecular 1,3-dipolar cycloaddition

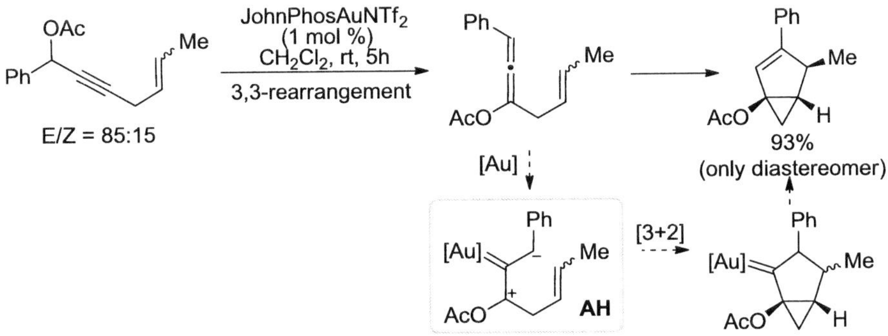

Scheme 16.21.

between a 1,3-dipolar gold carbene (i.e., **AH**) and an alkene, although a proposed stepwise mechanism is more consistent with the steric outcome (Scheme 16.21).[64]

16.4 CONCLUSION AND OUTLOOK

Gold carbenes or gold-stabilized carbocations have been proposed as critical reactive intermediates in a range of homogeneous gold-catalyzed reactions, and they are responsible for the variety and richness of gold chemistry. Density functional theory calculations have revealed that in these species there is significant back donation from the metal center to the α-carbocation and the resulting stability may be similar to that offered by a methoxy group.

These reactive intermediates can be generated from alkynes via a variety of intramolecular approaches including enyne cycloisomerization, 1,2-acyloxy migration from propargylic carboxylates and from intramolecular oxidation of alkynes, and intermolecular approaches including intermolecular alkyne oxidation and nitrene or carbene transfer to alkynes. The intramolecular approaches have been recognized to be highly versatile for the construction of a variety of cyclic structures, and the intermolecular counterpart offers desirable synthetic efficiency and flexibility that promises to circumvent the hazardous diazo compounds used in metal carbenoid chemistry.

The generation of 1,3-dipolar gold carbenes from allenes directly offers a rare but yet facile access to all-carbon 1,3-dipoles. Their concerted 1,3-dipolar cycloadditions permits rapid construction of ring structures with excellent stereoselectivities.

Gold carbenes are electrophilic in nature, the extent of which is affected by their substituents. With alkenyl groups or alkyl groups, they can undergo cyclopropanations and C–H insertion reactions, typical chemistry of carbenes and metal carbenoids, in addition to behaving as electrophiles. With electron-withdrawing substituents such as carbonyl or imino groups, they become highly electrophilic and chiefly react with nucleophiles in stepwise manners. With all the developments discussed earlier, the field of gold carbene chemistry is still much underdeveloped, and there are a range of challenges and opportunities: (i) α-oxo/imino gold carbenes are too reactive to be effectively trapped intermolecularly; this could be addressed through novel ligand design; (ii) oxidants and nitrene-transfer reagents are not atom economic and in many cases difficult to prepare; the development of cheap, atom-economic, and readily available reagents would make large-scale applications possible; (iii) the application of 1,3-dipolar gold carbenes is still limited and the development of intermolecular versions will substantially expand their synthetic utility; and (iv) the intermolecular enyne reaction, though challenging, promises to expand the gold carbene chemistry beyond cycloisomerization and deliver a variety of skeletons in multicomponent reactions.

SUGGESTED READING

Benitez, D., Shapiro, N. D., Tkatchouk, E., Wang, Y., Goddard III, W. A., Toste, F. D. "A Bonding Model for Gold(I) Carbene Complexes." *Nat. Chem.* **2009**, *1*, 482–486.

Echavarren, A. M., Nevado, C. "Non-Stabilized Transition Metal Carbenes as Intermediates in Intramolecular Reactions of Alkynes with Alkenes." *Chem. Soc. Rev.* **2004**, *33*, 431–436.

Fedorov, A., Moret, M.-E., Chen, P. "Gas-Phase Synthesis and Reactivity of a Gold Carbene Complex." *J. Am. Chem. Soc.* **2008**, *130*, 8880–8881.

Fürstner, A., Davies, P. W. "Catalytic Carbophilic Activation: Catalysis by Platinum and Gold Π Acids." *Angew. Chem. Int. Ed.* **2007**, *46*, 3410–3449.

Hashmi, A. S. K. "Gold-Catalyzed Organic Reactions." *Chem. Rev.* **2007**, *107*, 3180–3211.

Huang, X., Zhang, L. "Two-Step Formal [3+2] Cycloaddition of Enones/Enals and Allenyl Mom Ether: Gold-Catalyzed Highly Diastereoselective Synthesis of Cyclopentanone Enol Ether Containing an All-Carbon Quaternary Center." *J. Am. Chem. Soc.* **2007**, *129*, 6398–6399.

López-Carrillo, V. N., Echavarren, A. M. "Gold(I)-Catalyzed Intermolecular [2+2] Cycloaddition of Alkynes with Alkenes." *J. Am. Chem. Soc.* **2010**, *132*, 9292–9294.

Wang, S., Zhang, G., Zhang, L. "Gold-Catalyzed Reaction of Propargylic Carboxylates Via an Initial 3,3-Rearrangement." *Synlett* **2010**, 692–706.

Ye, L., Cui, L., Zhang, G., Zhang, L. "Alkynes as Equivalents of α-Diazo Ketones in Generating α-Oxo Metal Carbenes: A Gold-Catalyzed Expedient Synthesis of Dihydrofuran-3-Ones." *J. Am. Chem. Soc.* **2010**, *132*, 3258–3259.

Zhang, L., Kozmin, S. A. "Gold-Catalyzed Cycloisomerization of Siloxy Enynes to Cyclohexadienes." *J. Am. Chem. Soc.* **2004**, *126*, 11806–11807.

REFERENCES

1 (a) Abu Sohel, S. M., Liu, R.-S. *Chem. Soc. Rev.* **2009**, *38*, 2269–2281. (b) Arcadi, A. *Chem. Rev.* **2008**, *108*, 3266–3325. (c) Fürstner, A., Davies, P. W. *Angew. Chem. Int. Ed.* **2007**, *46*, 3410–3449. (d) Gorin, D. J., Sherry, B. D., Toste, F. D. *Chem. Rev.* **2008**, *108*, 3351–3378. (e) Hashmi, A. S. K. *Chem. Rev.* **2007**, *107*, 3180–3211. (f) Jiménez-Núñez, E., Echavarren, A. M. *Chem. Rev.* **2008**, *108*, 3326–3350. (g) Krause, N., Winter, C. *Chem. Rev.* **2011**, *111*, 1994–2009. (h) Li, Z., Brouwer, C., He, C. *Chem.*

Rev. **2008**, *108*, 3239–3265.(i) Patil, N. T., Yamamoto, Y. *Chem. Rev.* **2008**, *108*, 3395–3442.(j) Wang, S., Zhang, G., Zhang, L. *Synlett* **2010**, 692–706.

2 Marion, N., Nolan, S. P. *Chem. Soc. Rev.* **2008**, *37*, 1776–1782.

3 Seidel, G., Mynott, R., Fürstner, A. *Angew. Chem. Int. Ed.* **2009**, *48*, 2510–2513.

4 Fedorov, A., Moret, M.-E., Chen, P. J. *Am. Chem. Soc.* **2008**, *130*, 8880–8881.

5 Benitez, D., Shapiro, N. D., Tkatchouk, E., Wang, Y., Goddard III, W. A., Toste, F. D. *Nature Chemistry* **2009**, *1*, 482–486.

6 (a) Zhang, L., Sun, J., Kozmin, S. A. *Adv. Synth. Catal.* **2006**, *348*, 2271–2296. (b) Echavarren, A. M., Nevado, C. *Chem. Soc. Rev.* **2004**, *33*, 431–436. (c) Amijs, C. H. M., Lopez-Carrillo, V., Raducan, M., Perez-Galan, P., Ferrer, C., Echavarren, A. M. *J. Org. Chem.* **2008**, *73*, 7721–7730.

7 Nieto-Oberhuber, C., Muñoz, M. P., Buñuel, E., Nevado, C., Cárdenas, D. J., Echavarren, A. M., *Angew. Chem. Int. Ed.* **2004**, *43*, 2402–2406.

8 López, S., Herrero-Gómez, E., Pérez-Galán, P., Nieto-Oberhuber, C., Echavarren, A. M. *Angew. Chem. Int. Ed.* **2006**, *45*, 6029–6032.

9 (a) Mamane, V., Gress, T., Krause, H., Fürstner, A. *J. Am. Chem. Soc.* **2004**, *126*, 8654–8655. (b) Luzung, M. R., Markham, J. P., Toste, F. D. *J. Am. Chem. Soc.* **2004**, *126*, 10858–10859.

10 Harrak, Y., Blaszykowski, C., Bernard, M., Cariou, K., Mainetti, E., Mouriès, V., Dhimane, A.-L., Fensterbank, L., Malacria, M. *J. Am. Chem. Soc.* **2004**, *126*, 8656–8657.

11 Horino, Y., Yamamoto, T., Ueda, K., Kuroda, S., Toste, F. D. *J. Am. Chem. Soc.* **2009**, *131*, 2809–2811.

12 Sun, J., Conley, M. P., Zhang, L., Kozmin, S. A. *J. Am. Chem. Soc.* **2006**, *128*, 9705–9710.

13 Zhang, L., Kozmin, S. A. *J. Am. Chem. Soc.* **2004**, *126*, 11806–11807.

14 Yeom, H.-S., Koo, J., Park, H.-S., Wang, Y., Liang, Y., Yu, Z.-X., Shin, S. *J. Am. Chem. Soc.* **2011**, *134*, 208–211.

15 Correa, A., Marion, N., Fensterbank, L., Malacria, M., Nolan, S. P., Cavallo, L. *Angew. Chem. Int. Ed.* **2008**, *47*, 718–721.

16 Li, G., Zhang, G., Zhang, L. *J. Am. Chem. Soc.* **2008**, *130*, 3740–3741.

17 Miki, K., Ohe, K., Uemura, S. *J. Org. Chem.* **2003**, *68*, 8505–8513.

18 Johansson, M. J., Gorin, D. J., Staben, S. T., Toste, F. D. *J. Am. Chem. Soc.* **2005**, *127*, 18002–18003.

19 Shapiro, N. D., Toste, F. D. *J. Am. Chem. Soc.* **2008**, *130*, 9244–9245.

20 Bhunia, S., Liu, R.-S. *J. Am. Chem. Soc.* **2008**, *130*, 16488–16489.

21 (a) Solorio-Alvarado, C. s. R., Echavarren, A. M. *J. Am. Chem. Soc.* **2010**, *132*, 11881–11883. (b) Solorio-Alvarado, C. s. R., Wang, Y., Echavarren, A. M. *J. Am. Chem. Soc.* **2011**, null-null.

22 (a) Doyle, M. P., McKervey, M. A., Ye, T. *Modern Catalytic Methods for Organic Synthesis with Diazo Compounds: From Cyclopropanes to Ylides*. Wiley: New York, **1998**. (b) Davies, H. M. L., Beckwith, R. E. *J. Chem. Rev.* **2003**, *103*, 2861–2903. (c) Taber, D. F. In *Carbon–Carbon Σ-Bond Formation*. Pattenden, G., Ed., Vol. *3*. Pergamon Press: Oxford and New York, **1991**, p. 1045–1062.

23 Fructos, M. R., Belderrain, T. R., de Fremont, P., Scott, N. M., Nolan, S. P., Diaz-Requejo, M. M., Perez, P. *J. Angew. Chem. Int. Ed.* **2005**, *44*, 5284–5288.

24 Fructos, M. R., de Frémont, P., Nolan, S. P., Díaz-Requejo, M. M., Pérez, P. *J. Organometallics* **2006**, *25*, 2237–2241.

25 Noey, E. L., Luo, Y., Zhang, L., Houk, K. N. *J. Am. Chem. Soc.* **2011**, *134*, 1078–1084.

26 Shapiro, N. D., Toste, F. D. *J. Am. Chem. Soc.* **2007**, *129*, 4160–4161.
27 Li, G., Zhang, L. *Angew. Chem. Int. Ed.* **2007**, *46*, 5156–5159.
28 Cuenca, A. B., Montserrai, S., Hossain, K. M., Mancha, G., Lledos, A., Medio-Simon, M., Ujaque, G., Asensio, G. *Org. Lett.* **2009**, *11*, 4906–4909.
29 (a) Davies, P. W., Albrecht, S. J. C. *Angew. Chem. Int. Ed.* **2009**, *48*, 8372–8375. (b) Davies, P. W. *Pure Appl. Chem.* **2010**, *82*, 1537–1544.
30 Yeom, H. S., Lee, J. E., Shin, S. *Angew. Chem. Int. Ed.* **2008**, *47*, 7040–7043.
31 Cui, L., Peng, Y., Zhang, L. *J. Am. Chem. Soc.* **2009**, *131*, 8394–8395.
32 Cui, L., Ye, L., Zhang, L. *Chem. Commun.* **2010**, *46*, 3351–3353.
33 Hashmi, A. S., Bührle, M., Salathé, R., Bats, *J. Adv. Synth. Catal.* **2008**, *350*, 2059–2064.
34 Lin, G.-Y., Li, C.-W., Hung, S.-H., Liu, R.-S. *Org. Lett.* **2008**, *10*, 5059–5062.
35 Li, C.-W., Lin, G.-Y., Liu, R.-S. *Chem. Eur. J.* **2010**, *16*, 5803–5811.
36 Jadhav, A. M., Bhunia, S., Liao, H.-Y., Liu, R.-S. *J. Am. Chem. Soc.* **2011**, 1769–1771.
37 (a) Witham, C. A., Mauleon, P., Shapiro, N. D., Sherry, B. D., Toste, F. D. *J. Am. Chem. Soc.* **2007**, *129*, 5838–5839. (b) Li, Z. G., Ding, X. Y., He, C. *J. Org. Chem.* **2006**, *71*, 5876–5880. (c) Ricard, L., Gagosz, F. *Organometallics* **2007**, *26*, 4704–4707. (d) Prieto, A., Fructos, M. R., Mar Díaz-Requejo, M., Pérez, P. J., Pérez-Galán, P., Delpont, N., Echavarren, A. M. *Tetrahedron* **2009**, *65*, 1790–1793.
38 Li, C.-W., Pati, K., Lin, G.-Y., Sohel, S. M. A., Hung, H.-H., Liu, R.-S. *Angew. Chem. Int. Ed.* **2010**, *49*, 9891–9894.
39 Ye, L., Cui, L., Zhang, G., Zhang, L. *J. Am. Chem. Soc.* **2010**, *132*, 3258–3259.
40 Ye, L., He, W., Zhang, L. *J. Am. Chem. Soc.* **2010**, *132*, 8550–8551.
41 (a) Robak, M. T., Herbage, M. A., Ellman, J. A. *Chem. Rev.* **2010**, *110*, 3600–3740. (b) Ellman, J. A., Owens, T. D., Tang, T. P. *Acc. Chem. Res.* **2002**, *35*, 984–995. (c) Zhou, P., Chen, B.-C., Davis, F. A. *Tetrahedron* **2004**, *60*, 8003–8030.
42 Vasu, D., Hung, H.-H., Bhunia, S., Gawade, S. A., Das, A., Liu, R.-S. *Angew. Chem. Int. Ed.* **2011**, *50*, 6911–6914.
43 Qian, D., Zhang, *J. Chem. Commun.* **2011**, *47*, 11152–11154.
44 Lu, B., Li, C., Zhang, L. *J. Am. Chem. Soc.* **2010**, *132*, 14070–14072.
45 Taber, D. F., Herr, R. J., Pack, S. K., Geremia, J. M. *J. Org. Chem.* **1996**, *61*, 2908–2910.
46 (a) Wang, Y., Ji, K., Lan, S., Zhang, L. *Angew. Chem. Int. Ed.* **2012**, *51*, 1915–1918. (b) Qian, D., Zhang, *J. Chem. Commun.* **2012**, *48*, 7082–7084.
47 Bhunia, S., Ghorpade, S., Huple, D. B., Liu, R.-S. *Angew. Chem. Int. Ed.* **2012**, *51*, 2939–2942.
48 He, W., Xie, L., Xu, Y., Xiang, J., Zhang, L. *Org. Biomol. Chem.* **2012**, *10*, 3168–3171.
49 (a) Zhang, H., Ryono, D. E., Devasthale, P., Wang, W., O'Malley, K., Farrelly, D., Gu, L., Harrity, T., Cap, M., Chu, C., Locke, K., Zhang, L., Lippy, J., Kunselman, L., Morgan, N., Flynn, N., Moore, L., Hosagrahara, V., Zhang, L., Kadiyala, P., Xu, C., Doweyko, A. M., Bell, A., Chang, C., Muckelbauer, J., Zahler, R., Hariharan, N., Cheng, P. T. W. *Bioorg. Med. Chem. Lett.* **2009**, *19*, 1451–1456. (b) Japelj, B., Recnik, S., Cebasek, P., Stanovnik, B., Svete, J. *J. Heterocycl. Chem.* **2005**, *42*, 1167–1173.
50 (a) Spearing, P. K., Lambert, M. H., Ray, J. A., Laudeman, C. P., Szewczyk, J. R., Banker, P. US *Application 2007-US77365* (**2008**). (b) Sharma, P., Kumar, A., Sahu, V., Upadhyay, S., Singh, *J. Med. Chem. Res.* **2009**, *18*, 383–395. (c) Levy, A., Pogodin, S., Cohen, S., Agranat, I. *Eur. J. Org. Chem.* **2007**, 5198–5211.
51 (a) Chuprakov, S., Kwok, S. W., Zhang, L., Lercher, L., Fokin, V. V. *J. Am. Chem. Soc.* **2009**, *131*, 18034–18035. (b) Horneff, T., Chuprakov, S., Chernyak, N.,

Gevorgyan, V., Fokin, V. V. *J. Am. Chem. Soc.* **2008**, *130*, 14972–14974. (c) Chuprakov, S., Malik, J. A., Zibinsky, M., Fokin, V. V. *J. Am. Chem. Soc.* **2011**, 10352–10355.
52 Li, C., Zhang, L. *Org. Lett.* **2011**, *13*, 1738–1741.
53 Davies, P. W., Cremonesi, A., Dumitrescu, L. *Angew. Chem. Int. Ed.* **2011**, *50*, 8931–8935.
54 Gorin, D. J., Davis, N. R., Toste, F. D. *J. Am. Chem. Soc.* **2005**, *127*, 11260–11261.
55 Hiroya, K., Matsumoto, S., Ashikawa, M., Ogiwara, K., Sakamoto, T. *Org. Lett.* **2006**, *8*, 5349–5352.
56 Wetzel, A., Gagosz, F. *Angew. Chem. Int. Ed.* **2011**, *50*, 7354–7358.
57 Lu, B., Luo, Y., Liu, L., Ye, L., Wang, Y., Zhang, L. *Angew. Chem. Int. Ed.* **2011**, 8358–8362.
58 Yan, Z.-Y., Xiao, Y., Zhang, L. *Angew. Chem. Int. Ed.* **2012**, *51*, 8624–8627.
59 Kramer, S., Skrydstrup, T. *Angew. Chem. Int. Ed.* **2012**, *51*, 4681–4684.
60 Huang, X., Zhang, L. *J. Am. Chem. Soc.* **2007**, *129*, 6398–6399.
61 Trillo, B., Lopez, F., Gulias, M., Castedo, L., Mascarenas, J. L. *Angew. Chem. Int. Ed.* **2008**, *47*, 951–954.
62 Trillo, B., Lopez, F., Montserrat, S., Ujaque, G., Castedo, L., Lledos, A., Mascarenas, J. L. *Chem. Eur. J.* **2009**, *15*, 3336–3339.
63 Mauleon, P., Zeldin, R. M., Gonzalez, A. Z., Toste, F. D. *J. Am. Chem. Soc.* **2009**, *131*, 6348.
64 Buzas, A., Gagosz, F. *J. Am. Chem. Soc.* **2006**, *128*, 12614–12615.

INDEX

Note: Page numbers in *italics* refer to Figures; those in **bold** to Tables.

Contemporary Carbene Chemistry, First Edition. Edited by Robert A. Moss and Michael P. Doyle.